Handbook of Starch Science and Technology

Fifteen years have passed since the last major treatise on *starch* (third edition) was published. Since then, knowledge of the molecular and macromolecular structures of starch; exploration of new sources of commercial starch; modification of the properties of starches via chemical, enzymic, genetic, and physical means; and investigations into potential uses of new products have proliferated. The *Handbook of Starch Science and Technology* explores new developments in starch science and technologies to achieve new paradigms in the development of natural glucose polymers. New developments of starches with enhanced nutritional and health benefits and specialized starch derivatives are discussed in terms of novel applications for the design of functional products and recent developments for structuring starch that have not been covered in the previous literature. Further, it discusses the uses of starch in the manufacture of starch inclusion complexes and nanoparticles and as a key component in carrier delivery applications.

Features:

- Explores the genetics and physiology of starch biosynthesis
- Covers the source, isolation, structure, and properties of starches
- Identifies the structure and behavior of typical components in starch – amylose, amylopectin, and phytoglycogen
- Includes specific information on the modification and application of starch derivatives
- Presents current and emerging trends for starch science and technology

This timely guide is for scientists and technologists working in the fields of agriculture, biotechnology, food, pharmaceuticals, chemical engineering, nutrition, and human health.

Handbook of Starch Science and Technology

Edited by
Ming Miao, Long Chen, and James N. BeMiller

CRC Press
Taylor & Francis Group
Boca Raton London New York

CRC Press is an imprint of the
Taylor & Francis Group, an **informa** business

First edition published 2025
by CRC Press
2385 NW Executive Center Drive, Suite 320, Boca Raton FL 33431

and by CRC Press
4 Park Square, Milton Park, Abingdon, Oxon, OX14 4RN

CRC Press is an imprint of Taylor & Francis Group, LLC

ISBN: 978-1-032-68369-0 (hbk)
ISBN: 978-1-032-73474-3 (pbk)
ISBN: 978-1-003-46439-6 (ebk)

DOI: 10.1201/9781003464396

Typeset in Times
by codeMantra

Contents

Preface

Starch is a valuable, sustainable resource and is ubiquitous in green plants, especially in seeds and other storage organs. The starch used to make processed food, paper, and pharmaceutical products has most often been modified in at least one way and often in multiple ways. Many current and emerging developments have occurred in the world of starch science and technology to enhance the starch's desirable attributes, to minimize its undesirable characteristics, and/or to add new functionalities, since the third edition of *Starch: Chemistry and Technology* was published in 2009.

This book provides recent knowledge development in the fundamental and practical aspects, covering starch development, sources, isolation processes, molecular and macromolecular structures, properties, functionalities, modifications, and versatile uses of commonly used starches on the basis of the previous three editions. Chapter 1 deals with the history of starch science, global production, and consumption scenario, as well as applications of starches. Chapters 2 and 3 give in detail the genetics and physiology of starch development. Chapters 4–8 describe the source, extraction, and typical components in starch including amylose, amylopectin, and phytoglycogen. Chapters 9–11 focus on the chemical and physical structures, functional properties, and measurement techniques. Chapters 12–21 present specific information on the modification approaches and production and use of starch derivatives, including conversion products, dextrins, inclusion complexes, plastification, blends and composites, electrospun fiber, nanomaterials, **digestibility and glycaemic features,** and clean label starch. Chapter 22 discusses the future trends in the development of structuring starch with novel techniques.

This book aims to provide a state-of-the-art overview of the starch science and technology that will be useful for academic, government, and industrial scientists in fields such as foods, beverages, agriculture, cosmetics, textiles, pharmaceuticals, adhesives, papers, and chemicals.

Ming Miao
Long Chen
James N. BeMiller

Biographies

Ming Miao received his Ph.D. degree in food science from Jiangnan University in 2009. He is a professor at the State Key Laboratory of Food Science and Resources at Jiangnan University. His research interests focus on the processing–structure–property relation of starch. Through long-time efforts, he has successfully developed enzymic processes for novel starch derivatives, including starch-based delivery systems, low glycemic foods, and probiotic oligosaccharides, and transferred the technologies to food enterprises for industrialization. He published over 150 research papers in high-impact journals and more than 50 invention patents. He has also co-edited 3 books and authored more than 20 book chapters. He also got more than 10 scientific and technological awards from Jiangsu Province, Ministry of Education, and China Light Industry Council. He has actively engaged in international scientific collaboration in food science and technology. He was a major organizer of international programs including Purdue-Jiangnan faculty exchanges and International Joint Carbohydrate Research Laboratory since 2013, and a multilateral collaborative program including Purdue, UC-Davis, Cornell, Georgia, USDA, and AAFC for the establishment of the Joint International Research Laboratory in China. During the past 10 years, he has co-organized a great deal of conferences, seminars, workshops, and symposia to help his colleagues broaden their global view of important scientific progresses in carbohydrates.

Long Chen is an associate professor at the School of Food Science and Technology at Jiangnan University. His research focuses on food hydrocolloids and material science. He was a visiting scholar in the research group of Prof. David Julian McClements at University of Massachusetts Amherst, where he worked on the structural design, fabrication, characterization, and application of starch-based delivery systems for bioactive components. He received his Ph.D. degree in Food Science and Engineering from Jiangnan University in 2019. He has published over 50 scientific articles in well-known journals, 10 patents, and some chapters and conference proceedings. In addition, he has edited 1 book named *Bioactive Delivery Systems for Lipophilic Nutraceuticals: Formulation, Fabrication, and Application*. He has received funds like the National Key Research and Development Program of China – China, National Natural Science Foundation of China, Jiangsu Natural Science Foundation – Youth Foundation Project, and Youth Fund Project of Basic Scientific Research Program of Jiangnan University. He has also been invited to serve as a reviewer for more than 30 international SCI journals, including *Food Hydrocolloids*, *Carbohydrate Polymers*, *Microchimica Acta*, *Food Chemistry*, and *Journal of Agricultural and Food Chemistry*.

James N. BeMiller earned a Ph.D. degree in biochemistry at Purdue University. He began his independent career with a faculty position at Southern Illinois University at Carbondale. After 25 years and rising to positions of Professor in the Department of Chemistry and Biochemistry and Professor and Department Head of Medical Biochemistry in the School of Medicine, he left there to return to Purdue University as the founding Director of the Whistler Center for Carbohydrate Research and Professor in the Department of Food Science. Professor BeMiller has authored 170 research papers and more than 140 book chapters, reviews, encyclopedia articles, and methods and has written, co-authored, or co-edited 28 books – almost all contributions being about carbohydrates. He has received awards that include best teaching awards from both Southern Illinois University at Carbondale and Purdue University, the Alsberg-Schoch Memorial Award (from the Cereal and Grains Association), the Medal of the Japanese Society of Applied Glycoscience, the Melville L. Wolfrom Award of the Division of Carbohydrate Chemistry of the American Chemical Society, and the Spencer Award (for agricultural and food chemistry) of the Kansas City Section of the American Chemical Society, and designation as a Fellow by the American Association for the Advancement of Science (AAAS) and the Institute of Food Technologists.

Contributors

Edith Agama-Acevedo
Instituto Politecnico Nacional
CEPROBI
Yautepec, Morelos, México

Yongfeng Ai
Department of Food and Bioproduct Sciences
University of Saskatchewan
Saskatoon, SK, Canada

Abayomi Ajala
School of Food Technology and Natural Sciences
Massey University
Palmerston North, New Zealand
and
Riddet Institute
Massey University
Palmerston North, New Zealand

Touseef Amna
Faculty of Science, Department of Biology
Albaha University
Albaha, Saudi Arabia

Ardha Apriyanto
Biopolymer Analytics
Institute of Biochemistry and Biology
University of Potsdam
Potsdam-Golm, Germany
and
Research and Development
Kawasan Industri Pulogadung
Jakarta Timur, Indonesia

Marianne Ayumi Shirai
Department of Food Technology
Federal University of Technology– Paraná
Londrina, Brazil

Luis A. Bello-Perez
Instituto Politecnico Nacional
CEPROBI
Yautepec, Morelos, México

Andreas Blennow
Blennow Holding AB
Helsingborg, Sweden

Victoria Butler
Department of Molecular and Cellular Biology
College of Biological Science
University of Guelph
Guelph, ON, Canada

Yuanhui Chen
School of Food Science and Technology
Jiangnan University
Wuxi, Jiangsu Province, P.R. China

Hao Cheng
School of Food Science and Technology
Jiangnan University
Wuxi, Jiangsu Province, P.R. China

Rosana Colussi
Center for Pharmaceutical and Food Chemical Science
Federal University of Pelotas
Pelotas, Brazil

Les Copeland
School of Life and Environmental Sciences
The University of Sydney
Sydney, NSW, Australia

John R. Dutcher
Department of Physics
University of Guelph
Guelph, ON, Canada

Joerg Fettke
Biopolymer Analytics
Institute of Biochemistry and Biology
University of Potsdam
Potsdam-Golm, Germany

Robert G. Gilbert
Queensland Alliance for Agriculture and Food Innovation
Centre for Nutrition and Food Sciences
The University of Queensland
Brisbane, QLD, Australia
and
Key Laboratory of Plant Functional Genomics of the
    Ministry of Education/Jiangsu Key Laboratory of Crop
    Genetics and Physiology
Agricultural College of Yangzhou University
Yangzhou, P.R. China
and
Jiangsu Co-Innovation Center for Modern Production
    Technology of Grain Crops
Yangzhou University
Yangzhou, P.R. China

Gregory M. Glenn
USDA-ARS, Western Regional Research Center
Albany, California, USA

Anjum Hamid Rather
Nanostructures and Biomimetics Lab, Department
    of Nanotechnology
University of Kashmir
Srinagar, Jammu and Kashmir, India

Jovin Hasjim
Département Fonctionnalisation des Amidons, Sucres et
    Polyols
Roquette Frères
Lestrem, France

Jay-lin Jane
Department of Food Science and Human Nutrition
Iowa State University
Ames, IA, USA

Hongxin Jiang
College of Food Science and Technology
Henan University of Technology
Zhengzhou, Henan, P.R. China

Xing Jin
World Centric, Product Development Center
Stone Mountain, Georgia, USA

Zhengyu Jin
School of Food Science and Technology
Jiangnan University
Wuxi, Jiangsu Province, P.R. China
and
State Key Laboratory of Food Science and Resources
Jiangnan University
Wuxi, Jiangsu Province, P.R. China

Lovedeep Kaur
School of Food Technology and Natural Sciences
Massey University
Palmerston North, New Zealand
and
Riddet Institute
Massey University
Palmerston North, New Zealand

Nisar A. Khan
Department of Pharmaceutical Sciences
University of Kashmir
Srinagar, Jammu and Kashmir, India

Yong-Ro Kim
Department of Biosystems Engineering
Integrated Major in Global Smart Farm
Center for Food and Bioconvergence
Seoul National University
Seoul, Republic of Korea

Wenmeng Liu
School of Food Science and Technology
Jiangnan University
Wuxi, Jiangsu Province, P.R. China

David Julian McClements
Department of Food Science
University of Massachusetts
Amherst, Massachusetts, USA

Carley Miki
Department of Physics
University of Guelph
Guelph, ON, Canada

Marie Sofie Møller
Department of Biotechnology and Biomedicine
Technical University of Denmark
Kongens Lyngby, Denmark

Syed Naeim Raza
Department of Pharmaceutical Sciences
University of Kashmir
Srinagar, Jammu and Kashmir, India

Yasunori Nakamura
Akita Prefectural University
Akita City, Japan
and
Starch Technologies, Co., Ltd.
Akita, Japan

Shinjae Park
Center for Food and Bioconvergence
Seoul National University
Seoul, Republic of Korea

Muheeb Rafiq
Nanostructures and Biomimetics Lab, Department of
    Nanotechnology
University of Kashmir
Srinagar, Jammu and Kashmir, India

Razia Rehman
Department of Pharmaceutical Sciences
University of Kashmir
Srinagar, Jammu and Kashmir, India

Rumysa Saleem Khan
Nanostructures and Biomimetics Lab, Department of
    Nanotechnology
University of Kashmir
Srinagar, Jammu and Kashmir, India

M. Shamshi Hassan
Department of Chemistry, College of Science
Albaha University
Albaha, Saudi Arabia

Faheem A. Sheikh
Nanostructures and Biomimetics Lab, Department of
    Nanotechnology
University of Kashmir
Srinagar, Jammu and Kashmir, India

Randal L. Shogren
World Centric, Product Development Center
Stone Mountain, Georgia, USA

Jaspreet Singh
School of Food Technology and Natural Sciences
Massey University
Palmerston North, New Zealand
and
Riddet Institute
Massey University
Palmerston North, New Zealand

Birte Svensson
Technical University of Denmark
Kongens Lyngby, Denmark

Ian J. Tetlow
Department of Molecular and Cellular Biology
College of Biological Science
University of Guelph
Guelph, ON, Canada

Elif Turabi Yolacaner
Department of Food Engineering
Hacettepe University
Ankara, Turkey

Francisco Vilaplana
Division of Glycoscience, Department of Chemistry
School of Engineering Sciences in Chemistry, Biotechnology
    and Health
KTH Royal Institute of Technology
AlbaNova University Centre
Stockholm, Sweden

Kai Wang
College of Food Science
South China Agricultural University
Guangzhou, P.R. China

Yong Wang
Department of Food Science and Biotechnology
Sejong University
Seoul, South Korea
and
Carbohydrate Bioproduct Research Centre
Sejong University
Seoul, South Korea

Sang-Ho Yoo
Department of Food Science and Biotechnology
Sejong University
Seoul, South Korea
and
Carbohydrate Bioproduct Research Centre
Sejong University
Seoul, South Korea
and
Biopolymer Research Center for Advanced Materials
Sejong University
Seoul, South Korea

Juliano Zanela
Federal University of Technology—Paraná
Dois Vizinhos, Brazil

An Overview of Starch

1

Ming Miao and James N. BeMiller

1.1 INTRODUCTION

Starch is an invaluable, sustainable resource. Starches and products derived from starches play, and will continue to play, unending roles in meeting the dietary and other needs of humans. Starch is ubiquitous in green plants, especially in seeds and other storage organs, where its function is to store energy and carbon atoms for the biosynthesis of other compounds. Consumption of this starch by humans provides the D-glucose they need for energy and biosynthesis of other compounds (Chapter 20). And it functions in the same way for other organisms, including other animal life. The use of starch in fermentation processes provides potable alcohol, fuel ethanol, and other compounds.

Starches are generally a mixture of two individually unique, hydrophilic polysaccharides (amylopectin and amylose), which occur together in nature as unique, partially crystalline, hydrophilic granules that are insoluble in room-temperature water (Chapter 9) and can be easily isolated and dried to a powder (Chapter 5). Heating starch granules in water disrupts first their amorphous and then their crystalline structures, solubilizes the amylose and amylopectin molecules, and changes the starch-water slurry into a hot, viscous paste (Chapters 10 and 14) which, upon cooling, forms a gel.

Starches are central components of the diets of humans and are nutritionally important as suppliers of glucose (Chapter 10). They are versatile in that they can be easily modified chemically, enzymatically, and physically to make even more useful products (Chapter 12). Especially does slight chemical modification transform them into essential ingredients for many processed food products where they provide bulk, body, texture, stability, and other functional attributes (BeMiller & Huber 2015). Enzyme-catalyzed reactions convert starch into glucose and high-fructose syrups, maltodextrins (Chapter 13), and modified amylose and amylopectin products (Miao et al. 2018; Miao & BeMiller 2023). Certain physical and chemical modifications (Chapter 12) can transform normally highly digestible starches into dietary fiber components that are also important nutritionally (Chapter 20).

Major commercial sources of starches include cereal grains, such as maize/corn (including waxy maize and high-amylose corn (amylomaize)), wheat, and rice (BeMiller & Whistler 2009). Roots and tubers, such as potato and cassava/tapioca, and certain legumes, such as yellow pea, are other sources. Each starch from each plant source, including different varieties of the same species, is unique in terms of its characteristics; so, starch is a generic term, i.e., there are many different starches, each with its own unique characteristics. Research continues into ways to modify the properties of starches from different sources via plant breeding (Chapter 3), the use of mutagens, and genetic engineering techniques (Chen et al. 2021).

Isolated starches have many industrial uses other than in the production of processed food. Major non-food uses are the production of fuel ethanol and paper and related products. The starch used to make processed food, paper, and pharmaceutical products has most often been chemically modified in at least one way and often in multiple ways (Chapter 12). Modification is done to enhance the starch's desirable attributes, to minimize its undesirable characteristics, and/or to add new functionalities.

1.2 HISTORY OF STARCH SCIENCE

The story of starch has been tied to the history of human civilization since the Palaeolithic era (Copeland & Hardy 2018; BeMiller & Whistler 2009). Kernels of starchy grains have been identified in archaeological artifacts; an example is the identification of rice kernels in archaeological sites along the middle region of the Yangtze River in the Hubei and Hunan provinces of China which date back hundreds of thousands of years. Starch was obtained by partial fermentation of wheat kernels and used as an adhesive to bind together strips of papyrus as well as in cosmetics and medicines in Ancient Egypt (dating back to as early as 800 BCE). According to a Roman treatise in 184 BCE, the procedure for making starch was to steep wheat or barley kernels in water for 10 days, press the resulting swollen grain and mix it with fresh water, and let the extracts settle. The settled starch was rewashed with water and finally dried in the sun. Boiling freshly ground wheat flour with vinegar for preparing modified starch was documented in

DOI: 10.1201/9781003464396-1

the early Roman Empire (23–74 CE). The product was used for coating papyrus to create a smooth surface, to whiten cloth, to powder hair, and to thicken food sauces. Chinese documents of about the year 312 have been found to have been coated with a low-viscosity rice-starch paste to prevent ink penetration. In the Middle Ages, the Dutch established a starch manufacturing process from wheat, with the isolated wheat starch mainly used as a fabric stiffener. The German word stärke is the word for both starch and strength – the two being related because of starch's early use in the laundry for stiffening fabrics. The custom of powdering hair with starch became popular with wealthy Europeans during the 16th, 17th, and 18th centuries. Starch was introduced into England during the reign of Queen Elizabeth I (mid-1500s).

Although there is a long history of beer, malt, and maltose being made from starchy grains without any scientific research system, starch science is built upon testable hypotheses, In 1819, the starch granule architecture first began to be unraveled by the Dutch scientist Antonie van Leeuwenhoek using microscopes (Seetharaman & Bertoft 2013). van Leeuwenhoek was the first person to see starch granules in plant cells, and he reported the sizes and shapes of discrete granules with the detail of growth rings, as well as their swelling behaviors in hot water. Essentially nothing happened in terms of starch research for about 100 years following van Leeuwenhoek's report, and only a little happened for about 200 years after his report was published (Villwock & BeMiller 2022a, 2022b). Knowledge of starch science was slow to develop largely because the required conceptual thinking was outside current scientific knowledge and because the instruments and techniques required to examine starch structures were not available. Key milestones in the development of starch science and

some relatively recent publications are listed in Table 1.1. It is obvious from the table that, until about 1930, progress in understanding the natures of starch granules and molecules was quite slow, then grew more and more rapidly, but has slowed again in recent years. As the following chapters will show, substantial progress has been made toward understanding starch's structure, properties, and functionalities, but there are still important gaps in our knowledge of this important biopolymer (Bertoft 2017; Jane 2007; Miao & Hamaker 2021; Nakamura 2023; Tetlow & Bertoft 2020; Wang et al., 2014). Among the aspects requiring greater understanding are the fine structures of starch molecules, the relationship of amylopectin models and starch crystallinity (Chapter 7), the granule matrix, blocklet and fibrillar structures within granules, location of amylose and amylopectin molecules within granules (Chapter 6), and granule models (Chapter 9).

1.3 CURRENT STATUS OF PRODUCTION AND USE OF STARCHES

1.3.1 Global Starch Production

Starch is stored in many plants, including the kernels of cereal grains, tubers of potatoes, roots of cassava, and seeds of legumes (BeMiller & Whistler 2009). Starch is widely used in both the food and non-food sectors. The starch market has exhibited a steady growth in demand and production over many years, and it

TABLE 1.1 Timeline of the knowledge development of starch science

TIME PERIOD	MAJOR DISCOVERIES IN STARCH SCIENCE
1700–1900	Rings observed in granules (van Leeuwenhoek 1719)
	Observed that malt produces sugar (Irvine 1785)
	Acid hydrolysis of starch (Kirchhoff 1811)
	Starch–iodine complex (Colin et al. 1814)
	Outer layer of starch (Raspail 1825)
	Optical rotation of dextrin (Biot et al. 1833)
	Discovery of diastase (Payen et al. 1833)
	Theory of apposition for granule growth (Fritzche 1834)
	Intussusception theory of starch (Payen 1838)
	Determination of reducing sugars (Trommer 1841)
	Quantitative method for reducing sugars (Fehling 1849)
	Discovery of maltose (Dubrunfaut 1847)
	Micelle theory of granule structure (Nägeli 1858)
	Rediscovery of maltose (O'Sullivan 1872)
	Polymer theory of starch (Musculus et al. 1878)
	Spherocrystal theory of granule structure (Schimper 1881)
	Ring structure of glucose (Tollens 1883)
	Diastase is a protein (Linter 1886)
	Two enzymes in diastase (Wijsman 1889)
	The glycosidic linkage (Fischer 1893)
	Trichitic model of starch (Meyer 1895)

(Continued)

TABLE 1.1 (*Continued*) Timeline of the knowledge development of starch science

TIME PERIOD	MAJOR DISCOVERIES IN STARCH SCIENCE
1900–2000	Discovery of cyclodextrin (Schardinger 1903)
	Method of methylation (Purdie et al. 1903)
	Definition of amylose and amylopectin (Maquenne et al. 1905)
	Cyclodextrin theory of starch (Pringsheim et al. 1912)
	X-ray spectra of starch (Scherrer 1918)
	Defined conformation of diastase products (Kuhn 1924)
	Correct ring structure of glucose (Haworth 1925)
	Structure of maltose (Haworth et al. 1926)
	Three types of X-ray spectra of starch (Katz 1930)
	Blocklet structure of starch granules (Hanson et al. 1934)
	Proposed helical structure in starch (Hanes 1937)
	Granule ghost in starch (Alsberg 1938)
	Alpha-1,6 branches in starch (Freudenberg et al. 1940)
	Randomly branched amylopectin structure (Meyer et al. 1940)
	Biosynthesis of starch by potato phosphorylase (Hanes et al. 1940)
	Fractionation of starch using butanol precipitation (Schoch 1942)
	Conformation of helical structure (Rundle et al. 1943)
	Developing high-amylose corn starch (Whistler et al. 1946)
	Oxidation of starch (Whistler et al. 1951)
	Discovery of branched amylose (Peat et al. 1952)
	Definition of A, B, C chains (Peat et al. 1952)
	Introduction of transmission electron microscopy (Whistler et al. 1957)
	Unitary theory of starch granule (Nikuni 1957)
	Fibrillar structure of starch granules (Sterling et al. 1958)
	Discovery of substrate for synthesizing starch chains (Leloir et al. 1960)
	Action patterns for starch-related enzymes (Robyt et al. 1963)
	Unit chain distribution of amylopectin (Lee et al. 1968)
	Introduction of scanning electron microscopy (Hall et al. 1969)
	Cluster model of amylopectin (French 1972)
	Proposed double helical structure in starch (Kainuma et al. 1972)
	Alternating layers in starch granules (Yamaguchi et al. 1979)
	Hairy billiard ball model for starch granule (Lineback 1984)
	Location of amylose and amylopectin (Jane et al. 1986)
	Classification of resistant starch (Englyst et al. 1987)
	Crystal packing model in starch granules (Imberty et al. 1988)
	Macropores and channels in starch (Fannon et al. 1992)
	Development of SEC-MALLS (Sarazin et al. 1992)
2000-present	Side-chain liquid-crystalline model (Waigh et al. 2000)
	Proposed a three-component nanocomplex consisting of amylose, protein, and free fatty acids (Hamaker et al. 2003)
	Building block-backbone model (Bertoft 2004)
	Spherulitic crystallization for granule initiation (Ziegler et al. 2005)
	Structural basis for slowly digestible starch (Hamaker et al. 2006)
	De novo biosynthesis of starch without a primer (Mukerjea et al 2012)
	Enzymatic transformation of cellulose to starch (You et al. 2013)
	Chemoenzymatic starch synthesis from carbon dioxide (Cai et al. 2021)
	Developing artificial "CO_2-sugar" platform (Yang et al. 2023)
	Building hollow cage and sheet from starch (Kierulf et al. 2024)

Note: (BeMiller & Whistler 2009; Cai et al., 2021; Crini et al., 2021; Kierulf et al., 2024; Miao & Hamaker 2021; Miao & BeMiller 2023; Seetharaman & Bertoft 2013; Villwock & BeMiller 2022a, 2022b; Yang et al., 2023)

is expected to maintain its pace in the coming years. The world market for starch was an estimated 119.6 million metric tons with a value of USD 97.8 billion in the year 2020, and it is expected to reach 199.8 million metric tons by 2030, growing at a compound annual growth rate (CAGR) of 6.7% over the period 2020–2030 (Grand View Research 2021). The United States is the largest consumer of isolated starch, accounting for more than half of the global consumption, followed by the Asia-Pacific region and Europe. In industrialized nations, starches are mostly produced in large-scale factories (Chapter 5). In developing countries, the majority of starch is produced in small- and medium-sized mills, often from locally available sources.

Starch is chiefly isolated from normal maize, cassava, potato, sweet potato, and wheat (Chapter 5). In the North American region, maize (including waxy maize and high-amylose corn (amylomaize)) is the principal source of starch, accounting for

76% of the worldwide production (in terms of revenue). The United States is the leading producer of maize starch, accounting for 63% of the worldwide production (in terms of revenue). About 9% of total starch produced worldwide (in terms of revenue) is obtained from wheat – mainly in Europe. Potato accounts for about 6% of the total starch produced across the world (in terms of revenue), with Europe producing 71% of the total. In tropical regions, such as Southeast Asia, cassava is the cheapest source of starch, accounting for approximately 98% of starch production (in terms of revenue). Sweet potato, maize, and potato are the principal sources of starch in Northeast Asia.

Sweetener production is the largest consumer of starch worldwide and in developed countries, while processed food is the major consumer in developing countries. North America (home to various multinational food processing companies) is the largest consumer of starch worldwide, accounting for more than 45% of global consumption (in terms of revenue). Europe emerged as the second-largest regional market in terms of value in 2020. The expanding application of starch (specifically in convenience foods) is expected to drive the market in the region in the coming years. The Asia-Pacific region is expected to register the highest revenue-based CAGR from 2020 to 2030 because of the robust growth of the manufacturing sector in the region. Especially in China, the starch market has been booming, owing to a variety of industrial applications to meet the fast-growing demand of consumers. For example, developing environmentally friendly and recyclable bio-based materials from biological resources is of great significance in replacing fossil resources, developing the economy, and building a resource-conserving and environmentally friendly society.

1.3.2 Products Derived from Native Starches

1.3.2.1 Modified starches and specialty starches

Modified starch is obtained through chemical, physical, or enzymatic modification of raw starch. Starches are modified primarily to enhance the positive attributes of either their granular or cooked states, to minimize the defects of either their granular or cooked states, and/or to provide functionalities that native starches cannot provide (BeMiller & Whistler 2009; Miao & BeMiller 2023). Modified starch is used in a broad range of applications in food and non-food industries. Food applications include thickening, gelling, stabilization, emulsification, and adhesion. For example, in batter and breading applications, starches and modified starches offer a sticky coating for a crispier and less oily product. Also, modified starch is used to bind water and reduce purge in processed meat and seafood. As a food additive, modified starch must follow limitations in modification and usage that are described in agency publications such as the US Code of Federal Regulations of 21CFR 172.892 for food starch-modified. Modified starches are discussed in detail in Chapter 12.

Plant breeding can lead to the modification of native starch at the raw material stage to create new specialty starches – the most prominent example being the development of high-amylose corn/maize starch (Chen et al. 2021). In addition to amylose and waxy genes in maize, other genes (including dull, sugary 1, sugary 2, shrunken 1, shrunken 2, soft starch (horny), and floury 1) affect the properties of maize starch (BeMiller & Whistler 2009). This aspect is discussed in detail in Chapter 3. Induced mutation and genetic engineering offer possibilities of wider modifications, some of which have been realized in the creation of high-amylose/low-amylopectin starches.

1.3.2.2 Sweeteners and polyols from starch

Glucose and high-fructose syrups are sweeteners made from starch. Starch, being a polysaccharide, can be broken down into monomeric (i.e., D-glucose (also known commercially as dextrose)) and/or oligomeric (i.e., maltose, isomaltose, and maltodextrins) components that can be further used to produce other products, such as polyols, which may be available in liquid, crystalline, or dry powder forms (BeMiller & Whistler 2009). These products are discussed in detail in Chapters 13 and 14.

Polyols are obtained by hydrogenation of reducing sugars. Hydrogenation reduces aldehyde and keto groups to hydroxyl groups. Reduction of glucose leads to sorbitol, reduction of fructose to a mixture of mannitol and sorbitol, and reduction of maltose to maltitol. In the United States, these commercial products may be described as sugar alcohols. The United States Food and Drug Administration (FDA) classifies them as generally recognized as safe (GRAS) products. Polyols are widely used throughout the world because of either their combination of sweetness (which is generally less than that of sucrose) and low-calorie nature or their functional properties. Polyols do not contribute to dental caries.

1.3.2.3 Fermented products

Starch is broken down into syrups by acid- or enzyme-catalyzed methods (Chapter 13). The starch-derived carbohydrates in the syrups then act as substrates in a variety of fermentation processes. The development of novel technologies and the increasing industrial demand have made available many complex organic substances. Among the fermentation products made from starch hydrolysates are ethanol, organic acids, amino acids, enzyme preparations, antibiotics, and microbial polysaccharides.

Ethanol can be used as potable alcohol in beverages, as anhydrous ethanol in biofuels, and as a feedstock for the production of other chemicals. The production of ethanol from starch using a fermentation process by *Saccharomyces* yeast started most probably in beer-producing countries approximately in the 12th century. The production of "pure" ethanol (95% ethanol–5% water) by distillation has also been known for a long time. The development of more efficient distillation processes in the 19th century led to a large increase in the large-scale production of industrial ethanol. More recently, the process for making the 99.5% (bv) ethanol required for its use as a motor-fuel additive has been greatly improved. From 14.5 kg (32 lbs) of starch in a bushel of corn, about 9.5 L (2.5 gallons) of ethanol is produced (BeMiller & Whistler 2009).

Organic acids (including citric, lactic, fumaric, malic, and gluconic acids) have become large-scale food and industrial ingredients through ecofriendly processes. They are mainly produced from glucose via microbial fermentations. Major new plants for organic acid production were built by US starch producers in the 1980s and 1990s. Citric acid, which makes up almost 85% of the total volume of the organic acid market, is now used mainly in soft drinks, desserts, jams, jellies, candies, and frozen fruits. Lactic acid (made industrially in anaerobic fermentation processes using *Leuconostoc*, *Streptococcus*, and *Lactobacillus* bacteria) is used to make poly(lactic acid), food preservatives, and emulsifying agents.

Amino acids are attractive and promising bulk biochemicals with market capacity requirements constantly increasing. They are used in a wide range of industrial applications, including animal feed additives, flavor enhancers in human foods, and components of cosmetic, nutritional, and pharmaceutical products. Examples of amino acids produced at industrial scale are glutamate, lysine, threonine, tryptophan, methionine, and cysteine – all of which are made via microbial processes due to the economic and environmental advantages of such processes. Genetic engineering techniques maximize the yield of the target compounds. The most common bacteria used for amino acid production are *Corynebacterium glutamicum* and *Escherichia coli*.

Enzyme preparations are essential for biotechnological processes for the production of food and functional chemicals, and they can be made via fermentation of starch and starch hydrolyzates.

1.3.3 Prominent Starch-Producing Companies

Cargill (a family-owned company with 150 years of experience) provides food ingredients, agricultural solutions, and industrial products in 70 countries and regions, including Africa, Asia-Pacific, Europe, Latin America, Middle East, and North America.

Archer Daniels Midland Company (founded in 1902) is one of the world's leading processors and distributors of agricultural products including soybean, corn, and wheat for food and animal feed. ADM, which operates in three segments: agricultural services and oilseeds, carbohydrate solutions, and nutrition, has 247 processing plants in the United States and 121 overseas.

Ingredion (a global ingredients solutions company in about 120 countries) makes sweeteners, starches, starch-based materials, nutrition ingredients, and other biomaterials that are used in everyday products from foods and beverages to paper and pharmaceuticals.

Tate & Lyle PLC (in existence for more than 160 years) is another world leader in sweetening, providing texture, and fortification of foods. Among its products are starch-based sweeteners (including low- and no-calorie sweeteners), gut-friendly fibers, texturants, and stabilizers.

Grain Processing Corporation (a family-owned company founded in 1943) is among the world's major manufacturers of corn-based products, including alcohol, maltodextrins, corn syrup solids, and starches for the food, beverage, nutraceutical, personal care, and industrial markets.

AGRANA Beteiligungs-AG (an internationally oriented Austrian industrial company) is to add value to agricultural commodities to create industrial products for downstream industries.

Coöperatie Koninklijke Avebe U.A. (founded in 1919) operates production sites in The Netherlands, Germany and Sweden and is a main manufacturer of potato starch for the development of food products.

Roquette Frères is a French company that produces and sells ingredients for the food, nutrition, and health markets. It operates in over 100 countries and has more than 30 manufacturing sites.

Zhucheng Xingmao Corn Developing Co., Ltd. (established in 1993 in P.R. China) is a large-scale producer of corn-derived industrial and refined products, including 400+ varieties of corn starches, modified starches, sweeteners, dietary fibers, and organic acids.

BENEO GmbH (established in 2007 in Germany) provides natural ingredients for food, feed, and pharmaceutical applications in 80 countries. It operates six production sites in Germany, Belgium, The Netherlands, Italy, and Chile.

Tereos is France's largest sugar and ethanol producer. It also provides modified starch and starch-derived products.

Global Bio-chem Technology Group Company Limited is a Chinese, corn-based product company that produces a wide range of products for animal feed, foods, beverages, cosmetics, textiles, pharmaceuticals, and the chemical industry.

1.4 APPLICATIONS AND FUTURE PERSPECTIVES

Each year, starch becomes more and more important to human lives and economies because there are abundant sources of starches, because the sources are renewable, because starches are rather easily isolated (Chapter 5) and handled (making them relatively inexpensive), because new sources of starches continue to be developed, because of their ease of use, because of their nutritional value (Chapter 10), because of their multifunctionality, because of the ease of modification to improve functionalities and to add new functionalities, because of their versatility, because of their importance as ingredients in many processed foods, and because of their non-food applications (BeMiller & Huber 2015; Miao & BeMiller 2023).

Several attributes of starches and products derived from starches make them so valuable (BeMiller & Whistler 2009). Among these are that native starches are completely biodegradable (Chapter 17), and films, fibers, and thermo-plastic materials with varying degrees of hydrophobicity and biodegradability can be made from starches. Another advantage is that starches can be easily converted into glu-cose syrups, high-fructose syrups, high-maltose syrups, and maltodextrins, which have their own uses in food and other products (Chapter 13). Glucose syrups are used as feed-stocks for the production of other chemicals. Also, starch itself is used to produce potable, industrial, and fuel ethanol. Yet another is that starches can be easily and relatively inexpen-sively modified to improve their functional characteristics or to add new functional characteristics – in both cases, adding value to them. (Generally, only small degrees of modification produce significant changes in the attributes of the starch.) The growing application of modified starches as thickening, stabilizing, and gelling agents, binders, excipients, etc. is anticipated to continue to generate high demand for them (Miao & BeMiller 2023). The growth of the food and beverage industry is likely to be a major factor driving the demand. The food and beverage seg-ment accounted for the largest revenue share (more than 55%) in 2020 for starch-providing companies, and expanding demand for on-the-go food products and various beverages from the young and working population is expected to drive the segment.

Starches are used as binders to adjust pellet characteris-tics in animal feed, aquaculture feed, and pet food products. They can be used as milk alternatives. Functionalities of starch such as biodegradability (Chapter 17), biocompatibility, and gel-forming ability have resulted in its increased applications in the pharmaceutical industry. Starch-based polymers are used as an alternative to metals and ceramics in tissue engineering. Starches are used to enhance the strength of paper, as coating binders, as adhesives for bonding layers of corrugated board, and in various other applications such as in the production of organic chemicals, enzymes, and plastics. Starches also play an important role in the printing and finishing phases of the textile industry. The growing applications of starches in end-use indus-tries are expected to drive the growth of the market.

More complete understandings of the structures and prop-erties of starch molecules and granules, the differences in the structures and properties of starch molecules and granules in different starches, and how the structures and properties of starch molecules and granules can be altered to meet applica-tion needs are necessary for the expansion of starch uses (Chen et al. 2021; Jane 2007; Nakamura 2023; Tetlow & Bertoft 2020). The results of the efforts of scientists to understand starch and what it can be used for more fully are brought up to date in this handbook.

REFERENCES

BeMiller, JN & Huber, KC 2015. Physical modification of food starch functionalities. *Annual Review of Food Science and Technology*, 6, 19–69.

BeMiller, JN & Whistler, RL 2009. *Starch: Chemistry and Technology* (Third Edition). New York: Academic Press.

Bertoft, E 2017. Understanding starch structure: Recent progress. *Agronomy*, 7, 56.

Cai, T, Sun, H, Qiao, J, Zhu, L, Zhang, F, Zhang, J, Tang, Z, Wei, X, Yang, J, Yuan, Q, Wang, W, Yang, X, Chu, H, Wang, Q, You, C, Ma, H, Sun, Y, Li, Y, Li, C, Jiang, H, Wang, Q & Ma, Y 2021. Cell-free chemoenzymatic starch synthesis from carbon diox-ide. *Science*, 373, 1523–1527.

Chen, J, Hawkins, E & Seung, D 2021. Towards targeted starch modification in plants. *Current Opinion in Plant Biology*, 60, 102013.

Copeland, L & Hardy, KA 2018. Archaeological starch. *Agronomy*, 8, 4.

Crini, G, French, AD, Kainum, K, Jane, J-L & Szent, L 2021. Contributions of Dexter French (1918–1981) to cycloamylose/ cyclodextrin and starch science. *Carbohydrate Polymers*, 257, 117620.

Grand View Research 2021. Industrial starch market size, share & trends analysis report by source (corn, wheat, cassava, potato), by product (native, cationic, ethylated, acid modified, unmodified), by application, by region, and segment forecasts, 2020–2028 (Overview). San Francisco. https://www.grandviewresearch. com/industry-analysis/industrial-starch-market-report

Jane, J-L 2007. Structure of starch granules. *Journal of Applied Glycoscience*, 54, 31–36.

Kierulf, A, Mosleh, I, Li, J, Li, P, Zarei, A, Khazdooz, L, Smoot, J & Abbaspourrad, A 2024. Food LEGO: Building hollow cage and sheet superstructures from starch. *Science Advances*, 10, eadi7069.

Miao, M & BeMiller, JN 2023. Enzymatic approaches for structuring starch to improve functionality. *Annual Review of Food Science and Technology*, 14, 271–295.

Miao, M & Hamaker, BR 2021. Food matrix effects for modulating starch bioavailability. *Annual Review of Food Science and Technology*, 12, 169–191.

Miao, M, Jiang, B, Jin, Z & BeMiller, JN 2018. Microbial starch-converting enzymes: Recent insights and perspectives. *Comprehensive Reviews in Food Science and Food Safety*, 17, 1238–1260.

Mukerjea, R & Robyt, JF 2012. *De novo* biosynthesis of starch chains without a primer and the mechanism for its biosynthesis by potato starch-synthase. *Carbohydrate Research*, 352, 137–142.

Nakamura, Y 2023. A model for the reproduction of amylopectin cluster by coordinated actions of starch branching enzyme iso-forms. *Plant Molecular Biology*, 112, 199–212.

Seetharaman, K & Bertoft, E 2013. Perspectives on the history of research on starch. Part VI: Postscriptum. *Starch-Stärke*, 65, 107–111.

Tetlow, IJ & Bertoft, E 2020. A review of starch biosynthesis in relation to the building block-backbone model. *International Journal of Molecular Sciences*, 21, 7011.

Villwock, K & BeMiller, JN 2022a. The architecture, nature, and mys-tery of starch granules. Part 1: A concise history of early inves-tigations and certain granule parts. *Starch-Stärke*, 74, 2100183.

Villwock, K & BeMiller, JN 2022b. The architecture, nature, and mystery of starch granules. Part 2. *Starch-Stärke*, 74, 2100184.

Waigh, TA, Kato, KL, Donald, AM, Gidley, MJ, Clarke, CJ, & Riekel, C 2000. Side-chain liquid-crystalline model for starch. *Starch-Stärke, 52*, 450–460.

Wang, K, Henry, RJ & Gilbert, RG 2014. Causal relations among starch biosynthesis, structure, and properties. *Springer Science Reviews*, 2, 15–33.

Yang, J, Song, W, Cai, T, Wang, Y, Zhang, X, Wang, W, Chen, P, Zeng, Y, Li, C, Sun, Y & Ma, Y 2023. *De novo* artificial synthesis of hexoses from carbon dioxide. *Science Bulletin*, 68, 2370–2381.

Ziegler, GR, Creek, JA, & Runt, J 2005. Spherulitic crystalli-zation in starch as a model for starch granule initiation. *Biomacromolecules*, 6, 1547-54.

Biochemistry and Molecular Biology of Starch Synthesis

2

Ardha Apriyanto and Joerg Fettke

2.1 INTRODUCTION

All living organisms depend on the conversion of carbon and energy, whether they are autotrophic or heterotrophic. As a result, the main metabolism of almost all living organisms is centered on the metabolism of carbohydrates. The fact that polysaccharides are the most prevalent polymers in the biosphere is, therefore, not surprising. The majority of plants and algae and certain cyanobacteria can produce starch, which is known as the most prevalent storage carbohydrate on Earth (Apriyanto, Compart & Fettke 2022; Cenci et al. 2014). Starch often accumulates in water-insoluble particles known as *starch granules*. In prokaryotes, the major storage carbohydrate is hydro-soluble glycogen, which is comparable to starch in its chemical structure and has two kinds of interglycosyl linkages. Instead of glycogen, however, some prokaryotic algal strains form water-insoluble starch-like polysaccharides (Nakamura et al. 2005; Suzuki et al. 2013). In lower eukaryotic plants, especially algae, the evolution of a chloroplast-containing cell is complex (Ball et al. 2011; Ball, Colleoni & Arias 2015; Blaby-Haas & Merchant 2019; Facchinelli et al. 2013; Morell et al. 2018; Nowack & Weber 2018). Thus, starch metabolism is also variable; this includes the position of the starch granules within the cell, the glucosyl donor, and the starch turnover. In green algae (and their derivatives), starch is the most common carbon storage product, but some algae store reduced carbon compounds in other forms, e.g., as β-1,3-linked glucosyl moieties, as paramylon/paramylum, and as galactans. The latter are polysaccharides that are mainly or exclusively composed of galactosyl residues; galactans are typical storage carbohydrates

in red algae (Ciancia, Matulewicz & Tuvikene 2020; Usov 2011). Furthermore, paramylon/paramylum occurs in algae such as Euglena, which lacks all of the enzymes that synthesize and degrade starch. Instead, in the cytosol, Euglena forms hydro-insoluble associations of linear glucan chains composed of β-1,3-glucosyl residues that possess an average molecular weight of approximately $2 \times 10^{5\,Da}$ per molecule of β-1,3-glucan (Gissibl et al. 2019; Suzuki et al. 2018; Tanaka et al. 2017).

In many cases, higher plants form two types of starch: assimilatory starch (also known as transient starch) and reserve starch (also known as storage starch). These two types of starch differ in their quantities per plant, the size and structure of starch particles, as well as their metabolism, biochemistry, and biology. Reserve starch is far more abundant per plant than assimilatory starch, which is why the first is commonly used for agriculture. The overall shape and size distribution of starch granules varies significantly between the two kinds of starches. Usually, both particles exist as polymers consisting of numerous number of alpha-heterocyclic glucosyl residues. In both assimilatory and reserve starch (as well as in glycogen), there are only two types of bonds that connect glucosyl residues: α-1,4- and α-1,6-glycosidic bonds. In glycogen and starch, all glucosyl moieties are usually found in an alpha configuration, and the majority of them possess three free OH-groups that are located at carbon atoms 2, 3, and 6 of a glucosyl residue.

Typically, assimilatory starch (as well as reserve starch) is accumulated as hydro-insoluble particles; therefore, at least two different subcellular sites should be distinguished: one site is the surface or near-surface region and the other is the interior part of the particle. In terms of biochemistry, the two sites are largely different.

BOX 2.1

The biosynthesis of polysaccharides deviates strongly from that of most proteins: carbohydrate-based macromolecules are formed without any programmable system being involved. In (ribosomal) protein biosynthesis, the programmable system is the ribosomes that are located in the cytosol, plastids, and mitochondria. Each programmable system is instructed by the messenger RNA population in the compartment. By contrast, the kinetic properties of the synthesizing enzymes determine the generated polysaccharides. Therefore, the kinetic features—especially the glucan specificities—of the enzymes that synthesize amylopectin and/or amylose are essential for an understanding of the properties of starch. Likewise, the kinetic properties are of prime importance for the rate of starch degradation during the mobilization process.

In a formal sense, the biosynthesis of glycans that possess a precisely defined sequence of both sugar moieties and intersugar linkages (including the exact positions of the branching points) resembles that of non-ribosomal peptides, which are structurally and functionally diverse products of secondary metabolism. Non-ribosomal peptides are mainly formed by prokaryotes and fungi. They are synthesized by one or several large multidomain protein(s) collectively named non-ribosomal peptide synthase(s) (Miller & Gulick 2016; Süssmuth & Mainz 2017; Weissman 2015), but no messenger RNA and/or ribosome is/are involved. Each non-ribosomal peptide requires a distinct multidomain protein or, alternatively, a distinct set of proteins. Due to this mode of biosynthesis, the products are quite frequently cyclic, and the monomers incorporated into non-ribosomal peptides considerably exceed the 20 proteogenic amino acids (Reimer et al. 2019). Similarly, the formation of glycans having a strictly defined order of carbohydrates requires at least an equal number of distinct enzymes. Provided that the biosynthetic enzymes have a high degree of glycosyl acceptor specificity, various enzymes are functional only in one distinct order and, thereby, they synthesize a specific and clearly defined glycan.

2.2 INTRACELLULAR LOCATION OF STARCH GRANULES

Typically, assimilatory and reserve starches are hydro-insoluble particles, but their intracellular locations vary (Figure 2.1). In green algae and higher plants, assimilatory starch particles are deposited within the stroma of photosynthesis-competent chloroplasts. In green algae, such as *Chlamydomonas reinhardtii*, *Chlorella pyrenoidosa*, or *Scenedesmus quadricauda*, there are at least two plastidial sites at which assimilatory starch particles are formed: one is in close proximity to the pyrenoid (McKay & Gibbs 1991; Wang & Jonikas 2020; Zhan et al. 2018), while the other is independent of the position of the pyrenoid. This site frequently occurs as multiple copies and is similar to the chloroplasts of higher plants.

In the algae mentioned above, the pyrenoid is located inside the chloroplast, but in other species (primary plastids, with the exception of green algae and many secondary plastids), it is situated outside the chloroplast. In green algae, this non-membrane sub-microstructure is often surrounded by starch shields (Kuchitsu, Tsuzuki & Miyachi 1988; Mackinder et al. 2016; Ramazanov et al. 1994). Starch particles attached to a pyrenoid have a different shape and are often larger than those formed in the stroma independently of the position of the pyrenoid(s). Thus, the two types of assimilatory starch particles differ morphologically, and their precise metabolic functions also appear to be different (Ball et al. 2011).

Higher plants lack pyrenoids. In these plants, the chloroplast stroma is the principle site of assimilatory starch accumulation. Initial data exist that may indicate the exact position at which starch granule formation occurs in Arabidopsis mesophyll cells (Atkinson et al. 2024; Sharma et al. 2024).

Depending on the plant species, the number of assimilatory plastidial starch granules varies, but it is not known which factor(s) determine(s) this number. In the mesophyll cells of *Arabidopsis thaliana*, the starch granule numbers, in most cases, do not exceed approximately six to seven granules per chloroplast (Liu, Li & Fettke 2021; Malinova, Qasim et al. 2018).

2.3 PROPERTIES OF ASSIMILATORY STARCH GRANULES

Despite the wide variation in their size distribution and shape, the inner architecture in assimilatory starch granules seems to be evolutionarily conserved. Likewise, the inner structure of reserve starch granules appears to be similar. Under a light microscope, the internal structures of starch granules appear to possess an alternating sequence of light and dark rings with an approximate thickness of a few hundred to several hundred nanometers (Bertoft 2015 and the literature therein). The granular rings are also known as the 'growth rings' of starch granules. Although well known, the reason for their existence and their exact chemical and/or structural nature are not fully understood. The dark rings are thought to be areas enriched in α-1,6-bonds; therefore, they are more amorphous and less crystalline than the light rings. They seem to contain largely double helices of glucosyl chains that are relatively short. Amylose appears to be located mainly in the amorphous layers (thereby contributing to the amorphous nature of the dark rings), but some amylose may also be found in the more crystalline sections (Bertoft 2015). Thus, the more crystalline rings seem not to be structurally perfect. During incubation in diluted acids, the crystallinity of the remaining starch granules appears to be enhanced; therefore, it is concluded that the amorphous regions of the starch particles are eroded by this treatment but the more crystalline areas are not. Similarly, when cracked granules are treated with α-amylase, the rings appear thinner and more prevalent (Bertoft 2015; Fulton et al. 2002). Quite often, the more crystalline rings are named 'semi-crystalline' areas, as opposed to the 'amorphous' dark rings, but other designations for the two internal regions of the starch granule are also used (Bertoft 2015, 2017). Mutants or transgenic plants that strongly differ in their amylose content (so-called high-amylose and *waxy* starches) still possess the ring structure, indicating that amylose is not an essential component for the dark rings. However, it is plausible that the structure of the amylopectin in wild-type and *waxy* starches is non-identical.

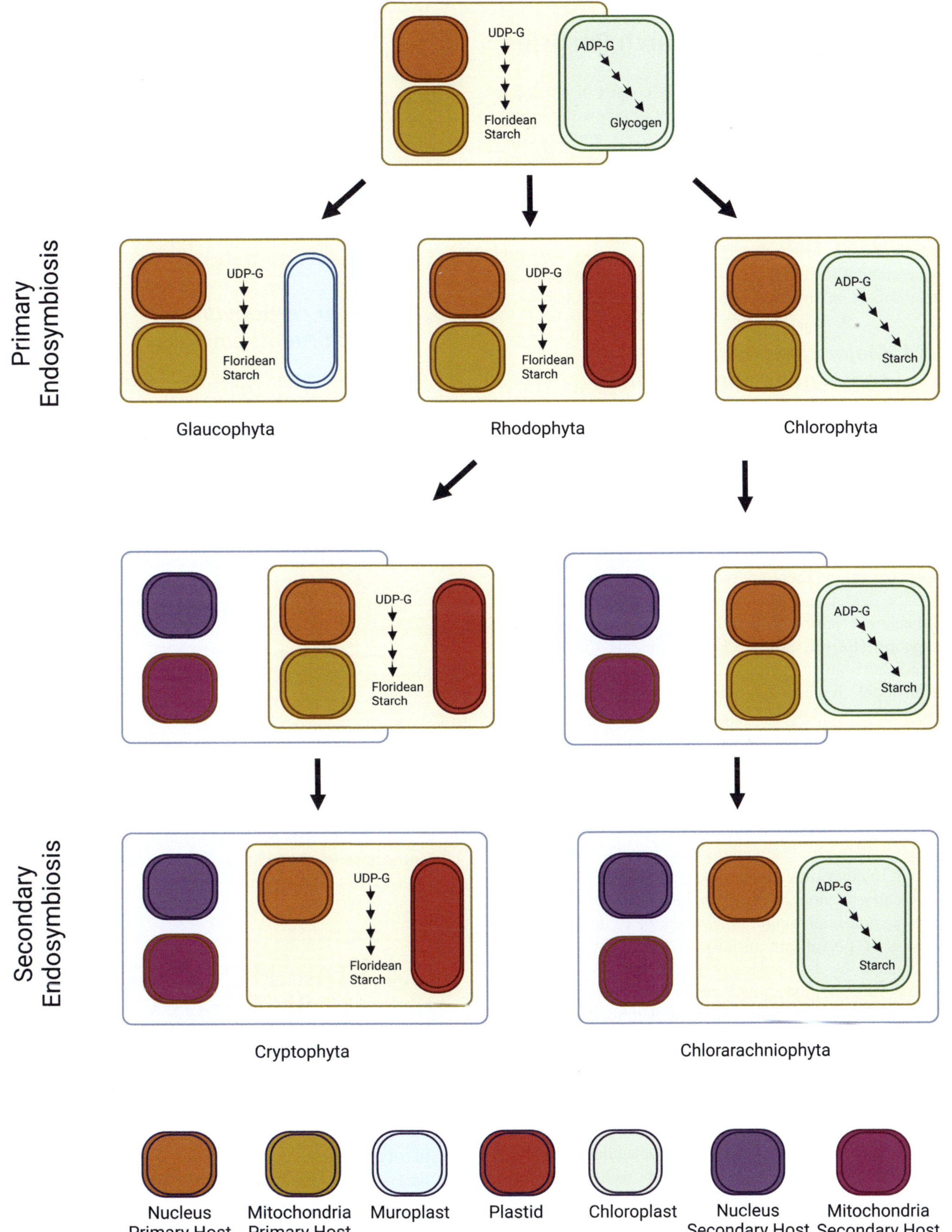

FIGURE 2.1 Schema of starch metabolism and granule formation in algae and green plants.

2.3.1 Main Carbohydrate Constituents of Starch Particles

Assimilatory starch particles essentially consist of amylopectin, and the amylose content is usually at or below the limit of detection. Similarly, in reserve starch particles, amylopectin is the major constituent. Depending on the plant species, this type of polyglucan accounts for at least 60% of the glucose-based carbohydrate content. By contrast, the reserve starch particles have reduced amylose amount. However, the composition of the starch granules also influences other parameters; thus, high amylose content results in more resistant starch, which has a positive impact on human health (Tiozon et al. 2023).

2.3.1.1 Major carbohydrate constituent: amylopectin

Amylopectin is a macromolecule type rather than a macromolecule. Several parameters, such as the molecular weight, are often available only as average values. Structurally, it may vary across the wild type and mutants (Frigård 2002).

A widely accepted structural model of amylopectin is still lacking. The same is true for the structure of the second type of polyglucan often found to be present in starch particles—amylose. Currently, there are two major amylopectin models: the cluster model and the more recently presented building block backbone notion, which describes the configuration of amylopectin (Bertoft 2015). The two models share the fact that the branching sites are packed, which promotes the creation of double helices. The two models also agree that the double helices are perpendicular to the exterior of the starch granulum. However, they differ significantly in the placements of the lengthy chains. In the cluster model, long chains are oriented similarly to double helices and can pierce two or more of them. In the building block backbone model, the lengthy chains are perpendicular to the double helices, forming a backbone within each amorphous lamella. At present, it is difficult to determine which model is correct.

Typically, amylopectin is a very large macromolecule with an average molecular weight in the order of 10^7 to 10^8 Daltons. This type of polyglucan usually exceeds that of amylose (if detectable) by 1–2 orders of magnitude (Bertoft 2015). Under natural conditions, amylopectin is a disperse molecule. As both aggregates and amylopectin degradation products occur (depending on the conditions of solubilization), precise molecular weight determinations are, to some extent, difficult to obtain. It has been repeatedly shown that in addition to the large amylopectin fraction, there are several other amylopectins that possess a significantly lower molecular weight. Depending on the plant species, the molecular weights of these smaller amylopectin molecules can be as low as approximately 11×10^4. The function(s) of the additional amylopectin molecule(s) is/are not yet clear. Currently, it is unclear if smaller molecules can continue to develop and eventually reach the size of 10^7 to 10^8 Daltons. In any case, the lower-molecular-weight amylopectins are minor constituents of the native starch granules (Bertoft 2015).

In amylopectin, clusters of branching points have been reported to occur. Likewise, it is known that the size distribution of amylopectin-derived chains tends to be considerably wider than that of glycogen. Therefore, the structure of amylopectin appears to differ from that of glycogen (Bertoft 2015 and the literature cited therein). Furthermore, in maize reserve starches, the ratio of small chains and long chains has been reported to differ in the A-type and B-type allomorphs. This implies that there are differences in the amylopectin structure between the starch particles of both allomorphs (Gérard et al. 2000). However, systematic studies are largely lacking (Bertoft 2015 and the literature cited therein).

2.3.1.2 Minor carbohydrate constituent: amylose

The other type component that frequently (but not always) constitutes starch particles is amylose. Similar to amylopectin, amylose represents a type of polyglucan macromolecule, rather than a polysaccharide molecule.

The average molecular weight of amylose is at least one order of magnitude lower than that of amylopectin. In reserve starch, amylose accounts for approximately 20%–30% of the dry weight of the starch particles (Hanashiro 2015). Amylose is a complicated combination of strictly linear and slightly branched molecules, with a much lower average degree of branching than amylopectin. Moreover, the branching points are not clustered; therefore, no double helices made up of vicinal glucan chains can be detected.

Similar to amylopectin, different types of chains can be distinguished in amylose, and they are designated as A- to C-type chains: type A and type B are strictly linear and branched side chains, respectively. In contrast to amylopectin, C-type chains (which contain the reducing end) can occur as both linear and branched chains. In amylose, they are significantly more abundant than in amylopectin. In the latter, only one C-type chain exists per amylopectin molecule.

Unfortunately, the exact position of amylose within the assimilatory and reserve starch granules is unknown.

2.4 STARCH DEVELOPMENT IN TISSUES

2.4.1 Assimilatory and Reserve Starch Turnover

The two types of starches differ not only in the mentioned parameters but also in their turnover (Figure 2.2).

Assimilatory starch reveals a diurnal turnover with synthesis in the light phase and degradation in the dark phase. This turnover is very well adjusted; thus, the amount of starch that is formed depends on the duration of the light phase. In addition, the degradation is precisely controlled; thus, the starch is typically almost completely degraded at the end of

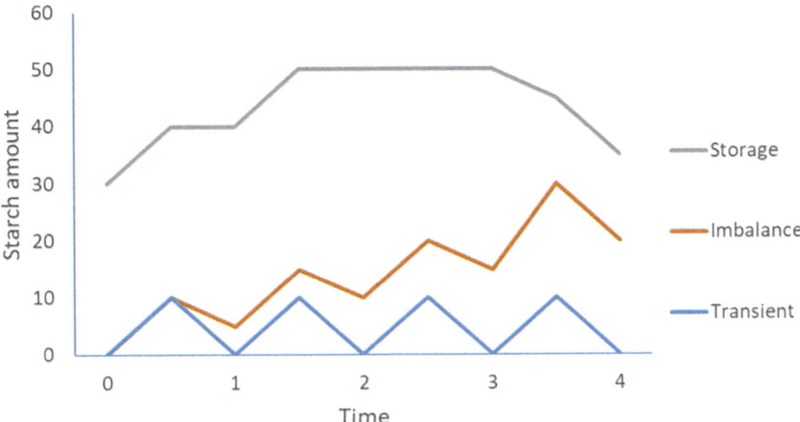

FIGURE 2.2   Starch turnover of transient/assimilatory, imbalanced, and storage/reserve starch.

the night, protecting the plant from carbon and energy starvation. The means by which this is sensed and regulated is largely unknown at present. However, several mutants have been described that show disturbances in this regulation. One example is included in Figure 2.2. Thus, increasing amounts of starch are accumulated over time. In principle, this is possible via either the additional synthesis of starch followed by degradation or vice versa.

By contrast, reserve starch formation typically takes place only once over a longer period, even weeks or months, followed by a phase in which the net starch turnover is null (rest phase) and a final phase that results in starch degradation to provide energy and carbon, typically for germination or sprouting.

2.4.2 Starch Formation in Photosynthetic Tissue of Higher Plants

In photosynthetic tissue, such as leaves and green stems, starch serves as an essential substrate for the production of energy. Starch biosynthesis is orchestrated through a complex sequence of biochemical pathways and molecular interactions, meticulously regulated to optimize the synthesis efficiency. The initiation and synthesis of starch occurs in the chloroplasts, facilitated by the absorption of solar radiation by chlorophyll pigments. In the subsequent darkness, the assimilatory starch usually is degraded. Other external parameters, such as the day/night durations and the temperature, are often not constant but vary throughout the year. In general, shorter nights lead to higher rates of assimilatory starch degradation. However, the system is able to compensate for a variety of external conditions.

Usually, the chloroplasts involved in carbon fixation store a part of the carbon fixed as starch particles. Both the biosynthesis and degradation of assimilatory starch have, therefore, frequently been studied with preparations of isolated chloroplasts or with protein extracts of isolated organelles (Fünfgeld et al. 2021; Hostettler et al. 2011; Kaiser & Bassham 1979; Neuhaus & Schulte 1996; Peavey, Steup & Gibbs 1977; Steup, Peavey & Gibbs 1976; Stitt & Heldt 1981; Stitt & Rees 1980).

Taken together, the data clearly show that photosynthesis-competent chloroplasts accumulate and degrade water-insoluble assimilatory starch particles. They possess the enzyme activity that is essential to synthesize and degrade starch—even that for functional starch turnover. The cytosolic steps are also important. Furthermore, they permanently possess the apparatus for the activation of proteins. All proteins involved in these paths are encoded by the nuclear DNA but not by the plastome, and proteins are imported post-transcriptionally from the cytosol into the chloroplasts.

In the photosynthetic tissue, starch synthesis is dependent on the diurnal cycle (Stitt & Zeeman 2012). The captured solar energy is transduced into chemical energy, predominantly in the form of adenosine triphosphate (ATP), via the photosynthetic processes. ATP subsequently functions as the primary energy molecule driving the synthesis of starch (Nakamura 2015).

The enzymatic pathway of starch synthesis starts with the conversion of glucose-1-phosphate to ADP-glucose, catalyzed by ADP-glucose pyrophosphorylase, a multimeric enzyme (Baris et al. 2009; Hädrich et al. 2011). ADP-glucose pyrophosphorylase, also known as AGPase, is a heterotetramer composed of two small and two large subunits. It has been discovered that small and large subunits have diverse roles in enzyme functioning. Small subunits have been proven to have both catalytic and regulatory activities, whereas large subunits are primarily responsible for controlling the allosteric characteristics of small subunits. It was demonstrated that the large subunit was incapable of assembling into a catalytically active oligomeric structure, but the small subunit could form a homotetramer with catalytic characteristics (Baris et al. 2009).

ADP-glucose acts as an activated glucosyl donor, pivotal for the subsequent polymerization reactions. Starch synthase (SS) then facilitates the incorporation of glucose units from ADP-glucose to the nascent polysaccharide chain, establishing α-1,4-glycosidic bonds that form the primary linear structure of starch. This enzyme exists in multiple isoforms, each contributing distinctively to the polymerization and modification of starch. The further diversification of the starch structure is accomplished by the action of the starch branching enzyme (SBE), which introduces α-1,6-glycosidic linkages by cleaving α-1,4 linkages and transferring the resultant oligosaccharide chains to create branch points. This branching is crucial in modulating the physicochemical properties of starch, such as its solubility and digestibility.

The regulatory mechanisms of starch synthesis and degradation, involving hormonal signals, sugar sensing, and enzymatic activity coordination, govern the dynamic process. These regulatory processes ensure that the starch biosynthesis in photosynthetic tissue aligns with the metabolic demands of the plant, providing a critical energy reservoir to support growth and survival during periods of diminished light or nutrient supplies (MacNeill et al. 2017).

Interestingly, the starch granule size in photosynthetic tissue is usually smaller than that in non-photosynthetic tissue. Regarding the starch structure, the predominant allomorphic form of starch within this tissue is the B type, characterized by a less dense structure compared to the A type (Imberty et al. 1988). This structural characteristic is advantageous for rapid starch turnover, facilitating swift degradation when required by the plant's metabolic needs.

Elucidating the biochemical and molecular mechanisms underpinning starch biosynthesis in photosynthetic tissue yields profound insights into plants' physiology, offering potential strategies to enhance agricultural productivity and resilience against environmental stresses.

2.4.3 Starch Formation in Non-photosynthetic Tissue

In non-photosynthetic tissue, starch also serves as energy storage. While starch synthesis is commonly associated with photosynthetic tissue, such as the leaves, it is fascinating to explore its development in non-photosynthetic tissue as well. Non-photosynthetic tissue, such as the tubers, seeds, and fruits, also undergoes starch synthesis to serve as a readily available energy source. In these types of tissue, starch synthesis also occurs in the plastids; the specialized plastid organelles that accumulate starch are known as amyloplasts. These organelles are responsible for the creation and storage of starches and, in most cases, its degradation (Nakamura 2015).

The process of starch synthesis in non-photosynthetic tissue follows a similar pathway to that in photosynthetic tissue, but with some unique characteristics. The carbon source for starch synthesis is mainly represented by sucrose (MacNeill et al. 2017). The first step in starch synthesis is the formation of glucose-1-phosphate from glucose-6-phosphate, catalyzed by the enzyme phosphoglucomutase. This glucose-1-phosphate then undergoes conversion to ADP-glucose, the key precursor for starch synthesis. The enzyme ADP-glucose pyrophosphorylase is responsible for catalyzing this conversion, and its activity is tightly regulated to ensure proper starch synthesis. Furthermore, the localization of ADP-glucose pyrophosphorylase can differ. For example, in potato, the activity is localized in the amyloplasts, whereas in cereals, the most activity is localized in the cytosol; therefore, here, a transporter of ADP-glucose in the plastidial envelope is needed (Soliman, Ayele & Daayf 2014). Once ADP-glucose is formed, it serves

as a substrate for SSs, which elongate the growing starch chain. Several isoforms of SSs exist, contributing to the diversity of the starch structures found in different types of tissue (Apriyanto et al. 2022; Mérida & Fettke 2021).

Moreover, in non-photosynthetic tissue, the balance between the activities of different isoforms plays a vital role in defining starch composition and characteristics. Another enzyme involved in starch synthesis is the SBE. SBE introduces branch points into the growing starch chain, leading to the formation of amylopectin, a highly branched starch molecule. The degree and pattern of branching are crucial for the physicochemical properties of starch in non-photosynthetic tissue. The regulation of starch synthesis in non-photosynthetic tissue is complex and involves various factors. End-product inhibition, metabolite sensing, and hormonal regulation are some of the mechanisms that contribute to the fine-tuning of starch synthesis in these types of tissue.

Compared to those found in photosynthetic tissue, the starch granules in non-photosynthetic tissue are usually larger and more spherical. The starch allomorphs inside the non-photosynthetic tissue are usually of the A type or mixed with the B type (referred to as the C-type allomorph) (MacNeill et al. 2017). The A-type allomorph is usually generated by shorter glucan chains that are evenly and tightly packed, hence excluding more water (Imberty et al. 1988). This is due to the low turnover of the starch degradation process in this tissue or the fact that it acts as a long-term reserve starch.

Understanding the biochemistry and molecular biology of starch synthesis in non-photosynthetic tissue provides useful knowledge about the mechanisms behind energy storage and utilization in plants. Moreover, this knowledge can have practical applications in crop improvement and the development of starch-based products with tailored properties.

The general starch biosynthetic pathway in photosynthetic and non-photosynthetic tissue can be seen in Figure 2.3. As an example of non-photosynthetic starch metabolism, the situation in potato tubers is shown. The differences in the biosynthetic pathways of starch metabolism in photosynthetic and non-photosynthetic tissue can be seen in Table 2.1.

2.5 ENZYME-CATALYZED REACTIONS OF STARCH SYNTHESIS

Starch metabolism in plants is essentially divided into two steps: synthesis and degradation. As previously mentioned, starch has two types of linkages: α-1,4- and α-1,6-glycosidic bonds. Therefore, most enzymes related to starch metabolism act on these glycosidic bonds. In fact, the synthesis and degradation processes cannot be separated, as it seems that at least some proteins/enzymes have overlapping functions. Furthermore, the action of the enzymes is influenced by several proteins, with currently unknown or missing catalytical functions.

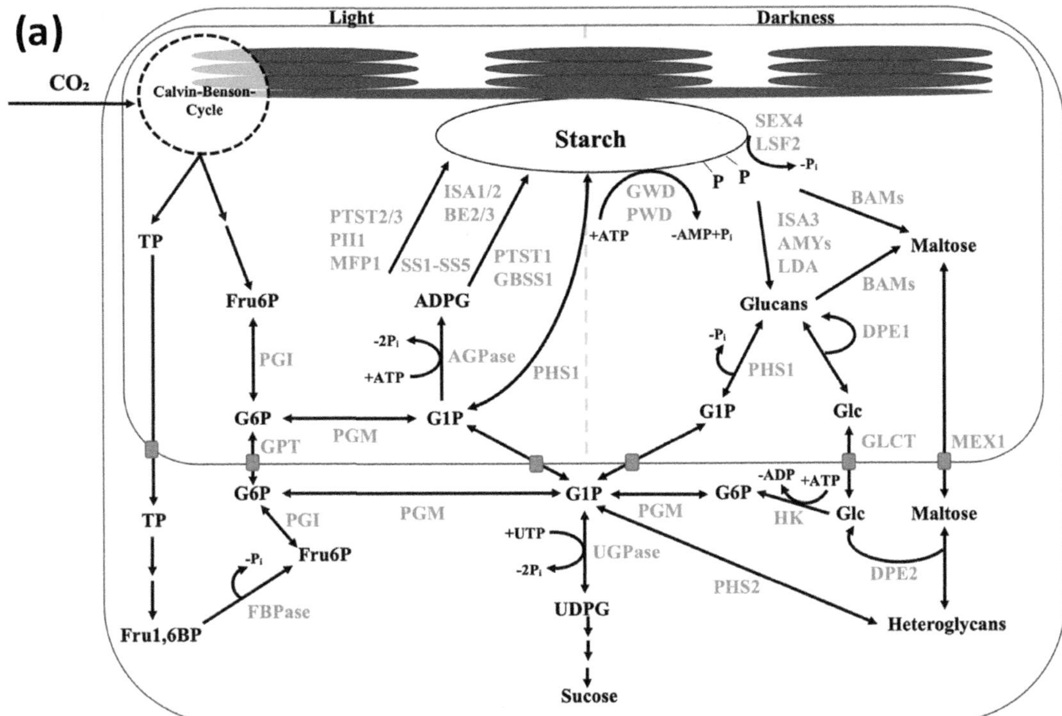

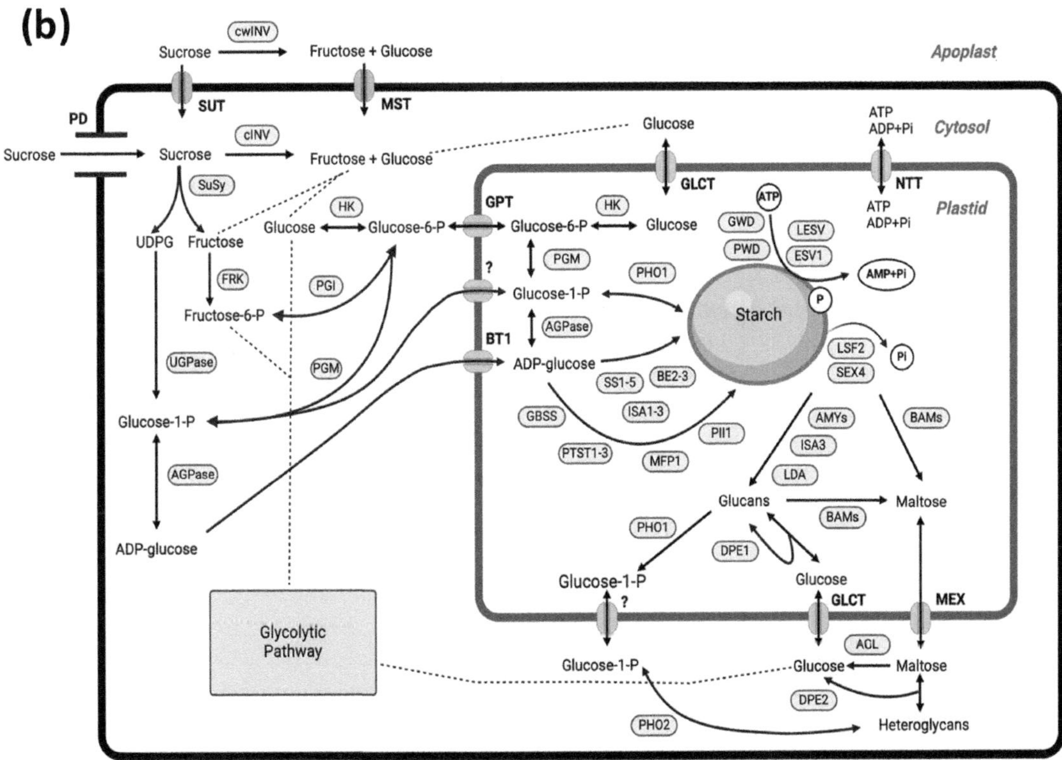

FIGURE 2.3 Schematic illustration of the starch biosynthetic pathway in photosynthetic and non-photosynthetic tissue. (a) The synthesis of transitory starch in the leaf tissue of *Arabidopsis thaliana*. (b) The synthesis of reserve starch in the storage tissue (tuber) of *Solanum tuberosum*. TP, triose phosphate; Fru1,6BP, fructose 1,6-bisphosphate; Fru6P, fructose 6-phosphate; G6P, glucose 6-phosphate; G1P, glucose 1-phosphate; ADPG, adenosine 5′-diphosphate glucose; UDPG, uridine 5′-diphosphate glucose; PGI, phosphoglucoisomerase; PGM, phosphoglucomutase; FBPase, fructose 1,6-bisphosphatase; UGPase, UDPG pyrophosphorylase; AGPase, ADPG pyrophosphorylase; SS, starch synthase; BE, branching enzyme; GBSS, granule-bound SS; PTST, protein targeting starch; PII, protein involved in starch initiation; MFP1, MAR-binding filament-like protein 1; ISA, isoamylase; LDA, limit dextrinase (debranching enzymes); GWD, α-glucan, water dikinase; PWD, phosphoglucan, water dikinase; SEX4, starch excess 4 and LSF2, like sex four (phosphatases); BAM, β-amylase; AMY, α-amylase; DPE, disproportionating enzyme; PHS/PHO, α-glucan phosphorylase; HK, hexokinase; GPT, glucose phosphate transporter; MEX, maltose exporter; GLCT, glucose transporter. (The figure was adapted from Apriyanto et al. 2022 and Flores-Castellanos & Fettke 2023.)

TABLE 2.1 Comparison of starch biosynthetic pathways in photosynthetic and non-photosynthetic tissue

STARCH-RELATED COMPARISON	PHOTOSYNTHETIC TISSUE	NON-PHOTOSYNTHETIC TISSUE
Primary function	Main energy	Reserve energy
Type of starch	Transient	Storage
Organ	Leaf, stem, and other photosynthetic active organs and tissue	Root, inner parts of fruit, seed, stem, and other non-photosynthetic active organs and tissue
Cell type	Autotroph	Heterotroph
Synthesis location	Chloroplast	Plastid (amyloplast)
Time of accumulation	Day	Light-independent
Time of degradation	Night	Light-independent
Regulation	Diurnal cycle	Demand-driven
Carbon donor	Photosynthate	Sucrose
Starch granule size/shape	Usually smaller/flat Conserved	Usually larger/more spherical Highly variable, including combined granules
Starch allomorph	Mostly B type	Mostly A and C type

Thus, the generation of the surface structures of starch granules, where the processes of synthesis and degradation mainly take place, is also dependent on non-catalytically active proteins and phosphorylation, the only known covalent modification of starch (see below).

2.5.1 Starch Synthesis

Generally, starch synthesis requires a subset of enzymes (Fettke & Fernie 2015; Streb & Zeeman 2012), including SSs (EC 2.4.1.2), plastidial phosphorylases (PHS1; EC 2.4.1.1), SBEs (EC 2.4.1.18), and starch debranching enzymes (DBEs), which are isoamylases (ISAs; EC 3.2.1.68) and limit dextrinases (LDAs; EC 3.1.2.41).

Recently, it was also found that starch granule initiation requires a subset of proteins, such as protein targeting starch 1–3 (PTST1–3) (Seung et al. 2015, 2017), protein involved in starch initiation 1 (PII1) (Vandromme et al. 2019), and MAR-binding filament-like protein 1 (MFP1) (Seung et al. 2018) and a glucan primer. The origin of the glucan primer is currently unknown. Furthermore, it seems that there is not only one method of starch initiation; this was found for both transitory starch (e.g., Arabidopsis (Li et al. 2022; Malinova & Fettke 2017; Mérida & Fettke 2021)) and storage starch (e.g., wheat (Kamble et al. 2023), oil palm (Apriyanto et al. 2023)).

The α-1,4 linkages between the glucose units during the elongation of the primer glucans are generated by SSs and, in part, by the plastidial phosphorylase's (PHS1) action. The starting point for starch synthesis is the production of ADP-glucose, a substrate for all SSs, via the ADP-glucose pyrophosphorylase (AGPase; EC 2.7.7.27) (Fettke & Fernie 2015; Mérida & Fettke 2021; Smith & Zeeman 2020; Streb & Zeeman 2012). SSs catalyze the glucosyl moieties' transfer from ADP-glucose to the non-reducing end of a glucan primer connected by 1,4-glucosidic bonds. The soluble SSs are involved in the formation of amylopectin, whereas amylose synthesis depends on granule-bound starch synthases (GBSS; EC 2.4.1.242).

Recently, it was found that at least six isoforms of SSs exist in plants (Nougué et al. 2014). These include five isoforms of SSs (SSI–SSV) and one isoform of GBSS. These genes have been widely characterized regarding their roles during starch synthesis (Pfister & Zeeman 2016). Four isoforms of SSs (SSI–SSIV) and GBSS have glycosyltransferase catalytic activity for the elongation of glucan chains (Lombard et al. 2014; Mérida & Fettke 2021). However, SSV seems to have no catalytic activity, but it is involved in starch granule initiation (Abt et al. 2020). Another SS isoform, SSIV, influences the initiation of starch granules and appears crucial in maintaining a consistent number of granules per chloroplast (Malinova, Qasim et al. 2018; Mérida & Fettke 2021). There is some evidence that SSI, SSII, and SSIII preferentially elongate the shortest, medium, and longest glucan chains of the amylopectin molecule, respectively (Fujita et al. 2006, 2012; Nakamura 2015; Zhang et al. 2008). In addition, PHS1 seems to elongate starch-related glucan chains (Flores-Castellanos & Fettke 2023; Malinova & Fettke 2017; Mérida & Fettke 2021) by catalyzing a reversible reaction that involves utilizing glucose-1-phosphate (G1P) as a substrate and transferring a glucose unit to the non-reducing ends of α-1,4-glucan chains, resulting in the release of inorganic phosphates (Malinova et al. 2014; Shoaib et al. 2021).

The SBEs establish the α-1,6-branches and are an important determinant of the starch structure. SBE catalyzes chain transfer by means of the cleavage of an α-1,4 linkage following the condensation of an α-1,6-linkage (Streb & Zeeman 2012).

During the synthesis of starch, degrading enzyme activity is essential—especially that of DBEs—for the generation of the clustered distribution of α-1,6-glycosidic bonds. DBEs include two types of enzymes, ISAs and LDAs (Streb & Zeeman 2012). In most cases, ISAs seem to be more important than LDAs for proper starch processing (Delatte et al. 2006). However, it is currently not well understood how the branching enzymes and the DBEs are regulated and targeted during starch synthesis, ensuring that the correct amylopectin molecules are formed.

2.5.2 Starch Degradation

Two plastidial dikinases, α-glucan water dikinase (GWD; EC 2.7.9.4) and phosphoglucan water dikinase (PWD; EC 2.7.9.5) (Mahlow et al. 2014; Ritte et al. 2002, 2006), as well as two dual-specificity phosphatases, starch excess four (SEX4; EC 3.1.3.48) (Kötting et al. 2009) and like starch excess four 2 (LSF2; EC 3.1.3.48) (Santelia et al. 2011), were identified as having important roles in the starch degradation process. Thus, a starch phosphorylation–dephosphorylation cycle takes place at the starch granule's surface, e.g., amylopectin glucan chains (Compart, Apriyanto & Fettke 2023; Mahlow, Orzechowski & Fettke 2016; Smith & Zeeman 2020). This is not only relevant for the daily turnover of transitory starch but also occurs in the case of storage starch, as phosphorylation was also detected here (Lloyd & Kossmann 2015). In the case of storage starch, this occurred during synthesis; however, for transitory starch, phosphorylation during synthesis has also been reported (Hejazi, Mahlow & Fettke 2014).

Recently, it was also found that starch degradation was influenced by protein early starvation 1 (ESV1; At1g42430) and its homolog, like early starvation 1 (LESV; At3g55760) (Feike et al. 2016; Malinova, Mahto et al. 2018; Singh et al. 2022). GWD and PWD selectively catalyze the phosphorylation of the C6 and C3 positions, respectively (Hejazi et al. 2014; Kötting et al. 2005; Ritte et al. 2002, 2006). Both enzymes are required for the normal metabolism of transitory starch and for the development of the entire plant (Hejazi et al. 2014). The action of GWD promotes the phase transition of glucan chains on the surface from a highly ordered state to a more solubilized state, rendering them more susceptible to the action of the downstream degradative enzymes (Hejazi et al. 2008; Mahlow et al. 2016). PWD prefers to phosphorylate mainly glucan structures formed by the action of GWD and therefore acts downstream of GWD (Hejazi, Steup & Fettke 2012; Kötting et al. 2005; Ritte et al. 2006). However, the action of PWD is not restricted to phosphoglucans (Fettke et al. 2009; Hejazi et al. 2009). This phosphorylation of starch was found to be influenced by ESV1 and LESV (Malinova, Mahto et al. 2018; Singh et al. 2022).

In order to catalyze the degradation of α-1,4 and α-1,6 linkages, the dephosphorylation of starch is also strongly required. SEX4 and LSF2 remove the phosphorylation at C6 and C3, respectively (Kötting et al. 2009; Santelia et al. 2011). This dephosphorylation is required to allow the progressive degradation of linear chains by beta-amylase (BAM; EC 3.2.1.2) because it can only continue its processive enzymatic activity until it comes across a phosphorylated glucosyl unit (Edner et al. 2007; Hejazi et al. 2008, 2009). The SEX4 and LSF2 enzymes (glucan phosphatase family) remove covalently bound phosphate esters, allowing BAM to continue its activity and accomplish the breakdown process (Silver, Kötting & Moorhead 2014).

In addition, the hydrolases ISA (EC 3.2.1.68), LDA (EC 3.1.2.41), and alpha-amylases (AMY; EC 3.2.1.1) release soluble phosphorylated glucans that cannot be completely converted into maltose by BAM (Smith & Zeeman 2020). In contrast to beta-amylases that act on α-1,4-glycosidic bonds from the non-reducing ends of polysaccharides, alpha-amylase cleaves the α-1,4-glycosidic linkages in polysaccharides containing three or more α-1,4-linked D-glucose units. Furthermore, during synthesis, the branches (α-1,6 linkages) are cleaved by DBE, including the two aforementioned types of enzymes, limited dextrinases (LDAs; EC 3.1.2.41) or pullulanase and ISAs (EC 3.2.1.68). ISAs can debranch amylopectin and glycogen, but not pullulan. By contrast, LDAs can debranch amylopectin and pullulan, but not glycogen (Hii et al. 2012).

Disproportionating enzymes, also known as 4-α-glucanotransferase, are another important type of enzyme for starch degradation. Disproportionating enzymes (DPE; EC 2.4.1.25) transfer 1,4-α-D-glucan units to other 1,4-α-D-glucan units, and this reaction is reversible. Maltooligosaccharides are the preferred substrates *in vitro*. Two isoforms of DPE were identified in plants—DPE1 is a plastidial enzyme and DPE2 is a cytosolic enzyme (Fettke et al. 2009; Smith & Zeeman 2020). One role of DPE1 is to create longer maltooligosaccharides from shorter ones so that the longer ones can become substrates for other starch degradation enzymes, such as beta-amylase (EC 3.2.1.2) and plastidial phosphorylase (EC 2.4.1.1). DPE2 is involved in metabolizing maltose exported from the chloroplast during starch degradation. As a 4-α-glucanotransferase, DPE2 transfers a donor glucosyl unit to an acceptor, releasing glucose in the process (Fettke et al. 2006; Smith & Seeman 2020).

Overall, several isoforms exist in planta for most of the mentioned enzymes' activity, making the processes of starch synthesis and degradation much more complex. Furthermore, these isoforms show different substrate affinity, kinetics, and interactions with other proteins/enzymes, which modulate their activity.

2.5.3 Non-Catalytical Proteins Involved in Starch Turnover

In the last decade, new starch-related proteins have been discovered that do not form interglucosyl bonds but are clearly starch-relevant. These proteins, with no enzymatic activity, are able to specifically alter the activity of the synthesizing and degrading enzymes, including ESV1 (Feike et al. 2016; Malinova, Mahto et al. 2018), like early starvation 1 (LESV) (Feike et al. 2016), PTST1 (Seung et al. 2015), like sex four 1 (LSF1) (Schreier et al. 2019), MAR-binding filament-like protein 1 (MFP1) (Sharma et al. 2024), protein involved in starch initiation 1 (PII1) (Vandromme et al. 2019), and β-amylase 4 (BAM4) (Fulton et al. 2008). The overall function can vary but is currently largely unknown.

2.6 CHARACTERIZATION OF STARCH-SYNTHESIZING ENZYMES

As previously mentioned, starch synthesis is a complex process that is also influenced by the degradation processes and involves the coordinated action of several enzymes.

TABLE 2.2   *In vitro* structural characterization of starch-related enzymes

NO.	ENZYME	SPECIES	METHOD	REFERENCE
1	GBSSI	*Oryza sativa*	X-ray	Momma and Fujimoto (2012)
2	SSI	*Hordeum vulgare*	X-ray	Cuesta-Seijo et al. (2013)
3	SSIV	*Arabidopsis thaliana*	X-ray	Nielsen et al. (2018)
4	SEX4	*Arabidopsis thaliana*	X-ray	Vander Kooi et al. (2010)
5	LSF2	*Arabidopsis thaliana*	X-ray	Meekins et al. (2013)
6	SBEI	*Oryza sativa*	X-ray	Gavgani et al. (2022)

These enzymes play crucial roles in the biosynthesis of starch, ensuring its proper structure and functionality. Understanding the characterization of starch-synthesizing enzymes is key to unraveling the intricate mechanisms behind starch synthesis.

Characterizing starch-synthesizing enzymes involves the observation of their biochemical properties, their substrate specificity, their activity under different conditions, and their roles in the biosynthesis and modification of starch. Overall, there are three major approaches to the characterization of starch-synthesizing enzymes. These include *in vitro* characterization, *in vivo* characterization, and *in silico* characterization.

2.6.1 *In Vitro* Characterization

In vitro characterization is the analysis of proteins outside a living organism and thus in a controlled laboratory environment. *In vitro* studies involve purified proteins or proteins in cell extracts and study the biochemical and biophysical properties of such proteins. This approach allows for detailed examination under various conditions, without the complexity of a living organism.

The ability to conduct experiments under controlled conditions allows for the precise manipulation of variables. This is crucial in isolating specific interactions or functions without the complexity of cellular or organismal contexts. Additionally, *in vitro* methods often support high-throughput screening, facilitating the rapid analysis of a large number of conditions or compounds. However, there are limitations to these techniques. They may not fully replicate the physiological conditions found in living organisms, potentially leading to differences in protein structure and function that could be vital in a natural setting. Furthermore, the absence of a cellular context means that important regulatory mechanisms, modifications, and protein–protein interactions may be overlooked, which could skew the understanding of a protein's role *in vivo*.

Several common methods enable this type of characterization, such as the following.

2.6.1.1 *Structural characterization*

The structural characterization of starch-synthesizing enzymes can be performed using several methods, such as X-ray crystallography, NMR spectroscopy, or cryo-electron microscopy (cryo-EM). X-ray crystallography is a powerful technique that provides high-resolution 3D structures of proteins in crystalline form, which is crucial in understanding molecular interactions at the atomic level. Complementing this, NMR spectroscopy offers excellent information into the structure and dynamics of proteins in solution, enabling researchers to explore their functional states in environments that closely mimic physiological conditions. Additionally, cryo-EM has become an indispensable tool in determining the structures of large protein complexes that are difficult to crystallize. This method has significantly advanced our understanding of the molecular machinery involved in complex biological processes, filling the gaps left by other structural determination methods. Together, these techniques form a comprehensive toolkit for the elucidation of the complex architectures and functionalities of proteins, pivotal in fields ranging from drug design to enzymology. At present, X-ray diffraction is the most commonly used tool for the 3D structural characterization of starch-related enzymes. However, NMR and cryo-EM are expected to be more widely used in the future. Examples of the *in vitro* structural characterization of starch-related enzymes can be seen in Table 2.2.

2.6.1.2 *Functional characterization*

2.6.1.2.1 *Enzyme assays*
Enzyme assays are crucial in measuring enzyme activity, kinetics, and substrate specificity, providing detailed insights into the catalytic abilities of enzymes under various conditions. The glucan microarray represents an alternative choice for the enzymatic assaying of starch. Complementing this, ligand binding studies employ techniques such as isothermal titration calorimetry (ITC) and surface plasmon resonance (SPR) to assess the binding affinity and kinetics between proteins and their ligands. In addition, affinity gel electrophoresis (AGE) can be used; this technique modifies the standard electrophoresis process to include a specific ligand within the gel matrix. It can give information about the binding affinity and specificity of proteins under non-denaturing conditions. Examples of functional characterization using these methods on starch-related enzymes can be seen in Table 2.3.

2.6.1.2.2 *Post-translational modifications*
Post-translational modifications (PTMs) play an essential part in modulating the function, stability, and localization of proteins within cells. Recently, it has been found that the PTMs of starch-related enzymes improve the starch quality in rice grains (Adegoke et al. 2021).

Mass spectrometry (MS) is a powerful analytical technique used extensively to identify and characterize these modifications, such as phosphorylation and succinylation. By providing

TABLE 2.3 *In vitro* functional characterization of starch-related enzymes

NO.	ENZYME	SPECIES	SUBSTRATE	METHOD	REFERENCES
1	GBSSI	*Hordeum vulgare*	Maltooligosaccharide	Glucan microarray	Cuesta-Seijo et al. (2015)
2	SSI	*Hordeum vulgare*	Maltoheptaose, β-Cyclodextrin	SPR	Wilkens et al. (2014)
3	SSIIa	*Zea mays*	Amylopectin	AGE	Liu et al. (2012)
4	SSIII	*Arabidopsis thaliana*	Starch	Adsorption assay	Valdez et al. (2008)
5	SSIV	*Hordeum vulgare*	Maltooligosaccharide	Glucan microarray	Cuesta-Seijo et al. (2015)
6	SEX4	*Arabidopsis thaliana*	Amylopectin Amylose β-Cyclodextrin	AGE SPR	Wilkens et al. (2016)
7	LSF2	*Arabidopsis thaliana*	Amylopectin Amylose β-Cyclodextrin	AGE SPR	Wilkens et al. (2016)
8	SBEI	*Solanum tuberosum*	Amylopectin Amylose Glycogen	AGE	Mouille et al. (1996)
9	GWD1	*Solanum tuberosum* *Elaeis guineensis*	Cyclodextrin Starch Starch	SPR ITC AGE Radioactivity Radioactivity	Compart, Apriyanto and Fettke (2023); Glaring et al. (2011); Hejazi et al. (2012); Mikkelsen, Suszkiewicz and Blennow (2006)
10	PWD	*Arabidopsis thaliana*	Cyclodextrin Starch	SPR Radioactivity	Christiansen et al. (2009); Hejazi et al. (2012)

TABLE 2.4 *In vitro* functional characterization related to PTMs of starch-related enzymes

NO.	PTM	ENZYME	SPECIES	METHOD	REFERENCE
1	Phosphorylation	SSI, GBSSI, SBEI, SBEII-a, and SSII-a	*Triticum aestivum*	LC-MS/MS	Chen et al. (2016)
2	Succinylation	AGPS, AGPL, SBEI, SBEII-b, PUL, and PHO	*Oryza sativa*	LC-MS/MS	Meng et al. (2019)

precise mass measurements and fragmentation patterns, MS can reveal the presence of chemical modifications on specific amino acid residues, providing insights into the molecular mechanisms that control proteins' properties. This technique enables scientists to explore how modifications such as phosphorylation alter enzyme activity or how glycosylation impacts proteins' stability and distribution within a cell. The most common method used for this type of characterization is liquid chromatography coupled with tandem mass spectrometry (LC-MS/MS).

Examples of *in vitro* functional characterization related to PTMs on starch-related enzymes can be seen in Table 2.4.

2.6.1.2.3 Protein–protein interactions

Additionally, protein–protein interaction assays, including pull-down assays, co-immunoprecipitation, and yeast two-hybrid assays, are employed to study the interactions between proteins. These methods help to elucidate the networks and pathways that proteins are involved in, contributing significantly to our understanding of cellular processes and mechanisms. Together, these assays provide a comprehensive toolkit for the exploration of the multifaceted roles that proteins play in biological systems, offering valuable insights that drive advancements in both basic and applied research.

Examples of functional characterization related to protein–protein interactions in starch-related enzymes can be seen in Table 2.5.

2.6.2 *In Vivo* Characterization

In vivo characterization involves studying proteins within their natural context in living organisms or cells, namely plants and algae, providing insights into their physiological roles, regulation, and interactions in a more complex biological setting.

This type of characterization offers distinct advantages but faces certain limitations with regard to studying proteins within plant biological systems. The primary advantage of this approach is that it allows researchers to observe proteins in the complex environment of a living plant/algae cell or entire organism. This context is critical, as it encompasses the full range of physiological interactions and modifications that proteins undergo, providing a comprehensive understanding of their functions and interactions as they occur naturally.

However, the *in vivo* situation also introduces significant challenges. It is more difficult to control and manipulate specific variables within the complex interplay of living systems, which can affect the clarity and interpretability of experimental results. Additionally, genetic redundancy—where multiple genes perform the same function—and compensatory mechanisms can obscure the roles of individual proteins, making it challenging to determine their unique contributions to an organism's physiology. These limitations necessitate careful experimental design and often require complementary *in vitro* studies to discern specific protein functions and interactions.

TABLE 2.5   *In vitro* functional characterization related to protein–protein interactions in starch-related enzymes

NO.	SPECIES	PROTEIN INTERACTION	METHOD	REFERENCES
1	*Arabidopsis thaliana*	GBSS/PTST2	Pull-down	Seung et al. (2015)
2	*Hordeum vulgare*	SSI/SSIIa/SBEIIa/SBEIIb and SBEIIa/SBEIIb/SBEI/PHO	Co-immunoprecipitation	Ahmed et al. (2015)
3	*Arabidopsis thaliana*	SSIV/PII1	Yeast two-hybrid	Vandromme et al. (2019)

TABLE 2.6   *In vivo* characterization performed via genetic manipulation in starch-related enzymes

NO.	ENZYME	SPECIES	METHOD	EFFECT	REFERENCES
1	PHS1/SSIV	*Arabidopsis thaliana*	Knockdown	Starch granule number and morphology	Malinova et al. (2017)
2	PHS1	*Triticum turgidum*	Tilling	Starch granule size	Kamble et al. (2023)
3	SSIII	*Solanum lycopersicum*	Overexpression	Alteration of amylopectin content	Miao et al. (2017)
4	SSIV	*Arabidopsis thaliana*	Knockdown	Starch initiation, granule number and morphology	Bürgy et al. (2021); Roldán et al. (2007)
5	GBSS	*Oryza sativa*	Site-directed mutagenesis (CRISPR-Cas9)	Alteration of amylose content	Pérez et al. (2019)
6	ISAI-3	*Arabidopsis thaliana*	Knockdown	Starch morphology	Wattebled et al. (2005)
7	ESV1 and LESV	*Arabidopsis thaliana*	Knockdown	Starch synthesis and degradation	Feike et al. (2016); Liu et al. (2023)

TABLE 2.7   *In vivo* characterization regarding the localization and interaction of proteins in starch-related enzymes

NO.	ENZYME	SPECIES	METHOD	LOCALIZATION	REFERENCES
1	GBSS1	*Arabidopsis thaliana*	Fluorescent tagging (GFP)	Chloroplast	Szydlowski et al. (2009)
2	SSIV	*Arabidopsis thaliana*	BiFC	Thylakoid membrane	Gámez-Arjona et al. (2014)
3	PTST1	*Hordeum vulgare*	FRET	Chloroplast	Zhong et al. (2019)

2.6.2.1 Genetic manipulation

Gene knockout/knockdown, overexpression, and site-directed mutagenesis are key genetic manipulation techniques used to elucidate protein functions within a plant or algae. By observing the phenotypic effects of removing or reducing a protein through gene knockout or knockdown, researchers can infer the protein's essential roles and contributions to biological pathways. Conversely, overexpression of a protein can enhance our understanding of its functions by revealing the biological consequences of increased protein levels, which might include changes in cell properties, signaling pathways, or physiological responses. Additionally, site-directed mutagenesis allows scientists to alter specific amino acids within a protein, thereby identifying functional domains or critical residues that are crucial for its activity or interaction with other biomolecules. This method provides insights into a protein's structural–functional relationships by observing how specific alterations affect the overall phenotype of the organism. In the future, genetic manipulation tools such as CRISPR-Cas will dominate the genetic manipulation field. Collectively, all of these techniques provide a robust framework for dissecting the complex roles that proteins play in living systems, offering valuable insights into their biological significance. See Apriyanto et al. (2022) and Compart, Singh et al. (2023) for a detailed review of the *in vivo* characterization of starch-synthesizing enzymes via genetic manipulation. Moreover, the impacts of *in vivo* genetic manipulation on starch parameters can be seen in Malinova, Qasim et al. (2018).

Examples of *in vivo* characterization via genetic manipulation in starch-related enzymes can be seen in Table 2.6.

2.6.2.2 Localization and interaction studies

Fluorescent tagging, bimolecular fluorescence complementation (BiFC), and Förster resonance energy transfer (FRET) are advanced techniques that leverage the power of fluorescence to study protein characteristics *in vivo*. Fluorescent tagging involves fusing proteins with fluorescent markers such as Green Fluorescent Protein (GFP), which allows researchers to track the localization and dynamics of these proteins within live cells. This method provides the real-time visualization of protein movements, offering insights into cellular processes. BiFC is particularly useful in observing protein–protein interactions, as it involves splitting a fluorescent protein into two non-fluorescent halves that only emit fluorescence when brought together by the interacting protein partners. This technique can pinpoint where and when proteins interact within the plant cell. Meanwhile, FRET measures the distance and interaction dynamics between two proteins by detecting the energy transfer between two different fluorescent molecules attached to each protein. When the proteins are close enough—typically within 10 nm—the energy emitted by one fluorescent tag can excite the other, emitting a detectable light signal. These methods collectively enhance our understanding of proteins' functions and interactions and their roles in cellular contexts.

Some examples of *in vivo* characterization regarding the localization and interaction of proteins in starch-related enzymes can be seen in Table 2.7.

2.6.3 *In Silico* Characterization

In silico protein characterization refers to the computational analysis and prediction of a protein's structure, function, interactions, and evolution using various bioinformatics tools and databases. This approach has become increasingly important in the field of molecular biology, offering a cost-effective and rapid means to gain insights into proteins of interest, especially when experimental data are limited or difficult to obtain. In addition, the advances and breakthroughs in computational technology, such as machine learning and quantum computing, will speed up the acquisition of new findings in the future (Pal et al. 2024).

While *in silico* methods are powerful and continue to improve, they have limitations. Predictive models are based on existing data and assumptions, which means that they may not capture novel folds or functions. Therefore, computational predictions often require experimental validation. Despite these challenges, *in silico* protein characterization is an indispensable component of modern molecular biology, complementing and guiding experimental efforts.

The most commonly applied approaches for *in silico* protein characterization are as follows.

2.6.3.1 Sequence analysis

Primary structure analysis primarily involves examining the fundamental aspects of a protein sequence. By utilizing tools such as ExPASy ProtParam, researchers can predict essential properties such as the molecular weight, isoelectric point, amino acid composition, and specific sequence motifs (Gasteiger et al. 2003). This level of analysis provides a foundational understanding of a protein's potential functional and structural characteristics based solely on its sequence.

Sequence alignment is a crucial step in comparative protein studies, leveraging tools such as BLAST (Basic Local Alignment Search Tool) and Clustal Omega (McGinnis & Madden 2004; Sievers & Higgins 2018). These tools help in identifying homologous proteins across different species, indicating possible functional and evolutionary links. Such alignments often reveal conserved domains and motifs, which are critical in understanding the roles that these proteins play in various biological processes and their structural conservation over time.

Gene Ontology (GO) annotation enriches the understanding of protein functions by associating specific sequences with known biological processes, functions, and cellular components. Using databases such as UniProt (UniProt Consortium 2018) and InterPro (Hunter et al. 2009), researchers can obtain GO terms linked to particular protein sequences. This annotation aids in systematically categorizing proteins' functions and understanding their roles within cellular contexts, enhancing both the depth and breadth of genomic and proteomic research.

2.6.3.2 Structural prediction

Secondary structure prediction utilizes sophisticated algorithms such as PSIPRED to map out the arrangement of alpha helices, beta strands, and coils within a protein sequence (McGuffin, Bryson & Jones 2000). This step is critical in understanding the basic structural elements that compose the overall architecture of the protein, providing insights into its potential folding patterns and functional regions based purely on its amino acid sequence.

Tertiary structure modeling is essential in visualizing the three-dimensional shapes of proteins, which is crucial in understanding their functions. Various computational techniques are employed based on the availability of similar known structures. Homology modeling, facilitated by tools such as SWISS-MODEL, is effective for proteins that have close structural relatives (Schwede et al. 2003). For proteins lacking identifiable structural analogs, alternative methods such as threading (e.g., I-TASSER) (Yang & Zhang 2015) and ab initio modeling (e.g., Rosetta) are used (Rohl et al. 2004). Recently, software such as AlphaFold has gained prominence due to its ability to accurately predict protein structures by leveraging deep learning techniques (Jumper et al. 2021).

Quaternary structure prediction involves predicting how protein subunits assemble into larger functional complexes, a process that is critical in understanding a protein's functionality in a cellular context. Tools such as PISA (Protein Interfaces, Surfaces, and Assemblies) are instrumental in this regard, allowing researchers to explore how individual protein structures interact and link together based on known structural data (Krissinel & Henrick 2007). This predictive capability is vital in understanding molecular biology at a systemic level.

2.6.3.3 Functional prediction

Active site prediction is a crucial aspect of functional protein analysis. Tools such as COACH (Yang, Roy & Zhang 2013) and MetaPocket (Huang 2009) utilize the known tertiary structures of proteins to identify potential active sites and ligand-binding regions. These predictions are pivotal in understanding how proteins interact with various molecules, which is essential in elucidating molecular functions within biological systems.

A starch-binding domain (SBD) is a specific type of carbohydrate-binding domain. Generally, these domains are independent protein modules that, while lacking enzymatic activity themselves, play a crucial role in directing the catalytic domain toward carbohydrate substrates, especially starch. This positioning allows the enzyme to effectively bind and process the substrate at its active site. SBDs, along with other carbohydrate-binding modules (CBMs), are categorized as distinct families within the comprehensive CAZy database (http://www.cazy.org/), highlighting their integral role in carbohydrate-related enzyme functions (Lombard et al. 2014).

Protein–protein interaction prediction is facilitated by the combination of computational methods that simulate docking (e.g., HADDOCK, PlaPPISite) and searching through databases for known interaction motifs (Dominguez, Boelens & Bonvin 2003; Yang et al. 2020). These techniques help in identifying potential interaction partners and mapping out the interaction surfaces, which are crucial in understanding the complex web of protein interactions that drives cellular processes.

Pathway analysis involves integrating proteins into broader biological contexts to elucidate their roles in metabolic and signaling pathways. Using databases such as KEGG (Kyoto Encyclopedia of Genes and Genomes), researchers can associate proteins with specific pathways based on a network of known interactions (Kanehisa et al. 2017). This analysis helps in understanding how proteins influence various biological processes, offering insights into the biological impacts of genetic variations and potential application targets.

2.6.3.4 Evolutionary analysis

Phylogenetic trees are a fundamental tool in evolutionary biology, used to illustrate the evolutionary relationships among various species or genes. Tools such as MEGA (Molecular Evolutionary Genetics Analysis) enable researchers to construct these trees from protein sequences, providing insights into how different organisms are related and how specific proteins have evolved over time (Tamura, Stecher & Kumar 2021). By examining these trees, scientists can infer patterns of evolutionary divergence and convergence, often shedding light on the functional evolution of proteins across different species.

Conservation analysis plays a critical role in identifying functionally important regions within proteins. Tools such as ConSurf (Ashkenazy et al. 2016) and the MEME suite (Bailey et al. 2015) analyze the degree of conservation across a protein family by examining multiple sequence alignments. By identifying regions that have remained highly conserved throughout evolution, researchers can pinpoint areas that are likely crucial for the protein's function. This type of analysis is particularly useful in highlighting active sites, binding sites, and other functional domains that are key to a protein's role in an organism. Such insights are invaluable in guiding experimental research and in understanding the fundamental mechanisms of proteins' functions.

Examples of the *in silico* characterization of starch-related enzymes can be seen in Table 2.8.

In addition to the above, there are many other methods that can be used to characterize starch-synthesizing enzymes, but they are not mentioned here due to space limitations. Practically, all of these characterization approaches (either *in vitro*, *in vivo*, or *in silico*) are complementary to each other and cannot be separated in order to characterize starch-related enzymes. Their application depends on many factors, such as the research question, protein target, researcher preferences, funding, and availability of instruments.

TABLE 2.8   *In silico* characterization of starch-related enzymes

NO.	ORGANISM	CHARACTERIZATION METHOD	REFERENCES
1	*Solanum tuberosum*	Sequence analysis	Kaur et al. (2024)
		• Primary sequence analysis • Sequence alignment • Gene ontology	
		Structural prediction	
		• Tertiary structure	
		Functional prediction	
		• Protein–protein interaction • Pathway analysis	
		Evolutionary analysis	
		• Phylogenetic analysis • Conservation analysis	
2	*Higher plants*	Sequence analysis	Qu et al. (2018)
		• Sequence alignment	
		Structural prediction	
		• Secondary structure • Tertiary structure	
		Functional prediction	
		• Active site prediction • Protein–protein interaction	
		Evolutionary analysis	
		• Phylogenetic analysis • Conservation analysis	

2.7 CONCLUSION

During the last decade, there has been significant progress in the field of starch research. Thus far, the most modern technologies have been used to solve the problems of starch research. However, there are still a number of research questions that need to be addressed in the future in order to fully understand starch metabolism. This chapter reviewed the most recent insights, including the formation of starch in photosynthetic and non-photosynthetic tissue, enzyme-catalyzed starch synthesis processes, and enzyme characterization using *in vitro*, *in vivo*, or *in silico* methods. Based on this chapter, it is expected that there will be many combinations of methods and multidisciplinary research collaborations that will accelerate the acquisition of new findings in starch research in the future.

REFERENCES

Abt, MR, Pfister, B, Sharma, M, Eicke, S, Bürgy, L, Neale, I, Seung, D, & Zeeman, SC, 2020, STARCH SYNTHASE5, a noncanonical starch synthase-like protein, promotes starch granule initiation in Arabidopsis, *The Plant Cell*, 32, 2543–2565.

Adegoke, TV, Wang, Y, Chen, L, Wang, H, Liu, W, Liu, X, Cheng, Y-C, Tong, X, Ying, J, & Zhang, J, 2021, Posttranslational modification of waxy to genetically improve starch quality in rice grain, *International Journal of Molecular Sciences*, 22(9), art. 4845.

Ahmed, Z, Tetlow, IJ, Ahmed, R, Morell, MK, & Emes, MJ, 2015, Protein-protein interactions among enzymes of starch biosynthesis in high-amylose barley genotypes reveal differential roles of heteromeric enzyme complexes in the synthesis of A and B granules, *Plant Science: An International Journal of Experimental Plant Biology*, 233, 95–106.

Apriyanto, A, Compart, J, & Fettke, J, 2022, A review of starch, a unique biopolymer - Structure, metabolism and in planta modifications, *Plant Science: An International Journal of Experimental Plant Biology*, 318, art. 111223.

Apriyanto, A, Compart, J, & Fettke, J, 2023, Transcriptomic analysis of mesocarp tissue during fruit development of the oil palm revealed specific isozymes related to starch metabolism that control oil yield, *Frontiers in Plant Science*, 14, art. 1220237.

Ashkenazy, H, Abadi, S, Martz, E, Chay, O, Mayrose, I, Pupko, T, & Ben-Tal, N, 2016, ConSurf 2016: An improved methodology to estimate and visualize evolutionary conservation in macromolecules, *Nucleic Acids Research*, 44, W344–W350.

Atkinson, N, Stringer, R, Mitchell, SR, Seung, D, & McCormick, AJ, 2024, SAGA1 and SAGA2 promote starch formation around proto-pyrenoids in Arabidopsis chloroplasts, *Proceedings of the National Academy of Sciences*, 121, art. e2311013121.

Bailey, TL, Johnson, J, Grant, CE, & Noble, WS, 2015, The MEME suite, *Nucleic Acids Research*, 43, W39–W49.

Ball, S, Colleoni, C, & Arias, MC 2015, The transition from glycogen to starch metabolism in cyanobacteria and eukaryotes, in *Starch: Metabolism and Structure*, 1st Ed, eds Y Nakamura, Springer, Japan, Tokyo, pp. 93–158.

Ball, S, Colleoni, C, Cenci, U, Raj, JN, & Tirtiaux, C, 2011, The evolution of glycogen and starch metabolism in eukaryotes gives molecular clues to understand the establishment of plastid endosymbiosis, *Journal of Experimental Botany*, 62, 1775–1801.

Baris, I, Tuncel, A, Ozber, N, Keskin, O, & Kavakli, IH, 2009, Investigation of the interaction between the large and small subunits of potato ADP-glucose pyrophosphorylase, *PLoS Computational Biology*, 5, art. e1000546.

Bertoft, E, 2015, Fine structure of Amylopectin, in Starch: Metabolism and Structure, 1st Ed, eds Y Nakamura, Springer, Japan, Tokyo, pp. 3–40.

Bertoft, E, 2017, Understanding starch structure: Recent progress, *Agronomy*, 7, art. 56.

Blaby-Haas, CE, & Merchant, SS, 2019, Comparative and functional algal genomics, *Annual Review of Plant Biology*, 70, 605–638.

Bürgy, L, Eicke, S, Kopp, C, Jenny, C, Lu, KJ, Escrig, S, Meibom, A, & Zeeman, SC, 2021, Coalescence and directed anisotropic growth of starch granule initials in subdomains of Arabidopsis thaliana chloroplasts, *Nature Communications*, 12, art. 6944.

Cenci, U, Nitschke, F, Steup, M, Minassian, BA, Colleoni, C, & Ball, SG, 2014, Transition from glycogen to starch metabolism in Archaeplastida, *Trends in Plant Science*, 19, 18–28.

Chen, G-X, Zhou, J-W, Liu, Y-L, Lu, X-B, Han, C-X, Zhang, W-Y, Xu, Y-H, & Yan, Y-M, 2016, Biosynthesis and regulation of wheat amylose and amylopectin from proteomic and phosphoproteomic characterization of granule-binding proteins, *Scientific Reports*, 6, art. 33111.

Christiansen, C, Hachem, MA, Glaring, MA, Viksø-Nielsen, A, Sigurskjold, BW, Svensson, B, & Blennow, A, 2009, A CBM20 low-affinity starch-binding domain from glucan, water dikinase, *FEBS Letters*, 583, 1159–1163.

Ciancia, M, Matulewicz, MC, & Tuvikene, R, 2020, Structural diversity in galactans from red seaweeds and its influence on rheological properties, *Frontiers in Plant Science*, 11, art. 559986.

Compart, J, Apriyanto, A, & Fettke, J, 2023, Glucan, water dikinase (GWD) penetrates the starch granule surface and introduces C6 phosphate in the vicinity of branching points, *Carbohydrate Polymers*, 321, art. 121321.

Compart, J, Singh, A, Fettke, J, & Apriyanto, A, 2023, Customizing starch properties: A review of starch modifications and their applications, *Polymers*, 15(16), art. 3491.

Cuesta-Seijo, JA, Nielsen, MM, Marri, L, Tanaka, H, Beeren, SR, & Palcic, MM, 2013, Structure of starch synthase I from barley: Insight into regulatory mechanisms of starch synthase activity, *Acta Crystallographica. Section D, Biological Crystallography*, 69, 1013–1025.

Cuesta-Seijo, JA, Nielsen, MM, Ruzanski, C, Krucewicz, K, Beeren, SR, Rydhal, MG, Yoshimura, Y, Striebeck, A, Motawia, MS, Willats, WGT, & Palcic, MM, 2015, In vitro biochemical characterization of all barley endosperm starch synthases, *Frontiers in Plant Science*, 6, art. 1265.

Delatte, T, Umhang, M, Trevisan, M, Eicke, S, Thorneycroft, D, Smith, SM, & Zeeman, SC, 2006, Evidence for distinct mechanisms of starch granule breakdown in plants, *The Journal of Biological Chemistry*, 281, 12050–12059.

Dominguez, C, Boelens, R, & Bonvin, AMJJ, 2003, HADDOCK: A protein-protein docking approach based on biochemical or biophysical information, *Journal of the American Chemical Society*, 125, 1731–1737.

Edner, C, Li, J, Albrecht, T, Mahlow, S, Hejazi, M, Hussain, H, Kaplan, F, Guy, C, Smith, SM, Steup, M, & Ritte, G, 2007, Glucan, water dikinase activity stimulates breakdown of starch granules by plastidial beta-amylases, *Plant Physiology*, 145, 17–28.

Facchinelli, F, Colleoni, C, Ball, SG, & Weber, APM, 2013, Chlamydia, cyanobiont, or host: Who was on top in the ménage à trois? *Trends in Plant Science, 18,* 673–679.

Feike, D, Seung, D, Graf, A, Bischof, S, Ellick, T, Coiro, M, Soyk, S, Eicke, S, Mettler-Altmann, T, Lu, KJ, Trick, M, Zeeman, SC, & Smith, AM, 2016, The starch granule-associated protein EARLY STARVATION1 is required for the control of starch degradation in Arabidopsis thaliana leaves, *The Plant Cell, 28,* 1472–1489.

Fettke, J, Chia, T, Eckermann, N, Smith, A, & Steup, M, 2006, A transglucosidase necessary for starch degradation and maltose metabolism in leaves at night acts on cytosolic heteroglycans (SHG), *The Plant Journal: For Cell and Molecular Biology, 46,* 668–684.

Fettke, J, & Fernie, AR, 2015, Intracellular and cell-to-apoplast compartmentation of carbohydrate metabolism, *Trends in Plant Science, 20,* 490–497.

Fettke, J, Hejazi, M, Smirnova, J, Höchel, E, Stage, M, & Steup, M, 2009, Eukaryotic starch degradation: Integration of plastidial and cytosolic pathways, *Journal of Experimental Botany, 60,* 2907–2922.

Flores-Castellanos, J, & Fettke, J, 2023, The plastidial glucan phosphorylase affects the maltooligosaccharide metabolism in parenchyma cells of potato (Solanum tuberosum L.) tuber discs, *Plant & Cell Physiology, 64,* 422–432.

Frigård, T, 2002, Gradual enzymatic modification of barley and potato amylopectin, *Carbohydrate Polymers, 47,* 169–179.

Fujita, N, Hanashiro, I, Suzuki, S, Higuchi, T, Toyosawa, Y, Utsumi, Y, Itoh, R, Aihara, S, & Nakamura, Y, 2012, Elongated phytoglycogen chain length in transgenic rice endosperm expressing active starch synthase IIa affects the altered solubility and crystallinity of the storage α-glucan, *Journal of Experimental Botany, 63,* 5859–5872.

Fujita, N, Yoshida, M, Asakura, N, Ohdan, T, Miyao, A, Hirochika, H, & Nakamura, Y, 2006, Function and characterization of starch synthase I using mutants in rice, *Plant Physiology, 140,* 1070–1084.

Fulton, DC, Edwards, A, Pilling, E, Robinson, HL, Fahy, B, Seale, R, Kato, L, Donald, AM, Geigenberger, P, Martin, C, & Smith, AM, 2002, Role of granule-bound starch synthase in determination of amylopectin structure and starch granule morphology in potato, *The Journal of Biological Chemistry, 277,* 10834–10841.

Fulton, DC, Stettler, M, Mettler, T, Vaughan, CK, Li, J, Francisco, P, Gil, M, Reinhold, H, Eicke, S, Messerli, G, Dorken, G, Halliday, K, Smith, AM, Smith, SM, & Zeeman, SC, 2008, Beta-AMYLASE4, a noncatalytic protein required for starch breakdown, acts upstream of three active beta-amylases in Arabidopsis chloroplasts, *The Plant Cell, 20,* 1040–1058.

Fünfgeld, MMFF, Wang, W, Ishihara, H, Arrivault, S, Feil, R, Smith, AM, Stitt, M, Lunn, JE, & Niittylä, T, 2021, The pathway of starch synthesis in Arabidopsis thaliana leaves, *bioRxiv,* art. 2021.01.11.426159.

Gámez-Arjona, FM, Raynaud, S, Ragel, P, & Mérida, A, 2014, Starch synthase 4 is located in the thylakoid membrane and interacts with plastoglobule-associated proteins in Arabidopsis, *The Plant Journal: For Cell and Molecular Biology, 80,* 305–316.

Gasteiger, E, Gattiker, A, Hoogland, C, Ivanyi, I, Appel, RD, & Bairoch, A, 2003, ExPASy: The proteomics server for in-depth protein knowledge and analysis, *Nucleic Acids Research, 31,* 3784–3788.

Gavgani, HN, Fawaz, R, Ehyaei, N, Walls, D, Pawlowski, K, Fulgos, R, Park, S, Assar, Z, Ghanbarpour, A, & Geiger, JH, 2022, A structural explanation for the mechanism and specificity of plant branching enzymes I and IIb. *The Journal of Biological Chemistry, 298(1),* art. 101395.

Gérard, C, Planchot, V, Colonna, P, & Bertoft, E, 2000, Relationship between branching density and crystalline structure of A- and B-type maize mutant starches, *Carbohydrate Research, 326,* 130–144.

Gissibl, A, Sun, A, Care, A, Nevalainen, H, & Sunna, A, 2019, Bioproducts from Euglena gracilis: Synthesis and applications, *Frontiers in Bioengineering and Biotechnology, 7,* art. 108.

Glaring, MA, Baumann, MJ, Abou Hachem, M, Nakai, H, Nakai, N, Santelia, D, Sigurskjold, BW, Zeeman, SC, Blennow, A, & Svensson, B, 2011, Starch-binding domains in the CBM45 family--low-affinity domains from glucan, water dikinase and α-amylase involved in plastidial starch metabolism, *The FEBS Journal, 278,* 1175–1185.

Hädrich, N, Gibon, Y, Schudoma, C, Altmann, T, Lunn, JE, & Stitt, M, 2011, Use of TILLING and robotised enzyme assays to generate an allelic series of Arabidopsis thaliana mutants with altered ADP-glucose pyrophosphorylase activity, *Journal of Plant Physiology, 168,* 1395–1405.

Hanashiro, I, 2015, Fine structure of Amylose, in Starch: Metabolism and Structure, 1st Ed, eds Y Nakamura, Springer, Japan, Tokyo, pp. 41–60.

Hejazi, M, Fettke, J, Haebel, S, Edner, C, Paris, O, Frohberg, C, Steup, M, & Ritte, G, 2008, Glucan, water dikinase phosphorylates crystalline maltodextrins and thereby initiates solubilization, *The Plant Journal, 55(2),* 323–334.

Hejazi, M, Fettke, J, Paris, O, & Steup, M, 2009, The two plastidial starch-related dikinases sequentially phosphorylate glucosyl residues at the surface of both the A- and B-type allomorphs of crystallized maltodextrins but the mode of action differs, *Plant Physiology, 150,* 962–976.

Hejazi, M, Mahlow, S, & Fettke, J, 2014, The glucan phosphorylation mediated by α-glucan, water dikinase (GWD) is also essential in the light phase for a functional transitory starch turn-over, *Plant Signaling & Behavior, 9,* art. e28892.

Hejazi, M, Steup, M, & Fettke, J, 2012, The plastidial glucan, water dikinase (GWD) catalyses multiple phosphotransfer reactions, *The FEBS Journal, 279,* 1953–1966.

Hii, SL, Tan, JS, Ling, TC, & Ariff, AB, 2012, Pullulanase: Role in starch hydrolysis and potential industrial applications, *Enzyme Research, 2012,* art. 921362.

Hostettler, C, Kölling, K, Santelia, D, Streb, S, Kötting, O, & Zeeman, SC, 2011, Analysis of starch metabolism in chloroplasts, *Methods in Molecular Biology, 775,* 387–410.

Huang, B, 2009, MetaPocket: A Meta approach to improve protein ligand binding site prediction, *Omics: A Journal of Integrative Biology, 13,* 325–330.

Hunter, S, Apweiler, R, Attwood, TK, Bairoch, A, Bateman, A, Binns, D, Bork, P, Das, U, Daugherty, L, Duquenne, L, Finn, RD, Gough, J, Haft, D, Hulo, N, Kahn, D, Kelly, E, Laugraud, A, Letunic, I, Lonsdale, D, Lopez, R, Madera, M, Maslen, J, McAnulla, C, McDowall, J, Mistry, J, Mitchell, A, Mulder, N, Natale, D, Orengo, C, Quinn, AF, Selengut, JD, Sigrist, CJA, Thimma, M, Thomas, PD, Valentin, F, Wilson, D, Wu, CH, & Yeats, C, 2009, InterPro: The integrative protein signature database, *Nucleic Acids Research, 37,* D211–D215.

Imberty, A, Chanzy, H, Pérez, S, Buléon, A, & Tran, V, 1988, The double-helical nature of the crystalline part of A-starch, *Journal of Molecular Biology, 201,* 365–378.

Jumper, J, Evans, R, Pritzel, A, Green, T, Figurnov, M, Ronneberger, O, Tunyasuvunakool, K, Bates, R, Žídek, A, Potapenko, A, Bridgland, A, Meyer, C, Kohl, SAA, Ballard, AJ, Cowie, A, Romera-Paredes, B, Nikolov, S, Jain, R, Adler, J, Back, T, Petersen, S, Reiman, D, Clancy, E, Zielinski, M, Steinegger, M, Pacholska, M, Berghammer, T, Bodenstein, S, Silver, D, Vinyals, O, Senior, AW, Kavukcuoglu, K, Kohli, P, & Hassabis, D, 2021, Highly accurate protein structure prediction with AlphaFold, *Nature, 596,* 583–589.

Kaiser, WM, & Bassham, JA, 1979, Light-dark regulation of starch metabolism in chloroplasts: I. Levels of metabolites in chloroplasts and medium during light-dark transition, *Plant Physiology, 63,* 105–108.

Kamble, NU, Makhamadjonov, F, Fahy, B, Martins, C, Saalbach, G, & Seung, D, 2023, Initiation of B-type starch granules in wheat endosperm requires the plastidial α-glucan phosphorylase PHS1, *The Plant Cell, 35,* 4091–4110.

Kanehisa, M, Furumichi, M, Tanabe, M, Sato, Y, & Morishima, K, 2017, KEGG: New perspectives on genomes, pathways, diseases and drugs, *Nucleic Acids Research, 45,* D353–D361.

Kaur, J, Manchanda, P, Kaur, H, Kumar, P, Kalia, A, Sharma, SP, & Taggar, MS, 2024, In-silico identification, characterization and expression analysis of genes involved in resistant starch biosynthesis in potato (Solanum tuberosum L.) varieties, *Molecular Biotechnology,* art. DOI:10.1007/s12033-024-01121-w

Kötting, O, Pusch, K, Tiessen, A, Geigenberger, P, Steup, M, & Ritte, G, 2005, Identification of a novel enzyme required for starch metabolism in Arabidopsis leaves. The phosphoglucan, water dikinase, *Plant Physiology, 137,* 242–252.

Kötting, O, Santelia, D, Edner, C, Eicke, S, Marthaler, T, Gentry, MS, Comparot-Moss, S, Chen, J, Smith, AM, Steup, M, Ritte, G, & Zeeman, SC, 2009, STARCH-EXCESS4 is a laforin-like Phosphoglucan phosphatase required for starch degradation in Arabidopsis thaliana, *The Plant Cell, 21,* 334–346.

Krissinel, E, & Henrick, K, 2007, Inference of macromolecular assemblies from crystalline state, *Journal of Molecular Biology, 372,* 774–797.

Kuchitsu, K, Tsuzuki, M, & Miyachi, S, 1988, Changes of starch localization within the chloroplast induced by changes in CO_2 concentration during growth of Chlamydomonas reinhardtii: Independent regulation of pyrenoid starch and stroma starch, *Plant and Cell Physiology, 29(8),* 1269–1278.

Li, X, Apriyanto, A, Castellanos, JF, Compart, J, Muntaha, SN, & Fettke, J, 2022, Dpe2/phs1 revealed unique starch metabolism with three distinct phases characterized by different starch granule numbers per chloroplast, allowing insights into the control mechanism of granule number regulation by gene co-regulation and metabolic profiling, *Frontiers in Plant Science, 13,* art. 1039534.

Liu, C, Pfister, B, Osman, R, Ritter, M, Heutinck, A, Sharma, M, Eicke, S, Fischer-Stettler, M, Seung, D, Bompard, C, Abt, MR, & Zeeman, SC, 2023, LIKE EARLY STARVATION 1 and EARLY STARVATION 1 promote and stabilize amylopectin phase transition in starch biosynthesis, *Science Advances, 9,* art. eadg7448.

Liu, F, Romanova, N, Lee, EA, Ahmed, R, Evans, M, Gilbert, EP, Morell, MK, Emes, MJ, & Tetlow, IJ, 2012, Glucan affinity of starch synthase IIa determines binding of starch synthase I and starch-branching enzyme IIb to starch granules, *The Biochemical Journal, 448,* 373–387.

Liu, Q, Li, X, & Fettke, J, 2021, Starch granules in Arabidopsis thaliana mesophyll and guard cells show similar morphology but differences in size and number, *International Journal of Molecular Sciences, 22(11),* art. 5666.

Lloyd, JR, & Kossmann, J, 2015, Transitory and storage starch metabolism: Two sides of the same coin? *Current Opinion in Biotechnology, 32,* 143–148.

Lombard, V, Golaconda Ramulu, H, Drula, E, Coutinho, PM, & Henrissat, B, 2014, The carbohydrate-active enzymes database (CAZy) in 2013, *Nucleic Acids Research, 42,* D490–D495.

Mackinder, LCM, Meyer, MT, Mettler-Altmann, T, Chen, VK, Mitchell, MC, Caspari, O, Freeman Rosenzweig, ES, Pallesen, L, Reeves, G, Itakura, A, Roth, R, Sommer, F, Geimer, S,

Mühlhaus, T, Schroda, M, Goodenough, U, Stitt, M, Griffiths, H, & Jonikas, MC, 2016, A repeat protein links Rubisco to form the eukaryotic carbon-concentrating organelle, *Proceedings of the National Academy of Sciences, 113,* 5958–5963.

MacNeill, GJ, Mehrpouyan, S, Minow, MAA, Patterson, JA, Tetlow, IJ, & Emes, MJ, 2017, Starch as a source, starch as a sink: The bifunctional role of starch in carbon allocation, *Journal of Experimental Botany, 68,* 4433–4453.

Mahlow, S, Hejazi, M, Kuhnert, F, Garz, A, Brust, H, Baumann, O, & Fettke, J, 2014, Phosphorylation of transitory starch by α-glucan, water dikinase during starch turnover affects the surface properties and morphology of starch granules, *The New Phytologist, 203,* 495–507.

Mahlow, S, Orzechowski, S, & Fettke, J, 2016, Starch phosphorylation: Insights and perspectives, *Cellular and Molecular Life Sciences: CMLS, 73,* 2753–2764.

Malinova, I, Alseekh, S, Feil, R, Fernie, AR, Baumann, O, Schöttler, MA, Lunn, JE, & Fettke, J, 2017, Starch synthase 4 and plastidal phosphorylase differentially affect starch granule number and morphology, *Plant Physiology, 174,* 73–85.

Malinova, I, & Fettke, J, 2017, Reduced starch granule number per chloroplast in the dpe2/phs1 mutant is dependent on initiation of starch degradation, *PloS One, 12,* art. e0187985.

Malinova, I, Mahlow, S, Alseekh, S, Orawetz, T, Fernie, AR, Baumann, O, Steup, M, & Fettke, J, 2014, Double knockout mutants of Arabidopsis grown under normal conditions reveal that the plastidial phosphorylase isozyme participates in transitory starch metabolism, *Plant Physiology, 164,* 907–921.

Malinova, I, Mahto, H, Brandt, F, Al-Rawi, S, Qasim, H, Brust, H, Hejazi, M, & Fettke, J, 2018, EARLY STARVATION1 specifically affects the phosphorylation action of starch-related dikinases, *The Plant Journal: For Cell and Molecular Biology, 95,* 126–137.

Malinova, I, Qasim, HM, Brust, H, & Fettke, J, 2018, Parameters of starch granule genesis in chloroplasts of Arabidopsis thaliana, *Frontiers in Plant Science, 9,* art. 761.

McGinnis, S, & Madden, TL, 2004, BLAST: At the core of a powerful and diverse set of sequence analysis tools, *Nucleic Acids Research, 32,* W20–W25.

McGuffin, LJ, Bryson, K, & Jones, DT, 2000, The PSIPRED protein structure prediction server, *Bioinformatics (Oxford, England), 16,* 404–405.

McKay, RML, & Gibbs, SP, 1991, Composition and function of pyrenoids: Cytochemical and immunocytochemical approaches, *Canadian Journal of Botany, 69,* 1040–1052.

Meekins, DA, Guo, H-F, Husodo, S, Paasch, BC, Bridges, TM, Santelia, D, Kötting, O, Vander Kooi, CW, & Gentry, MS, 2013, Structure of the Arabidopsis glucan phosphatase like sex four2 reveals a unique mechanism for starch dephosphorylation, *The Plant Cell, 25,* 2302–2314.

Meng, X, Mujahid, H, Zhang, Y, Peng, X, Redoña, ED, Wang, C, & Peng, Z, 2019, Comprehensive analysis of the lysine succinylome and protein co-modifications in developing rice seeds, *Molecular & Cellular Proteomics: MCP, 18,* 2359–2372.

Mérida, A, & Fettke, J, 2021, Starch granule initiation in Arabidopsis thaliana chloroplasts, *The Plant Journal: For Cell and Molecular Biology, 107,* 688–697.

Miao, H, Sun, P, Liu, Q, Jia, C, Liu, J, Hu, W, Jin, Z, & Xu, B, 2017, Soluble starch synthase III-1 in amylopectin metabolism of banana fruit: Characterization, expression, enzyme activity, and functional analyses, *Frontiers in Plant Science, 8,* art. 454.

Mikkelsen, R, Suszkiewicz, K, & Blennow, A, 2006, A novel type carbohydrate-binding module identified in alpha-glucan, water dikinases is specific for regulated plastidial starch metabolism, *Biochemistry, 45,* 4674–4682.

Miller, BR, & Gulick, AM, 2016, Structural biology of nonribosomal peptide synthetases, Methods in Molecular Biology, 1401, 3–29.

Momma, M, & Fujimoto, Z, 2012, Interdomain disulfide bridge in the rice granule bound starch synthase I catalytic domain as elucidated by X-ray structure analysis, *Bioscience, Biotechnology, and Biochemistry*, 76, 1591–1595.

Morell, MK, Li, Z, Regina, A, Rahman, S, D'Hulst, C, & Ball, SG 2018, Control of starch biosynthesis in vascular plants and algae, in *Annual Plant Reviews Online*, eds JA Roberts, Wiley, pp. 258–289. Hoboken, New Jersey, USA.

Mouille, G, Maddelein, ML, Libessart, N, Talaga, P, Decq, A, Delrue, B, & Ball, S, 1996, Preamylopectin processing: A mandatory step for starch biosynthesis in plants, *The Plant Cell*, 8, 1353–1366.

Nakamura, Y, 2015, Biosynthesis of reserve starch, in Starch: Metabolism and Structure, 1st Ed, eds Y Nakamura, Springer, Japan, Tokyo, pp. 161–210.

Nakamura, Y, Takahashi, J-I, Sakurai, A, Inaba, Y, Suzuki, E, Nihei, S, Fujiwara, S, Tsuzuki, M, Miyashita, H, Ikemoto, H, Kawachi, M, Sekiguchi, H, & Kurano, N, 2005, Some Cyanobacteria synthesize semi-amylopectin type alpha-polyglucans instead of glycogen, *Plant & Cell Physiology*, 46, 539–545.

Neuhaus, HE, & Schulte, N, 1996, Starch degradation in chloroplasts isolated from C3 or CAM (crassulacean acid metabolism)-induced Mesembryanthemum crystallinum L, *The Biochemical Journal*, 318 (Pt 3), 945–953.

Nielsen, MM, Ruzanski, C, Krucewicz, K, Striebeck, A, Cenci, U, Ball, SG, Palcic, MM, & Cuesta-Seijo, JA, 2018, Crystal structures of the catalytic domain of Arabidopsis thaliana starch synthase IV, of granule bound starch synthase from CLg1 and of granule bound starch synthase I of Cyanophora paradoxa illustrate substrate recognition in starch synthases, *Frontiers in Plant Science*, 9, art. 1138.

Nougué, O, Corbi, J, Ball, SG, Manicacci, D, & Tenaillon, MI, 2014, Molecular evolution accompanying functional divergence of duplicated genes along the plant starch biosynthesis pathway, *BMC Evolutionary Biology*, 14, art. 103.

Nowack, ECM, & Weber, APM, 2018, Genomics-informed insights into endosymbiotic organelle evolution in photosynthetic eukaryotes, *Annual Review of Plant Biology*, 69, 51–84.

Pal, S, Bhattacharya, M, Lee, S-S, & Chakraborty, C, 2024, Quantum computing in the next-generation computational biology landscape: From protein folding to molecular dynamics, *Molecular Biotechnology*, 66, 163–178.

Peavey, DG, Steup, M, & Gibbs, M, 1977, Characterization of starch breakdown in the intact spinach chloroplast, *Plant Physiology*, 60, 305–308.

Pérez, L, Soto, E, Farré, G, Juanos, J, Villorbina, G, Bassie, L, Medina, V, Serrato, AJ, Sahrawy, M, Rojas, JA, Romagosa, I, Muñoz, P, Zhu, C, & Christou, P, 2019, CRISPR/Cas9 mutations in the rice Waxy/GBSSI gene induce allele-specific and zygosity-dependent feedback effects on endosperm starch biosynthesis, *Plant Cell Reports*, 38, 417–433.

Pfister, B, & Zeeman, SC, 2016, Formation of starch in plant cells, *Cellular and Molecular Life Sciences: CMLS*, 73, 2781–2807.

Qu, J, Xu, S, Zhang, Z, Chen, G, Zhong, Y, Liu, L, Zhang, R, Xue, J, & Guo, D, 2018, Evolutionary, structural and expression analysis of core genes involved in starch synthesis, *Scientific Reports*, 8, art. 12736.

Ramazanov, Z, Rawat, M, Henk, M, Mason, C, Matthews, S, & Moroney, J, 1994, The induction of the CO_2-concentrating mechanism is correlated with the formation of the starch sheath around the pyrenoid of Chlamydomonas reinhardtii, *Planta*, 195, 210–216.

Reimer, JM, Eivaskhani, M, Harb, I, Guarné, A, Weigt, M, & Schmeing, TM, 2019, Structures of a dimodular nonribosomal peptide synthetase reveal conformational flexibility, *Science*, 366(6466), art. eaaw4388.

Ritte, G, Heydenreich, M, Mahlow, S, Haebel, S, Kötting, O, & Steup, M, 2006, Phosphorylation of C6- and C3-positions of glucosyl residues in starch is catalysed by distinct dikinases, *FEBS Letters*, 580, 4872–4876.

Ritte, G, Lloyd, JR, Eckermann, N, Rottmann, A, Kossmann, J, & Steup, M, 2002, The starch-related R1 protein is an alpha-glucan, water dikinase, *Proceedings of the National Academy of Sciences*, 99, 7166–7171.

Rohl, CA, Strauss, CEM, Misura, KMS, & Baker, D, 2004, Protein structure prediction using Rosetta, *Methods in Enzymology*, 383, 66–93.

Roldán, I, Wattebled, F, Mercedes Lucas, M, Delvallé, D, Planchot, V, Jiménez, S, Pérez, R, Ball, S, D'Hulst, C, & Mérida, A, 2007, The phenotype of soluble starch synthase IV defective mutants of Arabidopsis thaliana suggests a novel function of elongation enzymes in the control of starch granule formation, *The Plant Journal: For Cell and Molecular Biology*, 49, 492–504.

Santelia, D, Kötting, O, Seung, D, Schubert, M, Thalmann, M, Bischof, S, Meekins, DA, Lutz, A, Patron, N, Gentry, MS, Allain, FH-T, & Zeeman, SC, 2011, The phosphoglucan phosphatase like sex Four2 dephosphorylates starch at the C3-position in Arabidopsis, *The Plant Cell*, 23, 4096–4111.

Schreier, TB, Umhang, M, Lee, S-K, Lue, W-L, Shen, Z, Silver, D, Graf, A, Müller, A, Eicke, S, Stadler-Waibel, M, Seung, D, Bischof, S, Briggs, SP, Kötting, O, Moorhead, GBG, Chen, J, & Zeeman, SC, 2019, LIKE SEX4 1 acts as a β-amylase-binding scaffold on starch granules during starch degradation, *The Plant Cell*, 31, 2169–2186.

Schwede, T, Kopp, J, Guex, N, & Peitsch, MC, 2003, SWISS-MODEL: An automated protein homology-modeling server, *Nucleic Acids Research*, 31, 3381–3385.

Seung, D, Boudet, J, Monroe, J, Schreier, TB, David, LC, Abt, M, Lu, K-J, Zanella, M, & Zeeman, SC, 2017, Homologs of PROTEIN TARGETING TO STARCH control starch granule initiation in Arabidopsis leaves, *The Plant Cell*, 29, 1657–1677.

Seung, D, Schreier, TB, Bürgy, L, Eicke, S, & Zeeman, SC, 2018, Two plastidial coiled-coil proteins are essential for normal starch granule initiation in Arabidopsis, *The Plant Cell*, 30, 1523–1542.

Seung, D, Soyk, S, Coiro, M, Maier, BA, Eicke, S, & Zeeman, SC, 2015, PROTEIN TARGETING TO STARCH is required for localising GRANULE-BOUND STARCH SYNTHASE to starch granules and for normal amylose synthesis in Arabidopsis, *PLoS Biology*, 13, art. e1002080.

Sharma, M, Abt, MR, Eicke, S, Ilse, TE, Liu, C, Lucas, MS, Pfister, B, & Zeeman, SC, 2024, MFP1 defines the subchloroplast location of starch granule initiation, *Proceedings of the National Academy of Sciences*, 121, art. e2309666121.

Shoaib, N, Liu, L, Ali, A, Mughal, N, Yu, G, & Huang, Y, 2021, Molecular functions and pathways of plastidial starch phosphorylase (PHO1) in starch metabolism: Current and future perspectives, *International Journal of Molecular Sciences*, 22(19), art. 10450.

Sievers, F, & Higgins, DG, 2018, Clustal Omega for making accurate alignments of many protein sequences, Protein Science: A Publication of the Protein Society, 27, 135–145.

Silver, DM, Kötting, O, & Moorhead, GBG, 2014, Phosphoglucan phosphatase function sheds light on starch degradation, *Trends in Plant Science*, 19, 471–478.

Singh, A, Compart, J, Al-Rawi, SA, Mahto, H, Ahmad, AM, & Fettke, J, 2022, LIKE EARLY STARVATION1 alters the glucan structures at the starch granule surface and thereby influences the action of both starch-synthesizing and starch-degrading enzymes, *The Plant Journal: For Cell and Molecular Biology*, *111*, 819–835.

Smith, AM, & Zeeman, SC, 2020, Starch: A flexible, adaptable carbon store coupled to plant growth, *Annual Review of Plant Biology*, *71*, 217–245.

Soliman, A, Ayele, BT, & Daayf, F, 2014, Biochemical and molecular characterization of barley plastidial ADP-glucose transporter (HvBT1), *PloS One*, *9*, art. e98524.

Steup, M, Peavey, DG, & Gibbs, M, 1976, The regulation of starch metabolism by inorganic phosphate, *Biochemical and Biophysical Research Communications*, *72*, 1554–1561.

Stitt, M, & Heldt, HW, 1981, Physiological rates of starch breakdown in isolated intact spinach chloroplasts, *Plant Physiology*, *68*, 755–761.

Stitt, M, & Rees, TA, 1980, Carbohydrate breakdown by chloroplasts of Pisum sativum, *Biochimica et Biophysica Acta*, *627*, 131–143.

Stitt, M, & Zeeman, SC, 2012, Starch turnover: Pathways, regulation and role in growth, *Current Opinion in Plant Biology*, *15*, 282–292.

Streb, S, & Zeeman, SC, 2012, Starch metabolism in Arabidopsis, *The Arabidopsis Book*, *10*, art. e0160.

Süssmuth, RD, & Mainz, A, 2017, Nonribosomal peptide synthesis-principles and prospects, *Angewandte Chemie (*International ed. *in English)*, *56*, 3770–3821.

Suzuki, E, Onoda, M, Colleoni, C, Ball, S, Fujita, N, & Nakamura, Y, 2013, Physicochemical variation of cyanobacterial starch, the insoluble α-Glucans in cyanobacteria, *Plant & Cell Physiology*, *54*, 465–473.

Suzuki, K, Nakashima, A, Igarashi, M, Saito, K, Konno, M, Yamazaki, N, & Takimoto, H, 2018, Euglena gracilis Z and its carbohydrate storage substance relieve arthritis symptoms by modulating Th17 immunity, *PLoS One*, *13*, art. e0191462.

Szydlowski, N, Ragel, P, Raynaud, S, Lucas, MM, Roldán, I, Montero, M, Muñoz, FJ, Ovecka, M, Bahaji, A, Planchot, V, Pozueta-Romero, J, D'Hulst, C, & Mérida, A, 2009, Starch granule initiation in Arabidopsis requires the presence of either class IV or class III starch synthases, *The Plant Cell*, *21*, 2443–2457.

Tamura, K, Stecher, G, & Kumar, S, 2021, MEGA11: Molecular evolutionary genetics analysis version 11, *Molecular Biology and Evolution*, *38*, 3022–3027.

Tanaka, Y, Ogawa, T, Maruta, T, Yoshida, Y, Arakawa, K, & Ishikawa, T, 2017, Glucan synthase-like 2 is indispensable for paramylon synthesis in Euglena gracilis, *FEBS Letters*, *591*, 1360–1370.

Tiozon, RN, Jr, Fettke, J, Sreenivasulu, N, & Fernie, AR, 2023, More than the main structural genes. Regulation of resistant starch formation in rice endosperm and its potential application, *Journal of Plant Physiology*, *285*, art. 153980.

UniProt Consortium, T, 2018, UniProt: The universal protein knowledgebase, *Nucleic Acids Research*, *46*, art. 2699.

Usov, AI, 2011, Polysaccharides of the red algae, *Advances in Carbohydrate Chemistry and Biochemistry*, *65*, 115–217.

Valdez, HA, Busi, MV, Wayllace, NZ, Parisi, G, Ugalde, RA, & Gomez-Casati, DF, 2008, Role of the N-terminal starch-binding domains in the kinetic properties of starch synthase III from Arabidopsis thaliana, *Biochemistry*, *47*, 3026–3032.

Vander Kooi, CW, Taylor, AO, Pace, RM, Meekins, DA, Guo, H-F, Kim, Y, & Gentry, MS, 2010, Structural basis for the glucan phosphatase activity of Starch Excess4, *Proceedings of the National Academy of Sciences*, *107*, 15379–15384.

Vandromme, C, Spriet, C, Dauvillée, D, Courseaux, A, Putaux, J-L, Wychowski, A, Krzewinski, F, Facon, M, D'Hulst, C, & Wattebled, F, 2019, PII1: A protein involved in starch initiation that determines granule number and size in Arabidopsis chloroplast, *The New Phytologist*, *221*, 356–370.

Wang, L, & Jonikas, MC, 2020, The pyrenoid, *Current Biology: CB*, *30*, R456–R458.

Wattebled, F, Dong, Y, Dumez, S, Delvallé, D, Planchot, V, Berbezy, P, Vyas, D, Colonna, P, Chatterjee, M, Ball, S, & D'Hulst, C, 2005, Mutants of Arabidopsis lacking a chloroplastic isoamylase accumulate phytoglycogen and an abnormal form of amylopectin, *Plant Physiology*, *138*, 184–195.

Weissman, KJ, 2015, The structural biology of biosynthetic megaenzymes, *Nature Chemical Biology*, *11*, 660–670.

Wilkens, C, Auger, KD, Anderson, NT, Meekins, DA, Raththagala, M, Abou Hachem, M, Payne, CM, Gentry, MS, & Svensson, B, 2016, Plant α-glucan phosphatases SEX4 and LSF2 display different affinity for amylopectin and amylose, *FEBS Letters*, *590*, 118–128.

Wilkens, C, Cuesta-Seijo, JA, Palcic, M, & Svensson, B, 2014, Selectivity of the surface binding site (SBS) on barley starch synthase I, *Biologia*, *69*, 1118–1121.

Yang, J, Roy, A, & Zhang, Y, 2013, Protein-ligand binding site recognition using complementary binding-specific substructure comparison and sequence profile alignment, Bioinformatics *(*Oxford, England*)*, *29*, 2588–2595.

Yang, J, & Zhang, Y, 2015, Protein structure and function prediction using I-TASSER, *Current Protocols in Bioinformatics*, *52*, 5.8.1–5.8.15.

Yang, X, Yang, S, Qi, H, Wang, T, Li, H, & Zhang, Z, 2020, PlaPPISite: A comprehensive resource for plant protein-protein interaction sites, *BMC Plant Biology*, *20*, art. 61.

Zhan, Y, Marchand, CH, Maes, A, Mauries, A, Sun, Y, Dhaliwal, JS, Uniacke, J, Arragain, S, Jiang, H, Gold, ND, Martin, VJJ, Lemaire, SD, & Zerges, W, 2018, Pyrenoid functions revealed by proteomics in Chlamydomonas reinhardtii, *PLoS One*, *13*, art. e0185039.

Zhang, X, Szydlowski, N, Delvallé, D, D'Hulst, C, James, MG, & Myers, AM, 2008, Overlapping functions of the starch synthases SSII and SSIII in amylopectin biosynthesis in Arabidopsis, *BMC Plant Biology*, *8*, art. 96.

Zhong, Y, Blennow, A, Kofoed-Enevoldsen, O, Jiang, D, & Hebelstrup, KH, 2019, Protein targeting to Starch 1 is essential for starchy endosperm development in barley, *Journal of Experimental Botany*, *70*, 485–496.

Starch Biosynthesis in Cereal Endosperms

3

Ian J. Tetlow and Victoria Butler

3.1 INTRODUCTION

Cereals are domesticated grasses of the Poaceae (or Gramineae) family, so-named after *Ceres*, the Roman goddess of harvest and fertility. Cereals are cultivated primarily for harvesting their mature seeds which have starch and protein-rich endosperms. Archaeological studies indicate that humans and their ancestors have forged a long-term relationship with the progenitors of domesticated grasses, utilizing their starchy seeds as a source of food many millennia before the advent of agriculture (Fellows et al. 2021; Hardy et al. 2012; Piperno et al. 2004). For example, evidence from African cave excavations suggests processing of starch from a wild ancestor of sorghum (*Sorghum bicolor* L.) more than 120,000 years before present (BP) (Mercader 2009; Prasad et al. 2005), and more recently evidence of cooking and ingestion of wild Triticeae grasses (cf. *Hordeum*) by Neanderthals 36–46,000 years BP (Henry, Brooks & Piperno 2011). The domestication of crop plants beginning at the end of the Pleistocene and continuing into the Holocene (11,700 years BP) precipitated profound cultural and political changes within human societies which laid the foundation of civilizations across the globe (Diamond 2002). Cereals have played a central role in early agriculture, probably across multiple geographic locations (Abbo, Lev-Yadun & Gopher 2010; Fuller, Willcox & Allaby 2011) and continue to be of paramount importance as cereal yield governs modern global agricultural productivity and food/ feed supply (Awika 2011). Cereals currently represent the major source of calories for most of the world's human population, and in developing countries, they may account for more than 60% of daily caloric intake, whereas in developed countries, this figure is close to 30% (Anon 2003; Awika 2011). Cereals represent six of the seven most grown crops grown worldwide (Langer & Hill 1982), with considerable proportions of global yields supplying the animal and livestock feed market, depending on forage grass yields (FAO 2021; Hatfield & Walthall 2014).

The major cereals in the Poaceae family are broadly split into two physiologically distinct groups: the PACMAD clade representing the Panicoideae, Aristidoideae, Chloridoideae, Micrairoideae, Arundinoideae, and Danthonioideae, and the BEP clade representing the Bambusoideae, Ehrhartoideae, and Pooideae (Tetlow & Emes 2017). Within the PACMAD clade are the panicoid grasses (subfamily Panicoideae), and the major cereals within this group are tropical and sub-tropical C4 grasses, which include maize (*Zea mays* L.), sorghum (*Sorghum bicolor* L.), pearl millet (*Pennisetum glaucum* L.), common millet (*Panicum miliaceum* L.), and foxtail millet (*Setaria italica* L.). The major cereals of the BEP clade are C3 grasses, which include sub-tropical rice (*Oryza sativa* L.) from the Ehrhartoideae subfamily and temperate cereals of the Pooideae subfamily (formerly Festucoideae) such as wheat (*Triticum aestivum* L.), barley (*Hordeum vulgare* L.), and rye (*Secale cereale* L.) of the Triticeae tribe and oats (*Avena sativa* L.) of the Aveneae tribe. In terms of global production, the most widely grown cereal crops are maize, rice, wheat, barley, and sorghum (Awika 2011). Cereal production is therefore a central component of modern agricultural production, and it makes a major contribution to meeting global food demand, which is predicted to increase up to 56%, as global population is expected to reach 9.6 billion people by 2050 (van Dijk et al. 2021). In addition to the pressures of exponential population growth, increased demand for meat in the daily diet of the rising middle class also places significant demands on cereal production, which is expected to increase 50% by 2030 (Kharas 2010; Parry & Hawkesford 2010). Problems in meeting projected global food demand are compounded by the myriad effects of climate change, which contribute to declines in productivity of arable land. Data suggests that global warming caused by anthropogenic production of greenhouse gases such as CO_2 in key agricultural regions will cause yield decline in major cereals (IPCC 2014; Russo et al. 2014) and increase the population at risk of hunger (Battisti & Naylor 2009; Parry, Rosenzweig & Livermore 2005; van Dijk et al. 2021). Elevated temperatures in cereal growing regions, particularly when they occur at early stages of floral development and grain filling, cause reduced seed set and reduced levels of grain filling leading to yield reduction. The development of reproductive structures in all cereals

DOI: 10.1201/9781003464396-3

is sensitive to heat stress (Barnabás, Jager & Fehér 2008; Beckles & Thitisaksakul 2014; Boehlein et al. 2019; Olsen 2004; Sabelli & Larkins 2009), but temperate cereals, whose members include major crops such as wheat, barley, rye, and oats, are particularly susceptible to heat stress throughout their development (Chowdhury & Wardlaw 1978; Harris et al. 2023; Wallwork et al. 1998).

Modern cereal grains are the product of domestication as well as active attempts at plant breeding resulting in a number of important physiological traits which have high general agronomic value (Smýkal et al. 2018). These traits include indehiscence of the spike (non-shattering seed that is unable to be shed at maturity) leading to dependence on humans for dispersal, increased seed size such that in major cereals (e.g., wheat and barley) as much as 50% of the above-ground dry matter is in mature grains (Austin et al. 1980), loss of out-crossing, reduced seed dormancy, and synchronous tillering and maturation. These important agronomic traits are collectively known as the *domestication syndrome* (Hammer 1984), and this phenotypic integration is likely the result of co-selection of traits (Darwin 1859; Milla et al. 2015) arising from either unconscious selection by farmers or artificial breeding often within human-constructed environments. The physiological basis for the high yields observed in modern cereals is therefore complex (Preece et al. 2017), but it has resulted in larger seed (endosperm) tissue in comparison with wild progenitor species (Evans & Dunstone 1970).

An understanding of the biology of cereals is essential if we are to meet future challenges in food demand by producing new higher-yielding plant varieties that are resilient to both biotic and abiotic stresses. The principal determinant of cereal yield is grain weight, over 70% of which is storage starch made inside specialized plastids (amyloplasts) of the endosperm cells. The aim of this chapter is to outline the current state of knowledge of the biochemistry and molecular biology of storage starch biosynthesis occurring in amyloplasts of cereals. Temporary reserves of starch produced in the chloroplasts of vegetative (autotrophic) tissues serve an important function in the fitness of the whole plant during its lifecycle (MacNeill et al. 2017; Stitt & Zeeman 2012). The pathway involved in the synthesis of chloroplast starch is distinct from that occurring in cereal amyloplasts and is outlined in detail in Chapter 2.

3.2 CEREAL ENDOSPERM

The storage starch produced by cereals is made inside the endosperm, a nutritive storage tissue that is a major component of the grain (caryopsis) and supports germination of the next generation of plants. Endosperm cells arise through the process of double fertilization resulting in a triploid (3n, 3C) tissue, which is peculiar to angiosperms (flowering plants). The endosperm of cereals and other grasses undergoes multiple rounds of division in the absence of cytokinesis and cell wall synthesis, which results in syncytium (coenocyte) formation (Lopes & Larkins

1993; Olsen 2004). Various plant hormones and gene networks govern endosperm coenocyte division (Jameson & Song 2016) as well as subsequent endosperm development (Basunia & Nonhebel 2019; Wu et al. 2024a). A detailed discussion of endosperm development in the Poaceae can be found in a recent general review (Olsen 2020) and articles that describe the species-specific differences in endosperm development between the important cereals such as wheat (Briarty, Hughes & Evers 1979), rice (Brown, Lemmon & Olsen 1996), maize (Dai, Ma & Song 2021; Leroux et al. 2014), barley (Bosnes, Weidemen & Olsen 1992), sorghum (Shull et al. 1990), and forage grasses (Hands 2012; Sabelli & Larkins 2009). In general, following coenocyte formation endosperm development is characterized by cellularization which occurs approximately 2–4 days after pollination (DAP), followed by cell differentiation around 4–5 DAP (Olsen 2004). These crucial stages of development dictate yield potential as they govern viable endosperm cell number and, ultimately, starch granule number (Brocklehurst 1977; Chojecki, Bayliss & Gale 1986a; Chojecki, Gale & Baylis 1986b; Jones, Schrieber & Roessler 1996), and they are also highly susceptible to environmental stresses (Beckles & Thitisaksakul 2014; Commuri & Jones 1999; Reddy & Daynard 1983).

The endosperm is a non-photosynthetic (heterotrophic) tissue and therefore dependent upon the import of carbon from the phloem in the form of sucrose (Suc) to sustain growth and development. Synthesis of storage starch within the endosperm cells takes place inside amyloplasts, which are derived from proplastids (Lopez-Juez & Pyke 2005). The complex process of starch synthesis as it relates to cereals is discussed in detail in the following sections of this chapter. Also discussed are the physiological peculiarities of endosperm tissue, which dictate carbon metabolism in the cells and the transport of precursors for starch biosynthesis from the cytosol to the amyloplast.

3.3 CARBON FLOW TO THE ENDOSPERM AND ITS METABOLISM

Multiple factors control the provision of carbon, specifically Suc, from photosynthetic (source) leaf tissue to the developing endosperm (sink), and essentially underpin cereal yields. A detailed exposition of whole-plant carbon partitioning and source–sink relationships is beyond the scope of this chapter, but nevertheless plays a pivotal role in controlling rates of starch deposition within the endosperm. We refer the reader to several recent reviews dealing with the control of whole-plant carbon partitioning in cereals (Bihmidine et al. 2013; Shen et al. 2023) and other crop plants (Bennett, Roberts & Wagstaff 2012; Dong & Beckles 2019; Fernie et al. 2020; Lawlor & Paul 2014; Lemoine et al. 2013; Yu, Lo & Ho 2015).

Following the cell differentiation stage, endosperm cells begin to synthesize storage starch inside amyloplasts using phloem-derived Suc as a carbon source. In the Poaceae, Suc synthesized from source leaves is loaded into the phloem *via* an apoplastic route (Liesche & Patrick 2017; van Bel 1993). Suc arriving at the developing spike in grasses and cereals must take a mandatory apoplastic route through the cell wall prior to direct transport into endosperm transfer cells by energy-dependent sucrose/H$^+$ transporters (members of the SUT family, Weschke et al. 2000) because the embryo and endosperm are physically separated from maternal tissues. In cereals of the subfamily Pooideae (e.g., wheat and barley), passive apoplastic transfer of Suc occurs from maternal to growing seed tissues *via* the central crease vein region of the developing seed (Cochrane 1983; Fisher & Gifford 1987; Thorne 1985; Wang, Offler & Patrick 1995). Whereas in maize and other related Panicoideae such as sorghum, the site of phloem unloading is the pedicel and placenta-chalaza (Bihmidine et al. 2013; Shannon & Dougherty 1972), and in rice, it is the vascular system on the ventral side of the ovary (Krishnan & Dayanandan 2003). Although most carbon uptake into cereal endosperm cells is in the form of Suc *via* SUT-mediated transport, a minor proportion is in the form of hexose sugars produced by apoplastic cleavage of Suc by cell wall invertases, particularly during the early stages of endosperm development (Weschke et al. 2003). Various SUT genes are differentially expressed during cereal development (Aoki et al. 2002) and appear to be particularly sensitive to heat stress (Harris et al. 2023; Mangelsen et al. 2011; Phann et al. 2013; Qin et al. 2008; Xu et al. 2018).

Once in the endosperm cells, Suc is metabolized *via* two possible, and not mutually exclusive, routes; either *via* hydrolytic cleavage by invertases (INV, EC 3.2.1.26) to glucose (Glc) and fructose (Fru), or by sucrose synthase (SuSy, EC 2.4.1.13) into UDP-glucose (UDPGlc) and Fru. Data from multiple cereals indicates that INV-mediated Suc cleavage predominates at early (cell division) stages of endosperm development, approximately 3–4 DAP (Ou-Lee & Setter 1985), where its activities increase sink strength and water influx by providing enhanced turgor gradients (Turgeon 2010) and regulate development-specific sugar signalling (Koch 1996, 2004; Ruan et al. 2010). As the metabolism of endosperm cells shifts from cell division and expansion to storage starch formation and deposition, INV-mediated Suc cleavage declines and SuSy becomes the more dominant mode of Suc breakdown (Figure 3.1a), providing precursors for starch synthesis and becoming a major determinant of sink strength (Black et al. 1995; Chourey et al. 1998; Kato 1995; Stein & Granot 2019; Thévenot et al. 2005). The relative dominance of each mode of Suc cleavage during cereal endosperm development is likely an adaptive response to the increasingly hypoxic cellular conditions prevailing in the growing sink tissues. A number of studies have established that cellular O$_2$ levels sharply decline as the endosperm tissue grows (Radchuk & Borisjuk 2014; Rolletschek et al. 2004, 2005, 2011; Van Dongen et al. 2004), and this hypoxic environment in turn influences Suc metabolism to promote starch synthesis, as well as enhances viability and longevity of the mature seed (Bailly & Kranner 2011; Hendry 1993). The provision of carbon precursors for starch synthesis *via* SuSy reduces the cellular ATP demand relative to that of INV-mediated Suc cleavage within the increasingly hypoxic environment of the

endosperm during grain filling. Hypoxia represses INV gene expression (Zeng et al. 1999) and upregulates the expression of SuSy (Ricard et al. 1991; Zeng et al. 1998).

All of the biochemical reactions involved in building the starch polymer are confined to plastids in higher plants, and in the case of the cereal endosperm, these are the amyloplasts. In the majority of plants, the plastids are also the site of synthesis of the immediate soluble precursor of starch, ADP-glucose (ADPGlc), by the key enzyme and first committed step of the starch synthesis pathway ADPGlc pyrophosphorylase (AGPase, EC 2.7.7.27) (Ghosh & Preiss 1966). However, a singular distinguishing feature of storage starch biosynthesis in the non-photosynthetic cells of developing endosperms in cereals and other Poaceae is the presence of a major form of AGPase located in the cytosol (discussed below) in addition to a minor plastidial AGPase activity (Comparot-Moss & Denyer 2009).

3.4 FORMATION OF ADPGLC BY CYTOSOLIC AGPASE

AGPase catalyzes a reversible (equilibrium) reaction that, *in vivo*, produces ADPGlc and inorganic pyrophosphate (PPi) from glucose-1-phosphate (Glc1P) and ATP. Metabolically, the reaction is pushed in favor of starch synthesis inside plastids by the rapid utilization of ADPGlc by various isoforms of starch synthase (SS, EC 2.4.1.21) and the cleavage of PPi to inorganic phosphate (Pi) by alkaline pyrophosphatase (EC 3.6.1.1, Gross & ap Rees 1986). ADPGlc formation by the cytosolic AGPase of cereal endosperm is likely brought about by the combined actions of PPi-consuming reactions in the cytosol (Kleczkowski 1994; Tiessen et al. 2012) and the rapid transport of ADPGlc across the amyloplast envelope membrane *via* the ADPGlc/ADP antiporter, also known as Brittle1 (Bowsher et al. 2007; Cakir et al. 2016; Kirchberger, Tjaden & Neuhaus 2008; Shannon et al. 1998). The enzyme UDPGlc (UGPase, EC 2.7.7.9) is catalytically very active in endosperm tissue, metabolizing UDPGlc to Glc1P for starch synthesis by using cytosolic PPi (Kleczkowski 1994, 1996) (Figure 3.1b). The provision of PPi by cytosolic AGPase for other PPi-consuming reactions, like UGPase reactions, essentially couples the AGPase reaction in the cereal endosperm (sink) tissue to Suc breakdown. Biochemical analysis of various cereals and wild grasses shows that the majority (70%–90%) of ADPGlc formation in the endosperm is *via* the cytosolic AGPase (Beckles, Smith & ap Rees 2001; Denyer et al. 1996b; Sikka et al. 2001; Thorbjørnsen et al. 1996). The amount of AGPase activity that is extra-plastidial appears to be species specific, as well as dependent on the developmental stage of the endosperm (Tetlow et al. 2003), and in maize, it is as much as 90% (Table 3.1). Studies of AGPase mutants in cereals indicate that cytosolic AGPase is responsible for providing the majority of carbon for storage starch synthesis in amyloplasts (Huang, Hennen-Bierwagen & Myers 2014; Johnson et al. 2003; Lee et al. 2007; Sakulsingharoj et al. 2007).

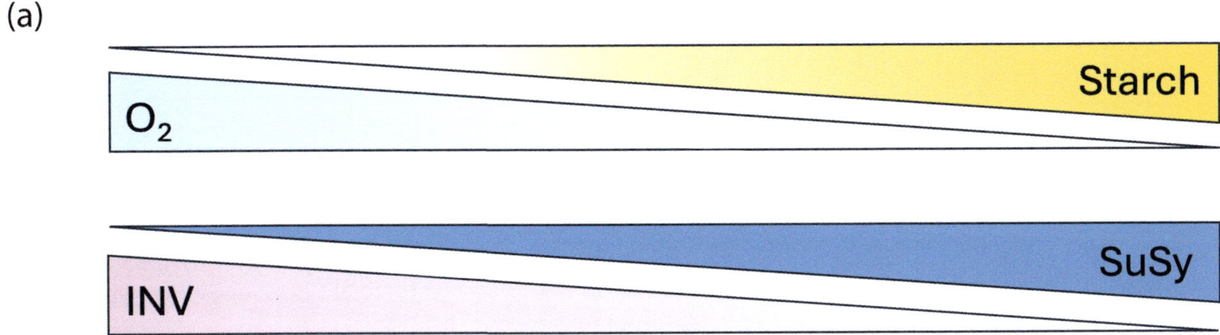

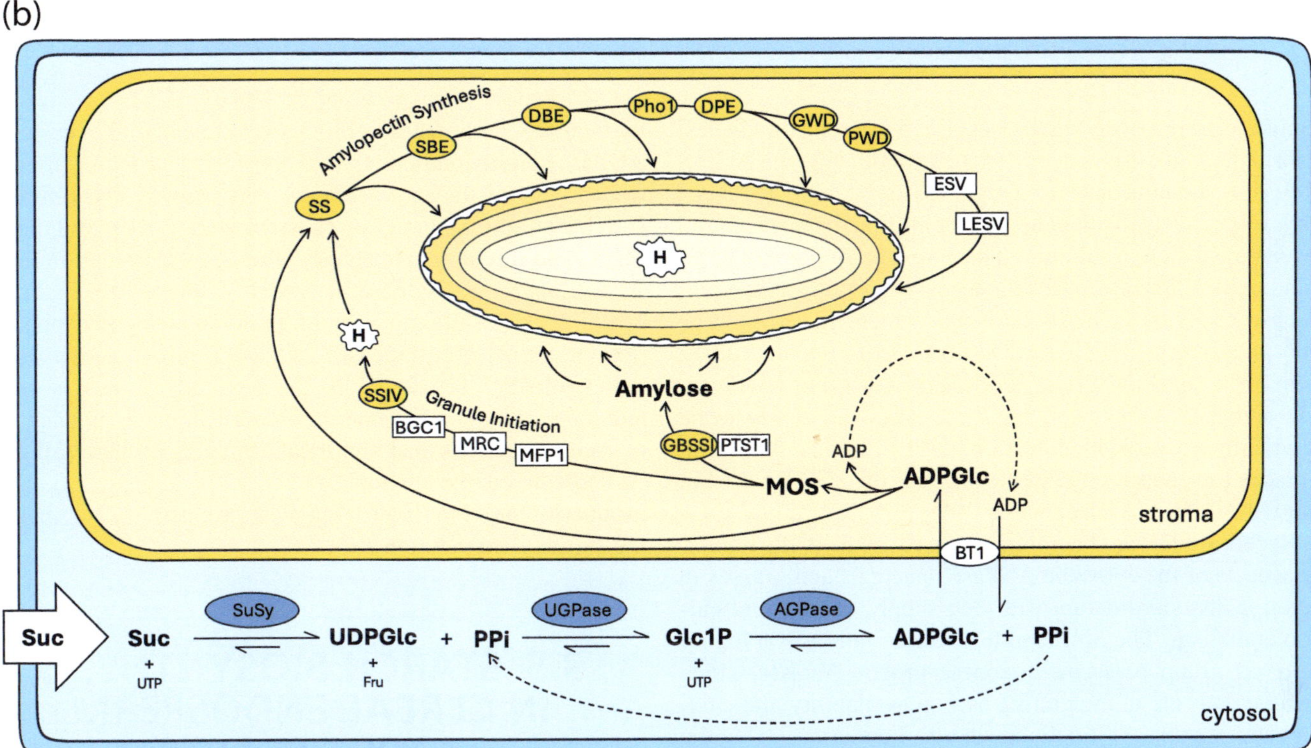

FIGURE 3.1   Starch synthesis in amyloplasts of developing cereal endosperm. (a) Schematic overview of changes in Suc metabolism during cereal endosperm development. O_2 levels rapidly decline as the tissue grows, and starch deposition steadily increases after cell differentiation until maturity. The predominant mode of Suc breakdown in the endosperm switches from INV-mediated to SuSy-mediated Suc cleavage in response to changing metabolic demands and O_2 availability. (b) Simplified diagram of starch synthesis in a cereal endosperm cell. During storage starch deposition SuSy is the predominant mode of Suc breakdown, providing UDPGlc for UGPase, which in turn provides a source of Glc1P for ADPGlc synthesis by AGPase in the cytosol. In the scheme shown, production of ADPGlc by AGPase is coupled to Suc breakdown *via* utilization of cytosolic PPi. Transport of ADPGlc into the amyloplast by the BT1 antiporter is coupled to export of ADP produced by SSs inside the plastid. ADPGlc and MOS are used as substrates for granule initiation and the synthesis of amylose, the latter is deposited within the amylopectin matrix. The amylopectin granule structure formed from the initial hilum (H) is formed *via* various classes of enzymes shown (some with multiple isoforms, not shown) and likely coordinated through protein–protein interactions. Note that the enzyme classes do not work in a sequential manner. Non-enzymatic proteins involved at the various stages of granule formation or amylopectin stabilization are depicted in square boxes. Dashed lines indicate cycling of enzyme products.

Plant AGPases are α2/β2 heterotetramers consisting of two small (α2) subunits (termed AGP-S) and two large (β2) subunits (AGP-L) encoded by multiple genes, unlike the homotetrameric (α4) enzyme found in bacteria (Cifuente et al. 2016; Preiss & Romero 1990). Molecular dynamics simulations of rice AGP-S and AGP-L subunits have suggested that the AGPase heterotetramer can take on multiple conformations (Maharana et al. 2024). The AGP-L and AGP-S subunits are also referred to as SH2 and BT2, respectively, in reference to *shrunken2* and *brittle2* maize mutations of the same subunits (Preiss 2004). The formation of the 200–240 kDa heterotetrameric AGPase protein complex relies on multiple hydrogen bonds formed between the AGP-S and AGP-L subunits (Pandey et al. 2016), with size differences between these

TABLE 3.1　Estimates of cytosolic AGPase activity (%) in cereal endosperms at varying stages of development determined through cell fractionation studies

CEREAL ENDOSPERM	% CYTOSOLIC ACTIVITY (DAP)	REFERENCES
Maize	95% (11–17)	Denyer et al. (1996b)
	≥90% (20)	Shannon et al. (1998)
Rice	90% (10–12)	Sikka et al. (2001)
Wheat	60%–93% (8–30)	Tetlow et al. (2003)
	93%–98% (10–12)	Burton et al. (2002)
Barley	70%–90% (11–13)	Thorbjørnsen et al. (1996)
	80%–90% (14)	Tiessen et al. (2012)
Wild barley (*Hordeum murinum*)	81% (11)	Beckles et al. (2001)
Early cultivated barley (*Hordeum spontaneum*)	76% (13)	Beckles et al. (2001)

DAP, days after pollination.

subunits being relatively small (1–5 kDa, depending on species) (Okita et al. 1990; Preiss et al. 1990). Traditionally, the AGP-S and AGP-L subunits have been regarded as catalytic and regulatory, respectively, but it is likely that all functions are shared between the different subunits. The cytosolic AGPase of cereals and grasses appears to lack the sensitivity to allosteric effectors observed with the plastidial forms, which are activated by 3-phosphoglycerate (3PGA) and inhibited by Pi (Doan, Rudi & Olsen 1999; Ghosh & Preiss 1966; Kleczkowski et al. 1993a, b). In addition, the cytosolic AGPase lacks the ability to be catalytically activated by reduction, unlike the plastidial form, as it lacks the conserved cysteine residues on AGP-S found in plastidial AGPases (Geigenberger 2011).

Recent studies are beginning to reveal some of the regulatory aspects of the cytosolic AGPase in cereals and grasses in relation to its coordination with Suc supply and general carbon metabolism. The SNF1 protein kinase family is a highly conserved group of serine/threonine protein kinases which play a pivotal role in integrating carbon availability and environmental signals with cellular energy homeostasis (Hardie 2007). Transgenic overexpression of potato, sweet potato, and wheat sucrose non-fermenting-1 (SNF1)-related protein kinase 1 (SnRK1a) in tobacco significantly increases AGPase activity (Jiang et al. 2013; Kumar et al. 2024; Wang et al. 2017). Furthermore, cytosolic AGP-L and SnRK1a interact, potentially facilitating this increase in AGPase activity (Kumar et al. 2024). Recently, expression levels of AGPase genes have been found to be regulated by transcription factors such as ZIP28 in wheat (Song et al. 2020) and TCP7 which binds to the AGP-L promoter in maize, regulating expression levels of AGP-L, AGP-S, and several SS isoforms (Ajayo et al. 2022). Table 3.2 comprises a comprehensive list of transcription factors that are directly or indirectly involved in the regulation of various starch synthesis genes in the cereals.

The possible role of protein phosphorylation as a means of post-translational regulation of AGPase activity and protein complex formation has led to several interesting and recent findings. Phosphorylation sites have been identified in AGP-S and AGP-L isoforms of cultivated wheat (Dong et al. 2015; Ferrero et al. 2020), AGP-L phosphorylation was found to increase with wheat grain development and measurable AGPase activity, indicating that phosphorylation may regulate the activity of the large subunit, and possibly the timing of starch synthesis in the grain (Ferrero et al. 2020). Mutation of serine and threonine phosphorylation sites identified in maize AGP-S decreases AGPase activity by almost 50% (Yu et al. 2019; 2023a), while mutation of a known serine phosphorylation site on AGP-L increases AGPase catalytic activity more than twofold (Yu et al. 2023b). Further study of the post-translational regulation of AGPase for modification or enhancement of AGPase activity had significant impacts on increasing the ADPGlc flux to the amyloplast for starch synthesis, making it a productive target for improving plant productivity and increasing starch yield (Section 3.6).

3.5　STARCH BIOSYNTHESIS IN CEREAL ENDOSPERM AMYLOPLASTS

3.5.1　Starch Granule Morphology in Cereal Endosperms

Cereals and wild grasses show a wide variation in temporal patterns of granule initiation and resultant granule morphology (Matsushima et al. 2013). Some of the major cereal crops of the Triticeae tribes (wheat, rye, and barley) and Aveneae (oats) tribes show a characteristic bimodal size distribution of starch grains in their mature seeds. Many forage grasses from the aforementioned tribes, as well as the Poeae tribe, which includes agriculturally important forage species such as ryegrass (*Lolium* sp.), meadow grass (*Poa* sp.), and cocksfoot (*Dactylis glomerata* L.), also show a bimodal distribution of starch granules in their endosperms (Shapter, Henry & Lee 2008; Stoddard & Sarker 2000). The bimodal starch granule size distribution in this largely temperate group of plants comprises large flattened (A-type) granules 10–35 μm in diameter, and small, round B-type granules

TABLE 3.2   Transcription factors shown to modulate expression of genes involved in starch biosynthesis

CEREAL	TRANSCRIPTION FACTOR	UP/DOWNREGULATED GENES (↑/↓)	REFERENCES
Maize	EREB94	↓: SSI	Li et al. (2017)
	EREB156	↑: AGPL(C), SSIIIa	Huang et al. (2016)
	TCP7	↑: GBSSI, BT1, SSI, SSIIa, SSIIIa ↓: AGPL(C), AGPS(C), SSIV	Ajayo et al. (2022)
	O2	↑: SuSy1, SuSy2, SSIII, SSIIa, SSI, GBSSI ↓: SBEI, SBEIIb, BT2, SBEIIb	Jia et al. (2013); Zhang et al. (2016b); Deng et al. (2020)
	PFB1	↑: AGPL(C), AGPS(C) ↓: SBEI, ISA1	Zhang et al. (2016a)
	O11	↑: SSV	Feng et al. (2018)
	NAC128 and NAC130	↑: AGPS(C), GBSSI, SSI, SSIIb, SBEI, PUL, SSIIa, ISA1	Zhang et al. (2019); Chen et al. (2023b)
	NKD1 and NDK2	↑: SuSy2, AGPS(C), AGPL(C), GBSSI, SSI, ISA2, BEIIb	Gontarek et al. (2016)
	NAC36	↑: AGPL2, AGPS2(C), SSI, GBSSII	Zhang et al. (2014)
	bZIP91	↑: AGPS1, SSI, ISA1	Chen et al. (2016)
	ABI4	↑: SSI	Hu et al. (2012)
	ABI19	↑: SuSy1, SuSy2, AGPS(C), SH1, GBSSI, SSI, SSII, SSIV, SSV, PUL	Yang et al. (2021)
	NAC34 *overexpressed in rice	↑: AGPS2b, GBSSI, SSIIa, SSIVa, BEI, BEIIb, ISA3, PUL ↓(3DAP), ↑ (after 3DAP): AGPL1, AGPL2, AGPS1, SSI, SSIIIa, BEIIa, ISA1, ISA2 ↓: GBSSII	Peng et al. (2019)
	PLATZ2	↑: GBSSI, SSI, SSIIIa, BEIIb ↓: AGPS(C), SH2, SSIIa	Li et al. (2021a)
	MADS1a	↑: AGPL2, AGPS1a, GBSSIa, SSIIa, BEIIb, ISA1 ↓: AGPL2, BEI, BEIIa, PUL	Dong et al. (2019)
Wheat	bZIP28	↑: APGL(C), AGPS(C)	Song et al. (2020)
	SPA-B	↓: AGPases, SuSys, GBSSI, SBE, SSI, SSIIa, Pho1	Guo et al. (2020)
	NAC019	↑: SuSy2, SSIIa, SBEIIb	Gao et al. (2021)
	NAC100	↑: SuSy2, GBSSI	Li et al. (2021b)
	NAC-A18	↑: AGPS2a ↓: SuSy1, AGPL2, GBSSI, SSIIIa, PUL, BEIIb	Wang et al. (2023)
	B3-2A1	↑: SuSy1, SuSy5, AGPL3, AGPS2, GBSSI	Xie et al. (2023)
Rice	RSR1	↓: AGPL1, AGPL2, AGPS1, AGPS2b, GBSSI, SSI, SSIIa, SSIIIa, BEI, BEIIb, ISA1, ISA2, ISA3, PUL, Pho1	Fu & Xue (2010)
	SERF1	↑: AGPL2, GBSSI, SSI, SSIIIa	Schmidt et al. (2014)
	RISBZ1/bZIP58	↑: AGPL3, GBSS1, SBEI, Pho1 ↓: AGPS2b, AGPL2, SSI, SSIIIa, SSIVb, BEIIb, ISA2	Wang et al. (2013)
	RPBF	↑: GBSSI, SSIIa	Kawakatsu et al. (2009)
	bZIP60	↑: AGPL2, GBSSI, SBEI, ISA2	Cao et al. (2022)
	NAC20/NAC26	↑: SSI, PUL, DPE1	Wang et al. (2020b)
	APP6	↑: AGPL3, AGPL4, AGPS1, AGPS2a, GBSSa, GBSSb, SSIIb, SSIIc, SSIVa, ISA1, PUL ↓: BE1, BEIIa, BEIIb	Peng et al. (2014a)
	FLO2	↑: AGPL1, AGPL2, AGPL3, AGPL4, AGPS1, AGPS2a, AGPS2b, GBSSI, SSI, SSIIa, SSIIc, SSIIIa, SSIVb, BEI, BEIIb, ISA1, ISA2, PUL	She et al. (2010)
	MADS1	↑: AGPS1, AGPS2, AGPL1, AGPL2, BT1	Liu et al. (2023b)
	Nhd1	↑: SSIIa, SSIIIa ↓: GBSSI, SBE3	Zhang et al. (2023)

Gene regulation was determined by qRT-PCR or yeast-2-hybrid (Y2H) analysis of overexpressed or knocked out transcription factors. AGPL2, AGPL4, AGPS1a, and AGPS2b are cytosolic, while AGPL1, AGPL3, AGPS1, AGPS1b, and AGPS2a are plastidial. Unspecified AGPase isoforms are denoted as "C" for cytosolic localization.

of 2–9 μm. Initiation of A-type and B-type granules is both spatially and temporally separated, and in amyloplasts, A-type granules are observed approximately 4 DAP, followed by initiation of smaller, more numerous B-type granules around 15–20 DAP (Dengate & Meredith 1984; Esch et al. 2023; Evers 1973; Howard et al. 2011; Parker 1985). B-type granules appear to be formed within outgrowths of the amyloplast envelope membrane known as stromules (Esch et al. 2023; Langeveld et al. 2000). Studies in wild wheat (*Aegilops peregrina* Hack.), or goatgrass, identified the B-GRANULE CONTENT1 (BGC1) gene, which controls B-type granule formation in a dose-dependent manner (Chia et al. 2017; Howard et al. 2011). BGC1 is orthologous to FLOURY ENDOSPERM6 (FLO6) (Table 3.3) in rice and barley (Peng et al. 2014b; Saito et al. 2017) and PROTEIN TARGETING TO STARCH2 (PTST2) in the model cereal/ grass Brachypodium (*Brachypodium distachyon* L.) and model dicotyledonous plant Arabidopsis (*Arabidopsis thaliana* L.) (Seung et al. 2017; Watson-Lazowski et al. 2022). In contrast to the A- and B-type starch granules formed in the temperate cereals of the Pooideae, storage starch granules of the major cereals of the Panicoideae (maize, millet, and sorghum) have simple, discrete granules of the order of 2–30 μm, which can be spherical, polyhedral, or lenticular (Buléon et al. 1998; Jane et al. 1994).

Rice is a member of a group of the grasses (including the bamboos, *Bambusa* sp. and *Dendrocalamus* sp.) that forms compound starch granules in the endosperm, which are comprised of tightly packed polyhedral granules arranged in spherical-like structures with membranous cross-walls called septa (Yun & Kawagoe 2010). During compound granule formation, multiple granule initiation events occur within an amyloplast, followed by granule fusion (Matsushima et al. 2013). Compound starch granules may represent the ancestral starch granule morphology, as primitive Panicoid grasses such as *Sorghum italica*, the earliest diverging lineage within the PACMAD clade, produce compound granules in the endosperm (Shapter, Henry & Lee 2008; Tateoka 1962). A group of rice mutants generated through ethyl methanesulfonate (EMS) mutagenesis, which possess the aforementioned FLOURY ENDOSPERM or opaque grain phenotype, have loosely packed starch granules that differ greatly from the typical compound granule structure observed in wild-type rice lines. Numerous FLO mutants that have been characterized are presented in Table 3.3, most of which represent compromised carbohydrate metabolism and, in some cases, abnormal mitochondrial function, which ultimately affects biosynthesis and accumulation of starch.

TABLE 3.3   List of characterized floury (FLO) endosperm mutants from rice and their mutant phenotype

GENE	PROTEIN	CELLULAR LOCALIZATION	MUTANT PHENOTYPE	REFERENCES
flo(a)	TPR domain-containing protein	?	Grain breaks easily, ↓ grain yield, ↓ crystallization, ↓ amylose, ↓ crude protein	Kim et al. (1993); Qiao et al. (2010)
flo1	?	?	Round loosely packed starch granules	Satoh & Omura (1981)
flo2	TPR domain-containing protein	Nucleus	↓ Grain quality, unusual seeds, ↑ amino acid content (lysine and histidine), downregulation of starch biosynthesis and storage protein genes *interacts with FLOC1	Kumamaru, Sato & Satoh (1997); She et al. (2010); Wu et al. (2015); Suzuki et al. (2020)
flo3	16-kDa globulin	Protein bodies	Floury phenotype grain center, ↑ storage proteins (not 16-kDa), delayed growth	Nishio and Iida (1993)
flo4	PPDKB	Cytosol	Loosely packed starch granules, ↑ protein and lipid content	Kang et al. (2005)
flo5	SSIIIa	Plastid	↓ Crystallinity, fewer >DP30 chains	Ryoo et al. (2007)
flo6	PTST2/BGC1	Plastid	Defects in compound granule formation, ↓ starch content	Peng et al. (2014b)
flo7	? (contains DUF1338)	Plastid	Thin and light grains, ↑ soluble sugars, irregular loosely packed starch granules, floury endosperm phenotype on endosperm periphery where FLO7 also strongly localizes	Sheng et al. (2015); L Zhang et al. (2016b)
flo8	UGP1	Cytosol	↓ Starch content, downregulation of starch biosynthesis genes	Long et al. (2017)
flo9	LESV	Plastid	↓ Starch content, ↓ amylopectin content, abnormal amylopectin biosynthesis	Yan et al. (2024)
flo10	P-type PPR protein	Mitochondria	Abnormal aleurone layer cells, abnormal mitochondria, ↓ ATP production, smaller starch grains	Wu et al. (2019)
flo11	Heat shock protein 70	Plastid	Abnormal amyloplast development, ↓ import of proteins into the amyloplast *Hsp70-cp2 associates with TIC on the amyloplast envelope	Zhu et al. (2018)

(Continued)

TABLE 3.3 (Continued)   List of characterized floury (FLO) endosperm mutants from rice and their mutant phenotype

GENE	PROTEIN	CELLULAR LOCALIZATION	MUTANT PHENOTYPE	REFERENCES
flo12	Alanine aminotransferase 1	Cytosol	Loosely packed starch granules, ↓ grain weight, ↓ amylose, ↑ protein content	Zhong et al. (2019)
flo13	Mitochondrial complex I subunit	Mitochondria	Embryonic lethal, abnormal mitochondria, ↓ grain weight	Hu et al. (2018)
flo14	P-type PPR protein	Nucleolus	Seed lethal, abnormal starch biosynthesis	Xue et al. (2019)
flo15	Glyoxalase I protein	Plastid	Defective compound granule formation, ↓ starch content, downregulation of starch synthesis genes	You et al. (2019)
flo16	NAD-dependent malate dehydrogenase	Cytosol	↓ Activity of starch synthesis enzymes, slower grain filling, ↓ grain weight, abnormal compound granules, ↓ starch and amylose content	Teng et al. (2019)
flo18	P-type PPR protein	Mitochondria	Abnormal mitochondria, small starch granules	Yu et al. (2021)
flo19	Glutamine amidotransferase	?	↓ Grain filling, ↓ Rubisco activity, altered expression of starch biosynthesis genes	Lou et al. (2021)
flo20	Serine hydroxymethyltransferase	Nucleus	↓ Starch and protein content, ↑ lipid content, loosely packed starch granules	Yan et al (2022)
flo22	P-type PPR protein	Mitochondria	Downregulation of starch and sucrose metabolism genes, delayed amyloplast development, abnormal mitochondria	Yang et al. (2023)
flo23	Fructose-2,6-bisphosphatase	Cytosol	↓ Starch, lipid, and protein content, irregular compound granules	Chen et al. (2023a)
flo24	Heat shock protein	Plastid	Floury endosperm phenotype exaggerated under heat stress compared to normal temperatures FLO24 interacts with AGPL1 (plastidial), AGPL3 (plastidial), and Pho1	Wu et al. (2024b)
flo26	Single-stranded DNA-binding protein	Mitochondria	Abnormal starch grains, altered transcription levels of mitochondrial genes, ↓ mitochondrial complex I activity and ATP production	Teng et al. (2024)

DUF, domain of unknown function; PPR, pentatricopeptide repeat; TPR, tetratricopeptide repeat.

3.5.2 Initiation of Storage Starch Granules

The initiation of water-insoluble starch granules from simple soluble precursors such as ADPGlc and maltooligosaccharides (MOS) inside plastids is a complex process involving enzymes of the starch metabolic pathway and a number of non-catalytic regulatory proteins. We are still at the early stages of understanding the details of the processes that govern granule initiation, as well as factors controlling starch granule number and morphology, resulting in wide variation of granule number and shape in the plant kingdom (Lindeboom, Chang & Tyler 2004). Much of our initial understanding of this process is derived from studies of the initiation of transient starch granules in leaf chloroplasts of the model plant Arabidopsis, which can be further explored in recent reviews on this subject and Chapter 2 of this volume (Mérida & Fettke 2021; Seung & Smith 2019). Studies on other plant species suggest that the mechanism of starch granule initiation likely differs according to species and tissue type. In amyloplasts of cereal endosperm and other sink organs such as potato tuber, there are typically less granule initiation events (Kram, Oostergetel & van Bruggen 1993), and there is no rebuilding of granules each day from the remnants of nocturnal starch degradation that occurs in leaves (Zhu et al. 2015).

All starch granules grow from a central, structurally disorganized zone referred to as the hilum (Ziegler, Creek & Runt 2005). From Arabidopsis research, it has been determined that STARCH SYNTHASE4 (SSIV, SS4) is required to initiate and form the correct number and size of starch granules in chloroplasts, and that SSIII also contributes to this process in some unknown way (Leterrier et al. 2008; Lu et al. 2018; Szydlowski et al. 2009). Recent studies in wheat (*Triticum* and *Aegilops* species) show that the 4 gene is required for normal A-type starch granule formation, and the loss of SS4 causes the replacement of A-type granules for compound granules (Hawkins et al. 2021). As mentioned above, BGC1 is important for B-type granule formation in the temperate cereals, and its downregulation does not appear to have any effect on the formation of A-type starch granules. However, complete loss of BGC1 causes structural defects in A-type granules, including compound granule formation, which is likely arising from unregulated initiation events (Chia et al. 2020; Saccomanno et al. 2022). In Brachypodium, which normally produces simple granules, the loss of BGC1 causes compound granules to form instead (Watson-Lazowski et al. 2022). Studies in Arabidopsis have shown that SSIV and PTST2 (BGC1) interact *via* coiled-coil domains (Lu et al. 2018; Seung et al. 2017). BGC1 possesses a family 48 carbohydrate binding domain (CBM48) (Janeček, Svensson & MacGregor 2011), which may enable it and the proteins it is associated with to localize to α-glucans present in the nascent starch granule/hilum.

It appears that SSIV is able to synthesize α-glucan structures that are resistant to amylolytic degradation by the plastidial α-amylase (EC 3.2.1.1) AMY3 (Seung et al. 2016). Recent studies have shown that, in addition to BGC1, plastidial starch phosphorylase (EC 2.4.1.1, Pho1) also plays a role in B-granule formation in wheat endosperm but is not necessary for the initiation of A-type granules (Kamble et al. 2023). The role of Pho1 in granule initiation appears to be different in other plant species such as Arabidopsis (Malinova et al. 2014). Both Pho1 and SSIII are capable of unprimed α-glucan synthesis using Glc1P and ADPGlc, respectively, and are therefore plausible donors of α-glucan substrate for SSIV during granule initiation (Malinova et al. 2017; Nakamura et al. 2017). Pho1 is able to form a protein complex with plastidial disproportionating enzyme (DPE1, EC 2.4.1.25), an enzyme involved in α-glucan chain length modification (Hwang et al. 2016b), and this protein complex may provide optimal length α-glucan chains as substrates to drive the granule initiation pathway in the amyloplast.

In addition to BGC1, a number of other non-enzymatic proteins aid the process of granule initiation *via* protein scaffolding functions and localize the granule synthesis machinery to specific sub-organellar locations. Recent studies suggest that a long coiled-coil protein termed MYOSIN-RESEMBLING CHLOROPLAST PROTEIN (MRC, also known as PROTEIN INVOLVED IN STARCH INITIATION1, or PII1) appears to regulate the timing of B-granule formation during wheat endosperm development (Chen et al. 2024). MRC is expressed early in endosperm development (before B-type granules form) and may suppress B-type starch granule initiation through protein–protein interactions with BGC1 (Chen et al. 2024). This study highlights distinctions between the pathways of starch granule initiation in cereals and dicotyledonous leaves of Arabidopsis. In Arabidopsis, MRC is thought to control leaf starch granule initiation through interactions with SSIV and PTST2 (BCG1) since *mrc⁻* mutants produce a single granule as opposed to multiple granules found in the wild type (Seung et al. 2018; Vandromme et al. 2019) Recent studies in potato (*Solanum tuberosum L.*) indicate a paralog of PTST2 (PTST2b), which unlike PTST2a does not interact with SS4, but appears to modulate granule morphology (Hochmuth et al. 2024). In chloroplasts, PTST2 (BGC1) and MRC interact with the thylakoid-associated MAR-BINDING FILAMENT-LIKE PROTEIN1 (MFP1) and with the non-enzymatic domain of SSIV localize granule initiation to specific areas in the plastid (thylakoid membranes in the case of chloroplasts) (Bürgy et al. 2021; Seung et al. 2018; Sharma et al. 2024). Studies in Arabidopsis show that MRC can also interact with the non-enzymic SS isoform SSV (Abt et al. 2020). Amyloplast proteins corresponding to MFP1 may serve a similar role in the process of starch granule initiation in heterotrophic plastids such that initiation occurs in the locality of internal plastid membranes. Ultrastructural studies with young amyloplasts from barley and maize show internal membrane systems that may act as sites of granule initiation in an analogous way to the thylakoids in chloroplasts (Buttrose 1960; Shannon, Garwood & Boyer 2009). There is still much yet to understand about how the starch granule initiation process is regulated in plants. For example, how and when are α-1,6-branch points introduced into the emerging polyglucan structure? How are the various protein–protein interactions described above regulated? More studies are needed in other plant systems and tissues, particularly given the idiosyncrasies observed in the granule initiation processes of the few model species studied to date.

3.5.3 Amylopectin Synthesis

Following its import into the plastidial stroma, ADPGlc produced from cytosolic AGPase is utilized for MOS synthesis for granule initiation, amylose synthesis, and amylopectin synthesis (Figure 3.1b). Starch is a composite of two polyglucans, amylose and amylopectin, the latter making up approximately 70%–80% of the granule (Table 3.4). Following granule initiation, various enzyme classes contribute to the growth of the granule outward from the central hilum, while non-catalytic proteins stabilize the nascent amylopectin complex. The structure of amylopectin is complex, based on α-1,4-linked glucan chains from 3 to >80 DP with α-1,6 branch points, which are arranged in alternating layers to create a higher structural order of repeating amorphous and semi-crystalline regions (Bertoft 2017). The organization of amylopectin clusters in the semi-crystalline regions of different starches is classified into structural allomorphs. The amylopectin of cereal endosperm starches that have mostly tightly packed short α-glucan chains adopt an "A" allomorph, while leaf and tuberous starches with longer, more loosely packed chains tend to adopt the "B" allomorph or a combination "C" type allomorph (Butler & Tetlow 2024; Damager et al. 2010). The cluster (Hizukuri 1986) and building block-backbone (Bertoft 2013) models postulate the organization of the bimodal chain-length distribution (CLD) of amylopectin, which makes up the starch granule. These models are discussed in Chapter 7 and in recent reviews by Bertoft, Blennow, and Hamaker (2024), Tetlow and Bertoft (2020), and Nakamura and Kainuma (2022).

3.5.3.1 α-Glucan chain elongation of amylopectin by starch synthases

Amylopectin synthesis commences with the elongation of α-glucan chains from the initiated starch granule by several SS isoforms. Most SS isoforms employ a distributive mechanism to elongate α-glucan chains by adding a single glucosyl unit from an ADPGlc donor molecule to a preexisting α-glucan chain with an α-1,4-glycosidic bond. SSIV (and likely the non-catalytic SSV) is involved in granule initiation (Section 3.5.2), while SSI, SSII, and SSIII elongate α-glucan chains of increasing lengths on the growing starch granule. Unique to the cereals, each SS isoform is encoded by two separate genes that are differently expressed in the endosperm and leaf, referred to as "a" and "b" (i.e., SSIIa, SSIIb), respectively. These unique tissue-specific forms of SS possess varying kinetic properties and substrate specificities, with expression changing throughout development (Hirose & Terao 2004; Imparl-Radosevich et al. 1999, 2003). This chapter will focus on SS "a" isoforms, which contribute to the synthesis of endosperm starch.

SSI is the most highly expressed of the SS isoforms during early endosperm development (Fujita & Nakamura 2012). SSI elongates short chains of 6–7 glucose units (DP), producing chains of 8–12 DP (Commuri & Keeling 2001), but

TABLE 3.4   Properties of endosperm starches from commonly grown cereal crops

CEREAL CROP	STARCH (%)	GRANULE TYPE	AMYLOSE (%)	AMYLOPECTIN (%)	PEAK CHAIN LENGTH (DP) OF AMYLOPECTIN	BRANCHING DEGREE (%)	STARCH ALLOMORPH
Maize	70–76[a, b]	Simple	22–30[a, c, d, e]	70–78[a, c, d, e]	13, 48[d] 15, 41[f]	4.3[g]	A[h]
Sorghum	70–85[b, i, j]	Simple	11–28[i, k]	89–72[i, k]	12, 18–20[j] 13, 43[l]	—	A[m]
Proso millet	54[n]	Simple	28[n]	72[n]	Cattail Millet	—	A[o]
Finger millet	52–63[n]		21–34[n]	66–79[n]	13, 45[d]		
Foxtail millet	69[n]		28[n]	72[n]			
Rice	72[a]	Compound	15–20[a, d]	80–85[a, d]	12, 46[d] 16, 43[f]	3.4–4.5[p]	A[f]
Oat	58–62[a, b]	Bimodal	19–30[a, q, r]	70–81[a, q, r]	13, 44[s] 15, 23, 46[r]	—	A[t]
Wheat	60–80[a, b, u]	Bimodal	25–28[a, d, u, v]	72–75[a, d, u, v]	12, 41[d] 19, 38[f] 12, 18, 47[v]	5.5–6.8[w]	A[x]
Barley	56–64[a, b]	Bimodal	20–29[a, d, e, v]	71–80[a, d, e, v]	12, 43[d] 12, 18, 47[v]	4.4–5.7[y]	A[z]
Rye	62–73[a, u]	Bimodal	15–30[a, u, v]	70–85[a, u, v]	11, 19, 48[v]	—	A[aa]

Peak chain lengths are chains with the highest proportions per peak on a CLD curve, for amylopectin with bi- or tri-modal distributions.

[a] Simsek, Whitney & Ohm (2012); [b] Herrera-Saldana, Huber & Poore (1990); [c] Wang et al. (1993); [d] Jane et al. (1999); [e] Schirmer et al. (2013); [f] Hizukuri (1985); [g] Mua & Jackson (1997); [h] Cheetham & Tao (1998); [i] Udachan, Sahoo & Hend (2012); [j] Sang et al. (2008); [k] Chanapamokkhot & Thongngam (2007); [l] Gaffa et al. (2004); [m] Zobel (1988); [n] Mahajan et al. (2021); [o] Annor et al. (2014); [p] Van Leeuwen et al. (2021); [q] Hoover & Senanayake (1996); [r] Tester & Karkalas (1996); [s] Xu et al. (2017); [t] Hoover et al. (2003); [u] Verwimp et al. (2004); [v] Fredriksson et al. (1998); [w] Liu et al. (2024); [x] Zhang et al. (2016a); [y] Regina et al. (2010); [z] Hanashiro, Abe and Hizukuri (1996); aa Katz & Van Itallie (1930).

has no specificity for elongation of shorter chains (3–4 DP) (Cuesta-Seijo et al. 2016). SSI rice mutants display an altered CLD profile, with reduced proportions of 6–12 DP chains in endosperm starch, but no alterations to amylopectin crystallinity or granule/grain structure (Fujita et al. 2006). Due to its short chain specificity, SSI will also elongate chains transferred by the endosperm specific SBEIIb, therefore stimulating one another's activities (Abe et al. 2014; Nakamura et al. 2014).

The product α-glucan chain of SSI is elongated by SSII, generating medium length chains of 12–24 DP (Morell et al. 2003). SSIIa expression is typically low in the endosperm, but is critical for proper amylopectin structure and granule formation (Cao et al. 1999; Imparl-Radosevich et al. 2003). *sugary2* mutants in maize, which lack SSII activity (termed *sex6* in barley), show severe changes in amylopectin CLD, starch crystallinity, and granule morphology, with overall reduced starch yield (Imparl-Radosevich et al. 2003; Morell et al. 2003). SSIIa activity in maize endosperm is increased significantly by protein phosphorylation, a process that recruits SSIIa, SSI, and SBEIIb into at least one protein complex (Mehrpouyan et al. 2021). In maize, wheat, and barley, decreased SSIIa activity is marked by an increase in amylose content (Morell et al. 2003; Schoen et al. 2021; Takeda and Preiss 1993; Yamamori et al. 2000; Zhang et al. 2004). Typically, in waxy maize lacking GBSS expression, an increase in gene expression levels is observed for the SS and SBE isoforms in the endosperm from 10 to 15 DAP, followed by a decrease of expression levels by 20 DAP, with the exception of SSIIa in which an increase in expression is observed at 20 DAP onward (Zhang et al. 2024). However, rice with low activities of SSIIa (*japonica* variety) coincides with low expression of GBSSI and reduced amylose content (Sano 1984).

The biological role of SSIII appears to be more complex. Along with its ability to elongate amylopectin chains to >30 DP, SSIII also plays an additional role in granule initiation. The SSIII isoform is unique in that it contains several functional domains and regions that are not found in other SSs, such as the N-terminal SSIII homology domain, which includes two coiled-coil domains for protein–protein interactions and substrate binding (Hennen-Bierwagen et al. 2008, 2009; Lin et al. 2012), as well as a C-terminal 14-3-3 binding site. SSIII also shows preference toward amylose rather than amylopectin or glycogen, indicating that it extends longer chains (Cao, James & Myers 2000). SSIII has been found to form protein complexes with other plastidial enzymes such as the small and large subunits of AGPase and pyruvate phosphate dikinase (EC 2.7.9.1), suggesting that it may directly utilize the relatively low amounts of ADPGlc synthesized within the plastid for unprimed synthesis and subsequent elongation of α-glucan chains (Hennen-Bierwagen et al. 2009; Lappe et al. 2018; Mérida & Fettke 2021; Szydlowski et al. 2009). SSIII has the second highest measurable catalytic activity of the SSs, *ssiii* mutants (*du1*) display clear altered granule morphology and crystalline structure, lacking chains that are >30 DP (Fujita et al. 2007; Ryoo et al. 2007). *ssiii* mutants of hexaploid wheat show alterations in CLD and granule morphology in a dosage-dependent manner (Fahy et al. 2022). This suggests that SSIII may be responsible for the formation of the relatively long interconnecting α-glucan chains bridging the alternating semi-crystalline and amorphous layers of the granule, as proposed in the amylopectin cluster model (James, Denyer & Myers 2003). The loss of SSIII alters activities of other starch biosynthetic enzymes, increasing and reducing SS and SBE activities, respectively (Boyer & Preiss 1981; Cao et al. 1999).

3.5.3.2 Formation of branch points on α-glucan chains by starch branching enzymes

A critical component of starch biosynthesis is starch branching enzymes (SBEs), whose actions play a crucial role in determining starch granule structure. SBEs rearrange linear α-glucan chains by hydrolytic cleavage of an α-1,4 bond and reintroduction of that chain as an α-1,6 bond (branch point). Reattachment of a chain can be either on the same chain (intrachain transfer) or a new chain (interchain transfer) (Borovsky et al. 1979; Drummond, Smith & Whelan 1972; Hawker et al. 1974). SBE activity increases the proportion of α-glucan chains, which in turn stimulates the action of SSs to elongate the non-reducing ends of short-medium length branched chains (Brust et al. 2014). The SS and SBE enzyme classes can form functional heteromeric protein complexes, which cooperatively build the amylopectin molecule (Crofts et al. 2015; Hennen-Bierwagen et al. 2008; Liu et al. 2012a; Tetlow et al. 2008). There are multiple SBE isoforms, each of which have different chain length transfer preferences, and are differentially expressed in various tissue types which could determine the unique structural characteristics of starch from various organs and plants (Nakamura et al. 2010) (Table 3.5). SBE genes are sorted into class I or II, with class I preferentially branching amylose and class II preferentially branching amylopectin *in vitro* (Gao et al. 1997; Takeda, Guan & Preiss 1993). Like SSs, the SBEs are distinguished into tissue-specific "a" and "b" isoforms, typically "a" being more highly expressed in the leaf and "b" more highly expressed in the endosperm, as observed in maize (Gao et al. 1997). In barley endosperm, SBEIIa and SBEIIb genes are expressed at approximately the same level, whereas in wheat, SBEIIa is predominantly expressed in the endosperm (Regina et al. 2005; Sun et al. 1998). In the cereal endosperm, SBEII is the most abundant class of BEs, unlike some dicots such as potato tubers, which predominantly express SBEI (Rydberg et al. 2001).

SBEs consist of a N-terminal domain that controls substrate specificity and chain length transfer preference (Jo et al. 2015; Palomo et al. 2009), a central domain responsible for catalytic activity, α-1,4 bond cleavage and substrate binding (Funane et al. 1998; Kuriki et al. 1996; Libessart & Preiss 1998; Wychowski et al. 2017), and a C-terminal domain thought to determine substrate affinity and catalytic activity (Hong & Preiss 2000; Kuriki, Stewart & Preiss 1997). Studies with recombinant rice SBEI demonstrate that its chain length transfer preference is determined by two amino acid loops, and that these can be interchanged to alter chain length specificity (Gavgani et al. 2022). SBEs are a critical part of starch biosynthesis, as mutation or knockout of these enzymes, specifically the SBEII class, causes atypical amylopectin structure and granule morphology, or lack of starch formation altogether. The *amylose extender (ae⁻)* maize mutant lacking SBEIIb reduces branching frequency of amylopectin and increases the average chain length, increasing the "apparent amylose" content of the starch (Liu et al. 2012b). In wheat, silencing of SBEIIa has been used to increase endosperm amylose content, as SBEIIb expression is low (Sestili et al. 2010). Double knockout mutants of class II BEs in Arabidopsis (*be2/be3* also termed *sbe2.1/sbe2.2*) result in a loss of starch biosynthesis almost entirely (Dumez et al. 2006). Conversely, overexpression of one or more SBEs in rice results in increased branching frequency, which reduces amylopectin crystallinity and increases the production of water-soluble α-glucans (Tanaka et al. 2004). SBEII plays a clearly important role in granule formation of cereal endosperm and leaf starch, while SBEI appears to play a role in determining optimal amylopectin structures for starch breakdown in the endosperm and subsequent seed germination (Pan et al. 2018; Xia et al. 2011). In maize, SBEI is almost exclusively expressed in the endosperm (Blauth et al. 2002), and is suggested to be important in the formation of the amorphous regions of amylopectin, while SBEIIb activity forms the crystalline lamellae (Nakamura 2023). There is a fine balance between amylopectin branching frequency and starch granule structure; thus, SBE expression and activity should be fine-tuned to modify branching frequency without compromising overall granule structure and crystallinity. For a more detailed understanding of SBEs and their contribution to starch biosynthesis, the reader is referred to reviews by Tetlow and Emes (2014) and Nakamura (2023).

TABLE 3.5   Substrate preference of cereal branching enzymes, their chain length transfer preference, and tissue of highest expression

CEREAL CROP	BE ISOFORM	SUBSTRATE PREFERENCE	SUBSTRATE MINIMUM CHAIN LENGTH (DP)	CHAIN LENGTH TRANSFER PREFERENCE (DP)	TISSUE EXPRESSION
Maize	SBEI	Amylose	15[a]	10–13 (up to 30)[b, c]	Endosperm[d]
	SBEIIa	Amylopectin	12[e]	6–7 (up to 11)[e] 7, 11, 23[f]	Leaf, endosperm[g]
	SBEIIb	Amylopectin	12[a]	6–7 (3 up to 9)[c]	Endosperm[d]
Rice	SBEI	Amylose	13, 39[h]	6–12, 29–38[h] 6–12, 26–39[i]	Endosperm[j]
	SBEIIa	Amylopectin	15[i] 12[h]	6–11[h] 6–7 (up to 12)[i]	Leaf[k]
	SBEIIb	Amylopectin	12[h] 13[i]	6–7 (up to 11)[h] 6–7 (up to 8)[i]	Endosperm[j]

[a] Guan et al. (1997); [b] Kuriki et al. (1997); [c] Takeda et al. (1993); [d] Gao et al. (1997); [e] Li et al. (2015); [f] Hernández et al. (2021); [g] Dang and Boyer (1989); [h] Sawada et al. (2009); [i] Nakamura et al. (2010); [j] Mizuno et al. (2001); [k] Ohdan et al. (2005).

3.5.3.3 Pre-amylopectin trimming by starch debranching enzymes and stabilization of the amylopectin complex

Debranching enzymes (DBEs), specifically isoamylases (ISA, EC 3.2.1.68), are key players in the formation of the organized semi-crystalline regions of starch and in promoting the water-insoluble properties of the granule. In plants, ISA hydrolyzes excess α-1,6 glycosidic branches generated by SBEs, "trimming" amylopectin chains, promoting crystallinity and allowing for continuous growth of the starch granule. The α-glucan chains released from the growing granule by ISA activity may be turned over by Pho1 and DPE or used for amylose synthesis by GBSS. There are two types of DBEs in plants, ISA, which debranch amylopectin and glycogen, and pullulanase (PUL, EC 3.2.1.41), which can debranch pullulan, a fungal α-glucan not found in plants. ISA1 and ISA2 are important for amylopectin synthesis and will be the focus of this section, while ISA3 and PUL play roles in starch turnover and can be further explored in reviews on Arabidopsis leaf starch degradation (Smith & Zeeman 2020; Zeeman, Kossmann & Smith 2010). ISA1 is responsible for α-1,6 bond hydrolysis of loosely packed α-glucan chains (Sim et al. 2014), while ISA2 is catalytically inactive but directs ISA1 to the granule surface (Facon et al. 2013; Sundberg et al. 2013). In both monocots and dicots, ISA1 and ISA2 form a functional heterodimeric protein complex for debranching unorganized α-glucan chains on the periphery of the granule. Unique to cereals is a ISA1 homodimer that is sufficient for normal starch granule formation in the endosperm, even in the absence of ISA2 (Kubo et al. 2005; Utsumi et al. 2011). Despite the seemingly redundant function of the ISA1/2 heterodimer and ISA1 homodimer, the ISA1/2 complex appears to be more advantageous under heat stress conditions as observed in rice endosperm (Utsumi & Nakamura, 2006). Complete loss of ISA1 produces the wrinkled seed *sugary1* phenotype, resulting in an accumulation of sucrose and phytoglycogen in the endosperm (Burton et al. 2002; Dinges et al. 2001; James, Robertson & Myers 1995; Kubo et al. 2005). While, a non-functional mutation in ISA1 results in failure to interact with ISA2 and mass accumulation of large phytoglycogen particles in maize endosperm (Inouchi et al. 1983; Takahashi et al. 2020), highlighting the critical role of ISA1 in establishing normal amylopectin structure. Reduced expression of SSIIa and GBSSI is observed in *sugary1* plants, the wrinkled seed phenotype of which can be restored by overexpression of SSIIa, but not GBSSI (Crofts et al. 2022). The *sugary1* phenotype was found to be further exacerbated in *flo6/isa1* mutants of barley, which led to the discovery that ISA1 can also be directed to the starch granule by FLO6 (PTST2) (Matsushima et al. 2023; Peng et al. 2014b). Furthermore, the non-catalytic protein LIKE EARLY STARVATION (LESV) has been found to interact with FLO6 to recruit ISA1 (Yan et al. 2024) or GBSSI (Dong et al. 2024) to the starch granule in rice endosperm, highlighting the importance of both FLO6 and ISA1 in the formation of normal starch.

In addition to physical self-assembly mechanisms that promote semi-crystalline amylopectin structures (Donald 2001; Waigh et al. 1999), recent evidence suggests that non-catalytic proteins EARLY STARVATION (ESV) and LESV may play a role in the organization of these regions (Liu et al. 2023a). These tryptophan-rich granule-bound proteins appear to have distinct functions in stabilizing the surface organization of the starch granule (Feike et al. 2016), LESV1 initiating the alignment and packing of α-glucan double helices within the semi-crystalline region, and ESV1 organizing peripheral α-glucan helices (Liu et al. 2023a). These hypotheses are further supported from studies by Osman et al. (2024) in which it was determined that ESV and LESV show affinity toward amylopectin and not amylose, and that LESV has affinity for glycogen, which possesses structural characteristics of untrimmed amylopectin, supporting its role as a stabilizer in the organization of otherwise disordered α-glucan chains.

3.5.3.4 α-glucan metabolism by plastidial starch phosphorylase and disproportionating enzyme

Plastidial starch phosphorylase (Pho1) is an α-glucan phosphorylase catalyzing a reversible reaction of either addition of Glc1P to a linear α-glucan chain, producing Pi (synthetic direction), or the removal of a Glc unit from an α-glucan chain and formation of Glc1P (degradative direction). Studies in rice endosperm suggest that Pho1 participates in an α-glucan-synthesizing direction, despite high stromal Pi levels which would be expected to promote starch degradation in the amyloplast (Hwang et al. 2010). A second starch phosphorylase gene (Pho2) encodes a cytosolic enzyme whose function in leaves is thought to be heteroglycan metabolism of the nocturnal products of starch degradation (Chia et al. 2004; Lu & Sharkey 2004; Steup, Robenek & Melkonian 1983). Pho1 is most highly expressed in maize endosperm, with peak levels of expression at 12DAP, followed by steadily decreasing expression as the seed matures, while Pho2 expression is highest in the embryo and relatively consistent throughout endosperm development (Yu et al. 2022). For further information on Pho2, the reader is referred to a review by Rathore et al. (2009). Expression levels of Pho1 mirror starch accumulation rates, and the incorporation of Pho1 into various SS and SBE protein complexes or functional interactions (Crofts et al. 2015; Liu et al. 2009; Subasinghe et al. 2014; Tetlow et al. 2004) strongly suggests its involvement in starch biosynthesis. However, studies of *pho1⁻* mutants in rice and Arabidopsis indicate that there is little change in plant phenotype or starch yield unless they are exposed to higher temperatures or lower humidity (Satoh et al. 2008; Zeeman et al. 2004), pointing to a possible role of Pho1 in providing an alternate pathway for starch biosynthesis under abiotic stress. Abscisic acid (ABA) plays a key role in plant response to abiotic stress, and exposure of maize seed to ABA for 12 hours showed increased levels of Pho1 and Pho2 transcription (Yu et al. 2022). The role of Pho1 as a modulator during abiotic stress is further supported by studies in rice, which indicate that Pho1 activity is higher in colder temperatures, and that an intrinsically disordered L80 region plays a role in Pho1 activity and stability at higher temperatures (Hwang et al. 2016a). Complementation of *pho1⁻* rice with Pho1 lacking the L80 region enhances plant growth and carbon assimilation compared to wild-type Pho1. Pho1 may play a wider role in controlling carbon flux in chloroplasts since it has been shown to interact with the PsaC subunit of photosystem I (Koper et al. 2021). For more detailed reviews on Pho1, the reader is referred to reviews by Hwang, Koper & Okita (2020), Rathore et al. (2009), and Shoaib et al. (2021).

The precise role of DPE is still being elucidated. DPE has both a plastidial (DPE1) and cytosolic isoform (DPE2), which, like the respective isoforms of starch phosphorylase, are considered to be involved in plastidial starch metabolism and cytosolic heteroglycan cycling, respectively. DPE1 cleaves maltose from maltotriose and attaches it to longer linear α-glucan chains, producing Glc for carbon metabolism (Niittylä et al. 2004; Weber, Schwacke & Flügge 2005). In wheat endosperm, DPE1 expression increases throughout grain development as well as germination, whereas in barley endosperm, DPE1 expression peaks between 10 and 15 DAP, while DPE2 expression remains consistent though much lower compared to its plastidial counterpart (Bresolin et al. 2006; Radchuk et al. 2009; Vinje et al. 2022). In rice endosperm, DPE1 and DPE2 expression is highest during early grain development, around 3–5 DAP (Akdogan et al. 2011; Dong et al. 2015). Starch degradation is slowed in *dpe1⁻* mutants of Arabidopsis (Critchley et al. 2001), supporting a role for this enzyme in efficient starch degradation and subsequent carbohydrate utilization. DPE1 overexpression and suppression in rice, as well as the characterization of wheat endosperm recombinant DPE1, indicate that DPE1 is also capable of extending chains of amylopectin and glycogen (Bresolin et al. 2006; Dong et al. 2015). Interestingly, Pho1 and DPE1 are found to form a protein complex in rice; when one of these enzymes is knocked down or overexpressed, there is a subsequent change in the activity of the other enzyme (Hwang et al. 2016b). The reader is directed to Mérida and Fettke (2021) for further information on both DPE isoforms.

3.5.3.5 Reversible phosphorylation of amylopectin by α-glucan water dikinases and α-glucan phosphatases

Amylopectin possesses varying amounts of covalently linked phosphate monoesters, which is dependent on the plant species and organ (Hizukuri et al. 1970). Typically, leaf starch contains less glucan phosphate than tubers and storage roots, but considerably more than that found in cereal endosperm storage starch (Blennow et al. 2000a, b; Butler & Tetlow 2024; Nitschke & Schmieder 2021). The mechanism for the phosphorylation of amylopectin is *via* the action of α-glucan water dikinase (GWD, EC 2.7.9.4, also known as R1 protein) or phosphoglucan water dikinase (PWD, EC 2.7.9.5), whereby the β-phosphate of ATP is transferred onto a glucosyl residue of an α-glucan chain (Ritte et al. 2002). Incorporation of phosphate onto amylopectin occurs during polymer synthesis (Nielsen et al. 1994), with the deposition of C6 phosphate being near an α-1,6 branch

point (Compart, Apriyanto & Fettke 2023). Phosphate monoesters are incorporated on glucosyl residues at two positions: the C6 position, which is catalyzed by GWD1, or the C3 position by PWD (GWD3) (Baunsgaard et al. 2005; Ritte et al. 2002, 2004). GWD2 is a cytosolic enzyme that is relatively unstudied and will not be discussed here, though work in Arabidopsis has suggested that GWD2 plays a role in seed development and germination (Pirone et al. 2017). The deposition of phosphate at the C6 carbon is thought to disrupt the interaction between double helices, while C3 phosphate breaks the α-glucan double helix, exposing otherwise hydrophobic areas (Blennow & Engelsen 2010; Hansen et al. 2009). Phosphorylation of amylopectin chains by GWD and PWD is thought to facilitate starch turnover by making the α-glucan chains at the granule surface more accessible to hydrolytic enzymes (Baunsgaard et al. 2005; Kötting et al. 2005; Ritte et al. 2004). The glucan phosphate content in cereal starch is extremely low, typically <0.01% (Blennow et al. 2000b) and is perhaps consistent with its function as a long-term store of carbon. Interestingly, downregulation of GWD in wheat reduces C6 phosphate levels in endosperm starch, and surprisingly causes an increase in vegetative biomass and grain yield (Bowerman et al. 2015; Ral et al. 2012), pointing to GWD as a potential target enzyme for cereal crop improvement (Section 3.6). For a recent review on starch phosphorylation, see You et al. (2020).

The removal of C6 and C3 phosphate residues is catalyzed by a specialized dual-specificity phosphatase termed STARCH EXCESS4 (SEX4, EC 3.1.3.48), named after the corresponding mutation in Arabidopsis that showed a starch excess phenotype due to inefficient breakdown of leaf starch (Kerk et al. 2006; Stitt & Zeeman 2012). SEX4 operates with other phosphatases termed LIKE SEX4-1 (LSF1) and LSF2 (Gentry, Brewer & Vander Kooi 2016; Kötting et al. 2009; Santelia et al. 2011). Specifically, SEX4 removes C6 and C3 phosphate residues, while LSF2 removes only C3 phosphate residues. LSF1 is a non-enzymatic protein that essentially functions as a scaffold to facilitate amylolytic attack for β-amylases (such as BAM1 and BAM3) during starch degradation (Comparot-Moss et al. 2010; Schreier et al. 2019). The precise function of α-glucan phosphatase activity during storage starch synthesis in the developing endosperm is not clear. The reader is further directed to a review on plant α-glucan phosphatases by Meekins, Vander Kooi and Gentry (2016).

3.5.4 Amylose Biosynthesis

Amylose is an essentially linear α-glucan chain made up of α-1,4-glycosidic linkages of variable lengths (Perez-Moral et al. 2018; Takeda, Maruta & Hizukuri 1992) and is sparsely branched (Hanashiro et al. 2008; Hanashiro, Sakaguchi & Yamashita 2013). The chain length or DP of amylose is typically a few hundred to thousand, with an estimated mass of 10^6 Da and therefore much longer than α-glucan chains found in amylopectin (Bertoft 2017; Donald 2001). In cereal endosperms, the amylose content of the starch granule is approximately 25%–30%, but this figure varies widely in other plant species (see Seung 2020 for a review on amylose). The presence of amylose in starches suggests an important biological function, but its precise biological role is not understood (Denyer et al. 2001; Seung et al. 2020). Physical analysis of starch granules using X-ray scattering suggests that amylose is synthesized within the loosely packed regions of amylopectin clusters or blocklets near the granule surface (Jenkins & Donald 1995; Pan & Jane 2000). Amylose is synthesized within the preexisting amylopectin structure by a specialized SS termed granule-bound starch synthase I (GBSSI) in the endosperm which, unlike other SS isoforms, elongates the α-1,4-linked chains using ADPGlc in a processive manner (Denyer et al. 1999). GBSSI activity is stimulated by the presence of MOS (Denyer et al. 1996a; Zeeman, Smith & Smith 2002) and has been shown to be phosphorylated in cereals, though the function of GBSS phosphorylation is undetermined (Ahmed et al. 2015; Grimaud et al. 2008; Liu et al. 2009). GBSS is encoded by the *Waxy* locus, its catalytic activity is responsible for amylose biosynthesis, and its loss in cereals is responsible for amylose-free or *waxy* starches (Shure, Wessler & Fedoroff 1983; Swinkels 1985). The catalytic activity essential for amylose synthesis is only possible through the assistance of a specialized, non-catalytic scaffolding protein termed PROTEIN TARGETING TO STARCH1 (PTST1) (Lohmeier-Vogel et al. 2008; Seung et al. 2015; Wang et al. 2020a). PTST1 is a member of the PTST family (whose members are involved in granule initiation, see above), and it interacts with GBSSI *via* conserved coiled-coil domains and appears to direct the GBSS1/PTST1 complex to the starch granule *via* a conserved CBM48 domain (Seung et al. 2015, 2017). Recent studies show that GBSSI (and GBSSII in chloroplasts) can also be targeted to the starch granule by ESVI and LESV (Dong et al. 2024). The essential role played by PTST1 in amylose synthesis is evidenced by the *waxy* amylose-free phenotype in the *ptst1⁻* Arabidopsis mutant (Seung et al. 2015). The loss of PTST1 in cereal endosperms is, however, less severe, suggesting that the role of PTST1 may be accomplished by other proteins (Wang et al. 2020a; Zhong et al. 2019). Following its targeting to the amylopectin matrix by PTST1, GBSSI becomes entrapped in the granules, and it is solely a starch granule-associated protein (Grimaud et al. 2008; Mu-Forster et al. 1996). GBSSI synthesis of amylose within the amylopectin matrix may prevent its premature degradation by hydrolytic enzymes in the amyloplast.

3.6 IMPROVEMENT OF STARCH BIOSYNTHESIS IN CEREAL ENDOSPERM, CHALLENGES, AND FUTURE PERSPECTIVES

New crop cultivars are selected through conventional plant breeding and through cross breeding of parent plants with valuable traits to produce offspring which share desirable traits (Ahmar et al. 2020). Conventional breeding techniques produce more resilient varieties of crops, but these practices are time consuming, reduce genetic diversity, and rely heavily on plant phenotype as a method of determining crop usefulness (Singh et al. 2023). Aside from the pressure of the ever-expanding global population, abiotic stress due to land atrophy (desertification, saline soils, and land compaction), climate change (temperature stress, drought, and flooding), and biotic stress (herbivory and weed competition) place even more strain on crops and often have significant negative effects on crop yield (Beckles & Thitisaksakul 2014; Thitisaksakul et al. 2012). To address these threats to crop productivity, conventional breeding methods as well as genetic manipulation of cereal crops aim to maintain and elevate cereal yields. For reviews on modern breeding of cereal crops, the reader is directed to Singh et al. (2023) for maize, Hernández-Soto et al. (2021) for rice, Alam, Baenziger and Freis (2024) and Alotaibi et al. (2021) for wheat, as well as a review on the value of wild germplasm in improving modern crops (Mondal, Kumar & Gnanesh 2023).

As described throughout this chapter, knockout or overexpression of individual genes has been useful in determining their function within a plant, particularly within the starch biosynthetic pathway (Sections 3.4 and 3.5). Altering expression levels of specific genes involved in starch biosynthesis can cause a myriad of effects on plant phenotype, yield, and stress resilience. Particularly, modification or overexpression of the key enzyme AGPase has led to significant increases in cereal yield; many of these transgenic events are reviewed in Tuncel and Okita (2013). Some recent biotechnological advances aimed at increasing yield in cereals by targeting AGPase and other enzymes involved in starch biosynthesis are detailed in Table 3.6.

Aside from biotechnological improvements to increase yield, modification of starch biosynthetic machinery can also improve cereal grain quality by increasing nutritional qualities such as resistant starch (Venn & Mann 2004). Levels of resistant starch can be modified by knockdown of SSIIa or SBE isoforms in various cereal crops by increasing average chain lengths of amylopectin, ultimately producing a higher "apparent amylose" content (Table 3.6). Wang et al. (2017) provide a review of SBE knockdown in the cereals, which is the primary method for increasing amylose content in cereal crops. High amylose starches are sought after as they can lower the glycemic load of carbohydrate heavy foods, and thus, the reader is referred to several reviews on cereal grain improvement and the beneficial impacts of resistant starch for human health (Baptista et al. 2024; Lafiandra, Riccardi & Shewry 2014). Further information on high amylose starches of maize or wheat and their uses are further detailed in Obadi, Qi and Xu (2023) and Li, Dhital and Gidley (2023), respectively.

Ultimately, there are many challenges to be faced when considering biotechnological improvement of crops, which prevents or slows crops with novel traits from reaching the market. While manipulation of endogenous gene expression or transgene expression can be a quick, more targeted, and potentially a more productive alternative to conventional breeding methods, approval of genetically modified organisms for human consumption can take 3–5 years in addition to the time already invested developing these lines (De Faria & Wieck 2016). As the incidence of abiotic stress faced by crops increases, these non-conventional methods of increasing crop productivity and nutrition have been more widely

TABLE 3.6   Biotechnological improvement of cereal crops to increase starch/grain yield and improve nutritional quality of the grain by modification of starch biosynthetic machinery

PLANT	MODIFICATION	EFFECT	REFERENCES
Yield Improvement			
Maize	CW-INV constitutive expression	↑ Grain yield (145%), ↑ starch content (20%)	Li et al. (2013)
	SuSy1 overexpression	↑ Amylose content (41%–69%), ↑ starch content (5%–13%), ↑ grain weight (19%–26%), ↑ starch granule diameter	Li et al. (2023)
Rice (*japonica*)	AGPase overexpression	↑ Vegetative biomass (61%), ↑ seed number (51%)	Oiestad et al. (2016)
Rice	*Zm*CBM48-1 transgenic overexpression	↑ Starch and protein content, ↓ soluble sugars, upregulated expression of starch biosynthesis genes	Peng et al. (2022)
Wheat	AGPS1b (plastidial) overexpression	↑ Spikelet weight (9%–11%), ↑ starch content (11%–14%)	Yang et al. (2017)
	GWD downregulation	↓ C6 phosphate, ↑ vegetative biomass, ↑ grain yield (26%–29%)	Ral et al. (2012)
Quality Improvement			
Barley	SBEI, SBEIIa, SBEIIb RNAi silencing	↑ Amylose (36%)	Carciofi et al. (2012)
Barley	SSIIa mutation	↑ Amylose (46%), ↓ grain weight (30%)	Morell et al. (2003)
Wheat	SBEIIa, SBEIIb RNAi silencing	↑ Amylose (50%)	Regina et al. (2006)
Wheat	SSIIa triple mutation	↑ Amylose (4%)	Schoen et al. (2021)

CW, cell wall; SDM, site-directed mutagenesis; RNAi, RNA interference; ↑, increased; ↓, reduced.

accepted. Hefferon (2015) outlines the increased use of GMOs in developed and developing countries, and Huesing et al. (2016) describe the benefits that adoption of GMO crops has provided for farmers and consumers as the global demand for food continues to rise.

Understanding starch biosynthesis in cereal endosperm is therefore key to effective design and implementation of biotechnological strategies that provide significant improvements to crop yield and grain nutritional quality. As pressures on the global food supply continue to rise and arable land diminishes, we must rationally consider and accept the use of biotechnological tools to sustain human life. Thus, it is important that we also understand the societal, ethical, and environmental challenges that must be faced on a global scale in order to promote plant biotechnology as an important tool for crop improvement for the future of mankind.

REFERENCES

Abe, N, Asai, H, Yago, H, Oitome, NF, Itoh, R, Crofts, N, Nakamura, Y & Fujita, N 2014. Relationships between starch synthase I and branching enzyme isozymes determined using double mutant rice lines. *BMC Plant Biology, 14*, 80.

Abbo, S, Lev-Yadun, S & Gopher, A 2010. Agricultural origins: centers and noncenters; a near eastern reappraisal. *Critical Reviews in Plant Science, 29*, 317–328.

Abt, MR, Pfister, B, Sharma, M, Eicke, S, Bürgy, L, Neale, I, Seung, D & Zeeman, SC 2020. STARCH SYNTHASE5, a noncanonical starch synthase-like protein, promotes starch granule initiation in Arabidopsis. *The Plant Cell, 32*, 2543-2565.

Ahmar, S, Gill, RA, Jung, K-H, Faheem, A, Qasim, MU, Mubeen, M & Zhou, W 2020. Conventional and molecular techniques from simple breeding to speed breeding in crop plants: recent advances and future outlook. *International Journal of Molecular Science, 21*, art. 2590.

Ahmed, Z, Tetlow, IJ, Ahmed, R, Morell, MK & Emes, MJ 2015. Protein-protein interactions among enzymes of starch biosynthesis in high-amylose barley genotypes reveal differential roles of heteromeric enzyme complexes in the synthesis of A and B granules. *Plant Science, 233*, 95–106.

Ajayo, BS, Li, Y, Wang, Y, Dai, C, Gao, L, Liu, H, Yu, G, Zhang, J, Huang, Y & Hu, Y 2022. The novel ZmTCP7 transcription factor targets AGPase-encoding gene *ZmBT2* to regulate storage starch accumulation in maize. *Frontiers in Plant Science, 13*, art. 943050.

Akdogan, G, Kubota, J, Kubo, A, Takaha, T & Kitamura, S 2011. Expression and characterization of rice disproportionating enzymes. *Journal of Applied Glycoscience, 58*, 99–105.

Alam, M, Baenziger, PS & Frels, K 2024. Emerging trends in wheat (*Triticum* spp.) breeding: implications for the future. *IMR Press, 16*, art. 2.

Alotaibi, F, Alharbi, S, Alotaibi, M, Al Mosallam, M, Motawei, M & Alrajhi, A 2021. Wheat omics: classical breeding to new breeding technologies. *Saudi Journal of Biological Sciences, 28*, 1433–1444.

Annor, GA, Marcone, M, Bertoft, E & Seetharaman, K 2014. Physical and molecular characterization of millet starches. *Cereal Chemistry, 91*, 286–292.

Anon 2003. Diet, nutrition and the prevention of chronic diseases. In *WHO Technical Report Series*, Geneva, Vol. 916, pp. 1–150.

Aoki, N, Whitfeld, P, Hoeren, F, Scofield, G, Newell, K, Patrick, J, Offler, C, Clarke, Rahman, S & Furbank, RT 2002. Three sucrose transporter genes are expressed in the developing grain of hexaploid wheat. *Plant Molecular Biology, 50*, 453-462.

Austin, R, Bingham, J, Blackwell, R, Evans, L, Ford, M, Morgan, C & Taylor, M 1980. Genetic improvements in winter wheat yields since 1900 and associated physiological changes. *Journal of Agricultural Science, 94,* 52–59.

Awika, J 2011. Major cereal grains production and use around the world. In: Awika, JM, Piironen, V & Bean, S (eds.), *Advances in Cereal Science: Implications to Food Processing and Health Promotion.* American Chemical Society, Atlantic City, NJ, Washington, DC. pp. 1–13.

Bailly, C & Kranner, I 2011. Analyses of reactive oxygen species and antioxidants in relation to seed longevity and germination. *Methods in Molecular Biology, 773,* 343–367.

Baptista, NT, Dessalles, R, Illner, A-K, Ville, P, Ribet, L, Anton, PM & Durand-Dubief, M 2024. Harnessing the power of resistant starch: a narrative review of its health impact and processing challenges. *Frontiers in Nutrition, 11,* art. 1369950.

Barnabás, B, Jager, K & Fehér, A 2008. The effect of drought and heat stress on reproductive processes in cereals. *Plant Cell Environment, 31,* 11–38.

Basunia, MA & Nonhebel, HM 2019. Hormonal regulation of cereal endosperm development with a focus on rice (*Oryza sativa*). *Functional Plant Biology, 46,* 493–506.

Battisti, DS & Naylor, RL 2009. Historical warnings of future food insecurity with unprecedented seasonal heat. *Science, 323,* 240–244.

Baunsgaard, L, Lütken, H, Mikkelsen, R, Glaring, MA, Pham, TT & Blennow, A 2005. A novel isoform of glucan, water dikinase phosphorylates pre-phosphorylated alpha-glucans and is involved in starch degradation in Arabidopsis. *Plant Journal, 41,* 595–605.

Beckles, D, Smith, A & ap Rees, T 2001. A cytosolic ADP-glucose pyrophosphorylase is a feature of graminaceous endosperms, but not of other starch storing organs. *Plant Physiology, 125,* 818–827.

Beckles, DM & Thitisaksakul, M 2014. How environmental stress affects starch composition and functionality in cereal endosperm. *Starch/Stäerke, 66,* 58–71.

Bennett, E, Roberts, JA & Wagstaff, C 2012. Manipulating resource allocation in plants. *Journal of Experimental Botany, 63,* 3391–3400.

Bertoft, E 2013. On the building block and backbone concepts of amylopectin structure. *Cereal Chemistry, 90,* 294–311.

Bertoft, E 2017. Understanding starch structure: recent progress. *Agronomy, 7,* 56.

Bertoft, E, Blennow, A & Hamaker, BR 2024. Perspectives on starch structure, function, and synthesis in relation to the backbone model of amylopectin. *Biomacromolecules, 25,* 5389-5401.

Bihmidine, S, Hunter, C, Johns, C, Koch, K & Braun, D 2013. Regulation of assimilate import into sink organs: update on molecular drivers of sink strength. *Frontiers in Plant Science, 4,* 177.

Black CC, Lobodia, T, Chen, JQ & Sung, S JS 1995. Can sucrose cleavage enzymes serve as markers for sink strength and is sucrose a signal molecule during plant sink development? *American Society of International Symposium on Sucrose Metabolism, American Society of Plant Physiologists.* Mar del Plata, Argentina, pp. 49–64.

Blauth, SL, Kim, KN, Klucinec, J, Shannon, JC, Thompson, DB & Guiltinan, M 2002. Identification of Mutator insertional mutants of starch-branching enzyme 1 (sbe1) in *Zea* mays L. *Plant Molecular Biology, 48,* 287–297.

Blennow, A, Bay-Smidt, AM, Olsen, CE & Møller, BL 2000a. The distribution of covalently bound phosphate in the starch granule in relation to starch crystallinity. *International Journal of Biological Macromolecules, 27,* 211–218.

Blennow, A & Engelsen, SB 2010. Helix-breaking news: fighting crystalline starch energy deposits in the cell. *Trends in Plant Science, 15,* 236–240.

Blennow, A, Engelsen, SB, Munck, L & Møller, BL 2000b. Starch molecular structure and phosphorylation investigated by a combined chromatographic and chemometric approach. *Carbohydrate Polymers, 41,* 163–174.

Boehlein, SK, Liu, P, Webster, A, Ribeiro, C, Suzuki, M, Wu, S, Guan, J-C, Stewart, JD, Tracy, WF, Settles, AM, McCarty, DR, Koch, KE, Hannah, LC, Hennen-Bierwagen, TA & Myers, AM 2019. Effects of long-term exposure to elevated temperature on *Zea* mays endosperm development during grain fill. *Plant Journal, 99,* 23–40.

Borovsky, D, Smith, EE, Whelan, WJ, French, D & Kikumoto, S 1979. The mechanism of Q-enzyme action and its influence on the structure of amylopectin. *Archives of Biochemistry and Biophysics, 198,* 627–631.

Bosnes, M, Weideman, F & Olsen, O-A 1992. Endosperm differentiation in barley wild-type and sex mutants. *Plant Journal, 2,* 661–674.

Bowerman, AF, Newberry, M, Dielen, A-S, Whan, A, Larroque, O, Pritchard, J, Gubler, F, Howitt, CA, Pogson, BJ, Morell, MK & Ral, J-P 2015. Suppression of glucan, water dikinase in the endosperm alters wheat grain properties, germination and coleoptile growth. *Plant Biotechnology Journal, 14,* 398–408.

Bowsher, CG, Scrase-Field, EFAL, Esposito, S, Emes, MJ & Tetlow, IJ 2007. Characterization of ADP-glucose transport across the cereal endosperm amyloplast envelope. *Journal of Experimental Botany, 58,* 1321–1332.

Boyer, CD & Preiss, J 1981. Evidence for independent genetic control of the multiple forms of maize endosperm branching enzymes and starch synthases. *Plant Physiology, 67,* 1141–1145.

Bresolin, NS, Li, Z, Kosar-Hashemi, B, Tetlow, IJ, Chatterjee, M, Rahman, S, Morell, MK & Howitt, CA 2006. Characterisation of disproportionating enzyme from wheat endosperm. *Planta, 224,* 20–31.

Briarty, LG, Hughes, CE & Evers, AD 1979. The developing endosperm of wheat—a stereological analysis. *Annals of Botany, 44,* 641–658.

Brocklehurst, P 1977. Factors controlling grain weight in wheat. *Nature, 266,* 348–349.

Brown, R, Lemmon, B & Olsen, O 1996. Development of the endosperm in rice (*Oryza sativa* L.): cellularization. *Journal of Plant Research, 109,* 301–313.

Brust, H, Lehmann, T, D'Hulst, C & Fettke J 2014. Analysis of the functional interaction of Arabidopsis starch synthase and branching enzyme isoforms reveals that the cooperative action of SSI and BEs results in glucans with polymodal chain length distribution similar to amylopectin. *PLoS One, 9,* art. 102364.

Buléon, A, Colonna, P, Planchot, V & Ball, S 1998. Starch granules: structure and biosynthesis. *International Journal of Biological Macromolecules, 23,* 85–112.

Bürgy, L, Eicke, S, Kopp, C, Jenny, C, Lu, KJ, Escrig, S, Meibom, A & Zeeman, SC. 2021. Coalescence and directed anisotropic growth of starch granule initials in subdomains of *Arabidopsis thaliana* chloroplasts. *Nature Communications, 12,* art. 6944.

Burton, RA, Jenner, H, Carrangis, L, Fahy, B, Fincher, GB, Hylton, C, Laurie, DA, Parker, M, Waite, D, Van Wegen, S, Verhoeven, T & Denyer, K 2002. Starch granule initiation and growth are altered in barley mutants that lack isoamylase activity. *Plant Journal, 31,* 97–112.

Butler, V & Tetlow, IJ 2024. Starch synthesis in plants. In: Nilsson, L (ed.), *Starch in Food. Structure, Function, and Applications,* Elsevier: Massachusetts, United States, pp. 1–34.

Buttrose, MS 1960. Submicroscopic development and structure of starch granules in cereal endosperms. *Journal of Ultrastructure Research, 4,* 231–257.

Cakir, B, Shiraishi, S, Tuncel, A, Matsusaka, H, Satoh, R, Singh, S, Crofts, N, Hosaka, Y, Fujita, N, Hwang, SK, Satoh, H & Okita, T 2016. Analysis of the rice ADP-glucose transporter (OsBT1) indicates the presence of regulatory processes in the amyloplast stroma that control ADP-glucose flux into starch. *Plant Physiology,* *170*, 1271–1283.

Cao, H, Imparl-Radosevich, J, Guan, H, Keeling, PL, James, MG & Myers, AM 1999. Identification of the soluble starch synthase activities of maize endosperm. *Plant Physiology, 120*, 205–215.

Cao, H, James, MG & Myers, AM 2000. Purification and characterization of soluble starch synthases from maize endosperm. *Archives of Biochemistry and Biophysics, 373*, 135–146.

Cao, R, Zhao, S, Jiao, G, Duan, Y, Ma, L, Dong, N, Lu, F, Zhu, M & Shao, G 2022. *OPAQUE3*, encoding a transmembrane bZIP transcription factor, regulates endosperm storage protein and starch biosynthesis in rice. *Plant Communications, 3*, art. 100463.

Carciofi, M, Blennow, A, Jensen, SL, Shaik, SS, Henriksen, A, Buléon, A, Holm, PB & Hebelstrup, KH 2012. Concerted suppression of all starch branching enzyme genes in barley produces amylose-only starch granules. *BMC Plant Biology, 12*, art. 223.

Chanapamokkhot, H & Thongngam, M 2007. The chemical and physico-chemical properties of sorghum starch and flour. *Agriculture and Natural Resources, 41*, 343–349.

Cheetham, NWH & Tao, L 1998. Variation in crystalline type with amylose content in maize starch granules: an X-ray powder diffraction study. *Carbohydrate Polymers, 36*, 277–284.

Chen, E, Yu, H, He, J, Peng, D, Zhu, P, Pan, S, Wu, X, Wang, J, Ji, C, Chao, Z, Xu, Z, Wu, Y, Chao, D, Wu, Y & Zhang, Z 2023b. The transcription factors ZmNAC128 and ZmNAC130 coordinate with Opaque2 to promote endosperm filling in maize. *The Plant Cell, 35*, 4066–4090.

Chen, J, Chen, Y, Watson-Lazowski, A, Hawkins, E, Barclay, JE, Fahy, B, Bowers, RD, Corbin, K, Warren, FJ, Blennow, A, Uauy, C & Seung, D 2024. Wheat MYOSIN-RESEMBLING CHLOROPLAST PROTEIN controls B-type starch granule initiation timing during endosperm development. *Plant Physiology, 196*, 1980-1196.

Chen, J, Yi, Q, Cao, Y, Wei, B, Zheng, L, Xiao, Q, Xie, Y, Gu, Y, Li, Y & Huang, H 2016. ZmZIP91 regulates expression of starch synthesis-related genes by binding to ACTCAT elements in their promoters. *Journal of Experimental Botany, 67*, 1327–1338.

Chen, X, Ji, Y, Zhao, W, Niu, H, Yang, X, Jiang, X, Zhang, Y, Lei, J, Yang, H, Chen, R, Gu, C, Xu, H, Dong, H, Duan, E, Teng, X, Wang, Y, Zhang, Y, Zhang, W, Wang, Y & Wan, J 2023a. Fructose-6-phosphate-2-kinase/fructose-2,6-bisphosphatase regulates energy metabolism and synthesis of storage products in developing rice endosperm. *Plant Science, 326*, art. 111503.

Chia, T, Adamski, NM, Saccomanno, B, Greenland, A, Nash, A, Uauy, C & Trafford, K 2017. Transfer of a starch phenotype from wild wheat to bread wheat by deletion of a locus controlling B-type starch granule content. *Journal of Experimental Botany, 68*, 5497–5509.

Chia, T, Chirico, M, King, R, Ramirez-Gonzalez, R, Saccomanno, B, Seung, D, Simmonds, J, Trick, M, Uauy, C, Verhoeven, T & Trafford, K 2020. A carbohydrate-binding protein, B-GRANULE CONTENT 1, influences starch granule size distribution in a dose-dependent manner in polyploid wheat. *Journal of Experimental Botany, 71*, 105–115.

Chia, T, Thorneycroft, SM, Chapple, A, Messerli, G, Chen, J, Zeeman, SC, Smith, SM & Smith, AM 2004. A cytosolic glucosyltransferase is required for the conversion of starch to sucrose in Arabidopsis leaves at night. *Plant Journal, 37*, 853–863.

Chojecki, A, Bayliss, M & Gale, M 1986a. Cell production and DNA accumulation in the wheat endosperm, and their association with grain weight. *Annals of Botany (Lond), 58*, 809–817.

Chojecki, A, Gale, M & Bayliss, M 1986b. The number and sizes of starch granules in the wheat endosperm, and their association with grain weight. *Annals of Botany (Lond), 58*, 819–831.

Chourey, PS, Taliercio, EW, Carlson, SJ & Ruan, YL 1998. Genetic evidence that the two isozymes of sucrose synthase present in developing maize endosperm are critical, one for cell wall integrity and the other for starch biosynthesis. *Molecular Genomics and Genetics, 259*, 88–96.

Chowdhury, SI & Wardlaw, IF 1978. The effect of temperature on kernel development in cereals. *Australian Journal of Agricultural Research, 29*, 205–223.

Cifuente, JO, Comino, N, Madriaga-Marcos, J, López-Fernández, S, García-Alija, M, Agirre, J, Albesa-Jové, D & Guerin, ME 2016. Structural basis of glycogen biosynthesis regulation in bacteria. *Structure, 24*, 1613–1622.

Cochrane, MP 1983. Morphology of the crease region in relation to assimilate uptake and water loss during caryopsis development in barley and wheat. *Australian Journal of Plant Physiology, 10*, 473–491.

Commuri, P & Jones, R 1999. Ultrastructural characterization of maize (*Zea mays* L.) kernels exposed to high temperature during endosperm cell division. *Plant Cell Environment, 22*, 375–385.

Commuri, PD & Keeling, PL 2001. Chain-length specificities of maize starch synthase I enzyme: studies of glucan affinity and catalytic properties. *Plant Journal, 25*, 475–486.

Comparot-Moss, S & Denyer, K 2009. The evolution of the starch biosynthetic pathway in cereals and other grasses. *Journal of Experimental Botany, 60*, 2481–2492.

Comparot-Moss, S, Kotting, O, Stettler, M, Edner, C, Graf, A, Weise, SE & Smith, AM 2010. A putative phosphatase, LSF1, is required for normal starch turnover in Arabidopsis leaves. *Plant Physiology, 152*, 685–697.

Compart, J, Apriyanto, A & Fettke, J 2023. Glucan, water dikinase (GWD) penetrates the starch granule surface and introduces C6 phosphate in the vicinity of branching points. *Carbohydrate Polymers, 321*, art. 121321.

Critchley, JH, Zeeman, SC, Tahaka, T, Smith, AM & Smith, SM 2001. A critical role for disproportionating enzyme in starch breakdown is revealed by a knock-out mutant in Arabidopsis. *The Plant Journal 26*, 89–100.

Crofts, N, Abe, N, Oitome, NF, Matsushima, R, Hayashi, M, Tetlow, IJ, Emes, MJ, Nakamura, Y & Fujita, N 2015. Amylopectin biosynthetic enzymes from developing rice seed form enzymatically active protein complexes. *Journal of Experimental Botany, 66*, 4469–4482.

Crofts, N, Domon, A, Miura, S, Hosaka, Y, Oitome, NF, Itoh, A, Noge, K & Fujita, N 2022. Starch synthases SSIIa and GBSSI control starch structure but do not determine starch granule morphology in the absence of SSIIIa and SSIVb. *Plant Molecular Biology, 108*, 379–398.

Cuesta-Seijo, JA, Nielsen, MM, Ruzanski, C, Krucewicz, K, Beeren, SR, Rydhal, MG, Yoshimura, Y, Striebeck, A, Motawia, MS, Willats, WGT & Palcic, MM 2016. *In vitro* biochemical characterization of all barley endosperm starch synthases. *Frontiers in Plant Science, 6*, 1–17.

Dai, D, Ma, Z & Song, R 2021. Maize endosperm development. *Journal of Integrative Plant Biology, 63*, 613–627.

Damager, I, Engelsen, SB, Blennow, A, Møller, BL & Motawia, MS 2010. First principles insight into the a-glucan structures of starch: their synthesis, conformation, and hydration. *Chemical Reviews, 110*, 2049–2080.

Dang, PL & Boyer, CD 1989. Comparison of soluble starch synthases and branching enzymes from leaves and kernels of normal and amylose-extender maize. *Biochemical Genetics, 27*, 521-532.

Darwin, C 1859. *The Origin of Species*. Wordsworth Editions Ltd, Hertfordshire, UK.

De Faria, RN & Wieck, C 2016. Regulatory differences in the approval of GMOs: extent and development over time. *World Trade Review, 15*, 85–108.

Deng, Y, Wang, J, Zhang, Z & Wu, Y 2020. Transactivation of *Sus1* and *Sus2* by Opaque2 is an essential supplement to sucrose synthase-mediated endosperm filling in maize. *Plant Biotechnology Journal, 18*, 1897–1907.

Dengate, H & Meredith, P 1984. Variation in size distribution of starch granules from wheat grain. *Journal of Cereal Science, 2*, 83–90.

Denyer, K, Clarke, B, Hylton, C, Tatge, H & Smith, AM 1996a. The elongation of amylose and amylopectin chains in isolated starch granules. *The Plant Journal, 10*, 1135–1143.

Denyer, K, Dunlap, F, Thorbjornsen, T, Keeling, P & Smith, A 1996b. The major form of ADP-glucose pyrophosphorylase in maize endosperm is extra-plastidial. *Plant Physiology, 112*, 779–785.

Denyer, K, Johnson, P, Zeeman, S & Smith, AM 2001. The control of amylose synthesis. *Journal of Plant Physiology, 158*, 479–487.

Denyer, K, Waite, D, Motawia, S, Møller, BL & Smith, AM 1999. Granule-bound starch synthase I in isolated starch granules elongates malto-oligosaccharides processively. *Biochemical Journal, 340*, 183–191.

Diamond, J 2002. Evolution, consequences and future of plant and animal domestication. *Nature, 418*, 700–707.

Dinges, JR, Colleoni, C, Myers, AM & James, MG 2001. Molecular structure of three mutations at the maize *sugary1* locus and their allele-specific phenotypic effects. *Plant Physiology, 125*, 1406–1418.

Doan, DNP, Rudi, H & Olsen, OA 1999. The allosterically unregulated isoform of ADP-glucose pyrophosphorylase from barley endosperm is the most likely source of ADP-glucose incorporated into endosperm starch. *Plant Physiology, 121*, 965–975.

Donald, AM 2001. Plasticization and self-assembly in the starch granule. *Cereal Chemistry, 78*, 307–314.

Dong, N, Jiao, G, Cao, R, Li, S, Zhao, S, Duan, Y, Ma, L, Li, X, Lu, F, Wang, H, Wang, S, Shao, G, Sheng, Z, Hu, S, Tang, S, Wei, X & Hu, P 2024. OsLESV and OsESV1 promote transitory and storage starch biosynthesis to determine rice grain quality and yield. *Plant Communications, 5*, art. 100893.

Dong, Q, Wang, F, Kong, J, Xu, Q, Li, T, Chen, L, Chen, H, Jiang, H, Li, C & Cheng, B 2019. Functional analysis of *ZmMADS1a* reveals its role in regulating starch biosynthesis in maize endosperm. *Scientific Reports, 9*, art. 3253.

Dong, S & Beckles, DM 2019. Dynamic changes in the starch-sugar interconversion within plant source and sink tissues promote a better abiotic stress response. *Journal of Plant Physiology, 234–235*, 80–93.

Dong, X, Zhang, D, Liu, J, Liu, QQ, Liu, H, Tian, L, Jiang, L & Qu, LQ 2015. Plastidial disproportionating enzyme participates in starch synthesis in rice endosperm by transferring maltooligosyl groups from amylose and amylopectin to amylopectin. *Plant Physiology, 169*, 2496–2512.

Drummond, GS, Smith, EE & Whelan, WJ 1972. Purification and properties of potato: α-1,4-glucan 6-glycosyltransferase (Q-enzyme). *European Journal of Biochemistry, 26*, 168–176.

Dumez, S, Wattebled, F, Dauvillée, D, Delvalle, D, Planchot, V, Ball, SG & D'Hulst, C 2006. Mutants of Arabidopsis lacking starch branching enzyme II substitute plastidial starch synthesis by cytoplasmic maltose accumulation. *The Plant Cell, 18*, 2694–2709.

Esch, L, Ngai, QY, Barclay, JE, McNelly, R, Hayta, S, Smedley, MA, Smith, AM & Seung, D 2023. Increasing amyloplast size in wheat endosperm through mutation of PARC6 affects starch granule morphology. *New Phytologist, 240*, 224–241.

Evans, LT & Dunstone, RL 1970. Some physiological aspects of evolution in wheat. *Australian Journal of Biological Science, 23*, 725–741.

Evers, A 1973. The size distribution among starch granules in wheat endosperm. *Starch-Stärke, 25*, 303–304.

Facon, M, Lin, Q, Azzaz, A, Hennen-Bierwagen, TA, Myers, AM, Putaux, J-L, Roussel, X, D'Hulst, C & Wattebled, F 2013. Distinct functional properties of isoamylase-type starch debranching enzymes in monocot and dicot leaves. *Plant Physiology, 163*, 1363–1375.

Fahy, B, Gonzalez, O, Savva, GM, Ahn-Jarvis, JH, Warren, FJ, Dunn, J, Lovegrove, A & Hazard, BA 2022. Loss of starch synthase IIIa changes starch molecular structure and granule morphology in grains of hexaploid bread wheat. *Scientific Reports, 12*, art. 10806.

FAO 2021. *Food Outlook: Biannual Report on Global Food Markets*. FAO, Rome.

Feike, D, Seung, D, Graf, A, Bischof, S, Ellick, T, Coiro, M, Soyk, S, Eicke, S, Mettler-Altmann, T, Lu, KJ, Trick, M, Zeeman, SC & Smith, AM 2016. The starch granule-associated protein EARLY STARVATION1 is required for the control of starch degradation in *Arabidopsis thaliana* Leaves. *The Plant Cell, 28*, 1472–1489.

Fellows Yates, JA, Velsko, IM, Aron, F, Posth, C, Hofman, CA, Austin, RM, Parker, CE, Mann, AE, Nägele, K, Arthur, KW, Arthur, JW, Bauer, CC, Crevecoeur, I, Cupillard, C, Curtis, MC et al. 2021. The evolution and changing ecology of the African hominid oral microbiome. *Proceedings of the National Academy of Sciences of the United States of America, 118*, art. e2021655118.

Feng, F, Qi, W, Lu, Y, Yan, S, Xu, L, Yang, W, Yuan, Y, Chen, Y, Zhao, H & Song, R 2018. OPAQUE11 is a central hub of the regulatory network for maize endosperm development and nutrient metabolism. *The Plant Cell, 30*, 375–396.

Fernie, AR, Bachem, CWB, Helariutta, Y, Neuhaus, HE, Prat, S, Ruan, Y-L, Stitt, M, Sweetlove, LJ, Tegeder, M, Wahl, V, Sonnewald, S & Sonnewald, U 2020. Synchronization of developmental, molecular and physical aspects of source-sink interactions. *Nature Plants, 6*, 55–66.

Ferrero, DML, Piattoni, CV, Asencion Diez, MD, Rojas, BE, Hartman, MD, Ballicora, MA & Iglesias, AA 2020. Phosphorylation of ADP-glucose pyrophosphorylase during wheat seeds development. *Frontiers in Plant Science, 11*, art. 1058.

Fisher, DB & Gifford, RM 1987. Accumulation and conversion of sugars by developing wheat grains. *Plant Physiology, 84*, 341–347.

Fredriksson, H, Silverio, J, Andersson, R, Eliasson, A-C & Åman, P 1998. The influence of amylose and amylopectin characteristics on gelatinization and retrogradation properties of different starches. *Carbohydrate Polymers, 35*, 119–134.

Fu, F-F & Xue, H-W 2010. Co-expression analysis identifies Rice Starch Regulator1, a rice AP2/EREBP family transcription factor, as a novel rice starch biosynthesis regulator. *Plant Physiology, 154*, 927–938.

Fujita, N & Nakamura, Y 2012. Distinct and overlapping functions of starch synthase isoforms. In: Tetlow, IJ (ed.), *Essential Reviews in Experimental Biology, The Synthesis and Breakdown of Starch*, Vol. 5, The Society for Experimental Biology, London. pp. 115–140.

Fujita, N, Yoshida, M, Asakura, N, Ohdan, T, Miyao, A, Hirochika, H & Nakamura, Y 2006. Function and characterization of starch synthase I using mutants in rice. *Plant Physiology, 140*, 1070–1084.

Fujita, N, Yoshida, M, Kondo, T, Saito, K, Utsumi, Y, Tokunaga, T, Nishi, A, Satoh, H, Park, JH, Jane, JL, Miyao, A, Hirochika, H & Nakamura, Y 2007. Characterization of SSIIIa-deficient mutants of rice: the function of SSIIIa and pleiotropic effects by SSIIIa deficiency in the rice endosperm. *Plant Physiology, 144*, 2009–2023.

Fuller, DQ, Willcox, G & Allaby, RG 2011. Cultivation and domestication had multiple origins: arguments against the core area hypothesis for the origins of agriculture in the Near East. *World Archaeology, 43*, 628–652.

Funane, K, Libessart, N, Stewart, D, Michishita, T & Preiss, J 1998. Analysis of essential histidine residues of maize branching enzymes by chemical modification and site-directed mutagenesis. *Journal of Protein Chemistry, 17*, 579–590.

Gaffa, T, Yoshimoto, Y, Hanashiro, I, Honda, O, Kawasaki, S & Takeda, Y 2004. Physicochemical properties and molecular structures of starches from Millet (*Pennisetum typhoides*) and Sorghum (*Sorghum bicolor* L. Moench) cultivars in Nigeria. *Cereal Chemistry, 81*, 255–260.

Gao, M, Fisher, DK, Kim, KN, Shannon, JC & Guiltinan, MJ 1997. Independent genetic control of maize starch-branching enzymes IIa and IIb. Isolation and characterization of a sbe2a cDNA. *Plant Physiology, 114*, 69–78.

Gao, Y, An, K, Guo, W, Chen, Y, Zhang, R, Zhang, X, Chang, S, Rossi, V, Jin, F, Cao, X, Xin, M, Peng, H, Hu, Z, Guo, W, Du, J, Ni, Z, Sun, Q & Yao, Y 2021. The endosperm-specific transcription factor TaNAC019 regulates glutenin and starch accumulation and its elite allele improves wheat grain quality. *The Plant Cell, 33*, 603–622.

Gavgani, HN, Fawaz, R, Ehyaei, N, Walls, D, Pawlowski, K, Fulgos, R, Park, S, Assar, Z, Ghanbarpour, A & Geiger, JH 2022. A structural explanation for the mechanism and specificity of plant branching enzymes I and IIb. *Journal of Biological Chemistry, 298*, art. 101395.

Geigenberger, P 2011. Regulation of starch biosynthesis in response to a fluctuating environment. *Plant Physiology, 155*, 1566–1577.

Gentry, MS, Brewer, MK & Vander Kooi, CW 2016. Structural biology of glucan phosphatases from humans to plants. *Current Opinion in Structural Biology, 40*, 62–69.

Ghosh, HP & Preiss, J 1966. Adenosine diphosphate glucose pyrophosphorylase: a regulatory enzyme in the biosynthesis of starch in spinach leaf chloroplasts. *Journal of Biological Chemistry, 241*, 4491–4504.

Gontarek, BC, Neelakandan, AK, Wu, H & Becraft, PW 2016. NKD transcription factors are central regulators of maize endosperm development. *The Plant Cell, 28*, 2916–2936.

Grimaud, F, Rogniaux, H, James, MG, Myers, AM & Planchot, V 2008. Proteome and phosphoproteome analysis of starch granule-associated proteins from normal maize and mutants affected in starch biosynthesis. *Journal of Experimental Botany, 59*, 3395–3406.

Gross, P & ap Rees, T 1986. Alkaline inorganic pyrophosphatase and starch synthesis in amyloplasts. *Planta, 167*, 140–145.

Guan, H, Li, P, Imparl-Radosevich, J, Preiss, J & Keeling, P 1997. Comparing the properties of *Escherichia coli* branching enzyme and maize branching enzyme. *Archives of Biochemistry and Biophysics, 342*, 92-98.

Guo, D, Hou, Q, Zhang, R, Lou, H, Li, Y, Zhang, Y, You, M, Xie, C, Liang, R & Li, B 2020. Over-expressing *TaSPA-B* reduces prolamin and starch accumulation in wheat (*Triticum aestivum* L.) grains. *International Journal of Molecular Sciences, 21*, art. 3257.

Hammer, K 1984. The domestication syndrome. *Kulturpflanze, 32*, 11–34.

Hanashiro, I, Abe, J & Hizukuri, S 1996. A periodic distribution of the chain length of amylopectin as revealed by high-performance anion-exchange chromatography. *Carbohydrate Research, 283*, 151–159.

Hanashiro, I, Itoh, K, Kuratomi, Y, Yamazaki, M, Igarashi, T, Matsugasako, J & Takeda, Y 2008. Granule-bound starch synthase I is responsible for biosynthesis of extra-long unit chains of amylopectin in rice. *Plant and Cell Physiology, 49*, 925–933.

Hanashiro, I, Sakaguchi, I & Yamashita, H 2013. Branched structures of rice amylose examined by differential fluorescence detection of side-chain distribution. *Journal of Applied Glycoscience, 60*, 79–85.

Hands, P 2012. Analysis of grain characters in temperate grasses reveals distinctive patterns of endosperm organisation associated with grain shape. *Journal of Experimental Botany, 63*, 6253–6266.

Hansen, PI, Spraul, M, Dvortsak, P, Larsen, FH, Blennow, A, Motawia, MS & Engelsen, SB 2009. Starch phosphorylation-maltosidic restrains upon 3′- and 6′-phosphorylation investigated by chemical synthesis, molecular dynamics and NMR spectroscopy. *Biopolymers, 91*, 179–193.

Hardie, DG 2007. AMP-activated/SNF1 protein kinases: conserved guardians of cellular energy. *Nature Reviews. Molecular Cell Biology, 8*, 774–785.

Hardy, K, Buckley, S, Collins, M, Estalrrich, A, Brothwell, D, Copeland, L, García-Tabernero, A, García-Vargas, S, de la Rasilla, M, Lalueza-Fox, C, Huguet, R, Bastir, M, Santamaría, D, Madella, M, Wilson, J, Cortés, ÁF & Rosas, A 2012. Neanderthal medics? Evidence for food, cooking, and medicinal plants entrapped in dental calculus. *Naturwissenschaften, 99*, 617–626.

Harris, PJ, Burrell, MM, Emes, MJ & Tetlow, IJ 2023. Effects of post-anthesis high-temperature stress on carbon partitioning and starch biosynthesis in a spring wheat (*Triticum aestivum* L.) adapted to moderate growth temperatures. *Plant Cell Physiology, 64*, 729–745.

Hatfield, JL & Walthall, CL 2014. Climate change: cropping system changes and adaptations. In: Van Alfen, NK (ed.), Encyclopedia of Agriculture and Food Systems, Academic Press, Cambridge, MA. pp. 256–265.

Hawker, JS, Ozbun, JL, Ozaki, H, Greenberg, E & Preiss, J 1974. Interaction of spinach leaf adenosine diphosphate glucose α-1,4-glucan α-4-glucosyl transferase and α-1,4-glucan, α-1,4-glucan-6-glycosyl transferase in synthesis of branched α-glucan. *Archives of Biochemistry and Biophysics, 160*, 530–551.

Hawkins, E, Chen, J, Watson-Lazowski, A, Ahn-Jarvis, J, Barclay, JE, Fahy, B, Hartley, M, Warren, FJ & Seung, D 2021. STARCH SYNTHASE 4 is required for normal starch granule initiation in amyloplasts of wheat endosperm. *New Phytologist, 230*, 2371–2386.

Hefferon, KL 2015. Nutritionally enhanced food crops; progress and perspectives. *International Journal of Molecular Sciences, 16*, 3895–3914.

Hendry, G 1993. Oxygen, free radical processes and seed longevity. *Seed Science Research, 3*, 141–153.

Hennen-Bierwagen, TA, Lin, Q, Grimaud, F, Planchot, V, Keeling, PL, James, MG & Myers, AM 2009. Proteins from multiple metabolic pathways associate with starch biosynthetic enzymes in high molecular weight complexes: a model for regulation of carbon allocation in maize amyloplasts. *Plant Physiology, 149*, 1541–1559.

Hennen-Bierwagen, TA, Liu, F, Marsh, RS, Kim, S, Gan, Q, Tetlow, IJ, Emes, MJ, James, MG & Myers, AM 2008. Starch biosynthetic enzymes from developing *Zea mays* endosperm associate in multisubunit complexes. *Plant Physiology, 146*, 1892–1908.

Henry, AG, Brooks, AS & Piperno, DR 2011. Microfossils in calculus demonstrate consumption of plants and cooked foods in Neanderthal diets (Shanidar III, Iraq; Spy I and II, Belgium). *PNAS, 108*, 486-491.

Hernández-Soto, A, Echeverría-Beirute, F, Abdelnour-Esquivel, A, Valdez-Melara, M, Boch, J & Gatica-Arias, A 2021. Rice breeding in the new era: comparison of useful agronomic traits. *Current Plant Biology, 27,* art. 100211.

Herrera-Saldana, RE, Huber, JT & Poore, MH 1990. Dry matter, crude protein, and starch degradability of five cereal grains. *Journal of Dairy Science, 73,* 2386–2393.

Hirose, T & Terao, T 2004. A comprehensive expression analysis of the starch synthase gene family in rice (*Oryza sativa* L.). *Planta, 220,* 9–16.

Hizukuri, S 1985. Relationship between the distribution of the chain length of amylopectin and the crystalline structure of starch granules. *Carbohydrate Research, 141,* 295–306.

Hizukuri, S 1986. Polymodal distribution of the chain lengths of amylopectin, and its significance. *Carbohydrate Research, 147,* 342–347.

Hizukuri, S, Tabata, S, Kagoshima & Nikuni, Z 1970. Studies on starch phosphate part 1. Estimation of glucose-6-phosphate residues in starch and the presence of other bound phosphate(s). *Starch/Stärke, 22,* 338–343.

Hochmuth, A, Carswell, M, Rowland, A, Scarbrough, D, Esch, L, Kamble, NU, Habig, JW, Seung, D 2024. Distinct effects of PTST2b and MRC on starch granule morphogenesis in potato tubers. *Plant Biotechnology Journal,* 1–18.

Hong, S & Preiss, J 2000. Localization of C-terminal domains required for the maximal activity or for determination of substrate preference of maize branching enzymes. *Archives of Biochemistry and Biophysics, 378,* 349–355.

Hoover, R & Senanayake, SPJN 1996. Composition and physicochemical properties of oat starches. *Food Research International, 29,* 15–26.

Hoover, R, Smith, C, Zhou, Y & Ratnayake, RMWS 2003. Physico-chemical properties of Canadian oat starches. *Carbohydrate Polymers, 52,* 253–261.

Howard, T, Rejab, NA, Griffiths, S, Leigh, F, Leverington-Waite, M, Simmonds, J, Uauy, C & Trafford, K 2011. Identification of a major QTL controlling the content of B-type starch granules in *Aegilops. Journal of Experimental Botany, 62,* 2217–2228.

Hu, T, Tian, Y, Zhu, J, Wang, Y, Jing, R, Lei, J, Sun, Y, Yu, Y, Li, J, Chen, X, Zhu, X, Hao, Y, Liu, L, Wang, Y & Wan, J 2018. OsNDUFA9 encoding a mitochondrial complex I subunit is essential for embryo development and starch synthesis in rice. *Plant Cell Reports, 37,* 1667–1679.

Hu, YF, Li, YP, Zhang, J, Liu, H, Tian, M & Huang, Y 2012. Binding of ABI4 to a CACCG motif mediates the ABA-induced expression of the ZmSSI gene in maize (*Zea mays* L.) endosperm. *Journal of Experimental Botany, 63,* 5979–5989.

Huang, B, Hennen-Bierwagen, T & Myers, AM 2014. Functions of multiple genes encoding ADP-glucose pyrophosphorylase subunits in maize endosperm, embryo and leaf. *Plant Physiology, 164,* 596–611.

Huang, H, Xie, S, Xiao, Q, Wei, B, Zheng, L & Wang, Y 2016. Sucrose and ABA regulate starch biosynthesis in maize through a novel transcription factor, ZmEREB156. *Scientific Reports, 6,* art. 27590.

Huesing, JE, Andres, D, Braverman, MP, Burns, A, Felsot, AS, Harrigan, GG, Hellmich, RL, Reynolds, A, Shelton, AM, van Rijssen, WJ, Morris, EJ & Eloff, JN 2016. Global adoption of genetically modified (GM) crops: challenges for the public sector. *Journal of Agricultural and Food Chemistry, 64,* 394–402.

Hwang, SK, Koper, K & Okita, TW 2020. The plastid phosphorylase as a multiple-role player in plant metabolism. *Plant Science, 290,* art. 110303.

Hwang, SK, Koper, K, Satoh, H & Okita, TW 2016b. Rice endosperm starch phosphorylase (Pho1) assembles with disproportionating enzyme (Dpe1) to form a protein complex that enhances synthesis of malto-oligosaccharides. *Journal of Biological Chemistry, 291,* 19994–20007.

Hwang, SK, Nishi, A, Satoh, H & Okita, TW 2010. Rice endosperm-specific plastidial alpha-glucan phosphorylase is important for synthesis of short chain malto-oligosaccharides. *Archives of Biochemistry and Biophysics, 495,* 82–92.

Hwang, SK, Singh, S, Cakir, B, Satoh, H & Okita, TW 2016a. The plastidial starch phosphorylase from rice endosperm: catalytic properties at low temperature. *Planta, 243,* 999–1009.

Imparl-Radosevich, JM, Gameon, JR, McKean, A, Wetterberg, D, Keeling, P & Guan, H 2003. Understanding catalytic properties and functions of maize starch synthase isozymes. *Journal of Applied Glycoscience, 50,* 177–182.

Imparl-Radosevich, JM, Nichols, DJ, Li, P, McKean, AL, Keeling, P & Guan, H 1999. Analysis of purified maize starch synthases IIa and IIb: SS isoforms can be distinguished based on their kinetic properties. *Archives of Biochemistry and Biophysics, 362,* 131–138.

Inouchi, N, Glover, DV, Takaya, T & Fuwa, H 1983. Development changes in fine structure of starches of several endosperm mutants of maize. *Starch/Stärke, 35,* 371–376.

IPCC 2014. *Climate Change 2014: Impacts, Adaptation, and Vulnerability. Organization & Environment. Working Group III Contribution to the Fifth Assessment Report of the Intergovernmental Panel on Climate Change.* Cambridge University Press, Cambridge and New York, NY.

James, MG, Denyer, K & Myers, AM 2003. Starch synthesis in the cereal endosperm. *Current Opinion in Plant Biology, 6,* 215–222.

James, MG, Robertson, DS & Myers, AM 1995. Characterization of the maize gene *sugary1,* a determinant of starch composition in kernels. *The Plant Cell, 7,* 417–429.

Jameson, PE & Song, J 2016. Cytokinin: a key driver of seed yield. *Journal of Experimental Botany, 67,* 593–606.

Jane, J, Chen, YY, Lee, LF, McPherson, AE, Wong, KS, Radosavljevic, M & Kasemsuwan, T 1999. Effects of amylopectin branch chain length and amylose content on the gelatinization and pasting properties of starch. *Cereal Chemistry, 76,* 629–637.

Jane, JL, Kasemsuwan, T, Leas, S, Zobel, H & Robyt, JF 1994. Anthology of starch granule morphology by scanning electron microscopy. *Starch/Stärke, 46,* 121–129.

Janeček, Š, Svensson, B & MacGregor, EA 2011. Structural and evolutionary aspects of two families of non-catalytic domains present in starch and glycogen binding proteins from microbes, plants and animals. *Enzymes, Microbiology, Technology, 49,* 429–440.

Jenkins, PJ & Donald, AM 1995. The influence of amylose on starch granule structure. *International Journal of Biomacromolecules, 17,* 315–321.

Jia, M, Wu, H, Clay, KL, Jung, R, Larkins, BA & Gibbon, BC 2013. Identification and characterization of lysine-rich proteins and starch biosynthesis genes in the opaque2 mutant by transcriptional and proteomic analysis. *BMC Plant Biology, 13,* art. 60.

Jiang, T, Zhai, H, Wang, F, Yang, N, Wang, B, He, S & Liu, Q 2013. Cloning and characterization of a carbohydrate metabolism-associated gene *IbSnRK1* from sweet potato. *Scientia Horticulturae, 158,* 22–32.

Jo, HJ, Park, S, Jeong, HG, Kim, JW & Park, JT 2015. *Vibrio vulnificus* glycogen branching enzyme preferentially transfers very short chains: N1 domain determines the chain length transferred. *FEBS Letters, 589,* 1089–1094.

Johnson, P, Patron, N, Bottrill, A, Dinges, J, Fahy, B, Parker, M, Waite, D & Denyer, K 2003. A low-starch barley mutant, Risø 16, lacking the cytosolic small subunit of ADP-glucose pyrophosphorylase, reveals the importance of the cytosolic isoform and the identity of the plastidial small subunit. *Plant Physiology, 131,* 684–696.

Jones, RJ, Schreiber, BMN & Roessler, JA 1996. Kernel sink capacity in maize: genotypic and maternal regulation. *Crop Science, 36,* 301–306.

Kamble, NU, Makhamadjonov, F, Fahy, B, Martins, C, Saalbach, G & Seung, D 2023. Initiation of B-type starch granules in wheat endosperm requires the plastidial á-glucan phosphorylase PHS1. *The Plant Cell, 35*, 4091–4110.

Kang, HG, Park, S, Matsuoka, M & An, G 2005. White-core endosperm floury endosperm-4 in rice is generated by knockout mutations in the C-type pyruvate orthophosphate dikinase gene (OsPPDKB). *The Plant Journal, 42*, 901–911.

Kato, T 1995. Change of sucrose synthase activity in developing endosperm of rice cultivars. *Crop Science, 35*, 827–831.

Katz, JR & van Itallie, TB 1930. Abhandlungen zur physikalischen Chemie der Stärke und der Brotbereitung: V. Alle Stärkearten haben das gleiche Retrogradationsspektrum. *Zeitschrift für Physikalische Chemie, 150A*, 90–99.

Kawakatsu, T, Yamamoto, MP, Touno, SM, Yasuda, H & Takaiwa, F 2009. Compensation and interaction between RISBZ1 and RPBF during grain filling in rice. *The Plant Journal, 59*, 908–920.

Kerk, D, Conley, TR, Rodriguez, FA, Tran, HT, Nimick, M, Muench, DG & Moorhead, GBG 2006. A chloroplast-localized dual-specificity protein phosphatase in Arabidopsis contains a phylogenetically dispersed and ancient carbohydrate-binding domain, which binds the polysaccharide starch. *The Plant Journal, 46*, 400–413.

Kharas, H 2010. *The emerging middle class in developing countries.* OECD Development Center Working Paper No. 285. OECD, Paris.

Kim, KH, Koh, HJ, Lee, JH, Park, SZ & Heu, MH 1993. Diversity of rice quality for processing: physicochemical characteristics and inheritance of floury endosperm mutants. *Korean Journal of Crop Science, 38*, 264–274.

Kirchberger, S, Tjaden, J & Neuhaus, H 2008. Characterization of the Arabidopsis Brittle1 transport protein and impact of reduced activity on plant metabolism. *Plant Journal, 56*, 51–63.

Kleczkowski, L 1996. Back to the drawing board: redefining starch synthesis in cereals. *Trends in Plant Science, 1*, 363–364.

Kleczkowski, LA 1994. Glucose activation and metabolism through UDP-glucose pyrophosphorylase in plants. *Phytochemistry, 37*, 1507–1515.

Kleczkowski, LA, Villand, P, Lüthi, E, Olsen, OA & Preiss, J 1993a. Insensitivity of barley endosperm ADP-glucose pyrophosphorylase to 3-phosphoglycerate and orthophosphate regulation. *Plant Physiology*, *101*, 179–186.

Kleczkowski, LA, Villand, P, Preiss, J & Olsen, OA 1993b. Kinetic mechanism and regulation of ADP-glucose pyrophosphorylase from barley (*Hordeum vulgare*) leaves. *Journal of Biological Chemistry, 268*, 6228–6233.

Koch, K 1996. Carbohydrate-modulated gene expression in plants. *Annual Review of Plant Physiology and Plant Molecular Biology, 47*, 509–540.

Koch, K 2004. Sucrose metabolism: regulatory mechanisms and pivotal roles in sugar sensing and plant development. *Current Opinion in Plant Biology, 7*, 235–246.

Koper, K, Hwang, S-K, Wood, M, Singh, S, Cousins, A, Kirchhoff, H & Okita, TW 2021. The rice plastidial phosphorylase participates directly in both sink and source processes. *Plant and Cell Physiology, 62*, 125–142.

Kötting, O, Pusch, K, Tiessen, A, Geigenberger, P, Steup, M & Ritte, G 2005. Identification of a novel enzyme required for starch metabolism in Arabidopsis leaves. The phosphoglucan, water dikinase. *Plant Physiology, 137*, 242–252.

Kötting, O, Santelia, D, Edner, C, Eicke, S, Marthaler, T, Gentry, MS, Comparot-Moss, S, Chen, J, Smith, AM, Steup, M, Ritte, G & Zeeman, SC 2009. STARCH-EXCESS4 is a Laforin-like phosphoglucan phosphatase required for starch degradation in *Arabidopsis thaliana. Plant Cell, 21*, 334–346.

Kram, AM, Oostergetel, GT & van Bruggen, E 1993. Localization of branching enzyme in potato tuber cells with the use of immuno-electron microscopy. *Plant Physiology, 101*, 237–243.

Krishnan, S & Dayanandan, P 2003. Structural and histochemical studies on grain-filling in the caryopsis of rice (*Oryza sativa* L.). *Journal of Biosciences, 28*, 455–469.

Kubo, A, Rahman, S, Utsumi, Y, Li, Z, Mukai, Y, Yamamoto, M, Ugaki, M, Harada, K, Satoh, H, Konik-Rose, C, Morell, M & Nakamura, Y 2005. Complementation of *sugary-1* phenotype in rice endosperm with the wheat isoamylase1 gene supports a direct role for isoamylase1 in amylopectin biosynthesis. *Plant Physiology, 137*, 43–56.

Kumamaru, T, Sato, H & Satoh, H 1997. High-lysine mutants of rice, *Oryza sativa* L. *Plant Breeding, 116*, 245–249.

Kumar, P, Madhawan, A, Sharma, A, Sharma, V, Das, D, Parveen, A, Fandade, V, Sharma, D & Roy, J 2024. A sucrose non-fermenting-1-related protein kinase 1 gene from wheat, *TaSNRK1a* regulates starch biosynthesis by modulating AGPase activity. *Plant Physiology and Biochemistry, 207*, art. 108407.

Kuriki, T, Guan, H, Sivak, M & Preiss, J 1996. Analysis of the active center of branching enzyme II from maize endosperm. *Journal of Protein Chemistry, 15*, 305–313.

Kuriki, T, Stewart, DC & Preiss, J 1997. Construction of chimeric enzymes out of maize endosperm branching enzymes I and II: activity and properties. *Journal of Biological Chemistry, 272*, 28999–29004.

Lafiandra, D, Riccardi, G & Shewry, PR 2014. Improving cereal grain carbohydrates for diet and health. *Journal of Cereal Science, 59*, 312–326.

Langer, RHM & Hill, GD 1982. *Agricultural Plants.* Cambridge University Press, Cambridge.

Langeveld, SMJ, Van wijk, R, Stuurman, N, Kijne, JW & de Pater, S 2000. B-type granule containing protrusions and interconnections between amyloplasts in developing wheat endosperm revealed by transmission electron microscopy and GFP expression. *Journal of Experimental Botany, 51*, 1357–1361.

Lappe, RR, Baier, JW, Boehlein, SK, Huffman, R, Lin, Q, Wattebled, F, Settles, AM, Hannah, LC, Borisjuk, L, Rolletschek, H, Stewart, JD, Scott, MP, Hennen-Bierwagen, TA & Myers, AM 2018. Functions of maize genes encoding pyruvate phosphate dikinase in developing endosperm. *Proceedings of the National Academy of Sciences USA, 115*, E24–E33.

Lawlor, DW & Paul, MJ 2014. Source/sink interactions underpin crop yield: the case for trehalose 6-phosphate/SnRK1 in improvement of wheat. *Frontiers in Plant Science, 5*, 418.

Lee, S, Hwang, S-K, Han, M, Eom, J-S, Kang, H-G, Han, Y, Choi, S-B, Cho, M-H, Bhoo, S, An, G, Hahn, T-R, Okita, T & Jeon, J-S 2007. Identification of the ADP-glucose pyrophosphorylase isoforms essential for starch synthesis in the leaf and seed endosperm of rice (*Oryza sativa* L.). *Plant Molecular Biology, 65*, 531–546.

Lemoine, R, La Camera, S, Atanassova, R, Dedaldechamp, F, Allario, T, Pourtau, N, Bonnemain, J-L, Laloi, M, Coutos-Thévenot, P, Maurosset, L, Faucher, M, Girousse, C, Lemonnier, P, Parrilla, J & Durand, M 2013. Source-to-sink transport of sugar and regulation by environmental factors. *Frontiers in Plant Science, 4*, 272.

Leroux, BM, Goodyke, AJ, Schumacher, KI, Abbott, CP, Clore, AM, Yadegari, R, Larkins, BA & Dannenhoffer, JM 2014. Maize early endosperm growth and development: from fertilization through cell type. *American Journal of Botany, 101*, 1259–1274.

Leterrier, M, Holappa, L, Broglie, K & Beckles, D 2008. Cloning, characterization and comparative analysis of a starch synthase IV gene in wheat: functional and evolutionary implications. *BMC Plant Biology, 8*, 98.

Li, B, Liu, H, Zhang, Y, Kang, T, Zhang, L, Tong, J, Xiao, L & Zhang, H 2013. Constitutive expression of cell wall invertase genes increases grain yield and starch content in maize. *Plant Biotechnology Journal, 11*, 1080-1091.

Li, C, Dhital, S & Gidley, MJ 2023. High amylose wheat foods: a new opportunity to improve human health. *Trends in Food Science & Technology, 135,* 93–101.

Li, C, Wu, AC, Go, RM, Malouf, J, Turner, MS, Malde, AK, Mark, AE & Gilbert, RG 2015. The characterization of modified starch branching enzymes: toward the control of starch chain-length distributions. *PLoS ONE, 10*, art. e0125507.

Li, H, Wang, Y, Xiao, Q, Luo, L, Zhang, C, Mao, C, Du, J, Long, T, Cao, Y, Yi, Q, Wang, Y, Li, Y, Huang, H, Liu, H, Hu, Y, Yu, G, Liu, Y, Zhang, J & Huang, Y 2021a. Transcription factor ZmPLATZ2 positively regulate the starch synthesis in maize. *Plant Growth Regulation, 93*, 291–302.

Li, H, Xiao, Q, Zhang, C, Du, J, Li, X, Huang, H, Wei, B, Li, Y, Yu, G, Liu, H, Hu, Y, Liu, Y, Zhang, J & Huang, Y 2017. Identification and characterization of transcription factor ZmEREB94 involved in starch synthesis in maize. *Journal of Plant Physiology, 216,* 11–16.

Li, J, Xie, L, Tian, X, Liu, S, Xu, D, Jin, H, Song, J, Dong, Y, Zhao, D, Li, G, Li, Y, Zhang, Y, Xia, X, He, Z & Cao, S 2021b. TaNAC100 acts as an integrator of seed protein and starch synthesis exerting pleiotropic effects on agronomic traits in wheat. *The Plant Journal, 108*, 829–840.

Libessart, N & Preiss, J 1998. Arginine residue 384 at the catalytic center is important for branching enzyme II from maize endosperm. *Archives of Biochemistry and Biophysics, 360*, 135–141.

Liesche, J & Patrick, J 2017. An update on phloem transport: a simple bulk flow under complex regulation. *F1000Research, 6,* art. 2096.

Lin, Q, Huang, B, Zhang, M, Zhang, X, Rivenbark, J, Lappe, RL, James, MG, Myers, AM & Hennen-Bierwagen, TA 2012. Functional interactions between starch synthase III and isoamylase-type starch-debranching enzyme in maize endosperm. *Plant Physiology, 158*, 679–692.

Lindeboom, N, Chang, PR & Tyler, RT 2004. Analytical, biochemical and physicochemical aspects of starch granule size, with emphasis on small granule starches: A review. *Starch/Stärke, 56,* 89–99.

Liu, C, Pfister, B, Osman, R, Ritter, M, Heutinck, A, Sharma, M, Eicke, S, Stettler, MF, Seung, D, Bompard, C, Abt, MR & Zeeman, SC 2023a. LIKE EARLY STARVATION 1 and EARLY STARVATION 1 promote and stabilize amylopectin phase transition in starch biosynthesis. *Science Advances, 9*, art. eadg7448.

Liu, F, Ahmed, Z, Lee, EA, Donner, E, Liu, Q, Ahmed, R, Morell, MK, Emes, MJ & Tetlow, IJ 2012b. Allelic variants of the amylose extender mutation of maize demonstrate phenotypic variation in starch structure resulting from modified protein-protein interactions. *Journal of Experimental Botany, 63*, 1183.

Liu, F, Makhmoudova, A, Lee, EA, Wait, R, Emes, MJ & Tetlow, IJ 2009. The *amylose extender* mutant of maize conditions novel protein-protein interactions between starch biosynthetic enzymes in amyloplasts. *Journal of Experimental Botany, 60*, 4423–4440.

Liu, F, Romanova, N, Lee, EA, Ahmed, R, Evans, M, Gilbert, EP, Morell, MK, Emes, MJ & Tetlow, IJ 2012a. Glucan affinity of starch synthase IIa determines binding of starch synthase I and starch branching enzyme IIb to starch granules. *Biochemical Journal, 448*, 373–387.

Liu, X, Qiao, L, Kong, Y, Wang, H, Yang, B 2024. Characterization of the starch molecular structure of wheat varying in the content of resistant starch. *Food Chemistry: X, 21*, art. 101103.

Liu, Z, Li, P, Yu, L, Hu, Y, Du, A, Fu, X, Wu, C, Luo, D, Hu, B, Dong, H, Jiang, H, Ma, X, Huang, W, Yang, X, Tu, S & Li, H 2023b. *OsMADS1* regulates grain quality, gene expressions, and regulatory networks of starch and storage protein metabolisms in rice. *International Journal of Molecular Sciences, 24*, art. 8017.

Lohmeier-Vogel, EM, Kerk, D, Nimick, M, Wrobel, S, Vickerman, L, Muench, DG & Moorhead, GB 2008. Arabidopsis At5g39790 encodes a chloroplast-localized, carbohydrate- binding, coiled-coil domain-containing putative scaffold protein. *BMC Plant Biology, 8*, 1–14.

Long, W, Dong, B, Wang, Y, Pan, P, Wang, Y, Liu, L, Chen, X, Liu, X, Liu, S, Tian, Y, Chen, L & Wan, J 2017. *FLOURY ENDOSPERM8*, encoding the UDP-glucose pyrophosphorylase 1, affects the synthesis and structure of starch in rice endosperm. *Journal of Plant Biology, 60*, 513–522.

Lopes, MA & Larkins, BA 1993. Endosperm origin, development, and function. *Plant Cell, 5,* 1383–1399.

Lopez-Juez, E & Pyke, KA 2005. Plastids unleashed: their development and their integration in plant development. The *International Journal of Developmental Biology, 49,* 557–577.

Lou, G, Chen, P, Zhou, H, Li, P, Xiong, J, Wan, S, Zheng, Y, Alam, M, Liu, R, Zhou, Y, Yan, H, Tian, Y, Bai, J, Rao, W, Tan, X, Gao, H, Li, Y, Gao, G, Zhang, Q, Li, X, Liu, C & He, Y 2021. FLOURY ENDOSPERM19 encoding a class I glutamine amidotransferase affects grain quality in rice. *Molecular Breeding, 41*, art. 36.

Lu, K-J, Pfister, B, Jenny, C, Eicke, S & Zeeman, SC 2018. Distinct functions of STARCH SYNTHASE 4 domains in starch granule formation. *Plant Physiology, 176*, 566–581.

Lu, Y & Sharkey, TD 2004. The role of amylomaltase in maltose metabolism in the cytosol of photosynthetic cells. *Planta, 207*, 271–274.

MacNeill, GJ, Mehrpouyan, S, Minow, MAA, Patterson, JA, Tetlow, IJ & Emes, MJ 2017. Starch as a source, starch as a sink: the bifunctional role of starch in carbon allocation. *Journal of Experimental Botany, 68,* 4433–4453.

Mahajan, P, Bera, MB, Panesar, PS & Chauhan, A 2021. Millet starch: A review. *International Journal of Biological Macromolecules, 180,* 61–79.

Maharana, J, Hwang, S-K, Singha, DL, Panda, D, Singh, S, Okita, TW & Modi, MK 2024. Exploring the structural assembly of rice ADP-glucose pyrophosphorylase subunits using MD simulation. *Journal of Molecular Graphics and Modelling, 129*, art. 108761.

Malinova, I, Alseekh, S, Feil, R, Fernie, AR, Baumann, O, Schöttler, MA, Lunn, JE & Fettke, J 2017. Starch synthase 4 and plastidal phosphorylase differentially affect starch granule number and morphology. *Plant Physiology, 174*, 73–85.

Malinova, I, Mahlow, S, Alseekh, S, Orawetz, T, Fernie, AR, Baumann, O, Steup, M & Fettke, J 2014. Double knockout mutants of Arabidopsis grown under normal conditions reveal that the plastidial phosphorylase isozyme participates in transitory starch metabolism. *Plant Physiology, 164*, 907–921.

Mangelsen, E, Kilian, J, Harter, K, Jansson, C, Wanke, D & Sundberg, E 2011. Transcriptome analysis of high-temperature stress in developing barley caryopses: early stress responses and effects on storage compound biosynthesis. *Molecular Plant, 4,* 97–115.

Matsushima, R, Hisano, H, Galis, I, Miura, S, Crofts, N, Takenake, Y, Oitome, NF, Ishimizu, T, Fujita, N & Sato, K 2023. FLOURY ENDOSPERM 6 mutations enhance the sugary phenotype caused by the loss of ISOAMYLASE1 in barley. *Theoretical and Applied Genetics, 136*, art. 94.

Matsushima, R, Yamashita, J, Kariyama, S, Enomoto, T & Sakamoto, W 2013. A phylogenetic re-evaluation of morphological variations of starch grains among Poaceae species. *Journal of Applied Glycoscience, 60,* 37–44.

Meekins, DA, Vander Kooi, CW & Gentry, MS 2016. Structural mechanisms of plant glucan phosphatases in starch metabolism. *The FEBS Journal, 283,* 2427–2447.

Mercader, J 2009. Mozambican grass seed consumption during the middle stone age. *Science, 326,* 1680–1683.

Mérida, A & Fettke, J 2021. Starch granule initiation in *Arabidopsis thaliana* chloroplasts. *Plant Journal, 107,* 688–697.

Mehrpouyan, S, Menon, U, Tetlow, IJ & Emes, MJ 2021. Protein phosphorylation regulates maize endosperm starch synthase IIa activity and protein-protein interactions. *The Plant Journal, 105,* 1098–1112.

Milla, R, Osborne, CP, Turcotte, MM & Violle, C 2015. Plant domestication through an ecological lens. *Trends in Ecology & Evolution, 30,* 463–469.

Mizuno, K, Kawasaki, T, Shimada, H, Satoh, H, Kobayashi, E, Okumura, S, Arai, Y & Baba T 1993. Alteration of the structural properties of starch composition by the lack of an isoform of starch branching enzyme in rice seeds. *Journal of Biological Chemistry, 268,* 19084-19091.

Mondal, R, Kumar, A & Gnanesh, BN 2023. Crop germplasm: current challenges, physiological-molecular perspective, and advance strategies towards development of climate-resilient crops. *Heliyon, 9,* art. e12973.

Morell, MK, Kosar-Hashemi, B, Cmiel, M, Samuel, MS, Chandler, P, Rahman, S, Buleon, A, Batey, IL & Li, Z 2003. Barley *sex6* mutants lack starch synthase IIa activity and contain a starch with novel properties. *Plant Journal, 34,* 173–185.

Mua, JP & Jackson, DS 1997. Fine structure of corn amylose and amylopectin fractions with various molecular weights. *Journal of Agricultural and Food Chemistry, 45,* 3840–3847.

Mu-Forster, C, Huang, R, Powers, JR, Harriman, RW, Knight, M, Singletary, GW, Keeling, PL & Wasserman, BP 1996. Physical association of starch biosynthetic enzymes with starch granules of maize endosperm. Granule-associated forms of starch synthase I and starch branching enzyme II. *Plant Physiology, 111,* 821–829.

Nakamura, Y 2023. A model for the reproduction of amylopectin cluster by coordinated actions of starch branching enzyme isoforms. *Plant Molecular Biology, 112,* 199–212.

Nakamura, Y, Aihara, S, Crofts, N, Sawada, T & Fujita, N 2014. *In vitro* studies of enzymatic properties of starch synthases and interactions between starch synthase I and starch branching enzymes from rice. *Plant Science, 224,* 1–8.

Nakamura, Y & Kainuma, K 2022. On the cluster structure of amylopectin. *Plant Molecular Biology, 108,* 291–306.

Nakamura, Y, Ono, M, Sawada, T, Crofts, N, Fujita, N & Steup, M 2017. Characterization of the functional interactions of plastidial starch phosphorylase and starch branching enzymes from rice endosperm during reserve starch biosynthesis. *Plant Science, 264,* 83–95.

Nakamura, Y, Utsumi, Y, Sawada, T, Aihara, S, Utsumi, C, Yoshida, M & Kitamura, S 2010. Characterization of the reactions of starch branching enzymes from rice endosperm. *Plant Cell Physiology, 51,* 776–794.

Nielsen, TH, Wischmann, B, Enevoldsen, K & Moller, BL 1994. Starch phosphorylation in potato-tubers proceeds concurrently with *de-novo* biosynthesis of starch. *Plant Physiology, 105,* 111–117.

Niittylä, T, Messerli, G, Trevisan, M, Chen, J, Smith, AM & Zeeman, SC 2004. A previously unknown maltose transporter essential for starch degradation in leaves. *Science, 303,* 87–89.

Nishio, T & Iida, S 1993. Mutant having a low content of 16-kDa allergenic protein in rice (*Oryza sativa* L.). *Theoretical and Applied Genetics, 86,* 317–321.

Nitschke, F & Schmieder, P 2021. Analysis of covalent modifications in α-glucans. In: Nitschke, F (ed.), *Enzymology of Complex Alpha-Glucans,* CRC Press: Boca Raton, FL, pp. 34–60.

Obadi, M, Qi, Y & Xu, B 2023. High-amylose maize starch: structure, properties, modifications and industrial applications. *Carbohydrate Polymers, 299,* art. 120185.

Ohdan, T, Francisco, P, Sawada, T, Hirose, T, Terao, T, Satoh, H & Nakamura, Y 2005. Expression profiling of genes involved in starch synthesis in sink and source organs of rice. *Journal of Experimental Botany, 56,* 3229–3244.

Oiestad, AJ, Martin, JM & Giroux, MJ 2016. Overexpression of ADP-glucose pyrophosphorylase in both leaf and seed tissue synergistically increase biomass and seed number in rice (*Oryza sativa ssp.* Japonica). *Functional Plant Biology, 43,* 1194-1204.

Okita, TW, Nakata, PA, Anderson, JM, Sowokinos, J, Morell, M & Preiss, J 1990. The subunit structure of potato tuber ADPglucose pyrophosphorylase. *Plant Physiology, 93,* 785-790.

Olsen, O-A 2004. Nuclear endosperm development in cereals and *Arabidopsis thaliana. Plant Cell, 16,* S214–S227.

Olsen, O-A 2020. The modular control of cereal endosperm development. *Trends in Plant Science, 25,* 275–290.

Osman, R, Bossu, M, Dauvillée, D, Spriet, C, Liu, C, Zeeman, SC, D'Hulst, C & Bompard, C 2024. LIKE EARLY STARVATION 1 interacts with amylopectin during starch biosynthesis. *Plant Physiology, 195,* 1851–1865.

Ou-Lee, T & Setter, T 1985. Enzyme activities of starch and sucrose pathways and growth of apical and basal maize kernels. *Plant Physiology, 79,* 848–851.

Palomo, M, Kralj, S, van der Maarel, MJEC & Dijkhuizen, L 2009. The unique branching pattern of *Deinococcus* glycogen branching enzymes are determined by their N-terminal domains. *Applied Environmental Microbiology, 75,* 1355–1362.

Pan, DD & Jane, J 2000. Internal structure of normal maize starch granules revealed by chemical surface gelatinization. *Biomacromolecules, 1,* 126–132.

Pan, T, Lin, L, Wang, J, Liu, Q & Wei, C 2018. Long branch-chains of amylopectin with B-type crystallinity in rice seed with inhibition of starch branching enzyme I and IIb resist in situ degradation and inhibit plant growth during seedling development. *BMC Plant Biology, 18,* 1–11.

Pandey, B, Tyagi, C, Chakraborty, O, Mishra, AK, Kumar, A & Jain, AK 2016. Heterodimeric interaction of the ADP-glucose pyrophosphorylase (AGPase) enzyme in *Hordeum vulgare. Indian Journal of Biotechnology, 15,* 334-342.

Parker, M 1985. The relationship between A-type and B-type starch granules in the developing endosperm of wheat. *Journal of Cereal Science, 3,* 271–278.

Parry, M, Rosenzweig, C & Livermore, M 2005. Climate change, global food supply and risk of hunger. *Philosophical Transactions of the Royal Society B, 360,* 2125–2138.

Parry, MAJ & Hawkesford, MJ 2010. Food security: increasing yield and improving resource efficiency. *Proceedings of the Nutrition Society, 69,* 592–600.

Peng, B, Kong, H, Li, Y, Wang, L, Zhong, M, Sun, L, Gao, G & Zhang, Q 2014a. *OsAAP6* functions as an important regulator of grain protein content and nutritional quality in rice. *Nature Communications, 5,* art. 4847.

Peng, C, Wang, Y, Liu, F, Ren, Y, Zhou, K, Lu, J, Zheng, M, Zhao, S, Zhang, L, Wang, C, Jiang, L, Zhang, X, Guo, X, Bao, Y & Wan, J 2014b. FLOURY ENDOSPERM6 encodes a CBM48 domain-containing protein involved in compound granule formation and starch synthesis in rice endosperm. *Plant Journal, 77,* 917–930.

Peng, X, Wang, Q, Wang, Y, Cheng, B, Zhao, Y & Zhu, S 2019. A maize NAC transcription factor, *ZmNAC34,* negatively regulates starch synthesis in rice. *Plant Cell Reports, 38,* 1473–1484.

Peng, X, Yu, W, Chen, Y, Jiang, Y, Ji, Y, Chen, L, Cheng, B & Wu, J 2022. A maize CBM domain containing the protein *ZmCBM48-1* positively regulates starch synthesis in the rice endosperm. *International Journal of Molecular Science, 23,* art. 6598.

Perez-Moral, N, Plankeele, J-M, Domoney, C & Warren, FJ 2018. Ultra-high performance liquid chromatography-size exclusion chromatography (UPLC-SEC) as an efficient tool for the rapid and highly informative characterisation of biopolymers. *Carbohydrate Polymers*, *196*, 422-426.

Phann, TTT, Ishibashi, Y, Miyazaki, M, Tran, HT, Okamura, K, Tanaka, S, Nakamura, J, Yuasa, T & Iwaya-Inoue, M 2013. High temperature-induced repression of the rice sucrose transporter (OsSUT1) and starch synthesis-related genes in sinks and source organs at milky ripening stage causes chalky grains. *Journal of Agronomy and Crop Science*, *199*, 178–188.

Piperno, D, Weiss, E, Holst, I & Nadel, D 2004. Processing of wild cereal grains in the Upper Palaeolithic revealed by starch grain analysis. *Nature*, *430*, 670–673.

Pirone, C, Gurrieri, L, Gaiba, I, Adamiano, A, Valle, F, Trost, P & Sparla, F 2017. The analysis of the different functions of starch-phosphorylating enzymes during the development of *Arabidopsis thaliana* plants discloses an unexpected role for the cytosolic isoform GWD2. *Physiologia Plantarum*, *160*, 447–457.

Prasad, V, Strömberg, C, Alimohammadian, A & Sahni, A 2005. Dinosaur coprolites and the early evolution of grasses and grazers. *Science, 310*, 1177–1180.

Preece, C, Livarda, A, Christin, P-A, Wallace, M, Martin, G, Charles, M, Jones, G, Rees, M & Osborne, CP 2017. How did the domestication of Fertile Crescent grain crops increase their yields? *Functional Ecology*, *31*, 387–397.

Preiss, J 2004. Plant starch synthesis. In: Eliasson, AC (ed.), *Starch in Food: Structure, Function and Applications*, CRC Press: Boca Raton, FL, pp. 3–56.

Preiss, J, Danner, S, Summers, PS, Morell, M, Barton, CR, Yang, L & Nieder, M 1990. Molecular characterization of the brittle-2 gene effect on maize endosperm ADP-glucose pyrophosphorylase subunits. *Plant Physiology*, *92*, 881-885.Preiss, J & Romero, T 1990. Physiology, biochemistry and genetics of bacterial glycogen synthesis. *Advances in Microbial Physiology*, *30*, 183–238.

Qiao, Y, Lee, S-I, Piao, R, Jiang, W, Ham, T-H, Chin, J-H, Piao, Z, Han, L & Kang, S-Y 2010. Fine mapping and candidate gene analysis of the floury endosperm gene, *FLO(a)*, in rice. *Molecules and Cells*, *29*, 167–174.

Qin, D, Wu, H, Peng, H, Yao, Y, Ni, Z, Li, Z, Zhou, C & Sun, Q 2008. Heat stress-responsive transcriptome analysis in heat susceptible and tolerant wheat (*Triticum aestivum* L.) by using Wheat Genome Array. *BMC Genomics, 9*, 432.

Radchuk, V & Borisjuk, L 2014. Physical, metabolic and developmental functions of the seed coat. *Frontiers in Plant Science*, 5, 510.

Radchuk, VV, Borijuk, L, Sreenivasulu, N, Merx, K, Mock, HP, Rolletschek, H, Wobus, U & Weschke, W 2009. Spatiotemporal profiling of starch biosynthesis and degradation in the developing barley grain. *Plant Physiology, 150*, 190–204.

Rathore, RS, Garg, N, Garg, S & Kumar, A 2009. Starch phosphorylase: role in starch metabolism and biotechnological applications. *Critical Reviews in Biotechnology, 29*, 214–224.

Ral, J-P, Bowerman, AF, Li, Z, Sirault, X, Furbank, R, Pritchard, JR, Bloemsma, M, Cavanagh, CR, Howitt, CA & Morell, MK 2012. Down-regulation of glucan, water-dikinase activity in wheat endosperm increases vegetative biomass and yield. *Plant Biotechnology Journal, 10*, 871-882.

Reddy, VM & Daynard, TB 1983. Endosperm characteristics associated with rate of grain filling and kernel size in corn. *Maydica, 28*, 339–355.

Regina, A, Bird, A, Topping, D, Bowden, S, Freeman, J, Barsby, T, Kosar-Hashemi, B, Li, Z, Rahman, S & Morell, M 2006. High-amylose wheat generated by RNA interference improves indices of large-bowel health in rats. *PNAS, 103*, art. 10.

Regina, A, Kosar-Hashemi, B, Li, Z, Pedler, A, Mukai, Y, Yamamoto, M, Gale, K, Sharp, PJ, Morell, MK & Rahman, S 2005. Starch branching enzymes IIb in wheat is expressed at low levels in the endosperm compared to other cereals and encoded at a non-syntenic locus. *Planta, 222*, 899–909.

Regina, A, Kosar-Hashemi, B, Ling, S, Li, Z, Rahman, S & Morell, M 2010. Control of starch branching in barley defined through differential RNAi suppression of starch branching enzyme IIa and IIb. *Journal of Experimental Biology, 61*, 1469–1482.

Ricard, B, Rivoal, J, Spiteri, A & Pradet, A 1991. Anaerobic stress induces the transcription and translation of sucrose synthase in rice. *Plant Physiology, 95*, 669–674.

Ritte, G, Lloyd, JR, Eckermann, N, Rotmann, A, Kossmann, J & Steup, M 2002. The starch related R1 protein is an α-glucan, water dikinase. *Proceedings of the National Academy of Science USA, 99*, 1766–1771.

Ritte, G, Scharf, A, Eckermann, N, Haebel, S & Steup, M 2004. Phosphorylation of transitory starch is increased during degradation. *Plant Physiology, 135*, 2068–2077.

Rolletschek, H, Koch, K, Wobus, U & Borisjuk, L 2005. Positional cues for the starch/lipid balance in maize kernels and resource partitioning to the embryo. *The Plant Journal, 42*, 69–83.

Rolletschek, H, Melkus, G, Grafahrend-Belau, E, Fuchs, J, Heinzel, N, Schreiber, F, Jakob, P & Borisjuk, L 2011. Combined noninvasive imaging and modeling approaches reveal metabolic compartmentation in the barley endosperm. *Plant Cell, 23*, 3041–3054.

Rolletschek, H, Weschke, W, Weber, H, Wobus, U & Borisjuk, L 2004. Energy state and its control on seed development: starch accumulation is associated with high ATP and steep oxygen gradients within barley grains. *Journal of Experimental Botany, 55*, 1351–1359.

Ruan, Y-L, Jin, Y, Yang, Y-J, Li, G-J & Boyer, JS 2010. Sugar input, metabolism, and signaling mediated by invertase: roles in development, yield potential, and response to drought and heat. *Molecular Plant, 3*, 942–955.

Russo, S, Dosio, A, Graverson, RG, Sillman, J, Carrao, H, Dunbar, MB, Singleton, A, Montagna, P, Barbola, P & Vogt, JV 2014. Magnitude of extreme heat waves in present climate and their projection in a warming world. *Journal of Geophysical Research: Atmospheres, 119*, 12500–12512.

Rydberg, U, Andersson, L, Andersson, R, Åman, P & Larsson, H 2001. Comparison of starch branching enzyme I and II from potato. *European Journal of Biochemistry, 268*, 6140–6145.

Ryoo, N, Yu, C, Park, C-S, Baik, Y, Park, IM, Cho, M-H, Bhoo, SH, An, G, Hahn, FR & Jeon, J-S 2007. Knockout of a starch synthase gene OsSSIIIa/Flo5 causes white-core floury endosperm in rice (*Oryza sativa* L.). *Plant Cell Reports, 26*, 1083–1095.

Sabelli, PA & Larkins, BA 2009. The development of endosperm in grasses. *Plant Physiology, 149*, 14–26.

Saccomanno, B, Berbezy, P, Findlay, K, Shoesmith, J, Uauy, C, Viallis, B & Trafford, K 2022. Characterization of wheat lacking B-type starch granules. *Journal of Cereal Science, 104*, 103398.

Saito, M, Tanaka, T, Sato, K, Vrinten, P & Nakamura, T 2017. A single nucleotide polymorphism in the "Fra" gene results in fractured starch granules in barley. *Theoretical and Applied Genetics, 131*, 353–364.

Sakulsingharoj, C, Choi, S-B, Hwang, S-K, Edwards, GE, Bork, J, Meyer, CR, Preiss, J & Okita, TW 2004. Engineering starch biosynthesis for increasing rice seed weight: the role of the cytoplasmic ADPglucose pyrophosphorylase. *Plant Science, 167*, 1323–1333.

Sang, Y, Bean, S, Seib, PA, Pedersen, J & Shi, YC 2008. Structure and functional properties of sorghum starches differing in amylose content. *Journal of Agricultural and Food Chemistry, 56*, 6680–6685.

Sano, Y 1984. Differential regulation of *waxy* gene expression in rice endosperm. *Theoretical and Applied Genetics, 68*, 467–473.

Santelia, D, Kötting, O, Seung, D, Schubert, M, Thalmann, M, Bischof, S, Meekins, DA, Lutz, A, Patron, N, Gentry, MS, Allain, FH-T & Zeeman, SC 2011. The phosphoglucan phosphatase like sex Four2 dephosphorylates starch at the C3-position in Arabidopsis. *Plant Cell, 23*, 4096–4111.

Satoh, H & Omura, T 1981. New endosperm mutations induced by chemical mutagens in rice, *Oryza sativa* L. *Japanese Journal of Breeding, 31*, 316–326.

Satoh, H, Shibahara, K, Tokunaga, T, Nishi, A, Tasaki, M, Hwang, S-K, Okita, TW, Kaneko, N, Fujita, N, Yoshida, M, Hosaka, Y, Sato, A, Utsumi, Y, Ohdan, T & Nakamura, Y 2008. Mutation of the plastidial α-glucan phosphorylase gene in rice affects the synthesis and structure of starch in the endosperm. *Plant Cell, 20*, 1833–1849.

Sawada, T, Francisco, PB, Aihara, S, Utsumi, Y, Yoshia, M, Oyama, Y, Tsuzuki, M, Satoh, H & Nakamura, Y 2009. *Chlorella* starch branching enzyme II (BEII) can complement the function of BEIIb in rice endosperm. *Plant Cell Physiology, 50*, 1062-1074.

Schirmer, M, Höchstötter, A, Jekle, M, Arendt, E & Becker, T 2013. Physiological and morphological characterization of different starches with variable amylose/amylopectin ratio. *Food Hydrocolloids, 32*, 52–63.

Schmidt, R, Schippers, JHM, Mieulet, D, Watanabe, M, Hoefgen, R, Guiderdoni, E & Mueller-Roeber, B 2014. SALT-RESPONSIVE ERF1 is a negative regulator of grain filling and gibberellin-mediated seedling establishment in rice. *Molecular Plant, 7*, 404–421.

Schoen, A, Joshi, A, Tiwari, V, Gill, B & Rawat, N 2021. Triple null mutations in starch synthase SSIIa gene homeologs lead to high amylose and resistant starch in hexaploid wheat. *BMC Plant Biology, 21*, art. 74.

Schreier, TB, Umhang, M, Lee, SK, Lue, WL, Shen, Z, Silver, D & Zeeman, SC 2019. LIKE SEX4 1 acts as a β-amylase-binding scaffold on starch granules during starch degradation. *Plant Cell, 31*, 2169–2186.

Sestili, F, Janni, M, Doherty, A, Botticella, E, D'Ovidio, R, Masci, S, Jones, HD & Lafiandra, D 2010. Increasing the amylose content of durum wheat through silencing of the SBEIIa genes. *BMC Plant Biology, 10*, 144.

Seung, D 2020. Amylose in starch: towards an understanding of biosynthesis, structure and function. *New Phytologist, 228*, 1490–1504.

Seung, D, Boudet, J, Monroe, JD, Schreier, TB, David, LC, Abt, M, Lu, K-J, Zanella, M & Zeeman, SC 2017. Homologs of PROTEIN TARGETING TO STARCH control starch granule initiation in Arabidopsis leaves. *Plant Cell, 29*, 1657–1677.

Seung, D, Echevarría-Poza, A, Steuernagel, B & Smith, AM 2020. Natural polymorphisms in arabidopsis result in wide variation or loss of the amylose component. *Plant Physiology, 182*, 870–881.

Seung, D, Lu, K-J, Stettler, M, Streb, S & Zeeman, SC 2016. Degradation of glucan primers in the absence of starch synthase 4 disrupts starch granule initiation in Arabidopsis. *The Journal of Biological Chemistry, 192*, 20718–20728.

Seung, D, Schreier, TB, Bürgy, L, Eicke, S & Zeeman, SC 2018. Two plastidial coiled-coil proteins are essential for normal starch granule initiation in Arabidopsis. *Plant Cell, 30*, 1523–1542.

Seung, D & Smith, AM 2019. Starch granule initiation and morphogenesis – progress in Arabidopsis and cereals. *Journal of Experimental Botany, 70*, 771–784.

Seung, D, Soyk, S, Coiro, M, Maier, BA, Eicke, S & Zeeman, SC 2015. Protein targeting to starch is required for localising granule-bound starch synthase to starch granules and for normal amylose synthesis in Arabidopsis. *PLoS Biology, 13*, e1002080.

Shannon, J & Dougherty, C 1972. Movement of ^{14}C-labelled assimilate into kernels of *Zea* mays L. II. Invertase activity of the pedicel and placenta-chalazal tissues. *Plant Physiology, 49*, 203–206.

Shannon, JC, Garwood, DL & Boyer, CD 2009. Genetics and physiology of starch development. In: BeMiller, JN & Whistler, RL (eds.), *Starch: Chemistry and Technology*, Elsevier Inc: Massachusetts, USA, pp. 24–82.

Shannon, JC, Pien, FM, Cao, H & Liu, KC 1998. Brittle-1, an adenylate translocator, facilitates transfer of extraplastidial synthesized ADP-glucose into amyloplasts of maize endosperms. *Plant Physiology, 117*, 1235–1252.

Shapter, F, Henry, R & Lee, L 2008. Endosperm and starch granule morphology in wild cereal relatives. *Plant Genetic Resources, 6*, 85–97.

Sharma, M, Abt, MR, Eicke, S, Ilse, TE, Liu, C, Lucas, MS, Pfister, B & Zeeman, SC 2024. MFP1 defines the subchloroplast location of starch granule initiation. *PNAS, 121*, e2309666121.

She, K-C, Kusano, H, Koizumi, K, Yamakawa, H, Hakata, M, Imamura, T, Fukuda, M & Naito, N 2010. A novel factor *FLOURY ENDOSPERM2* is involved in regulation of rice grain size and starch quality. *The Plant Cell, 22*, 3280–3294.

Shen, S, Ma, S, Wu, L, Zhou, S-L & Ruan, Y-L 2023. Winners take all: competition for carbon resource determines grain fate. *Trends in Plant Science, 28*, 893–901.

Sheng, Z, Fang, P, Li, S, Jiao, G, Xie, L, Hu, P, Tang, S & Wei, X 2015. Phenotype of rice floury endosperm mutant *flo7* and fine mapping of mutated gene. *Rice Science, 22*, 162–170.

Shoaib, S, Liu, L, Ali, A, Mughal, N, Yu, G & Huang, Y 2021. Molecular functions and pathways of plastidial starch phosphorylase (PHO1) in starch metabolism: current and future perspectives. *International Journal of Molecular Sciences, 22*, 10450.

Shull, J, Chandrashekar, A, Kirleis, A & Ejeta, G 1990. Development of Sorghum (*Sorghum bicolor*) endosperm in varieties of varying hardness. *Food Structure, 9*, 253–267.

Shure, M, Wessler, S & Fedoroff, N 1983. Molecular identification and isolation of the *Waxy* locus in maize. *Cell, 35*, 225–233.

Sikka, V, Choi, B, Kavakli, I, Sakulsingharoj, C, Gupta, S, Ito, H & Okita, T 2001. Subcellular compartmentation and allosteric regulation of the rice endosperm ADPglucose pyrophosphorylase. *Plant Science, 161*, 461–468.

Sim, L, Beeren, SR, Findinier, J, Dauvillée, D, Ball, SG, Henriksen, A & Palcic, MM 2014. Crystal structure of the Chlamydomonas starch debranching enzyme isoamylase ISA1 reveals insights into the mechanism of branch trimming and complex assembly. *Journal of Biological Chemistry, 289*, 22991–23003.

Simsek, S, Whitney, K & Ohm, J-B 2012. Analysis of cereal starches by high-performance size exclusion chromatography. *Food Analytical Methods, 6*, 181–190.

Singh, S, Singh, N, Isono, N & Noda, T 2010. Relationship of granule size distribution and amylopectin structure with pasting, thermal, and retrogradation properties in wheat starch. *Journal of Agricultural and Food Chemistry, 58*, 1180–1188.

Singh, VK, Pundir, S, Chaturvedi, D, Kaur, A, Pandey, A, Mandal, S, Kumar, R, Singh, RK, Bhati, HP, Dhanda, PS, Yadav, A, Kole, C & Khauisk, P 2023. Enhancing maize (Zea mays L.) crop through advanced techniques: a comprehensive approach. In: Kaushik, P (ed), *New Prospects of Maize*. Intech Open. pp. 1–25.

Smith, AM & Zeeman, SC 2020. Starch: A flexible, adaptable carbon store coupled to plant growth. *Annual Review of Plant Biology, 71*, 217–245.

Smýkal, P, Nelson, MN, Berger, JD & von Wettberg, EJB 2018. The impact of genetic changes during crop domestication. *Agronomy, 8*, 119.

Song, Y, Luo, G, Shen, L, Yu, K, Yang, W, Li, X, Sun, J, Zhan, K, Cui, D, Liu, D & Zhang, A 2020. *TubZIP28*, a novel bZIP family transcription factor from *Triticum urartu*, and *TabZIP28*, its homologue from *Triticum aestivum*, enhance starch synthesis in wheat. *New Phytologist, 226*, 1384–1398.

Stein, O & Granot, D 2019. An overview of sucrose synthase in plants. *Frontiers in Plant Science, 10,* 95.

Steup, M, Robenek, H & Melkonian, M 1983. *In-vitro* degradation of starch granules isolated from spinach chloroplasts. *Planta, 158,* 428–436.

Stitt, M & Zeeman, SC 2012. Starch turnover: pathways, regulation and role in growth. *Current Opinion in Plant Biology, 15,* 282–292.

Stoddard, F & Sarker, R 2000. Characterization of starch in *Aegilops* species. *Cereal Chemistry, 77,* 445–447.

Subasinghe, RM, Liu, F, Polack, UC, Lee, EA, Emes, MJ & Tetlow, IJ 2014. Multimeric states of starch phosphorylase determine protein-protein interactions with starch biosynthetic enzymes in amyloplasts. *Plant Physiology and Biochemistry, 83,* 168–179.

Sun, C, Sathish, P, Ahlandsberg, S & Jansson, C 1998. The two genes encoding starch branching enzymes IIa and IIb are differentially expressed in barley. *Plant Physiology, 118,* 37–49.

Sundberg, M, Pfister, B, Fulton, D, Bischof, S, Delatte, T, Eicke, S, Stettler, M, Smith, SM, Streb, S & Zeeman, SC 2013. The heteromultimeric debranching enzyme involved in starch synthesis in Arabidopsis requires both isoamylase1 and isoamylase2 subunits for complex stability and activity. *PLoS One, 8,* art. e75223.

Suzuki, R, Imamura, T, Nonaga, Y, Kusano, H, Teramura, H, Sekine, K-T, Yamashita, T & Shimada, H 2020. A novel FLOURY ENDOSPERM2 (FLO2)-interacting protein, is involved in maintaining fertility and seed quality in rice. *Plant Biotechnology (Tokyo), 37,* 47-55.

Swinkels, JJM 1985. Composition and properties of commercial native starches. *Starch/Stärke, 37,* 1–5.

Szydlowski, N, Ragel, P, Raynaud, S, Lucas, M, Roldán, I, Montero, M, Munoz, F, Ovecka, M, Bahaji, A, Planchot, V, Pozueta-Romero, J, D'Hulst, C & Mérida, A 2009. Starch granule initiation in Arabidopsis requires the presence of either class IV or class III starch synthases. *Plant Cell, 21,* 2443–2457.

Takahashi, S, Kumagai, Y, Igarashi, H, Horimai, K, Ito, H, Shimada, T, Kato, Y & Hamada, S 2020. Biochemical analysis of a new sugary-type rice mutant, Hemisugary1, carrying a novel allele of the *sugary-1* gene. *Planta, 251,* 29.

Takeda, Y, Guan, HP & Preiss, J 1993. Branching of amylose by the branching isoenzymes of maize endosperm. *Carbohydrate Research, 240,* 253–263.

Takeda, Y, Maruta, N & Hizukuri, S 1992. Structures of amylose subfractions with different molecular sizes. *Carbohydrate Research, 226,* 279-285.

Takeda, Y & Preiss, J 1993. Structures of B90 (sugary) and W64A (normal) maize starches. *Carbohydrate Research, 240,* 265–275.

Tanaka, N, Fujita, N, Nishi, A, Satoh, H, Hosaka, Y, Ugaki, M, Kawasaki, S & Nakamura, Y 2004. The structure of starch can be manipulated by changing the expression levels of starch branching enzyme IIb in rice endosperm. *Plant Biotechnology Journal, 2,* 507–516.

Tateoka, T 1962. Starch grains of endosperm in grass systematics. *Botanical Magazine Tokyo, 75,* 377–383.

Teng, X, Wang, Y, Liu, L, Yang, H, Wu, M, Chen, X, Ren, Y, Wang, Y, Duan, E, Dong, H, Jiang, L, Zhang, Y, Zhang, W, Chen, R, Liu, S, Liu, X, Tian, Y, Chen, L, Wang, Y & Wan, J 2024. Rice *floury endosperm26* encoding a mitochondrial single-stranded DNA-binding protein is essential for RNA-splicing of mitochondrial genes and endosperm development. *Plant Science, 346,* art. 112151.

Teng, X, Zhong, M, Zhu, X, Wang, C, Ren, Y, Wang, Y, Zhang, H, Jiang, L, Wang, D, Hao, Y, Wu, M, Zhu, J, Zhang, X, Guo, X, Wang, Y & Wan, J 2019. *FLOURY ENDOSPERM16* encoding a NAD-dependent cytosolic malate dehydrogenase plays an important role in starch synthesis and seed development in rice. *Plant Biotechnology Journal, 17,* 1914–1927.

Tester, RF & Karkalas, J 1996. Swelling and gelatinization of oat starches. *Cereal Chemistry, 73,* 271–277.

Tetlow, IJ, Beisel, KG, Cameron, S, Makhmoudova, A, Liu, F, Bresolin, NS, Wait, R, Morell, MK & Emes, MJ 2008. Analysis of protein complexes in amyloplasts reveals functional interactions among starch biosynthetic enzymes. *Plant Physiology, 146,* 1878–1891.

Tetlow, IJ & Bertoft, E 2020. A review of starch biosynthesis in relation to the building block-backbone model. *International Journal of Molecular Sciences, 21,* 7011.

Tetlow, IJ, Davies, EJ, Vardy, KA, Bowsher, CG, Burrell, MM & Emes, MJ 2003. Subcellular localization of ADPglucose pyrophosphorylase in developing wheat endosperm and analysis of a plastidial isoform. *Journal of Experimental Botany, 54,* 715–725.

Tetlow, IJ & Emes, MJ 2014. A review of starch-branching enzymes and their role in amylopectin biosynthesis. *IUBMB Life, 66,* 546–558.

Tetlow, IJ & Emes, MJ 2017. Starch biosynthesis in the developing endosperms of grasses and cereals. *Agronomy, 7,* 81.

Tetlow, IJ, Wait, R, Lu, Z, Akkasaeng, R, Bowsher, CG, Esposito, S, Kosar-Hashemi, B, Morell, MK & Emes, MJ 2004. Protein phosphorylation in amyloplasts regulates starch branching enzyme activity and protein-protein interactions. *The Plant Cell, 16,* 694–708.

Thévenot, C, Simond-Côte, E, Reyss, A, Manicacci, D, Trouverie, J, Le Guilloux, M, Ginhoux, V, Sidicina, F & Prioul, J-L 2005. QTLs for enzyme activities and soluble carbohydrates involved in starch accumulation during grain filling in maize. *Journal of Experimental Botany, 56,* 945–958.

Thitisaksakul, M, Jiménez, RC, Arias, MC & Beckles, DM 2012. Effects of environmental factors on cereal starch biosynthesis and composition. *Journal of Cereal Science, 56,* 67-80.

Thorne, JH 1985. Developing seeds. *Annual Reviews in Plant Physiology, 36,* 317–343.

Thorbjørnsen, T, Villand, P, Denyer, K, Olsen, O & Smith, A 1996. Distinct isoforms of ADPglucose pyrophosphorylase occur inside and outside the amyloplasts in barley endosperm, *Plant Journal, 10,* 243–250.

Tiessen, A, Nerlich, A, Faix, B, Hümmer, C, Fox, S, Trafford, K, Weber, H, Weschke, W & Geigenberger, P 2012. Subcellular analysis of starch metabolism in developing barley seeds using a non-aqueous fractionation method. *Journal of Experimental Botany, 63,* 2071–2087.

Tuncel, A & Okita, TW 2013. Improving starch yield in cereals by over-expression of ADPglucose pyrophosphorylase: expectations and unanticipated outcomes. *Plant Science, 211,* 52-60.

Turgeon, R 2010. The role of phloem loading reconsidered. *Plant Physiology, 152,* 1817–1823.

Udachan, IS, Sahoo, AK & Hend, GM 2012. Extraction and characterization of sorghum (*Sorghum bicolor L. Moench*) starch. *International Food Research Journal, 19,* 315–319.

Utsumi, Y & Nakamura, Y 2006. Structural and enzymatic characterization of the isoamylase1 homo-oligomer and the isoamylase1-isoamylase2 hetero-oligomer from rice endosperm. *Planta, 225,* 75–87.

Utsumi, Y, Utsumi, C, Sawada, T, Fujita, N & Nakamura, Y 2011. Functional diversity of isoamylase oligomers: the ISA1 homo-oligomer is essential for amylopectin biosynthesis in rice endosperm. *Plant Physiology, 156,* 61–77.

van Bel, A 1993. Strategies of phloem loading. *Annual Review of Plant Physiology and Plant Molecular Biology, 44,* 253–281.

van Dijk, M, Morley, T, Rau, ML & Saghai, Y 2021. A meta-analysis of projected food demand and population at risk of hunger for the period 2010–2050. *Nature Food, 2,* 494–501.

van Dongen, J, Roeb, GW, Dautzenberg, M, Froehlich, A, Vigeolas, H, Minchin, PEH & Geigenberger, P 2004. Phloem import and storage metabolism are highly coordinated by the low oxygen concentrations within developing wheat seeds. *Plant Physiology, 135,* 1809–1821.

Van Leeuwen, MP, Toutounji, MR, Mata, J, Ward, R, Gilbert, EP, Castignolles, P & Gaborieau, M 2021. Assessment of starch branching and lamellar structure in rice flours. *Food Structure, 29,* art. 100201.

Vandromme, C, Spriet, C, Dauvillée, D, Courseaux, A, Putaux, J-L, Wychowski, A, Krzewinski, F, Facon, M, D'Hulst, C & Wattebled, F 2019. PII1: a protein involved in starch initiation that determines granule number and size in Arabidopsis chloroplast. *New Phytologist, 221,* 356–370.

Venn, BJ & Mann, JI 2004. Cereal grains, legumes and diabetes. *European Journal of Clinical Nutrition, 58,* 1443–1461.

Verwimp, T, Vandeputte, GE, Marrant, K & Delcour, JA 2004. Isolation and characterisation of rye starch. *Journal of Cereal Science, 39,* 85–90.

Vinje, MA, Walling, JG, Henson, CA & Duke, SH 2022. Temporal expression analysis of barley disproportionating enzyme 1 (DPE1) during grain development and malting. *Journal of the American Society of Brewing Chemists, 81,* 396–403.

Waigh, TA, Donald, AM, Heidelbach, F, Riekel, C & Gidley, MJ 1999. Analysis of the native structure of starch granules with small angle X-ray microfocus scattering. *Biopolymers, 49,* 91–105.

Wallwork, MAB, Jenner, CF, Logue, SJ & Sedgley, M 1998. Effect of high temperature during grain filling on starch synthesis in the developing barley grain. *Functional Plant Biology, 25,* 173–181.

Wang, HL, Offler, CE & Patrick, JW 1995. The cellular pathway of photosynthate transfer in the developing wheat grain. 3. A structural analysis and physiological studies of the pathway from the endosperm cavity to the starchy endosperm. *Plant Cell and Environment, 18,* 389–407.

Wang, J, Chen, Z, Zhang, Q, Meng, S & Wei, C 2020b. The NAC transcription factors OsNAC20 and OsNAC26 regulate starch and storage protein synthesis. *Plant Physiology, 184,* 1775–1791.

Wang, J, Hu, P, Chen, Z, Liu, Q & Wei, C 2017. Progress in high-amylose cereal crops through inactivation of starch branching enzymes. *Frontiers in Plant Science, 8,* art. 469.

Wang, J-C, Xu, H, Zhu, Y, Liu, Q-Q & Cai, X-L 2013. OsbZIP58, a basic leucine zipper transcription factor, regulates starch biosynthesis in rice endosperm. *Journal of Experimental Botany, 64,* 3453–3466.

Wang, W, Wei, X, Jiao, G, Chen, W, Wu, Y, Sheng, Z, Hu, S, Xie, L, Wang, J, Tang, S & Hu, P 2020a. GBSS-BINDING PROTEIN, encoding a CBM48 domain-containing protein, affects rice quality and yield. *Journal of Integrative Plant Biology, 62,* 948–966.

Wang, X, Liu, Y, Hao, C, Li, T, Majeed, U, Liu, H, Li, H, Hou, J & Zhang, X 2023. Wheat *NAC-A18* regulates grain starch and storage proteins synthesis and affects grain weight. *Theoretical and Applied Genetics, 136,* art. 123.

Wang, Y-J, White, P, Pollack, L & Jane, J 1993. Characterization of starch structures of 17 maize endosperm mutant genotypes with Oh43 inbred line background. *Cereal Chemistry, 70,* 171–179.

Watson-Lazowski, A, Raven, E, Feike, D, Hill, L, Elaine Barclay, J, Smith, AM & Seung, D 2022. Loss of PROTEIN TARGETING TO STARCH 2 has variable effects on starch synthesis across organs and species. *Journal of Experimental Botany, 73,* 6367–6379.

Weber, APM, Schwacke, R & Flügge, U-I 2005. Solute transporters of the plastid envelope membrane. *Annual Review of Plant Biology, 56,* 133–164.

Weschke, W, Panitz, R, Gubatz, S, Wang, Q, Radchuk, Weber, H & Wobus, U 2003. The role of invertases and hexose transporters in controlling sugar ratios in maternal and filial tissues of barley caryopses during early development. The *Plant Journal, 33,* 395–411.

Weschke, W, Panitz, R, Sauer, N, Wang, Q, Neubohn, B, Weber, H & Wobus, U 2000. Sucrose transport into barley seeds: molecular characterization of two transporters and implications for seed development and starch accumulation. *The Plant Journal, 21,* 455–467.

Wu, H, Galli, M, Spears, CJ, Zhan, J, Liu, P, Yadegari, R, Dannenhoffer, JM, Gallavotti, A & Becraft, PW 2024a. NAKED ENDOSPERM1, NAKED ENDOSPERM2, and OPAQUE2 interact to regulate gene networks in maize endosperm development. *Plant Cell, 36,* 19–39.

Wu, H, Ren, Y, Dong, H, Xie, C, Zhao, L, Wang, X, Zhang, F, Zhang, B, Jiang, X, Huang, Y, Jing, R, Wang, J, Miao, R, Bao, X, Yu, M, Nguyen, T, Mou, C, Wang, Y, Wang, Y, Lei, C, Cheng, Z, Jiang, L & Wan, J 2024b. FLOURY ENDOSPERM24, a heat shock protein 101 (HSP101), is required for starch biosynthesis and endosperm development in rice. *New Phytologist, 242,* 2635–2651.

Wu, M, Ren, Y, Cai, M, Wang, Y, Zhu, S, Zhu, J, Hao, Y, Teng, X, Zhu, X, Jing, R, Zhang, H, Zhong, M, Wang, Y, Lei, C, Zhang, X, Guo, X, Cheng, Z, Lin, Q, Wang, J, Jiang, L, Bao, Y, Wang, Y & Wan, J 2019. Rice FLOURY ENDOSPERM10 encodes a pentatricopeptide repeat protein that is essential for the trans-splicing of mitochondrial nad1 intron 1 and endosperm development. *New Phytologist, 223,* 736–750.

Wu, Y-P, Pu, C-H, Lin, H-Y, Huang, H-Y, Huang, Y-C, Hong, C-Y, Chang, M-C & Lin, Y-R 2015. Three novel alleles of FLOURY ENDOSPERM2 (FLO2) confer dull grains with low amylose content in rice. *Plant Science, 233,* 44-52.

Wychowski, A, Bompard, C, Grimaud, F, Potocki-Vronse, G, D'Hulst, CD, Wattebled, F & Roussel, X 2017. Biochemical characterization of *Arabidopsis thaliana* starch branching enzyme 2.2 reveals an enzymatic positive cooperativity. *Biochimie, 140,* 146–158.

Xia, H, Yandeau-Nelson, M, Thompson, DB & Guiltinan, MJ 2011. Deficiency of maize starch-branching enzyme I results in altered starch fine structure, decreased digestibility and reduced coleoptile growth during germination. *BMC Plant Biology, 11,* art. 95.

Xie, L, Liu, S, Zhang, Y, Tian, W, Xu, D, Li, J, Luo, X, Li, L & Bian, Y 2023. Efficient proteome-wide identification of transcription factors targeting *Glu-1*: a case study for functional validation of TaB3-2A1 in wheat. *Plant Biotechnology Journal, 21,* 1952–1965.

Xu, J, Kuang, Q, Wang, K, Zhou, S, Wang, S, Liu, X & Wang, S 2017. Insights into molecular structure and digestion rate of oat starch. *Food Chemistry, 220,* 25–30.

Xu, Q, Chen, S, Yunjuan, R, Chen, S & Liesche, J 2018. Regulation of sucrose transporters and phloem loading in response to environmental cues. *Plant Physiology, 176,* 930–945.

Xue, M, Liu, L, Yu, Y, Zhu, J, Gao, H, Wang, Y & Wan, J 2019. Lose-of-function of a rice nucleolus-localized pentatricopeptide repeat protein is responsible for the *floury endosperm14* mutant phenotypes. *Rice, 12,* art. 100.

Yamamori, M, Fujita, S, Hayakawa, K, Matsuki, J & Yasui, T 2000. Genetic elimination of a starch granule protein, SGP-1, of wheat generates an altered starch with apparent high amylose. *Theoretical and Applied Genetics, 101,* 21–29.

Yan, H, Zhang, W, Wang, Y, Jin, J, Xu, H, Fu, Y, Shan, Z, Wang, X, Teng, X, Li, X, Wang, Y, Hu, X, Zhang, W, Zhu, C, Zhang, X, Zhang, Y, Wang, R, Zhang, J, Cai, Y, You, X, Chen, J, Ge, X, Wang, L, Xu, J, Jiang, L, Liu, S, Lei, C, Zhang, X, Wang, H, Ren, Y & Wan, J 2024. Rice LIKE EARLY STARVATION1 cooperates with FLOURY ENDOSPERM6 to modulate starch biosynthesis and endosperm development. *The Plant Cell, 36,* 1892–1912.

Yan, M, Pan, T, Zhu, Y, Jiang, X, Yu, M, Wang, R, Zhang, B, Jing, R, Cheng, Z, Zhang, X, Lei, C, Lin, Q, Zhu, S, Guo, X, Ren, Y & Wan, J 2022. *FLOURY ENDOSPERM20* encoding SHMT4 is required for rice endosperm development. *Plant Biotechnology Journal, 20,* 1438–1440.

Yang, H, Wang, Y, Tian, Y, Teng, X, Lv, Z, Lei, J, Duan, E, Dong, H, Yang, X, Zhang, Y, Sun, Y, Chen, X, Bao, X, Chen, R, Gu, C, Zhang, Y, Jiang, X, Ma, W, Zhang, P, Ji, Y, Zhang, Y, Wang, Y & Wan, J 2023. Rice *FLOURY ENDOSPERM22*, encoding a pentatricopeptide repeat protein, is involved in both mitochondrial RNA splicing and editing and is crucial for endosperm development. *Journal of Integrative Plant Biology, 65,* 755–771.

Yang, T, Guo, L, Ji, C, Wang, H, Wang, J, Zheng, X, Xiao, Q & Wu, Y 2021. The B3 domain-containing transcription factor ZmABI19 coordinates expression of key factors required for maize seed development and grain filling. *The Plant Cell, 33,* 104–128.

Yang, Y, Gao, T, Xu, M, Dong, J, Li, H, Wang, P, Li, G, Gao, T, Kang, G & Wang, Y 2017. Functional analysis of a wheat AGPase plastidial small subunit with a truncated transit peptide. *Molecules, 22,* art. 386.

You, X, Zhang, W, Hu, J, Jing, R, Cai, Y, Feng, Z, Kong, F, Zhang, J, Yan, H, Chen, W, Chen, X, Ma, J, Tang, X, Wang, P, Zhu, S, Liu, L, Jiang, L & Wan, J 2019. FLOURY ENDOSPERM15 encodes a glyoxalase I involved in compound granule formation and starch synthesis in rice endosperm. *Plant Cell Reports, 38,* 345–359.

You, Y, Zhang, M, Yang, W, Li, C, Liu, Y, Li, C, He, J & Wu, W 2020. Starch phosphorylation and the *in vivo* regulation of starch metabolism and characteristics. *International Journal of Biological Macromolecules, 159,* 823–831.

Yu, G, Lv, Y, Shen, L, Wang, Y, Qing, Y, Wu, N Li, Y, Huang, H, Zhang, N, Liu, Y, Hu, Y, Liu, H, Zhang, J & Huang, Y 2019. The proteomic analysis of maize endosperm protein enriched by Phos-tag™ reveals the phosphorylation of Brittle-2 subunit of ADP-glc pyrophosphorylase in starch biosynthesis process. *International Journal of Molecular Sciences, 20,* art. 986.

Yu, G, Mou, Y, Shoaib, N, He, X, Liu, L, Di, R, Mughal, N, Zhang, N & Huang, Y 2023b. Serine 31 phosphorylation-driven regulation of AGPase activity: potential implications for enhanced starch yields in crops. *International Journal for Molecular Sciences, 24,* art. 15283.

Yu, G, Shoaib, N, Xie, Y, Liu, L, Mughal, N, Li, Y, Huang, H, Zhang, N, Zhang, J, Liu, Y, Hu, Y, Liu, H & Huang, Y 2022. Comparative study of starch phosphorylase genes and encoded proteins in various monocots and dicots with emphasis on maize. *International Journal of Molecular Sciences, 23,* art. 4518.

Yu, G, Shoaib, N, Yang, Y, Liu, L, Mughal, N, Mou, Y & Huang, Y 2023a. Effect of phosphorylation sites mutations on the subcellular localization and activity of AGPase Bt2 subunit: implications for improved starch biosynthesis in maize. *Agronomy, 13,* art. 2119.

Yu, M, Wu, M, Ren, Y, Wang, Y, Li, J, Lei, C, Sun, Y, Bao, X, Wu, H, Yang, H, Pan, T, Wang, Y, Jing, R, Yan, M, Zhang, H, Zhao, L, Zhao, Z, Zhang, X, Guo, X, Cheng, Z, Yang, B, Jiang, L & Wang, J 2021. Rice *FLOURY ENDOSPERM 18* encodes a pentatricopeptide repeat protein required for 5' processing of mitochondrial *nad5* messenger RNA and endosperm development. *Journal of Integrative Plant Biology, 63,* 854–847.

Yu, S-M, Lo, S-F & Ho, D 2015. Source-sink communication: regulated by hormone, nutrient, and stress cross-signalling. *Trends in Plant Science, 20,* 844–857.

Yun, M & Kawagoe, Y 2010. Septum formation in amyloplasts produces compound granules in the rice endosperm and is regulated by plastid division proteins. *Plant Cell Physiology, 51,* 1469–1479.

Zhang, H, Qin, F, Xu, G, Geng, S, Yuan, Y, Wang, M, Jiao, F & Chen, J 2024. The relationship between starch synthesis enzyme activity, gene expression, and amylopectin fine structure in waxy maize. *Cereal Research Communications.* https://doi.org/10.1007/s42976-024-00509-3

Zhang, J, Chen, J, Yi, Q, Hu, Y, Liu, H, Liu, Y & Huang, Y 2014. Novel role of ZmaNAC36 in co-expression of starch synthetic genes in maize endosperm. *Plant Molecular Biology, 84,* 359–369.

Zhang, L, Ren, Y, Lu, B, Yang, C, Feng, Z, Liu, Z, Chen, J, Ma, W, Wang, Y, Yu, X, Wang, Y, Zhang, W, Wang, Y, Liu, S, Wu, F, Zhang, X, Guo, X, Bao, Y, Jiang, L & Wan, J 2016b. FLOURY ENDOSPERM7 encodes a regulator of starch synthesis and amyloplast development essential for peripheral endosperm development in rice. *Journal of Experimental Botany, 67,* 633–647.

Zhang, X, Colleoni, C, Ratushna, V, Sirghie-Colleoni, M, James, MG & Myers, AM 2004. Molecular characterization demonstrates that the *Zea mays* gene *sugary2* codes for the starch synthase isoform SSIIa. *Plant Molecular Biology, 54,* 865–879.

Zhang, Y, Guo, Q, Feng, N, Wang, J, Wang, S & He, Z 2016a. Characterization of A- and B-type starch granules in Chinese wheat cultivars. *Journal of Integrative Agriculture, 15,* 2203–2214.

Zhang, Y, Zhang, S, Zhang, J, Wei, W, Zhu, T, Qu, H, Liu, Y & Xu, G 2023. Improving rice eating and cooking quality by enhancing endogenous expression of a nitrogen-dependent floral regulator. *Plant Biotechnology Journal, 21,* 2654–2670.

Zhang, Z, Dong, J, Ji, C, Wu, Y & Messing, J 2019. NAC-type transcription factors regulate accumulation of starch and protein in maize seeds. *PNAS, 116,* 11223–11228.

Zhang, Z, Zheng, X, Yang, J, Messing, J & Wu, Y 2016b. Maize endosperm-specific transcription factors O2 and PBF network the regulation of protein and starch synthesis. *PNAS, 113,* 39.

Zhong, M, Liu, X, Liu, F, Ren, Y, Wang, Y, Zhu, J, Teng, X, Duan, E, Wang, F, Zhang, H, Mining, W, Hao, Y, Zhu, X, Jing, R, Guo, X, Jiang, L, Wang, Y & Wan, J 2019. *FLOURY ENDOSPERM12* encoding alanine aminotransferase 1 regulates carbon and nitrogen metabolism in rice. *Journal of Plant Biology, 62,* 61–73.

Zhu, X, Teng, X, Wang, Y, Hao, Y, Jing, R, Wang, Y, Liu, Y, Zhu, J, Wu, M, Zhong, M, Chen, X, Zhang, Y, Zhang, W, Wang, C, Wang, Y & Wan, J 2018. FLOURY ENDOSPERM11 encoding a plastid heat shock protein 70 is essential for amyloplast development in rice. *Plant Science, 277,* 89–99.

Zeeman, SC, Kossmann, J & Smith, AM 2010. Starch: its metabolism, evolution, and biotechnological modification in plants. *Annual Review of Plant Biology, 61,* 209–234.

Zeeman, SC, Smith, SM & Smith, AM 2002. The priming of amylose synthesis in Arabidopsis leaves. *Plant Physiology, 128,* 1069–1076.

Zeeman, SC, Thorneycroft, D, Schupp, N, Chapple, A, Weck, M, Dunstan, H, Haldimann, P, Bechtold, N, Smith, AM & Smith, SM 2004. Plastidial alpha-glucan phosphorylase is not required for starch degradation in Arabidopsis leaves but has a role in the tolerance of abiotic stress. *Plant Physiology, 135,* 849–858.

Zeng, Y, Wu, Y, Avigne, WT & Koch, KE 1998. Differential regulation of sugar-sensitive sucrose synthases by hypoxia and anoxia indicate complementary transcriptional and posttranscriptional responses. *Plant Physiology, 116,* 1573–1583.

Zeng, Y, Wu, Y, Avigne, WT & Koch, KE 1999. Rapid repression of maize invertase by low oxygen. Invertase/sucrose synthase balance, sugar signaling potential, and seedling survival. *Plant Physiology, 121,* 599–608.

Zhong, Y, Blennow, A, Kofoed-Enevoldsen, O, Jiang, D & Hebelstrup, K 2019. Protein Targeting to Starch 1 is essential for starchy endosperm development in barley. *Journal of Experimental Botany, 70,* 485–496.

Zhu, F, Bertoft, E, Wang, Y, Emes, M, Tetlow, I & Seetharaman, K 2015. Structure of Arabidopsis leaf starch is markedly altered following nocturnal degradation. *Carbohydrate Polymers, 117,* 1002–1013.

Ziegler, GR, Creek, JA & Runt, J 2005. Spherulitic crystallization in starch as a model for starch granule initiation. *Biomacromolecules, 6,* 1547–1554.

Zobel, HF 1988. Molecule to granules: a comprehensive starch review. *Starch, 40,* 44–50.

Source of Starch

4

Andreas Blennow

4.1 DIFFERENT SOURCES OF STARCH AND PRODUCTION

4.1.1 The Major Sources of Starch

Of the top 10 crops worldwide by quantity, six are pivotal starch crops, specifically maize, wheat, rice, potato, cassava, and barley (Table 4.1). Among these, maize, cassava, wheat, and potato are the primary contributors to starch production, collectively accounting for nearly 100 million metric tons annually (Table 4.1). Maize leads starch production with a substantial 75%, followed by cassava at 14%, wheat at 7%, and potato at 4% (Vilpoux & Junior 2023) (Table 4.1). Additionally, various other botanical sources, though produced in smaller quantities, play a crucial role in starch production, including rice, barley, yams, sweet potatoes, and sorghum (Table 4.1). The combined annual value of starch extracted from these main crops approximates 100 billion USD, demonstrating a Compound Annual Growth Rate (CAGR) of 8.3% (Global Starch Market: Industrial Starch Market Global Report 2024). This underscores the strategic importance and tremendous value of starch as a natural resource.

The quality and functionality of these different major sources vary considerably. Their granular structures are highly variable with respect to size (1–100 μm), shape (surface smoothness, presence of pores, form-factors, etc.), branching pattern, amylose content, and crystalline, thermal, and visco-elastic properties (e.g., Sanderson et al. 2006; Gismondi et al. 2019). While the amylose content typically ranges from 15% to 35% for starches in general, it is on the lower end for potato and rice starches, typically about 20%–25% for potato and cassava starch (Hoover 2001), and 14%–26% for normal rice starch (Williams et al. 1958). In particular, potato starch has a high level of bound phosphate monoesters, providing very transparent and viscous paste systems, a direct effect of the starch-phosphate esters (Viksø et al. 2001). Tuberous starches generally have moderately low amylose, protein, and lipid contents (Hoover 2001), with granules that are typically large and smooth. Yams and starch from rhizomes such as arrowroot can, in some cases, have relatively high amylose contents (Hoover 2001). Generally, the crystalline polymorphs for cereal and seed starches are of the A-type, while for tuberous, rhizome, and root starch types, the crystalline polymorphs vary and are often of the B-type, corresponding to the longer chain lengths of amylopectin in B-type starches (e.g., Sanderson et al. 2006; Blennow et al. 2000).

4.1.2 Additional Sources of Starch

Interesting and relevant sources of starch include pulses such as pea and fava bean (*Vicia faba*) (Robinson, Balk & Domoney 2019), mung bean (*Vigna radiata*) (Hoover et al. 1997), and common bean (*Phaseolus vulgaris*) (Zhang et al. 2022). These starches typically provide increased amylose content, resulting in higher levels of resistant starch. Additionally, starches from pulses often exhibit C-type crystalline polymorphs, which are a mixture of A- and B-type polymorphs. Sago palm (*Metroxylon sagu*) is an extremely hardy plant and a staple food crop in lowland Papua New Guinea. Notably, the sago palm has the highest productivity of all starch crops globally, with starch content in the pith ranging from 19% to 40% and a moderate amylose content of around 25%–30% (Singhal et al. 2008).

Amaranth comprises several species with starches that have diverse properties (Zhu 2017). Originally regarded as an ancient Mexican crop, amaranth was recently reintroduced. Its starch is characterized by very small granules (1–2 μm) and variable amylose content, typically low but occasionally exceeding 30%. Amaranth starch is also noted for its excellent freeze-thaw stability. Millet (*Panicum miliaceum*) starch occurs in numerous species, with amylose content ranging from 11% to 34%. Its starch granules are typically rough and polygonal, varying significantly in size depending on the species (1–50 μm) (Mahajan et al. 2021). Quinoa (*Chenopodium quinoa*) is considered an emerging "superfood." Like amaranth, quinoa seeds contain very small starch granules (1–3 μm) with consistently low amylose content (Li & Zhu 2018). Due to their small granular size, both amaranth and quinoa starches are suitable for special applications such as Pickering emulsions

DOI: 10.1201/9781003464396-4

TABLE 4.1 Global top 25 ranking crops in quantity of fruit, vegetable, grain, roots and tubers 2022 (FAO 2023) and starch 2020 (Vilpoux & Junior, 2023). Starch crops in light gray.

RANK	CROP	QUANTITY (MILL METRIC TONNES)	STARCH PRODUCED (MILL METRIC TONNES)
1	Sugar cane	1950	
2	Maize	1200	65.8
3	Wheat	808	6.3
4	Rice	776	
5	Oil palm	425	
6	Potatoes	375	3.9
7	Soya beans	349	
8	Cassava	335	12.1
9	Tomatoes	186	
10	Barley	155	
11	Bananas	135	
12	Onions	111	
13	Watermelons	100	
14	Yams	95	
15	Sweet potatoes	95	
16	Cucumbers	95	
17	Apples	96	
18	Rapeseed	87	
19	Seed cotton	70	
20	Oranges	76	
21	Grapes	75	
22	Cabbages	73	
23	Coconuts	62	
24	Eggplants	59	
25	Sorghum	58	

(Li & Zhu 2018). Starch extracted from ginger species, including *Curcuma zedoaria* and *Curcuma longa*, typically displays flat and smooth granules with relatively high amylose content and occasionally high starch-phosphate esters (Blennow et al. 2000; Rasmi et al. 2024).

More unconventional and minor crop sources of starch, particularly from roots and tubers, have been recommended to extend functionalities and avoid additional post-harvest modifications (Junior & de Francisco 2020). Some of these include *Canna edulis* (canna), *Cyperus esculentus* (tiger nut), *Dioscorea bulbifera* (yam), *Hedychium coronarium* (march lily), and *Xanthosoma sagittifolium* (cocoyam). We are observing how these resources may gain importance for specialized purposes.

The gelatinization, solubility, swelling, digestion, and other properties of native starches are highly dependent on their amylose and starch phosphate contents. Regarding these structures and their resulting functionalities, I refer to reviews in this area and other chapters in this volume. Importantly, all these starch resources offer a wide range of diverse qualities, which are significantly expanded through rapidly advancing breeding technologies (Section 4.3).

4.2 PLANT STARCH BIOSYNTHESIS

4.2.1 Biosynthesis of a Solid Granule

The biosynthesis of starch granules occurs in the plastids of storage organs and is, in general, an exceptionally coordinated metabolic system. Its complexity arises from the involvement of various enzymes, including enzyme and non-enzyme homologs, diurnal and other regulations, and the inherent interfacial catalytic mechanisms of biosynthesis at starch granular surfaces, which generate organized molecular structures (Tetlow & Bertoft 2020; Li, Zheng et al. 2021). This complexity increases when considering the initiation of starch granule growth and its morphogenesis during growth (Seung & Smith 2019). The schematic outline for the biosynthesis of amylose, the primarily linear molecules in the starch granule, and amylopectin, the branched molecules of the starch granule, is shown in Figure 4.1, with the main enzyme activities

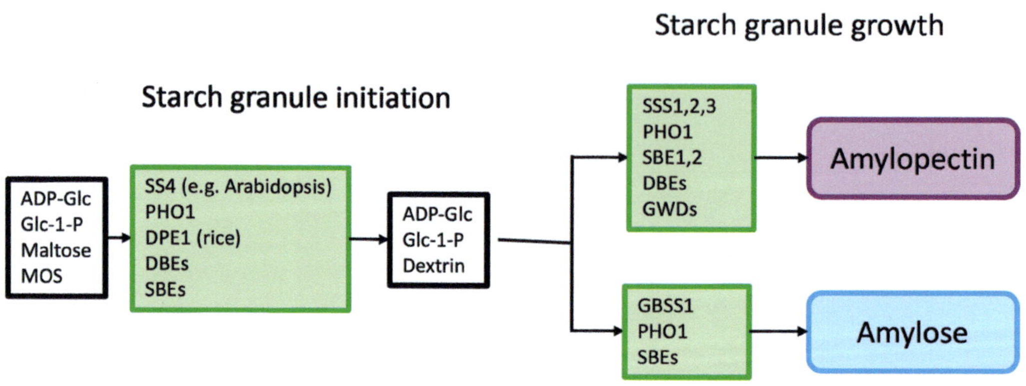

Starch granule growth

Starch granule initiation

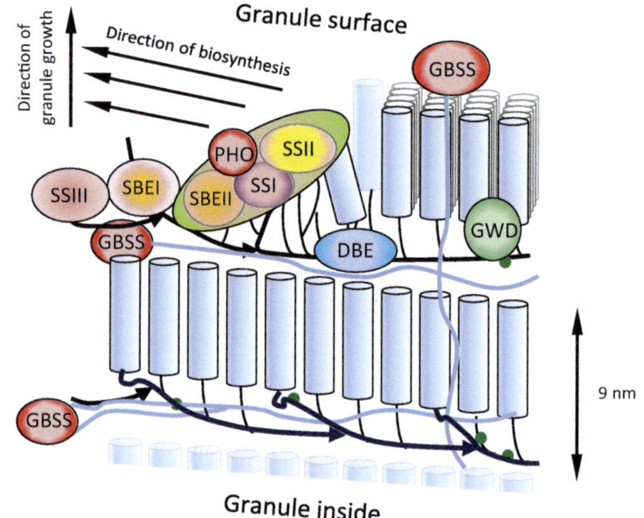

FIGURE 4.1 General view of the two main phases of starch granule biosynthesis: initiation and growth. Initiation of starch granules entails ADP-glucose (ADP-Glc), glucose-1-phosphate (Glc-1-P), maltose and maltooligosaccharides (MOS) serving as substrates for the enzymes starch synthase IV (SSIV), DPE1, and phosphorylase 1 (PHO1), starch branching enzymes (SBEs), and debranching enzymes (DBEs). Once initiated, amylopectin is synthesized from ADP-Gluc, Glc-1-P, and dextrin (diverse MOS) serving as substrates for SSS1,2,3, (PHO1), SBE1,2, DPE1, DBEs, and glucan water dikinases 1 and 3 (GWD1,3). Amylose biosynthesis requires the granule-bound starch synthase 1 (GBSS1), PHO1, and a minor activity of SBEs. There is a tremendous diversity between species of enzyme homologs; hence, this view is very general. Non-enzymatic factors described in the text are not included. White squares: metabolites; light gray squares:enzymes; dark gray: final product

annotated, many of which are found as complexes (Tetlow & Bertoft 2020; Hennen-Bierwagen et al. 2009; Shoaib et al. 2021). Due to the solid state of native starch granules, the complexity is not primarily chemical but physical. The significance of starch phosphorylation for partially solubilizing the granular structure in both starch degradation (Blennow & Engelsen 2010) and biosynthesis (Skeffington et al. 2014; Xu et al. 2017) highlights the cellular challenges in synthesizing and degrading such a solid and recalcitrant carbohydrate. To this end, limited information is available on how such catalysis occurs, and only recently was interfacial catalysis described for starch-acting hydrolases (Tian, Wang, Zhong et al. 2023), which may shed some light on these biosynthetic mechanisms.

This section generally describes the storage starch biosynthetic grid of starch crops in general, bearing in mind the tremendous difference between different crops and crop organs, i.e., seeds, grain, tubers, and roots.

4.2.2 Activation of Sugars and the Overall Biosynthetic Capacity

The initial stages of starch biosynthesis involve ADP-glucose pyrophosphorylases (AGPases) (Hannah 2005) and, alternatively, phosphorylase (Pho) (Shoaib et al. 2021; Sharma et al. 2023). These enzymes provide activated glucose moieties (ADP-glucose and glucose-1-phosphate, respectively) for the biosynthetic metabolic pathway. The primary route is generally considered to be AGPase, which is tightly regulated by allosteric mechanisms (Boehlein et al. 2010) and redox control (affecting many other enzymes involved in starch biosynthesis) (Skryhan et al. 2018). Pho has been shown to play specific roles, such as in stress situations, and to provide maltooligosaccharide primers for starch biosynthesis (Shoaib et al. 2021). Recent documentation shows that the knockout of Pho generates rounded starch granules in potato tubers (Sharma et al.

FIGURE 4.2 Enzymes of starch biosynthesis act in concert on the granule surface and inside the granule matrix, interacting with each other and with the granule. GBSS is found tightly bound in the granule matrix, while others are differently bound to various starch motifs. The granule representation and localization of enzymes shown are schematic, and they collect data from a wide number of data on starch structure and enzyme activities and interactions (Hwang et al. 2016; Hennen-Bierwagen et al. 2009; Tetlow & Bertoft 2020). Hence, for SSSs, homologs elongating long chains are suggested to be located at backbone structures, while homologs extending shorter chains are co-localized in crystalline areas together with, e.g., SBEs. Non-enzymatic factors described in the text are not included. Abbreviations as in Figure 4.1.

2023). Further steps of starch biosynthesis, including the plausible semi-crystalline structures on the starch granule surface and the positions and actions of key enzymes, are outlined in Figure 4.2.

The subsequent step in starch formation takes place specifically in the stroma of plant plastids. This process involves the sequential addition of glucosyl residues to form α-1,4 linkages and α-1,4 - 1,6 transfer thereby elongating and branching the polysaccharide, followed by the trimming of branch linkages and phosphorylation of amylopectin. These reactions are catalyzed by starch synthases, including granule-bound starch synthases (GBSSs), soluble starch synthases (SSs), phosphorylase (Pho), starch branching enzymes (SBEs), starch debranching enzymes (DBEs), and glucan water dikinases (GWD). The formation of amylopectin involves SSs and SBEs working together to create highly branched precursor segments, referred to as "pre-amylopectin." These segments are partially debranched by DBEs, leading to the production of mature, long-chained amylopectin (Lin et al. 2012). The various starch synthase isoforms are crucial as they transfer the glucosyl moiety of ADP-glucose to the non-reducing end of preexisting α-1,4-glucan of different lengths. However, the biosynthetic pathway for the formation of the initial primer remains somewhat unclear.

4.2.3 Elongation of Starch Chains

Multiple isoforms of starch synthases are present, including GBSS, which is primarily located within the granule matrix, and soluble SSs, which are distributed between granular and stromal fractions depending on isoforms, species, tissues, and developmental stages. GBSS consists of two main isoforms: GBSSI, which is predominant in amylose biosynthesis in the endosperm, and GBSSII, which is primarily active in autotrophic tissues (Jeon et al. 2010). Various forms of SSs are identified across different plant sources, exhibiting partly overlapping and synergistic functions (Fujita et al. 2007; Zhang et al. 2008). These isoforms are generally denoted as SSI, II, III, and IV. SSI, II, and III generally elongate longer chains sequentially, while SSIV (Seung et al. 2018), along with phosphorylase (Mérida & Fettke 2021), has been shown to be involved in the initiation of starch granules in Arabidopsis. In Arabidopsis, SSIV plays a critical role in starch granule initiation, as evidenced by a significant reduction in granule numbers upon mutation (Burgy et al. 2021). In maize, SSI, SSIIa, SSIIb, and SSIIIa are all identified in the endosperm (Qu et al. 2018). In rice, SSI primarily generates shorter chains (DP 6–15), while SSIIa favors longer chains (DP > 24) (Nakamura 2002). In maize, SSIIb appears to play a minor role, possibly serving as a secondary or complementary gene (Liu et al. 2014). SSIIIa mainly acts as a facilitator, coordinating the activities of enzymes involved in starch biosynthesis (Lin et al. 2012).

4.2.4 Branching and Branch Trimming

SBEs play a critical role in the formation of branch points within glucan chains. They achieve this by breaking α-1,4 linkages and attaching the released oligosaccharides to a linear glucan chain segment, thereby creating an α-1,6 linkage. The SBE family is generally divided into three isoforms: SBEI, SBEIIa, and SBEIIb (Qu et al. 2018). SBEI preferentially acts on longer amylopectin chains, whereas SBEII, on the other hand, targets shorter chains with a degree of polymerization (DP) usually less than 12 (Jeon et al. 2010). Despite the significant homology between SBEIIa

and SBEIIb (Li, Dhital et al. 2019), their expression patterns are distinct. In maize endosperms, SBEIIb is more highly expressed compared to SBEIIa, whereas SBEIIa is more prevalent in leaf chloroplasts (Tetlow & Emes 2014). SBEs efficiently produce highly branched soluble glucans, while DBEs are responsible for removing excess α-d-1,6 linkages from randomly branched water-soluble polysaccharides. This "glucan trimming" process is crucial for the proper and efficient crystallization of the amylopectin molecule (Delatte et al. 2005; Wattebled et al. 2005), resulting in the formation of small linear α-1,4-glucans. These glucans are subsequently recycled into the biosynthetic pathway and incorporated into starch polysaccharides through a series of reactions primarily mediated by disproportionating enzymes (D-enzymes) as demonstrated in rice (Dong et al. 2015). DBEs can be classified as either isoamylases or pullulanases based on their substrate specificity. Typically, isoamylases target distantly spaced branches of amylopectin, whereas pullulanases are more active on tightly spaced branches within the glucan polymer (Lin et al. 2012).

4.2.5 Starch Phosphorylation

Phosphorylation of amylopectin during starch biosynthesis is crucial for both physiological processes and the functional properties of starch. It can enhance both starch synthesis and degradation, as well as contribute to the development of viscous starches (Blennow & Engelsen 2010; Mahlow et al. 2016). Notably, potato starch contains a significant amount of phosphate monoesters, with concentrations varying from 0 to 60 nmol/mg starch across different crops, and it is particularly abundant in potatoes (Blennow et al. 2000; Noda et al. 2007). The phosphorylation of the C-6 carbon in the glucosyl ring is mediated by glucan water dikinase 1 (GWD1), while the C-3 position is phosphorylated by glucan water dikinase 3/phosphoglucan water dikinase (GWD3/PWD) (Baunsgaard et al. 2005; Mahlow et al. 2016). Research using force-field models has shown that the phosphate group at C-3 is highly exposed, causing structural conflicts and strain within the starch double helix (Engelsen et al. 2003). This observation has been confirmed for various starches, indicating that high phosphate content can lead to structural disarray (Ding et al. 2023). For practical applications, the esters of starch phosphate are essential for starch functionality, influencing properties like viscosity and paste clarity.

4.2.6 Non-Enzymatic Factors in Starch Biosynthesis

Recently, several non-enzyme factors have been discovered, many of which were identified using Arabidopsis as a model and their function in some cases later verified in starch crops. These non-enzymatic proteins, alongside enzymatic proteins, play critical roles in starch biosynthesis. One notable factor is a non-enzymatic domain of the SS homolog SSIV, which guides the localization and growth of starch granules in specific regions (Burgy et al. 2021). Another example is the Protein Targeting To

Starch (PTST) proteins, which possess a carbohydrate-binding module of the CBM48 type and coiled-coil folding, facilitating the transient binding and transportation of proteins. PTST1, for instance, binds and transports the amylose synthesizing enzyme GBSS to the starch granular surface. Disruption of PTST1 leads to amylose-deficiency, as demonstrated by a decrease in amylose content in cassava roots (Bull et al. 2018), due to the disrupted starch granular binding dependent on PTST. Similarly, rice mutants lacking a GBSS-Binding Protein, homologous to barley PTST, exhibit significantly reduced starch and amylose content in grains and an amylose-free phenotype in leaves (Wang et al. 2020). Surprisingly, in barley, mutation of this protein resulted in a starch-less endosperm, suggesting a broader role (Zhong et al. 2019). Additionally, the starch synthase isoform SSV, along with Mar-Binding Filament-Like Protein 1 (MFP1) and Myosin-Related Chloroplast Protein (MRC), both containing coiled coils, can independently bind to and assist in the localization of PTST2 and SSIV (Seung et al. 2018; Chen et al. 2022). Recently, new proteinuous factors, Early Starvation 1 (ESV1) and Like Early Starvation 1 (LESV1), were identified in Arabidopsis using a starch biosynthesis reconstituting yeast system (Liu et al. 2023). These non-enzymatic proteins likely participate in protein-mediated phase transition, facilitating the crystallization of starch granule chains during biosynthesis, with LESV promoting and ESV1 stabilizing this process.

For temperate cereals like wheat and barley, the formation of small B-type granules during endosperm development is significant but less understood. Recently, non-enzymatic proteins, such as B-Granule Content1 (BGC1) in wheat and goatgrass (Aegilops), have been identified and proposed to play a role in this process. This protein has counterparts in barley and rice known as Floury Endosperm6 (FLO6), and in Arabidopsis and Brachypodium, it is referred to as Protein Targeting To Starch 2 (PTST2) (Watson-Lazowski et al. 2022). In wheat, the suppression of both SSIV and BGC1 results in compound starch granules similar to those found in rice. Additionally, recent findings indicate that a double BGC1/phosphorylase1 mutation can enhance the formation of B-type granules in wheat (Kamble et al. 2023). It's important to note that in Arabidopsis, these mutations have different outcomes; for instance, the loss of phosphorylase1 does not impact the number of starch granules formed (Kamble et al. 2023), suggesting that these proteins serve different functions across various species.

4.3 PLANT GENETIC ENGINEERING AND BREEDING-CLASSICAL BREEDING AND GENOME EDITING

4.3.1 New Breeding Technologies

For thousands of years, natural mutants of starch crops, such as sugary maize and high amylopectin glutinous rice, have been gathered and selected. The early and mid-20th century marked the initial efforts to create and isolate mutants specifically

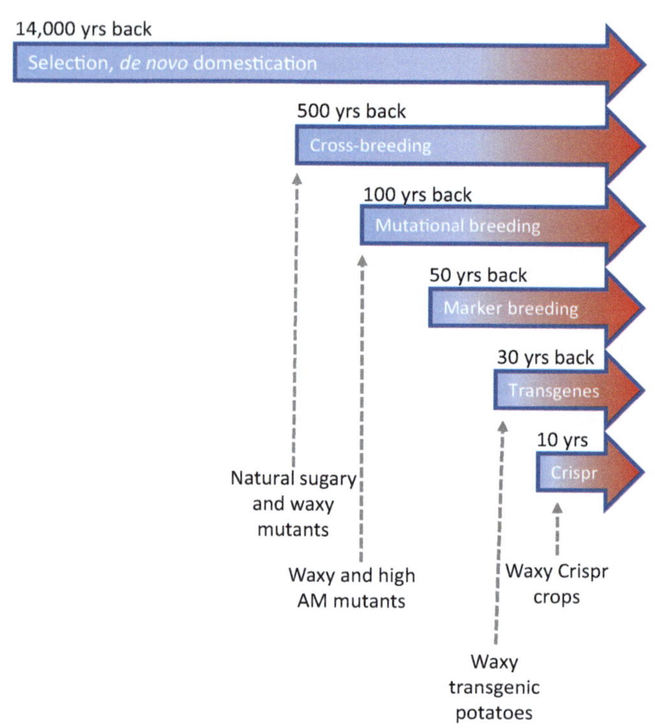

FIGURE 4.3 Schematic timeline for crop domestication, breeding, and biotechnological improvements. Approximate timepoints for the introduction of important improved starch crops indicated as detailed in the text.

for starch quality, resulting in the development of waxy and high-amylose maize genotypes (BeMiller 2009). The first transgenic starch crops to receive licenses were amylopectin potatoes in the mid-1990s, namely "NewLeaf Plus" from Monsanto and "Amflora" from BASF (www.isaaa.org). However, due to public concerns and the lengthy licensing processes for genetically modified (GM) crops, these products have not seen significant market success. With the advent of Crispr-Cas9-based genome editing techniques (Van Vu et al. 2022), researchers developed amylose-free potatoes (Andersson et al. 2017), although they have yet to enter the market. Figure 4.3 provides a schematic timeline of progress in crop improvement for novel starch qualities. The subsequent section will outline the milestones in starch crop breeding development.

Breeding crops involves four primary strategies: cross-breeding, mutation breeding, transgenic breeding, and genome editing. The advent of CRISPR/Cas9 has revolutionized plant biotechnology by offering a rapid and precise method for mutagenesis (Gao 2021). However, public skepticism toward genetically engineered crops remains a significant challenge to the commercialization of transgenic crops (Blennow et al. 2013; Hebelstrup, Sagnelli & Blennow, 2015). Classical chemical and radiation mutagenesis for starch crop bioengineering, especially for increasing amylose content, is generally accepted (Compart et al. 2023; Zhong et al. 2022). Regulatory landscapes are gradually embracing more precise engineering methods, such as transient genome editing using preassembled ribonucleoproteins or mRNA, which ensure DNA editing without foreign DNA integration (Buchholzer & Frommer 2023). This technique employs protoplasts or biolistic protocols for the delivery of CRISPR/Cas9 (Gao 2021).

Despite these advancements, regulations for genome editing vary globally. In many countries, guidelines allow the use of edited lines in agriculture similar to conventionally bred lines, provided they lack transgenes. Approval decisions for edited crops are typically made on a case-by-case basis. However, in the EU, genome editing is currently classified under GMO regulations, although this is under re-evaluation (Buchholzer & Frommer 2023).

4.3.2 The Classical Maize Mutants

Starch quality improvement through breeding has a rich history, primarily targeting genes responsible for enzymes and factors that influence amylose content in starch, such as GBSS, SSs, SBEs, DBEs, and GWDs, resulting in a variety of starch structures and functionalities (Zhong et al. 2022). Technically and practically, significant insights into starch structure engineering have come from maize mutants, including amylose-extender (*ae*), waxy (*wx*), sugary1 (*su1*), sugary2 (*su2*), and dull1 (*du1*) mutations.

The *ae* mutants, deficient in the SBEIIb homolog, produce starch with long-chain amylopectin and high apparent amylose content (Wang et al. 2017). The *wx* mutants lack GBSS1 and have very low or no amylose content (Huang et al. 2010; Šárka & Dvořáček 2017). The *su1* mutants are characterized by a deficiency in debranching enzyme activity, particularly the isoamylase-type homolog, leading to the accumulation of highly branched and soluble phytoglycogen instead of starch and elevated sugar levels (Dinges et al. 2001).

In other plant species, similar mutations produce analogous effects. The *su2* mutants have defects in starch synthase II (SSII), resulting in an increased number of short amylopectin chains and a decrease in intermediate chain length, a phenotype observed in SSII-deficient species (Zhang et al. 2004). The *du1* mutants, affected by SSII deficiency and secondary effects on SBEIIa, exhibit elevated apparent amylose content and a mix of amylopectin and intermediate material with higher branching due to impaired elongation by SSII (Gao et al. 1998).

"Amylose-extender" (*ae*) maize varieties with increased amylose levels are deficient in the SBEIIb isoform (Wang et al.

2017; Zhong et al. 2022). Classical mutagenesis and breeding have recently produced wheat lines with amylose content as high as 93% (Li et al. 2019).

To increase the amylose content in starch, three primary strategies can be identified as outlined formally by Zhong, Qu, et al. (2022): (1) Enhancing AM production through overexpression of GBSSs, which usually results in only modest increases in amylose; (2) creating "AM-like" structures by suppressing SBEs, a commonly effective approach; and (3) boosting the relative amylose proportion by reducing amylopectin biosynthesis through the suppression of SS, DBE, or GWD. It is important to note that, while the amylose content in starch is positively correlated with resistant starch, this correlation tends to diminish when amylose content exceeds approximately 50%, and significant variability in resistant starch is observed among different high amylose starch types (Li et al. 2024; Tian et al. 2024).

4.3.3 Recent Advances in Starch Bioengineering; the Gene Scissor; CRISPR Approaches Targeting GBSS Generating Low Amylose Starches

In addition to classical mutagenesis protocols, which are complemented by advanced molecular techniques and high-throughput genome screening (e.g., Knudsen et al. 2022), the development of starch bioengineering has been significantly accelerated by insights gained from starch mutants and other species, as well as advancements in breeding technologies. Transgene technologies emerged several decades ago, leading to the creation of a high-yielding pure amylose barley line via RNA interference, which silenced all starch branching enzyme genes (SBEI, SBEIIa, and SBEIIb) directly in barley grains (Carciofi et al. 2012). The general effects on starch granule morphology in dependence of amylose content are illustrated in Figure 4.4.

Recently, CRISPR-Cas9 genome editing techniques, also known as "gene scissors," have made rapid advancements.

(a) (b) (c)

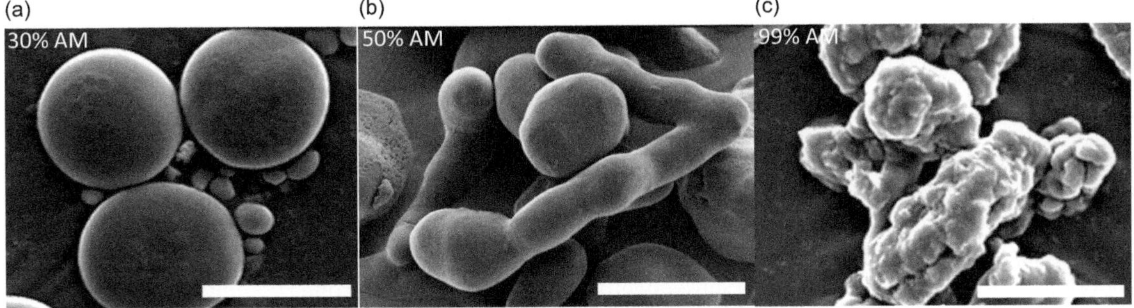

FIGURE 4.4 The effects of amylose on starch granule morphology. (a) Starch granules from normal barley showing rounded, larger A-type granules and smaller B-type granules. (b) High amylose starch granules from maize demonstrating irregular, compounded and elongated structures. (c) Granules from the amylose-only barley line displaying very distorted and compounded granules. Scale bars: 10 μm.

This approach involves an RNA–protein complex that recognizes a specific DNA sequence, which is then cleaved by the Cas9 nuclease, resulting in mutation-prone self-repair mechanisms within the plant. Since the initial publications on CRISPR-Cas9 for starch engineering in 2016, the field has expanded significantly, with approximately 150 articles published, including over 100 from 2022 alone (ISI Web of Science). The GBSS gene is a common target, resulting in amylose-free (waxy) phenotypes. For tetraploid crops like potato, complete GBSS knockout (KO) plants have been successfully generated by various laboratories (Andersson et al. 2017; Johansen et al. 2019; Veillet et al. 2019; Toinga-Villafuerte et al. 2022; Abeuova et al. 2023).

Recent advancements in CRISPR technologies are noteworthy. For example, transient CRISPR-Cas9 expression has been performed in single-cell protoplasts (Andersson et al. 2017), and ribonucleoprotein (RNP) Cas9 protein strategies have been developed to enhance efficiency and offer a DNA-free approach (Johansen et al. 2019). Additionally, base editing has been employed to achieve more precise nucleotide mutations at specific sites (Veillet et al. 2019; Abeuova et al. 2023). In sweet potato, GBSSI suppression has resulted in low amylose (5.75%) lines with minimal yield loss (Wang et al. 2019). Moreover, simultaneous mutation of GBSS and PTST (which targets GBSS to the starch granule) in cassava has produced lines with either complete amylose removal (GBSS1 KO) or low amylose content (PTST KO), allowing for fine-tuning of amylose levels for crop improvement (Bull et al. 2018). A recent study in potato demonstrated that complete GBSS mutation combined with partial SBE mutation can restore the aberrant granule morphology caused by SBE gene knockout (Jayarathna et al. 2024). The following sections will review recent efforts in engineering starch crops with novel functionalities using genome editing techniques such as CRISPR-Cas9 and Zinc Finger Nuclease (ZFN) approaches.

4.3.4 CRISPR Approaches Targeting SBEs Generating High Amylose Starch: Potato

A prevalent method for producing high amylose crops involves the suppression of SBEs, which results in higher amylose content and the formation of "amylose-like" material—amylopectin with reduced molecular size and minimal branching (Zhong, Tai et al. 2022). CRISPR-Cas9 editing has been applied to a range of crops including potato, sweet potato, rice, cassava, and brassica, with notable results.

Due to its ease of genetic transformation and plant regeneration, potato is a primary focus for this approach. Initial attempts to target the SBEI and SBEII homologs in potato led to varieties with increased amylose content, along with altered amylopectin chain lengths and branching patterns, and changes in granule development (Tuncel et al. 2019). Subsequently, Zhao et al. (2021) achieved full knockout of both SBEI and SBEII using CRISPR-Cas9, resulting in almost complete inhibition of starch biosynthesis and the production of pure amylose in tubers, though with very low yield. This was accomplished using a ribonucleoprotein (RNP) Cas9 protein strategy, which avoids transfer and possible integration of foreign DNA.

An intriguing study on a potato mutant lacking the SBE3 homolog, which is analogous to the rice BEI homolog, revealed an unexpected decrease of 8%–10% in amylose content compared to controls, indicating a unique role for this SBE variant (Takeuchi et al. 2021). Unlike rice SBEI mutants, the potato SBE3 mutant exhibited increased iodine binding, reflecting longer chains within the amylopectin (Satoh et al. 2003).

4.3.5 Crispr Approaches Targeting SBEs Generating High Amylose Starch: Rice, Wheat, Cassava, and Brassica

As highlighted earlier, rice stands out as a critical starch crop globally. To enhance the functional properties of rice starch by increasing amylose content, CRISPR/Cas9 was employed to target the SBEI and SBEIIb genes, resulting in a rise in amylose content to 25% and resistant starch content to 9.8% (Sun et al. 2017). More recently, a multiplexing (targeting several genes) approach using CRISPR to target all four SBE homologs in rice led to an increase in amylose content from 17% to 42%, and resistant starch content was elevated by 15% (Biswas et al. 2023). However, achieving homozygous lines proved challenging due to the complexity of the multiplexing strategy. Similar to rice, wheat is a major cereal crop with potential health benefits from high amylose varieties. CRISPR/Cas9 technology has been successfully used to develop both winter and spring wheat lines with partial or triple-null alleles of the SBEIIa gene, reaching amylose levels of 45% in winter wheat and 47% in spring wheat (Li et al. 2021). It is important to note that classical mutagenesis has achieved amylose levels up to 93% (Li, Gidley & Dhital 2019).

Cassava, known for its resilience in arid regions, has also been targeted for high amylose content through CRISPR/Cas9-mediated mutagenesis of the SBEII gene. This has resulted in cassava lines with significantly increased amylose (up to 56%) and resistant starch content (up to 35%) (Luo et al. 2022). This modification has transformed the crystalline structure of the starch from A-type to B-type. In sweet potato, CRISPR/Cas9 targeting of the SBEII gene has produced high amylose lines (40%) with minimal impact on yield (Wang et al. 2019). Additionally, canola/rapeseed (Brassica napus) was edited for six different SBE homologs to explore the effects of multi-gene engineering and gene-dosage effects on oil production and crop performance. This approach was not aimed at starch production, but rather at understanding how starch metabolism influences oil yield. The edited plants exhibited traits such as thicker main stems (Wang, Wang et al. 2023).

It is crucial to recognize that enhancing amylose content can significantly impact the overall primary metabolism of the crop. Targeting one or a few genes often leads to changes in the expression of other starch-related genes (Zhong et al. 2020), and environmental factors such as drought, flooding, and temperature during crop growth play a vital role in starch functionality (Guo et al. 2023).

4.3.6 Additional Genes Targeted Using Genome Editing Protocols: Soluble Starch Synthases (SSSs), Phosphorylase (Pho), and the FtsZ1 Gene

The key targets for advancing starch crop bioengineering through gene editing are the enzymes and proteins directly involved in starch biosynthesis. Soluble Starch Synthases (SSSs) play a crucial role in amylopectin formation. Modifying these enzymes often results in a decreased amylopectin content and a corresponding increase in amylose, although this can also lead to reduced crop yields (Zhong, Qu et al. 2022). In rice, CRISPR-Cas9 was used to knock out SSS homologs SSSII-1, SSSII-2, and SSSII-3, achieving a significant increase in amylose content from 23% in controls to 63% (Jameel et al. 2022). However, the claims of high levels of resistant starch and improved grain yield were not substantiated with data. Another important SSS homolog, SSIV, which is involved in the initiation of starch granules, was targeted using ZFN technology in rice (Jung et al. 2018). This approach resulted in altered starch content and dwarf plant phenotypes, indicating that the SSIV target was not effective for creating high-yield starch crops.

The plastidial Pho1 (Pho1) plays a crucial role in starch biosynthesis, serving as an alternative pathway for SSS-assisted chain elongation. Hence, to investigate the role of this enzyme in potato, CRISPR-Cas9 technology was utilized to introduce specific point mutations in the Pho1a gene, resulting in a G261V mutation. This mutation led to altered morphology and starch content in tubers and leaves, particularly under cold stress (Nezhdanova et al. 2022). This genetic modification also impacted the expression of starch catabolism-related genes, such as β-amylases, amylase inhibitors, and stress-responsive MADS-domain transcription factors in response to cold stress.

In a separate study, another CRISPR-targeted mutation in the plastidial Pho gene in potato (Sharma et al. 2023) underscored its essential role in normal starch granule biosynthesis in tubers. Knockout mutants exhibited an increased number of rounded starch granules and a decrease in amylose content, although the functionality of this altered starch type has yet to be fully assessed.

Additionally, a different approach involved targeting enzymes related to plastid biogenesis using CRISPR/Cas9 to edit the FtsZ1 gene (Pfotenhauer et al. 2023). The FtsZ1 protein, a tubulin-like GTPase, regulates plastid size. The suppression of FtsZ1 in potato tubers resulted in a doubling of starch granule size without affecting the overall plant phenotype.

4.4 CONCLUSION AND PERSPECTIVES

Ensuring a diverse range of starch resources is vital for addressing the combined challenges of climate change and a growing global population. Recent advancements in breeding, particularly through genome editing, present significant opportunities to enhance starch yield, quality, and resilience, although climate factors also influence quality improvements (Guo et al. 2023). Major starch crops, including maize, wheat, rice, potato, cassava, and barley, are central to global production, with maize contributing to 75% of annual starch output. Alternative sources such as pulses, sago palm, amaranth, millet, quinoa, and various ginger species provide unique characteristics and applications, broadening the use of starch in food, biomaterials, and other areas. A deeper understanding of starch biosynthesis and granule formation is essential for optimizing starch production and ensuring its future importance as a key bioresource.

Recent advances in plant genetic engineering have shifted from traditional breeding methods to cutting-edge genome editing techniques like CRISPR/Cas9, which have revolutionized starch crop improvement. While early breeding of natural starch mutants laid the necessary foundation, the commercial success of transgenic crops faced setbacks due to public and regulatory challenges in the 1990s. Classical mutagenesis in maize has provided critical insights into starch biosynthesis, especially regarding strategies to increase amylose content. CRISPR/Cas9 editing has proven effective in targeting genes such as GBSS and SBEs in crops like potato, rice, cassava, and wheat, leading to significant improvements in starch properties.

Looking ahead, combining classical mutagenesis with advanced genome editing techniques offers promising prospects for starch crop bioengineering. The development of DNA-free genome editing methods may address regulatory and public acceptance issues, facilitating broader commercial application. While traditional breeding remains fundamental, modern genome editing technologies, especially CRISPR/Cas9, are poised to transform agricultural biotechnology, offering more precise, efficient, and publicly accepted approaches to enhancing starch crops and expanding starch sources.

REFERENCES

Abeuova, L, Kali, B, Tussipkan, D, Akhmetollayeva, A, Ramankulov, Y & Manabayeva, S 2023. CRISPR/Cas9-mediated multiple guide RNA-targeted mutagenesis in the potato. *Transgenic Research*, 32(5), 383–397.

Abt, MR, Pfister, B, Sharma, M, Eicke, S, Burgy, L, Neale, I, Seung, D & Zeeman, SC 2020. STARCH SYNTHASE5, a noncanonical starch synthase-like protein, promotes starch granule initiation in Arabidopsis. *The Plant Cell*, 32(8), 2543–2565.

Andersson, M, Turesson, H, Nicolia, A, Fält, AS, Samuelsson, M & Hofvander, P 2017. Efficient targeted multiallelic mutagenesis in tetraploid potato (*Solanum tuberosum*) by transient CRISPR-Cas9 expression in protoplasts. *Plant Cell Reports*, 36, 117–128.

Baunsgaard, L, Mogensen, HL, Mikkelsen, R, Glaring, MA, Pham, TT & Blennow, A 2005. A novel isoform of glucan water dikinase phosphorylates prephosphorylated α-glucans and is involved in starch degradation in Arabidopsis. *The Plant Journal*, 41(4), 595–605.

BeMiller, JN 2009. One hundred years of commercial food carbohydrates in the United States. *Journal of Agricultural and Food Chemistry*, 57(18), 8125–8129.

Biswas, S, Ibarra, O, Shaphek, M, Molina-Risco, M, Faion-Molina, M, Bellinatti-Della Gracia, M, Thomson, MJ & Septiningsih, EM 2023. Increasing the level of resistant starch in 'Presidio' rice through multiplex CRISPR-Cas9 gene editing of starch branching enzyme genes. *Plant Genome, 16*(2), art. e20225.

Blennow, A 2018, Chapter 4, Starch bioengineering. In: M Sjoo & L Nilsson (eds.), *Starch in Food, Structure, Function and Applications* 2nd Ed, Woodhead Publishing, Elsevier, ISBN: 9780081008966, pp. 179–122.

Blennow, A & Engelsen, SB 2010. Helix-breaking news: Fighting crystalline starch energy deposits in the cell. *Trends in Plant Science*, 15(4), 236–240.

Blennow, A, Engelsen, SB, Munck, L & Møller, BL 2000. Starch molecular structure and phosphorylation investigated by a combined chromatographic and chemometric approach. *Carbohydrate Polymers*, 41, 163–174.

Blennow, A, Jensen, SL, Shaik, SS, Skryhan, K, Carciofi, M, Holm, PB, Hebelstrup, KH & Tanackovic, V 2013. Future cereal starch bioengineering — Cereal ancestors encounter gene-tech and designer enzymes. *Cereal Chemistry*, 90(4), 274–287.

Boehlein, SK, Shaw, JR, Stewart, JD & Hannah, LC 2010. Studies of the kinetic mechanism of maize endosperm ADP-glucose pyrophosphorylase uncovered complex regulatory properties. *Plant Physiology*, 152, 1056–1064.

Buchholzer, M & Frommer, WB 2023. An increasing number of countries regulate genome editing in crops. *New Phytologist*, 237, 12–15.

Bull, SE, Seung, D, Chanez, C, Mehta, D, Kuon, J-E, Truernit, E, Hochmuth, A, Zurkirchen, I, Zeeman, SC, Gruissem, W & Vanderschuren, H 2018. Accelerated ex situ breeding of GBSS- and PTST1-edited cassava for modified starch. *Science Advances*, 4(9), art. eaat6086.

Burgy, L, Eicke, S, Kopp, C, Jenny, C, Lu, KJ, Escrig, S, Meibom, A & Zeeman, SC 2021. Coalescence and directed anisotropic growth of starch granule initials in subdomains of Arabidopsis thaliana chloroplasts. *Nature Communications*, 12(1), art. 6944.

Carciofi, M, Blennow A, Jensen, SL, Shaik, SS, Henriksen, A, Buléon, A., Holm, PB, Hebelstrup, KH 2012. Concerted suppression of all starch branching enzyme genes in barley produces amylose-only starch granules. *BMC Plant Biology* 12, 223, 1–16.

Chen, J, Chen, Y, Watson-Lazowski, A, Hawkins, E, Barclay, JE, Fahy, B, Bowers, RD, Corbin, K, Warren, FJ, Blennow, A, Uauy, C & Seung, D 2022. Contrasting roles of the wheat MRC in promoting starch granule initiation in leaves and repressing the onset of B-type granule initiation in the endosperm. *bioRxiv*, art. 2022.10.07.511297.

Compart, J, Singh, A, Fettke, J & Apriyanto, A 2023. Customizing starch properties: A review of starch modifications and their applications. *Polymers*, 15, art. 3491.

Delatte, T, Trevisan, M, Parker, ML & Zeeman, SC 2005. Arabidopsis mutants Atisa1 and Atisa2 have identical phenotypes and lack the same multimeric isoamylase, which influences the branch point distribution of amylopectin during starch synthesis. *Plant Journal*, 41, 815–830.

Ding, L, Liag, W, Qu, J, Persson, S, Li, X, Herburger, K, Kirkensgaard, JJK, Khakimov, B, Enemark-Rasmussen, K, Blennow, A & Zhong, Y 2023. Effects of natural starch-phosphate monoester content on the multi-scale structures of potato starches. *Carbohydrate Polymers, 310*, 120740.

Dinges, JR, Colleoni, C, Myers, AM & James, MG 2001. Molecular structure of three mutations at the maize sugary1 locus and their allele-specific phenotypic effects. *Plant Physiology*, 125, 1406–1418.

Dong, X, Zhang, D, Liu, J, Liu, QQ, Liu, H, Tian, L, Jiang, L & Qu, LQ 2015. Plastidial disproportionating enzyme participates in starch synthesis in rice endosperm by transferring maltooligosyl groups from amylose and amylopectin to amylopectin. *Plant Physiology*, 169(4), 2496–512.

Engelsen, SB, Madsen, MAO, Blennow, A, Motawia, S, Møller, BL & Larsen, S 2003. The phosphorylation site in double helical amylopectin as investigated by a combined approach using chemical synthesis, crystallography and molecular modeling. *FEBS Letters*, 541(1–3), 137–144.

FAO 2023. https://www.fao.org/3/cc9205en/cc9205en.pdf

Fujita, N, Yoshida, M, Kondo, T, Saito, K, Utsumi, Y, Tokunaga, T, Nishi, A, Satoh, H, Park, JH, Jane, JL, Miyao, A, Hirochika, H & Nakamura, Y 2007. Characterization of SSIIIa-deficient mutants of rice: The function of SSIIIa and pleiotropic effects by SSIIIa deficiency in the rice endosperm. *Plant Physiology*, 144(4), 2009–2023.

Gao, C 2021. Genome engineering for crop improvement and future agriculture. *Cell*, 184, 1622–1635.

Gao, M, Wanat, J, Stinard, PS, James, MG & Myers, AM 1998. Characterization of dull1, a maize gene coding for a novel starch synthase. *The Plant Cell,* 10(3), 399–412.

Gismondi, A, D'Agostino, A, Canuti, L, Di Marco, G, Basoli, F & Canini, A 2019. Starch granules: A data collection of 40 food species. *Plant Biosystems*, 153(2), 273–279.

Global starch market: Industrial Starch Market Global Report 2024. https://www.thebusinessresearchcompany.com/report/industrial-starch-global-market-report

Guo, K, Liang, W, Wang, S, Guo, D, Liu, F, Persson, S, Herburger, K, Petersen, BL, Liu, X, Blennow, A & Zhong, Y 2023. Strategies for starch customization: Agricultural modification. *Carbohydrate Polymers,* 312, art. 121336.

Hannah, LC 2005. Starch synthesis in the maize endosperm. *Maydica*, 50, 497–506.

Hebelstrup, KH, Sagnelli, D & Blennow, A 2015. The future of starch bioengineering: GM microorganisms or GM plants? *Frontiers in Plant Science*, 6, art. 247.

Hennen-Bierwagen, TA, Lin, Q, Grimaud, F, Planchot, V, Keeling, PL, James, MG & Myers, AM 2009. Proteins from multiple metabolic pathways associate with starch biosynthetic enzymes in high molecular weight complexes: A model for regulation of carbon allocation in maize amyloplasts. *Plant Physiology,* 149, 1541–1559.

Hoover, R 2001. Composition, molecular structure, and physicochemical properties of tuber and root starches: A review. *Carbohydrate Polymers*, 45, 253–267.

Hoover, R, Li, YX, Hynes, G & Senanayake, N 1997. Physicochemical characterization of mung bean starch. *Food Hydrocolloids*, 11(4), 401–408.

Huang, BQ, Tian, ML, Zhang, JJ & Huang, YB 2010. Waxy locus and its mutant types in maize Zea mays L. *Agricultural Sciences in China*, 9(1), 1–10.

Hwang, S-K, Koper, K, Satoh, H & Okita, TW 2016. Rice endosperm starch phosphorylase (Pho1) assembles with disproportionating enzyme (Dpe1) to form a protein complex that enhances synthesis of malto-oligosaccharides. *Journal of Biological Chemistry*, 201, 19994–20007.

Jameel, MR, Ansari, Z, Al-Huqail, AA, Naaz, S & Qureshi, MI 2022. CRISPR/Cas9-mediated genome editing of soluble starch synthesis enzyme in rice for low glycemic index. *Agronomy*, 12, art. 2206.

Jayarathna, S, Hofvander, P, Péter-Szabó, Z, Andersson, M & Andersson, R 2024. GBSS mutations in an SBE mutated background restore the potato starch granule morphology and produce ordered granules despite differences to native molecular structure. *Carbohydrate Polymers*, 331, 121860.

Jeon, J-S, Ryoo, N, Hahn, T-R, Walia, H & Nakamura, Y 2010. Starch biosynthesis in cereal endosperm. *Plant Physiology and Biochemistry*, 48(6), 383–392.

Johansen, IE, Liu, Y, Jørgensen, B, Bennett, EP, Andreasson, E, Nielsen, KL, Blennow, A & Petersen, BL 2019. High efficacy full allelic CRISPR/Cas9 gene editing in tetraploid potato. *Scientific Reports*, 9, art. 17715.

Jung, Y-J, Nogoy, FM, Lee, S-K, Cho, Y-G & Kang, K-K 2018. Application of ZFN for site directed mutagenesis of rice SSIVa gene. *Biotechnology and Bioprocess Engineering*, 23(1), 108–115.

Junior, JFSS & de Francisco, A 2020. Unconventional food plants as an alternative in starch production. *Cereal Foods World*, 65, March/April. https://doi.org/10.1094/CFW-65-2-0018

Kamble, NU, Makhamadjonov, F, Fahy, B, Martins, C, Saalbach, G & Seung, D 2023. Initiation of B-type starch granules in wheat endosperm requires the plastidial α-glucan phosphorylase PHS1. *The Plant Cell*, 35, 4091–4110.

Knudsen, S et al. 2022. FIND-IT: Accelerated trait development for a green evolution. *Science Advances*, 8, art. eabq2266.

Li, G & Zhu, F 2018. Quinoa starch: Structure, properties, and applications. *Carbohydrate Polymers*, 181, 851–861.

Li, H et al. 2024. Resistant starch intake facilitates weight loss in humans by reshaping the gut microbiota. *Nature Metabolism*, 6, 578–597.

Li, H, Dhital, S, Slade, AJ, Yu, W, Gilbert, RG & Gidley, MJ 2019. Altering starch branching enzymes in wheat generates high-amylose starch with novel molecular structure and functional properties. *Food Hydrocolloids*, 92, 51–59.

Li, H, Gidley, MJ & Dhital, S 2019. High-amylose starches to bridge the "fiber gap": Development, structure, and nutritional functionality. *Comprehensive Reviews in Food Science and Food Safety*, 18(2), 362–379.

Li, J, Jiao, G, Sun, Y, Chen, J, Zhong, Y, Yan, L, Jiang, D, Ma, Y & Xia, L 2021. Modification of starch composition, structure and properties through editing of TaSBEIIa in both winter and spring wheat varieties by CRISPR/Cas9. *Plant Biotechnology Journal*, 19(5), 937–951.

Li, R, Zheng, W, Jiang, M & Zhang, H 2021. A review of starch biosynthesis in cereal crops and its potential breeding applications in rice (Oryza Sativa L.). *Peer Journal*, 9, art. e12678.

Lin, Q, Huang, B, Zhang, M, Zhang, X, Rivenbark, J, Lappe, RL, James, MG, Myers, AM & Hennen-Bierwagen, TA 2012. Functional interactions between starch synthase III and isoamylase-type starch-debranching enzyme in maize endosperm. *Plant Physiology*, 158(2), 679–692.

Liu, C, Pfister, B, Osman, R, Ritter, M, Heutinck, A, Sharma, M, Eicke, S, Fischer-Stettler, M, Seung, D, Bompard, C, Abt, MR & Zeeman, SC 2023. LIKE EARLY STARVATION1 and EARLY STARVATION1 promote and stabilize amylopectin phase transition in starch biosynthesis. *Science Advances*, 9, art. eadg7448.

Liu, H, Liu, H, Wei, L, Yang & Lin, Z 2014. A transposable element insertion disturbed starch synthase gene SSIIb in maize. *Molecular Breeding*, 34, 1159–1171.

Luo, S, Ma, Q, Zhong, Y, Jing, J, Wei, Z, Zhou, W, Lu, X, Tian, Y & Zhang, P 2022. Editing of the starch branching enzyme gene SBE2 generates high-amylose storage roots in cassava. *Plant Molecular Biology*, 108, 429–442.

Mahajan, P, Bera, MB, Panesar, PS & Chauhan, A 2021. Millet starch: A review. *International Journal of Biological Macromolecules*, 180, 61–79.

Mahlow, S, Orzechowski, S & Fettke, J 2016. Starch phosphorylation: Insights and perspectives. *Cellular and Molecular Life Sciences*, 73(14), 2753–2764.

Mérida, A & Fettke, J 2021. Starch granule initiation in Arabidopsis thaliana chloroplasts. *Plant Journal*, 107(3), 688–697.

Nakamura, Y 2002. Towards a better understanding of the metabolic system for amylopectin biosynthesis in plants: Rice endosperm as a model tissue. *Plant and Cell Physiology*, 43(7), 718–725.

Nezhdanova, AV, Efremov, GI, Slugina, MA, Kamionskaya, AM, Kochieva, EZ & Shchennikova, AV 2022. Effect of a radical mutation in plastidic starch phosphorylase PHO1a on potato growth and cold stress response. *Horticulturae*, 8(8), art. 730.

Noda, T, Kottearachchi, S, Tsuda, S, Mori, M, Takigawa, S, Matsuura-Endo, C & Yamauchi, H 2007. Starch phosphorus content in potato (*Solanum tuberosum* L.) cultivars and its effect on other starch properties. *Carbohydrate Polymers*, 68(4), 793–796.

Pfotenhauer, AC, Occhialini, A, Harbison, SA, Li, L, Piatek, AA, Luckett, CR, Yang, Y, Stewart, CN & Lenaghan, SC 2023. Genome-editing of FtsZ1 for alteration of starch granule size in potato tubers. *Plants-Basel*, 12(9), art. 1878.

Qu, J, Xu, S, Zhang, Z, Chen, G, Zhong, Y, Liu, L, Zhang, R, Xue, J & Guo, D 2018. Evolutionary, structural and expression analysis of core genes involved in starch synthesis. *Scientific Reports*, 8(1), art. 12736.

Rasmi, Y, Kırboğa, KK, Tekin, B & Demir, M 2024, Chapter 13-Turmeric starch: Structure, functionality, and applications. In: JM Lorenzo & SP Bangar (eds.), *Non-Conventional Starch Sources*, Academic Press, pp. 377–405.

Robinson, GHJ, Balk, J & Domoney, C 2019. Improving pulse crops as a source of protein, starch and micronutrients. *Nutrition Bulletin*, 44(3), 202–215.

Sanderson, JS, Daniels, RD, Donald, AM, Blennow, A & Engelsen, SB 2006. Exploratory SAXS and HPAEC-PAD studies of starches from diverse plant genotypes. *Carbohydrate Polymers* 64, 433-443.

Šárka, E & Dvořáček, V 2017. Biosynthesis of waxy starch – A review. *Plant Soil and Environment*, 63(8), 335–341.

Satoh, H, Nishi, A, Yamashita, K, Takemoto, Y, Tanaka, Y, Hosaka, Y, Sakurai, A, Fujita, N & Nakamura, Y 2003. Starch-branching enzyme I-deficient mutation specifically affects the structure and properties of starch in rice endosperm. *Plant Physiology*, 133, 1111–1121.

Seung, D, Schreier, TB, Burgy, L, Eicke, S & Zeeman, SC 2018. Two plastidial coiled-coil proteins are essential for normal starch granule initiation in Arabidopsis. *Plant Cell*, 30(7), 1523–1542.

Seung, D & Smith, AM 2019. Starch granule initiation and morphogenesis—Progress in Arabidopsis and cereals. *Journal of Experimental Botany*, 70, 771–784.

Sharma, S, Friberg, M, Vogel, P, Turesson, H, Olsson, N, Andersson, M & Hofvander, P 2023. Pho1a (plastid starch phosphorylase) is duplicated and essential for normal starch granule phenotype in tubers of *Solanum tuberosum* L. *Frontiers in Plant Science*, 14, art. 1220973.

Shoaib, N, Liu, L, Ali, A, Mughal, N, Yu, G & Huang, Y 2021. Molecular functions and pathways of plastidial starch phosphorylase (PHO1) in starch metabolism: Current and future perspectives. *International Journal of Molecular Sciences*, 22(19), art. 10450.

Singhal, RS, Kennedy, JF, Gopalakrishnan, SM, Kaczmarek, A, Knill, CJ & Akmar, PF 2008. Industrial production, processing, and utilization of sago palm-derived products. *Carbohydrate Polymers*, 72, 1–20.

Skeffington, AW, Graf, A, Duxbury, Z, Gruissem, W & Smith, AM 2014. Glucan, water dikinase exerts little control over starch degradation in Arabidopsis leaves at night. *Plant Physiology*, 165, 866–879.

Skryhan, K, Gurrieri, L, Sparla, F, Trost, P & Blennow, A 2018. Redox regulation of starch metabolism. *Frontiers in Plant Science*, 9, 1344.

Sun, Y, Jiao, G, Liu, Z, Zhang, X, Li, J, Guo, X, Du, W, Du, J, Francis, F, Zhao, Y & Xia, L 2017. Generation of high-amylose rice through Crispr/Cas9-mediated targeted mutagenesis of starch branching enzymes. *Frontiers in Plant Science*, 8, 298.

Takeuchi, A, Ohnuma, M, Teramura, H, Asano, K, Noda, T, Kusano, H, Tamura, K & Shimada, H 2021. Creation of a potato mutant lacking the starch branching enzyme gene StSBE3 that was generated by genome editing using the CRISPR/dMac3-Cas9 system. *Plant Biotechnology*, 38(3), 345–353.

Tetlow, IJ & Bertoft, E 2020. A review of starch biosynthesis in relation to the building block-backbone model. *International Journal of Molecular Sciences*, 21, art. 7011.

Tetlow, IJ & Emes, M 2014. A review of starch-branching enzymes and their role in amylopectin biosynthesis. *IUBMB Life*, 66(8), 521–585.

Tian, Y, Petersen, BL, Liu, X, Li, H, Kirgensgaard, JJK, Enemark-Rasmussen, K, Khakimov, B, Hebelstrup, KH, Zhong, Y & Blennow, A 2024. Characterization of different high amylose starch granules. Part II: Structure evolution during digestion and distinct digestion mechanisms. *Food Hydrocolloids*, 149, art. 109593.

Tian, Y, Wang, Y, Liu, X, Herburger, K, Westh, P, Møller, MS, Svensson, B, Zhong, Y & Blennow, A 2023. Interfacial catalysis during amylolytic degradation of starch granules: Current understanding and kinetic approaches. *Molecules*, 28, art. 3799.

Tian, Y, Wang, Y, Zhong, Y, Møller, MS, Westh, P, Svensson, B & Blennow, A 2023. Interfacial enzyme kinetics reveals degradation mechanisms behind resistant starch. *Food Hydrocolloids*, 140, 108621.

Toinga-Villafuerte, S, Vales, MI, Awika, JM & Rathore, KS 2022. CRISPR/Cas9-mediated mutagenesis of the granule-bound starch synthase gene in the potato variety Yukon Gold to obtain amylose-free starch in tubers. *International Journal of Molecular Sciences,* 23(9), art. 4640.

Tuncel, A, Corbin, KR, Ahn-Jarvis, J, Harris, S, Hawkins, E, Smedley, MA, Harwood, W, Warren, FJ, Patron, NJ & Smith, AM 2019. Cas9-mediated mutagenesis of potato starch-branching enzymes generates a range of tuber starch phenotypes. *Plant Biotechnology Journal,* 17, 2259–2271.

Van Vu, T, Das, S, Hensel, G & Kim, J-Y 2022. Genome editing and beyond: What does it mean for the future of plant breeding? *Planta,* 255, art. 130.

Veillet, F, Chauvin, L, Kermarrec, M-P, Sevestre, F, Merrer, M, Terret, Z, Szydlowski, N, Devaux, P, Gallois, J-L & Chauvin, J-E 2019. The *solanum tuberosum* GBSSI gene: A target for assessing gene and base editing in tetraploid potato. *Plant Cell Reports,* 38(9), 1065–1080.

Viksø-Nielsen, A. Blennow, A, Kristensen, KH, Jensen, A & Møller, BL 2001. Structural, physicochemical, and pasting properties of starches from potato plants with repressed r1-gene. *Biomacromolecules* 3, 836-841.

Vilpoux, OF & Silveira Junior, JFS 2023. Chapter 3: Starchy crops morphology, extraction, properties and applications. In: FV Olivier & PC Marney (eds.), *Global Production and Use of Starch*, pp. 43–66, https://doi.org/10.1016/B978-0-323-90058-4.00014-1

Wang, H, Wu, Y, Zhang, Y, Yang, J, Fan, W, Zhang, H, Zhao, S, Yuan, L & Zhang, P 2019. CRISPR/Cas9-based mutagenesis of starch biosynthetic genes in sweet potato (*Ipomoea batatas*) for the improvement of starch quality. *International Journal of Molecular Sciences*, 20(19), art. 4702.

Wang, J, Hu, P, Chen, Z, Liu, Q & Wei, C 2017. Progress in high-amylose cereal crops through inactivation of starch branching enzymes. *Frontiers in Plant Science*, 8, art. 469.

Wang, L, Wang, Y, Makhmoudova, A, Nitschke, F, Tetlow, IJ & Emes, MJ 2023. CRISPR–Cas9-mediated editing of starch branching enzymes results in altered starch structure in Brassica napus. *Plant Physiology,* 188, 1866–1886.

Wang, W, Wei, X, Jiao, G, Chen, W, Wu, Y, Sheng, Z, Hu, S, Xie, L, Wang, J, Tang, S & Hu, P 2020. GBSS-BINDING PROTEIN, encoding a CBM48 domain-containing protein, affects rice quality and yield. *Journal of Integrated Plant Biology*, 62(7), 948–966.

Wang, Y, Tian, Y, Christensen, SJ, Blennow, A & Svensson, B, Møller, MS 2023. An enzymatic approach to quantify branching on the surface of starch granules by interfacial catalysis. *Food Hydrocolloids*, 146, art. 109162.

Watson-Lazowski, A, Raven, E, Feike, D, Hill, L, Elaine Barclay, J, Smith, AM & Seung, D 2022. Loss of PROTEIN TARGETING TO STARCH 2 has variable effects on starch synthesis across organs and species. *Journal of Experimental Botany*, 73(18), 6367–6379.

Wattebled, F, Dong, Y, Dumez, S, Delvallé, D, Planchot, V, Berbezy, P, Vyas, D, Colonna, P, Chatterjee, M & Ball, S 2005. Mutants of Arabidopsis lacking a chloroplastic isoamylase accumulate phytoglycogen and an abnormal form of amylopectin. *Plant Physiology,* 138, 184–195.

Williams, VR, Wu, WT, T'sai, HY & Bates, HG 1958. Rice starch, varietal differences in amylose content of rice starch. *Journal of Agricultural and Food Chemistry*, 6(1), 47–48.

Xu, X, Dees, D, Dechesne, A, Huang, X-F, Visser, RGF & Trindade, LM 2017. Starch phosphorylation plays an important role in starch biosynthesis. *Carbohydrate Polymers*, 157, 1628–1637.

Zhang, C, Narayanamoorthy, S, Ming, S, Li, K, Cantre, D, Sui, Z & Corke, H 2022. Rheological properties, structure and digestibility of starches isolated from common bean (*Phaseolus vulgaris* L.) varieties from Europe and Asia. *LWT - Food Science and Technology*, 161, art. 11335.

Zhang, X, Colleoni, C, Ratushna, V, Sirghie-Colleoni, M, James, MG & Myers, AM 2004. Molecular characterization demonstrates that the Zea mays gene sugary2 codes for the starch synthase isoform SSIIa. *Plant Molecular Biology*, 54(6), 865–879.

Zhang, X, Szydlowski, N, Delvalle, D, D'Hulst, C, James, MG & Myers, AM 2008. Overlapping functions of the starch synthases SSII and SSIII in amylopectin biosynthesis in Arabidopsis. *BMC Plant Biology*, 8, art. 96.

Zhao, X, Jayarathna, S, Turesson, H, Fält, AS, Nestor, G, González, MN, Olsson, N, Beganovic, M, Hofvander, P, Andersson, R & Andersson, M 2021. Amylose starch with no detectable branching developed through DNA-free CRISPR-Cas9 mediated mutagenesis of two starch branching enzymes in potato. *Scientific Reports*, 11, art. 4311.

Zhong, Y, Blennow, A, Kofoed-Enevoldsen, O, Jiang, D & Hebelstrup, KH 2019. Protein Targeting to Starch 1 (PTST1) is essential for starchy endosperm development in barley. *Journal of Experimental Botany*, 70, 485–496.

Zhong, Y, Liu, L, Qu, J, Li, S, Blennow, A, Seytahmetovna, SA, Liu, X & Guo, D 2020. The relationship between the expression pattern of starch biosynthesis enzymes and molecular structure of high amylose maize starch. *Carbohydrate Polymers*, 247, art. 116681.

Zhong, Y, Qu, JZ, Blennow, A, Liu, XX & Guo, DW 2021. Expression pattern of starch biosynthesis genes in relation to the starch molecular structure in high-amylose maize. *Journal of Agriculture and Food Chemistry*, 69, 2805–2815.

Zhong, Y, Qu, JZ, Liu, X, Ding, LY, Liu, Bertoft, E, Petersen, BL, Hamaker, BR, Hebelstrup, KH & Blennow, A 2022. Different genetic strategies to generate high amylose starch mutants by engineering the starch biosynthetic pathways. *Carbohydrate Polymers*, 287, art. 119327.

Zhong, Y, Tai, L, Blennow, A, Ding, L, Herburger, K, Qu, J, Xin, A, Guo, D, Hebelstrup, KH & Liu, X 2022. High-amylose starch: Structure, functionality and applications. *Critical Reviews in Food Science and Nutrition*, 63(27), 8568–8590.

Zhu, F 2017. Structures, physicochemical properties, and applications of amaranth starch. *Critical Reviews in Food Science and Nutrition*, 57(2), 313–325.

Methods for Starch Extraction

5

Les Copeland

5.1 INTRODUCTION

Starch is a biopolymer of considerable importance for humans. It has many applications in food and non-food industries, and it is a major source of the energy intake in the modern human diet. Starch occurs widely in plant tissues as an energy reserve, but is most abundant in seeds and storage organs, from which it is extracted on a laboratory scale for structure-function studies and on a large scale for commercial applications. Starch occurs naturally as semi-crystalline granules that vary in shape and size between and within botanical species. This variability is due to inherent differences in plant genetics and environmental factors that influence starch deposition during plant growth. Starch granules are made up of two polymers of glucose, the lightly branched amylose and more highly branched amylopectin, both of which have linear glucan chains joined by α-1,4 linkages substituted with α-1,6 branching links. The extent of crystallinity ranges from about 15% in high-amylose starches to 45%–50% for waxy starches, which are made up of essentially only amylopectin. Starch granules in their native state are inherently stable but are susceptible to hydrolysis by acid or amylolytic enzymes and are partially broken down in alkaline conditions. Starch structure, function and technological applications have long been areas of intensive research that has produced a vast body of published literature, including many reviews on diverse aspects of the starch science (for recent reviews, see Wang et al. 2015, 2020; Bertoft 2017; Seung 2020; Apriyanto, Compart & Fettke 2022; Li et al. 2023).

Starch extraction is the first step for studies on structure and function and for technical uses of this important biopolymer. Commercially, around 85 million tonnes of starch are extracted annually worldwide for multiple purposes. In foods, starch and its modified forms are used as texture modifiers, emulsifiers, stabilizers, fat replacers and as a source of dextrins and glucose syrups. Starch is also used in pharmaceuticals and as a fermentation substrate for production of a range of biologicals such as amino acids, organic acids, vitamins, antibiotics and hormones. Non-food applications of starch include cosmetics, paper, packaging materials, textiles, and building products, as well as the production of bioethanol (Burrell 2003; Batey 2017; Shevkani et al. 2017). Maize is the largest source of commercial starch (about 80%), but significant amounts are also obtained from wheat, potato and cassava, with smaller amounts from various crops for specialty uses (Waterschoot et al. 2015). Maize is the main crop used for starch extraction in North America, whereas wheat and potato are major sources in Europe. Wheat is used for starch production in Australia, while cassava and rice are used to a lesser extent, mainly in Asia.

In addition to these main sources of commercial starch, other grains provide smaller amounts of starch particularly suited to specific end-uses. For example, barley starch has a lower gelatinization temperature and water absorption potential than maize or wheat starch (Vasanthan & Hoover 2009; Zhu 2017), whereas sorghum starch has a higher gelatinization temperature, retrogrades to a greater degree and is less susceptible to *in vitro* enzymatic digestion than starches from other cereals (Khoddami et al. 2023). Oat starch granules are small and irregularly shaped with relatively high lipid content and are useful for certain food and non-food applications (Punia et al. 2020). Starch from pulse grains generally has higher amylose content and stronger gelling properties than cereal starches (Li et al. 2019).

Due to the variability of native starches, most extraction methods are designed to suit the botanical source and the purpose and scale of the work. Other relevant considerations may be the equipment available, cost factors and analytical methods for evaluating the isolated product. In some cases, there may also be an intention to recover co-products of value, such as proteins, non-starch polysaccharides and fibre. The purpose of this chapter is to describe the general principles for extracting starch, not to catalogue the myriad of specific extraction methods used. Illustrative examples are given for extraction methods from cereal and pulse grains, roots and tubers, and from less conventional sources, including recovery of starch from agricultural and food wastes. Due to space limitations, the literature cited in this chapter is only a small selection from the vast resource of publications describing methods for starch extraction. Many other excellent publications could have been cited; these can be found easily nowadays with contemporary tools for searching the literature.

5.2 GENERAL PRINCIPLES FOR EXTRACTING STARCH

The earliest recorded history of starch use by humans dates back to around 800 BCE in Ancient Egypt, where starch was obtained from partially fermented wheat grains for use as a fabric stiffener and adhesive, as well as in cosmetics and medicines (Copeland 2020). A Roman treatise dating back to 184 BCE describes a procedure for preparing starch by steeping wheat or barley grains in water for 10 days. The soaked material was pressed and mixed with fresh water, and the suspension was filtered through a linen cloth. The filtered slurry was allowed to settle and the sediment was washed with water and dried in the sun. Subsequently, Pliny the Elder (23–74 CE) in the Early Roman Empire documented extracting starch by boiling freshly ground wheat flour with vinegar. The extracted starch was used to coat papyrus, whiten cloth, powder hair and thicken food sauces.

Two seminal discoveries in the 18th century opened the way for the vast amount of research on starch that has followed. These were the first microscopic examination of starch morphology by the Dutch scientist Antonie van Leeuwenhoek in 1716, and the separation of wheat flour into gluten and starch in 1745 by Beccari of Bologna (Copeland 2020). As in ancient times, the fundamental principle for isolating starch and separating it from other plant cellular constituents is the higher density of starch (1.5–1.6 g/cm^3; Dengate, Baruch & Meredith 1978) compared with proteins and non-starch polysaccharides (1.3 g/cm^3; Fischer, Polikarpov & Craievich 2004).

The optimum method for extracting starch will depend on the botanical source and the purpose and purity sought for the preparation. For studies on starch structure and function, the material obtained should be as representative as possible of the native state and free from contaminants. The extraction method should cause minimal damage to the surface and internal molecular structures of starch granules, as well as change the size distribution of the starch molecules and their chain-length distributions as little as possible. The extraction should not favour small over large molecules. An extraction method aimed at optimizing one parameter, such as yield of granules, may differ from a method seeking to obtain starch with the least damage to glucan chains (Zhao et al. 2020). As the examples discussed subsequently show, the functional properties of starch, such as water absorption, swelling, gelatinization, pasting and susceptibility to amylolysis, can be affected by the extraction procedure.

5.3 EXTRACTION OF STARCH FROM GRAINS

Dry grains from cereals, pulses and pseudocereals, which contain about 10%–15% moisture, are the most common source for extracting starch. For laboratory-scale extractions, grain that has been cleaned to remove impurities is ground in a small laboratory mill or coffee grinder and sieved to obtain material of uniform particle size. For large-scale extractions, flour is obtained by dry or wet milling, as discussed in depth in several monographs (e.g., Delcour & Hoseney 2010; Whistler & BeMiller 2010). A detailed discussion of the milling of wheat and peas is illustrative of methods used for grains and pulses, respectively (Abdel-Aal 2024). For dry milling, the grain is first softened by tempering to a moisture content of 15%–16% and then broken down with a series of different types of rollers. Flours obtained by dry milling may be collected in different fractions or they may be fractionated by air classifications or with electrostatic separation technology (Schutyser et al. 2015; Pulivarthi et al. 2023; Abdel-Aal 2024). Air classification separates particles based on size and density, with starch granules being typically in the coarse fractions, whereas protein is enriched in the finer particles. Li et al. (2019) discuss how high purity starch with low damage can be isolated from dehulled pea, lentil and faba bean flours after fractionation by air classification. For wet milling, which is the main method for extracting starch from maize, the grain is first softened by steeping in water at about 50°C, usually with bisulphite or dilute alkali added to reduce microbial growth and help break down interactions between the starch, proteins and fibre.

The milling process of cereals separates the outer bran layer and germ from the endosperm, which typically comprises about 60%–80% starch, 9%–15% protein and 5%–15% non-starch polysaccharides. The bran layer, which makes up about 10% of the cereal grain, contains little starch, whereas the germ (5%–10% of the cereal grain) contains proteins, lipids and starch in different proportions and composition to the endosperm. In wholemeal flour, the bran and germ are not separated from the endosperm, and a small amount of the starch extracted will originate from the germ. The higher amounts of lipid in the germ can cause rancidity in wholemeal flour on prolonged storage. Dry milling has advantages over wet milling in convenience, cost-effectiveness and in producing less aqueous residues, but it can result in a higher amount of starch damage, which may affect the physical and chemical properties. Starch obtained after wet milling tends to have higher purity compared to that from dry milling (Kringel et al. 2020; Abdel-Aal 2024).

To extract the starch, flour is mixed well with a roughly equal weight of water, and the mixture is sieved and allowed to settle under gravity or with gentle centrifugation. The starch granules precipitate as a dense, white layer covered with a layer containing residual cell debris. The soluble cellular contents are removed by decanting the supernatant, and the cell debris can be scraped off from the starch. In larger scale extractions, hydrocyclones are often used to recover the starch. The isolated starch is washed repeatedly by resuspending and decanting, and then dried. An alcohol washing step may be included to reduce the lipid content. If relevant to the purpose for the extraction, the amount of damaged starch in the preparation should be measured. Starch granules from the endosperm of wheat, barley, rye and the wheat–rye hybrid triticale have a bimodal size distribution, with large and small granules differing in structural and functional properties (Li et al. 2023).

Starch granules in cereal endosperm are usually associated *in situ* with proteins and other cellular constituents such as non-starch polysaccharides (e.g., β-glucans in barley and oats, arabinoxylans in wheat and rye); these associations may also occur during the extraction procedure. Starch from soft wheat varieties has the lipid-binding puroindoline proteins bound on the surface of granules through specific interactions (Bhave & Morris 2008), whereas other endosperm proteins may be associated more loosely with the granule surface (Wang et al. 2013). Starch granules also contain internal proteins, such as entrapped enzymes and proteins of starch biosynthesis (Wang et al. 2013; Seung 2020). Lipids associated with the granule surface are mainly triglycerides and, to a lesser extent, free fatty acids, glycolipids and phospholipids, whereas internal lipids are predominantly monoacyl lipids that may be part of inclusion complexes with amylose (Wang et al. 2020).

On isolation from cereal endosperm, starch granules are usually associated with about 1%–2% lipid and 0.5%–1% protein; non-starch polysaccharides may also be present. As all of these contaminants may affect the properties of starch, additions are often included in the extraction medium to release starch from other cellular constituents associated with the granule surface. The protein content of the starch may be reduced by extracting starch under alkaline conditions (e.g., 0.1%–0.3% NaOH) or with proteases in the medium. Alkaline conditions also suppress the activity of amylases which may be present in the flour. Sodium bisulphite or sodium metabisulphite can inhibit enzymatic browning and microbial growth, whereas surfactants (e.g., sodium lauryl sulphate and dodecylbenzene sulfonate) and glycanase enzymes can improve extraction yield and enhance the purity of the isolated starch. An example of extracting high quality starch from rye grain using a xylanase–protease mixture in an enzyme-assisted extraction is described by Buksa (2018).

Wheat flour may be processed differently compared to other starch sources if the intention is to recover its two products of value, starch and gluten. Accordingly, wheat flour is processed to first separate starch from the gluten. The wheat flour–water mixture is kneaded so that the gluten proteins interact and aggregate into an insoluble, rubbery mass from which the starch can be washed out and collected by sedimentation or centrifugation. Additives are generally not used in these types of wheat starch extractions to avoid damage to the gluten. Wheat starch that is certified as suitable for use in products for coeliacs needs to meet the Codex Alimentarius guideline of containing less than 20 ppm gluten; this standard may be achieved by using gluten specific peptidases in the extraction protocol (Walter, Wieser & Kohler 2014).

Numerous studies comparing results of starch extractions using different media have been described, of which only a few will be covered here. A study by Nie et al. (2023) compared various techniques for extracting starch from highland barley. Starch extracted with ultrasound assistance had significantly more damage and lower molecular weight than starch extracted with combinations of added enzymes (cellulase and xylanase or cellulase, xylanase and papain), which had the highest molecular weight, relative crystallinity and resistance to enzymatic digestion. Alkali extraction gave the highest yield, but the starch had greater susceptibility to amylolysis and lower thermal stability (Nie et al. 2023). Three extraction methods were compared for starch from quinoa seeds: steeping flour in an alkaline medium (0.3% NaOH), milling after soaking overnight in sodium metabisulphite and protease-assisted extraction (Junejo et al. 2022). The extraction with protease treatment was best for retaining the intact structure of starch granules, whereas the alkali method was highly effective for extraction yield and purity of the starch. Similar findings for extracting starch have been compared for millet varieties (Kumar, Tangsrianugul & Suphantharika 2023; Singh et al. 2023), high amylose maize (Tian et al. 2021) and vetch (Zhang et al. 2022). An in-depth study by Zhao et al. (2020) assessed eight different techniques for extracting starch from rice and chickpeas, comparing the methods according to the greatest number of larger molecules and longer chains that were extracted, assuming that these are the properties most susceptible to degradation and loss during extraction. The optimal method for starch extraction from rice was proposed to use 0.45% sodium metabisulfite at 4°C for 0.5 hour with protease addition, whereas for chickpea, starch extraction using 0.45% sodium bisulphite for 6 hours and washing the starch repeatedly with water gave the best results.

Many studies have shown that extracting starch under mild alkaline conditions from a diverse range of sources results in higher yield of granules of greater purity and with lower protein content, compared to using only water or water with sodium bisulphite or metabisulphite added (El Halal et al. 2019; Felisberto et al. 2021; Junejo et al. 2022; Kumar et al. 2023; Singh et al. 2023). However, alkaline steeping has also been shown to alter the functional properties of starch. Alkaline treatment removes surface proteins and lipids from the granules, but it also causes leaching of amylose (Wang et al. 2014). The resulting changes in the internal granular structure may be relatively small but can lead to substantial changes in starch swelling power, thermal stability, gelatinization and pasting profile, and increased susceptibility to *in vitro* enzymatic attack (Cardoso, Samios & da Silveira 2006; Cardoso et al. 2007; Karim et al. 2008; Nadiha et al. 2010; Thys et al. 2008; Wang & Copeland 2012; Sun et al. 2015). The effects of alkali on starch can occur during a relatively short exposure. Starch isolated from pea flour after steeping in 0.02% NaOH at 30°C for 18 hours had different physicochemical properties and was not suitable for making vermicelli compared to starch isolated by other methods (Sun et al. 2015).

5.4 EXTRACTING STARCH FROM ROOTS AND TUBERS

Isolation of starch from roots and tubers follows the same principles as for extracting starch from grains. The process begins with soaking the washed, peeled and macerated raw material in water, usually containing up to 5% added sodium metabisulphite. The slurry is strained through cloth mesh, and the starch in the filtrate is collected by sedimentation or centrifugation. The layer of cell debris can easily be scraped off the thick white layer of starch before it is washed several times by resuspension in water, settling and decanting, and finally dried. Roots and tubers contain low amounts of proteins and fats, and separation of the starch from non-starch components is relatively simple (Phogat et al. 2020; Al Maqtari et al. 2024).

A comparison of different extraction methods of potato starch showed that a combination of treatments with 0.25% of NaOH, a 2% w/v mix of SDS-mercaptoethanol and cellulase, followed by 5.25% sodium hypochlorite to avoid browning, was needed to produce starch in the highest yield having the desired physicochemical and functional properties (Phogat et al. 2020; Neeraj et al. 2021). A low-cost, small-scale method suitable for on-farm isolation of starch from potato has been described by Singh, Kaur and Sachdev (2021).

5.5 EXTRACTING STARCH FROM NOVEL SOURCES

Extracting starch from novel sources, such as plant material that is normally discarded during industrial processing, is an area of increasing activity for identifying starch with different functionalities and due to the interest in sustainability and reducing waste. Reducing the loss of carbohydrates, especially starch in agricultural and food industry wastes, can add value through generating co-products. Methods for starch recovery, mostly for laboratory research, from plant parts without commercial value and discarded waste from fruit processing have been the subject of recent reviews (Makroo et al. 2021; Chacon et al. 2024). Materials such as seeds from durian, jackfruit, loquat, lychee, mango, tamarind and pine, as well as other unconventional sources, have been the subject of studies. In general, seeds are washed, dried and ground into a flour and then starch is extracted along the lines already described. In some cases, fibrous non-starch polysaccharides, proteins and lipids may also be recovered. The extraction of starch from cassava industrial waste has also been described by Thuppahige et al. (2023).

Ultrasound-assisted extraction of starch has been applied in the extraction of starch from various sources as a means of overcoming some of the disadvantages of traditional methods, such as time required and amount of waste produced (Bernardo et al. 2018; Buksa 2018). Ultrasound-assisted extraction has also been applied starch extraction from corn (Flores-Silva et al. 2017), oats (Ünlü & Aykaç 2023), yam (Bernardo et al. 2018) and microalgae (Di Caprio et al. 2022). This technology decreases extraction time and increases the yield of starch obtained, but in most studies, the shear forces generated by ultrasound treatment induced significant changes in the structure and physicochemical properties of starch granules, including disruption of granule crystallinity and depolymerization of starch chains. For example, the structure of starch granules

isolated with sonication from corn (Flores-Silva et al. 2017) and the seed of lychee fruit (Morales-Trejo et al. 2022) was disrupted in a way that affected the properties of the starch. The starch granules from the lychee fruit seeds had cracks on the surface and increased apparent amylose content due to fragmentation of the starch chains.

5.6 CONCLUDING REMARKS

The optimum method for extracting starch will depend on the source of the starch, the scale and reason for the extraction, and the purity of the preparation sought. For studies on starch structure and function, which are usually conducted on a laboratory scale, the material obtained should be as representative as possible of the native state, free from contaminants and with as little as possible structural change to the whole molecule. Methods for evaluating the isolated product are often relevant factors, and in this regard, caution should be applied when interpreting the assessment of functional properties of starch isolated under alkaline conditions. For larger scale technical and industrial applications, cost and convenience and ensuring no loss of desired end-use properties will be important. An intention to also recover co-products, such as proteins, non-starch polysaccharides and fibre, may influence the method of extraction. The optimal method for extracting starch will depend on the objective. Given the variability of native starch due to genetic and environmental influences, applying an extraction method designed and tested for the particular purpose is always recommended.

REFERENCES

Abdel-Aal, E-SM 2024. Insights into grain milling and fractionation practices for improved food sustainability with emphasis on wheat and peas. *Foods 13*, 1532. https://doi.org/10.3390/foods13101532

Al-Maqtari, QA, Li, B, He, H-J, Mahdi, AA, Al-Ansi, W, & Saeed, A 2024. An overview of the isolation, modification, physicochemical properties, and applications of sweet potato starch. *Food and Bioprocess Technology 17*, 1–32. https://doi.org/10.1007/s11947-023-03086-1

Apriyanto, A, Compart, J, & Fettke, J 2022. A review of starch, a unique biopolymer – Structure, metabolism and in planta modifications. *Plant Science 318*, art. 111223. https://doi.org/10.1016/j.plantsci.2022.111223

Batey, IL 2017. The diversity of uses for cereal grains. In: CW Wrigley & IL Batey (eds.), *Cereal Grains. Assessing and Managing Quality*, Woodhead, Cambridge UK, pp. 45–56. https://doi.org/10.1016/B978-0-08-100719-8.00003-6

BeMiller, JN & Whistler, RL 2009. *Starch: Chemistry and Technology* (Third Edition). New York: Academic Press.

Bernardo, CO, Ascheri, JLR, Chávez, DWH, & Carvalho, CWP 2018. Ultrasound assisted extraction of yam (*Dioscorea bulbífera*) starch: Effect on morphology and functional properties. *Starch/Stärke 70*, art.1700185. https://doi.org/10.1002/star.201700185

Bertoft, E 2017. Understanding starch structure: Recent progress. *Agronomy 7*, art. 56. https://doi.org/10.3390/agronomy7030056

Bhave, M, & Morris, CF 2008. Molecular genetics of puroindolines and related genes: Allelic diversity in wheat and other grasses. *Plant Molecular Biology 66*, 205–219. https://doi.org/10.1007/s11103-007-9263-7

Buksa, K 2018. Extraction and characterization of rye grain starch and its susceptibility to resistant starch formation. *Carbohydrate Polymers 194*, 184–192. https://doi.org/10.1016/j.carbpol.2018.04.024

Burrell, MM 2003. Starch: The need for improved quality and quantity – An overview. *Journal of Experimental Botany 54*, 451–456. https://doi.org/10.1093/jxb/erg049

Cardoso, MB, Putaux, J-L, Samios, D, & da Silveira, NP 2007. Influence of alkali concentration on the deproteinization and/or gelatinization of rice starch. *Carbohydrate Polymers 70*, 160–165. https://doi.org/10.1016/j.carbpol.2007.03.014

Cardoso, MB, Samios, D, & da Silveira, NP 2006. Study of protein detection and ultrastructure of Brazilian rice starch during alkaline extraction. *Starch/Stärke 58*, 345–352. https://doi.org/10.1002/star.200600495

Chacon, WDC, Gaviria, YAR, de Melo, APZ, Verruck, S, Monteiro, AR, & Valencia, GA 2024. Physicochemical properties and potential food applications of starches isolated from unconventional seeds: A review. *Starch/Stärke 76*, 2200228. https://doi.org/10.1002/star.202200228

Copeland, L 2020. History of starch research. In: S Wang (ed.), *Starch Structure, Functionality and Application in Foods*, Springer Nature, Singapore, pp. 1–7. https://doi.org/10.1007/978-981-15-0622-2_1

Delcour, JA, & Hoseney, RC 2010. *Principles of Cereal Science and Technology*, 3rd Ed, AACC International, pp. 139–148.

Dengate, H, Baruch, D, & Meredith, P 1978. The density of wheat starch granules: A tracer dilution procedure for determining the density of an immiscible dispersed phase. *Starch/Stärke 30*, 80–84. https://doi.org/10.1002/star.19780300304

Di Caprio, F, Chelucci, R, Francolini, I, Altimari, P, & Pagnanelli, F 2022. Extraction of microalgal starch and pigments by using different cell disruption methods and aqueous two-phase system. *Journal of Chemical Technology and Biotechnology 97*, 67–78. https://doi.org/10.1002/jctb.6910

El Halal, SLM, Kringel, DH, Zavareze, ED, & Dias, ARG 2019. Methods for extracting cereal starches from different sources: A review. *Starch/Stärke 71*,1900128. https://doi.org/10.1002/star.201900128

Felisberto, MHF, Sanches, EA, Campelo, PH, & Clerici, MTPS 2021. Non-conventional starch sources. *Current Opinion in Food Science 39*, 93–102. https://doi.org/10.1016/j.cofs.2020.11.011

Fischer, H, Polikarpov, I, & Craievich, A 2004. Average protein density is a molecular-weight- dependent function. *Protein Science 13*, 2825–2828. https://doi.org/10.1110/ps.04688204

Flores-Silva, PC, Roldan-Cruz, CA, Chavez-Esquivel, G, Vernon-Carter, EJ, Bello- Perez, LA, & Alvarez-Ramirez, J 2017. *In vitro* digestibility of ultrasound-treated corn starch. *Starch/Stärke 69*, art.1700040. https://doi.org/10.1002/star.201700040

Junejo, SA, Wang, J, Liu, Y, Jia, R, Zhou, Y, & Li, S 2022. Multi-scale structures and functional properties of quinoa starch extracted by alkali, wet-milling, and enzymatic methods. *Foods 11*, art. 2625. https://doi.org/10.3390/foods11172625

Karim, AA, Nadiha, MZ, Chen, FK, Phuah, YP, Chui, YM, & Fazilah, A 2008. Pasting and retrogradation properties of alkali-treated sago (*Metroxylon sagu*) starch. *Food Hydrocolloids 22*, 1044–1053. https://doi.org/10.1016/j.foodhyd.2007.05.011

Khoddami, A, Messina, V, Venkata, KA, Farahnaky, A, Blanchard, CL, & Roberts, TH 2023. Sorghum in foods: Functionality and potential in innovative products. *Critical Reviews in Food Science and Nutrition 63*, 1170–1186. https://doi.org/10.1080/10408398.2021.1960793

Kringel, DH, El Halal, SLM, Zavareze, ER, & Dias, ARG 2020. Methods for the extraction of roots, tubers, pulses, pseudocereals, and other unconventional starches sources: A review. *Starch/Stärke 72*, art. 1900234. https://doi.org/10.1002/star.201900234

Kumar, SR, Tangsrianugul, N, & Suphantharika, MA 2023. Review on isolation, characterization, modification, and applications of proso millet starch. *Foods 12*, art. 2413. https://doi.org/10.3390/foods12122413

Li, L, Yuan, TZ, Setia, R, Raja, RB, Zhang, B, & Ai, Y 2019. Characteristics of pea, lentil and faba bean starches isolated from air classified flours in comparison with commercial starches. *Food Chemistry 299*, 599–607. https://doi.org/10.1016/j.foodchem.2018.10.064

Li, M, Daygon, VD, Solah, V, & Dhital, S 2023. Starch granule size: Does it matter? *Critical Reviews in Food Science and Nutrition 63*, 3683–3703. https://doi.org/10.1080/10408398.2021.1992607

Makroo, HA, Naqash, S, Saxena, J, Sharma, S, Majid, D, & Dar, BN 2021 Recovery and characteristics of starches from unconventional sources and their potential applications: A review. *Applied Food Research 1*, art. 100001. https://doi.org/10.1016/j.afres.2021.100001

Morales-Trejo, F, Trujillo-Ramirez, D, Aguirre-Mandujano, E, Lobato-Calleros, C, Vernon-Carter, EJ, & Alvarez-Ramirez, J 2022. Ultrasound-assisted extraction of Lychee (Litchi chinensis Sonn.) seed starch: Physicochemical and functional properties. *Starch/Stärke 74*, art. 100092. https://doi.org/10.1002/star.202100092

Nadiha, MZN, Fazilah, A, Bhat, R, & Karim, AA 2010. Comparative susceptibilities of sago, potato and corn starches to alkaline treatment. *Food Chemistry 121*, 1053–1059. https://doi.org/10.1016/j.foodchem.2010.01.048

Neeraj, Siddiqui, S, Dalal, N, Srivastva, A, & Pathera, AK 2021. Physicochemical, morphological, functional, and pasting properties of potato starch as a function of extraction methods. *Journal of Food Measurement and Characterization 15*, 2805–2820. https://doi.org/10.1007/s11694-021-00862-5

Nie, M, Piao, C, Wang, A, Xi, H, Chen, Z, He, Y, Wang, L, Liu, L, Huang, Y, Wang, F, & Tong, L-T 2023. Physicochemical properties and *in vitro* digestibility of highland barley starch with different extraction methods. *Carbohydrate Polymers 303*, 120458. https://doi.org/10.1016/j.carbpol.2022.120458

Phogat, N, Siddiqui, S, Dalal, N, Srivastva1, A, & Bindu, B 2020. Effects of varieties, curing of tubers and extraction methods on functional characteristics of potato starch. *Journal of Food Measurement and Characterization 14*, 3434–3444. https://doi.org/10.1007/s11694-020-00579-x

Pulivarthi, MK, Buenavista, RM, Bangar, SP, Li, Y, Pordesimo, LO, Bean, SR, & Siliveru, K 2023. Dry fractionation process operations in the production of protein concentrates: A review. *Comprehensive Reviews in Food Science and Food Safety 22*, 4670–4697. https://doi.org/10.1111/1541-4337

Punia, S, Sandhu, KS, Dhull, SG, Siroha, AK, Purewal, SS, Kaur, M, & Kidwai, MK 2020. Oat starch: Physicochemical, morphological, rheological characteristics and its application - A review. *International Journal of Biological Macromolecules 154*, 493–498. https://doi.org/10.1016/j.ijbiomac.2020.03.083

Schutyser, MAI, Pelgrom, PJM, van der Goot, AJ, & Boom, RM 2015. Dry fractionation for sustainable production of functional legume protein concentrates. *Trends in Food Science and Technology 45*, 327–335. https://doi.org/10.1016/j.tifs.2015.04.013

Seung, D 2020. Amylose in starch: Towards an understanding of biosynthesis, structure and function. *New Phytologist 228*, 1490–1504. https://doi.org/10.1111/nph.16858

Shevkani, K, Singh, N, Bajaj, R, & Kaur, A 2017. Wheat starch production, structure, functionality and applications - A review. *International Journal of Food Science and Technology 52*, 38–58. https://doi.org/10.1111/ijfs.13266

Singh, H, Singh, R, Kumar, N, Kaur, BP, & Upadhyay, A 2023. Effects of wet-milling extraction methods on nutritional, functional, and structural properties of barnyard millet starch. *Starch/Stärke* art. 2300136. https://doi.org/10.1002/star.202300136

Singh, R, Kaur, S, & Sachdev, PS 2021. A cost effective technology for isolation of potato starch and its utilization in formulation of ready to cook, non cereal, and non glutinous soup mix. *Journal of Food Measurement and Characterization 15*, 3168–3181. https://doi.org/10.1007/s11694-021-00887-w

Sun, Q, Chu, L., Xiong, L, & Si, F 2015. Effects of different isolation methods on the physicochemical properties of pea starch and textural properties of vermicelli. *Journal of Food Science and Technology 52*, 327–334. https://doi.org/10.1007/s13197-013-0980-4

Thuppahige, VT, Moghaddam, L, Welsh, ZG, Wang, T, Xiao, H-W, & Karim, A 2023. Extraction and characterisation of starch from cassava (*Manihot esculenta*) agro-industrial wastes. *LWT - Food Science and Technology 182*, 114787. https://doi.org/10.1016/j.lwt.2023.114787

Thys, RCS, Westfahl, H, Norena, CPZ, Marczak, LDF, Silveira, NP, & Cardoso, MB 2008. Effect of alkaline treatment on ultrastructure of C-type starch granules. *Biomacromolecules 9*, 1894–1901. https://doi.org/10.1021/bm800143w

Tian, Y, Qin, F, Zhong, Y, Cui, Y, Yang, Y, Liu, YS, & Liu, X 2021. Optimal extraction methods for high amylose maize starch studied by size-exclusion chromatography (SEC) and small-angle X-ray scattering (SAXS). *ACS Food Science and Technology 1*, 1920–1927. https://doi.org/10.1021/acsfoodscitech.1c00247

Ünlü, EB, & Aykaç, C 2023. Effect of ultrasound on isolation and properties of oat starch. *Czech Journal of Food Science 41*, 111–117. https://doi.org/10.17221/94/2022-CJFS

Vasanthan, T, & Hoover, R 2009. Barley starch: Production, properties, modification and uses. In: J BeMiller & R Whistler (eds.), *Starch: Chemistry and Technology,* 3rd Ed, Academic Press, San Diego, pp. 601–628. ISBN: 978-0-12–746275-2.

Walter, T, Wieser, H, & Koehler, P 2014. Production of gluten-free wheat starch by peptidase treatment. *Journal of Cereal Science 60*, 202–209. https://doi.org/10.1016/j.jcs.2014.02.012

Wang, S, Chen Chao, C, Cai, J, Niu, B, Copeland, L, & Wang, S 2020. Starch–lipid and starch–lipid–protein complexes: A comprehensive review. *Comprehensive Reviews in Food Science and Food Safety 20*, 1056–1079. https://doi.org/10.1111/1541-4337.12550

Wang, S, & Copeland, L 2012. Effect of alkali treatment on structure and function of pea starch granules. *Food Chemistry 135*, 1635–1642. https://doi.org/10.1016/j.foodchem.2012.06.003

Wang, S, Hassani, ME, Crossett, B, & Copeland, L 2013. Extraction and identification of internal granule proteins from waxy wheat starch. *Starch/Stärke 65*, 186–190. https://doi.org/10.1002/star.201200093

Wang, S, Li, C, Copeland, L, Niu, Q, & Wang, S 2015. Starch retrogradation: A comprehensive review. *Comprehensive Reviews in Food Science and Food Safety 14*, 568–585. https://doi.org/10.1111/1541-4337.12143

Wang, S, Luo, H, Zhang, J, Zhang, Y, He, Z, & Wang, S 2014. Alkali-induced changes in functional properties and in vitro digestibility of wheat starch: The role of surface proteins and lipids. *Journal of Agricultural and Food Chemistry 62*, 3636–3643. https://doi.org/10.1021/jf500249w

Waterschoot, J, Gomand, SV, Fierens, E, & Delcour, JA 2015. Production, structure, physicochemical and functional properties of maize, cassava, wheat, potato and rice starches. *Starch/Stärke 67*, 14–29. https://doi.org/10.1002/star.201300238

Zhang, X, Cheng, Y, Jia, X, Geng, D, Bian, X, & Tang, N 2022. Effects of extraction methods on physicochemical and structural properties of common vetch starch. *Foods 11*, art. 2920. https://doi.org/10.3390/foods11182920

Zhao, Y, Tan, X, Wu, G, & Gilbert, RG 2020. Using molecular fine structure to identify optimal methods of extracting starch. *Starch/Stärke 72*, 1900214. https://doi.org/10.1002/star.201900214

Zhu, F 2017. Barley starch: Composition, structure, properties, and modifications. *Comprehensive Reviews in Food Science and Food Safety 16*, 558–579. https://doi.org/10.1016/j.tifs.2015.02.0

Amylose

6

Jovin Hasjim, Kai Wang, Francisco Vilaplana, and Robert G. Gilbert

6.1 INTRODUCTION

Amylose is one of the two major molecules in starch, the other being amylopectin. Both molecules contain only glucose monomers with $\alpha(1\rightarrow4)$ and $\alpha(1\rightarrow6)$ glycosidic linkages. The linear part of starch molecules is linked together by $\alpha(1\rightarrow4)$ glycosidic linkages, whereas the branching points are made up of $\alpha(1\rightarrow6)$ glycosidic linkages. The amylose molecules are on average smaller than the amylopectin molecules, having a molecular weight in the range of 10^4–10^6; they are primarily linear with a few long chain branches (Takeda, Shitaozono, & Hizukuri 1990; Vilaplana et al. 2014; Wang et al. 2019). On the other hand, amylopectin has a larger molecular weight (10^7–10^9) with a highly branched structure. In the native starch granules, amylopectin branches form double helices that create alternating crystalline and amorphous lamellae, while amylose molecules are in an amorphous conformation interspersed with amylopectin double helices (Kasemsuwan & Jane 1994).

Having a predominantly linear structure, amylose can spontaneously form single-helical complexes with alcohols and some lipids, including fatty acids, monoglycerides, and phospholipids. The presence of amylose–lipid complexes reduces the swelling of the starch granules during heating in water, resulting in a lower viscosity of the cooked paste (Jane et al. 1999; Tester & Morrison 1990). Amylose can also form a double helical structure that crystalizes into a heat-stable, more ordered structure. The crystalline structure of amylose double helices has a high melting temperature above 120°C (Gruchala & Pomeranz 1993; Sievert & Pomeranz 1990), which is not easily dissociated by conventional cooking. This has a considerable impact on the texture of food product and its stability during storage. Therefore, amylose content is an important parameter in determining the properties and end-uses of a starch.

Based on the amylose content, starch can be grouped into waxy starch (0%–8% amylose), normal starch (15%–35% amylose), and high-amylose starch (>45% amylose). For normal starches, legume starches are known to have higher amylose contents than cereal starches (Fredriksson et al. 1998; Jane et al. 1999). The amylose contents of some starches from different botanical origins are shown in Table 6.1 as examples. Waxy starches are preferred as thickening agents, because they have better stability on the consistency of the paste over time than normal and high-amylose varieties. Some plant species have natural waxy (or glutinous) varieties that have been cultivated for many decades, such as waxy maize, waxy rice, and waxy wheat (Graybosch 1998; Tian et al. 2009; Zhang et al. 2019). Other waxy starches, including waxy potato and waxy tapioca varieties, have been developed more recently (Hsieh et al. 2019; Svegmark et al. 2002) due to the demand for "clean label" starch, where the consumers try to avoid chemically modified starches, although they have been recognized as safe for human consumption and have been used in food industry for decades. The development of high-amylose starches was mainly driven by their nutritional benefits, because they contain high amounts of resistant starch (RS), i.e., a portion of starch that cannot be digested in the small intestine, but can be partially fermented in the colon (King et al. 2008; Li, Gidley, & Dhital 2019; Li et al. 2008; Slade et al. 2012). RS prevents the blood-sugar spikes caused by the rapid digestion of starch to glucose in the upper gastrointestinal tract, which increases the propensity to diabetes. High-amylose starches also form strong films, which are suitable for bioplastics and pharmaceutical excipients.

6.2 MOLECULAR STRUCTURE OF AMYLOSE

6.2.1 Whole-Molecular Size

The whole (or branched) molecular structure of starch is commonly characterized using size-exclusion chromatography (SEC, also known as gel permeation chromatography) (Gilbert, Witt, & Hasjim 2013). Starch SEC chromatograms normally exhibit bimodal distributions, corresponding to amylopectin molecules, which have a large hydrodynamic size, and amylose molecules, which are somewhat smaller. However, waxy starches only show unimodal distributions as they are devoid of amylose (e.g., Witt, Gidley, & Gilbert 2010). Examples of the chromatograms of waxy and normal maize starches is given in Figure 6.1. Amylopectin, having larger average hydrodynamic size, elutes

TABLE 6.1 Amylose contents of various starches (extracted from Jane et al. 1999) based on iodine potentiometry (see Section 6.7.2)

STARCH SOURCE	APPARENT AMYLOSE CONTENT (%)	ABSOLUTE AMYLOSE CONTENT (%)
Waxy maize[a]	0.0	/
Normal maize	29.4	22.5
Waxy rice[a]	0.0	/
Sweet rice[a]	2.1	/
Rice	25.0	20.5
Wheat	28.8	25.8
Barley	25.5	23.6
Waxy amaranth[a]	3.4	/
Cattail millet	19.8	15.3
Mung bean	37.9	30.7
Chinese taro	13.8	13.8
Tapioca	25.3	17.8
Amylomaize V	52.0	27.3
Amylomaize VII	68.0	40.2
Potato	36.0	16.9
Green leaf canna	43.2	22.7
Water chestnut	29.0	16.0

[a] These starches do not contain amylose.

separation being outside the separation range of the column (Cave et al. 2009; Gidley et al. 2010). The characterization of amylose molecules, on the other hand, is less affected by these problems. Other characterization techniques, including various field-flow fractionation methods (Wan et al. 2023), are less developed or less commonly implemented for the characterization of starch whole-molecular structure, but are seen as the best means of overcoming artifacts caused by shear scission in SEC.

The whole-molecular size of amylose depends on many factors, such as botanical origins, varieties, genetic mutations, and environmental conditions. For comparison purpose, the SEC chromatogram of amylose molecules can be reduced to the hydrodynamic size at the peak maximum and the average hydrodynamic size, which reflect the average whole-molecular size of amylose. The hydrodynamic radii at the peak maximum of whole amylose molecules from various rice varieties in dimethyl sulfoxide (DMSO) solution containing 5% w/w LiBr range between 19 and 41 nm in Li et al. (2016) and between 10 and 19 nm in Syahariza et al. (2013), while their weight-average hydrodynamic radii range between 17 and 25 nm and between 8 and 11 nm, respectively. The peak area of amylose in the SEC chromatogram, in comparison to the peak area of amylopectin or the whole peak area, cannot be used to determine the amylose content of starch because shear scission of amylopectin molecules means that some degraded amylopectin molecules elute in what is actually the size range of amylose region. As stated, the problem of shear scission of amylopectin molecules during SEC elution can be overcome using techniques with lower shear during size separation, such as various field-flow fractionation methods. These have not been applied to any significant extent to starch, although they have been proven to give good results for glycogen, a "sister-molecule" to starch, and this is seen as an important area for future development.

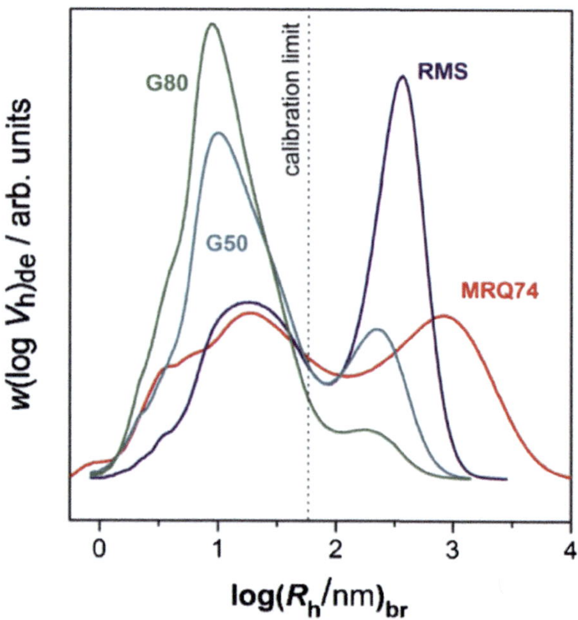

FIGURE 6.1 SEC whole-molecular (fully branched) size distributions of rice starch (MRQ74), normal maize starch (RMS), and high-amylose maize starches (G50 and G80) (reprint from Vilaplana, Hasjim, & Gilbert 2012 with permission). The distributions are normalized to the same total area under the curves (AUC). The peak on the left is amylose molecules, which have smaller hydrodyanamic radii, while the peak on the right is amylopectin molecules, which have larger hydrodynamic radii.

before amylose since smaller molecules have longer pathway than the larger molecules in the SEC column. Problems characterizing amylopectin molecules using SEC due to its large molecular size include shear scission, low recovery, and poor

6.2.2 Branch Chain Length

The chain-length distribution (CLD) of the branches from amylose molecules can be characterized using SEC (Gong et al. 2019; Li et al. 2013; Li et al. 2016; Syahariza et al. 2013; Tao et al. 2019; Vilaplana et al. 2014; Wang et al. 2014). Prior to size characterization, the individual branches are obtained by enzymatic debranching of starch whole molecules, such as using isoamylase. In most cases, it is not necessary to separate amylose and amylopectin molecules for this characterization since there is a clear separation between amylopectin branches and amylose branches at degree of polymerization (DP) ~100 (Vilaplana et al. 2014). This has been confirmed as the peaks associated to amylose branches with DP > 100 are undetectable in the CLDs of waxy starches (Gong et al. 2019; Teng et al. 2016; Witt, Gidley, & Gilbert 2010). Three peak maxima can usually be observed in the CLD of amylose at DP ~300, 1,000–5,000, and >5,000, although often the peak maximum at DP > 5,000 is less prominent than the other two (Vilaplana et al. 2014). The CLDs of amylopectin and amylose from various botanical origins are shown in Figure 6.2. It is not possible to analyze the chain length of amylose using fluorophore-assisted capillary electrophoresis (FACE) or high-performance anion-exchange chromatography (HPAEC): the lengths of the amylose chains go beyond the detection limit of FACE, which is up to DP ~150 (Wu, Li, & Gilbert 2014), well below the DP of all but the shortest amylose chains; the range for HPAEC is even less, up to DP 65–70.

Similar to whole-molecular size distributions, the CLDs of amylose branches vary with botanical origins, varieties, and genetic mutations. Wang et al. (2014) compared the CLDs of amylose samples from different botanical origins. The first peak with maximum at DP ~300 was quite similar for all samples although it was not apparent in the CLDs of barley and potato amyloses. On the other hand, the second peak, with maximum between DP 1,500 and 3,400, did vary with botanical origin. The maxima of the second peak from the

amylose CLDs of barley, cow pea, potato, and sweet potato starches appeared at larger sizes (DP > 2,500) than those of the other starches examined in this study. In addition, the DPs at the peak maxima of amylose branches were correlated with the DPs at the peak maxima of amylopectin branches, but not with the amylose content. This is consistent with the inference that the enzymes involved in the biosynthesis of amylopectin may also play a role in the biosynthesis of amylose (Zhu et al. 2020), which is further explored in Section 6.5.2.

Variations in the amylose CLD have been reported among different varieties of the same species, not only in the peak area, which is related to the amylose content, but also in the DP and the height of the peak maxima (Gong et al. 2019; Li et al. 2016; Syahariza et al. 2013; Tao et al. 2019). However, no relationships have been found between the peak area and the DP or the height of the peak maxima. Genetic mutation can alter both amylose content and its CLD, such as in high-amylose maize starches (Vilaplana et al. 2014). Besides the higher amylose content, the height ratio of the peak maximum at DP ~300 to that at DP ~1,000 is higher for high-amylose maize starches than for normal maize starch.

The effects of growing temperature on amylose CLD of sorghum starch were found to be less pronounced than those on amylopectin CLD (Li et al. 2013). However, the difference in the two growing temperatures (38°C/21°C and 32°C/22°C for day/night temperature) in this study might not be large enough to impose any effects on amylose CLD and amylose content. Furthermore, sorghum is known to be more drought-resistant than most crops, and it can withstand harsh growing conditions. On the other hand, several rice varieties grown at three locations in Australia displayed small, but significant, differences in their amylose CLDs (Li & Liu 2019). Another study on sorghum starch (Kaufman et al. 2017) also showed a significant, but small, difference in the amylose contents of the crops grown in two consecutive years. There are currently not enough data to understand the effects of growing environment on amylose CLD.

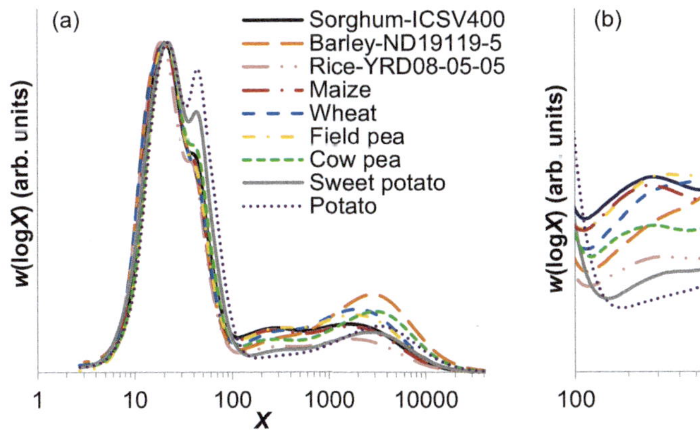

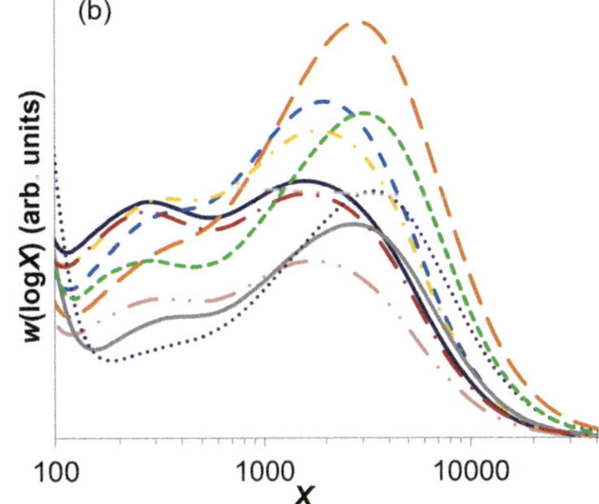

FIGURE 6.2   CLDs of (a) debranched starch and (b) amylose from various botanical sources. (Reprint from Wang et al. 2014 with permission.) All distributions are normalized to yield the same height of the first (amylopectin) peak to avoid the effect of different sample concentrations.

6.2.3 Average Number of Branches

The average number of branches per amylose molecule can be determined by comparing the average size of the whole molecules and that of the individual branches (Wang et al. 2019). Two-dimensional SEC can further characterize the average number of branches per molecule based on the whole-molecular size. This technique was introduced by Vilaplana and Gilbert (2011). The whole starch molecules are first separated by their molecular size into several fractions using preparative SEC. Each collected fraction is then enzymatically debranched and analyzed using analytical SEC to obtain its CLD. The resulting CLDs are plotted against the average hydrodynamic volume/radius of their whole molecules to obtain a two-dimensional molecular size distribution. Thus, it is possible to understand how the average number of branches per amylose molecule varies with the size of amylose molecules. The two-dimensional SEC technique is further discussed in Section 6.7.7.

In general, amylose has less than five branches per molecule (Takeda, Hizukuri, & Juliano 1986; Wang et al. 2019). Higher average numbers of branches per amylose molecule have been reported in the past (Takeda, Shirasaka, & Hizukuri 1984), which could be due to either the botanical specificity or the contamination from residual amylopectin branches resulting from imperfect separation of the amylopectin and amylose components. Moreover, size fractionation of amylose polymers revealed that the length of the amylose branches was dependent on the whole-molecular size. Larger molecules have longer branches, while smaller molecules have shorter branches (Vilaplana et al. 2014; Wang et al. 2019). The average number of branches per amylose molecule, on the other hand, is not, or much less, dependent on the whole-molecular size. The amylopectin CLD is less dependent on the whole-molecular size, which could be due to the constraints imposed by the crystallinity requirement in starch granules (Wu & Gilbert 2010).

6.3 HELICAL AND CRYSTALLINE STRUCTURES OF AMYLOSE

6.3.1 Amylose Double Helices in Native Starch Granules

Amylose is present in an amorphous conformation inside native starch granules as amylose content correlates negatively with the degree of crystallinity (Cheetham & Tao 1998). Regardless of its amorphous conformation, amylose intersperses with the amylopectin double helices in the crystalline lamellae (Kasemsuwan & Jane 1994). Amylose molecules leach out more easily from the swollen granules during heating than amylopectin molecules, because of the former's amorphous conformation (Srichuwong et al. 2005). On the other hand, the semi-crystalline structure of amylopectin needs to be disrupted before the amylopectin molecules become soluble. Furthermore, amylopectin, being

the larger molecules with a high number of branches, has more entanglement in the granules than the predominantly linear amylose molecules.

However, it has been proposed that double helices of amylose are present in native high-amylose starch granules, especially when the amylose content is above 70% (Jiang et al. 2010a, 2010d). The conclusion gelatinization temperature of native high-amylose maize starch increases with amylose content, and the area (or the enthalpy change) of the endotherm above 100°C becomes larger with higher amylose content (Li et al. 2008), which is further discussed in Section 6.4.1. The melting temperature of retrograded amylose, having a double-helical crystalline structure (see Section 6.3.2), has also been reported to be above 120°C (Gruchala & Pomeranz 1993; Sievert & Pomeranz 1990). Therefore, the high conclusion gelatinization temperature of native high-amylose starch granules, above 100°C, which is not observed from waxy and normal starches, seems to be associated with the presence of crystalline amylose double helices. Furthermore, both native high-amylose starch granules and retrograded amylose have the hexagonal B-type crystalline pattern with the X-ray diffraction peaks at 2θ angles of 5.5°, 15°, 17°, 22°, and 24°. Jiang et al. (2010a) collected the components with melting temperatures above 100°C from native high-amylose maize starch granules by removing the other part of the starch using a thermostable α-amylase in a boiling water bath. Structural analysis revealed that a fraction of these components consisted of long-chain double helices with an average DP of 840–951, which most likely originated from amylose branches (DP > 100) (see Section 6.2.2). Furthermore, the amount of these thermally stable double helices in native high-amylose maize starch granules increased with amylose content. Interaction among amylose molecules is possible in high-amylose starches, considering the high concentration of amylose molecules in the granules. It has been proposed that the interaction between amylose molecules begins during the synthesis of starch granules in amyloplast, resulting in the fusing of two or more starch granules into irregular shapes, such as rod, filament, triangle, and L-shape (Jiang et al. 2010b). Multiple hila (or centers of the starch granule) have been reported within a single granule of high-amylose maize starch.

6.3.2 Double Helices of Retrograded Amylose

Although amylose molecules are present in an amorphous conformation in native starch granules, they recrystallize into a double helical structure (also known as retrogradation) faster than amylopectin molecules after gelatinization (less than a day vs. several days, respectively). The longer, more linear structure gives amylose molecules more mobility, less entanglement, and less steric hindrance than amylopectin molecules with a high number of short branches (Sievert & Wüsch 1993). An exothermic transition below 70°C, related to amylose chain reassociation, can be observed during cooling after a pure amylose has been heated to 180°C. Due to its structure, retrograded amylose can form a stronger, more perfect crystalline structure than retrograded amylopectin. The melting temperature of retrograded amylose is above 120°C (Gruchala & Pomeranz 1993; Sievert & Pomeranz 1990), while that of retrograded amylopectin generally appears between 40°C and 60°C (Jane et al. 1999). Retrograded amylose has the hexagonal B-type crystalline pattern (Leloup et al. 1992; Shi, Mao, & Shi 2024), but the monoclinic A-type crystalline pattern, with the X-ray diffraction peaks at 2θ angles of 15°, 17°, 18°, and 23°, can be achieved if the retrogradation takes place under certain conditions, such as low moisture or high temperature (Cai et al. 2010; Hasjim & Jane 2009). Retrograded amylose can also self-assemble into spherulites with birefringence, having diameters up to several microns (Nordmark & Ziegler 2002; Shi, Mao, & Shi 2024).

There are several ways to increase the amount of retrograded amylose. One way is through the repeated heating and cooling treatments, which accelerate the nucleation and propagation processes of starch crystallites (Gruchala & Pomeranz 1993; Sievert & Pomeranz 1990; Slade & Levine 1987). The rate of nucleation is higher at low temperature, whereas the propagation is more rapid at higher temperature, closer to the melting temperature. Another way to increase the amount of retrograded amylose is by enzymatically debranching starch to produce small linear glucans from amylopectin that have more mobility than amylose molecules (Chang et al. 2019; Huong, Hoa, & Van Hung 2021; Shi, Mao, & Shi 2024). Therefore, starch with a high amount of retrograded amylose crystals can be obtained by combining enzymatic debranching and heating/cooling cycles. It is also possible to obtain crystalline starch similar to retrograded amylose from a debranched waxy starch (Cai & Shi 2010). However, although these small linear glucans have higher mobility than amylose molecules, they might not have the length to form perfect crystalline structure of double helices, and thus, their crystallinity might be lower than that of retrograded amylose.

6.3.3 Single-Helical Inclusion Complex

Amylose molecules can form single-helical inclusion complexes with hydrophobic molecules, including polyiodide, alcohols, and some lipids, such as fatty acids, monoglycerides, and phospholipids. The single helices of amylose are left-handed single helices with a hydrophobic inner cavity, while the hydroxyl groups protrude outward creating a hydrophilic periphery (Immel & Lichtenthaler 2000). The hydrophobic molecules fill the inner cavity through various interactions, including hydrophobic interactions and van der Waal forces. This is the basis of iodine colorimetry and iodine potentiometry for amylose content analysis (see Sections 6.7.1 and 6.7.2, respectively). The single-helical inclusion complexes of amylose give the V-type crystalline pattern with the X-ray diffraction peaks at 2θ angles of 8°, 13°, and 20°. The number of glucose units per helical turn can range from six to eight glucose units, depending on the size of the complexing compounds (Jane & Robyt 1984). The dissociation temperature of the amylose–lipid complexes is also dependent on the complexing compounds, such as the length of hydrocarbon chains and the presence of *cis* double bonds (Raphaelides & Karkalas 1988; Tufvesson, Wahlgren & Eliasson 2003a; Tufvesson,

Wahlgren & Eliasson 2003b). Due to the ability of amylose to complex with alcohols, amylose can be separated from amylopectin and intermediate components using butan-1-ol (see Section 6.6).

Amylose–lipid complexes have two forms, namely less ordered and more ordered complexes (Biliaderis & Seneviratne 1990; Biliaderis & Galloway 1989; Tufvesson, Wahlgren & Eliasson 2003a). They are also known as form I complex and form II complex, respectively. The former has a lower dissociation temperature (<100°C), while the latter has a higher dissociation temperature (>100°C). The form I complex forms spontaneously between amylose and lipids. The form II complex can be obtained by incubating the form I complex at a temperature above its dissociation temperature. Similar to retrograded amylose, amylose–lipid complexes can also form spherulites (Bhosale & Ziegler 2010; Fanta et al. 2008).

6.4 EFFECTS OF AMYLOSE ON STARCH PROPERTIES

6.4.1 Gelatinization Properties

Waxy and normal starches from the same species (or subspecies) have similar gelatinization properties, except that waxy starch has a higher enthalpy change when measured by differential scanning calorimetry (DSC) (Jane et al. 1999). Gelatinization is the melting of starch native crystalline structure, which is formed by amylopectin double helices, and hence, amylose does not have a strong influence on starch gelatinization properties. However, the presence of amylose dilutes the amount of amylopectin in the starch granules, resulting in lower crystallinity (Cheetham & Tao 1998), which is reflected by the lower enthalpy change during gelatinization.

On the other hand, the gelatinization temperature, especially the conclusion temperature, of high-amylose maize starch increases with amylose content (Li et al. 2008). It has been postulated that crystalline amylose double helices are present when the concentration of amylose in starch granules is high (see Section 6.3.1). Thus, the high conclusion temperature of high-amylose starches (above 100°C) is due to the melting of crystalline amylose, similar to the melting temperature of retrograded amylose, which is above 120°C (Gruchala & Pomeranz 1993; Sievert & Pomeranz 1990).

6.4.2 Pasting Properties and Gelation

Pasting properties involve viscosity evolution when a starch slurry is heated to a certain temperature, followed by a cooling step. They are usually determined using devices, such as the Brabender Viscograph and Amylograph or the Rapid Visco Analyzer (RVA). Amylose can affect the pasting properties of starch in several ways. Although peak viscosity is a property of amylopectin, the presence of amylose in non-waxy cereal starches can retard the swelling of starch granules during heating. This is because amylose complexes with lipids (see Section 6.3.3) on the surface of starch granule, creating a stronger shell. Amylose–lipid complexes have been associated with "ghost granules", which are the remnants of starch granules after being heated in excess water under little or no shear (Li & Wei 2020; Zhang et al. 2014). Amylose–lipid complexes cause not only a decrease in peak viscosity, but also an increase in pasting temperature, an increase in trough viscosity (the lowest viscosity during heating), and a decrease in breakdown viscosity (the difference between peak and trough viscosities) (Jane et al. 1999). These phenomena are not observed with non-waxy starches that contain little or no lipids, such as mung bean and banana starches, although their amylose contents can be higher than those of non-waxy cereal starches. In addition, the defatting of non-waxy cereal starches results in higher peak viscosity and larger breakdown viscosity (Chinma et al. 2012; Lorenz 1976) because it removes the effects of amylose–lipid complexes.

Amylose is known to be the main contributor to final viscosity, which is the viscosity at the end of cooling step, causing a large setback viscosity, which is the difference between trough and final viscosities. Waxy starches have low final viscosity and small setback viscosity due to the lack of amylose, while non-waxy starches show high final viscosity and large setback viscosity that correlate positively with amylose content (Varavinit et al. 2003; Yanagisawa et al. 2006; Zeng et al. 1997). During the cooling step, leached amylose molecules form a network in the aqueous phase that traps swollen starch granules (Wang, Liu, & Ai 2022). Since it is not related to amylose–lipid complex, the defatting of non-waxy cereal starches does not decrease the final viscosity. However, the defatting process increases the amount of free (uncomplexed) amylose molecules, and thus, it further increases the final viscosity, giving a larger setback viscosity (Chinma et al. 2012; Lorenz 1976).

Gel is normally obtained from the paste of a gelatinized non-waxy starch after being left at room temperature or being stored at a cold temperature. The presence of swollen granules filling the space with leached amylose molecules acting as a glue is crucial for gel formation (Wang, Liu, & Ai 2022). On the other hand, the paste from a gelatinized waxy starch does not form gel, not only because it does not contain amylose, but also because its weak granular structure is easily disrupted by agitation during gelatinization. Therefore, waxy starch is more suitable as a thickening agent where gelling is not desired, while non-waxy starch can be a good gelling agent. Gel texture characteristics, such as hardness and gumminess, correlates positively with amylose content (Bao et al. 2006; Sandhu & Singh 2007). When the granular structure is completely disrupted, such as by heating in DMSO solution, only high-amylose starches show gel formation, while normal starches fail to form gel (Wang, Liu, & Ai 2022).

6.4.3 Starch Retrogradation

Starch retrogradation, which is the recrystallization of starch molecules after gelatinization, has been associated with textural changes of food products during storage, such as bread staling and gel syneresis. Amorphous starch molecules can bind more water molecules than starch molecules in a double helical confirmation. Therefore, when starch reverts back to the crystalline structure with double helices, it loses its water binding capacity, releasing the water molecules from the food system. Water molecules play a big role in food texture, and hence, losing water molecules during storage is not desirable for foods with long shelf lives. As mentioned in Section 6.3.2, amylose molecules, due to their primarily linear structure, retrograde faster than amylopectin molecules (Sievert & Wüsch 1993). The freeze–thaw stability of starch gel, determined as the amount of water retained in the gel after cycles of freezing and thawing, has been found to be negatively correlated with amylose content (Srichuwong et al. 2012). Thus, starch with a high amylose content is less preferred for foods targeted to have long shelf lives. The hardness of cooked rice is also positively correlated with amylose content, which is due to the faster rate of starch retrogradation after cooking (Yu, Ma, & Sun 2009). However, there are some advantages from amylose retrogradation. For example, retrograded amylose is resistant to hydrolysis by α-amylase (Jane & Robyt 1984) and has been regarded as a type of dietary fiber (see Section 6.4.4). Furthermore, the mechanical and thermal properties of starch film are better with higher amylose content, including the film strength and toughness (Cano et al. 2014; Li et al. 2011; Wang et al. 2017), which can be attributed to the presence of crystalline structure from retrograded amylose.

6.4.4 Starch Digestibility and Resistant Starch

Starch digestibility, as observed using cooked rice grain or native starch granules, is negatively correlated with amylose content (Syahariza et al. 2013; Teng et al. 2016). This could be associated with amylose molecules, located at the periphery of starch granules (see Section 6.5.3), forming a strong shell by entanglement with amylopectin molecules, proteins, and lipids. Ghost granules are more evident after cooking of starches containing amylose because the shell is more difficult to be disintegrated by normal cooking methods (Debet & Gidley 2007; Zhang et al. 2014). Thus, amylose may form a barrier to prevent the digestive enzymes from hydrolyzing the starch molecules inside the granules.

Another possible mechanism of how amylose reduces starch digestibility has been revealed following the evolution of starch molecules during *in vitro* and *in vivo* digestions of native starch granules, as well as *in vitro* digestion of starch extrudate. A new species of linear glucans having an average DP of ~50 becomes more apparent at the end of the digestion process (Hasjim et al. 2010a; Teng et al. 2016; Witt, Gidley, & Gilbert 2010). The amount of these molecules is positively correlated with amylose content, and thus, they originate from amylose molecules. Linear glucans with an average DP of ~50 were also obtained when retrograded amylose was hydrolyzed by α-amylases, because the enzymes cannot hydrolyze the crystalline part of retrograded amylose (Jane & Robyt 1984). Therefore, it can be concluded that amylose molecules retrograde during digestion, becoming more resistant to the digestive enzymes, and eventually, the crystalline part of amylose having the average DP of ~50 remains in the digesta, as the enzymes cannot hydrolyze it.

RS is the portion of starch that resists the hydrolysis by the digestive enzymes in the gastrointestinal tract. Hence, it can reach the colon where it is partially fermented by the gut microflora. Because of the similar health benefits, it has been included as a type of dietary fiber (Jones 2014). RS has been grouped into five types based on its crystalline structure and interactions with non-starch components (Eerlingen & Delcour 1995; Englyst, Kingman, & Cummings 1992; Hasjim, Ai, & Jane 2013). Amylose is the main contributor of the enzyme resistance in three types of RS.

RS type 2 is found in native starch granules with B- and C-type crystalline patterns that cannot be completely hydrolyzed by the digestive enzymes. The C-type crystalline pattern is the combination of the A- and the B-type crystalline patterns. Some examples of RS type 2 are native starches from potato, banana, and high-amylose maize. Although potato and banana starches contain high amounts of RS, they are not heat resistant, and their enzyme resistance disappears after being gelatinized. On the other hand, high-amylose maize starch has a high gelatinization temperature due to the presence of crystalline amylose double helices (see Section 6.3.1), and its enzyme resistance remains after normal cooking process (Jiang et al. 2010a; Li et al. 2008). Therefore, high-amylose maize starch has been commonly used as the source of RS in many food products, such as in baked goods. Among high-amylose maize starches, the amount of RS type 2 after food processing is positively correlated with amylose content.

RS type 3 is retrograded amylose (see Section 6.3.2), whose crystalline structure is highly resistant to enzyme hydrolysis (Jane & Robyt 1984). This crystalline structure has a high melting temperature above 120°C, and thus, it is not affected by common food processing (Gruchala & Pomeranz 1993; Sievert & Pomeranz 1990). A small amount of RS type 3 can be found in common food products, such as stale bread, cold leftover rice, and cooled leftover potato. The residues of retrograded amylose after enzyme hydrolysis have an average DP of ~50 (Jane & Robyt 1984), which is also found in native and processed non-waxy starches after enzyme hydrolysis (Hasjim et al. 2010a; Teng et al. 2016; Witt, Gidley, & Gilbert 2010), indicating that crystalline amylose double helices can form during the digestion of non-waxy starches. The amount of RS type 3 increases with amylose content. Certain types of processing, such as repeated heating and cooling cycles, can increase the amount of retrograded amylose (Gruchala & Pomeranz 1993; Sievert & Pomeranz 1990). Enzymatic debranching of starch can increase the amount of linear chains that behave similarly to amylose and thus has been used to enhance the RS content (González-Soto et al. 2004; Pongjanta et al. 2009). It is also possible to produce RS type 3 from waxy starches using enzymatic debranching (Cai et al. 2010; Cai & Shi 2010).

RS type 5 comprises amylose–lipid complexes (see Section 6.3.3). Both forms of amylose–lipid complex, form I and form II complexes, show some resistance against enzymatic hydrolysis with the form II, due to its more ordered structure, more resistant than the form I. Starch can be enzymatically debranched before complexing with lipids to increase the amount of amylose–lipid complexes (Hasjim, Ai, & Jane 2013; Hasjim & Jane 2009; Zhang et al. 2012). This method has been reported to increase the RS of a high-amylose maize starch from 37% to 75% (Hasjim, Ai, & Jane 2013). The result from a human trial revealed that the glycemic and insulinemic indices of a bread product were only 55% and 43%, respectively, of those of the control bread when two-thirds of the wheat flour was substituted by RS type 5 (Hasjim et al. 2010b). Some prebiotic effects, such as the production of short-chain fatty acids and some prevention of colon cancer cell development were observed when rats were fed with RS type 5 (Zhao et al. 2011).

6.5 AMYLOSE BIOSYNTHESIS

6.5.1 Granule-Bound Starch Synthase

In developing grain, granule-bound starch synthase (GBSS) is the main enzyme for amylose biosynthesis (Ahuja et al. 2013; Ball, van de Wal, & Visser 1998; Denyer et al. 2001). GBSSI is present in storage tissues, while GBSSII is found in non-storage tissues, such as leaves. GBSS is exclusively bound to starch granules and uses ADP-glucose as the substrate to progressively elongate the linear amylose chains through the formation of new $\alpha(1{\rightarrow}4)$ glycosidic linkages. GBSSI requires a primer, i.e., a maltooligosaccharide, which might be a product from the debranching process during amylopectin biosynthesis or from the degradation of long starch chains (Denyer et al. 1996, 1999). Short maltooligosaccharides can penetrate into a starch granule, where they are captured by *GBSSI* and elongated to form long amylose chains. The amylose content of starch is closely related to the expression of *GBSSI* in the grain. A complete knockout of *GBSSI* gene in the endosperm results in a waxy variety that does not contain any amylose, while reducing the activity of GBSSI produces a partially waxy variety with lower amylose content. Mutation of the *GBSSI* gene, however, can only affect the amylose content without altering the CLD of amylose (Wang et al. 2015). Increasing GBSS activity, on the other hand, does not always produce high-amylose varieties. For example, overexpression of *GBSSI* did not increase the amylose content of wheat and potato starches (Flipse et al. 1996; Sestili et al. 2012). This suggests that there might be an upper limit of amylose biosynthesis, which could be regulated by other factors in addition to the activity of GBSSI, such as the availability of primers and ADP-glucose substrates and the space constraint in the starch granule. The role of GBSSI in amylose biosynthesis cannot be replaced by other starch synthases in the grain, i.e., soluble starch synthases (SSS), which can only elongate amylopectin chains. However, it has been proposed that GBSSI also participates in the synthesis of extra-long chains of amylopectin, as the absence of GBSSI results in the loss of these extra-long chains (Hanashiro et al. 2008).

6.5.2 Starch Branching Enzymes

The main purpose of starch branching enzymes (SBEs) is to create new branches on amylopectin molecules through $\alpha(1\rightarrow6)$ glycosidic linkages (Tetlow & Emes 2014). There are two classes of SBE. In an *in vitro* study (Guan & Preiss 1993), SBEI showed a higher activity for branching amylose molecules than for branching amylopectin molecules, while SBEII showed the opposite, indicating that SBEI and SBEII are responsible for the synthesis of long and short branches, respectively. However, it is not yet proven *in vivo* if SBEI is also responsible for the number of branches and the CLD of amylose molecules. It has been postulated that the branching of amylose is due to a random interaction between a growing amylose chain and a branching enzyme in starch granules (Ball, van de Wal, & Visser 1998).

The mutation of *SBE*, on the other hand, can affect the biosynthesis of amylose in developing grain and the amylose content of the starch. Knocking out the enzyme(s) responsible for the biosynthesis of amylopectin, including SBE, reduces the amount of amylopectin in starch granules, which also means an increase of amylose content. In addition, the synthesis of amylopectin is competing with the synthesis of amylose for the ADP-glucose substrates (Han et al. 2019). During amylopectin biosynthesis, SBE creates new branches, and each branch carries one non-reducing end that can be elongated by SSS through the formation of new $\alpha(1\rightarrow4)$ glycosidic linkages. The reduction of SBEIIb activity or the knockout of *SBEIIb* gene in maize significantly decreases the formation of new short branches and thus slows down the synthesis of amylopectin (Han et al. 2022; Zhao et al. 2015). This leads to more ADP-glucose substrates for amylose biosynthesis. The biosynthesis of amylopectin in maize can be further suppressed to increase amylose content through the knock-out of more than one SBE. In addition, the resulting amylopectin molecules have an increased number of long branches that behave similarly to amylose branches, increasing the apparent amylose content of the starch. High-amylose maize lines with apparent amylose content above 70% were developed through the suppression of both *SBEIIb* and *SBEIa* (Li et al. 2008; Yao, Thompson, & Guiltinan 2004). However, the mutation also alters the starch granular structure and creates new hybrid molecules (Jiang et al. 2010a; Li et al. 2008; Vilaplana et al. 2014). The starch granules of waxy and normal maize are polygonal in shape, while high-amylose maize starch is composed of both spherical and irregularly-shaped granules, such as rod, filament, triangle, and L shapes, where some granules contain multiple hila. The hybrid molecules have the characteristics of both amylose and amylopectin, such as intermediate components, having the molecular size of amylose with amylopectin CLDs, and amylopectin with extra-long branches that overlap with amylose CLDs (Figure 6.3). The appearance of hybrid molecules affects the amylose content analysis as there is no clear separation between amylose and amylopectin, as discussed in Sections 6.7.6 and 6.7.7.

High-amylose rice lines, similar to those of maize, were developed by suppressing *SBEIIb* (Liu et al. 2021). On the other hand, high-amylose wheat and high-amylose barley lines were developed by the mutation of *SBEIIa* (Regina et al. 2010; Sestili et al. 2010; Slade et al. 2012). This is mainly due to the fact that SBEIIb is the dominant branching enzyme in maize and rice, while SBEIIa is the dominant in wheat and barley. High-amylose potato lines with amylose content up to 94% were developed by downregulating both *SBEI* and *SBEII* genes (Hofvander et al. 2004; Zhao, Andersson, & Andersson 2018). The high-amylose potato lines have lower starch yield, and the granules are smaller in size and more irregular in shape than in the

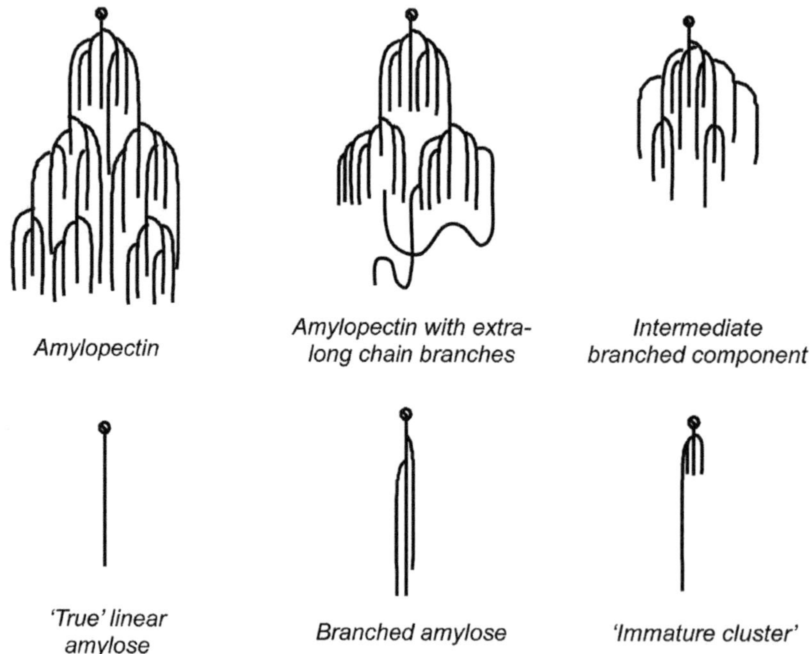

FIGURE 6.3 Glucan molecules found in high-amylose maize starches. (Reprint from Vilaplana et al. 2014 with permission.)

normal lines. Overexpression of *GBSSI* and suppression of *SSS* can also be used to increase the amylose content of starch, but they are less effective than the mutation of *SBE*.

6.5.3 Amylose Accumulation in Developing Grain

The rate of amylose biosynthesis in developing grain is not constant being accumulated more at the later stage of grain development. Li, Blanco, and Jane (2007) reported that the amylose content in normal maize starch increased from 9.2% at the 12th day after pollination to 24.2% at the 30th day after pollination, and then remained at the same level until grain maturation. Jiang et al. (2010c) also reported that the amylose content of high-amylose maize line GEMS-0067 increased from 55% at the 15th day after pollination to 88% at the 40th day after pollination, and then remained constant until maturation. The amylose accumulation in the developing grain is related to the activity of GBSSI, which increases up to the 25th day after pollination before declining when the grain reaches maturity (Guo et al. 2006).

6.5.4 Location of Amylose in Starch Granules

In native starch granules, more amylose molecules are located closer to the periphery than to the hilum, as revealed through the surface gelatinization of starch granules (Kuakpetoon & Wang 2007; Pan & Jane 2000). Using 13 M LiCl solution, starch granules were gelatinized or solubilized slowly at room temperature from the surface to the hilum. The solubilized parts at successive times were then collected and analyzed to understand the composition at different layers of starch granules. The concentration of amylose molecules was found to be higher for the layer that is closer to the periphery of starch granules. This confirms that the biosynthetic rate of amylose molecules is higher at the later stage of grain development (see Section 6.5.3), as starch granules are synthesized outward from the hilum.

6.6 SEPARATION OF AMYLOSE AND AMYLOPECTIN MOLECULES

Amylose molecules can be separated from amylopectin molecules, and from intermediate components if present, using the fact that the long linear amylose branch chains can form inclusion complexes with alcohols having four carbon units or more, while the branch chains of amylopectin and intermediate components are too short and can only form inclusion complexes with small alcohol molecules, i.e., methanol and ethanol. The separation of amylose molecules from amylopectin molecules and intermediate components is normally performed using butan-1-ol (Jane & Chen 1992). The starch sample has to be solubilized and defatted prior to the separation step, which can be performed simultaneously by dissolving the sample in DMSO, followed by precipitation using ethanol. This step is important to remove preexisting crystalline and helical structures, which can affect the complexation between amylose and butan-1-ol and the precipitation of the complex. The starch-ethanol precipitate is redissolved in hot deionized water under reflux in a boiling water bath, and then butan-1-ol is added to complex with amylose molecules in the solution. The amylose-butan-1-ol complexation is allowed to continue as the temperature cools down slowly to room temperature, for example, 30 in a Dewar flask. The amylose-butan-1-ol complex is collected as precipitate using a centrifuge, which can be stored in methanol or ethanol. It is crucial to remove the butan-1-ol from the amylose sample before usage, such as by the defatting process described above, as it can affect the functionality of amylose, including forming double helices and complexing with iodine and lipids.

The amylopectin molecules and intermediate components that remain in the supernatant, as they cannot complex with butan-1-ol, can be further purified by repeating the addition of butan-1-ol to complex with the residual amylose in the supernatant until no precipitate is observed. The amylopectin molecules and intermediate components can be collected through ethanol precipitation, followed by centrifugation. The precipitate is washed with ethanol and dried in a convection oven at 40°C to remove water and ethanol. The isolated amylopectin and intermediate components can be used in the determination of absolute amylose content by iodine potentiometry (see Section 6.7.2).

6.7 ANALYTICAL METHODS FOR AMYLOSE CONTENT

Due to the significant influence of amylose on the functional properties of starch, it is important to determine the amylose content of a starch sample for its end-uses. However, the measured amylose content of starch depends on the method used, as some are based on the properties of amylose, while others are based on the structure (Vilaplana, Hasjim, & Gilbert 2012). The most commonly used method is based on the complexation of amylose and iodine, followed by quantification using colorimetric or potentiometric methods. Amylose is also known to complex with lipids, including phospholipids. The amount of amylose–lipid complexes can be determined based on the enthalpy change of the complex formation or dissociation, which can be measured using DSC. There is also a commercial kit to quantify amylose content using concanavalin A, a lectin that binds amylopectin molecules at the multiple non-reducing ends. Amylose can be easily (but imperfectly) separated from amylopectin based on the differences in their molecular structures, since amylose has a smaller average whole-molecular size and longer branch chains than amylopectin. Thus, amylose content can be obtained from either the whole-molecular size distribution or the CLD as the ratio of the area under the curve (AUC) of amylose molecules or branches to the total AUC.

The two-dimensional plot of whole-molecular size distribution and CLD provides a more accurate amylose content than either whole-molecular size distribution or CLD alone. Although the two-dimensional SEC is labor - and time-intensive, it can be useful when the separation between amylose and amylopectin is not clear, such as in the presence of extra-long amylopectin branches and intermediate components (Figure 6.3) that are commonly found in high-amylose starches.

6.7.1 Iodine Colorimetry

Starch forms colored complexes in a solution containing iodine and iodide ions (Pesek & Silaghi-Dumitrescu 2024; Thoma & French 1960). These are a single-helical complexes where the hydrophobic inner part of the starch helices allows the iodine and iodide ions to form polyiodide (Yu, Houtman, & Atalla 1996). Amylopectin–iodine complexes give a red to purple color, whereas amylose–iodine complexes give a blue color. This is attributed to longer polyiodide molecules that can fit inside the single helices of long amylose branches. Hence, the amount of amylose can be quantified based on the blue color of the complexes using spectrophotometry at a wavelength between 590 and 720 nm. Some methods use a single wavelength for the absorbance of blue color, while others use two wavelengths (Hoover & Ratnayake 2001; Juliano et al. 1981; Zhu et al. 2008). The dual wavelength method is so as to remove the interference from amylopectin–iodine complexes that can cause the overestimation of amylose–iodine complexes. A standard curve of the absorbance is required for the quantification of amylose content, and it is normally obtained using commercially available (potato) amylose alone or mixed with amylopectin (or a waxy starch) at different concentrations. Starch or flour standards with known amylose contents have also been used to generate the standard curve. Iodine colorimetry is commonly used to obtain amylose content both for research and in the food industry as, it is easy and quick. There are various official methods of iodine colorimetry, particularly ISO 6647-2:2020 and AACC 61-03.01. The variations in the iodine colorimetric methods have, however, caused many problems when comparing amylose content results among different labs (Fitzgerald et al. 2009).

The intensity and the stability of the blue color obtained from amylose–iodine complexes depends on several factors, such as pH, time, temperature, and the concentrations of starch sample, iodine, and iodide ions. It was found that the amylose content obtained by iodine colorimetry at pH 4.5 was higher than that at pH 10.2 (Perez & Juliano 1978). Amylose–iodine complexes are less stable at alkaline pH, probably because starch molecules are more hydrophilic above pH 7. The absorbance of the blue color developed from the complexes also declines with time (Murdoch 1992). In addition, amylose–iodine complexes are more stable below 15°C, where the absorbance remains almost independent of temperature (Fonslick & Khan 1989; Morrison & Laignelet 1983). Above 15°C, the absorbance decreases with the increase in temperature due to the dissociation of the complexes. It is important to have sufficient iodine and iodide ions to complex with all of the amylose in the starch sample. However, high concentrations of starch, iodine, and iodide ions may produce black

specks due to the precipitation of the amylose–iodine complexes that can affect the reading of the absorbance (Morrison & Laignelet 1983). To allow the iodine to complex with amylose, preexisting double and single helical structures need to be disrupted, which in turn solubilizes the starch molecules. There are several methods to solubilize starch molecules, including using KOH, NaOH, DMSO, and urea solutions, and this may impact the amylose–iodine complex formation (Morrison & Laignelet 1983). Moreover, and most importantly, KOH and NaOH both can break glycosidic bonds (depending on the alkali concentration). Therefore, the analysis needs to be carefully performed to obtain an accurate amylose content.

The presence of endogenous lipids, which also form single-helical complexes with amylose, prevents amylose–iodine complex formation, thereby causing an underestimation of the actual amylose content. A defatting step is required to remove the endogenous lipids before the analysis in order to obtain an accurate amylose content (Hoover & Ratnayake 2001; Juliano et al. 1981; Morrison & Laignelet 1983). Some lipids form strong complexes with amylose, which are not easily dissociated. Furthermore, the solvent used for defatting may form single-helical complexes with amylose. Thus, it is imperative to choose the right solvent and extraction conditions for a complete removal of the lipids and solvent before iodine calorimetry.

Choosing the right amylose and amylopectin references to establish the standard curve is also important, because the structural differences among starches from different botanical origins can affect the complexation with iodine. It is better to use the same botanical origin as the sample. In addition, commercially available amylose is extracted through the complexation of amylose and butan-1-ol (see Section 6.6), and residual butan-1-ol may be present, which can prevent the formation of amylose–iodine complexes. Thus, the amylose and amylopectin standards should be defatted beforehand.

6.7.2 Iodine Potentiometry

Similar to iodine colorimetry, iodine potentiometry is also based on the complexation of amylose and iodine. However, instead of measuring the blue color developed by the complexes using spectrophotometry, iodine potentiometry is a potentiometric titration of solubilized starch molecules with iodine solution (Bates, French, & Rundle 1943). A potentiometer is used to monitor the potential during the titration, which is plotted as the potential on the y-axis vs. the volume of iodine solution added being on the x-axis. The potential remains constant at the beginning of the titration until the complexation between amylose and iodine is complete, since bound iodine has no effect on the potential; then the potential starts to rise due to the increasing amount of free iodide ions in the solution. Based on the iodine affinity of pure amylose, i.e., the amount of iodine/iodide that can be bound by pure amylose, the amylose content of a starch sample can be calculated from its iodine affinity.

It is possible to use iodine potentiometry to determine the actual amylose content (absolute amylose content) and the amylose content including the long branches of amylopectin and intermediate components that behave like amylose (apparent amylose content) (Jane et al. 1999; Kasemsuwan et al. 1995). Before the analysis, starch sample needs to be defatted to remove the effect of amylose–lipid complexes that can prevent amylose–iodine complexation. The defatting step is achieved simultaneously with solubilization, by dissolving starch sample in DMSO, followed by precipitation using ethanol. The iodine affinity of defatted starch gives the apparent amylose content. The absolute amylose content is obtained by correcting the iodine affinity of defatted starch with that of amylopectin and intermediate components. The amylopectin and intermediate components can be obtained by using the procedure mentioned in Section 6.6. The corrected iodine affinity is the iodine affinity of the amylose component alone without interference from the long branches of amylopectin and intermediate components. The apparent and absolute amylose contents of starches from various botanical origins are provided in Table 6.1.

6.7.3 Concanavalin A

There is a commercially available kit to determine the amylose content using concanavalin A (Gibson, Solah, & McCleary 1997; Matheson & Welsh 1988; Yun & Matheson 1990), which is a lectin or a carbohydrate - binding protein extracted and purified from jack beans. Concanavalin A specifically binds with carbohydrates at the non-reducing end of α-D-glucopyranosyl or α-D-mannopyranosyl unit. Amylopectin, having a high number of non-reducing ends, binds more concanavalin A than amylose, resulting in the precipitation of amylopectin. Amylose, which remains in the supernatant, is then collected and quantified enzymatically or chemically. Since this test is based on the solubility of amylose, the starch sample needs to be completely solubilized and defatted before the analysis. Lipids forming complexes with amylose can reduce the solubility of amylose, causing the underestimation of amylose content. Similar to iodine potentiometry (Section 6.7.2), the solubilization and the defatting step can be achieved simultaneously by dissolving the starch sample in DMSO, followed by precipitation using ethanol. The starch precipitate is finally redissolved in DMSO before adding concanavalin A. The amount of amylose in the supernatant, after the precipitation of amylopectin by concanavalin A, can be determined by hydrolyzing the amylose to glucose using amyloglucosidase and then colorimetrically quantifying the amount of glucose using the glucose oxidase and peroxidase (GOPOD) method (Gibson, Solah, & McCleary 1997). It can also be determined using a less specific method, such as the phenol-sulfuric acid method for the quantification of total carbohydrates (Yun & Matheson 1990).

6.7.4 Differential Scanning Calorimetry

DSC has been proposed as a method to determine the amylose content of starch (Mestres et al. 1996; Sievert & Holm 1993), although it is less commonly applied than the other methods discussed here. The DSC method is based on the complexation of amylose and lipids (see Section 6.3.3). Excess complexing lipids, such as L-α-lysophosphatidylcholine from egg yolk, are added to the starch sample to make sure that there are enough lipids to complex completely with the amylose in the starch sample. The starch sample with added lipids is then heated in a DSC followed by a cooling step (Mestres et al. 1996). During the heating step, starch gelatinizes and preexisting amylose–lipid complexes are dissociated. The final temperature of the heating step is important to make sure that there is no residual crystalline structure that can inhibit the complexation between amylose and lipids. For pure amylose and retrograded starch, a higher final temperature up to 170°C might be needed to dissociate preexisting crystalline amylose (Sievert & Holm 1993). Amylose–lipid complexes reform spontaneously during the cooling step, and the exotherm of the complex formation appearing between 65°C and 92°C is used to quantify the amount of amylose–lipid complexes. There is a linear relationship between the amylose content and the area (or enthalpy change) of the exotherm from the complex formation. Since the exotherm of the complex formation from pure amylose is known to be around 27–28.5 J/g, the amylose content of a starch sample can be estimated from the enthalpy change of the exotherm. Alternatively, a second heating step can be employed after the cooling step, and the amount of amylose–lipid complexes can be determined from the endotherm of amylose–lipid complex dissociation during the second heating step (Sievert & Holm 1993). Unreacted lipids do not affect the exotherm of complex formation or the endotherm of complex dissociation, and hence, a defatting step prior to the DSC analysis is not necessary. However, the endotherm and the exotherm peaks of amylose–lipid complexes are dependent on the botanical origins (i.e., molecular structure) and can be influenced by the presence of proteins (Zhu et al. 2008).

6.7.5 Whole-Molecular Size Distribution

The whole-molecular analysis of a starch sample (see Section 6.2.1), such as using SEC, yields two peaks associated with amylose, which has a smaller average hydrodynamic size, and amylopectin, which has a larger average hydrodynamic size (Figure 6.1). Therefore, the amylose content could in principle be estimated as the ratio of AUC of amylose to the total AUC of amylose and amylopectin (Vilaplana, Hasjim, & Gilbert 2012). However, since amylose and amylopectin have distinct branching structures and conformations in solution, different values of differential refractive indices, dn/dc, should be applied for amylose and amylopectin (Tizzotti et al. 2013). In addition, amylopectin, having a large molecular size, unavoidably suffers from shear scission during SEC analysis (Cave et al. 2009), resulting in the artifactual presence of a larger amount of smaller molecules that may overlap with the amylose peak. Poor separation of amylose and amylopectin has been reported for the SEC whole-molecular size analysis of starch (Gidley et al. 2010). Furthermore, some of the largest amylopectin

TABLE 6.2 Amylose contents (%, dry starch basis) of rice flour, normal maize starch, and high-amylose maize starches obtained by various methods, as provided by Vilaplana, Hasjim, and Gilbert (2012)

SAMPLE	IODINE COLORIMETRY[a]	CONCANAVALIN A[a]	WHOLE-MOLECULAR SIZE DISTRIBUTION	CHAIN-LENGTH DISTRIBUTION	TWO-DIMENSIONAL MOLECULAR SIZE DISTRIBUTION
Rice flour	22.8±0.1	31.8±0.3	40.1	29.9	28.1
Normal maize	30.2±0.3	28.7±0.6	43.6	30.2	34.1
High-amylose maize Gelose 50	48.2±1.3	46.1±1.1	77.0	64.0	50.7[b] 54.7[c]
High-amylose maize Gelose 80	64.5±5.4	58.9±1.2	93.3	83.7	53.1[b] 63.1[c]

[a] Mean±standard deviation. The iodine colorimetry was performed following ISO 6647-2:2011, and the absorbance was obtained at 720 nm and the correction factor of 85% was used for normal maize and high-amylose maize starches.
[b] The separation was performed according to Line 1 in panel (c) and (d) Figure 6.4.
[c] The separation was performed according to Line 2 in panel (c) and (d) Figure 6.4.

molecules might be blocked by the guard column that is placed before the SEC analytical columns, reducing the apparent concentration of amylopectin. Intermediate components, which are part of the hybrid molecules commonly found in high-amylose starches, have similar whole-molecular sizes to that of amylose but have CLDs similar to that of amylopectin. They elute at similar times as amylose (Vilaplana et al. 2014). Hence, the whole-molecular size distribution obtained using SEC is not a good method to determine amylose content based on molecular structure as it may underestimate the amount of amylopectin in the sample and the intermediate components are included as part of amylose. Other characterization techniques, such as various field-flow fractionation methods (Wan et al. 2023), may provide a more accurate whole-molecular size distribution of starch and thus a better estimation of amylose content, but they are less developed at this time. Table 6.2 compares the amylose contents obtained from SEC whole-molecular size distribution of starch to those obtained by other methods .

6.7.6 Chain-Length Distribution

Amylose and amylopectin have very distinct branching structures. Amylose molecules have only a few long branches, whereas amylopectin molecules have a large number of short branches. Based on the difference in the chain lengths of amylose and amylopectin molecules, it is possible to obtain the amylose content of a starch sample from the CLD, such as that obtained using SEC (see Section 6.2.2), as the ratio of the AUC of amylose branches to the total AUC of all branches (Vilaplana, Hasjim, & Gilbert 2012; Wang et al. 2014). One of the benefits of using CLD over whole-molecular size distribution is that all molecules are small linear glucans with only $\alpha(1\rightarrow4)$ linkages. Therefore, only one dn/dc is required for the analysis, and no molecules are sheared during the separation or blocked by the guard column. For normal starches, there is a clear separation between amylopectin branches (smaller hydrodynamic size) and amylose branches (larger hydrodynamic size) at DP ~100 (Figure 6.2). Therefore, this method can reveal the real amylose content of normal starches, which is based on the molecular structure rather than the functionality. The results

obtained are similar to, but not the same as, the absolute amylose content from iodine potentiometry. The sample preparation for analysis, on the other hand, takes a longer time than iodine colorimetry, concanavalin A, and DSC methods, as it involves not only the solubilization of starch molecules, such as in DMSO solution, but also an enzymatic debranching step. It is important that the starch sample is completely debranched to obtain an accurate result.

For the hybrid molecules found in high-amylose starches (Figure 6.3), this method can separate the intermediate components from amylose as the intermediate components have CLDs similar to that of amylopectin, whose branches are smaller than those of amylose. However, it is not possible to separate the extra-long branches of amylopectin from amylose branches because they elute at similar times and thus overestimating the amylose content of high-amylose starches as shown in Table 6.2. Although this method is not suitable for high-amylose starches, it still gives accurate amylose content for waxy and normal starches, and the results are not dependent on the botanical origins, i.e., based on the molecular structure.

6.7.7 Two-Dimensional Molecular Size Distribution

Two-dimensional molecular size distribution (Figure 6.4) is currently the only method that can structurally differentiate amylose from amylopectin and the hybrid molecules in high-amylose starches (Vilaplana et al. 2014; Vilaplana & Gilbert 2011; Vilaplana, Hasjim, & Gilbert 2012). Although this method is very laborious compared with the other methods discussed here, it provides the most accurate results of amylose content based on the molecular structure with the least ambiguity caused by the hybrid molecules. For this analysis, starch molecules, after being dissolved in DMSO solution, are firstly separated based on their whole-molecular size using a preparative SEC. Despite the poorer resolution of the preparative SEC column, it is needed so as to be able to collect sufficient amounts of samples separated by whole-molecular size, which undergo subsequent enzymatic debranching and size characterization. Fractions based on

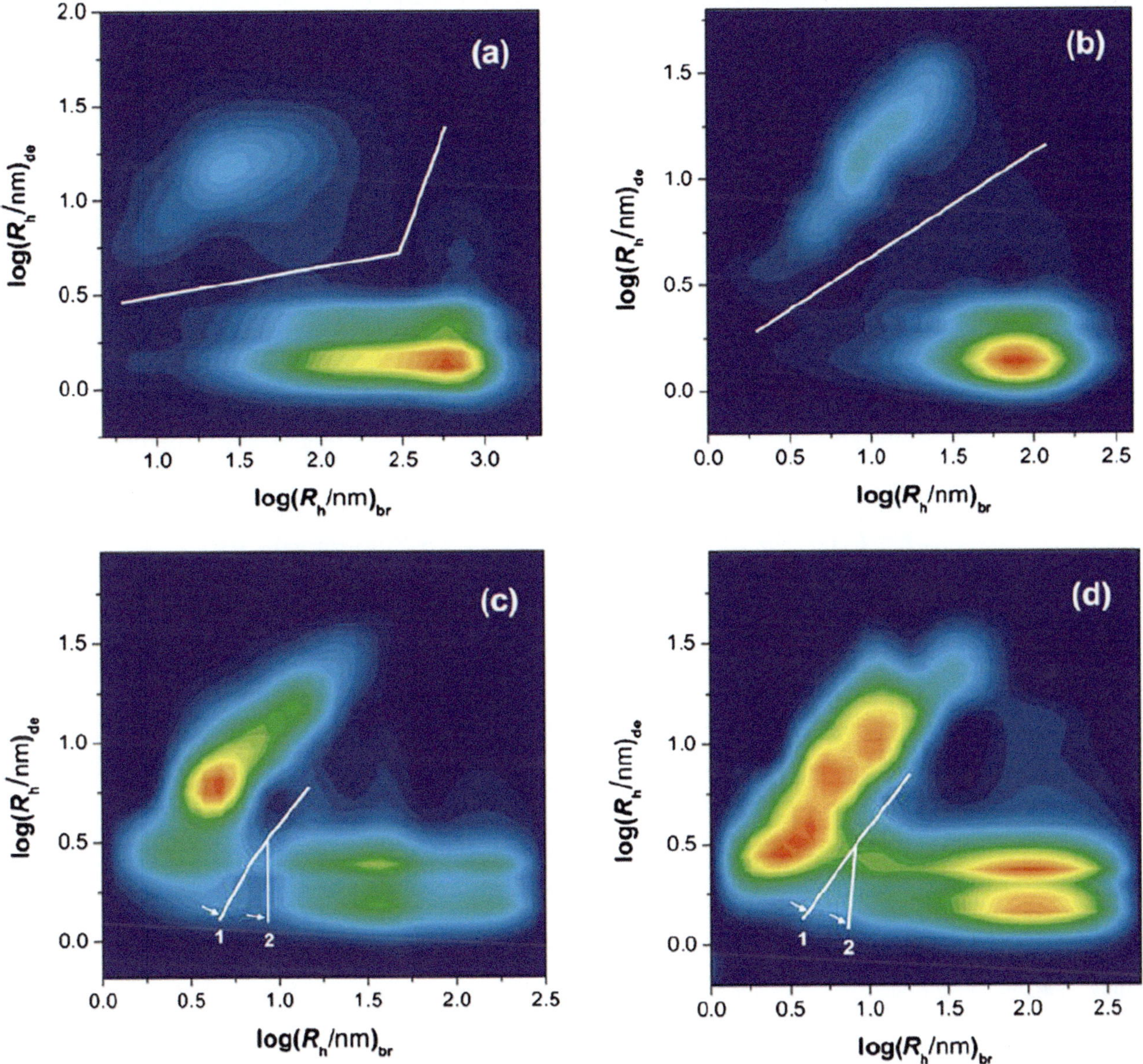

FIGURE 6.4 Two-dimensional molecular size distributions for (a) rice flour/starch, (b) normal maize starch, (c) high-amylose maize starch Gelose 50, and (d) high-amylose starch Gelose 80. (Reprint from Vilaplana, Hasjim, & Gilbert 2012 with permission.)

whole-molecular size are collected at different elution times and enzymatically debranched to convert the branches into individual linear chains. The fractions before and after debranching are each characterized using analytical SEC to obtain both whole-molecular size distribution and the CLD of each fraction, respectively. Then, the CLDs are plotted as a function of the weight-average molecular sizes (such as hydrodynamic volume or radius) of the corresponding whole molecules. Figure 6.5 summarizes the experimental methodology to obtain two-dimensional molecular size distribution of starch, an example of fraction division of rice starch based on the whole-molecular size, and the whole-molecular size distributions and the CLDs of the collected fractions.

From the two-dimensional molecular size distribution, one can identify the amylose region as the molecules that satisfy two criteria: (1) smaller whole-molecular size than amylopectin and (2) branches with DP > 100. The mass of a region in a two-dimensional plot can be defined as follows:

$$\iint_{\text{region}} w(\log R_{h,br}, R_{h,de})\, d\log R_{h,br}\, d\log R_{h,de} \qquad (6.1)$$

Hence, a clear separation of the amylose region is required for obtaining a reliable amylose content. Figure 6.4 shows examples of the two-dimensional molecular size distributions from rice starch, normal maize starch, and high-amylose maize starches. The white line separates the area associated with amylose branches from that with amylopectin branches and the branches from hybrid molecules. There is a clear separation between the amylose and amylopectin in rice starch and normal maize starch. However, as expected, the separation is not clear for high-amylose maize starches, and thus, two lines are proposed as illustrated by Line 1 and Line 2 (in Figure 6.4). The amylose contents obtained by this method are also included in Table 6.2 for comparison with those obtained by other methods, noting that the two-dimensional molecular size distribution method

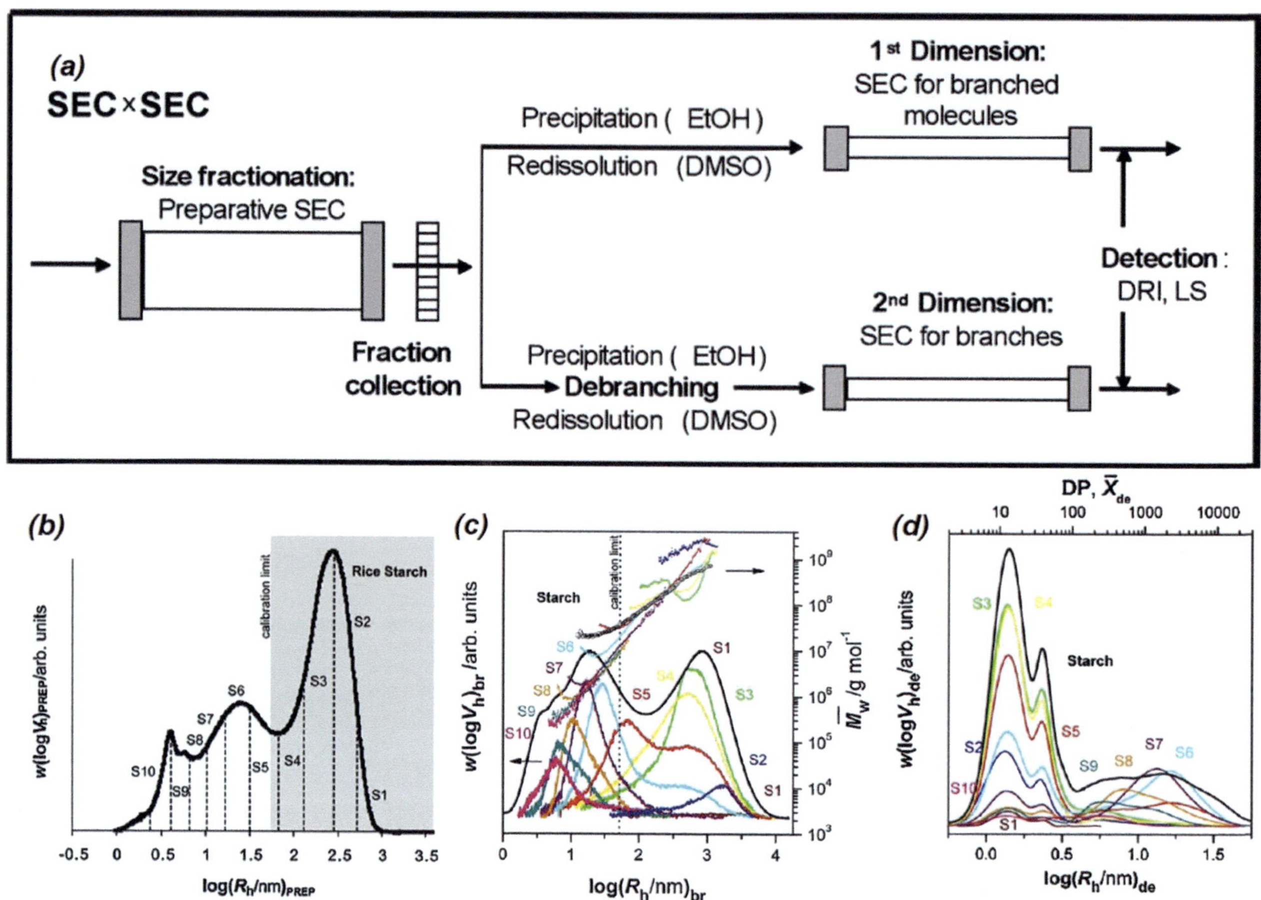

FIGURE 6.5   (a) Experimental methodology to obtain two-dimensional molecular size distribution of starch, combining preparative and analytical SEC with enzymatic debranching. (b) An example of fraction division of rice starch based on whole-molecular size. The fractions were collected using preparative SEC at different elution volumes. (c) SEC whole-molecular size distributions and (d) CLDs of the collected fractions of rice starch. (Reprint from Vilaplana & Gilbert 2011 with permission.)

is the only one suitable for such purposes for high-amylose starches. Although variations are present in the amylose content results obtained using the two dividing lines, the amylose contents are more accurate than those obtained by the other methods because interference from the intermediate components and the extra-long branch chains of amylopectin is mostly eliminated, which cannot be avoided when using the other methods.

6.8 CONCLUSION

Amylose is present in an amorphous conformation in native starch granules, whereas amylopectin branches are arranged in alternating crystalline and amorphous lamellae. However, after gelatinization, amylose retrogrades into crystalline double-helical structure faster than amylopectin. In addition, amylose can spontaneously complex with lipids. These properties of amylose affect the gelatinization, solubility, pasting, gelation, and retrogradation rate of starch, which, in turn, influence the texture, stability, and digestibility of food products. In the past, only amylose content

was determined and correlated with the properties of starch. The whole-molecular size and the CLD of amylose were less studied. It is, therefore, not well understood if the variation in the molecular structure of amylose affects the properties of starch. Recently, it has been revealed that the average chain length of amylose is highly dependent on its whole-molecular size, whereas the number of branches per molecule (which is less than five branches per molecules) is less dependent on its whole-molecular size. Although it is known that GBSS is responsible for the elongation of amylose chains, it is still not known which enzyme creates the branches on the amylose chains, and how the number of branch per amylose molecule is controlled during the biosynthesis. Furthermore, it is not well understood which factors cause or control the variations observed in the molecular structure and CLD of amylose among different botanical origins and among different varieties of the same species. To date, there are not enough data to comprehend the impact of growing environment on amylose content and its structure. In addition, it is not easy to compare the amylose contents from different studies, as there are different methods to determine the amylose content of starch; some are based on the properties of amylose, while others are based on the structure.

REFERENCES

Ahuja, G, Jaiswal, S, Hucl, P, and Chibbar, RN 2013. Genome-specific granule-bound starch synthase I (GBSSI) influences starch biochemical and functional characteristics in near-isogenic wheat (*Triticum aestivum* L.) lines. *Journal of Agricultural and Food Chemistry*, *61*, 12129–12138.

Ball, SG, van de Wal, MHBJ, and Visser, RGF 1998. Progress in understanding the biosynthesis of amylose. *Trends in Plant Science*, *3*, 462–467.

Bao, J, Shen, S, Sun, M, and Corke, H 2006. Analysis of genotypic diversity in the starch physicochemical properties of nonwaxy rice: Apparent amylose content, pasting viscosity and gel texture. *Starch/Stärke*, *58*, 259–267.

Bates, FL, French, D, and Rundle, RE 1943. Amylose and amylopectin content of starches determined by their iodine complex formation. *Journal of the American Chemical Society*, *65*, 142–148.

Bhosale, RG and Ziegler, GR 2010. Preparation of spherulites from amylose-palmitic acid complexes. *Carbohydrate Polymers*, *80*, 53–64.

Biliaderis, CG and Galloway, G 1989. Crystallization behavior of amylose-V complexes: Structure-property relationships. *Carbohydrate Research*, *189*, 31–48.

Biliaderis, CG and Seneviratne, HD 1990. On the supermolecular structure and metastability of glycerol monostearate-amylose complex. *Carbohydrate Polymers*, *13*, 185–206.

Cai, L and Shi, YC 2010. Structure and digestibility of crystalline short-chain amylose from debranched waxy wheat, waxy maize, and waxy potato starches. *Carbohydrate Polymers*, *79*, 1117–1123.

Cai, L, Shi, YC, Rong, L, and Hsiao, BS 2010. Debranching and crystallization of waxy maize starch in relation to enzyme digestibility. *Carbohydrate Polymers*, *81*, 385–393.

Cano, A, Jiménez, A, Cháfer, M, Gónzalez, C, and Chiralt, A 2014. Effect of amylose: Amylopectin ratio and rice bran addition on starch films properties. *Carbohydrate Polymers*, *111*, 543–555.

Cave, RA, Seabrook, SA, Gidley, MJ, and Gilbert, RG 2009. Characterization of starch by size-exclusion chromatography: The limitations imposed by shear scission. *Biomacromolecules*, *10*, 2245–2253.

Chang, R, Li, M, Wang, Y, Chen, H, Xiao, J, Xiong, L, Qiu, L, Bian, X, Sun, C, and Sun, Q 2019. Retrogradation behavior of debranched starch with different degrees of polymerization. *Food Chemistry*, *297*, 125001.

Cheetham, NWH and Tao, L 1998. Variation in crystalline type with amylose content in maize starch granules: An X-ray powder diffraction study. *Carbohydrate Polymers*, *36*, 277–284.

Chinma, CE, Abu, JO, James, S, and Iheanacho, M 2012. Chemical, functional and pasting properties of defatted starches from cowpea and soybean and application in stiff porridge preparation. *Nigerian Food Journal*, *30*, 80–88.

Debet, MR and Gidley, MJ 2007. Why do gelatinized starch granules not dissolve completely? Roles for amylose, protein, and lipid in granule "ghost" integrity. *Journal of Agricultural and Food Chemistry*, *55*, 4752–4760.

Denyer, K, Clarke, B, Hylton, C, Tatge, H, and Smith, AM 1996. The elongation of amylose and amylopectin chains in isolated starch granules. *The Plant Journal*, *10*, 1135–1143.

Denyer, K, Johnson, P, Zeeman, S, and Smith, AM 2001. The control of amylose synthesis. *Journal of Plant Physiology*, *158*, 479–487.

Denyer, K, Waite, D, Motawia, S, Møller, BL, and Smith, AM 1999. Granule-bound starch synthase I in isolated starch granules elongates malto-oligosaccharides processively. *Biochemical Journal*, *340*, 183–191.

Eerlingen, RC and Delcour, JA 1995. Formation, analysis, structure and properties of type III enzyme resistant starch. *Journal of Cereal Science*, *22*, 129–138.

Englyst, HN, Kingman, SM, and Cummings, JH 1992. Classification and measurement of nutritionally important starch fractions. *European Journal of Clinical Nutrition*, *46*(Suppl 2), S33–S50.

Fanta, GF, Felker, FC, Shogren, RL, and Salch, JH 2008. Preparation of spherulites from jet cooked mixtures of high amylose starch and fatty acids. Effect of preparative conditions on spherulite morphology and yield. *Carbohydrate Polymers*, *71*, 253–262.

Fitzgerald, MA, Bergman, CJ, Resurreccion, AP, Möller, J, Jimenez, R, Reinke, RF, Martin, M, Blanco, P, Molina, F, Chen, M-H, Kuri, V, Romero, MV, Habibi, F, Umemoto, T, Jongdee, S, Graterol, E, Reddy, KR, Bassinello, PZ, Sivakami, R, Rani, NS, Das, S, Wang, YJ, Indrasari, SD, Ramli, A, Ahmad, R, Dipti, SS, Xie, L, Lang, NT, Singh, P, Toro, DC, Tavasoli, F, and Mestres, C 2009. Addressing the dilemmas of measuring amylose in rice. *Cereal Chemistry*, *86*, 492–498.

Flipse, E, Keetels, CJAM, Jacobsen, E, and Visser, RGF 1996. The dosage effect of the wildtype GBSS allele is linear for GBSS activity but not for amylose content: Absence of amylose has a distinct influence on the physico-chemical properties of starch. *Theoretical and Applied Genetics*, *92*, 121–127.

Fonslick, J and Khan, A 1989. Thermal stability and composition of the amylose-iodine complex. *Journal of Polymer Science Part A: Polymer Chemistry*, *27*, 4161–4167.

Fredriksson, H, Silverio, J, Andersson, R, Eliasson, AC, and Åman, P 1998. The influence of amylose and amylopectin characteristics on gelatinization and retrogradation properties of different starches. *Carbohydrate Polymers*, *35*, 119–134.

Gibson, TS, Solah, VA, and McCleary, BV 1997. A procedure to measure amylose in cereal starches and flours with concanavalin A. *Journal of Cereal Science*, *25*, 111–119.

Gidley, MJ, Hanashiro, I, Hani, NM, Hill, SE, Huber, A, Jane, J, Liu, Q, Morris, GA, Rolland-Sabaté, A, Striegel, AM, and Gilbert, RG 2010. Reliable measurements of the size distributions of starch molecules in solution: Current dilemmas and recommendations. *Carbohydrate Polymers*, *79*, 255–261.

Gilbert, RG, Witt, T, and Hasjim, J 2013. What is being learned about starch properties from multiple-level characterization. *Cereal Chemistry*, *90*, 312–325.

Gong, B, Cheng, L, Gilbert, RG, and Li, C 2019. Distribution of short to medium amylose chains are major controllers of *in vitro* digestion of retrograded rice starch. *Food Hydrocolloids*, *96*, 634–643.

González-Soto, RA, Agama-Acevedo, E, Solorza-Feria, J, Rendón-Villalobos, R, and Bello-Pérez, LA 2004. Resistant starch made from banana starch by autoclaving and debranching. *Starch/Stärke*, *56*, 495–499.

Graybosch, RA 1998. Waxy wheats: Origin, properties, and prospects. *Trends in Food Science & Technology*, *9*, 135–142.

Gruchala, L and Pomeranz, Y 1993. Enzyme-resistant starch: Studies using differential scanning calorimetry. *Cereal Chemistry*, *70*, 163.

Guan, HP and Preiss, J 1993. Differentiation of the properties of the branching isozymes from maize (*Zea mays*). *Plant Physiology*, *102*, 1269–1273.

Guo, S-J, Li, J-R, Qiao, W-H, and Zhang, X-S 2006. Analysis of amylose accumulation during seed development in maize. *Acta Genetica Sinica*, *33*, 1014–1019.

Han, H, Yang, C, Zhu, J, Zhang, L, Bai, Y, Li, E, and Gilbert, RG 2019. Competition between granule bound starch synthase and starch branching enzyme in starch biosynthesis. *Rice*, *12*, 96.

Han, J, Guo, Z, Wang, M, Liu, S, Hao, Z, Zhang, D, Yong, H, Weng, J, Zhou, Z, Li, M, and Li, X 2022. Using the dominant mutation gene *Ae1–5180* (*amylose extender*) to develop high-amylose maize. *Molecular Breeding*, *42*, 57.

Hanashiro, I, Itoh, K, Kuratomi, Y, Yamazaki, M, Igarashi, T, Matsugasako, J, and Takeda, Y 2008. Granule-bound starch synthase I is responsible for biosynthesis of extra-long unit chains of amylopectin in rice. *Plant and Cell Physiology*, *49*, 925–933.

Hasjim, J, Ai, Y, and Jane, J 2013. Novel applications of amylose-lipid complex as resistant starch type 5. In: Y-C Shi & CC Maningat (eds.), *Resistant Starch: Sources, Applications and Health Benefits*, John Wiley & Sons, Ltd., West Sussex, pp. 79–94.

Hasjim, J and Jane, J 2009. Production of resistant starch by extrusion cooking of acid-modified normal-maize starch. *Journal of Food Science*, *74*, C556–C562.

Hasjim, J, Lavau, GC, Gidley, MJ, and Gilbert, RG 2010a. *In vivo* and *in vitro* starch digestion: Are current *in vitro* techniques adequate? *Biomacromolecules*, *11*, 3600–3608.

Hasjim, J, Lee, SO, Hendrich, S, Setiawan, S, Ai, Y, and Jane, J 2010b. Characterization of a novel resistant-starch and its effects on postprandial plasma-glucose and insulin responses. *Cereal Chemistry*, *87*, 257–262.

Hofvander, P, Andersson, M, Larsson, CT, and Larsson, HÃ 2004. Field performance and starch characteristics of high-amylose potatoes obtained by antisense gene targeting of two branching enzymes. *Plant Biotechnology Journal*, *2*, 311–320.

Hoover, R and Ratnayake, WS 2001. Determination of total amylose content of starch. *Current Protocols in Food Analytical Chemistry*, *00*, E2.3.1–E2.3.5.

Hsieh, CF, Liu, W, Whaley, JK, and Shi, YC 2019. Structure, properties, and potential applications of waxy tapioca starches - A review. *Trends in Food Science & Technology*, *83*, 225–234.

Huong, NTM, Hoa, PN, and Van Hung, P 2021. Effects of microwave treatments and retrogradation on molecular crystalline structure and *in vitro* digestibility of debranched mung-bean starches. *International Journal of Biological Macromolecules*, *190*, 904–910.

Immel, S and Lichtenthaler, FW 2000. The hydrophobic topographies of amylose and its blue iodine complex. *Starch/Stärke*, *52*, 1–8.

Jane, J and Chen, JF 1992. Effect of amylose molecular size and amylopectin branch chain length on paste properties of starch. *Cereal Chemistry*, *69*, 60–65.

Jane, J, Chen, YY, Lee, LF, McPherson, AE, Wong, KS, Radosavljevic, M, and Kasemsuwan, T 1999. Effects of amylopectin branch chain length and amylose content on the gelatinization and pasting properties of starch. *Cereal Chemistry*, *76*, 629–637.

Jane, J and Robyt, JF 1984. Structure studies of amylose-V complexes and retrograded amylose by action of alpha amylases, and a new method for preparing amylodextrins. *Carbohydrate Research*, *132*, 105–118.

Jiang, H, Campbell, M, Blanco, M, and Jane, J 2010a. Characterization of maize amylose-extender (*ae*) mutant starches: Part II. Structures and properties of starch residues remaining after enzymatic hydrolysis at boiling-water temperature. *Carbohydrate Polymers*, *80*, 1–12.

Jiang, H, Horner, HT, Pepper, TM, Blanco, M, Campbell, M, and Jane, J 2010b. Formation of elongated starch granules in high-amylose maize. *Carbohydrate Polymers*, *80*, 533–538.

Jiang, H, Lio, J, Blanco, M, Campbell, M, and Jane, J 2010c. Resistant-starch formation in high-amylose maize starch during kernel development. *Journal of Agricultural and Food Chemistry*, *58*, 8043–8047.

Jiang, H, Srichuwong, S, Campbell, M, and Jane, J 2010d. Characterization of maize amylose-extender (*ae*) mutant starches. Part III: Structures and properties of the Naegeli dextrins. *Carbohydrate Polymers*, *81*, 885–891.

Jones, JM 2014. CODEX-aligned dietary fiber definitions help to bridge the 'fiber gap'. *Nutrition Journal*, *13*, 34.

Juliano, BO, Perez, CM, Blakeney, AB, Castillo, T, Kongseree, N, Laignelet, B, Lapis, ET, Murty, VVS, Paule, CM, and Webb, BD 1981. International cooperative testing on the amylose content of milled rice. *Starch/Stärke*, *33*, 157–162.

Kasemsuwan, T and Jane, J 1994. Location of amylose in normal starch granule. II. Locations of phosphodiester cross-linking revealed by phosphorus-31 nuclear magnetic resonance. *Cereal Chemistry*, *71*, 282–287.

Kasemsuwan, T, Jane, J, Schnable, P, Stinard, P, and Robertson, D 1995. Characterization of the dominant mutant amylose-extender (*Ae1–5180*) maize starch. *Cereal Chemistry*, *72*, 457–464.

Kaufman, RC, Wilson, JD, Bean, SR, Xu, F, and Shi, YC 2017. Sorghum starch properties as affected by growing season, hybrid, and kernel maturity. *Journal of Cereal Science*, *74*, 127–135.

King, RA, Noakes, M, Bird, AR, Morell, MK, and Topping, DL 2008. An extruded breakfast cereal made from a high amylose barley cultivar has a low glycemic index and lower plasma insulin response than one made from a standard barley. *Journal of Cereal Science*, *48*, 526–530.

Kuakpetoon, D and Wang, YJ 2007. Internal structure and physicochemical properties of corn starches as revealed by chemical surface gelatinization. *Carbohydrate Research*, *342*, 2253–2263.

Leloup, VM, Colonna, P, Ring, SG, Roberts, K, and Wells, B 1992. Microstructure of amylose gels. *Carbohydrate Polymers*, *18*, 189–197.

Li, E, Hasjim, J, Singh, V, Tizzotti, M, Godwin, ID, and Gilbert, RG 2013. Insights into sorghum starch biosynthesis from structure changes induced by different growth temperatures. *Cereal Chemistry*, *90*, 223–230.

Li, H, Gidley, MJ, and Dhital, S 2019. High-amylose starches to bridge the "fiber gap": Development, structure, and nutritional functionality. *Comprehensive Reviews in Food Science and Food Safety*, *18*, 362–379.

Li, H and Liu, Y 2019. Effects of variety and growth location on the chain-length distribution of rice starches. *Journal of Cereal Science*, *85*, 77–83.

Li, H, Prakash, S, Nicholson, TM, Fitzgerald, MA, and Gilbert, RG 2016. The importance of amylose and amylopectin fine structure for textural properties of cooked rice grains. *Food Chemistry*, *196*, 702–711.

Li, L, Blanco, M, and Jane, J 2007. Physicochemical properties of endosperm and pericarp starches during maize development. *Carbohydrate Polymers*, *67*, 630–639.

Li, L, Jiang, H, Campbell, M, Blanco, M, and Jane, J 2008. Characterization of maize amylose-extender (*ae*) mutant starches. Part I: Relationship between resistant starch contents and molecular structures. *Carbohydrate Polymers*, *74*, 396–404.

Li, M, Liu, P, Zou, W, Yu, L, Xie, F, Pu, H, Liu, H, and Chen, L 2011. Extrusion processing and characterization of edible starch films with different amylose contents. *Journal of Food Engineering*, *106*, 95–101.

Li, Z and Wei, C 2020. Morphology, structure, properties and applications of starch ghost: A review. *International Journal of Biological Macromolecules*, *163*, 2084–2096.

Liu, ZD, Wang, J, Li, L, and Wu, P 2021. Mechanistic insights into the role of starch multi-level structures in functional properties of high-amylose rice cultivars. *Food Hydrocolloids*, *113*, 106441.

Lorenz, K 1976. Physico-chemical properties of lipid-free cereal starches. *Journal of Food Science*, *41*, 1357–1359.

Matheson, NK and Welsh, LA 1988. Estimation and fractionation of the essentially unbranched (amylose) and branched (amylopectin) components of starches with concanavalin A. *Carbohydrate Research*, *180*, 301–313.

Mestres, C, Matencio, F, Pons, B, Yajid, M, and Fliedel, G 1996. A rapid method for the determination of amylose content by using differential-scanning calorimetry. *Starch/Stärke*, *48*, 2–6.

Morrison, WR and Laignelet, B 1983. An improved colorimetric procedure for determining apparent and total amylose in cereal and other starches. *Journal of Cereal Science*, *1*, 9–20.

Murdoch, KA 1992. The amylose-iodine complex. *Carbohydrate Research*, *233*, 161–174.

Nordmark, TS and Ziegler, GR 2002. Spherulitic crystallization of gelatinized maize starch and its fractions. *Carbohydrate Polymers*, *49*, 439–448.

Pan, DD and Jane, J 2000. Internal structure of normal maize starch granules revealed by chemical surface gelatinization. *Biomacromolecules*, *1*, 126–132.

Perez, CM and Juliano, BO 1978. Modification of the simplified amylose test for milled rice. *Starch/Stärke*, *30*, 424–426.

Pesek, S and Silaghi-Dumitrescu, R 2024. The iodine/iodide/starch supramolecular complex. *Molecules*, *29*, 641.

Pongjanta, J, Utaipattanaceep, A, Naivikul, O, and Piyachomkwan, K 2009. Debranching enzyme concentration effected on physicochemical properties and α-amylase hydrolysis rate of resistant starch type III from amylose rice starch. *Carbohydrate Polymers*, *78*, 5–9.

Raphaelides, S and Karkalas, J 1988. Thermal dissociation of amylose-fatty acid complexes. *Carbohydrate Research*, *172*, 65–82.

Regina, A, Kosar-Hashemi, B, Ling, S, Li, Z, Rahman, S, and Morell, M 2010. Control of starch branching in barley defined through differential RNAi suppression of starch branching enzyme IIa and IIb. *Journal of Experimental Botany*, *61*, 1469–1482.

Sandhu, KS and Singh, N 2007. Some properties of corn starches II: Physicochemical, gelatinization, retrogradation, pasting and gel textural properties. *Food Chemistry*, *101*, 1499–1507.

Sestili, F, Botticella, E, Proietti, G, Janni, M, D'Ovidio, R, and Lafiandra, D 2012. Amylose content is not affected by overexpression of the *Wx-B1* gene in durum wheat. *Plant Breeding*, *131*, 700–706.

Sestili, F, Janni, M, Doherty, A, Botticella, E, D'Ovidio, R, Masci, S, Jones, HD, and Lafiandra, D 2010. Increasing the amylose content of durum wheat through silencing of the *SBEIIagenes*. *BMC Plant Biology*, *10*, 144.

Shi, J, Mao, Y, and Shi, YC 2024. Effects of crystallization temperature on structure and digestibility of spherulites formed from debranched high-amylose maize starch. *Carbohydrate Polymers*, *342*, 122332.

Sievert, D and Holm, J 1993. Determination of amylose by differential scanning calorimetry. *Starch/Stärke*, *45*, 136–139.

Sievert, D and Pomeranz, Y 1990. Enzyme-resistant starch. II. Differential scanning calorimetry studies on heat-treated starches and enzyme-resistant starch residues. *Cereal Chemistry*, *67*, 217–221.

Sievert, D and Wüsch, P 1993. Amylose chain association based on differential scanning calorimetry. *Journal of Food Science*, *58*, 1332–1335.

Slade, AJ, McGuire, C, Loeffler, D, Mullenberg, J, Skinner, W, Fazio, G, Holm, A, Brandt, KM, Steine, MN, Goodstal, JF, and Knauf, VC 2012. Development of high amylose wheat through TILLING. *BMC Plant Biology*, *12*, 69.

Slade, L and Levine, H 1987. Recent advances in starch retrogradation. In: SS Stivala, V Crescenzi & ICM Dea (eds.), *Industrial Polysaccharides: The Impact of Biotechnology and Advanced Methodologies*, Gordon and Breach Science Publishers, New York, pp. 387–430.

Srichuwong, S, Isono, N, Jiang, H, Mishima, T, and Hisamatsu, M 2012. Freeze-thaw stability of starches from different botanical sources: Correlation with structural features. *Carbohydrate Polymers*, *87*, 1275–1279.

Srichuwong, S, Sunarti, TC, Mishima, T, Isono, N, and Hisamatsu, M 2005. Starches from different botanical sources II: Contribution of starch structure to swelling and pasting properties. *Carbohydrate Polymers*, *62*, 25–34.

Svegmark, K, Helmersson, K, Nilsson, G, Nilsson, PO, Andersson, R, and Svensson, E 2002. Comparison of potato amylopectin starches and potato starches - Influence of year and variety. *Carbohydrate Polymers*, *47*, 331–340.

Syahariza, ZA, Sar, S, Hasjim, J, Tizzotti, MJ, and Gilbert, RG 2013. The importance of amylose and amylopectin fine structures for starch digestibility in cooked rice grains. *Food Chemistry*, *136*, 742–749.

Takeda, Y, Hizukuri, S, and Juliano, BO 1986. Purification and structure of amylose from rice starch. *Carbohydrate Research*, *148*, 299–308.

Takeda, Y, Shirasaka, K, and Hizukuri, S 1984. Examination of the purity and structure of amylose by gel-permeation chromatography. *Carbohydrate Research*, *132*, 83–92.

Takeda, Y, Shitaozono, T, and Hizukuri, S 1990. Structures of subfractions of corn amylose. *Carbohydrate Research*, *199*, 207–214.

Tao, K, Li, C, Yu, W, Gilbert, RG, and Li, E 2019. How amylose molecular fine structure of rice starch affects functional properties. *Carbohydrate Polymers*, *204*, 24–31.

Teng, A, Witt, T, Wang, K, Li, M, and Hasjim, J 2016. Molecular rearrangement of waxy and normal maize starch granules during *in vitro* digestion. *Carbohydrate Polymers*, *139*, 10–19.

Tester, RF and Morrison, WR 1990. Swelling and gelatinization of cereal starches. I. Effects of amylopectin, amylose, and lipids. *Cereal Chemistry*, *67*, 551–557.

Tetlow, IJ and Emes, MJ 2014. A review of starch-branching enzymes and their role in amylopectin biosynthesis. *IUBMB Life*, *66*, 546–558.

Thoma, JA and French, D 1960. The starch-iodine-iodide interaction. Part I. Spectrophotometric investigations. *Journal of the American Chemical Society*, *82*, 4144–4147.

Tian, M, Tan, G, Liu, Y, Rong, T, and Huang, Y 2009. Origin and evolution of Chinese waxy maize: Evidence from the *Globulin-1* gene. *Genetic Resources and Crop Evolution*, *56*, 247–255.

Tizzotti, MJ, Sweedman, MC, Schäfer, C, and Gilbert, RG 2013. The influence of macromolecular architecture on the critical aggregation concentration of large amphiphilic starch derivatives. *Food Hydrocolloids*, *31*, 365–374.

Tufvesson, F, Wahlgren, M, and Eliasson, AC 2003a. Formation of amylose-lipid complexes and effects of temperature treatment. Part 1. Monoglycerides. *Starch/Stärke*, *55*, 61–71.

Tufvesson, F, Wahlgren, M, and Eliasson, AC 2003b. Formation of amylose-lipid complexes and effects of temperature treatment. Part 2. Fatty acids. *Starch/Stärke*, *55*, 138–149.

Varavinit, S, Shobsngob, S, Varanyanond, W, Chinachoti, P, and Naivikul, O 2003. Effect of amylose content on gelatinization, retrogradation and pasting properties of flours from different cultivars of Thai rice. *Starch/Stärke*, *55*, 410–415.

Vilaplana, F and Gilbert, RG 2011. Analytical methodology for multidimensional size/branch-length distributions for branched glucose polymers using off-line 2-dimensional size-exclusion chromatography and enzymatic treatment. *Journal of Chromatography A*, 1218, 4434–4444.

Vilaplana, F, Hasjim, J, and Gilbert, RG 2012. Amylose content in starches: Toward optimal definition and validating experimental methods. *Carbohydrate Polymers*, *88*, 103–111.

Vilaplana, F, Meng, D, Hasjim, J, and Gilbert, RG 2014. Two-dimensional macromolecular distributions reveal detailed architectural features in high-amylose starches. *Carbohydrate Polymers*, *113*, 539–551.

Wan, Y, Chua, SMH, Yao, Y, Adler, L, Navarro, M, Roura, E, Tilley, RD, Li, C, Nilsson, L, Gilbert, RG, and Sullivan, MA 2023. Comparing different techniques for obtaining molecular size distributions of glycogen. *European Polymer Journal*, *201*, 112518.

Wang, K, Hasjim, J, Wu, AC, Henry, RJ, and Gilbert, RG 2014. Variation in amylose fine structure of starches from different botanical sources. *Journal of Agricultural and Food Chemistry*, *62*, 4443–4453.

Wang, K, Hasjim, J, Wu, AC, Li, E, Henry, RJ, and Gilbert, RG 2015. Roles of GBSSI and SSIIa in determining amylose fine structure. *Carbohydrate Polymers*, *127*, 264–274.

Wang, K, Vilaplana, F, Wu, A, Hasjim, J, and Gilbert, RG 2019. The size dependence of the average number of branches in amylose. *Carbohydrate Polymers*, *223*, 115134.

Wang, K, Wang, W, Ye, R, Xiao, J, Liu, Y, Ding, J, Zhang, S, and Liu, A 2017. Mechanical and barrier properties of maize starch-gelatin composite films: Effects of amylose content. *Journal of the Science of Food and Agriculture*, *97*, 3613–3622.

Wang, X, Liu, S, and Ai, Y 2022. Gelation mechanisms of granular and non-granular starches with variations in molecular structures. *Food Hydrocolloids*, *129*, 107658.

Witt, T, Gidley, MJ, and Gilbert, RG 2010. Starch digestion mechanistic information from the time evolution of molecular size distributions. *Journal of Agricultural and Food Chemistry*, *58*, 8444–8452.

Wu, AC and Gilbert, RG 2010. Molecular weight distributions of starch branches reveal genetic constraints on biosynthesis. *Biomacromolecules*, *11*, 3539–3547.

Wu, AC, Li, E, and Gilbert, RG 2014. Exploring extraction/dissolution procedures for analysis of starch chain-length distributions. *Carbohydrate Polymers*, *114*, 36–42.

Yanagisawa, T, Domon, E, Fujita, M, Kiribuchi-Otobe, C, and Takayama, T 2006. Starch pasting properties and amylose content from 17 waxy barley lines. *Cereal Chemistry*, *83*, 354–357.

Yao, Y, Thompson, DB, and Guiltinan, MJ 2004. Maize starch-branching enzyme isoforms and amylopectin structure. In the absence of starch-branching enzyme IIb, the further absence of starch-branching enzyme Ia leads to increased branching. *Plant Physiology*, *136*, 3515–3523.

Yu, S, Ma, Y, and Sun, DW 2009. Impact of amylose content on starch retrogradation and texture of cooked milled rice during storage. *Journal of Cereal Science*, *50*, 139–144.

Yu, X, Houtman, C, and Atalla, RH 1996. The complex of amylose and iodine. *Carbohydrate Research*, *292*, 129–141.

Yun, SH and Matheson, NK 1990. Estimation of amylose content of starches after precipitation of amylopectin by concanavalin-A. *Starch/Stärke*, *42*, 302–305.

Zeng, M, Morris, CF, Batey, IL, and Wrigley, CW 1997. Sources of variation for starch gelatinization, pasting, and gelation properties in wheat. *Cereal Chemistry*, *74*, 63–71.

Zhang, B, Dhital, S, Flanagan, BM, and Gidley, MJ 2014. Mechanism for starch granule ghost formation deduced from structural and enzyme digestion properties. *Journal of Agricultural and Food Chemistry*, *62*, 760–771.

Zhang, B, Huang, Q, Luo, F, and Fu, X 2012. Structural characterizations and digestibility of debranched high-amylose maize starch complexed with lauric acid. *Food Hydrocolloids*, *28*, 174–181.

Zhang, C, Zhu, J, Chen, S, Fan, X, Li, Q, Lu, Y, Wang, M, Yu, H, Yi, C, Tang, S, Gu, M, and Liu, Q 2019. *Wxlv*, the ancestral allele of rice waxy gene. *Molecular Plant*, *12*, 1157–1166.

Zhao, X, Andersson, M, and Andersson, R 2018. Resistant starch and other dietary fiber components in tubers from a high-amylose potato. *Food Chemistry*, *251*, 58–63.

Zhao, Y, Hasjim, J, Li, L, Jane, J, Hendrich, S, and Birt, DF 2011. Inhibition of azoxymethane-induced preneoplastic lesions in the rat colon by a cooked stearic acid complexed high-amylose cornstarch. *Journal of Agricultural and Food Chemistry*, *59*, 9700–9708.

Zhao, Y, Li, N, Li, B, Li, Z, Xie, G, and Zhang, J 2015. Reduced expression of starch branching enzyme IIa and IIb in maize endosperm by RNAi constructs greatly increases the amylose content in kernel with nearly normal morphology. *Planta*, *241*, 449–461.

Zhu, J, Yu, W, Zhang, C, Zhu, Y, Xu, J, Li, E, Gilbert, RG, and Liu, Q 2020. New insights into amylose and amylopectin biosynthesis in rice endosperm. *Carbohydrate Polymers*, *230*, 115656.

Zhu, T, Jackson, DS, Wehling, RL, and Geera, B 2008. Comparison of amylose determination methods and the development of a dual wavelength iodine binding technique. *Cereal Chemistry*, *85*, 51–58.

Amylopectin

7

Yasunori Nakamura

7.1 INTRODUCTION

Amylopectin, a major component of starch, is a highly branched glucan, and its fine structure is specific in terms of the non-random branching. Short chains having a degree of polymerization (DP) up to approximately 24–27 form a group (widely called "cluster"), and the group is interconnected to other group by long chains, although some long chains interconnect only two groups, but the others link three and more groups. The other structural feature of amylopectin molecules is a fact that adjacent external segments of two chains form a double helix when the segment lengths reach $DP \geq 10$ (Gidley & Bulpin 1987). Generation of the double helix gives rise to a dramatic effect on the starch properties. The double helices chemically interact with the adjacent double helices and form the semi-crystalline region horizontal to the surface of the starch granule (Perez et al. 2009; Yamaguchi, Kainuma & French 1979; Villwock & BeMiller 2022a, 2022b), as explained below. Due to the non-random nature of amylopectin branching, the chain-length distribution (CLD) of amylopectin side chains having different DP forms a specific polymodal pattern (Figure 7.1), in contrast to a single modal pattern exhibited in glycogen or phytoglycogen chain distribution (Figure 7.1b).

The first cluster model for representation of the structural features of amylopectin molecules was proposed by Nikuni (1969). Then, French depicted a model, which has been widely known as the "French model" (Figure 7.2a), based on the biochemical analysis of the uneven distribution of amylopectin branching (Kainuma & French 1970), structural characteristics and chemical behavior of Naegeli amylodextrin obtained from waxy maize starch (Kainuma & French 1971), the X-ray diffraction analysis of starch granules (Kainuma & French 1971, 1972), and the space filling model and computer modeling for amylopectin chains (Kainuma & French 1972). The French model explains that the external linear segments of chains are arranged in parallel to one another so that they can form double helices (Kainuma & French 1971) and multiple helices are densely arranged in a horizontal direction to the chain direction (Yamaguchi, Kainuma & French 1979). The model also suggests that a huge amylopectin molecule can be synthesized by simply increasing/reproducing the clusters to

the non-reducing ends of component chains. Later, several cluster models have been reported by different researchers, as described and discussed below (Figure 7.2).

In recent years, Bertoft and his colleagues have proposed a new model called the "building block backbone model" (Bertoft 2013, 2017; Tetlow & Bertoft 2020), although this chapter uses shortly the "backbone model". Surprisingly, the basic concept of the backbone model is quite different from that of the cluster model. In the cluster model, the cluster chains including mostly short chains are linked by long penetrating chains in the tandem fashion, whereas in the backbone model, long chains are completely separated from short chains by carrying many short chain groups, as discussed below.

In spite of accumulation of numerous analytical data, the presence of two contradictory models suggests that we are still far from a full understanding of the sophisticated fine structure of amylopectin, mainly due to the limitation of analytical methodologies. In this chapter, the present status of our understanding of the structural features of amylopectin and the regulation of its biosynthesis is mainly described. Some related topics are discussed instead of description of every aspect of starch science fields. For example, see review papers for the role of ADPglucose pyrophosphorylase and protein structures of starch biosynthetic enzymes (Preiss 2009; Suzuki & Suzuki 2016; Tetlow & Emes 2017). Among studies using various plant species, results with rice plants are frequently described since the most comprehensive studies have been conducted.

7.2 BIOCHEMICAL, PHYSICAL, AND MICROSCOPIC APPROACHES FOR ANALYSIS OF THE FINE STRUCTURE OF AMYLOPECTIN

The fine structure of amylopectin has been extensively examined by many approaches because its structure and variation are responsible for the physicochemical and functional properties of the starch. To analyze the fine structure of amylopectin, the most important but commonly used methodology is to measure the CLD of amylopectin after its branch

DOI: 10.1201/9781003464396-7

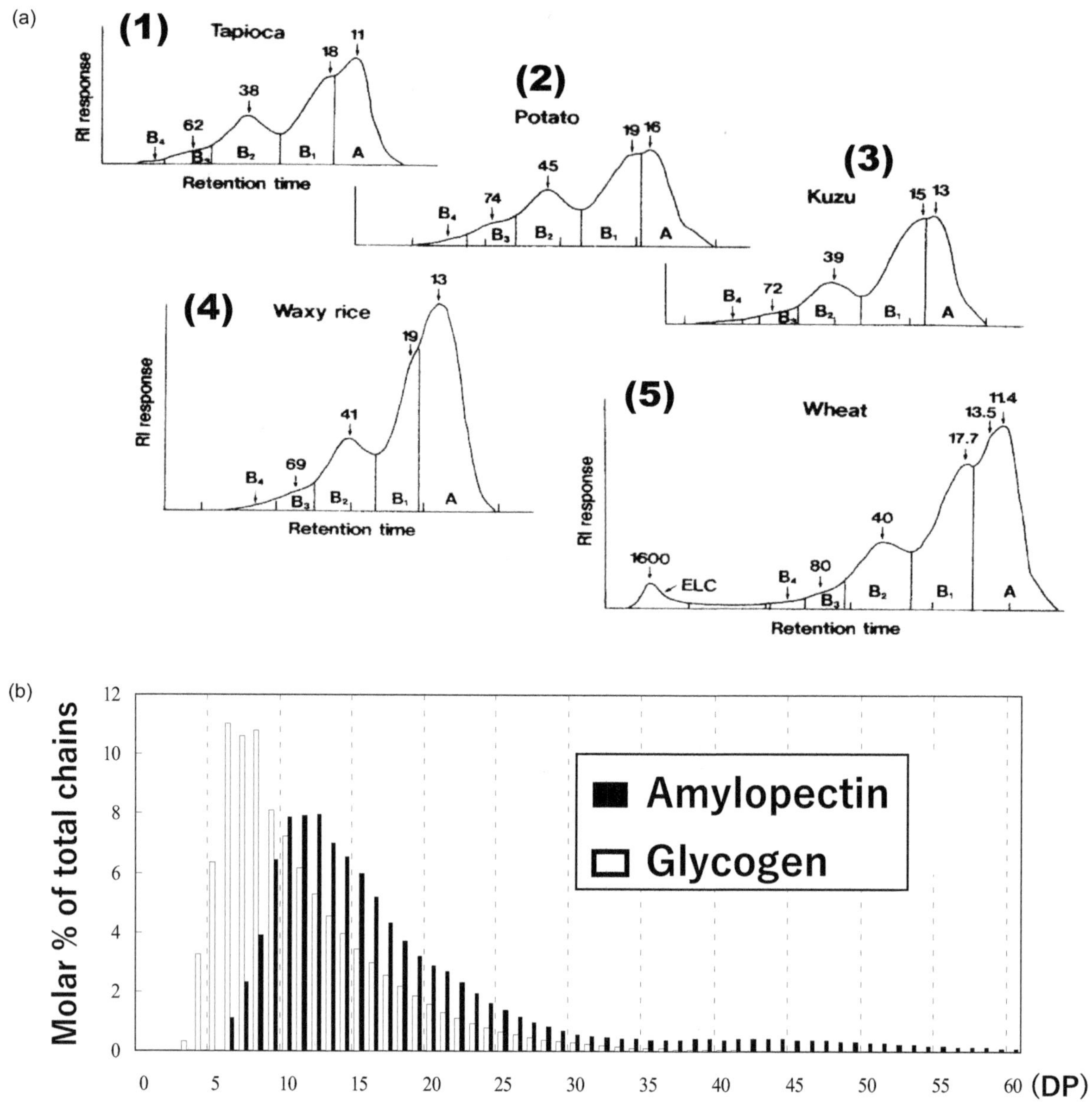

FIGURE 7.1 Chain-length distribution (CLD) pattern of amylopectin and glycogen. (a) Gel filtration chromatography of debranched amylopectin from various plant sources (Hizukuri 1996). 1, cassava tuberous root; 2, potato tuber; 3, kuzu root; 4, waxy rice endosperm; 5, wheat endosperm. (b) CLD analysis of debranched amylopectin from rice endosperm (closed bar) and glycogen from oyster (open bar) by the FACE method.

linkages are removed by treatment with debranching enzymes such as isoamylase (ISA) and pullulanase (PUL). The debranched linear chains were usually separated by conventional gel permeation chromatography (GPC)/size exclusion chromatography (SEC) or high-performance anion-exchange chromatography (HPAEC) (see Reviews by Hizukuri 1996; Jane 2009). As shown in Figure 7.1a, component chains of amylopectin show polymodal distributions. Quantification of these chains has been conducted by several ways. The amounts of fractionated dextrins and their reducing values were conventionally determined by a phenol-sulfuric method and a modified Park-Johnson method (Hizukuri, Takeda & Yasuda 1981). Currently, one of the most widely used methods is to automatically measure the carbohydrate amounts by using a pulsed amperometric detector (PAD), and this is usually called the HPAEC-PAD analysis. This method has many advantages. Researchers can obtain reliable and reproducible data in comparatively short period by using a rather smaller amount of starch samples. However, we need to be careful that the PAD response is not equivalent to all chains, and therefore, the obtained peak data including chains of different DP are not necessarily quantitative on a molar basis (Koizumi, Fukuda & Hizukuri 1991). To overcome these difficulties, Jane and her colleagues introduced an enzyme reactor in which separated amylodextrins are converted to glucose by amyloglucosidase into the HPAEC-PAD system, and they named this as the HPAEC-ENZ-PAD chromatography (Wong & Jane 1997). The other advanced method is

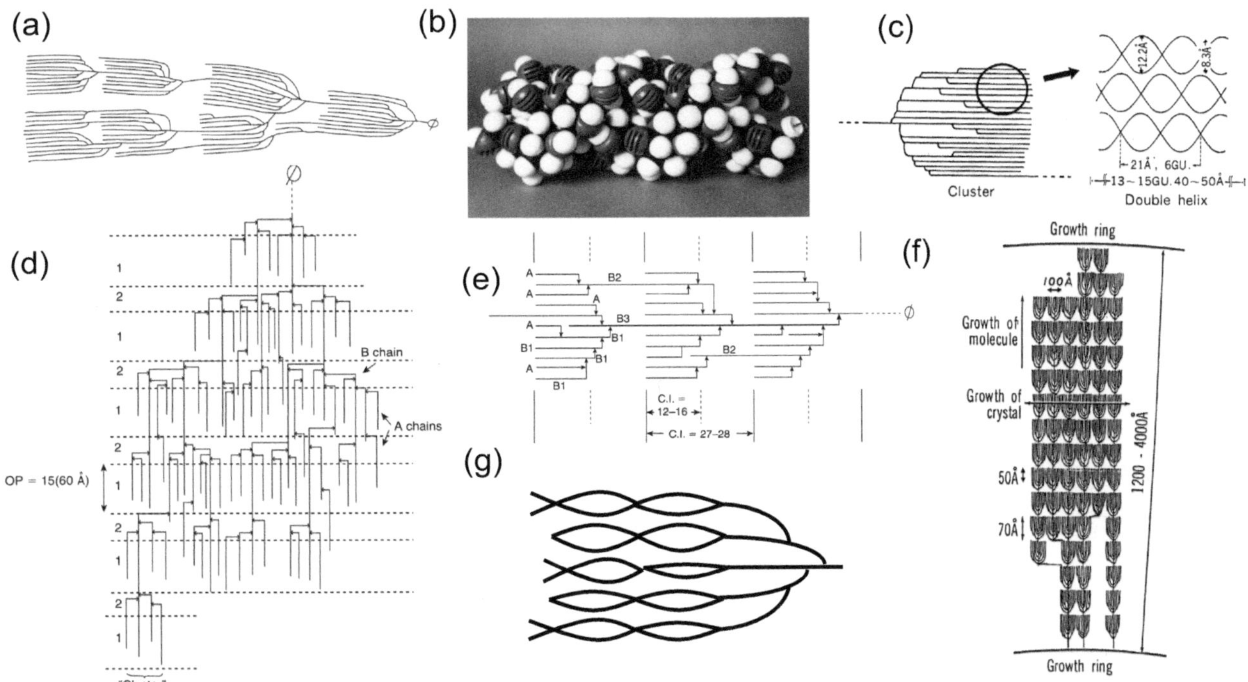

FIGURE 7.2   Cluster models for amylopectin structure. (a) The French model for amylopectin cluster (French 1972). (b) The Kainuma–French double helix model (Kainuma and French 1972). (c) Schematic representation of double helix unit of amylopectin (Kainuma 1977). (d) The Robin model (Robin et al. 1974). (e) The Hizukuri Model (Hizukuri 1986). (f) The Kainuma–French model (Kainuma 1980; French 1984). (a–f reproduced from Kainuma archives with permission of K. Kainuma.) (g) The Nakamura model showing the two types of branches (Nakamura 2002).

a fluorophore-assisted carbohydrate electrophoresis (FACE) method (O'Shea et al. 1998). Every debranched chain is labeled with a fluorophore 8-amino-1,3,6-pyrenesulfonic acid (APTS) at its reducing end, separated by capillary electrophoresis, and quantified by laser-induced fluorescent detector. The FACE method has a number of advantages that it can clearly separate linear dextrins of DP from 1 until approximately 100, quantifies each chain at molar basis, and needs only a small amount of sample. In addition, the results are satisfactorily reproducible.

Currently, it is actually impossible to determine the exact branch positions in B chains having multiple branch points. Although, in principle, the sequence analysis of maltooligosaccharides (MOS) with α-1,4 and α-1,6 linkages is possible by Q-TOF MS/MS method (Usui et al. 2009), it is practically difficult for the routine samples because they are composed of heterogeneous MOS. Thus, the average values for the lengths of external and internal segments were usually calculated (see reviews by Hizukuri 1996; Perez & Bertoft 2010). In these studies, amylopectin was often treated by hydrolyzing enzymes, and the resulting glucans, namely α-limit dextrins (α-LD), β-LD, phosphorylase-LD (ψ-LD), and ψ,β-LD, are used for analytical experiments (see reviews by Bertoft 2017; Hizukuri 1996).

Although it is difficult to determine the molecular size of native amylopectin molecules (Gidley et al. 2010), several attempts to separate multiple components of amylopectin have been successful (Suzuki, Hanashiro & Fujita 2023; Yoo & Jane 2002).

7.3 CLUSTER MODELS

Starch molecules consist of two types of chains (Peat, Whelan & Thomas 1952). The A-chains are linked only to other chains by their reducing ends at carbon 6 of a glucose unit of another chain. The B-chains carry at least one chain branched at their carbon 6. The C-chain is the only chain having a free reducing end in each amylopectin molecule, whereas it is a kind of B-chain. The cluster structure of amylopectin has been most widely accepted. The French model was proposed in 1972 (French 1972) (Figure 7.2a), based on several experimental results and observations. First, analysis of the products obtained after treatment of waxy maize starch with porcine pancreatic α-amylase showed that approximately 65%–70% of branches are singly branched, whereas the remaining about 30%–35% are localized in densely branched regions of the amylopectin (Kainuma & French 1970). The result clearly indicates the uneven localization of branches in amylopectin molecules. Second, Kainuma and French (1972) found that the crystalline starch is composed of the tightly packed amylopectin chains and proposed a left-handed double helix structure for B-type potato starch of Naegeli amylodextrin (ND) (Figure 7.2b and c), which was based on the following observations and ideas (Kainuma & French 1971, 1972):

1. ND was highly resistant to acids.
2. ND was mostly composed of linear chains with the similar chain length.

3. Once ND was dispersed in hot water, it was stained red-purple, which is a characteristic color of short amylose chains in iodine solution.
4. The left-handed double helical structure fits the space-filling models and the computer modeling.
5. The double helices can well explain the basic behavior of starch granules, which are insoluble and sedimented in water having a specific gravity of up to approximately 1.6.

Later, the dimension and spatial arrangement of double helices in A-type and B-type starches were confirmed (Imberty & Perez 1988; Imberty et al. 1991).

Up to now, several cluster models have been reported, although some representative ones are explained in this chapter. Robin et al. (1974) proposed a modified cluster model (Figure 7.2d). This model includes various modes of interconnection of clusters, as explained by Thompson (2000). Some long B chains penetrate into the clusters, and thus, clusters are arranged in a tandem fashion, whereas others stretch their chains to carry multiple clusters. In this way, the number of clusters rapidly increases toward the non-reducing direction by both series and parallel interconnections. Hizukuri (1986) proposed a new quantitative model, which is known as the "Hizukuri model" (Figure 7.2e), based on the detailed analysis of the chain-length profiles of debranched amylopectins as well as the density of branches and the average distance of branch-branch points by using a variety of methods. This model clearly shows the size of a single cluster and defines

B-chains, as B1 chains are in a single cluster, while B2, B3, and B4 chains extend into 2, 3, and 4 clusters.

Yamaguchi, Kainuma and French (1979) showed comprehensive scanning and transmission electron microscopic (SEM and TEM, respectively) observations by using starch samples from waxy maize, which were carefully designed and differently prepared (Figure 7.3). The image of waxy maize amylopectin dispersed in dimethylsulfoxide showed the extended clusters with rod shapes having approximately 10 nm in diameter and 12–400 nm in length. Then, they observed the fibrous structures in wet mashed waxy maize starch, which were interpreted as a consequence of the amylopectin molecules themselves (Figure 7.3b). It is noted that the fibers are oriented radially and perpendicular to the growth ring-like structures, consistent with the earlier observations by Sterling (1974). They also observed that dozens of approximately 5-nm-thick string-like structures were aligned in parallel with the growth rings and granule surface by using extensive acid hydrolysis of the starch (Figure 7.3a and c). The results strongly suggest that double helices from numerous clusters were densely packed side by side. From these observations, Kainuma and French (Kainuma 1980; French 1984) illustrated the amylopectin cluster arrangement model, in which the whole amylopectin molecule grows between growth rings and radially toward the granule surface (Figure 7.3d).

In recent years, biochemical studies of amylopectins having different fine structures indicated that there are two types of branches in amylopectin cluster. Jane and her colleagues (1997) found the difference in branch positions of amylopectin between A-type and B-type crystalline allomorphs

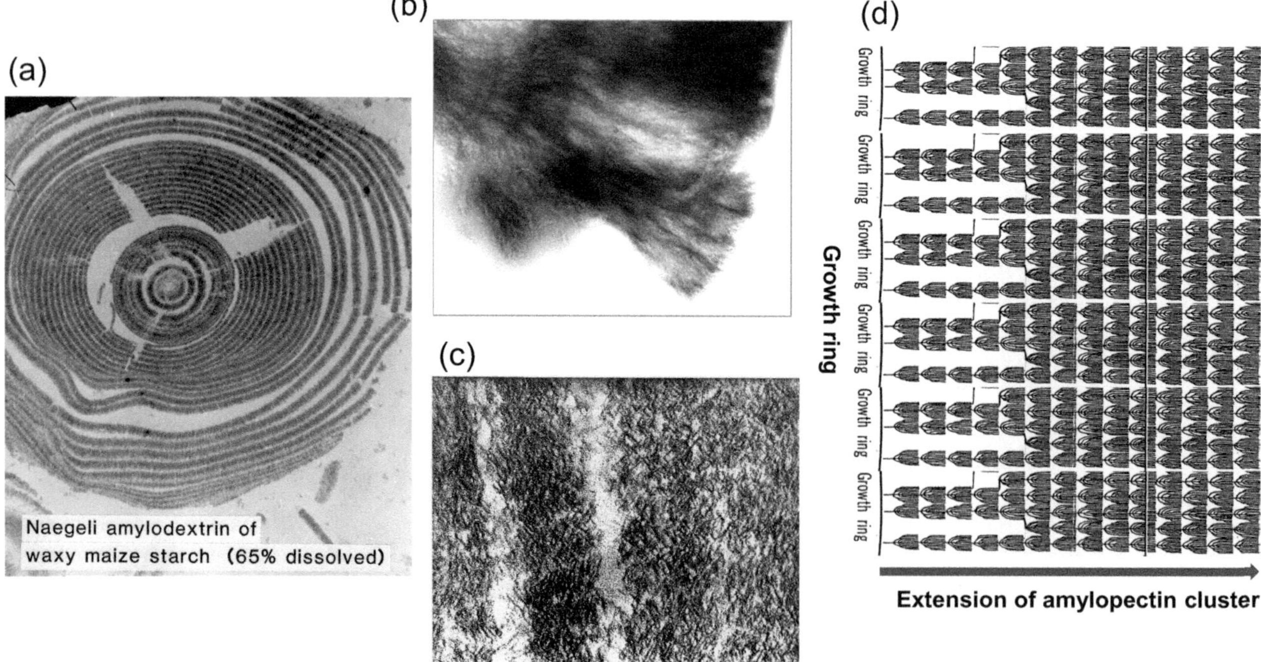

FIGURE 7.3   Amylopectin molecules in the starch granules. (See details by Yamaguchi et al. 1979.) (a) TEM image of central section of waxy maize starch Naegeli amylodextrin (ND) (65% dissolved). Note that growth rings are clearly seen. (b) TEM image of wet mashed native waxy maize starch granule. (c) TEM image of ND of waxy maize starch granule (enlarged Figure 7.3a). (d) Schematic representation of developing clusters in the starch granule in cereal endosperm. (Reproduced from Kainuma archives with permission of K. Kainuma.)

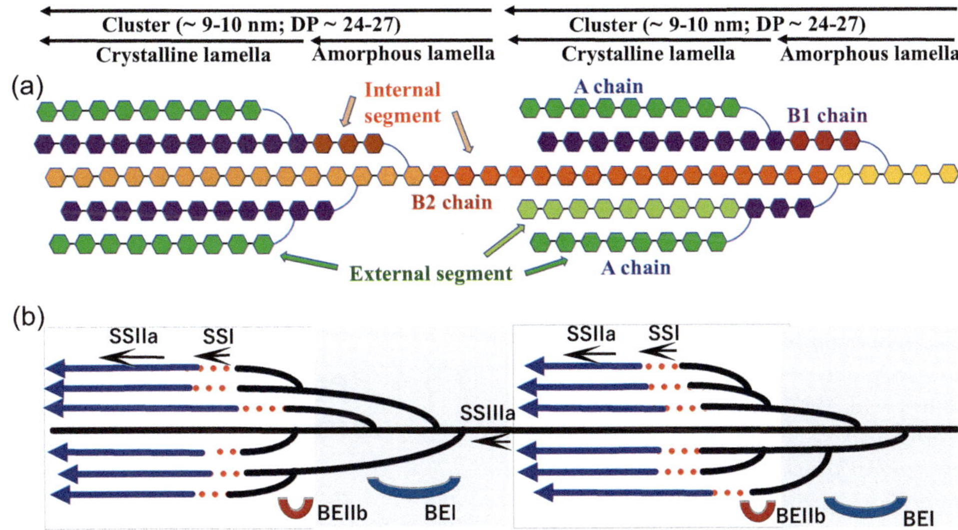

FIGURE 7.5   Schematic representation of the typical amylopectin cluster and roles of key BE and SS isoforms. (a) Nomenclature of cluster chains. (b) Roles of key isoforms. See text in detail.

chains, in particular B2 and B3 chains (Fujita et al. 2007). The *ss3a* (*dull*) mutants in cereals are known to appear dull in their endosperms (Fujita et al. 2007; Gao et al. 1998). The *ss3a* mutant amylopectin of rice has enriched chains of DP10–15 and depleted chains of DP30–60 (Fujita et al. 2007), indicating that SSIIIa plays an important role in the formation of B2 and B3 chains in addition to long B1 chains. Notably, loss of SSIIIa induces pleiotropic effects on activities of SSI and GBSSI (Fujita et al. 2007). The experimental observation that a rice endosperm *ss3a* amylopectin has a lower proportion of very short chains of DP6–10 compared with its wild-type one can be explained by an elevated SSI activity. The increase in the amylose content of the mutant starch is considered to be due to an elevated GBSSI activity.

7.5.2 Starch Branching Enzyme (BE)

Starch BE is the only enzyme which can form branches in amylopectin molecules. BE plays an essential role in determining the structural features of amylopectin. Plants have two types of BE isoforms, usually referred to as BEI and BEII. Cereals have further BEIIa and BEIIb isoforms. Cereal BEIIb plays a specific role in starch biosynthesis of the endosperm, because in its absence, the starch becomes highly resistant to thermal gelatinization and digestion by hydrolyzing enzymes, and the *be2b* mutant is often designated as *amylose-extender* (*ae*) because the mutant starches have the high amylose contents (Shannon, Garwood & Boyer 2009). It is noted that the *ae* mutation also causes a modified structure of amylopectin in rice endosperm with more long B chains and fewer short chains of DP of lower than approximately 13 (Nishi, Nakamura & Satoh 2001). The drastic change in amylopectin structure can be explained by a specific role of BEIIb, which is responsible for the synthesis of short chains by forming branches, mostly in the crystalline lamellae (Figures 7.5 and 7.6).

By contrast, the role of BEI is to synthesize longer chains by forming branches of amylopectin in the amorphous lamellae, although the role of BEI can be significantly complemented by other BE isoforms (Satoh et al. 2003) (Figure 7.5). The contribution of BEIIa is considered to support both BEIIb and BEI because the amylopectin structure and the starch content in the *be2a* mutant are not significantly affected (Nakamura 2002; Nakamura et al. 2022).

Direct evidence of the chain-length specificities of BE isoforms has been obtained from *in vitro* studies of the chain-length analysis of the products obtained from enzymatic reactions including purified enzyme and glucan substrates with known structural features (Guan et al. 1997; Takeda, Guan & Preiss 1993; Nakamura et al. 2010). For example, detailed analyses indicate that the rice BEIIb isoform is very sensitive to the length of the external chains of branched glucans because it almost exclusively forms short chains of DP7 and 6 by attacking the donor chains with external chain lengths of DP12–14, leaving residual chains of DP6–7 (Nakamura et al. 2010; Sawada et al. 2014; Nakamura 2023). Rice BEIIa can attack broader short external segments of DP approximately 12–17 to synthesize mainly short branch chains of DP6–8, and therefore, the chain length of the residual segments of DP is around 6–9 (Nakamura et al. 2010; Sawada et al. 2014; Nakamura 2023). On the other hand, rice BEI mainly reacts to the external segments of DP16–21 to produce intermediate chains of DP10–12, although this isoform can react to longer segments of DP up to about 24 (Nakamura et al. 2010; Sawada et al. 2014; Nakamura 2023). The capacities of branching reaction to the internal segments are markedly different among three BE isoforms, although it is most likely that these activities are limited compared with those to the external segments. Rice BEI can react to the internal segments to some extent (Nakamura et al. 2010). It is noted that, however, rice BEIIb is unable to attack the internal segments, whereas rice BEIIa seems to be only slightly reactive to the internal segments (Nakamura et al. 2010). All these reaction characteristics of BE isoforms should be consistent with their contributions to the fine structure of amylopectin in plant tissues when they are expressed.

7.5.3 Starch Debranching Enzyme (DBE)

The involvement of isoamylase 1 (ISA1), one of DBEs, in the synthesis of amylopectin was proved in maize endosperm for the first time (James, Robertson & Myers 1995). Then, the essential role of ISA or ISA1 has also been observed in endosperm of rice (Nakamura et al. 1996) and barley (Burton et al. 2002), the leaf of *Arabidopsis* (Zeeman et al. 1998), a green alga *Chlamydomonas reinhardtii* (Dauvillee et al. 2000), and *Cyanobacterium* strain CLg1 (Cenci et al. 2013) (see review by Hennen-Bierwagen, James & Myers 2012). The involvement of ISA1 in amylopectin biosynthesis was verified by molecular biology approaches. For example, introducing the wheat *ISA1* gene into a *sugary-1* mutant line EM914 of rice restored the capacity for the synthesis of starch (Kubo et al. 2005), while in rice endosperm, starch was replaced by phytoglycogen by the *ISA1* gene silencing (Utsumi et al. 2011).

It is likely that PUL, the other type of DBE, also plays a part in the synthesis of normal amylopectin at least to some extent. Fujita et al. (2009) found that the amount of phytoglycogen was increased by the double mutation between ISA1 and PUL genes in rice endosperm compared with the corresponding single mutants.

Of particular interest is that the *sugary* phenotypes are wholly plant species- and tissue-specific. For example, in some rice *sugary-1* mutant lines, starch is completely replaced by phytoglycogen in the endosperm (Nakamura et al. 1997), while all maize *isa1* mutants, both starch and phytoglycogen exist in the endosperm, although the relative amounts differ among mutant lines (Rahman et al. 1998). The functional form of ISA in rice endosperm is the ISA1 homomer (Utsumi et al. 2011). In contrast, in many other plant tissues/organs, the ISA1-ISA2 heteromer functions in amylopectin biosynthesis, e.g., in the maize endosperm (Kubo et al. 2010), potato tubers (Bustos et al. 2004), *Arabidopsis leaf* (Delatte et al. 2005), and *Chlamydomonas* (Dauvillee et al. 2001). It is interesting to note that both homomer and heteromer function in maize leaf (Lin et al. 2013).

What is the function of DBEs? Although no convincing evidence has been reported, the most accepted idea is that DBEs play an essential role in trimming the shape of amylopectin branch structure by removing improperly formed branches that are accidentally but occasionally synthesized at the position that interferes with the generation of double helices during the synthesis of clusters (Hennen-Bierwagen, James & Myers 2012; Nakamura 2002). In this connection, recently Nagamatsu et al. (2022) reported a notable observation in rice endosperm. The result suggests a possibility that ISA1 plays an essential role in removing improper branches formed by BEIIb to trim the cluster structure in developing rice endosperm. The idea seems to be consistent with observations that starch is accumulated in leaves and culms of rice where BEIIb is not expressed.

In summary, DBEs are responsible for the normal amylopectin structure in many plant species and tissues. It is known that ISA1 can remove branches having any chain lengths, whereas ISA3 can react to only short chains of DP2–4 (Kobayashi et al. 2016; Takashima et al. 2007), suggesting that ISA1 and ISA3 are involved in starch biosynthesis and degradation, respectively. It is rational to suppose that plants have differentiated ISA isoforms that can play a role in starch biosynthesis, whereas their ISA's role was originally designed to degrade starch molecules by debranching short chains.

7.5.4 Regulation of Amylopectin Biosynthesis

The major process of starch biosynthesis is to reproduce the cluster of amylopectin by increasing the number of clusters. The specific fine structure of amylopectin is constructed and regulated by coordinated actions and balance of activities among starch biosynthetic enzyme isoforms. However, when contributions of some isoforms to the amylopectin structure and starch synthesis are specific, changes in their activities greatly affect the starch-related phenotypes. For example, the downregulation of *BEIIb* gene expression in *japonica*-type rice greatly modifies the amylopectin chain-profile in the endosperm, with a great increase in the proportion of long and intermediate chains in contrast to a marked decrease in short chains (Butardo et al. 2011). They also found that the structural change in amylopectin profoundly alters the starch granule morphology and crystallinity (to B-type from A-type), as well as the distinct low digestibility of cooked grains. The same line of experimental results with rice plants was reported (see review by He & Wei 2020). The loss of BEIIb activity caused the similar starch phenotypes in the *ae* mutant starches in maize (Li et al. 2008). BEIIb activity levels can be changed in rice endosperm by introducing the wild-type *BEIIb* gene into the *be2b* mutant line from *japonica*-type rice (Tanaka et al. 2004). With the increase of BEIIb activity, the *ae* phenotypes are alleviated and finally similar to the wild-type one when the activity is close to the wild type. Notably, when BEIIb is overexpressed, the starch becomes water-soluble, and the glucan loses the cluster structure by excess branching. All these results are consistent with the conclusion that cereal BEIIb plays a specific role in amylopectin fine structure and synthesis by attacking a limited chain-length of DP almost 12–14 of the external segments and forming short chains of DP almost exclusively 7 and 6, and therefore, the proportion of short chains is high in the presence of BEIIb (Nakamura et al. 2010). Therefore, in the *ae* mutants, the average length of component chains of amylopectin becomes longer including the length of double helices (Figure 7.6).

How are many isoforms coordinated? According to the synthetic model (Figures 7.5–7.7), there are at least two interactions. The first interaction is between BEI and SSIIIa in the synthesis of the initial step for the new cluster, resulting in the formation of B2–4 chains. Recently, Nakamura (2023) proposed a new model for the enzymatic reactions for the cluster

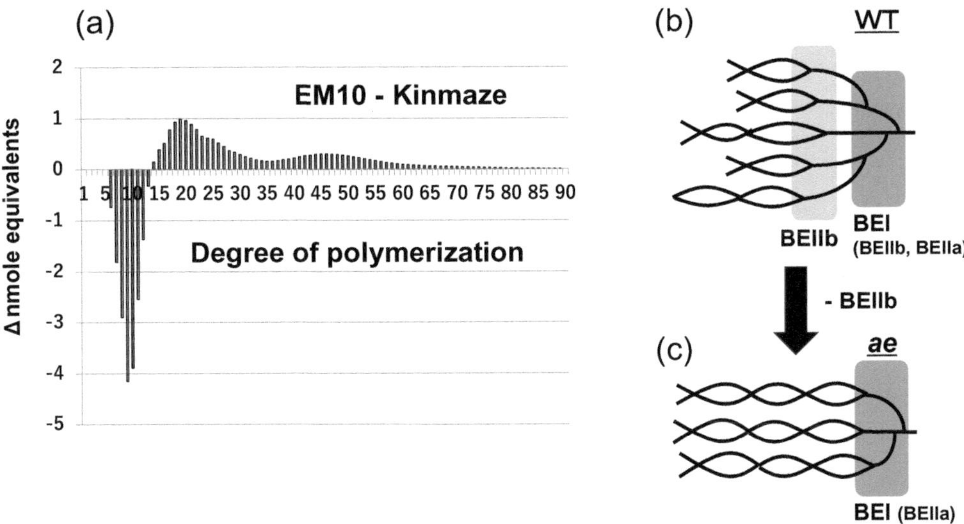

FIGURE 7.6 Specific roles of BEIIb in amylopectin biosynthesis in cereal endosperm. (a) Effect of *be2b* mutation on chain-length distribution (CLD) of amylopectin in rice endosperm. The figure was obtained by subtraction of CLD of wild-type cultivar Kinmaze from that of a *be2b* mutant line (EM10). (b) Schematic representation of the amylopectin cluster of a *japonica*-type rice cultivar Kinmaze. (c) Schematic representation of amylopectin cluster of a *be2b* (*ae*) mutant line EM10 generated from Kinmaze. See text in detail.

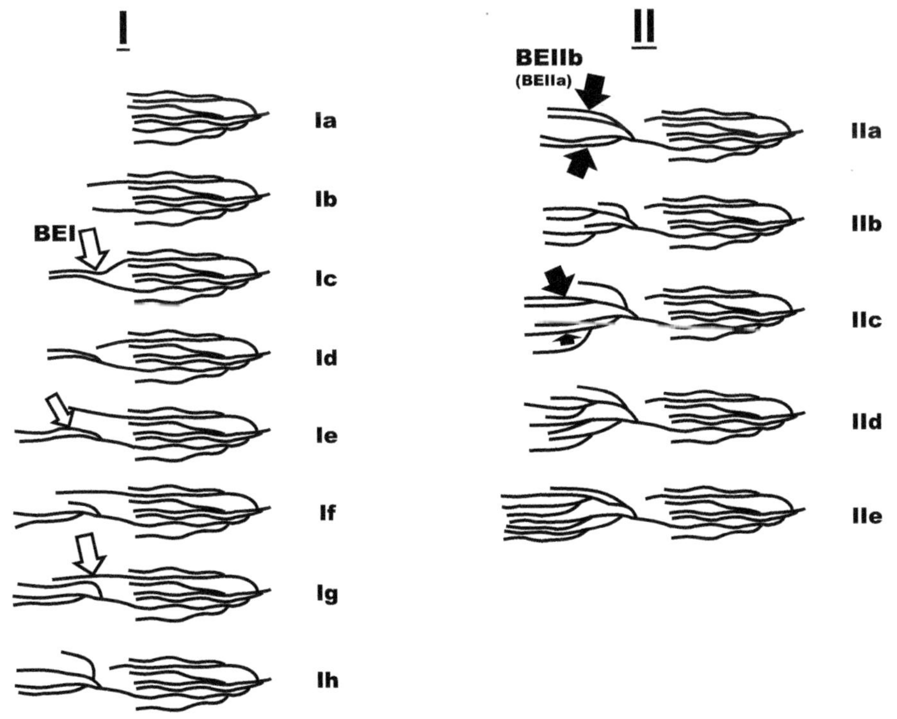

FIGURE 7.7 Initiation of the synthesis of a new cluster in the cluster reproduction process of amylopectin biosynthesis (Nakamura 2023). Open and closed arrows indicate the positions where BEI and BEIIb, respectively, play important roles in the formation of branched chains. See text in detail.

reproduction with emphasis on the major role of BEI in the initiation step of the cluster reproduction (Figure 7.7).

The second interaction is among BEIIb, SSI, SSIIa, and ISA1 to synthesize short chains to construct the crystalline lamellae in cereal endosperm amylopectin (Nakamura 2002, 2015a, 2018). On the other hand, there have been several observations suggesting the functional and/or physical coordination between isoforms. Recently, biochemical studies indicated that in developing cereal endosperms, all BE isoforms associate with other starch synthetic isoforms, forming various combinations of protein–protein complexes (Tetlow et al. 2004, 2008).

Since then, these observations have been accumulated in endosperm from various cereals (Crofts et al. 2017; Emes & Tetlow 2012). It is particularly interesting that a trimetric complex composed of BEIIb-SSI-SSIIa is found in developing endosperm of maize (Liu et al. 2009, 2012) and rice (Crofts et al. 2015, 2017). Further convincing evidence of the roles of these protein–protein complexes in starch biosynthesis remains to be accumulated.

7.5.5 Initiation Process of Amylopectin Biosynthesis

How is the first cluster glucan(s) formed from a simple compound such as maltose? Regrettably, however, only limited information on the molecular mechanism for this process has been available. The first evidence that plastidial glucan phosphorylase (Pho1) is involved in the initiation of glucan biosynthesis of developing endosperm was reported by Satoh et al. (2008). The rice *pho1* mutants exhibited several distinct phenotypes. The endosperm starch content or the seed weight greatly differed among seeds even in the same panicles; most of seeds had reduced starch contents (shriveled seeds), while some seeds contained no starch at all (empty seeds), and some others were comparable to the wild-type ones (plump seeds). The proportion of empty seeds or shriveled seeds was markedly higher when the plants were grown at lower temperatures (20°C) than at 30°C. These results suggest that Pho1 is involved in the initiation of amylopectin biosynthesis. Once the initiation step proceeds successfully, the seed can synthesize starch in the endosperm. It is also highly possible that the additional

factor(s) support(s) the function of Pho1 in the initiation step, but this support is inactive at low temperature.

Recently, it was found that rice Pho1 is capable of synthesizing linear glucans as well as MOS from maltose (Nakamura et al. 2019), whereas barley Pho1 can synthesize MOS from glucose-1-phosphate (G1P) without any sugars added (Cuesta-Seijo et al. 2017). Hwang et al. (2010) showed that Pho1 can elongate MOS even at a higher concentration of inorganic phosphate (Pi) than that of G1P. They also showed that the Pi activity is high even under low temperature (Hwang et al. 2016a). In addition, the synthesis of MOS and dextrins by Pho1 is accelerated by forming the protein complex between Pho1 and disproportionating enzyme (DPE1) (Hwang et al. 2016b). Moreover, Pho1 functionally interacts with BEs during the synthesis of branched maltodextrins (Nakamura et al. 2012), although it is likely that their physical association is not necessarily needed for the functional interaction in rice endosperm (Nakamura et al. 2017).

Then, in what way can Pho1 play an essential role in amylopectin initiation? Figure 7.8 illustrates one scenario. First, Pho1 synthesizes linear maltodextrins (LMDs) from maltose vis MOS. It is highly possible that the Pho1–DPE1 complex can efficiently synthesize long LMD. When LMD is fully elongated so that BE can attack, branched maltodextrins (BMDs) are produced. Then, the concerted actions among SS and BE isoforms might form the pre-amylopectin molecules with the cluster structure. The pre-amylopectin can be used as primer for the cluster reproduction reaction. Nakamura, Ono & Ozaki (2019) isolated fine grains having a diameter smaller than approximately 800 nm from rice endosperm at very early developmental stage. Interestingly, the amounts of glucans in these granules accounted for only about 0.11% of the total glucan amount in the

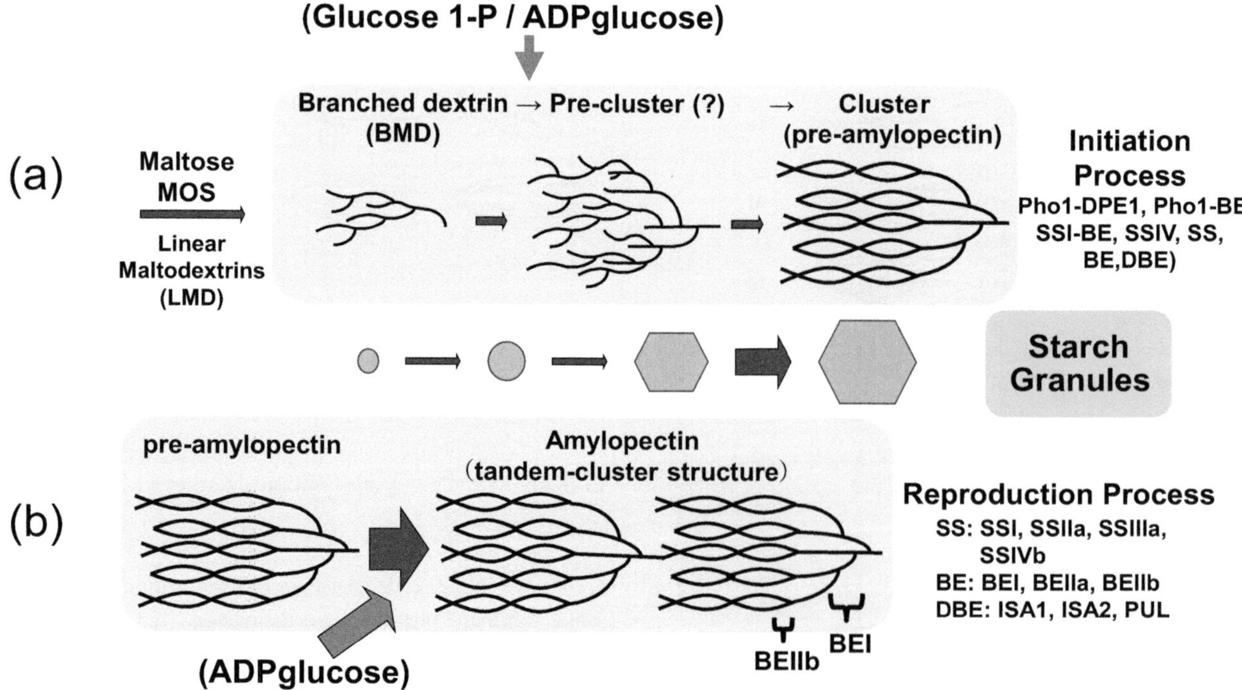

FIGURE 7.8 The reproduction stage and the initiation stage in amylopectin biosynthesis (Nakamura et al. 2019). (a & b) See text in detail.

endosperm. Recently, Nakamura & Steup (2025) isolated BMD and found that it is devoid of the cluster structure, having only a few long B chains, and that the amount of BMD is extremely lower than that of starch in the endosperm. These results indicate that the carbon flow from maltose and/or MOS to BMD is extremely lower than that from the precursor (pre-amylopectin/amylopectin) to mature amylopectin molecules.

Recently, Dong et al. (2023) observed that in rice *pho1* mutants, the levels of maltose and MOS markedly were higher than in the wild type, suggesting that Pho1 together with DPE1 plays an important role in the LMD. To establish the functional role of Pho1 in the starch initiation in developing cereal endosperm, more convincing evidence needs to be accumulated (Hwang et al. 2020).

Initiation of starch biosynthesis includes not only starch molecule initiation (Nakamura 2015b), but also starch granule initiation. Recently, several reports have shown that some specific plastid proteins such as PTST (PROTEIN TARGETING TO STARCH), MRC (MYOSIN-RESEMBLING CHLOROPLAST PROTEIN), and MFP (MAR BINDING FILAMENT-LIKE PROTEIN) play important roles in the starch granule initiation (Seung et al. 2018; Seung & Smith 2019).

How is the initiation of amylopectin biosynthesis and the reproduction of amylopectin molecules collaborated? There is a large body of literature in the presence of glucose, maltose, and MOS in plant tissues where starch is vigorously synthesized, while only a few attempts have examined the LMD and/or BMD, which are involved in the synthesis of pre-amylopectin and/or amylopectin. The pool size of BMD as well as LMD is kept extremely lower than that of amylopectin/pre-amylopectin having the cluster, as stated above. In other words, cereal endosperms efficiently synthesize amylopectin molecules using pre-amylopectin as the precursor, but the *de novo* synthesis of amylopectin synthesis from G1P, glucose, or maltose is very much limited, and thus only a low amount of BMD and/or LMD, a possible intermediate of the pathway for the *de novo* synthesis, is found during rapid starch biosynthesis.

7.6 AMYLOPECTIN IN THE SEMI-CRYSTALLINE STRUCTURE

Starch granules exhibit semi-crystalline nature, with hierarchical structural layers from the hilum to the surface of the granule. Amylopectin molecules are densely and radially arranged between growth ring by orientating their non-reducing ends toward the surface of the granule (Figure 7.3). Amylopectin molecules consist of alternating crystalline lamellae and amorphous lamellae with a constant repeat distance of 9–10 nm, usually called "9-nm repeat" (Jenkins et al. 1993) (Figure 7.5). The crystalline lamellae are mainly composed of double helices formed by amylopectin side chains, and numerous lamellae of neighboring amylopectin molecules are laterally aligned almost throughout the granule (Villwock & BeMiller 2022a, 2022b; Yamaguchi, Kainuma & French 1979) (Figure 7.3).

During amylopectin biosynthesis, each molecule develops toward the surface of the granule by increasing the number of clusters, in harmony both temporally and spatially with the development of neighboring molecules (Figure 7.3d). When are the double helices of amylopectin side chains formed, during its synthesis or after the synthesis? Based on the biochemical analysis of BE enzymatic reactions, it was proposed that the actual substrate of BE is a double helix, rather than a single glucan chain (Borovsky, Smith & Whelan 1975; Borovsky et al. 1979). On the contrary, recent X-ray crystalline analysis of BE proteins determined the location of both the donor chain binding site (Hayashi et al. 2017) and the acceptor chain binding site (Gavgani et al. 2022), and confirmed that there is no overlap between both sites on the protein surface and that these chain binding sites are separated by two distinct loops. These results show that a single glucan chain, but not a double helical strand, is bound to BE (Gavgani et al. 2022). It is rational to support the conclusion obtained from the crystallographic protein structural analysis. As described above (Section 7.5.2), BE is very sensitive to the chain length of external segment of glucan chain. In particular, BEIIb isoform can be reactive to the substrate chains having DP12–14 and form chains of DP7 and 6 (Figure 7.6a). If the double helical chains are the substrate, these chains must be separated after the BE reaction because the length of donor chain is shortened to be the DP6 or 7 chain. The activities of SSI and SSIIa isoforms are higher to shorter chains of DP smaller than 9 than longer chains of DP≥10 (Cuesta-Seijo et al. 2015). The result strongly suggests that SS efficiently attacks a single glucan chain. If BE can be reactive only to the double helix chains, but not a single chain, amylopectin side chains must frequently repeat the binding and dissociation among other chains for the formation of double helices by changing their counterparts. Is it possible that the double helix forms immediately after the external lengths of adjacent two chains reach DP≥10 and then provides BEIIb with it when the external segment(s) of component chains has (have) DP12–14? In agreement with an idea proposed by Myers et al. (2000), it is easily acceptable that chemical processes such as the formation of double helices and the crystallization of amylopectin molecules follow the enzymatic reactions. It is known that starch accounts for 70%–90% of total dry weight of the kernel, with a specific gravity of ≫1.0. At the mature dehydration stage, double helices of amylopectin are aligned in parallel to the growth rings (Figure 7.3). The synthesis and dense packing of amylopectin molecules must be controlled so that this can be the main driving force for the formation of liquid crystalline structures of starch granules in cereal endosperms (Waigh et al. 2000). O'Sullivan & Perez (1999) proposed that only limited combinations of internal chain-lengths match the parallel arrangement when the double helices chains are derived from the same chains. The alignment of double helix–double helix is not necessarily limited within the same clusters. In addition, the alignment is possible among the helices from adjacent different clusters. Therefore, it is unclear as to how and to what extent does this relationship affect the arrangement of double helices in starch granules.

7.7 FUTURE PERSPECTIVES

Despite various limitations in our thorough understanding of the fine structure of amylopectin, its structural features and the regulatory mechanism of amylopectin biosynthesis have been explained and discussed. At present, two representative models, the cluster model and the backbone model, have been proposed regarding the structural features of elements in amylopectin molecules, although critical points of both models are still in debate. In addition, the contribution of each starch biosynthetic enzyme isoform as well as functional interactions among them and the initiation of amylopectin biosynthesis from simple carbohydrates still remain to be elucidated. Although it has been reported that many isoforms exist by forming a variety of heteromeric protein–protein complexes in plant tissues, their physiological roles are not fully understood (see reviews by Crofts, Nakamura & Fujita 2017; Emes & Tetlow 2012). It is unclear if the formation of protein complex is required for the functional interactions between isoforms, and it is most likely that the extent or importance of functional interaction in starch biosynthesis is not proportional to the intensity of the physical association between them.

The significance of *in vitro* studies on the molecular mechanism of the physiological role of individual isoform in amylopectin biosynthesis will increase. The position of branches and the chain length of external segments and internal segments between the reducing sites and the first branches and between branches are reflected by characteristics of each BE and SS isoform in its branching reaction and chain elongation reaction, respectively. At present, there is a big gap in such information, and therefore, we must construct a model from available fragmentary results and observations. In that sense, it is very important for starch scientists to propose new models and discuss on each critical topic toward our better understanding of structures of starch molecules, starch biosynthesis, and its regulatory mechanism.

The accumulation of huge amounts of information on fine structure of starch molecules and internal structure of starch granules obtained from biochemical, genetic, and crystalline analysis approaches, as well as the rapid development of molecular biology methodology and spread of big data such as the whole genome sequences of many species during the last three decades, is of great help for our better understanding of several aspects of starch structure and properties. Thus, it is time for us to apply these results for food, medical, and the other industrial uses.

ACKNOWLEDGMENTS

The author thanks Professor Keiji Kainuma for critical reading of the manuscript. In particular, the author could learn the historical studies performed by Professor Dexter French and his colleagues from him. He also kindly provides the author with many figures and pictures used in Figures 7.2 and 7.3. The author also thanks Dr. Jay-lin Jane for critical reading of the manuscript with helpful comments.

REFERENCES

Bertoft, E 2013, On the building block and backbone concepts of amylopectin structure, *Cereal Chemistry, 90,* 294–311.

Bertoft, E 2017. Understanding starch structure: Recent progress, *Agronomy, 7,* 56.

Bertoft, E, Kallman, A, Koch, K, Andersson, R, & Âman, P 2011, The cluster structure of barley amylopectins of different genetic backgrounds, *International Journal of Biological Macromolecules, 49,* 441–453.

Bertoft, E, Koch, K, & Âman, P 2012, Building block organization of clusters in amylopectin from different structural types, *International Journal of Biological Macromolecules, 50,* 1212–1223.

Borovsky, D, Smith, EE, & Whelan, WJ 1975, Temperature dependence of the action of Q-enzyme and the nature of the substrate for Q-enzyme, *FEBS Letters, 54,* 201–205.

Borovsky, D, Smith, EE, Whelan, WJ, French, D, & Kikumoto, S 1979, The mechanism of Q-enzyme action and its influence on the structure of amylopectin, *Archives of Biochemistry & Biophysics, 198,* 627–631.

Burton, RA, Jenner, H, Carrangis, L, Fahy, B, Fincher, GB, Hylton, C, Laurie, DA, Parker, M, Waite, D, van Wegen, S, Verhoeven, T, & Denyer, K 2002, Starch granule initiation and growth are altered in barley mutants that lack isoamylase activity, *Plant Journal, 31,* 97–112.

Bustos, R, Fahy, B, Hylton, CM, Seale, R, Nebane, NM, Edwards, A, Martin, C, & Smith, AM 2004, Starch granule initiation is controlled by a heteromultimeric isoamylase in potato tubers. *Proceedings of National Academy of Science USA, 101,* 2215–2220.

Butardo, VM, Fitzgerald, MA, Bird, AR, Gidley, MJ, Flanagan, BM, Larroque, O, Resurreccion, AP, Laidlaw, HKC, Jobling, SA, Morell, MK, & Rahman, S 2011, Impact of down-regulation of starch branching enzyme IIb in rice by artificial microRNA- and hairpin RNA-mediated RNA silencing, *Journal of Experimental Botany, 62,* 4927–2941.

Cenci, U, Chabi, M, Ducatez, M, Tirtiaux, C, Nirmal-Raj, J, Utsumi, Y, Kobayashi, D, Sasaki, S, Suzuki, E, Nakamura, Y, Putaux, J, Roussel, X, Durand-Terrasson, A, Bhattacharya, D, Vercoutter-Edouart, AS, Maes, E, Arias, MC, Palcic, M, Sim, L, Ball, SG, & Colleoni, C 2013, Convergent evolution of polysaccharide debranching defines a common mechanism for starch accumulation in Cyanobacterium and plants, *Plant Cell, 25,* 3961–3975.

Commuri, P, & Keeling, PL 2001, Chain-length specificities of maize starch synthase I enzyme: Studies of glucan affinity and catalytic properties. *Plant Journal, 25,* 475–486.

Crini, G, French, AD, Kainuma, K, Jane, JL, & Szente, L 2021, Contributions of Dexter French (1918–1981) to cycloamylose/cyclodextrin and starch science, *Carbohydrate Polymers, 257,* 117620.

Crofts, N, Abe, N, Oitome, NF, Matsushima, R, Hayashi, M, Tetlow, IJ, Emes, MJ, Nakamura, Y, & Fujita, N 2015, Amylopectin biosynthetic enzymes from developing rice seed from enzymatically active protein complexes, *Journal of Experimental Botany, 66,* 4469–4482.

Crofts, N, Nakamura, Y, & Fujita, N 2017, Critical and speculative review of the roles of multi-protein complexes in starch biosynthesis in cereals, *Plant Science, 262,* 1–8.

Cuesta-Seijo, JA, Nielsen, MM, Ruzanski, C, Krucewicz, K, Beeren, SR, Rydhal, MG, Yoshimura, Y, Striebeck, A, Motawia, MS, Willats, WGT, & Palcic, MM 2015, *In vitro* biochemical characterization of all barley endosperm starch synthases. *Frontiers of Plant Science, 6,* 1265.

Cuesta-Seijo, JA, Ruzanski, C, Krucewicz, K, Meier, S, Hagglund, P, Svensson, B, & Palcic, MM 2017, Functional and structural characterization of plastidic starch phosphorylase during barley endosperm development. *PLoS One, 12,* e0175488.

Dauvillee, D, Colleoni, C, Mouille, G, Buleon, A, Gallant, DJ, Bouchet, B, Morell, MK, d'Hulst, C, Myers, AM, & Ball, SG 2001, Two loci control phytoglycogen production in the monocellular green alga *Chlamydomonas reinhardtii, Plant Physiology, 125,* 1710–1722.

Dauvillee, D, Mestre, VV, Colleoni, C, Slomianny, M, Mouille, G, Detrue, B, d'Hulst, C, Bliard, C, Nuzillard, J, & Ball, SG 2000, The debranching enzyme complex missing in glycogen accumulating mutants of *Chlamydomonas reinhardtii* displays an isoamylase-type specificity, *Plant Science, 157,* 145–156.

Delatte, T, Trevisan, M, Parker, ML, & Zeeman, SC 2005, *Arabidopsis* mutants Atisa1 and Atisa2 have identical phenotypes and lack the same multimeric isoamylase, which influences the branch point distribution of amylopectin during starch synthesis, *Plant Journal, 41,* 815–830.

Dong, X, Chen, L, Yang, H, Tian, L, Dong, F, Chai, Y, & Qu, L 2023, Pho1 cooperates with DPE1 to control short maltooligosaccharide mobilization during starch synthesis initiation in rice endosperm, *Theoretical and Applied Genetics, 136,* 47.

Emes, MJ, & Tetlow, IJ 2012, The role of heteromeric protein complexes in starch synthesis, In *Starch: Origins, Structure and Metabolism,* ed I Tetlow, SEB, London, pp. 255–278.

French, D 1972, Fine structure of starch and its relationship to the organization of starch granules, *Journal of Japanese Society of Starch Science (Denpun Kagaku, in Japanese), 19,* 8–25.

French, D 1984, Organization of starch granules, In *Starch: Chemistry and Technology*, 2nd Ed, ed RL Whistler, JN BeMiller, & E Paschall, Academic Press, New York, pp. 183–247.

Fujita, N, Toyosawa, Y, Utsumi, Y, Higuchi, T, Hanashiro, I, Ikegami, A, Akuzawa, S, Yoshida, M, Mori, A, Inomata, K, Itoh, R, Miyao, A, Hirochika, H, Satoh, H, & Nakamura, Y 2009, Characterization of pullulanase (PUL)-deficient mutants of rice (*Oryza sativa* L.) and the function of PUL on starch biosynthesis in the developing rice endosperm, *Journal of Experimental Botany, 60,* 1009–1023.

Fujita, N, Yoshida, M, Asakura, N, Ohdan, T, Miyao, A, Hirochika, H, & Nakamura, Y 2006, Function and characterization of starch synthase I using mutants in rice, *Plant Physiology, 150,* 1070–1084.

Fujita, N, Yoshida, M, Kondo, T, Saito, K, Utsumi, Y, Tokunaga, T, Nishi, A, Satoh, H, Park, J, Jane, J, Miyao, A, Hirochika, H, & Nakamura, Y 2007, Characterization of SSIIIa-deficient mutants of rice: The function of SSIIIa and pleiotropic effects by SSIIIa deficiency in the rice endosperm, *Plant Physiology, 144,* 2009–2023.

Gao, M, Wanat, J, Stinard, PS, & James, MG, 1998, Characterization of dull1, a maize gene encoding for a novel starch synthase, *Plant Cell, 10,* 399–412.

Gavgani, HN, Fawaz, R, Ehyaei, N, Walls, D, Pawlowsky, K, Fulgos, R, Park, S, Assar, Z, Ghanbarpour, A, & Geiger, JH 2022, A structural explanation for the mechanism and specificity of plant branching enzyme I and IIb. *Journal of Biological Chemistry, 298,* 101395.

Gidley, MJ, & Bulpin, PV 1987, Crystallisation of malto-oligosaccharides as models of the crystalline forms of starch: Minimum chain-length requirement for the formation of double helices, *Carbohydrate Research, 161,* 291–300.

Gidley, MJ, Hanashiro, I, Hani, NM, Hill, SE, Huber, A, Jane, J, Liu, Q, Morris, GA, Rolland-Sabate, A, Striegel, AM, & Gilbert, RG 2010, Reliable measurements of the size distributions of starch molecules in solution: Current dilemmas and recommendations. *Carbohydrate Polymers, 79,* 255–261.

Guan, HP, Li, P, Imparl-Radosevich, J, Preiss, J, & Keeling, P 1997, Comparing the properties of *Escherichia coli* branching enzyme and maize branching enzyme, *Archives of Biochemistry and Biophysics, 342,* 92–98.

Hanashiro, I, Higuchi, T, Aihara, S, Nakamura, Y, & Fujita, N 2011, Structures of starches from rice mutants deficient in the starch synthase isozyme SSI and SSIIIa, *Biomacromolecules, 12,* 1621–1628.

Hayashi, M, Suzuki, R, Colleoni, C, Ball, SG, Fujita, N, & Suzuki, E 2017, Bound substrate in the structure of cyanobacterial branching enzyme supports a new mechanistic model, *Journal of Biological Chemistry, 292,* 5465–5475.

He, W, & Wei, C 2020, A critical review on structural properties and formation mechanism of heterogeneous starch granules in cereal endosperm lacking starch branching enzyme, *Food Hydrocolloids, 100,* 105434.

Hennen-Bierwagen, TA, James, MG, & Myers, AM 2012, Involvement of debranching enzymes in starch biosynthesis, in *Starch: Origins, Structure and Metabolism,* ed I Tetlow, SEB, London, pp. 179–215.

Hizukuri, S 1986, Polymodal distribution of the chain lengths of amylopectins, and its significance. *Carbohydrate Research, 147,* 342–347.

Hizukuri, S 1996, Starch: Analytical aspects, in *Carbohydrates in Food*, ed A-C Eliasson, Marcel Dekker, Inc., New York, pp. 347–429.

Hizukuri, S, Takeda, Y, & Yasuda, M 1981, Multi-branched nature of amylose and the action of debranching enzymes, *Carbohydrate Research, 94,* 205–213.

Hwang, S, Koper, K, & Okita, TW 2020, Plastid phosphorylase as a multiple-role player in plant metabolism, *Plant Science, 290,* 110303.

Hwang, S, Koper, K, Satoh, H, & Okita, TW 2016b, Rice endosperm starch phosphorylase (Pho1) assembles with disproportionating enzyme (Dpe1) to form a protein complex that enhances synthesis of malto-oligosaccharides. *Journal of Biological Chemistry, 291,* 19994–20007.

Hwang, S, Nishi, A, Satoh, H, & Okita, TW 2010, Rice endosperm-specific plastidial α-glucan phosphorylase is important for synthesis of short-chain malto-oligosaccharides, *Archives of Biochemistry and Biophysics, 495,* 82–95.

Hwang, S, Singh, S, Cakir, B, Satoh, H, & Okita, TW 2016a, The plastidial starch phosphorylase from rice endosperm: Catalytic properties at low temperature. *Planta, 243,* 998–1009.

Imberty, A, Buleon, A, Tran, V, & Perez, S 1991, Recent advances in knowledge of starch structure. *Starch/Stärke, 43,* 375–384.

Imberty, A, Chanzy, H, Perez, S, Buleon, A, & Tran, V 1988, The double-helical nature of the crystalline part of A-starch, *Journal of Molecular Biology, 201,* 365–378.

Imberty, A, & Perez, S 1988, A revisit to the three-dimensional 3-D structure of B-type starch. *Biopolymers, 27,* 1205–1221.

James, MG, Robertson, DS, & Myers, AM 1995, Characterization of the maize gene *sugary1*, a determinant of starch composition in kernels, *Plant Cell, 7,* 417–429.

Jane, J, Wong, K, & McPherson, AE 1997, Branch-structure difference in starches of A- and B-type X-ray patterns revealed by their Naegeli dextrins, *Carbohydrate Research, 300,* 219–227.

Jane, J-L 2009, Structural features of starch granules II, in *Starch: Chemistry and Technology,* 3rd Ed, eds J BeMiller & RL Whistler, Academic Press, San Diego, CA, pp. 193–236.

Jenkins, PJ, Cameron, RE, & Donald, AM 1993, A universal feature in the starch granules from different botanical sources, *Starch/Stärke, 45,* 417–420.

Kainuma, K 1977, Fine structure of starch and maltohexaose forming amylase. *Journal of Japanese Society of Starch Science, 4,* 141–147.

Kainuma, K 1980, Fine structure of starch (in Japanese), *Journal of Cookery Society of Japan, 13,* 83–90.

Kainuma, K, & French, D 1970, Multiple branched oligosaccharides by the action of porcine pancreatic α-amylase on amylopectin, *Proceedings of Amylase Symposium, 5,* 35–37.

Kainuma, K, & French, D 1971, Naegeli amylodextrin and its relationship to starch granule structure. I. Preparation and properties of amylodextrins from various starch types, *Biopolymers, 10,* 1673–1680.

Kainuma, K, & French, D 1972, Naegeli amylodextrin and its relationship to starch granule structure. II. Role of water in crystallization of B-starch, *Biopolymers, 11,* 2241–2250.

Kobayashi, T, Sasaki, S, Utsumi, Y, Fujita, N, Umeda, K, Sawada, T, Kubo, A, Abe, J, Colleoni, C, Ball, S, & Nakamura, Y 2016, Comparison of chain-length preferences and glucan specificities of isoamylase-type α-glucan debranching enzymes from rice, cyanobacteria, and bacteria, *PLoS One, 11,* e0157020.

Koizumi, K, Fukuda, M, & Hizukuri, S 1991, Estimation of the distribution of chain length of amylopectins by high-performance liquid chromatography with pulsed amperometric detection, *Journal of Chromatography, 585,* 233–238.

Kubo, A, Colleoni, C, Dinges, JR, Lin, Q, Lappe, RR, Rivenbark, JG, Meyer, AJ, Ball, SG, James, MG, Hennen-Bierwagen, TA, & Myers, AM 2010, Functions of heteromeric and homomeric isoamylase-type starch debranching enzymes in developing maize endosperm, *Plant Physiology, 153,* 956–969.

Kubo, A, Rahman, S, Utsumi, Y, Li, Z, Mukai, Y, Yamamoto, M, Ugaki, M, Harada, K, Satoh, H, Konik-Rose, C, Morell, M, & Nakamura, Y 2005, Complementation of *sugary-1* phenotype in rice endosperm with the wheat *isoamylase1* gene supports a direct role for isoamylase-1 in amylopectin biosynthesis, *Plant Physiology, 137,* 43–56.

Li, L, Jiang, H, Campbell, M, Blanco, M, & Jane, J 2008, Characterization of maize amylose-extender (*ae*) mutant starches. Part I: Relationship between resistant starch contents and molecular structures, *Carbohydrate Polymers, 74,* 396–404.

Lin, Q, Facon, M, Putaux, JL, Dinges, JR, Wattebled, F, D'Hulst, C, Hennen-Bierwagen, TA, & Myers, AM 2013, Function of isoamylase-type starch debranching enzymes ISA1 and ISA2 in the *Zea mays* leaf, *New Phytologist, 200,* 1009–1021.

Liu, F, Makhmoudova, A, Lee, EA, Wait, R, Emes, MJ, & Tetlow, IJ 2009, *Amylose extender* mutant of maize conditions novel protein-protein interactions between starch biosynthetic enzymes in amyloplasts, *Journal of Experimental Botany, 60,* 4423–4440.

Liu, F, Romanova, N, Lee, EA, Ahmed, R, Evans, M, Gilbert, EP, Morell, MK, Emes, MJ, & Tetlow, IJ, 2012, Glucan affinity of starch synthase IIa determines binding of starch synthase I and starch branching enzyme IIb to starch granules, *Biochemical Journal, 448,* 373–387

Myers, A, Morell, MK, James, MG, & Ball, SG 2000, Recent progress toward understanding the biogenesis of the amylopectin crystal, *Plant Physiology, 122,* 989–998.

Nagamatsu, S, Wada, T, Matsushima, R, Fujita, N, Miura, S, Crofts, N, Hosaka, Y, Yamaguchi, O, & Kumamaru, T 2022, Mutation in BEIIb mitigates the negative effect of the mutation in ISA1 on grain filling and amylopectin formation in rice, *Plant Molecular Biology, 108,* 497–512.

Nakamura, Y 2002, Towards a better understanding of the metabolic system for amylopectin biosynthesis in plants: Rice endosperm as a model tissue, *Plant & Cell Physiology, 47,* 718–725.

Nakamura, Y 2015a, Biosynthesis of reserve starch, In *Starch: Metabolism and Structure,* ed Y Nakamura, Springer, Tokyo, pp. 161–209.

Nakamura, Y 2015b, Initiation process of starch biosynthesis, In *Starch: Metabolism and Structure,* ed Y Nakamura, Springer, Tokyo, pp. 315–332.

Nakamura, Y 2018, Rice starch biotechnology: Rice endosperm as a model of cereal endosperms. *Starch/Stärke, 70,* 1600376.

Nakamura, Y 2023, A model for the reproduction of amylopectin cluster by coordinated actions of starch branching isoforms, *Plant Molecular Biology, 112,* 199–212.

Nakamura, Y, Francisco, PB, Jr, Hosaka, Y, Sato, A, Sawada, T, Kubo, A, & Fujita, N 2005, Essential amino acids of starch synthase IIa differentiate amylopectin structure and starch quality between *japonica* and *indica* rice varieties, *Plant Molecular Biology, 58,* 213–227.

Nakamura, Y, & Kainuma, K 2022, On the cluster structure of amylopectin, *Plant Molecular Biology, 108,* 291–306.

Nakamura, Y, Kubo, A, Ono, M, Yashiro, K, Matsuba, G, Wang, Y, Matsubara, A, Mizutani, G, Matsuki, J, & Kainuma, K 2022, Changes in fine structure of amylopectin and internal structures of starch granules in developing endosperms and culms caused by starch branching enzyme mutations of rice endosperm, *Plant Molecular Biology, 108,* 481–496.

Nakamura, Y, Kubo, A, Shimamine, T, Matsuda, T, Harada, K, & Satoh, H 1997, Correlation between activities of starch debranching enzyme (R-enzyme or pullulanase) and α-glucan structure in endosperms of *sugary-1* mutants of rice. *Plant Journal, 12,* 143–153.

Nakamura, Y, Ono, M, & Ozaki, N 2019, Structural features of α-glucans in the very early developmental stage of rice endosperm, *Journal of Cereal Science, 89,* 102778.

Nakamura, Y, Ono, M, Sawada, T, Crofts, N, Fujita, N, & Steup, M 2017, Characterization of the functional interactions of plastidial starch phosphorylase and starch branching enzymes from rice endosperm during reserve starch biosynthesis, *Plant Science, 264,* 83–95.

Nakamura, Y, Ono, M, Utsumi, Y, & Steup, M 2012, Functional interaction between plastidial starch phosphorylase and starch branching enzymes from rice during the synthesis of branched maltodextrins. *Plant & Cell Physiology, 53,* 8969–878.

Nakamura, Y, & Steup, M 2025, Contents and structures of branched and linear maltodextrins, malto-oligosaccharides, and sugars in the early developmental stages of *phosphorylase1* mutant endosperm of rice, *Carbohydrate Polymers, 348,* 122923.

Nakamura, Y, Umemoto, T, Takahata, Y, Komae, K, Amano, E, & Satoh, H 1996, Changes in structure of starch and enzyme activities affected by *sugary* mutations. In developing rice endosperm. Possible role of starch debranching enzyme (R-enzyme) in amylopectin biosynthesis, *Physiologia Plantarum, 97,* 491–498.

Nakamura, Y, Utsumi, Y, Sawada, T, Aihara, S, Utsumi, C, Yoshida, M, & Kitamura, S 2010, Characterization of the reactions of starch branching enzyme from rice endosperm, *Plant & Cell Physiology, 51,* 776–794.

Nikuni, Z 1969, Starch and cooking. *Science for Cookery (Chori Kagaku, in Japanese), 2,* 6.

Nishi, A, Nakamura, Y, & Satoh, H 2001, Biochemical and genetic analysis of *amylose-extender* mutations of rice endosperm, *Plant Physiology, 127,* 459–472.

Ohdan, T, Francisco, PB, Jr, Sawada, T, Hirose, T, Terao, T, Satoh, H, & Nakamura, Y 2005, Expression profiling of genes involved in starch synthesis in sink and source organs of rice, *Journal of Experimental Botany, 56,* 3229–3244.

O'Shea, MG, Samuel, MS, Konik, CM, & Morell, MK 1998, Fluorophore-assisted carbohydrate electrophoresis (FACE) of oligosaccharides: Efficiency of labelling and high-resolution separation, *Carbohydrate Research, 307,* 1–12.

O'Sullivan, AC, & Perez, S 1999, The relationship between internal chain length of amylopectin and crystallinity in starch, *Biopolymers, 50,* 381–390.

Peat, S, Whelan, WJ, & Thomas, GJ 1952, Evidence of multiple branching in waxy maize starch, *Journal of Chemical Society,* 4546–4548.

Perez, S, Baldwin, PM, & Gallant, DJ 2009, Structural features of starch granules I, In *Starch: Chemistry and Technology,* 3rd Ed, ed J BeMiller, & R Whistler, Academic Press, New York, pp. 149–192.

Perez, S, & Bertoft, E 2010, The molecular structures of starch components and their contribution of to the architecture of starch granules: A comprehensive review, *Starch/Stärke, 62,* 389–420.

Preiss, J 2009, Biochemistry and molecular biology of starch biosynthesis, In *Starch: Chemistry and Technology,* 3rd Ed, ed J BeMiller, & R Whistler, Academic Press, New York, pp. 83–148.

Rahman, A, Wong, K, Jane, J, Myers, AM, & James, MG 1998, Characterization of SU1 isoamylase, a determinant of storage starch structure in maize, *Plant Physiology, 117,* 425–435.

Robin, JP, Mercier, C, Charbonnière, R, & Guillbot, A 1974, Lintnerized starches. Gel filtration and enzymatic studies of insoluble residues from prolonged acid treatment of potato starch, *Cereal Chemistry, 51,* 389–405.

Satoh, H, Nishi, A, Yamashita, K, Takemoto, Y, Tanaka, Y, Hosaka, Y, Sakurai, A, Fujita, N, & Nakamura, Y 2003, Starch-branching enzyme I-deficient mutation specifically affects the structure and properties of starch in rice endosperm, *Plant Physiology, 133,* 1111–1121.

Satoh, H, Shibahara, K, Tokunaga, T, Nishi, A, Tasaki, M, Okita, TW, Kaneko, N, Fujita, N, Yoshida, M, Hosaka, Y, Sato, A, Utsumi, Y, Ohdan, T, & Nakamura, Y 2008, Mutation of the plastidial α-glucan phosphorylase gene in rice dramatically affects the synthesis and structure of starch in the endosperm, *Plant Cell, 20,* 1833–1849.

Sawada, T, Nakamura, Y, Ohdan, T, Saitoh, A, Francisco, PB, Jr, Suzuki, E, Fujita, N, Shimonaga, T, Fujiwara, S, Tsuzuki, M, Colleoni, C, & Ball, S 2014, Diversity of reaction characteristics of glucan branching enzymes and the fine structure of a-glucan from various sources, *Archives of Biochemistry and Biophysics, 562,* 9–21.

Seung, D, Schreier, TB, Burgy, L, Eicke, S, & Zeeman, SC 2018, Two plastidial coiled-coil proteins are essential for normal starch granule initiation in Arabidopsis, *Plant Cell, 30,* 1523–1542.

Seung, D, & Smith, AM 2019, Starch granule initiation and morphogenesis – Progress in Arabidopsis and cereals, *Journal of Experimental Botany, 70,* 771–784.

Shannon, JC, Garwood, DL, & Boyer, CD 2009, Genetics and physiology of starch development, in *Starch: Chemistry and Technology,* 3rd Ed, ed J BeMiller, & R Whistler, Academic Press, New York, pp. 23–82.

Sterling, C 1974, Fibrillar structure of starch. Evidence for crossed fibrils from incipient gelatinization, *Starch/Stärke 26,* 105–144.

Suzuki, E, & Suzuki, R 2016, Distribution of glucan-branching enzymes among prokaryotes, *Cellular and Molecular Life Sciences, 73,* 2643–2660.

Suzuki, N, Hanashiro, I, & Fujita, N 2023, Molecular weight distribution of whole starch in rice endosperm by gel-permeation chromatography, *Journal of Applied Glycoscience, 70,* 23–32.

Takashima, Y, Senoura, T, Yoshizaki, T, Hamada, S, Ito, H, & Matsui, H 2007, Differential chain-length specificities of two isoamylase-type starch debranching enzymes from developing seeds of kidney bean, *Biosciences, Biotechnology & Biochemistry, 71,* 2308–2312.

Takeda, Y, Guan, H, & Preiss, J 1993, Branching of amylose by the branching isoenzymes of maize endosperm, *Carbohydrate Research, 240,* 253–263.

Tanaka, N, Fujita, N, Nishi, A, Satoh, H, Hosaka, Y, Ugaki, M, Kawasaki, S, & Nakamura, Y 2004, The structure of starch can be manipulated by changing expression levels of starch branching enzyme IIb in rice endosperm, *Plant Biotechnology Journal, 2,* 507–516.

Tetlow, I, & Emes, MJ 2017, Starch biosynthesis in the developing endosperms of grasses and cereals, *Agronomy, 7,* 81.

Tetlow, I, Wait, R, Lu, Z, Akkasaeng, R, Bowsher, CG, Esposito, S, Kosar-Hashemi, B, Morell, MK, & Emes, MJ 2004, Protein phosphorylation in amyloplasts regulates starch branching enzyme activity and protein-protein interactions, *Plant Cell, 16,* 694–708.

Tetlow, IJ, Beisel, KG, Cameron, S, Makhmoudove, A, Liu, F, Bresolin, NS, Wait, R, Morell, MK, & Emes, MJ 2008, Analysis of protein complexes in wheat amyloplasts reveals functional interactions among starch biosynthetic enzymes, *Plant Physiology, 146,* 1878–1891.

Tetlow, IJ, & Bertoft, E 2020, A review of starch biosynthesis in relation to the building block-backbone model, *International Journal of Molecular Sciences, 21,* 1–37.

Thompson, DB 2000, On the non-random nature of amylopectin branching, *Carbohydrate Polymers, 43,* 223–239.

Umemoto, T, Yano, M, Satoh, H, Shomura, A, & Nakamura, Y 2002, Mapping of a gene responsible for the difference in amylopectin structure between *japonica*-type and *indica*-type rice varieties, *Theoretical and Applied Genetics, 104,* 1–8.

Usui, T, Ogata, M, Murata, T, Ichikawa, K, Sakano, Y, & Nakamura, Y 2009, Sequential analysis of α-glucanooligosaccharides with α–(1–4) and α–(1–6) linkages by negative ion Q-TOF MS/MS spectrometry, *Journal of Carbohydrate Chemistry, 28,* 421–430.

Utsumi, Y, Utsumi, C, Sawada, T, Fujita, N, & Nakamura, Y 2011, Functional diversity of isoamylase oligomers: The ISA1 homo-oligomer is essential for amylopectin biosynthesis in rice endosperm, *Plant Physiology, 156,* 61–77.

Villwock, K, & BeMiller, JN 2022a, The architecture, nature, and mystery of starch granules. Part 1: A concise history of early investigations and certain granule parts, *Starch/Stärke, 74,* 2100183.

Villwock, K, & BeMiller, JN 2022b, The architecture, nature, and mystery of starch granules. Part 2, *Starch/Stärke, 74,* 2100184.

Waigh, TA, Gidley, MJ, Komanshek, BU, & Donald, AM 2000, The phase transformations in starch during gelatinization: A liquid crystalline approach, *Carbohydrate Research, 328,* 165–176.

Wong, K, & Jane, J 1997, Quantitative analysis of debranched amylopectin by HPAEC-PAD with a postcolumn enzyme reactor, *Journal of Chromatography, 297–310.*

Yamaguchi, M, Kainuma, K, & French, D 1979, Electron micrographic observations of waxy maize starch, *Journal of Ultrastructure, 69,* 249–261.

Yoo, S, & Jane, J 2002, Molecular weights and gyration radii of amylopectins determined by high-performance size-exclusion chromatography equipped with multi-angle lase-light scattering and refractive index detectors, *Carbohydrate Polymers, 49,* 307–314.

Zeeman, S, Umemoto, T, Lue, WL, Au-Yeung, P, Martin, C, & Smith, AM 1998, A mutant of Arabidopsis lacking a chloroplastic isoamylase accumulates both starch and phytoglycogen, *Plant Cell, 10,* 1699–1712.

Phytoglycogen

8

Carley Miki and John R. Dutcher

8.1 MOLECULAR ARCHITECTURE

In plants, one of the most common forms of glucose-based polysaccharide is starch, which consists of highly branched chains of amylopectin with a small amount of linear chains of amylose. In certain varieties of plants, e.g., the su1 maize mutant of sweet corn, a different glucose-based polysaccharide called phytoglycogen (PG) is formed (Morris & Morris 1939). Both amylopectin and PG consist of α-1,4-linked linear chains with α-1,6 branching. However, the average length of the linear chains between branch points in PG is 10–12 glucose units, which is about half that in amylopectin, with a degree of branching that is correspondingly larger by almost a factor of 2 (Inouchi, Glover & Fuwa 1987; Yun & Matheson 1993). The increased branching in PG, which results from a deficiency in an endo-acting debranching enzyme isoamylase (Ball & Morell 2003), has two main consequences. First, the shorter length of the linear chain segments in PG prevents crystallization of the chains. Second, the larger, regular degree of branching produces a dendrimer or tree-like architecture for the PG particles (Ball & Morell 2003; Yun & Matheson 1993), which results in compact, highly branched nanoparticles.

The architecture of dendrimers is characterized by several parameters: the number n of glucose units in the linear chains between branch points, the functionality f or number of chains emanating from each branch point, and the total number of repeats or generations G of the branching structure within the molecule (Rubinstein & Colby 2003). For PG, $n \sim 10-12$ glucose units; each branch point produces two new linear chains (trifunctional, $f = 3$); and each particle consists of $G = 11$ generations, as determined using small angle neutron scattering (SANS) from the measured average number of glucose units, $N = 66{,}600$, in each particle (Simmons et al. 2020). The standard schematic diagram of a dendrimer has rigid chains that extend away from the center of the particle (Figure 8.1a), producing an increase in density with radial distance. However, this simple picture does not apply to PG since the flexible glucose chains fold back into the interior of the particle, producing a large uniform density of glucose chains within the particle (Figure 8.1b), as demonstrated in computer simulations (Ballauff & Likos 2004; Lescanec & Muthukumar 1990;

Mansfield & Klushin 1993), self-consistent field theory calculations (Boris & Rubinstein 1996; Morling et al. 2024; Zook & Pickett 2003), pyrene excimer fluorescence experiments (Kim & Duhamel 2023), and SANS experiments (Nickels et al. 2016; Rosenfeldt et al. 2002; Simmons et al. 2020) of large generation dendrimer polymers. One of the implications of the highly branched, dense core of PG nanoparticles is that many bioactive molecules cannot be incorporated inside the particles. As a result, bioactive delivery applications of PG focus on the association of bioactive molecules with the outer surface of native and modified PG particles.

8.2 PHYSICAL PROPERTIES

In this section, we focus on recent studies of the physical properties of high-purity PG particles extracted and purified using the physical techniques of centrifugation and ultrafiltration (Baylis et al. 2021).

8.2.1 Structure, Hydration, and Morphology

The compact, highly branched structure of PG particles was first observed in high-resolution electron microscopy images of dried PG particles that were negatively stained with uranyl acetate and supported on carbon films (Figure 8.2a) (Putaux et al. 1999). These measurements showed the underlying "cauliflower-like" substructure of the particles related to their dendritic architecture. Recently, atomic force microscopy (AFM) force spectroscopy has been used to obtain high-resolution images of individual hydrated PG particles. For these measurements, the PG particles were covalently bound to an underlying flat gold substrate, mediated by a self-assembled monolayer of 4-mercaptophenylboronic acid (4-MPBA), and imaged in water (Baylis et al. 2021). In the AFM images, each pixel corresponds to a force–distance curve in which the AFM tip is pressed into and then pulled away from the PG particle, allowing the simultaneous measurement of the particle morphology and mechanical stiffness.

DOI: 10.1201/9781003464396-8

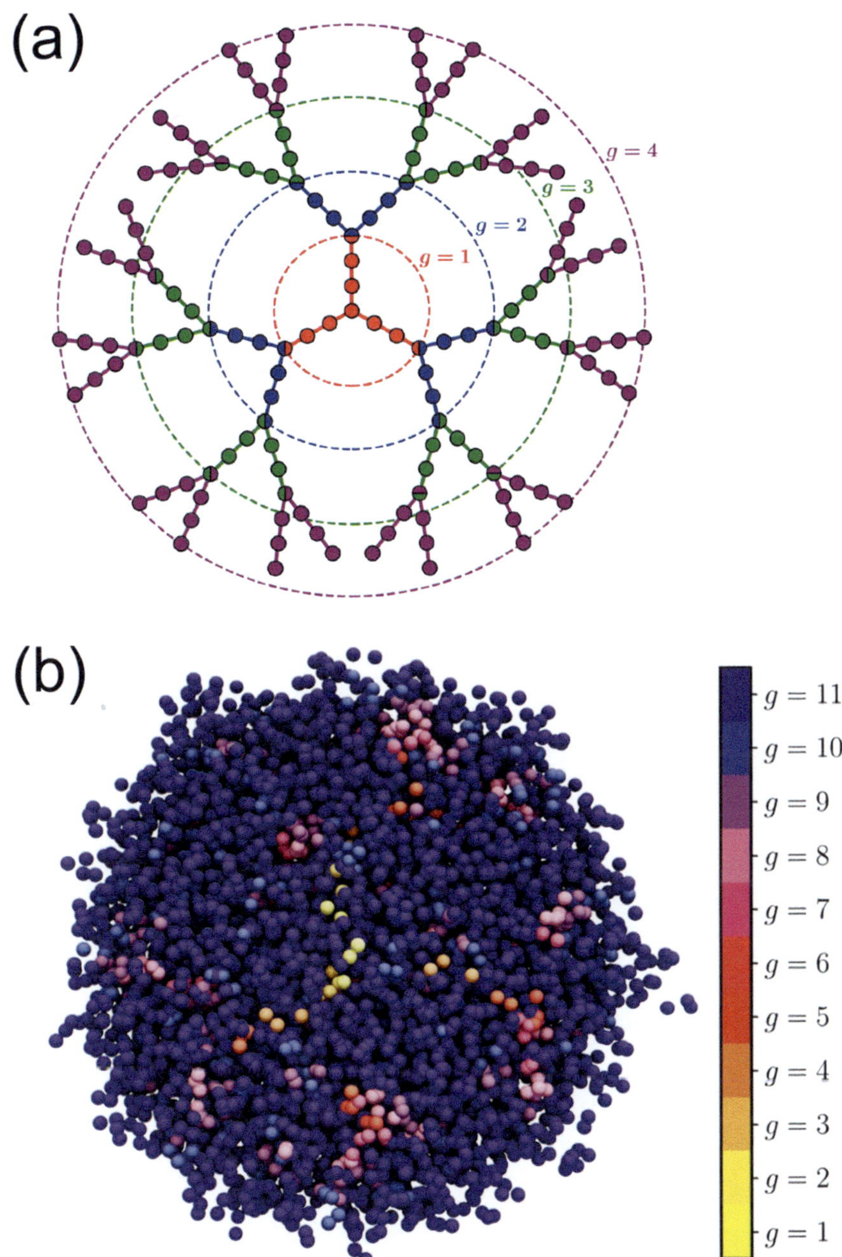

FIGURE 8.1   (a) Schematic diagram of the architecture of a trifunctional, $G = 4$, $n = 3$ dendrimer. Monomers in each generation g are shown in different shades of grey. (b) Dynamic self-consistent field theory image of a cross-section of an 11-generation dendrimer. Different generations are represented with different shades of grey, according to the greyscale code on the right-hand side. (The images were reprinted with permission from Morling et al. 2024. Copyright 2024 American Chemical Society.)

These measurements showed that PG particles are soft and deformable. For the AFM tip just contacting the PG particles, the particles were approximately circular, "fuzzy" objects (Figure 8.2b). For modest applied forces (several nN), the PG particles were highly deformed, revealing a distinctive branched structure within the particles (Figure 8.2c). The AFM force spectroscopy images also allowed the determination of the equivalent spherical particle radius $r_{AFM} = (3V/4\pi)^{1/3}$, where V is the measured volume of the PG particle on the substrate. In addition, the force–distance curves within each image were used to generate high-resolution maps of Young's modulus

E within individual PG particles, as discussed below. The effect of PG particle hydration was also measured by comparing the results for fully hydrated particles with those for dried particles: the dried particles were much smaller (r_{AFM} reduced by a factor of 1.7) and much less deformable (E increased by a factor of 5,200) (Baylis et al. 2021).

SANS is a powerful technique to study the structure and morphology of biological molecules such as PG. It allows the determination of the radial density profile of spherical molecules such as PG, as well as their molar mass and degree of hydration. In the SANS experiment, the scattering intensity I is

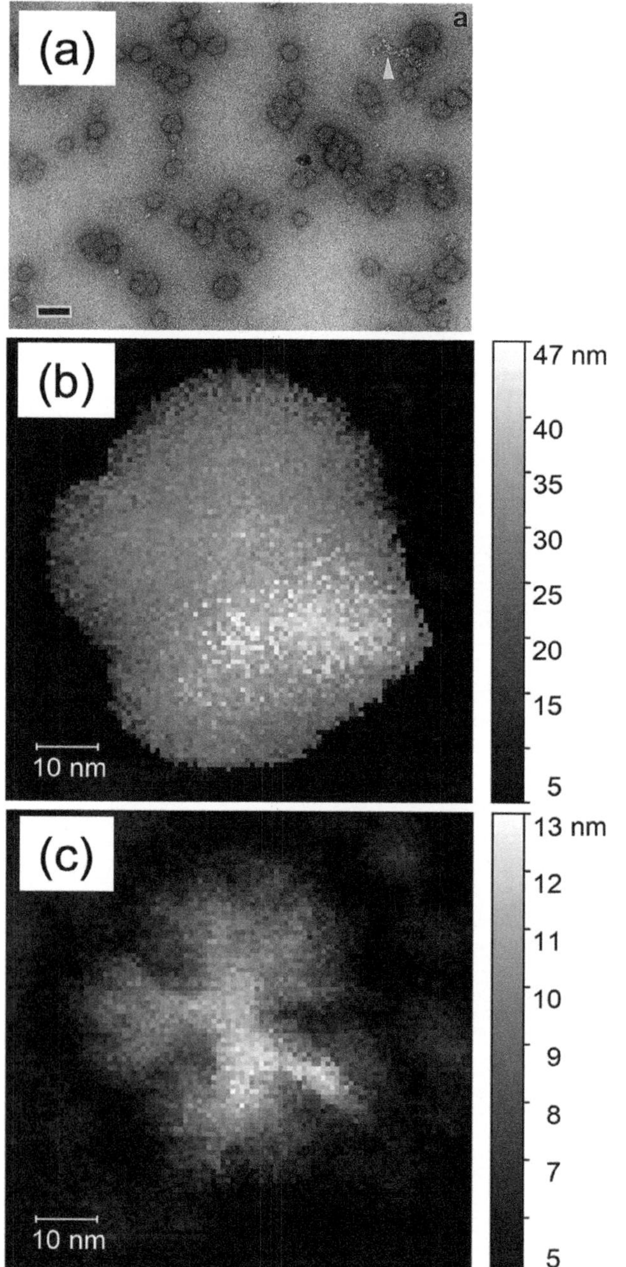

FIGURE 8.2 (a) Transmission electron microscopy image of PG particles negatively stained with 2% uranyl acetate on a carbon film. The scale bar corresponds to 100 nm. (Reprinted from Putaux et al. 1999, with permission from Elsevier.) (b) Atomic force microscopy (AFM) force spectroscopy height image of a PG particle that corresponds to the point of contact between the AFM tip and the particle when measured in water. (c) AFM force spectroscopy height image of the same PG particle as in (b) measured in water for an applied force of 2 nN. (Images in (b) and (c) reprinted with permission from Baylis et al. 2021. Copyright 2021 American Chemical Society.)

measured as a function of the scattering wavevector q, and the resulting scattering curve is fit to a model of the radial density profile. The simplest radial density profile for a spherical particle has uniform density and, for PG, this provides a reasonable fit to the measured SANS curve (Simmons et al. 2020).

However, there was a small amount of additional scattering intensity at large q values corresponding to a feature with a small length scale that was consistent with short glucose chains on the surface of the PG particles. To determine the morphology of the glucose chains, PG particles were modified with a hydrophobic compound, octenyl succinic anhydride (OSA). At large degrees of substitution (DS), an additional peak was observed in the SANS curve at large q, indicating an additional small length scale feature that was consistent with the collapse of the hydrophobically modified chains on the surface of the PG particles to form "seeds". The best fit of the scattering curve to the raspberry model (Larson-Smith, Jackson & Pozzo 2010) was obtained for an average length of the outer chains of 20 glucose units (Simmons et al. 2020). These measurements yielded a core radius for the PG particle of $r_{SANS} = 21$ nm, with an additional radial extent for the outer chains of 1 nm.

By varying the $D_2O:H_2O$ ratio in the surrounding solvent, it was possible to match the neutron scattering length density (NSLD) of the solvent to that of the hydrated PG particles, suppressing the scattering from the particles. This contrast match condition, which accounted for the exchange of the labile hydrogens on each glucose monomer, occurred for 51.2% D_2O (Simmons et al. 2020). Dividing the sum of the atomic scattering lengths for the atoms in a glucose monomer by the NSLD value at the contrast match condition yielded the glucose monomer volume. This value, together with the measured particle radius, allowed the determination of the average number of glucose units in the PG particles to be 66,600, which is consistent with an 11-generation dendrimer. A comparison of the NSLD at the contrast match condition with that measured in pure D_2O allowed the determination of the average number of water molecules (20 ± 2) associated with each glucose unit in the PG particles, which corresponds to each PG particle sorbing 220% of its own weight in water.

8.2.2 Water Ordering and Dynamics

The SANS measurements allow the determination of the total amount of water sorbed by each PG particle, which is confined to the very small distances between the tightly packed glucose chains within the particles. To learn about the nature of the ordering of this nano-confined water, attenuated total reflection-infrared (ATR-IR) spectroscopy was used to measure the OH-stretch band of water (3,000–3,700 cm^{-1}) in ultrathin films of PG on a ZnSe crystal as the relative humidity (RH) of the gas surrounding the films was changed from 4% to 90% (Grossutti & Dutcher 2016). By subtracting the IR spectrum collected at the lowest RH from those collected at higher values of RH, the contributions due to bonds within the polysaccharide were eliminated, leaving just those due to the water within the PG particles. The IR spectra were dominated by a large peak at the lowest OH-stretch wavenumber ~3,225 cm^{-1}, indicating that the water inside PG particles is highly ordered relative to that of bulk water and other polysaccharides (Grossutti & Dutcher 2016). The ordering of the water was quantified using a parameter $R_{network} = A_{3,225}/A_{3,400}$, where A_i refers to the IR absorbance measured at a wavenumber of $i\,\mathrm{cm}^{-1}$ $\left(\pm\,20\,\mathrm{cm}^{-1}\right)$. For native PG particles, the value of $R_{\mathrm{network}} \sim 2.1$ for RH > 70%, which is considerably larger than that for bulk water ($R_{\mathrm{network}} \sim 0.95$).

The high degree of ordering of water inside PG particles also leads to a significant slowing of its dynamics. In quasi-elastic neutron scattering (QENS) measurements, in which the diffusion of water can be measured on time scales of tens of ps to ~1 ns, it was shown that motion of water inside the PG particles (hydration water) was slowed by a factor of ~5.8 relative to that of bulk water (Nickels et al. 2016). These measurements also revealed that the retardation of the water dynamics depended on the scattering wavevector q or, in other words, the length scale of the measurement, ranging from 0.3 to 3 nm. The QENS measurement also provided an independent measure of the amount of hydration water in the PG particles, which was consistent with the value determined using SANS (Nickels et al. 2016; Simmons et al. 2020).

8.2.3 Mechanical Properties

8.2.3.1 Mechanical modulus

AFM force spectroscopy was used to measure high-resolution maps of Young's modulus E within individual particles (Baylis et al. 2021). These maps were interpreted in terms of an inner stiff region and an outer softer region within each PG particle that was consistent with the internal structure observed in high force AFM images (Figure 8.2). The average Young's modulus value for each PG particle, averaged over many particles ($n = 103$), was $E = 690 \pm 23$ kPa.

The median value of the mean Young's modulus $\bar{E}$ measured on many PG particles was compared to the ensemble averaged bulk modulus K measured using osmotic pressure techniques. Three different techniques – dialysis, gravimetric analysis and ellipsometry – were used to measure the dependence of the osmotic pressure Π on the concentration C of PG (Grossutti & Dutcher 2021). The $\Pi - C$ data was used to calculate the dependence of the bulk modulus K on the hydration of the PG particles, showing that K increased from 200 kPa for fully hydrated particles to ~7 GPa for dry particles.

The relationship between the bulk modulus K, for a three-dimensional compression of the particles, and Young's modulus E, for a one-dimensional compression of the particles as in the AFM experiments, is given by $K = E/[3(1-2\nu)]$ (Landau & Lifshitz 1986), where ν is Poisson's ratio that describes the lateral extension of a material when it is subjected to a vertical compression. The softness of PG particles, combined with their porous nature, makes their compressibility comparable to that of a sponge for which ν is close to zero (no lateral extension for a vertical compression). With this assumption of $\nu = 0$, $K = E/3$, and this relationship is satisfied quite closely for hydrated PG particles ($K \sim E/3.4$) (Baylis et al. 2021; Grossutti & Dutcher 2021).

The relationship between the bulk modulus K, measured using ellipsometry (Grossutti, Dutcher & Miki 2017; Grossutti et al. 2017), and the water ordering parameter R_{network}, measured using IR spectroscopy, was determined as a function of RH for thin films of native PG, as well as for cationically and anionically modified PG, dextran and hyaluronic acid (Grossutti & Dutcher 2020). The amount of sorbed water was considerably larger, and the values of K and R_{network} were considerably smaller, for both charge-modified versions of PG than for native PG. K increased linearly with R_{network}, showing that the hydration water ordering dictates the stiffness of the PG particles.

8.2.3.2 Rheological properties

The mechanical properties and interactions between PG particles can also be probed in rheology measurements of aqueous dispersions of PG particles. Unlike other polysaccharides, PG particles show a gradual increase in the zero-shear viscosity η_0 as the concentration C of PG particles is increased (Shamana et al. 2018). It is only for $C > 20\%$ w/w that η_0 increases dramatically. This distinctive $\eta_0(C)$ behavior is well described by the Vogel–Fulcher–Tammann (VFT) model of the glass transition (Shamana et al. 2018). For soft particles, the divergence in η_0 is shifted to larger values of C or, alternatively, larger effective volume fractions ϕ_{eff}, relative to that for hard particles, providing a direct indication of the softness of the PG particles. Shear stress σ versus shear strain rate $\dot{\gamma}$ measurements (flow curves) for highly concentrated dispersions showed that flow occurred for σ values below the apparent yield stress. This suggested an additional relaxation mechanism, which was interpreted in terms of the relaxation of short glucose chains decorating the outer surface of the particles, as observed for star polymers (Erwin et al. 2010) and as determined independently in SANS measurements of PG particles (Simmons et al. 2020).

8.2.4 Challenges in Measuring Complex Physical Properties of PG Particles

Taken together, the comprehensive results obtained for the physical properties of high-purity PG nanoparticles show that they are soft, deformable, hydrated and hairy. Given these complex characteristics, it is challenging to measure their fundamental properties. For example, the radius of the spherical particles can be measured using different scattering and microscopy techniques, but different techniques provide different measures of this quantity. Care must be taken to reconcile different values measured in different experiments. Dynamic light scattering (DLS) is a standard method to determine the hydrodynamic radius r_h, based on the diffusion of the small PG particles, for which the most commonly reported value of r_h is the intensity-averaged value, $[r_h]_I$. However, for small (radius $r < 100$ nm) nanoparticles, one needs to account for the strong, power law dependence of the scattering intensity I on the particle radius: $I \sim r^6$ (Dyuzheva, Kargu & Klyubin 2002). Because of this, the intensity-averaged values $[r_h]_I$ are skewed to larger values, and a more accurate measure of the particle radius is given by the number-averaged r_h value $[r_h]_N$. The transformation from the intensity-averaged to the number-averaged values is well defined for particles which have a relatively narrow, monomodal polydispersity distribution. Even for PG, a relatively monodisperse nanoparticle, the difference between $[r_h]_I (= 35\ \text{nm})$ and $[r_h]_I (= 22.5\ \text{nm})$ is considerable, and it is best to compare the value of $[r_h]_N$ with those measured using other techniques. AFM can be used to image the lateral and vertical extent of small soft particles, thereby determining their volume. However, it is important to realize that the particles are deformed because of their adhesion to an underlying substrate and to calculate an equivalent spherical radius r_{AFM}, which yields a value of $r_{AFM} = 23.0\ \text{nm}$ for PG particles. SANS allows the determination of the particle radius r_{SANS} by fitting the measured scattering curves $I(q)$ to a model of the radial particle density, which yields a value of $r_{SANS} = 21.0\ \text{nm}$. We can see that excellent agreement is obtained between the values of $[r_h]_N$, r_{AFM}, and r_{SANS} by carefully accounting for the physical nature of each measure of the PG particle radius. Similar considerations must be taken into account in reconciling different measurements of other physical parameters, e.g., the mechanical modulus of PG particles (bulk modulus K vs. Young's modulus E for 3D vs. 1D compression of the particles, respectively), as discussed above.

8.3 MODIFICATIONS AND DERIVATIZATIONS

The physical properties of PG particles can be tuned by modifying the particles in a variety of ways. The morphology of PG particles can be altered by hydrolyzing the linkages between glucose units, either nonspecifically using dilute acids or specifically using a variety of different enzymes. In addition, enzymes can also be used to add glucose units to the particles (Feng et al. 2023; Takata et al. 2014). Chemical groups can also be covalently attached to the particles, effectively creating new nanoparticles with different functionalities.

8.3.1 Enzymatic Modification

The enzymes that are responsible for growing and degrading PG particles in plants can also be used to modify their morphology. Because of their considerable size, enzyme molecules cannot penetrate the PG particles but rather are limited to act on the surface of the particles. Their action can be controlled through environmental changes such as pH and temperature. Enzymatic treatments are often used to help determine structural properties of phytoglycogen and other polysaccharides such as branching ratio and chain-length distribution because of their highly regulated action (Miao et al. 2014; Roman et al. 2022). For example, β-amylase cleaves α-1,4 but not

α-1,6 linkages, working inward from the non-reducing chain ends. This enzymatic action removes maltose units from the chain ends until a branch point is reached. As a result, the short chains on the surface of the particles are removed, with very little change in the particle radius and molar mass (Roman et al. 2022). These particles are referred to as β-limit dextrins. Other enzymes target other glucosidic linkages, such as pullulanase, a debranching enzyme that targets α-1,6-linkages, and glucoamylases, exoamylases that target both α-1,4 and α-1,6 linkages, resulting in digestion of entire PG particles. Different enzymes can also be combined so that they work cooperatively to break all the glucosidic linkages in the particles.

In addition to hydrolyzing the glucosidic bonds, enzymes such as phosphorylase in the presence of glucose-1-phosphate can catalyze the addition of glucose units to the glucose chains through α-1,4 linkages (Feng et al. 2023; Takata et al. 2014). In this case, the outer chains can be extended but do not branch. An interesting feature of this modification is the possibility of crystallization of the linear chains as they grow (Feng et al. 2023).

8.3.2 Acid Hydrolysis

Exposure of the PG particles to dilute acids at elevated temperatures causes hydrolysis of the glucosidic linkages. Unlike larger enzyme molecules, small acid molecules can penetrate PG particles and cause hydrolysis throughout the particles (Powell et al. 2015). This produces smaller PG particles which also have reduced densities. One consequence of this was shown dramatically in rheology measurements of densely packed acid-hydrolyzed PG particles (Shamana & Dutcher 2022), in which the increase in zero-shear viscosity η_0 with increasing effective volume fraction ϕ_{eff} transitioned from a VFT dependence for $\phi_{eff} < 1$ to an Arrhenius dependence for $\phi_{eff} > 1$. Counterintuitively, the particles could move more easily when they were closely packed together, and this behavior was interpreted in terms of a new relaxation mode due to the extension and retraction of longer glucose chains produced by the acid hydrolysis.

8.3.3 Covalent Attachment of Chemical Groups

The functionality of the outer surface of PG can be altered by covalently attaching different chemical groups, adding properties such as charge and hydrophobicity. These modifications can be used to improve the association of PG particles with bioactive molecules and their compatibility in complex formulations. PG can be chemically modified through reaction with the hydroxyl groups on the C2, C3, and C6 carbons in the glucose units. Many different chemistries have been used to attach chemical groups, such as esterification, oxidation, carboxymethylation, hydroxypropylation, and cationization (Besford, Cavalieri & Caruso 2020; Chen et al. 2023).

Carboxymethylation (CM) is an anionic modification commonly applied to starch and other glucose-based polysaccharides (Chen et al. 2020; Liu et al. 2020). In the case of PG, CM was shown to increase the uptake of water and decrease the water ordering and mechanical stiffness, relative to that for native PG (Grossutti & Dutcher 2020), with similar changes observed for a cationic modification with quaternary ammonium groups using glycidyltrimethylammonium chloride (GTAC). Both CM and GTAC modifications of PG can achieve relatively high degrees of substitution (DS ~ 1), where DS can have a maximum value of 3 if all three labile hydrogens are substituted on each glucose monomer.

Another important modification of PG particles is obtained by attaching hydrophobic octenyl succinic anhydride (OSA) groups, which is used to achieve a food-grade modification of starch (Din, Xiong & Fei 2017; Sweedman et al. 2013), for which DS values of up to 0.23 have been achieved (Simmons et al. 2020). For the largest DS value, SANS measurements showed that the hydrophobically modified outer chains on the PG particles collapsed in water to form uniform "seeds" on the particle surface (Simmons et al. 2020). OSA also contains a carboxyl group, which is negatively charged for $pH > pKa \sim 4.2$, so that interactions between PG particles and with other molecules are determined by the interplay between hydrophobicity and electrostatics.

8.4 APPLICATIONS

PG occurs as soft, compact, hydrated, hairy nanoparticles with an underlying dendritic architecture. These physical properties, combined with its non-toxicity and digestibility, make it ideal for applications involving human health and personal care. Because of the relatively high cost of small-scale production of PG, initial efforts have targeted high-end applications such as personal care, bioactive delivery, and immunomodulation, as described below.

8.4.1 Personal Care

The unique interaction of PG with water makes it attractive for use in personal care formulations as a moisturizer and skin rejuvenation agent. SANS studies showed that PG nanoparticles take up to 220% of their weight in water when fully hydrated (Simmons et al. 2020). ATR-FTIR spectroscopy studies revealed that the water taken up by thin films of PG was more highly ordered than for hyaluronic acid (HA), a polysaccharide commonly used as a moisturizing agent in cosmetic formulations (Grossutti & Dutcher 2016). These results suggest that water sorbed by PG is more strongly bound and could potentially provide controlled release of water as a moisturizer within personal care formulations (Grossutti & Dutcher 2016), (Grossutti, Dutcher & Miki 2017). The strong interaction of water with PG can also provide unique lubrication for health care applications: surface forces apparatus measurements showed that PG particles had a very low friction coefficient, comparable to currently used synthetic polymers but with the advantage of having a natural origin (Adibnia et al. 2021).

The unique rheological properties of PG are also desirable from a formulation perspective. PG has been found to maintain a low viscosity at concentrations up to ~20% w/w, much higher than other polysaccharides commonly used in personal care formulations (Shamana et al. 2018). The minimal effect of PG on the viscosity of formulations has advantages in terms of processing as well as consumer use, where viscosity can impact the spreading and tactile properties of a product.

PG was also found to perform well as an anti-aging agent for skincare products. *In vitro* studies on human skin cells found that PG promoted the production of hyaluronic acid and collagen, both important compounds for maintaining a youthful texture and appearance to skin (Miki et al. 2018). The cells were also found to proliferate more quickly, compared to cells that had not been exposed to PG. This may indicate a faster rate of cellular turnover – how quickly the cells of the epidermis are replenished – which helps improve skin complexion and could also be beneficial for wound healing applications (Miki et al. 2018). To test these effects in topically applied products, where active ingredients need to penetrate the outer barrier layer of the skin, *in vitro* skin penetration studies were performed. Using fluorescently labeled PG and human skin samples, it was found that PG could penetrate the outer layer of the skin to a depth of 150–170 μm, corresponding to the dermal layer of the skin where living cells reside, demonstrating their potential to affect skin cells when applied topically (Miki et al. 2018). In addition, clinical studies that compared the performance of PG in a skin cream formulation relative to that of a placebo also found that PG improved the appearance of participant's skin, evaluated by an expert grader on multiple metrics including appearance of fine lines, coarse wrinkles, and hyperpigmentation (Baumann, Wandrey & Zülli 2024; Miki et al. 2018).

8.4.2 Bioactive Delivery

The nanoparticle geometry and chemistry of PG allows it to associate with a variety of small molecules – the small, water-soluble PG particles have a large surface area-to-volume ratio with many hydroxyl groups on the outer surface, providing a large area with many sites for interaction with small molecules. Because penetration of bioactives into the PG particles is inhibited by the dense core of the particles, their association with PG is typically limited to the PG particle surface.

A variety of studies have demonstrated the binding and association of PG with small molecules, ranging from the interaction of sugar-binding molecules such as lectins with native PG (Charlesworth et al. 2022) to the interaction of lipophilic bioactive compounds with hydrophobically modified PG (Chen & Yao 2017a; Scheffler et al. 2010; Xue et al. 2021). Interestingly, several studies have demonstrated significant increases in the effective water solubility of hydrophobic compounds such as resveratrol (Shi et al. 2020) and astaxanthin (van Heijst, Whiting & Dutcher 2024) by promoting physical association of the compounds with native PG using solvent evaporation techniques. In addition, spray drying of PG with the antimicrobial rifampicin produced particles that, when incorporated into dry powder inhaler formulations, produced small particles with excellent aerosol performance (Tse et al. 2021). The bioavailability of bioactive compounds solubilized using PG has been demonstrated *in vitro* by measuring their permeability across monolayers of Caco-2 intestinal cells (Chen & Yao 2017a, 2017b; Shi et al. 2020; Xue et al. 2021). Recent reviews provide additional details on bioactive delivery applications of PG (Besford, Cavalieri & Caruso 2020; Chen et al. 2023; Shi et al. 2023).

8.4.3 Immunomodulation

Immunomodulation therapies combat disease by either stimulating or suppressing the immune system. These therapies are currently being developed to treat a wide variety of conditions including cancer, autoimmune disorders, and viral infections.

Complexes of cationically modified PG and a synthetic double-stranded RNA, polyinosinic cytidylic acid or poly(I:C), have been shown to induce cell death and enhance the innate immune response in two different ovarian cancer cell lines (Lewis et al. 2023). The cellular response was amplified by a factor of 2 compared to poly(I:C) alone, demonstrating the effectiveness of PG in enhancing dsRNA-mediated immune responses in ovarian cancer cells.

Poly(I:C)–cationic PG complexes were also found to protect human lung HEL-299 cells against infection from a variety of human coronaviruses (Semple et al. 2022). The poly(I:C)–PG complexes induced the production of type I interferon in HEL-299 cells, providing more effective protection against infection compared to that for poly(I:C) alone (Semple et al. 2022). Although the HEL-299 cell line did not propagate the SARS-COV-2 virus responsible for the recent COVID-19 pandemic, these studies serve as an important model to study innate immune stimulation and potential treatments for coronaviruses.

8.5 SUMMARY AND PERSPECTIVE

Recent research has led to an improved understanding of the complex properties of PG, a glucose-based polysaccharide with a unique dendritic architecture that is produced in the kernels of sweet corn as compact, highly branched nanoparticles. PG is remarkable because it is a relatively monodisperse natural nanoparticle that can be produced with high purity in very large quantities, which makes it suitable for a wide range of nanotechnology applications. These applications have been further broadened through chemical modification to achieve different functionalities. Additional promising applications of PG will continue to be identified and exploited as more insight is achieved into the relationship between the physical and chemical properties of PG nanoparticles and their functionality, and as the cost of production is reduced through further scaling up the production of this sustainable nanotechnology.

REFERENCES

Adibnia, V, Ma, Y, Halimi, I, Walker, GC, Banquy, X, & Kumacheva, E 2021. Phytoglycogen nanoparticles: nature-derived superlubricants. *ACS Nano*, *15(5)*, 8953–8964.

Ball, SG, & Morell, MK 2003. From bacterial glycogen to starch: understanding the biogenesis of the plant starch granule. *Annu. Rev. Plant Biol.*, *54*, 207–233.

Ballauff, M, & Likos, CN 2004. Dendrimers in solution: insight from theory and simulation. *Angew. Chem., Int. Ed.*, *43*, 2998–3020.

Baumann, J, Wandrey, F, & Zülli, F 2024. Energizing the skin with phytoglycogen. *Personal Care*, January 2024, 36–39.

Baylis, B, Shelton, E, Grossutti, M, & Dutcher, JR 2021. Force spectroscopy mapping of the effect of hydration on the stiffness and deformability of phytoglycogen nanoparticles. *Biomacromolecules*, *22*, 2985–2995.

Besford, QA, Cavalieri, F, & Caruso, F 2020. Glycogen as a building block for advanced biological materials. *Adv. Mater.*, *32*, 1904625.

Boris, D, & Rubinstein, M 1996. A self-consistent mean field model of a starburst dendrimer: dense core vs dense shell. *Macromolecules*, *29*, 7251–7260.

Charlesworth, K, van Heijst, N, Maxwell, A, Baylis, B, Grossutti, M, Leitch, JJ, & Dutcher, JR 2022. Binding affinity of concanavalin A to native and acid-hydrolyzed phytoglycogen nanoparticles. *Biomacromolecules*, *23*, 4778–4785.

Chen, H, & Yao, Y 2017a. Phytoglycogen to increase lutein solubility and its permeation through Caco-2 monolayer. *Food Res. Int.*, *97*, 258–264.

Chen, H, & Yao, Y 2017b. Phytoglycogen improves the water solubility and Caco-2 monolayer permeation of quercetin. *Food Chem.*, *221*, 248–257.

Chen, L, Zhao, N, McClements, DJ, Hamaker, BR, & Miao, M 2023. Advanced dendritic glucan-derived biomaterials: from molecular structure to versatile applications. *Compr. Rev. Food Sci. Food Saf.*, *22*, 4107–4146.

Chen, Y, Xue, J, Wusigale, Wang, T, Hu, Q, & Luo, Y 2020. Carboxymethylation of phytoglycogen and its interactions with caseinate for the preparation of nanocomplex. *Food Hydrocoll.*, *100*, 105390.

Din, Z-U, Xiong, H, & Fei, P 2017. Physical and chemical modification of starches: a review. *Crit. Rev. Food Sci. Nutr.*, *57*, 2691–2705.

Dyuzheva, MS, Kargu, OV, & Klyubin, VV 2002. The effect of polydispersity on the size of colloidal particles determined by the dynamic light scattering. *Colloid J.*, *64*, 33–38.

Erwin, BM, Cloitre, M, Gauthier, M, & Vlassopoulos, D 2010. Dynamics and rheology of colloidal star polymers. *Soft Matter*, *6*, 2825–2833.

Feng, W, Wang, Z, Campanella, OH, Zhang, T, & Miao, M 2023. Fabrication of phytoglycogen-derived core-shell nanoparticles: structure and characterizations, *Food Chem.*, *423*, 136317.

Grossutti, M, Bergmann, E, Baylis, B, & Dutcher JR 2017. Equilibrium swelling, interstitial forces, and water structuring in phytoglycogen nanoparticle films. *Langmuir, 33*, 2810–2816.

Grossutti, M, & Dutcher, JR 2016. Correlation between chain architecture and hydration water structure in polysaccharides. *Biomacromolecules, 17*, 1198–1204.

Grossutti, M, & Dutcher, JR 2020. Hydration water structure, hydration forces, and mechanical properties of polysaccharide films. *Biomacromolecules, 21*, 4871–4877.

Grossutti, M, & Dutcher, JR 2021. Correlation of mechanical and hydration properties of soft phytoglycogen nanoparticles. *Carbohydr. Polym., 251*, 116980.

Grossutti, M, Dutcher, JR, & Miki, C 2017. PG Nanoparticles: 1. Key properties relevant to its use as a natural moisturizing ingredient. *Household and Personal Care Today, 12(1)*, 47–51.

Inouchi, N, Glover, D.V, & Fuwa, H 1987. Chain length distribution of amylopectins of several single mutants and the normal counterpart, and sugary-1 phytoglycogen in maize (*Zea mays* L.). *Starch-Starke, 39*, 259–266.

Kim, D, & Duhamel, J 2023. Interior of glycogen probed by pyrene excimer fluorescence. *Carbohydr. Polym., 299*, 120205.

Landau, LD, & Lifshitz, EM 1986. *Theory of Elasticity*, 3rd ed., Pergamon Press: Oxford.

Larson-Smith, K, Jackson, A, & Pozzo, DC 2010. Small angle scattering model for Pickering emulsions and raspberry particles. *J. Colloid Interface Sci., 343*, 36–41.

Lescanec, RL, & Muthukumar, M 1990. Configurational characteristics and scaling behavior of starburst molecules: a computational study. *Macromolecules, 23*, 2280–2288.

Lewis, A, Tran, A, Aldor, NL, Jadaa, NA, Feng, T, Moore, E, DeWitte-Orr, SJ, & Poynter, SJ 2023. PG-DsRNA nanoparticles demonstrate differential cytotoxicity and immunostimulatory potential in two ovarian cancer cell lines. *J. Nanopart. Res., 25(6)*, 104.

Liu, Y, Lu, K, Hu, X, Jin, Z, & Miao, M 2020. Structure, properties and potential applications of phytoglycogen and waxy starch subjected to carboxymethylation. *Carbohydr. Polym., 234*, 115908.

Mansfield, ML, & Klushin, LI 1993. Monte Carlo studies of dendrimer macromolecules. *Macromolecules, 26*, 4262–4268.

Miao, M, Li, R, Jiang, B, Cui, SW, Lu, K, & Zhang, T 2014. Structure and digestibility of endosperm water-soluble a-glucans from different sugary maize mutants, *Food Chem., 143(143)*, 156–162.

Miki, C, DeWitte-Orr, S, Foldvari, M, & Moore, M 2018. Anti-ageing properties of PG. *Household Personal Care Today, 13(2)*, 26–28.

Morling, B, Luyben, S, Dutcher, JR, & Wickham, RA 2024. Efficient modeling of high-generation dendrimers in solution using dynamical self-consistent field theory. *Macromoleucles, 57(9)*, 4617–4628.

Morris, DL, & Morris, CT 1939. Glycogen in the seed of *Zea Mays* (variety Golden Bantam). *J. Biol. Chem., 130(2)*, 535–544.

Nickels, JD, Atkinson, J, Papp-Szabo, E, Stanley, C, Diallo, SO, Perticaroli, S, Baylis, B, Mahon, P, Ehlers, G, Katsaras, J, & Dutcher, JR 2016. Structure and hydration of highly-branched, monodisperse phytoglycogen nanoparticles. *Biomacromolecules, 17*, 735–743.

Powell, PO, Sullivan, MA, Sheehy, JJ, Schulz, BL, Warren, FJ, & Gilbert, RG 2015. Acid hydrolysis and molecular density of phytoglycogen and liver glycogen helps understand the bonding in glycogen α (composite) particles. *PLoS One, 10*, e0121337.

Putaux, J-L, Buléon, A, Borsali, R, & Chanzy, H 1999. Ultrastructural aspects of phytoglycogen from cryo-transmission electron microscopy and quasi-elastic light scattering data. *Int. J. Biol. Macromol., 26*, 145–150.

Roman, L, Baylis, B, Klinger, K, de Jong, J, Dutcher, JR, & Martinez, MM 2022. Changes to fine structure, size and mechanical modulus of phytoglycogen nanoparticles subjected to high-shear extrusion, *Carbohydr. Polym., 298*, 120080.

Rosenfeldt, S, Dingenouts, N, Ballauf, M, Werner, N, Vögtle, F, & Lindner, P 2002. Distribution of end groups within a dendritic structure: a SANS study including contrast variation. *Macromolecules, 35*, 8098–8105.

Rubinstein, M, & Colby, RH 2003, *Polymer Physics*, Oxford University Press: New York.

Scheffler, SL, Huang, L, Bi, L, & Yao, Y 2010. *In vitro* digestibility and emulsification properties of phytoglycogen octenyl succinate. *J. Agric. Food Chem., 58(8)*, 5140.

Semple, SL, Alkie, TN, Jenik, K, Warner, BM, Tailor, N, Kobasa, D, & DeWitte-Orr, SJ 2022. More tools for our toolkit: the application of HEL-299 cells and DsRNA-nanoparticles to study human Coronaviruses in vitro. *Virus Res., 321*, 198925.

Shamana, H, & Dutcher, JR 2022. Transition in the glassy dynamics of melts of acid-hydrolyzed phytoglycogen nanoparticles, *Biomacromolecules, 23*, 2040–2050.

Shamana, H, Grossutti, M, Papp-Szabo, E, Miki, C, & Dutcher, JR 2018. Unusual polysaccharide rheology of aqueous dispersions of soft phytoglycogen nanoparticles. *Soft Matter, 14*, 6496–6505.

Shi, Y, Chen, S, Bai, H, Chen, L, & Miao, M 2023, Phytoglycogen-based systems, in *Bioactive Delivery Systems for Lipophilic Nutraceuticals: Formulation, Fabrication, and Application*, eds. M Miao, L Chen, & DJ McClements, Royal Society of Chemistry, pp. 322–346.

Shi, Y, Ye, F, Lu, K, Hui, Q & Miao, M 2020. Characterizations and bioavailability of Dendrimer-like glucan nanoparticulate system containing resveratrol. *J. Agric. Food Chem., 68*, 6420–6429.

Simmons, J, Nickels, JD, Michalski, M, Grossutti, M, Shamana, H, Stanley, CB, Schwan, AL, Katsaras, J, & Dutcher, JR 2020. Structure, hydration, and interactions of native and hydrophobically modified phytoglycogen nanoparticles. *Biomacromolecules, 21*, 4053–4062.

Sweedman, MC, Tizzotti, MJ, Schäfer, C, & Gilbert, RG 2013. Structure and physicochemical properties of octenyl succinic anhydride modified starches: a review. *Carbohydr. Polym., 92*, 905–920.

Takata, Y, Shimohigoshi, R, Yamamoto, K, & Kadokawa, J-I 2014. Enzymatic synthesis of dendritic amphoteric α-glucans by thermostable phosphorylase catalysis. *Macromol. Biosci., 14*, 1437–1443.

Tse, JY, Kadota, K, Imakubo, Uchiyama, TH, & Tozuka, Y 2021. Enhancement of the extra-fine particle fraction of levofloxacin embedded in excipient matrix formulations for dry powder inhaler using response surface methodology. *Eur. J. Pharm. Sci., 156*, 105600.

van Heijst, N, Whiting, P, & Dutcher, JR 2024. Solubilization of hydrophobic astaxanthin in water by physical association with phytoglycogen nanoparticles. *Biomacromolecules, 25(7)*, 4110–4117.

Xue, J, Li, Z, Duan, H, He, J, & Luo, Y 2021. Chemically modified phytoglycogen: physicochemical characterizations and applications to encapsulate curcumin. *Colloids Surf. B, 205*, 111829.

Yun, SH, & Matheson, NK 1993. Structures of the amylopectins of waxy, normal, amylose-extender, and *wx: ae* genotypes and of the phytoglycogen of maize. *Carbohydr. Polym., 243*, 307–321.

Zook, TC, & Pickett, GT 2003. Hollow-core dendrimers revisited. *Phys. Rev. Lett., 90*, 015502–015504.

Chemical and Physical Structures of Starch

9

Hongxin Jiang and Jay-Lin Jane

9.1 INTRODUCTION

Starch is the major energy-storage glucan in higher plants, algae, and cyanobacteria (Ball, Colleoni & Arias 2015; Robyt 1998). Normal starch comprises amylose and amylopectin. Amylose consists of hundreds to thousands glucopyranosyl units linked by α-D-(1→4) glycosidic bonds and with a few branch chains linked by α-D-(1→6) glycosidic bonds (Refer to Chapter 6). Amylopectin is a highly branched glucopyranosyl gigantic polymer with degree of polymerization (DP) up to 10^9, consisting of linear chains linked by α-D-(1→4) glycosidic bonds and branched by ~5% α-D-(1→6) glycosidic bonds (Refer to Chapter 7). Maize *amylose extender* (*ae*) mutant starches contain a minor component known as the intermediate component, which has branched structures with branch chain length longer than that of amylopectin but molecular weight similar to that of amylose (Kasemsuwan et al. 1995; Li et al. 2008).

Amylopectin has semi-crystalline structures in native starch granules (Gallant, Bouchet & Baldwin 1997; Kainuma & French 1971, 1972), which maintain the integrity of starch granules and prevent them from dispersing in an aqueous medium at the ambient temperature. Amylose is present in an amorphous structure and oriented side by side, interspersed, and intertwined with amylopectin in normal native starch granules (Jane et al. 1992; Kasemsuwan & Jane 1994). Amylose molecules, however, form long-chain double helices in native high-amylose maize starch granules. The amylose double helices have high conclusion gelatinization temperatures, up to 130°C. Consequently, high-amylose maize starch after being cooked in excess water at boiling-water temperature is not fully gelatinized and exhibits resistance to enzymatic hydrolysis (Cheng et al. 2024; Jiang et al. 2010a, 2010b; Li et al. 2008). In cereal starches, amylose also forms helical complex with lipids (Morrison 1988; refer to Chapter 15).

Due to differences in branch chain lengths of amylopectin, packing of amylopectin double helices differs in starch granules of different botanical sources, resulting in A-, B-, or C-type polymorph (Cai et al. 2014a, c; Li et al. 2008; Imberty & Perez 1988; Sarko & Wu 1978; Takahashi, Kumano & Nishikawa 2004; Wu & Sarko 1978a, 1978b). The C-type polymorph is a combination of the A- and B-types. The V-type X-ray diffraction pattern is also observed for some high-amylose starches (Refer to Chapter 15). Starch granules isolated from different botanical sources exhibit great differences in size, granule shape, and internal structure. This chapter aims to discuss current understandings on starch helical, crystalline, and granular structures.

9.2 CONFORMATION AND HELICAL STRUCTURES OF STARCH

In contrary to the general perception that starch is a hydrophilic polymer because of the presence of three hydroxy groups on each anhydro-glucose unit, starch, like most other biopolymers, is an amphiphilic polymer, consisting of hydrophilic and hydrophobic groups. Carbons of the glucopyranosyl unit carry both hydroxyl groups and hydrogen atoms. Hydroxyl groups of starch are hydrophilic and interact favorably with water molecules in the aqueous dispersion, whereas the array of hydrocarbon groups of the starch chain is hydrophobic as evidenced by the hydrophobic cavity of cyclodextrins that are biosynthesized from starch chains by cyclodextrin glucosyltransferases (Szejtli 1998, 2004). Cyclodextrins are well known to encapsulate fatty acids by interacting and holding the hydrophobic tail of the fatty acid in the hydrophobic cavity (Harada et al. 1997; Szejtli 1998, 2004).

With the presence of these hydrophilic and hydrophobic groups, amylose in an aqueous dispersion is known to form helical complex instantly with compounds carrying hydrophobic moiety, such as polyiodide ions to develop blue color, *n*-butyl alcohol to form crystallites and precipitate, and fatty acids to form left-handed helical complexes (Calabrese & Khan 1999; Huang et al. 2020; Jane & Robyt 1984). Depending on the size of cross-section of the complexing agent, amylose can form helical complex of six, seven, or eight glucose units per turn (French & Murphy 1977; refer to Chapter 15). The hydrocarbon groups located outside the helix are likely to facilitate the crystallite

DOI: 10.1201/9781003464396-9

formation. Those amylose-helical complex crystallites have different morphologies (Choisnard, Wouessidjewe & Putaux 2018; Huang et al. 2020; Putseys, Lamberts & Delcour 2010) and display a thickness of 10 nm (Yamashita, Ryugo, & Monobe 1973), as confirmed by results obtained using enzymatic hydrolysis of amylose-helical complex crystallites (Jane & Robyt 1984). When there is no complexing agent present in the aqueous dispersion, a starch chain tends to interact with another chain, either parallel or antiparallel, to have the hydrophobic sides of the chains folded facing each other, away from the polar aqueous medium, to form double helices and reach a low energy state (Imberty et al. 1987, 1988; Imberty & Perez 1988; Imberty et al. 1991; Sarko & Wu 1978; Takahashi, Kumano & Nishikawa 2004; Wu & Sarko 1978a, 1978b). This process is known as retrogradation.

9.2.1 Conformational Changes of Starch Molecules

Linear starch chains in a diluted aqueous dispersion display a random-coil conformation because glucopyranosyl units are in a $^4C_1$ conformation (Banks & Greenwood 1971; Norisuye 1996; Robyt 1998), which differ from the straight chain of cellulose linked by β-D-(1→4) glycosidic bonds. The pH, temperature, the presence of salts, and types of solvent could affect the amylose conformation in the aqueous dispersion (Banks & Greenwood 1971; Everett & Foster 1959; Gidley & Bulpin 1987, 1989; Rao & Foster 1963). In an alkaline dispersion with high pH, the conformation of amylose is an extended helix due to negative-charge repulsion (French & Murphy 1977). Short linear chains (DP 10–12) in an aqueous dispersion tend to crystallize the A-type polymorph, whereas longer chains (DP > 12) tend to form the B-type polymorph (Gidley & Bulpin 1987; Pfannemüller 1987). At high concentrations, high temperatures, or the presence of some salts or organic compounds, amylose chains tend to develop the A-type polymorph (Ai & Jane 2017; Hizukuri, Fujii & Nikuni 1960; Cai et al. 2010; Cai & Shi 2013, 2014).

9.2.2 Double Helix

Branch chains of amylopectin form left-handed parallel-stranded double helix in native starch granules (Imberty et al. 1987, 1988; Imberty & Perez 1988; Imberty et al. 1991; Kainuma & French 1971, 1972). Double helices with pitches (12 glucose units per turn) of 2.138 nm and 2.08 nm–2.114 nm are observed for A- and B-type polymorphic crystallites, respectively (Buleon, Véronèse & Putaux 2007; Imberty et al 1988; Imberty & Perez 1988; Takahashi, Kumano & Nishikawa 2004). In addition to hydrophobic interaction, the parallel strands of double helix are stabilized through two interstrand hydrogen bonds between O2 and O6 atoms, and no intramolecular hydrogen bond is observed within one strand for either A- or B-type crystalline

amylose (Imberty & Perez 1988; Takahashi, Kumano & Nishikawa 2004; Wu & Sarko 1978a, 1978b).

After a systematic study on conformations of the α-D-(1→6) glycosidic bond between two strands of a double helix, either no deformation of the glucopyranose ring involving in the branching point (Imberty & Pérez 1989) or a small deformation of the glucopyranose ring adjacent to the α-D-(1→6) glycosidic bond of the short chain (Buleon & Tran 1990) is reported. The inherent flexibility of the α-D-(1→6) glycosidic bond allows a very short folding back of the lateral chain right after one glucose unit, which makes the propagation of the double helix possible from the second glucose unit after the branching point on each strand (Buleon & Tran 1990).

9.3 CRYSTALLINE STRUCTURE

9.3.1 Crystalline Structure of Amylopectin

Models of A- and B-type polymorphs were proposed by Kainuma and French (1972) and Wu and Sarko (1978a, 1978b) using X-ray diffraction data collected from Naegeli dextrin or stretched films of potato amylose (Refer to Chapter 6). Native starches exhibit A-, B-, or C-type polymorphs, resulting from parallel packing of left-handed double helices (Buleon, Véronèse & Putaux 2007; Gallant, Bouchet & Baldwin 1997; Imberty, Chanzy & Perez 1987; Imberty et al. 1987, 1988; Imberty & Perez 1988; Imberty et al. 1991; Takahashi, Kumano & Nishikawa 2004).

A- and B-type polymorphs differ in the packing of double helices and water content (Buléon, Véronèse & Putaux 2007; Imberty et al 1988; Imberty & Perez 1988; Takahashi, Kumano & Nishikawa 2004). In the A-type polymorph, double helices are packed with the B2 space group in a monoclinic unit cell (a = 2.124 nm, b = 1.172 nm, c axis = 1.069 nm, and γ = 123.5°) with four water molecules between the double helices (Imberty et al. 1988). The c axis is parallel to the helix axis. In the B-type polymorph, double helices are packed with the $P6_1$ space group in a hexagonal unit cell with lattice parameters of a = b = 1.85 nm, c axis = 1.04 nm, and γ = 120° and 36 water molecules in the center cavity of the unit cell (Imberty et al 1988; Imberty & Perez 1988; Takahashi, Kumano & Nishikawa 2004). The neighboring double helices are stabilized by four hydrogen bonds of O3—O3 and O2—O6 (Takahashi, Kumano & Nishikawa 2004).

The parallel packing of the double helices of amylopectin branch chains results in alternating amorphous lamella of branching-linkage rich area of amylopectin and crystalline lamella concentrated with linear chains (French 1984; Gallant, Bouchet & Baldwin 1997). Native starches consisting of more short-branch chains and fewer long-branch chains (Figure 9.1), such as normal and waxy rice, maize, taro, wheat, sorghum, and barley starches, display the A-type polymorph. Whereas, those consisting of more long-branch chains (Figure 9.1), such as *aewx*

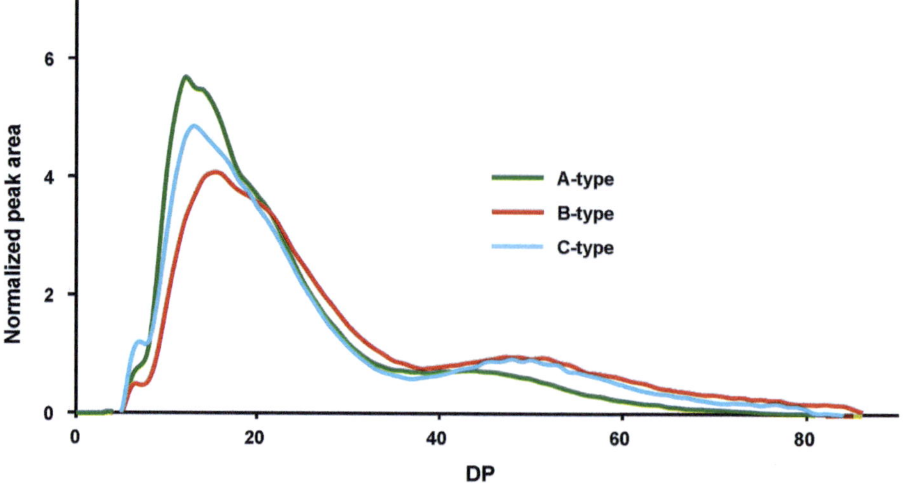

FIGURE 9.1 Averaged branch-chain length distributions of amylopectins from starches of A-, B-, and C-type polymorphs (Jane et al. 2006).

maize, high-amylose maize, potato, and canna starches, display the B-type polymorph. Starches with proportions of long- and short-branch chains between those of A- and B-polymorphs, such as banana, pulse, lotus rhizome, Chinese yam, and water chestnut starches, display the C-type polymorph (Cai et al. 2014a, c; Hizukuri 1985; Jane et al. 1999; Wang, Yu & Yu 2008). Averaged chain-length distributions of the A-, B-, and C-type polymorphic starches determined using a high-performance anion-exchange chromatography equipped with a glucoamylase reactor and a pulsed amperometric detector are shown in Figure 9.1. In native high-amylose maize starch, especially the amylose content higher than 50%, amylose chains also form double helices and display the B-type polymorph (Jiang et al. 2010a, 2010b, 2010c; Li et al. 2008). The crystalline structure of the B-type polymorphic starch can be converted to the A-type using heat-moisture treatments (Fonseca et al. 2021). The A-type polymorph, however, cannot be converted back to the B-type except using an ultra-high-pressure treatment in the presence of water (Katopo, Song & Jane 2002).

The relative crystallinity of native starch ranges from 15% to 45%, which are affected by the starch origin, starch hydration level, and analytical method (Buleon et al. 1987; Lemke et al. 2004; Lopez-Rubio et al. 2008; Paris et al. 1999; Zobel 1988). Dried starch shows little X-ray diffraction peaks, and the relative crystallinity of starch increases with increasing water contents up to ~30% (Buleon et al. 1987, 1998; Donald et al. 2001; Kainuma & French 1972).

9.3.2 Locations of Amylopectin Branch Linkages in Starch Crystalline Lamellae

Based on the results of branch chain lengths of amylopectin and locations of branch linkages, revealed from different proportions of the linear and branch chains of Naegeli dextrins of the A- and B-type polymorphic starches, Jane, Wong and

McPherson (1997) proposed that the amylopectin of the B-type polymorphic starch consists of more long-branch chains (B2, B3, and B4 chains), and the branch linkages are located mostly in the amorphous lamella (Figure 9.2). Therefore, the Naegeli dextrin of the B-type polymorphic starch comprises mostly linear chains. The amylopectin of the A-type polymorphic starch consists of more short-branch chains, and the branch linkages are located in both amorphous and crystalline lamellae (Figure 9.2). Similar findings on locations of branch linkages in the A- and B-type polymorphic rice starches are reported by Nakamura et al., while studying the reaction mechanism of branching enzymes I (BEI) and IIb (BEIIb) in wild type and *ae* mutant rice (Nakamura et al. 2022, refer to Chapter 7).

9.3.3 Distribution of A- and B-Type Polymorphs in the C-Type Polymorphic Starch Granule

C-type polymorphic starches isolated from different plant sources have the hilum at various locations within the granules, with the A- and B-type polymorphs located in different parts of the granule. Distributions of the A- and B-type polymorphic starch in naive starch granules have been intensively studied. Smooth pea starch has a centric hilum and exhibits C-type polymorph. It loses the birefringence at the central/inner part of the granule and transitions to the A-type polymorph after being heated to 66°C in excess water. These results indicate that starch of the B-type polymorph is located in the central/inner part of the granule, surrounded by starch of the A-type polymorph at the periphery (Bogracheva et al. 1998). A study using synchrotron microfocus mapping confirms that the smooth pea starch has 60% A-type polymorph located essentially in the outer part of the granule and 40% B-type located at the center (Buleon et al. 1998). A Chinese yam starch granule with eccentric hilum has the B-type polymorph near the hilum and surrounded by the A-type polymorph (Cai et al. 2014c; Wang et al. 2009).

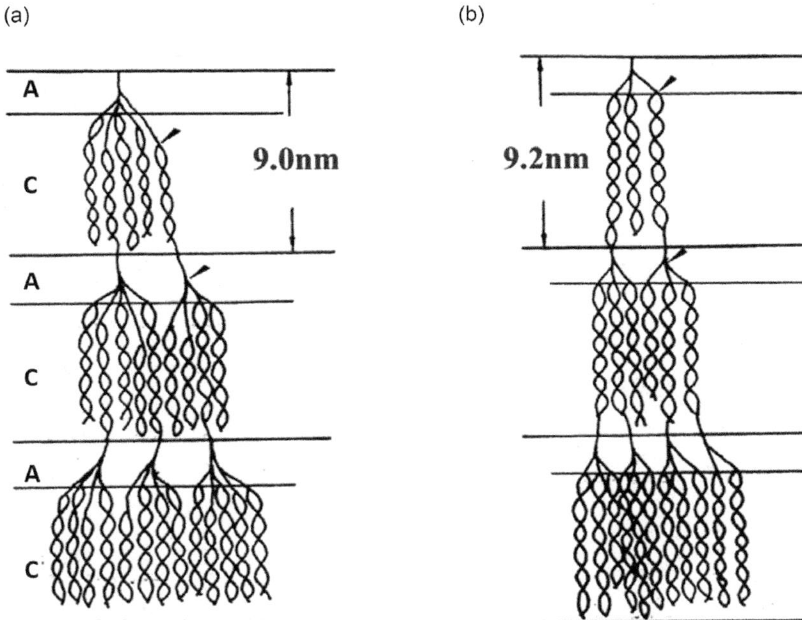

FIGURE 9.2 Proposed models for branching patterns of (a) waxy maize starch (A-type polymorph) and (b) potato starch (B-type polymorph) (Jane, Wong & Mcpherson 1997). A: Amorphous region; C: Crystalline region. The chain length between the arrows stands for the internal long B-chain.

A high-amylose rice starch granule consists of several sub-granules with their respective centric hila and displays the C-type polymorph, with the A-type polymorph located in the inner part of sub-granules and the B-type polymorph located at the periphery of the sub-granules and the starch granule (Wei et al. 2010). Lotus rhizome starch granule has eccentric hilum at one end of the granule and shows C-type polymorph, with A-type polymorph at the periphery of the hilum end, B-type polymorph in the distal region of the granule, and mixed A- and B-type polymorphs at the center of the granule (Cai et al. 2014a). Although there are various distributions of the A- and B-type polymorphs in C-type polymorphic starch granules, the mechanism underlying the differences in these distributions still remains to be understood. Starch from the tuber of Apios fortunei shows two groups of granules, with the B-type polymorphic granules gelatinizing at a low temperature (~68°C) and the A-type polymorphic granules gelatinizing at a high temperature (~76°C) (Fan et al. 2017).

9.4 GRANULAR STRUCTURE OF STARCH

9.4.1 Morphology of Starch Granule

Depending on the botanical origin, genetic mutation, locations in the plant, and the stage of granule development, starch granules vary from less than 1 μm to more than 100 μm in diameter and show various unique shapes (Chanzy et al.

2006; French 1984; Jane et al. 1994; Li et al. 2019; Seung & Smith 2019). In general, starch granules in leaves have a flat shape and are smaller in size than those in other parts of a plant because starch in chloroplasts is for transitory storage of energy (Seung & Smith 2019). Amaranth, Chinese taro, cow cockle, iliuaua dasheen, small pigweed, and quinoa starches have granule diameters between less than 1 and 4 μm (Jane et al. 1994). Starch from the pseudo-bulbs of *Phajus grandifolius,* however, has ogival shapes with the hilum located at the apex of the granule and a length of more than 100 μm (Chanzy et al. 2006).

Spherical, polygonal, oval, disk, and kidney shapes are commonly observed for normal and waxy starch granules (Jane et al. 1994). Wheat, rye, and barley starches have large disk-shaped A-granules with diameters ranging from 18 to 33 μm and small disk-shaped C-granules and round B-granules with sizes ranging from 2 to 8 μm (Dai 2009; Jane et al. 1994). Rice, waxy rice, and oat starches are known to have compound starch granules, having multiple sub-granules biosynthesized in a single amyloplast. Those sub-granules are tightly packed together and difficult to separate. After separation, those sub-granules display a polyhedral shape. The sub-granules of rice starch have sizes ranging from 3 to 8 μm (Jane et al. 1994; Ai & Jane 2017). Normal and waxy maize starches typically exhibit spherical and polygonal shapes with granule sizes of 5–20 μm. Potato starch has large granules with oval and spherical shapes and sizes ranging from 15 to 75 μm. Most pulse starches have kidney-shaped granules of 5–70 μm (Jane et al. 1994; Li et al. 2019). Tapioca starch consists of round and cup-shaped granules with sizes of 5–25 μm. The cup shape is a feature for tapioca, apple, and other fruit and tuber starches (Jane et al. 1994; Stevenson et al. 2006).

High-amylose maize, wheat, and barley starches, with the amylose contents of 50% or more, consist of elongated and other irregular shaped granules (Cai et al. 2014b, Jiang et al. 2010a, b, c, d; Li et al. 2008; Li et al. 2020; Regina et al. 2006, 2010). The filamentous granules present in high-amylose maize starch have diameters of 3–5 μm and lengths as long as more than 50 μm (Li et al. 2008; Jiang et al. 2010a, d). High-amylose wheat and barley starches also contain taco-shaped granules in addition to regular and elongated granules (Li et al. 2020; Regina et al. 2006, 2010).

9.4.2 Organization of Starch Granule

Depending on the botanical origin and genetic background, parallelly arranged double-helical crystalline lamellae are 5.9–6.8 nm in thickness, corresponding to about three pitches of a sixfold double helix (Imberty & Perez 1988; Imberty et al. 1988). The amorphous lamella has an average thickness of 2.2 nm, corresponding to one pitch (Imberty & Perez 1988; Imberty et al. 1988). Crystalline lamellae prepared from hydrochloric acid hydrolysis of native waxy maize starch remain the A-type polymorph and exist in the shape of parallelepipedal blocks with lengths ranging from 20 to 40 nm, widths from 15 to 30 nm, and thicknesses from 5 to 7 nm, which is consistent with DP 15–21 of a single strand of the double helix (Putaux et al. 2003).

Lamellae are effectively organized into blocklets with diameters ranging from 10 to 500 nm, depending on the botanic origin and genetic background of the starch and their locations within the granule (Baker, Miles & Helbert 2001; Gallant, Bouchet & Baldwin 1997; Ohtani et al. 2000; Tang, Mitsunaga & Kawamara 2006). Small blocklets of 10–50 nm are observed in wheat starch granule using atomic force microscopy, whereas large blocklets of 50–300 nm are found on the surface of native potato starch granules (Baldwin et al. 1998). Native maize starch granules have blocklets of 10–30 nm (Baker, Miles & Helbert 2001). Blocklets in the starch granule of A-type polymorph consist of around 210–6,300 chains, and that in the B-type polymorphic starch has around 140–4,400 chains (Tang, Mitsunaga & Kawamura 2006). A blocklet may consist of several amylopectin molecules (Tang, Mitsunaga & Kawamura 2006). Waxy and normal starch granules grow from the hilum toward the granule surface, with the growth rings (120–400 nm) characterized by alternating well-ordered semi-crystalline hard shells consisting of large blocklets and less-ordered semi-crystalline soft shells consisting of small blocklets (French 1984; Gallant, Bouchet & Baldwin 1997).

Pores and channels are found on rice, maize, sorghum, wheat, millet, rye, barley, and waxy rice and maize starch granules (Fannon, Hauber & BeMiller 1992; Fannon, Shull, & BeMiller 1993; Fannon et al. 2004; Huber & BeMiller 1997, 2000; Kim & Huber 2008). Potato, arrow-root, and canna starches, however, display smooth granule surfaces without pores and channels. Channels are from granule surface toward the center cavity at the hilum with different lengths (Huber & BeMiller 1997). Pores have diameters of about 0.1–0.3 μm, while internal channels have diameters of about 0.05–0.1 μm

(Fannon, Hauber & BeMiller 1992; Fannon, Shull, & BeMiller 1993; Gallant, Bouchet & Baldwin 1997). A large number of pores, however, are observed on normal maize starch granules isolated from the endosperm 45 days after pollination (DAP), but few are found on the surface of granules harvested on 30 DAP (Li, Blanco & Jane 2007). These results suggest that pores and channels could also be results of indigenous amylase hydrolysis of starch granules. They also could be produced by drying processes of starch granules, enzyme hydrolysis during wet milling, heat-moisture treatment, or mechanical force. Cavities at the hilum are commonly observed for starch granules. Maize and sorghum starch granules have cavities with irregular shape and often show a star shape (Huber & BeMiller 1997).

9.4.3 Internal Structure of Starch Granule

The biosynthesis of starch granules in the amyloplast and chloroplast initiates at the hilum. The birefringence pattern of a starch granule observed using a microscope with polarized light reflects the arrangement of double helices in the granule from the hilum to the surface of the granule (French 1984). The hilum is the center of the Maltese-cross birefringence. Normal and waxy maize starch granules display a symmetrical Maltese-cross birefringence pattern, reflecting that amylopectin double helices in the granule are aligned radially from the hilum to the periphery and are perpendicular to the surface of the granule (French 1984; Jane 2006, 2007).

The amylose content and amylopectin branch chain-length distribution vary from the hilum to the periphery during the growth of starch granule as revealed by chemical surface gelatinization of starch granules (Jane & Shen 1993; Pan & Jane 2000; Kuakpetoon & Wang 2007) and a developmental study of maize starch (Li et al. 2007, Jiang et al. 2010b). Surface gelatinization of starch granules using a saturated salt solution, such as lithium chloride or calcium chloride, enables removing of starch from the surface. The remaining starch is used for structural analyses to reveal the structural differences between the inner and outer parts of the granules. Results show that amylose is more concentrated at the periphery, and the branch chain length of amylopectin is shorter at the periphery (Jane & Shen 1993; Pan & Jane 2000; Kuakpetoon & Wang 2007). These results are consistent with those obtained from developmental studies of maize starch (Li et al. 2007).

Confocal laser-light scanning micrographs (CLSMs) of starch granules stained with Rhodamine B show that B-type polymorphic-starch granules possess solid internal structures, whereas A-type polymorphic-starch granules display porous internal structures (Jane et al. 2004). The porous internal structure of the A-type polymorphic granules and the solid internal structures of the B-type polymorphic granules are consistent with the scanning electron microscopic images of normal maize (Pan & Jane 2000) and potato starch granules (Jane & Shen 1993), respectively, after the starch molecules on the granule surface being removed using chemical surface gelatinization. These results are in line with small blocklets observed in the A-type polymorphic starch granules and large blocklets in the B-type polymorphic starch granules (Baker, Miles & Helbert 2001; Baldwin et al. 1998).

The formation of porous internal structure in A-type polymorphic-starch granules is proposed as a result of conversion from the hexagonal packing of the B-type polymorph, with 36 water molecules at the center of the packing unit, to monoclinic packing of the A-type polymorph that has an additional double helix at the center (Jane et al. 2004; Jane 2006, 2007; Pan & Jane 2000). Hexagonal packing is kinetically favored, whereas monoclinic packing is thermodynamically favored starch crystalline structure. Amylopectin of the A-type polymorphic starch comprises a large proportion of short A- and B1-chains with scattered branch linkages (Figure 9.2; Jane et al. 1997, 1999), which extend through a single lamella and possess free chain-ends. Those short-chain double helices with free chain-ends have freedom to rearrange into a thermodynamically favored A-type polymorph and, consequently, generate open space and result in a porous structure (Jane 2006). In general, starch granules with the A-type polymorph have porous internal structures. In contrast, the B- and C-type polymorphic starch granules exhibit uniform and solid internal structures (Jane 2006, 2007).

9.4.4 Structure of Elongated Starch Granules

High-amylose maize starch from a double mutant of *ae* and high-amylose modifier gene contains up to 32% elongated granules in addition to spherical and polygonal granules (Li et al. 2008). When viewed under a polarized-light microscope, elongated starch granules exhibit various birefringence patterns, such as several Maltese crosses overlapping in one granule (Figure 9.3, granule b), one or more Maltese crosses in one part of the granule and weak or no birefringence in the rest of the granule (Figure 9.3, granule c), and a granule with weak or no birefringence (Figure 9.3, granule d; Jiang et al. 2010d; Wolf et al. 1964). Variations in the birefringence

pattern of elongated starch granule indicate that the arrangements of double helices in the granule vary between elongated granules and even within an elongated granule. This is also confirmed by the CLSMs of starch granules labeled using 8-aminopyrene-1,3,6-trisulfonic acid, showing spherical granule with fluorescence at the center (Figure 9.4a) but multiple regions of different fluorescence intensities in one elongated granule (Figure 9.4b–g) (Cai et al. 2014b; Glaring et al. 2006; Jiang et al. 2010d).

To understand the formation mechanism of elongated starch granules during biosynthesis of starch, Jiang et al. (2010d) studied the high-amylose maize endosperm collected on 20 DAP using a transmission electron microscopy. The subaleurone layer of high-amylose maize endosperm shows that most of amyloplasts at an early developmental stage contain two or more small starch granules (Figure 9.5a), which fuse and grow into an elongated starch granule at a later development stage (Figure 9.5b–d; Jiang et al. 2010d). This fusion is likely a result of amylose molecules of two adjacent small granules forming double helices and connecting the two small granules because of the high amylose content (Figure 9.6b; Jiang et al. 2010d; Li et al. 2008), which prevents the amyloplast from dividing. Consequently, the adjacent small granules grow into one elongated granule (Figure 9.6c). A proposed model of elongated granule formation in the amyloplast of high-amylose maize endosperm is illustrated in Figure 9.6.

9.4.5 Structure of Disk-Shaped Starch Granules

Wheat, barley, and triticale are known to have large, disk-shaped A-granules and small, spherical B-granules (Ao & Jane 2007; Uttam et al. 2023; Esch et al. 2023). To understand how the chemical structures of amylopectin and amylose relate to the shape of starch granules, Ao and Jane (2007) characterize the structures of amylopectin molecules isolated from both A- and B-granules. The results show that amylopectin of the

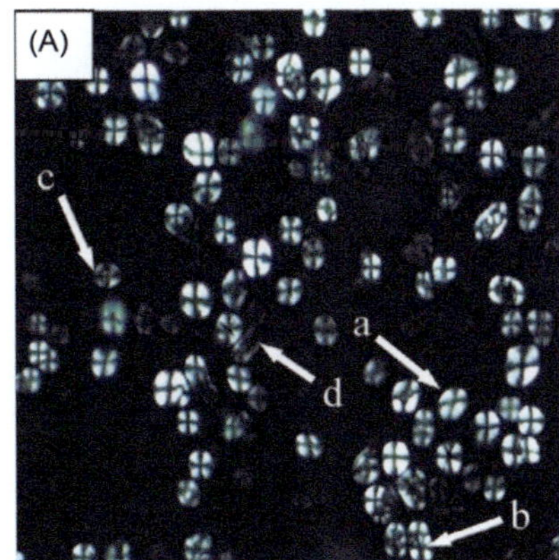
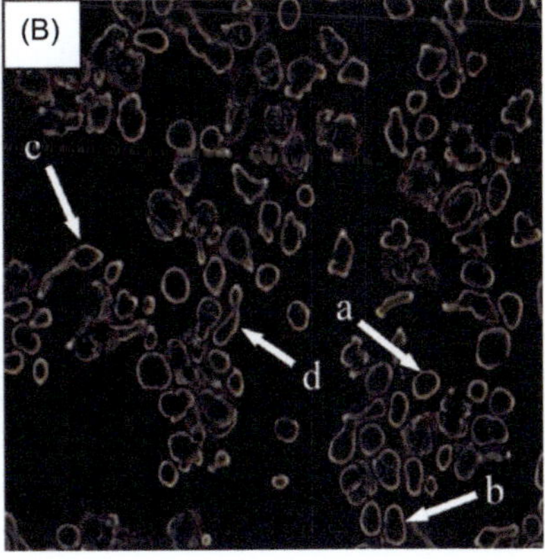

FIGURE 9.3   Polarized (A) and phase contrast (B) light micrographs of high-amylose maize starch (Jiang & Jane 2013). Arrows indicate birefringence patterns of (a) one granule displaying one Maltese cross, (b) Maltese crosses overlapping in one granule, (c) a single granule consisting of one or more Maltese crosses and weak or no birefringence on the remainder of the granule, (d) granule showing weak or no birefringence. Bar = 20 µm.

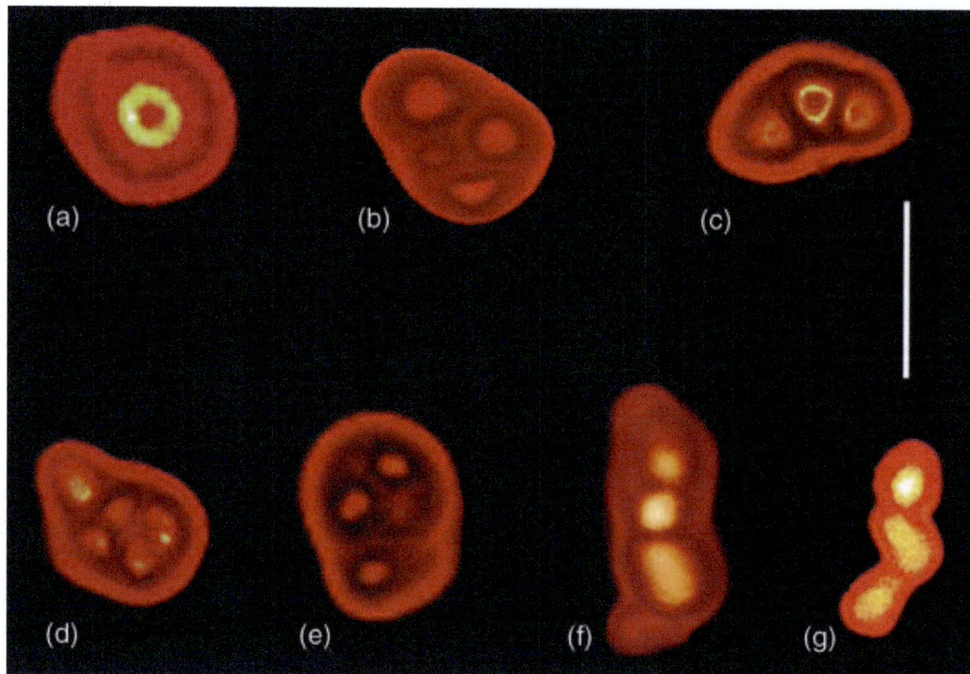

FIGURE 9.4 Confocal laser-scanning microscopic images of high-amylose maize starch granules (Jiang et al. 2010d). Starch granules are labeled using 8-aminopyrene-1,3,6-trisulfonic acid. (a) Spherical granule showing bright colour around the hilum; (b), (c), (d), (e), (f), and (g) show elongated starch granules, displaying multiple regions with intense fluorescence. Bar=10 μm.

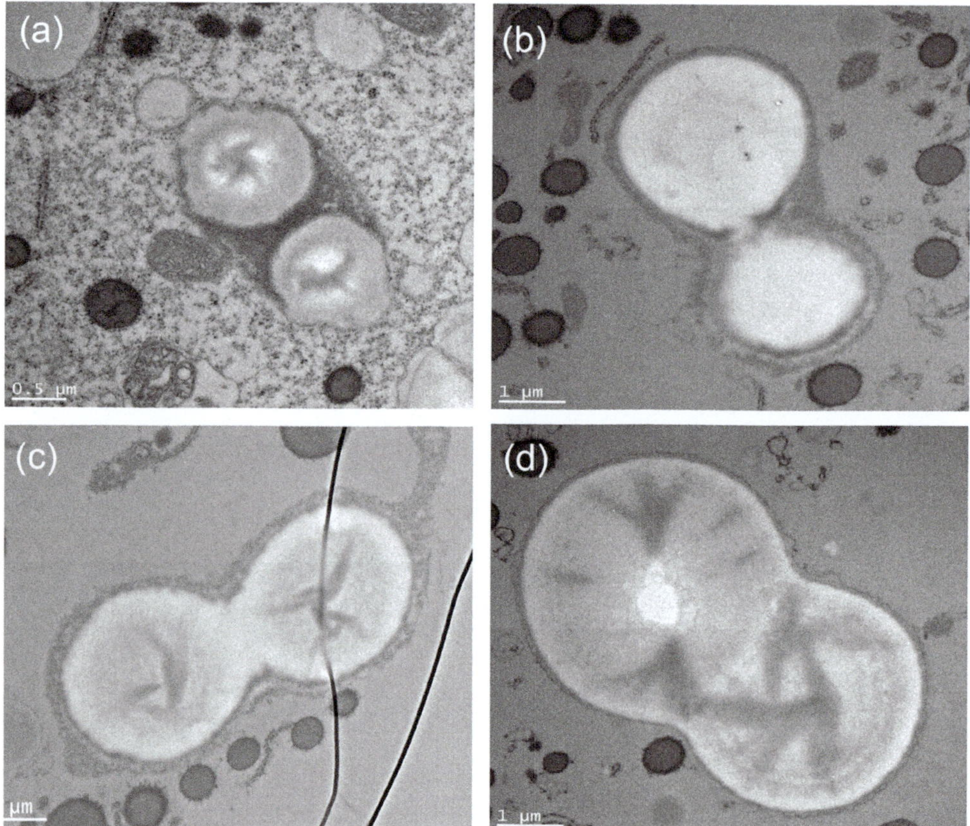

FIGURE 9.5 Transmission electron micrographs of high-amylose maize endosperm tissue harvested on 20 days after pollination (Jiang et al. 2010d). (a) Two starch granules initiated in one amyloplast at an early stage of starch granule development. (b&c) Initial fusion of two starch granules in one amyloplast. (d) Two fused starch granules forming an elongated starch granule.

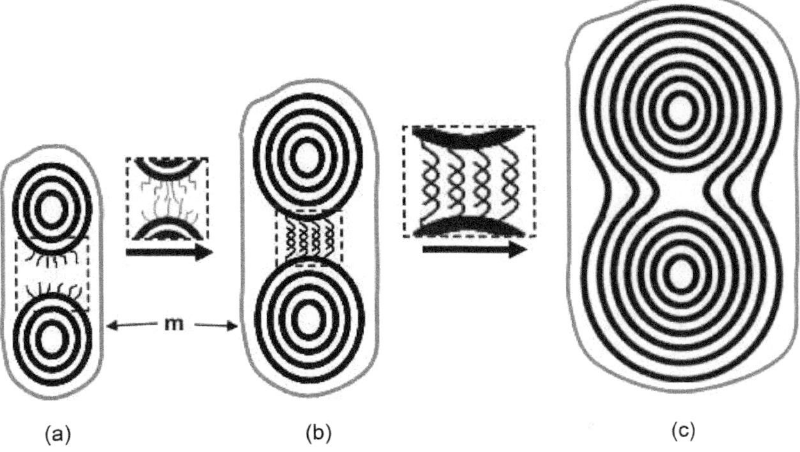

FIGURE 9.6 Proposed formation mechanism of elongated starch granule in the amyloplast during kernel development of high-amylose maize (Jiang et al. 2010d). (a) Two starch granules, each with hilum and growth rings, are initiated in amyloplast at an early stage of granule development. (b) Two granules begin fusion through amylose interaction by forming double helices, which prevent amyloplast division. (c) Fused granules have integrated outer layer growth rings. m: membrane boundary of amyloplast.

A-granules comprises substantially more long-branch chains (B2, B3, and B4 chains) and fewer short-branch chains (A and B1 chains) than the B-granules. The differences in branch chain-length distribution of amylopectin between the A- and the B-granules are substantially greater than that of the large and small granules of other starch varieties, such as maize starch. The large A-granule starch consists of more absolute amylose content than the small B-granule starch, which is consistent with other starches. Structural models of the disk-shaped A-granule and the spherical B-granule relating to their amylopectin branch chain-length distributions are shown in Figure 9.7 (Ao & Jane 2006).

9.4.6 Structure of Kidney-Shaped Legume Starch Granules

Legume starches, having the C-type polymorph, commonly display unique kidney-shaped starch granules, which resemble the shapes of beans and peas. It is puzzling how the kidney-shaped starch granule develops during biosynthesis and has an indentation on the granule. Numerous experimental results obtained using various techniques have been reported by researchers. Normal light micrographs (NLMs) of some pea starch granules (Figure 9.8a) show fissures on the granule surface (Cai et al. 2014c). Polarized light micrographs (PLMs) of the starch granules show a Maltese cross and cracks on the outer part of the starch granule (Figure 9.8b). Scanning electron micrographs of the same starch sample (Figure 9.8c) show indentations on starch granules (Cai et al. 2014c), which appear as the same features as the fissures shown on NLMs (Figure 9.8a). PLMs of pea starch before (Figure 9.8d) and after heating to 66°C (Figure 9.8e) show that starch in the inner part of the granule gelatinizes after heating to 66°C, whereas starch in the outer part of the granule remains crystalline structure

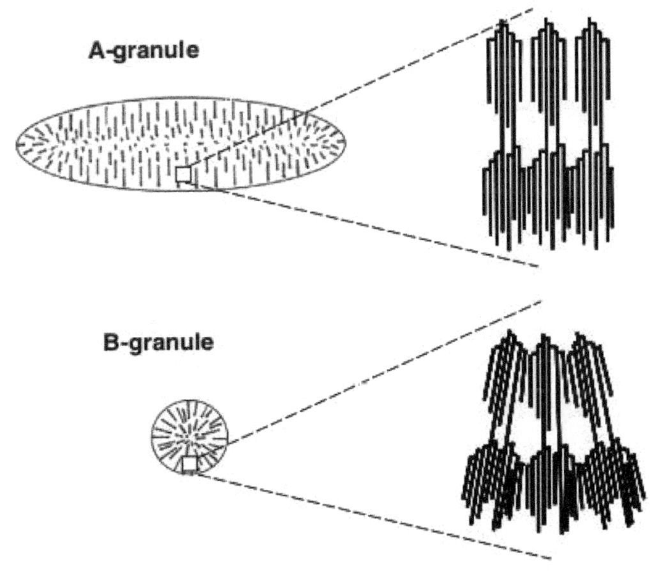

FIGURE 9.7 Proposed granular and molecular structures of the wheat A- and B-granules (Ao & Jane 2006). The amylopectin branch-structures of the A- and B-granules were constructed on the basis of the molar ratios of short chains (A and B1 chains) to long chains (B2 and longer chains). Amylopectin molecules of the A-granule consist of more long chains but lesser short chains, displaying a cylindrical shape and better aligned parallel into a disk-shaped A-granule. Amylopectin molecules of the B-granule consist of more short chains but lesser long chains, displaying a cone shape and fitting in a spherical B-granule.

and displays the A-type polymorph (Bogracheva et al. 1998), indicating the starch in the inner part of the granule possessing the B-type polymorph. The structure was confirmed by Buleon et al. (1998) using synchrotron (Figure 9.8f). CLSMs of various pulse starches stained with Rhodamine B show a homogeneous stain in the inner part of the granule, whereas

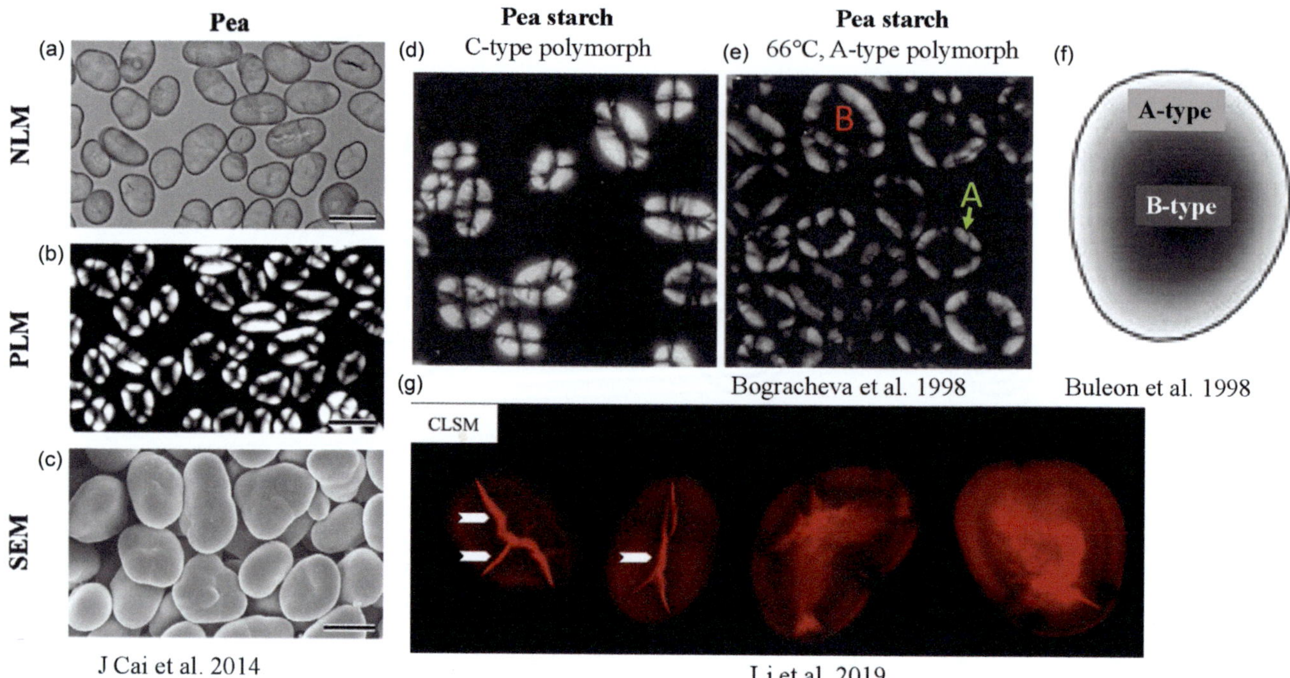

Pea (a)

NLM

PLM

SEM

Pea starch (d) C-type polymorph

Pea starch (e) 66°C, A-type polymorph

(f)

Bogracheva et al. 1998

Buleon et al. 1998

(g)

J Cai et al. 2014

Li et al. 2019

FIGURE 9.8   Normal-light (a), polarized-light (b), and scanning electron (c) micrographs of pea starch, polarized-light micrographs of pea starch before (d) and after heating to 66°C (e), and confocal laser-light scanning micrographs of various pulse starches stained with Rhodamine B (g). Reprinted with permissions.

the outer part shows a porous structure (Figure 9.8g, Li et al. 2019), which is consistent with the feature of the A-type polymorphic starch (Jane et al. 2004). All these results suggest that the inner part B-type polymorphic starch of the granule retains the uniform structure, whereas the outer part porous A-type polymorphic starch develops fissures and indentations, resulting in the kidney-shaped granule.

9.5 CONCLUSION

Starch chains form left-handed double helices in an aqueous medium because their hydrophobic sides tend to fold facing each other to achieve a low energy sate. In starch granules, starch chains are synthesized from the hilum to the periphery, perpendicular to the granule surface, and the packing of amylopectin branch chain double helices results in alternating amorphous lamellae, rich in branch linkages, and crystalline lamellae, having mostly linear chains. The semi-crystalline amylopectin molecules are further organized into blocklets and then into growth rings featured with alternating compact hard shells consisting of large blocklets and soft shells comprising small blocklets. Amylose remains amorphous and is interspersed and intertwined with amylopectin. The branch chain-length distribution of amylopectin determines the type of starch polymorph, that with more short-branch chains develops to the A-type polymorph and that with more

long-branch chains to the B-type. Branch linkages of the A-type starch amylopectin locate in both amorphous and crystalline lamellae, whereas those of the B-type locate in the amorphous lamella. Granules of the A-type polymorph display porous internal structure, whereas those of the B-type polymorph display uniform and solid internal structures. C-type polymorphic starch granules consist of both the A- and B-type polymorphic starch, and the locations of the A- and B-polymorphic starch vary with the botanical origin. High-amylose starch consists of elongated starch granules in addition to spherical and polygonal shapes.

REFERENCES

Ai, Y & Jane, JL 2017. Understanding starch structure and functionality, in *Starch in Food: Structure, Function and Applications*, 2nd Ed, eds M Sjöö & L Nilsson, Woodhead Publishing, Cambridge, UK, pp. 151–178.

Ao, Z, & Jane, JL 2006. Characterization and modeling of the A-and B-granule starches of wheat, triticale, and barley. *Carbohydrate Polymers, 67*, 46–55.

Baker, AA, Miles, MJ, & Helbert, W 2001. Internal structure of the starch granule revealed by AFM. *Carbohydrate Research, 330*, 249–256.

Baldwin, PM, Adler, J, Davies, MC, & Melia, CD 1998. High resolution imaging of starch granule surfaces by atomic force microscopy. *Journal of Cereal Science, 27*, 255–265.

Ball, S, Colleoni, C, & Arias, MC 2015. The transition from glycogen to starch metabolism in Cyanobacteria and Eukaryotes, in *Starch: Metabolism and structure*, ed Y Nakamura, Springer, New York, pp. 93–158.

Banks, W, & Greenwood, CT 1971. The conformation of amylose in dilute solution. *Starch-Stärke, 23*, 300–314.

Bogracheva, TY, Morris, VJ, Ring, SG, & Hedley, CL 1998. The granular structure of C-type pea starch and its role in gelatinization. *Biopolymers, 45(4)*, 323–332.

Buleon, A, Bizot, H, Delage, MM, & Pontoire, B 1987. Comparison of X-ray diffraction patterns and sorption properties of the hydrolyzed starches of potato, wrinkled and smooth pea, broad bean and wheat. *Carbohydrate Polymers, 7*, 461–482.

Buleon, A, Gérard, C, Riekel, C, Vuong, R, & Chanzy, H 1998. Details of the crystalline ultrastructure of C-starch granules revealed by synchrotron microfocus mapping. *Macromolecules, 31*, 6605–6610.

Buleon, A, & Tran, V 1990. Systematic conformational search for the branching point of amylopectin. *International Journal of Biological Macromolecules, 12*, 345–352.

Buleon, A, Véronèse, G, & Putaux, J-L 2007. Self-association and crystallization of amylose. *Australian Journal of Chemistry, 60*, 706–718.

Cai, C, Cai, J, Man, J, Yang, Y, Wang, Z, & Wei, C 2014a. Allomorph distribution and granule structure of lotus rhizome C-type starch during gelatinization. *Food Chemistry, 142*, 408–415.

Cai, C, Zhao, L, Huang, J, Chen, Y, & Wei, C 2014b. Morphology, structure, and gelatinization properties of heterogeneous starch granules from high-amylose maize. *Carbohydrate Polymers, 102*, 606–614.

Cai, J, Cai, C, Man, J, Zhou, W, & Wei, C 2014c. Structural and functional properties of C-type starches. *Carbohydrate Polymers, 101*, 289–300.

Cai, L, & Shi, YC 2013. Self-assembly of short linear chains to A-and B-type starch spherulites and their enzymatic digestibility. *Journal of Agricultural and Food Chemistry, 61(45)*, 10787–10797.

Cai, L, & Shi, YC 2014. Preparation, structure, and digestibility of crystalline A-and B-type aggregates from debranched waxy starches. *Carbohydrate Polymers, 105*, 341–350.

Cai, L, Shi, YC, Rong, L, & Hsiao, BS 2010. Debranching and crystallization of waxy maize starch in relation to enzyme digestibility. *Carbohydrate Polymers, 81(2)*, 385–393.

Calabrese, VT, & Khan, A 1999. Amylose-iodine complex formation without KI: Evidence for absence of iodide ions within the complex. *Journal of Polymer Science Part A: Polymer Chemistry, 37*, 2711–2717.

Chanzy, H, Putaux, JL, Dupeyre, D, Davies, R, Burghammer, M, Montanari, S, & Riekel, C 2006. Morphological and structural aspects of the giant starch granules from Phajus grandifolius. *Journal of Structural Biology, 154*, 100–110.

Cheng, G, Gu, Z, Yang, Y, Wang, X, Zhao, R, Feng, Y, … & Jiang, H 2024. Understanding resistant-starch formation during drying high-amylose maize kernels. *International Journal of Biological Macromolecules, 260*, art. 129419.

Choisnard, L, Wouessidjewe, D, & Putaux, JL 2018. Polymorphism of crystalline complexes of V-amylose with fatty acids. *International Journal of Biological Macromolecules, 119*, 555–564.

Dai, Z 2009. Starch granule size distribution in grains at different positions on the spike of wheat (*Triticum aestivum* L.). *Starch-Stärke, 61(10)*, 582–589.

Donald, AM, Kato, KL, Perry, PA, & Waigh, TA 2001. Scattering studies of the internal structure of starch granules. *Starch-Stärke, 53*, 504–512.

Esch, L, Ngai, QY, Barclay, JE, McNelly, R, Hayta, S, Smedley, MA, Smith, AM, & Sueng, D 2023. Increasing amyloplast size in wheat endosperm through mutation of PARC6 affects starch granule morphology. *New Phytologist, 240*, 224–241.

Everett, WW, & Foster, JF 1959. The conformation of amylose in solution. *Journal of the American Chemical Society, 81*, 3464–3469.

Fan, X, Zhao, L, Zhang, L, Xu, B, & Wei, C 2017. A new allomorph distribution of C-type starch from root tuber of Apios fortunei. *Food Hydrocolloids, 66*, 334–342.

Fannon, JE, Gray, JA, Gunawan, N, Huber, KC, & BeMiller, JN 2004. Heterogeneity of starch granules and the effect of granule channelization on starch modification. *Cellulose, 11*, 247–254.

Fannon, JE, Hauber, RJ, & BeMiller, JN 1992. Surface pores of starch granules. *Cereal Chemistry, 69(3)*, 284–288.

Fannon, JE, Shull, JM, & BeMiller, JN 1993. Interior channels of starch granules. *Cereal Chemistry, 70*, 611–611.

Fonseca, LM, El Halal, SLM, Dias, ARG, & da Rosa Zavareze, E 2021. Physical modification of starch by heat-moisture treatment and annealing and their applications: A review. *Carbohydrate Polymers, 274*, art. 118665.

French, A, & Murphy, VG 1977. Computer modeling in the study of starch. *Cereal Foods World, 22(2)*, 61–70.

French, D 1984. Organization of starch granules, in *Starch: Chemistry and Technology*, 2nd Ed, ed RL Whistler & JN BeMiller, Academic Press, San Diego, pp. 183–247.

Gallant, DJ, Bouchet, B, & Baldwin, PM 1997. Microscopy of starch: Evidence of a new level of granule organization. *Carbohydrate Polymers, 32*, 177–191.

Gidley, MJ, & Bulpin, PV 1987. Crystallisation of malto-oligosaccharides as models of the crystalline forms of starch: Minimum chain-length requirement for the formation of double helices. *Carbohydrate Research, 161*, 291–300.

Gidley, MJ, & Bulpin, PV 1989. Aggregation of amylose in aqueous systems: The effect of chain length on phase behavior and aggregation kinetics. *Macromolecules, 22*, 341–346.

Glaring, MA, Koch, CB, & Blennow, A 2006. Genotype-specific spatial distribution of starch molecules in the starch granule: A combined CLSM and SEM approach. *Biomacromolecules, 7(8)*, 2310–2320.

Harada, A, Adachi, H, Kawaguchi, Y, & Kamachi, M 1997. Recognition of alkyl groups on a polymer chain by cyclodextrins. *Macromolecules, 30*, 5181–5182.

Hizukuri, S 1985. Relationship between the distribution of the chain length of amylopectin and the crystalline structure of starch granules. *Carbohydrate Research, 141*, 285–306.

Hizukuri, S, Fujii, M, & Nikuni, Z 1960. The effect of inorganic ions on the crystallization of amylodextrin. *Biochimica Biophysica Acta, 40*, 346–348.

Huber, KC, & BeMiller, JN 1997. Visualization of channels and cavities of corn and sorghum starch granules. *Cereal Chemistry, 74*, 537–541.

Huber, KC, & BeMiller, JN 2000. Channels of maize and sorghum starch granules. *Carbohydrate Polymers, 41*, 269–276.

Huang, Q, Chen, X, Wang, S, & Zhu, J 2020. Amylose-lipid complex, in *Starch Structure Functionality and Application in Foods*, ed S Wang, Springer, New York, USA, pp. 57–76.

Imberty, A, Buléon, A, Tran, V, & Péerez, S 1991. Recent advances in knowledge of starch structure. *Starch-Stärke, 43*, 375–384.

Imberty, A, Chanzy, H, Perez, S, Buleon, A, & Tran, V 1987. New three-dimensional structure for A-type starch. *Macromolecules, 20,* 2634–2636.

Imberty, A, Chanzy, H, Pérez, S, Bulèon, A, & Tran, V 1988. The double-helical nature of the crystalline part of A-starch. *Journal of Molecular Biology, 201,* 365–378.

Imberty, A, & Perez, S 1988. A revisit to the three-dimensional structure of B-type starch. *Biopolymers, 27,* 1205–1221.

Imberty, A, & Pérez, S 1989. Conformational analysis and molecular modelling of the branching point of amylopectin. *International Journal of Biological Macromolecules, 11,* 177–185.

Jane, J 2006. Current understanding on starch granule structures. *Journal of Applied Glycoscience, 53,* 205–213.

Jane, J 2007. Structure of starch granules. *Journal of Applied Glycoscience, 54,* 31–36.

Jane, J, Atichokudomchai, N, & Suh, D 2004. Internal structures of starch granules revealed by confocal laser-light scanning microscopy, in *Starch: Progress in Structural Studies, Modification and Applications,* eds P Tomasik, VP Yuryev & E Bertoft, Polish Society of Food Technologists, Cracow, PP. 147-156.

Jane, J, Chen, YY, Lee, LF, McPherson, AE, Wong KS, Radosavljevic M, & Kasemsuwan, T 1999. Effects of amylopectin branch chain length and amylose content on the gelatinization and pasting properties of starch. *Cereal Chemistry, 76,* 629–637.

Jane, J, Kasemsuwan, T, Leas, S, Zobel, H, & Robyt, JF 1994. Anthology of starch granule morphology by scanning electron microscopy. *Starch-Stärke, 46,* 121–129.

Jane, J, & Robyt, JF 1984. Structure studies of amylose-V complexes and retrograded amylose by action of alpha amylases and a new method for preparing amylodextrins. *Carbohydrate Research, 132,* 105–118.

Jane, J, & Shen, JJ 1993. Internal structure of the potato starch granule revealed by chemical gelatinization. *Carbohydrate Research, 247,* 279–290.

Jane, J, Wong, KS, & McPherson, AE 1997. Branch-structure difference in starches of A-and B-type X-ray patterns revealed by their Naegeli dextrins. *Carbohydrate Research, 300,* 219–227.

Jane, J, Xu, A, Radosavljevic, M, & Seib, PA 1992. Location of amylose in normal starch granules. I. Susceptibility of amylose and amylopectin to cross-linking reagents. *Cereal Chemistry, 69,* 405–409.

Jiang, H, Campbell, M, Blanco, M, & Jane, JL 2010a. Characterization of maize amylose-extender (*ae*) mutant starches: Part II. Structures and properties of starch residues remaining after enzymatic hydrolysis at boiling-water temperature. *Carbohydrate Polymers, 80,* 1–12.

Jiang, H, Horner, HT, Pepper, TM, Blanco, M, Campbell, M, & Jane, JL 2010d. Formation of elongated starch granules in high-amylose maize. *Carbohydrate Polymers, 80,* 533–538.

Jiang, H, & Jane, JL 2013. Type 2 resistant starch in high-amylose maize starch and its development, in *Resistant Starch: Sources, Applications and Health Benefits,* eds CC Maningat & Y-C Shi, Wiley-Blackwell, Hoboken, NJ, USA, pp. 23–42.

Jiang, H, Lio, J, Blanco, M, Campbell, M, & Jane, JL 2010b. Resistant-starch formation in high-amylose maize starch during kernel development. *Journal of Agricultural and Food Chemistry, 58,* 8043–8047.

Jiang, H, Srichuwong, S, Campbell, M, & Jane, JL 2010c. Characterization of maize amylose-extender (*ae*) mutant starches. Part III: Structures and properties of the Naegeli dextrins. *Carbohydrate Polymers, 81,* 885–891.

Kainuma, K, & French, D 1971. Naegeli amylodextrin and its relationship to starch granule structure. I. Preparation and properties of amylodextrins from various starch types. *Biopolymers: Original Research on Biomolecules, 10,* 1673–1680.

Kainuma, K, & French, D 1972. Naegeli amylodextrin and its relationship to starch granule structure. II. Role of water in crystallization of B-starch. *Biopolymers, 11,* 2241–2250.

Kasemsuwan, T, & Jane, J 1994. Location of amylose in normal starch granules. II. Locations of phosphodiester cross-linking revealed by phosphorus-31 nuclear magnetic resonance. *Cereal Chemistry, 71,* 282–287.

Kasemsuwan, T, Jane, L, Schnable, P, Stinard, P, & Robertson, D 1995. Characterization of the dominant mutant amylose-extender (Ae1–5180) maize starch. *Cereal Chemistry, 72,* 457–464.

Katopo, H, Song, Y, & Jane, J 2002. Effect and mechanism of ultra-high hydrostatic pressure on the structure and properties of starches. *Carbohydrate Polymers, 47,* 233–244.

Kim, HS, & Huber, KC 2008. Channels within soft wheat starch A-and B-type granules. *Journal of Cereal Science, 48,* 159–172.

Kuakpetoon, D, & Wang, YJ 2007. Internal structure and physicochemical properties of corn starches as revealed by chemical surface gelatinization. *Carbohydrate Research, 342(15),* 2253–2263.

Lemke, H, Burghammer, M, Flot, D, Rössle, M, & Riekel, C 2004. Structural processes during starch granule hydration by synchrotron radiation microdiffraction. *Biomacromolecules, 5,* 1316–1324.

Li, C, Dhital, S, Gilbert, RG & Gidley, MJ 2020. High-amylose wheat starch: Structural basis for water absorption and pasting properties. *Carbohydrate Polymers, 245,* art. 116557.

Li, L, Blanco, M, & Jane, JL 2007. Physicochemical properties of endosperm and pericarp starches during maize development. *Carbohydrate Polymers, 67(4),* 630–639.

Li, L, Jiang, H, Campbell, M, Blanco, M, & Jane, JL 2008. Characterization of maize amylose-extender (*ae*) mutant starches. Part I: Relationship between resistant starch contents and molecular structures. *Carbohydrate Polymers, 74,* 396–404.

Li, L, Yuan, TZ, Setia, R, Raja, RB, Zhang, B, & Ai, Y 2019. Characteristics of pea lentil and faba bean starches isolated from air-classified flours in comparison with commercial starches. *Food Chemistry, 276,* 599–607.

Lopez-Rubio, A, Flanagan, BM, Gilbert, EP, & Gidley, MJ 2008. A novel approach for calculating starch crystallinity and its correlation with double helix content: A combined XRD and NMR study. *Biopolymers, 9,* 761–768.

Morrison, WR 1988. Lipids in cereal starches: A review. *Journal of Cereal Science, 8(1),* 1–15.

Nakamura, Y, Kubo, A, Ono, M, Yashiro, K, Mstsuba, G, Wang, Y, Matsubara, A, Mizutani, G, Matsuki, J, & Kainuma, K 2022. Changes in fine structure of amylopectin and internal structures of starch granules in developing endosperms and culms caused by starch branching enzyme mutations of rice endosperm. *Plant Molecular Biology, 108,* 481–496.

Norisuye, T 1996. Conformation and properties of amylose in dilute solution. *Food Hydrocolloids, 10,* 109–115.

Ohtani, T, Yoshino, T, Hagiwara, S, & Maekawa, T, 2000. High-resolution imaging of starch granule structures using atomic force microscopy. *Starch-Stärke, 52,* 150–153.

Pan, DD, & Jane, J 2000. Internal structure of normal maize starch granules revealed by chemical surface gelatinization. *Biomacromolecules, 1(1),* 126–132.

Paris, M, Bizot, H, Emery, J, Buzaré, JY, & Buléon, A 1999. Crystallinity and structuring role of water in native and recrystallized starches by 13C CP-MAS NMR spectroscopy 1: Spectral decomposition. *Carbohydrate Polymers, 39,* 327–339.

Pfannemüller, B 1987. Influence of chain length of short monodisperse amyloses on the formation of A-and B-type X-ray diffraction patterns. *International Journal of Biological Macromolecules, 9*, 105–108.

Putaux, J-L, Molina-Boisseau, S, Momaur, T, & Dufresne, A 2003. Platelet nanocrystals resulting from the disruption of waxy maize starch granules by acid hydrolysis. *Biomacromolecules, 4*, 1198–1202.

Putseys, JA, Lamberts, L, & Delcour, AJ 2010. Amylose-inclusion complexes: Formation, identity and physico-chemical properties. *Journal of Cereal Science*, 51(3), 238–247.

Rao, VSR, & Foster, JF 1963. Studies of the conformation of amylose in solution. *Biopolymers: Original Research on Biomolecules, 1*, 527–544.

Regina, A, Bird, A, Topping, D, Bowden, S, Freeman, J, Barsby, T, … & Morell, M 2006. High-amylose wheat generated by RNA interference improves indices of large-bowel health in rats. *Proceedings of the National Academy of Sciences, 103*(10), 3546–3551.

Regina, A, Kosar-Hashemi, B, Ling, S, Li, Z, Rahman, S, & Morell, M 2010. Control of starch branching in barley defined through differential RNAi suppression of starch branching enzyme IIa and IIb. *Journal of Experimental Botany, 61*(5), 1469–1482.

Robyt, JF (ed) 1998, *Essentials of Carbohydrate Chemistry*, Springer, New York.

Sarko, A, & Wu, HC 1978. The crystal structures of A-, B-, and C-polymorphs of amylose and starch. *Starch-Stärke, 30*, 73–78.

Seung, D, & Smith, AM 2019. Starch granule initiation and morphogenesis—Progress in Arabidopsis and cereals. *Journal of Experimental Botany, 70(3)*, 771–784.

Stevenson, DG, Domoto, PA, & Jane, JL 2006. Structures and functional properties of apple (Malus domestica Borkh) fruit starch. *Carbohydrate Polymers, 63(3)*, 432–441.

Szejtli, J 1998. Introduction and general overview of cyclodextrin chemistry. *Chemical Reviews, 98*, 1743–175

Szejtli, J 2004. Past, present, and future of cyclodextrin research. *Pure and Applied Chemistry, 76*, 1825–1845.

Takahashi, Y, Kumano, T, & Nishikawa, S 2004. Crystal structure of B-amylose. *Macromolecules, 37*, 6827–6832.

Tang, H, Mitsunaga, T, & Kawamura, Y 2006. Molecular arrangement in blocklets and starch granule architecture. *Carbohydrate Polymers, 63*, 555–560.

Uttam Kamble, N, Makhamadjonov, F, Fahy, B, Martins, C, Saalbach, G, & Seung, D 2023. Initiation of B-type starch granules in wheat endosperm requires the plastidial α-glucan phosphorylase PHS1. *Pant Cell, 35*, 4091–4110.

Wang, S, Yu, J, & Yu, J 2008. Conformation and location of amorphous and semi-crystalline regions in C-type starch granules revealed by SEM, NMR, and XRD. *Food Chemistry, 110*, 39–46.

Wang, S, Yu, J, Zhu, Q, Yu, J, & Jin, F 2009. Granular structure and allomorph position in C-type Chinese yam starch granule revealed by SEM, ^{13}C CP/MAS NMR, and XRD. *Food Hydrocolloids, 23*, 426–433.

Wei, C, Qin, F, Zhou, W, Yu, H, Xu, B, Chen, C, … & Liu, Q 2010. Granule structure and distribution of allomorphs in C-type high-amylose rice starch granule modified by antisense RNA inhibition of starch branching enzyme. *Journal of Agricultural and Food Chemistry, 58*, 11946–11954.

Wolf, MJ, Seckinger, HL, & Dimler, RJ 1964. Microscopic characteristics of high-amylose corn starches. *Starch-Stärke, 16*, 377–382.

Wu, HCH, & Sarko, A 1978a. The double-helical molecular structure of crystalline B-amylose. *Carbohydrate Research, 61*, 7–25.

Wu, HCH, & Sarko, A 1978b. The double-helical molecular structure of crystalline A-amylose. *Carbohydrate Research, 61*, 27–40.

Yamashita, Y, Ryugo, J, & Monobe, K 1973. An electron microscopic study on crystallites of amylose V-complexes. *Journal of Electron Microscopy, 22*, 19–26.

Zobel, HF 1988. Molecules to granules: A comprehensive starch review. *Starch-Stärke, 40*, 44–55.

Functional Properties of Starch

10

Wenmeng Liu, David Julian McClements, Zhengyu Jin, and Long Chen

10.1 INTRODUCTION

Starch is an abundant, natural, renewable, safe, cheap, and biodegradable polysaccharide that is found in cereals, tubers, legumes, and various other edible plants (Ma et al. 2017). It is mainly found in the form of starch granules, which mainly consist of the amylose and amylopectin, but also contain minor amounts of proteins, lipids, and trace elements (BeMiller 2011). Variations in starch granule size, morphology, composition, and molecular properties impact the functional properties of starch. These properties (such as pasting, thickening, and gelling) play a crucial role in guiding the application of starch in various fields including foods, cosmetics, biomedicines, textiles, paper, adhesives, and other materials (Hung & Lai 2024), as shown in Figure 10.1. For instance, starch with a high amylopectin content exhibits resistance to aging, resulting in soft breads that are rapidly digested in the human body (Wang et al. 2023a). Currently, researchers are focusing on gaining a better understanding of the functional properties of starch, with the aim of improving its performance and extending its range of applications. In this chapter, we summarize the functional properties of starch. Initially, the factors influencing the gelatinization, retrogradation, rheology, and nutritional attributes of starch are discussed. Then, the molecular and physicochemical mechanisms underlying these properties are reviewed. It is anticipated that this chapter will serve as a valuable reference for enhancing the utilization of starch within foods and other commercial products.

10.2 FUNCTIONAL PROPERTIES

10.2.1 Gelatinization

10.2.1.1 Gelatinization mechanism

The gelatinization of starch forms the basis for many of its applications in foods, as it leads to modifications in the appearance, texture, and stability of foods (Zhang et al. 2015a). Gelatinized starches that form high-viscosity pastes can be used for their thickening, binding, and stabilizing properties (Kumari, Kaur, & Thiruvalluvan 2024). A wide range of instruments are available for the assessment of starch gelatinization, with the most widely used being the rapid visco analyzer (RVA) due to its speed and simplicity (Cozzolino 2016). The gelatinization of starch involves the swelling and then disruption of starch granules as they are heated in water, which leads to the leaching of amylose molecules and the destruction of the ordered internal structure of the granules. During cooling, new hydrogen bonds are formed between the starch molecules, resulting in the formation of a paste. During the initial stage of gelatinization, water enters the amorphous growth rings inside the granules, which causes them to swell and amylose molecules to leach out. During this phase, only the amorphous region undergoes significant expansion, leading to breakdown of the crystalline structure. This is followed by helix-helix side chain dissociation and then a helix-coil transition, which is referred to as double helical dissociation (Li 2022; Punia 2020; Waigh et al. 2000).

Starch gelatinization parameters, such as peak viscosity (PV), final viscosity (FV), setback viscosity (SB), breakdown viscosity (BD), trough viscosity (TV), and pasting temperature (PT), can be obtained when a starch suspension is subjected to a standardized heating-holding-cooling procedure within an RVA instrument (Liu et al. 2024b). The peak viscosity is related to the ability of starch particles to absorb water, swell, and dissociate. Often, the larger the PV, the stronger the dissociation ability, which influences the quality of the final product. The breakdown viscosity is related to the shear resistance exhibited by the expanded starch particles, which is commonly employed for assessing the shear stability of starch-based products. The trough viscosity represents the lowest viscosity observed during the cooling stage, while the final viscosity denotes the viscosity of the starch paste after heating and cooling, resulting in a sticky consistency. The setback viscosity reflects both degradation and rearrangement processes occurring within amylose molecules, which is a crucial parameter for assessing starch aging. Lastly, the pasting temperature is the critical temperature at which water absorption and expansion take

DOI: 10.1201/9781003464396-10

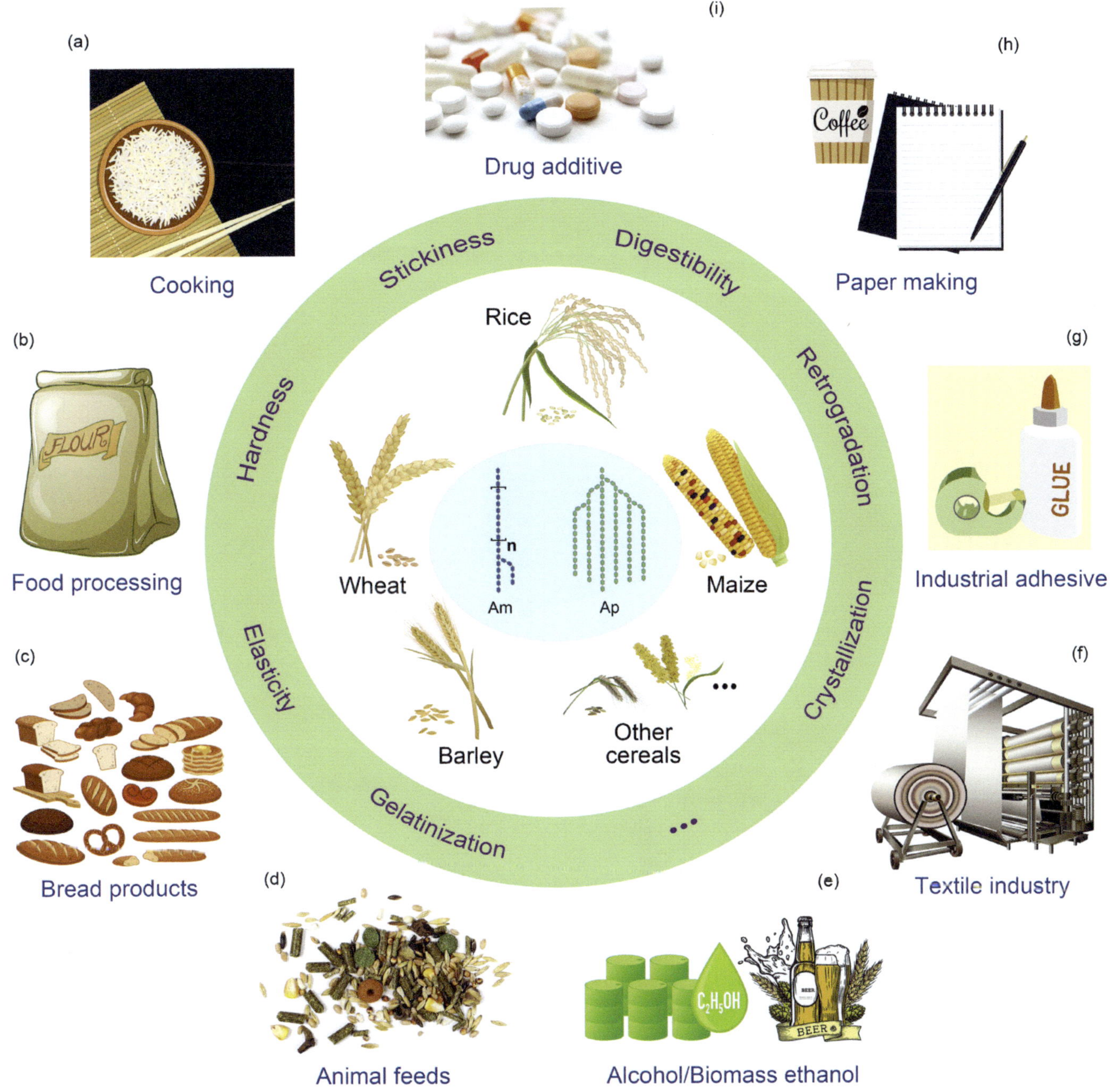

FIGURE 10.1 The dependence of starch properties on different factors influences the application of starch in foods and other products (Huang et al. 2021).

place during starch gelatinization, which serves as an indicator for the cooking behavior of starch. A higher PT necessitates a correspondingly higher cooking temperature (Zhang et al. 2021). These starch gelatinization characteristics are pivotal for determining the suitability of different starches for different applications. For instance, starches with low SB values are often used in frozen food products, whereas those with high PV values are used as gelling or thickening agents (Liu et al. 2024b). There are significant variations in the gelatinization characteristics of starches from different biological origins and that have received different processing treatments, which are attributed to differences in granule size, composition, physical state, molecular characteristics, moisture contents, and concentrations. To fully understand the gelatinization mechanisms of starch, it is important to have an appreciation of the major factors that impact gelatinization.

10.2.1.2 Factors affecting gelatinization

The gelatinization characteristics of a particular starch ingredient depends on numerous factors, including its botanical origin, maturity, and processing, as these factors impact the

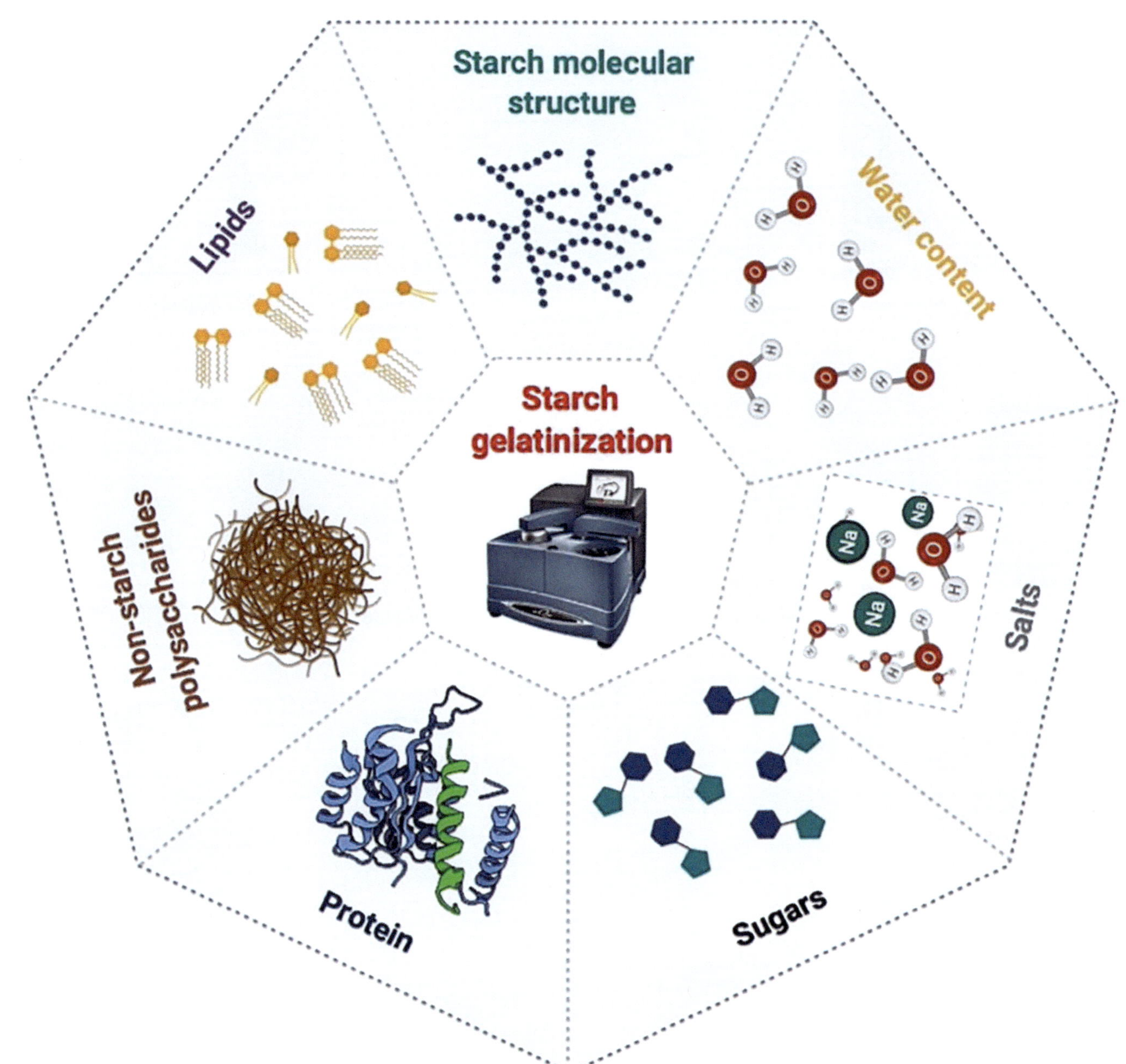

FIGURE 10.2 Factors affecting the gelatinization characteristics of starch (Li 2022).

composition and structural organization of the starch granules (Figure 10.2). In this section, various internal factors (such as origin and maturity) and external factors (such as processing conditions) that influence starch properties are reviewed.

Internal factors: The biological origin, maturity, and growing conditions of starch granules influence their composition, structure, size, and shape, which alter their gelatinization characteristics. Starch mainly consists of carbohydrates (amylose and amylopectin), as well as trace components of water, proteins, lipids, and minerals (K, Ca, P, Na, Mg, Si, Fe, Al, and Zn) (Bangar et al. 2021; Contreras-Jiménez et al. 2019). Studies have shown that the rheology of gelatinized starches is influenced by these trace components, such as their mineral

composition (Contreras-Jiménez et al. 2019). The water content of starch granules also impacts their gelatinization behavior. A small amount of water is required to promote starch gelatinization, while too much water prevents the formation of starch pastes.

The gelatinization of starch pastes is strongly influenced by the properties and behavior of amylose and amylopectin (Wang et al. 2018). For example, waxy, native, and high amylose corn starches exhibit different gelatinization degrees based on differences in their amylose and amylopectin contents, with the waxy corn starch having the highest paste viscosity due to its higher amylopectin content (Luo et al. 2020). The gelatinization temperature and retrogradation of

into an ordered structure, which leads to an increase in the viscoelasticity and hardness of starch gels. Rong et al. (2022) also reported that starch type influences the shear modulus of starch pastes, which was attributed to differences in amylose content. Indeed, the G' and G'' values of waxy starches were the lowest. However, Luo et al. (2020) reported that the viscoelasticity of high amylose corn starch was lower than that of waxy corn starch and ordinary corn starch, indicating that the amylose ratio alone does not solely determine this property.

The rheological properties of starch also depend on endogenous lipids and proteins. For instance, G' has been reported to depend on the endogenous protein concentration in starch, which was attributed to the protective effects of these proteins on starch granule integrity (Li et al. 2019). The presence of endogenous lipids in starch granules can inhibit their expansion during heating, thereby altering their viscoelasticity. Furthermore, the removal of endogenous lipids often increases the viscoelasticity of starches due to their ability to form V-type complexes with amylose (Ding et al. 2021). The influence of endogenous lipids on starch characteristics also depends on the starch type, extraction method, and composition (Zhang et al. 2019). It is important to note that the nature of the endogenous lipids in starch granules is influenced by the origin of the starch.

Before heating, the shear modulus of starch suspensions is relatively low because of the limited interactions between the raw starch granules. As the temperature is raised, starch granules absorb water and expand, which leads to an increase in the shear modulus because the starch granules become more tightly packed together. As the temperature continues to rise, amylose leaches from the starch granules and dissolves in the surrounding water. When the system is cooled, the amylose molecules undergo a coil-to-helix transition and then the helices form hydrogen-bonded cross-links with each other, leading to the formation of a three-dimensional network structure that increases the shear modulus (Wu et al. 2018). Upon prolonged shearing and heating, some amylopectin molecules may also form a continuous network structure with amylose. Notably, partially gelatinized starch has a higher order and denser structure than fully gelatinized starch, leading to higher G' and G'' values (Xiao et al. 2020).

External factors: Additives also play an important in regulating the viscoelastic properties of starch gels. The addition of hydrocolloids to starch suspensions significantly enhances the G' and G'' values, which has been attributed to the entanglement and cross-linking of different biopolymers in the composite system (Liu et al. 2022b). The presence of this three-dimensional network structure enhances the viscoelasticity of starch-hydrocolloid gels (Zhang et al. 2018). The addition of polyphenols (glycyrrhizic acid) has been reported to enhance the viscoelasticity of corn starch in a dose-dependent manner, thereby demonstrating their potential for modulating gel properties. Indeed, the rheological properties of starch-based edible inks can be optimized using these additives so as to increase

their suitability for 3D food printing applications (Liu et al. 2024a). However, the addition of tea polyphenols has been reported to hinder the formation of viscoelastic starch-based gels by impeding the formation of amylose double helix junction zones in the starch network (Huang, Wu, & Chen 2022; Li et al. 2023). Thus, additives play a crucial role in influencing the rheological characteristics of starch, which means they can be used to tailor its properties for specific applications.

The high viscosity and strong shear thinning properties of starch pastes are widely utilized in yogurts, puddings, sauces, dressings, and other semi-solid food products, whereas the gelling properties of starches are widely used in pastas, cookies, breads, and other solid food products. Consequently, starch suspensions with the appropriate rheological properties should be selected for each application.

10.2.3 Retrogradation

10.2.3.1 Retrogradation mechanism

The recrystallization of gelatinized starches during cooling and storage is referred to as retrogradation (Wang et al. 2015). Retrogradation involves the rearrangement and association of starch molecules, resulting in a transition from a disordered amorphous state to a more ordered semi-crystalline state. Retrogradation can be characterized through a series of molecular and physicochemical transformations that occur simultaneously and are often linked to each other, such as an increase in crystallinity, viscosity, gel strength, and turbidity, as well as water leakage, and a reduction in digestibility (Matignon & Tecante 2017; Patel et al. 2017). These changes often adversely affect the desirable quality attributes of foods.

The retrogradation of starch can be divided into short-term and long-term retrogradation processes. The former involves the rearrangement and recrystallization of amylose molecules after starch gelatinization, while the latter involves the recrystallization of the amylopectin molecules (Nasseri & Mohammadi 2014). Amylopectin is more prone to forming intramolecular and intermolecular bonds, thereby accelerating long-term degradation (Martinez et al. 2018; Matalanis, Campanella, & Hamaker 2009). The retrogradation of starch has several implications for food applications. Retrogradation can lead to the hardening of gel matrices, thereby reducing the chewability of bread, which is undesirable (Hayes et al. 2020). In contrast, retrogradation reduces the rate of enzymatic starch digestion in the gastrointestinal tract, which may be beneficial for health (Rolandelli, Rodríguez, & Buera 2024). Commonly, the degree of retrogradation in a starch-based product is determined by measuring changes in its texture (Liu et al. 2021), as these textural changes impact food processing, quality, and digestion. Therefore, a comprehensive understanding and regulation of starch retrogradation is essential for controlling product quality and improving production operations.

10.2.3.2 Factors affecting starch retrogradation

Starch retrogradation is impacted by various factors, which can be categorized as internal or external factors depending on their origin:

Internal factors: The retrogradation of starch is influenced by internal factors such as the origin, maturity, growth conditions, and processing of the starch granules, as these parameters influence the amylose to amylopectin ratio, the length of the amylopectin side chains, the degree of polymerization, and the granule dimensions (Xu et al. 2013; Zhu et al. 2020). The linear amylose chains in starch pastes have a high propensity to cross-link and form crystalline zones, whereas only the linear side chains in amylopectin molecules can form localized crystallization zones. Thus, high-amylose corn starch is more prone to retrograde than regular corn starch (Liu et al. 2017), which can be attributed to the faster rearrangement and association of the linear amylose chains. Additionally, amylose requires less space for the formation of double helices or crystalline regions than amylopectin. Due to its highly branched structure, amylopectin has a slower association time and consequently takes longer to retrograde than amylose (Chang et al. 2021).

The retrogradation of starch is also influenced by the branching degree and chain length of the amylopectin and amylose molecules. Studies on the retrogradation of sorghum, corn, and rice starch showed that amylopectin molecules with longer branches tend to have stronger intramolecular and intermolecular interactions, thereby promoting starch retrogradation (Matalanis et al. 2009). In another study, Chang et al. (2021) reported that the presence of longer branches in amylopectin influenced the formation of longer and more stable double helices in starch gels. Starch retrogradation is also influenced by the fine structure of amylose and amylopectin molecules, such as the distribution of chain lengths (Tao et al. 2019).

The extent of retrogradation has also been reported to depend on the level of polymerization of the starch molecules. For instance, Vandeputte et al. (2003) reported that a degree of polymerization (DP) of glucose units in amylopectin side chains below 6~9 or above 25 resulted in a lower starch aging rate. This effect was primarily attributed to the fact that a minimum degree of polymerization (DP=6) is required for double helices to form in starch. Consequently, if the DP is below this value, helices will not form, and so hydrogen-bonded cross-links between helices cannot occur. In contrast, if the DP is too high, the starch molecules move too slowly to effectively come into contact and form cross-links.

The degree of retrogradation varies among different types of starch, and compositional differences also play a significant role as an influencing factor. For instance, starch with higher lipid content can potentially retard the aging process due to the formation of amylose–lipid complexes during gelatinization. Higher lipid content delayed the degradation rate of oat starch, and the aging degree of oat starch was intensified after lipid removal. However, the aging degree varies among different types of starch. In comparison to wheat and corn starch, oat starch exhibits lower aging degree due to its high viscosity and low hardness (Hoover et al. 1994; Punia et al. 2020).

External factors: The retrogradation of starch is also influenced by various external factors, including emulsifiers, small sugar molecules, and food additives. Their impact on starch retrogradation is dependent on their specific species and properties, as well as the conditions in which retrogradation takes place. For instance, glucose, fructose, and maltose can affect starch degradation to varying degrees depending on the temperature at which it occurs (Guo & Du 2014). Moreover, differences in polysaccharide nature can also lead to variations in retrogradation degree. A study reported that low concentrations and low polymerization of synanthrin can inhibit short-term amylose degradation (Luo et al. 2017). This is achieved by facilitating water binding to alter the water distribution within the starch–hydrocolloid complex system, as well as engaging in hydrogen bonding interactions with starch to impede recrystallization and prevent starch aging (Chen et al. 2017). However, it is crucial to exercise precise control over the appropriate quantity of supplementation. In previous investigation, it can be observed that the incorporation of *bamboo shoot* polysaccharides within the range of 3%–6% facilitated the retrogradation process of starch gel, whereas an inclusion level between 9% and 15% impeded both short-term and long-term retrogradation of starch (Zheng et al. 2024). The impact of sugars on starch aging is primarily attributed to the following factors: (1) These sugars can form a starch-sugar-starch structure by binding with starch molecules, thereby impeding starch gelatinization. (2) Sugar has the ability to either promote or inhibit the dehydration contraction during starch retrogradation. (3) Both sugar and water can interact with starch, resulting in plasticization of starch chains and an increase in their fluidity.

Lipids can hinder the retrogradation process by forming starch–lipid complexes with amylose through hydrophobic interactions. However, the extent of retrogradation relies on the lipid's characteristics. Lipids, such as fatty acids, possess both hydrophilic and hydrophobic groups, and the length of the lipid chain also impacts the formation of lipid–starch complexes. Longer chain lipids exhibit lower hydrophilicity compared to shorter chain lipids, resulting in a more stable starch–lipid complex formation. Certain lipids can also form complexes with amylopectin, which further aids in inhibiting the degradation process. Additionally, it has been reported that the impact of lipids on starch retrogradation is influenced by factors such as starch type, lipid properties, and environmental conditions during degradation. For instance, Prakaywatchara, Wattanapairoj, and Thirathumthavorn (2018) discovered that different lipids' inhibitory effects on rice flour aging depend on their type and content. Conversely, Garcia and Franco (2015) found that only when added in appropriate amounts can starch retrogradation be inhibited.

10.2.4 Nutritional Aspects

Starch is a major source of energy in the human diet. The rate and extent of starch digestion and glucose absorption within the human gastrointestinal tract influences the nutritional and health properties of starchy foods. The gastrointestinal fate of starch is usually monitored using *in vitro* digestion methods that simulate the digestive characteristics of the human body, which provides a useful indication of its nutritional effects. Starch is primarily digested by α-amylase and glucosidases secreted in the mouth and small intestine, resulting in the production of glucose that can be absorbed by the human body (Chen et al. 2022). Based on its digestibility, starch can be categorized into three types: rapidly digestible starch (RDS), slowly digestible starch (SDS), and resistant starch (RS) (Teixeira et al. 2016). RDS leads to a large spike in blood sugar levels and insulin secretion, which can contribute to various health complications such as obesity, diabetes, and cardiovascular diseases. SDSs are primarily broken down at a slow rate within the small intestine, leading to a more moderate increase in blood sugar levels. The majority of resistant starch is not hydrolyzed in the digestive tract, but it can be fermented by intestinal flora, which are mainly located in the colon (Liu et al. 2022a). Fermentation of resistant starch generates short-chain fatty acids (SFCAs), which have beneficial effects on human health by regulating the intestinal flora.

10.2.4.1 Factors affecting starch digestibility

Like its other physicochemical and functional properties, the digestibility of starch is influenced by both internal and external factors:

Internal factors: The internal factors influencing starch digestibility are related to the composition and structure of the starch granules themselves, such as their amylose content, chain length, granular structure, crystal type, and endogenous components (Table 10.1). A positive correlation has been reported between the amylose content of starch and its resistance to enzymatic hydrolysis (Chi et al. 2021). Other researchers have also reported a positive correlation between amylose content and RS content (Yang et al. 2021c). If starch granules retain some of their semi-crystalline structure within the gastrointestinal tract, then the amorphous regions are digested more rapidly than the crystalline regions because it is easier for the enzymes to access the starch molecules in these regions. The distribution of amylose chain lengths has also been shown to impact starch digestibility. Starch digestibility tends to be lower for amylose molecules with shorter chains, which may be due to the formation of more numerous dense crystalline regions by these linear starch molecules (Gong et al. 2019). Conversely, Chi et al. (2021) reported that a higher proportion of short-chain amylose molecules led to a stronger resistance of the starch to digestive enzymes, thus inhibiting the rapid increase in postprandial blood sugar levels. These results suggest that different types of starch have different digestibilities, which may be important when selecting a type of starch for a particular food application.

TABLE 10.1 Possible factors and mechanisms affecting starch digestibility

INTERNAL FACTORS	DIGESTIBILITY	POSSIBLE MECHANISM	REFERENCES
Amylose content	Starch with high amylose content has low digestibility	Amylose is resistant to enzymatic hydrolysis	Luo et al. (2021)
Chain length	Amylopectin with more long chains has lower digestibility	The presence of larger amylopectin molecules is associated with an increased abundance of shorter branches and a decreased number of long chains, resulting in a deceleration of starch hydrolysis by α-amylase	Li et al. (2021a)
Degree of polymerization	Amylopectin with intermediate (DP 25–36) inhibited enzymatic digestion	The medium degree of polymerization promotes the formation of rigidity and crystalline structure	Kim et al. (2013)
Crystal type	A-type starch presented higher digestibility than B-type starch	The crystalline site of type starch is sensitive to digestive enzymes	Chi et al. (2021)
Starch type	The content of RS in pea starch is between that of corn starch and potato starch	Pea starch belongs to C-type crystal structure and its enzyme sensitivity is between A type and B type starch	Ouyang et al. (2021)
Moisture content	Reducing water content increased *in vitro* digestibility of extruded high-amylose jackfruit starch	Decreased water content presents a lower double helix structure and long/short-range molecular sequence	Zhang et al. (2022)
Particle size	The small-sized starch granules exhibited increased starch digestibility	The smaller the starch particles, the greater the specific surface area and the increased contact area between enzyme and substrate. This facilitates the digestion of small grains of starch	Dong et al. (2021)
Endogenous protein	The presence of endogenous proteins prevents the digestion of starch	Endogenous proteins impede digestion by limiting starch expansion and gelatinization and physically impeding enzyme–starch interactions	Wu et al. (2024)
Endogenous lipid	Removing non-starch components increased the digestibility of starch	The presence of lipids can form starch–lipid complexes with amylose	Yang et al. (2021a)

Endogenous proteins and lipids in starch granules can serve as physical barriers that impede the accessibility of enzymes to starch molecules, thereby diminishing their digestibility. For instance, the rate of digestion of oat starch increased when the proteins were removed (Tang et al. 2019). Furthermore, an increase in protein content corresponded to greater resistance against starch hydrolysis (Li et al. 2021b). Additionally, endogenous proteins can form hydrogen bonds with amylose, leading to the formation of complexes that reduce the ability of this enzyme to digest starch (Lu et al. 2021).

Endogenous lipids can form complexes with starch, which restricts the accessibility of digestive enzymes to the starch molecules. These starch–lipid complexes can regulate starch digestibility by inducing changes in the short-range ordered structure, crystalline structure, and helix content of the starch. However, studies have shown that the interactions of lipids with amylopectin are weaker than those with amylose and therefore have less effect (Lin et al. 2020; Oyeyinka, Singh, & Amonsou 2021).

The crystalline structure of starch is a crucial determinant of its digestibility (Wang et al. 2023b). Starch may adopt different crystalline structures, with type A starch exhibiting a more compact structure than type B starch and therefore being less digestible (Chi et al. 2021). The crystal structure of type C starch is between that of types A and B starch, resulting in an intermediate enzyme sensitivity (Ouyang et al. 2021). V-shaped crystalline structure has been reported to be more resistant to starch digestion due to its dense structure. Researchers have reported that the greater the relative crystallinity of starch, the lower its *in vitro* digestibility (Chi et al. 2021). Similarly, it has been observed that a higher degree of crystallinity in starch reduces enzyme accessibility, thereby leading to an increased resistance to digestion, which is important for the development of low glycemic index (GI) foods.

The digestibility of starch is also influenced by the structure of the starch granules and molecules. The presence of a short-range ordered structure can impede the interaction between enzymes and starch, thereby reducing the contact area and ultimately inhibiting starch hydrolysis (Chi et al. 2021). The surface morphology and particle size of starch granules also influence its digestibility. For instance, smaller starch granules are digested more rapidly, regardless of plant origin, which can be attributed to a higher surface area of starch exposed to the digestive enzymes (Dong et al. 2021). Other studies have also shown that starch granules with small sizes are digested more rapidly by amylase (Tawheed et al. 2018). The presence of small holes and grooves on the surfaces of starch granules facilitates the penetration of amylase into the interior of the granules, while granules with a smoother surface exhibit strong resistance to amylase activity.

External factors: The digestibility of starch is also significantly influenced by external factors such as processing operations, additives, and solution conditions (Lal et al. 2021). The presence of exogenous substances, such as salts, sugars, and hydrocolloids, can alter the digestibility of starch. Salt ions can penetrate into starch granules and disrupt the hydrogen bonds between the starch chains, thereby enhancing starch digestibility.

The presence of high levels of sugars can increase the packing of the starch chains, which reduces the contact between starch and digestive enzymes, thereby reducing starch digestion (Yang et al. 2023). The digestibility of starch can also be controlled by treating it with non-thermal processing technologies, such as microwave heating, pulsed electric field treatments, and ultrasound irradiation, which can be attributed to their ability to induce structural changes in the starch granules (Liu et al. 2023).

10.3 FUTURE & CHALLENGES

The properties of starch are influenced by internal and external factors, such as the composition and structure of starch granules and molecules, as well as processing operations, additives, and solution conditions. Starch is an extremely versatile ingredient that can be used in a broad range of industries because of its diverse functional attributes, including thickening, gelling, binding, and film-forming. A great deal of knowledge about the functionality of starch has already been gained, but further research is still required. More studies are required on the internal molecular structure and physicochemical properties of different kinds of starch granules. More research is needed on the development of physical, chemical, and biological methods to improve and extend the functional performance of starch. In addition, further research is required to understand the factors that impact the digestibility of different kinds of starch within the human body and to design starchy foods with specific digestibility profiles. Starches with lower digestibility rates could be used to create foods that reduce the risk of chronic diet-related diseases, such as diabetes, overweight, and heart disease.

REFERENCES

Agi, A., Junin, R., Gbadamosi, A., Abbas, A., Azli, N. B., & Oseh, J. 2019. Influence of nanoprecipitation on crystalline starch nanoparticle formed by ultrasonic assisted weak-acid hydrolysis of cassava starch and the rheology of their solutions. *Chemical Engineering and Processing - Process Intensification, 142*, art. 107556.

Ahmad, F. B., & Williams, P. A. 1999. Effect of salts on the gelatinization and rheological properties of sago starch. *Journal of Agricultural and Food Chemistry, 47*(8), 3359–3366.

Allan, M. C., Rajwa, B., & Mauer, L. J. 2018. Effects of sugars and sugar alcohols on the gelatinization temperature of wheat starch. *Food Hydrocolloids, 84*, 593–607.

Bangar, S. P., Ashogbon, A. O., Dhull, S. B., Thirumdas, R., Kumar, M., Hasan, M., Chaudhary, V., & Pathem, S. 2021. Proso-millet starch: Properties, functionality, and applications. *International Journal of Biological Macromolecules, 190*, 960–968.

BeMiller, J. N. 2011. Pasting, paste, and gel properties of starch-hydrocolloid combinations. *Carbohydrate Polymers, 86*(2), 386–423.

Chang, Q., Zheng, B., Zhang, Y., & Zeng, H. 2021. A comprehensive review of the factors influencing the formation of retrograded starch. *International Journal of Biological Macromolecules, 186*, 163–173.

Chen, J., Cui, Y., Shi, W., Ma, Y., & Zhang, S. 2023. The interaction between wheat starch and pectin with different esterification degree and its influence on the properties of wheat starch-pectin gel. *Food Hydrocolloids, 145*, art. 109062.

Chen, L., Tian, Y., Tong, Q., Zhang, Z., & Jin, Z. 2017. Effect of pullulan on the water distribution, microstructure and textural properties of rice starch gels during cold storage. *Food Chemistry, 214*, 702–709.

Chen, S., Qin, L., Chen, T., Yu, Q., Chen, Y., Xiao, W., Ji, X., & Xie, J. 2022. Modification of starch by polysaccharides in pasting, rheology, texture and in vitro digestion: A review. *International Journal of Biological Macromolecules, 207*, 81–89.

Chi, C., Li, X., Huang, S., Chen, L., Zhang, Y., Li, L., & Miao, S. 2021. Basic principles in starch multi-scale structuration to mitigate digestibility: A review. *Trends in Food Science & Technology, 109*, 154–168.

Contreras-Jiménez, B., Vázquez-Contreras, G., de los Ángeles Corncjo-Villegas, M., del Real-López, A., & Rodríguez-García, M. E. 2019. Structural, morphological, chemical, vibrational, pasting, rheological, and thermal characterization of isolated jicama (*Pachyrhizus spp.*) starch and jicama starch added with Ca(OH)₂. *Food Chemistry, 283*, 83–91.

Cozzolino, D. 2016. The use of the rapid visco analyser (RVA) in breeding and selection of cereals. *Journal of Cereal Science, 70*, 282–290.

Ding, Y., Cheng, J., Lin, Q., Wang, Q., Wang, J., & Yu, G. 2021. Effects of endogenous proteins and lipids on structural, thermal, rheological, and pasting properties and digestibility of adlay seed (Coix lacryma-jobi L.) starch. *Food Hydrocolloids, 111*, art. 106254.

Dong, S., Fang, G., Luo, Z., & Gao, Q. 2021. Effect of granule size on the structure and digestibility of jackfruit seed starch. *Food Hydrocolloids, 120*, art. 106964.

Garcia, M. C., & Franco, C. M. L. 2015. Effect of glycerol monostearate on the gelatinization behavior of maize starches with different amylose contents. *Starch-Starke, 67*(1–2), 107–116.

Gong, B., Cheng, L., Gilbert, R. G., & Li, C. 2019. Distribution of short to medium amylose chains are major controllers of in vitro digestion of retrograded rice starch. *Food Hydrocolloids, 96*, 634–643.

Guo, L., & Du, X. F. 2014. Retrogradation kinetics and glass transition temperatures of *Pueraria lobata* starch, and its mixtures with sugars and salt. *Starch-Starke, 66*(9–10), 887–894.

Hayes, A. M. R., Okoniewska, M., Martinez, M. M., Zhao, B., & Hamaker, B. R. 2020. Investigating the potential of slow-retrograding starches to reduce staling in soft savory bread and sweet cake model systems. *Food Research International, 138*, art. 109745.

Hoover, R. 2001. Composition, molecular structure, and physicochemical properties of tuber and root starches: A review. *Carbohydrate Polymers, 45*(3), 253–267.

Hoover, R., Vasanthan, T., Senanayake, N. J., & Martin, A. M. 1994. The effects of defatting and heat-moisture treatment on the retrogradation of starch gels from wheat, oat, potato, and lentil. *Carbohydrate Research, 261*(1), 13–24.

Huang, L., Tan, H., Zhang, C., Li, Q., & Liu, Q. 2021. Starch biosynthesis in cereal endosperms: An updated review over the last decade. *Plant Communications, 2*(5), art. 100237.

Huang, Y., Wu, P., & Chen, X. D. 2022. Mechanistic insights into the influence of flavonoids from dandelion on physicochemical properties and in vitro digestibility of cooked potato starch. *Food Hydrocolloids, 130*, art. 107714.

Hung, S.-H., & Lai, L.-S. 2024. Changes in the pasting and rheological properties of wheat, corn, water caltrop and lotus rhizome starches by the addition of Annona montana mucilage. *International Journal of Biological Macromolecules, 265*, art. 131009.

Kim, B.-S., Kim, H.-S., Hong, J.-S., Huber, K. C., Shim, J.-H., & Yoo, S.-H. 2013. Effects of amylosucrase treatment on molecular structure and digestion resistance of pre-gelatinised rice and barley starches. *Food Chemistry, 138*(2), 966–975.

Kishore, A., Patil, R. J., Singh, A., & Pati, K. 2024. Jicama (Pachyrhizus spp.) a nonconventional starch: A review on isolation, composition, structure, properties, modifications and its application. *International Journal of Biological Macromolecules, 258*, art. 129095.

Kumari, S., Kaur, B. P., & Thiruvalluvan, M. 2024. Ultrasound modified millet starch: Changes in functional, pasting, thermal, structural, in vitro digestibility properties, and potential food applications. *Food Hydrocolloids, 153*, art. 110008.

Lal, M. K., Singh, B., Sharma, S., Singh, M. P., & Kumar, A. 2021. Glycemic index of starchy crops and factors affecting its digestibility: A review. *Trends in Food Science & Technology, 111*, 741–755.

Li, C. 2022. Recent progress in understanding starch gelatinization - An important property determining food quality. *Carbohydrate Polymers, 293*, art. 119735.

Li, C., Chen, G., Ran, C., Liu, L., Wang, S., Xu, Y., Tan, Y., & Kan, J. 2019. Adlay starch-gluten composite gel: Effects of adlay starch on rheological and structural properties of gluten gel to molecular and physico-chemical characteristics. *Food Chemistry, 289*, 121–129.

Li, C., Gong, B., Huang, T., & Yu, W.-W. 2021a. In vitro digestion rate of fully gelatinized rice starches is driven by molecular size and amylopectin medium-long chains. *Carbohydrate Polymers, 254*, art. 117275.

Li, E., Cao, P., Cao, W., & Li, C. 2022. Relations between starch fine molecular structures with gelatinization property under different moisture content. *Carbohydrate Polymers, 278*, art. 118955.

Li, H.-T., Li, Z., Fox, G. P., Gidley, M. J., & Dhital, S. 2021b. Protein-starch matrix plays a key role in enzymic digestion of high-amylose wheat noodle. *Food Chemistry, 336*, art. 127719.

Li, J., Shen, M., Xiao, W., Li, Y., Pan, W., & Xie, J. 2023. Regulating the physicochemical and structural properties of different starches by complexation with tea polyphenols. *Food Hydrocolloids, 142*, art. 108836.

Lin, L., Yang, H., Chi, C., & Ma, X. 2020. Effect of protein types on structure and digestibility of starch-protein-lipids complexes. *LWT, 134*, art. 110175.

Liu, B., Zhao, Y., Li, Y., Tao, L., Pan, P., Bi, Y., Song, S., & Yu, L. 2024a. Investigation of the structure, rheology and 3D printing characteristics of corn starch regulated by glycyrrhizic acid. *International Journal of Biological Macromolecules, 263*, art. 130277.

Liu, R., Xu, C., Cong, X., Wu, T., Song, Y., & Zhang, M. 2017. Effects of oligomeric procyanidins on the retrogradation properties of maize starch with different amylose/amylopectin ratios. *Food Chemistry, 221*, 2010–2017.

Liu, W., Chen, L., McClements, D. J., Zou, Y., Chen, G., & Jin, Z. 2024b. Vanillin-assisted preparation of chitosan-betaine stabilized corn starch gel: Gel properties and microstructure characteristics. *Food Hydrocolloids, 148*, art. 109510.

Liu, W., Wang, R., Li, J., Xiao, W., Rong, L., Yang, J., Wen, H., & Xie, J. 2021. Effects of different hydrocolloids on gelatinization and gels structure of chestnut starch. *Food Hydrocolloids, 120*, art. 106925.

Liu, W., Zhang, Y., Wang, R., Li, J., Pan, W., Zhang, X., Xiao, W., Wen, H., & Xie, J. 2022a. Chestnut starch modification with dry heat treatment and addition of xanthan gum: Gelatinization, structural and functional properties. *Food Hydrocolloids, 124*, art. 107205.

Liu, W., Zhang, Y., Xu, Z., Pan, W., Shen, M., Han, J., Sun, X., Zhang, Y., Xie, J., Zhang, X., & Yu, L. 2022b. Cross-linked corn bran arabinoxylan improves the pasting, rheological, gelling properties of corn starch and reduces its in vitro digestibility. *Food Hydrocolloids, 126*, art. 107440.

Liu, W. M., McClements, D. J., Peng, X. W., Jin, Z. Y., & Chen, L. 2023. Recent progress in regulating starch digestibility using natural additives and sustainable processing operations. *Critical Reviews in Food Science and Nutrition, 7*, 1–15.

Lu, X., Ma, R., Qiu, H., Sun, C., & Tian, Y. 2021. Mechanism of effect of endogenous/exogenous rice protein and its hydrolysates on rice starch digestibility. *International Journal of Biological Macromolecules, 193*, 311–318.

Luo, D., Li, Y., Xu, B., Ren, G., Li, P., Li, X., Han, S., & Liu, J. 2017. Effects of inulin with different degree of polymerization on gelatinization and retrogradation of wheat starch. *Food Chemistry, 229*, 35–43.

Luo, Y., Han, X., Shen, M., Yang, J., Ren, Y., & Xie, J. 2021. Mesona chinensis polysaccharide on the thermal, structural and digestibility properties of waxy and normal maize starches. *Food Hydrocolloids, 112*, art. 106317.

Luo, Y., Shen, M., Li, E., Xiao, Y., Wen, H., Ren, Y., & Xie, J. 2020. Effect of *Mesona chinensis* polysaccharide on pasting, rheological and structural properties of corn starches varying in amylose contents. *Carbohydrate Polymers, 230*, art. 115713.

Ma, M., Wang, Y., Wang, M., Jane, J.-L., & Du, S.-K. 2017. Physicochemical properties and in vitro digestibility of legume starches. *Food Hydrocolloids, 63*, 249–255.

Madruga, M. S., de Albuquerque, F. S. M., Silva, I. R. A., do Amaral, D. S., Magnani, M., & Queiroga Neto, V. 2014. Chemical, morphological and functional properties of Brazilian jackfruit (*Artocarpus heterophyllus L.*) seeds starch. *Food Chemistry, 143*, 440–445.

Martinez, M. M., Li, C., Okoniewska, M., Mukherjee, I., Vellucci, D., & Hamaker, B. 2018. Slowly digestible starch in fully gelatinized material is structurally driven by molecular size and A and B1 chain lengths. *Carbohydrate Polymers, 197*, 531–539.

Matalanis, A. M., Campanella, O. H., & Hamaker, B. R. 2009. Storage retrogradation behavior of sorghum, maize and rice starch pastes related to amylopectin fine structure. *Journal of Cereal Science, 50*(1), 74–81.

Matignon, A., & Tecante, A. 2017. Starch retrogradation: From starch components to cereal products. *Food Hydrocolloids, 68*, 43–52.

Nasseri, R., & Mohammadi, N. 2014. Modeling of starch retrogradation onset in its aqueous solution using thermoreversible gelation concept. *Carbohydrate Polymers, 99*, 325–330.

Obadi, M., Qi, Y., & Xu, B. 2023. High-amylose maize starch: Structure, properties, modifications and industrial applications. *Carbohydrate Polymers, 299*, 120185.

Ouyang, Q., Wang, X., Xiao, Y., Luo, F., Lin, Q., & Ding, Y. 2021. Structural changes of A-, B- and C-type starches of corn, potato and pea as influenced by sonication temperature and their relationships with digestibility. *Food Chemistry, 358*, art. 129858.

Oyeyinka, S. A., Singh, S., & Amonsou, E. O. 2021. A review on structural, digestibility and physicochemical properties of legume starch-lipid complexes. *Food Hydrocolloids, 349*, art. 129165.

Patel, H., Royall, P. G., Gaisford, S., Williams, G. R., Edwards, C. H., Warren, F. J., Flanagan, B. M., Ellis, P. R., & Butterworth, P. J. 2017. Structural and enzyme kinetic studies of retrograded starch: Inhibition of α-amylase and consequences for intestinal digestion of starch. *Carbohydrate Polymers, 164*, 154–161.

Perry, P. A., & Donald, A. M. 2002. The effect of sugars on the gelatinisation of starch. *Carbohydrate Polymers, 49*(2), 155–165.

Prakaywatchara, P., Wattanapairoj, C., & Thirathumthavorn, D. 2018. Effects of emulsifier types and levels in combination with glycerol on the gelatinization and retrogradation properties of gluten-free rice-based wonton wraps. *Starch-Starke, 70*(5–6). art. 1700227.

Punia, S. 2020. Barley starch modifications: Physical, chemical and enzymatic - A review. *International Journal of Biological Macromolecules, 144*, 578–585.

Punia, S., Sandhu, K. S., Dhull, S. B., Siroha, A. K., Purewal, S. S., Kaur, M., & Kidwai, M. K. 2020. Oat starch: Physico-chemical, morphological, rheological characteristics and its applications - A review. *International Journal of Biological Macromolecules, 154*, 493–498.

Qiao, L., Li, Y., Chi, Y., Ji, Y., Gao, Y., Hwang, H., Aker, W. G., & Wang, P. 2016. Rheological properties, gelling behavior and texture characteristics of polysaccharide from Enteromorpha prolifera. *Carbohydrate Polymers, 136*, 1307–1314.

Renzetti, S., van den Hoek, I. A. F., & van der Sman, R. G. M. 2021. Mechanisms controlling wheat starch gelatinization and pasting behaviour in presence of sugars and sugar replacers: Role of hydrogen bonding and plasticizer molar volume. *Food Hydrocolloids, 119*, art. 106880.

Rolandelli, G., Rodríguez, S. D., & Buera, M. D. P. 2024. Modulation of the retrogradation kinetics of sweet potato starch by the addition of pectin, guar gum, and gallic acid. *Food Hydrocolloids, 146*, art. 109211.

Rong, L., Liu, W., Shen, M., Xiao, W., Chen, X., Yang, J., & Xie, J. 2022. The effects of Mesona chinensis Benth gum on the pasting, rheological, and microstructure properties of different types of starches. *Current Research in Food Science, 5*, 2287–2293.

Schirmer, M., Höchstötter, A., Jekle, M., Arendt, E., & Becker, T. 2013. Physicochemical and morphological characterization of different starches with variable amylose/amylopectin ratio. *Food Hydrocolloids, 32*(1), 52–63.

Singh, A., Geveke, D. J., & Yadav, M. P. 2017. Improvement of rheological, thermal and functional properties of tapioca starch by using gum arabic. *LWT, 80*, 155–162.

Szwengiel, A., Lewandowicz, G., Górecki, A. R., & Błaszczak, W. 2018. The effect of high hydrostatic pressure treatment on the molecular structure of starches with different amylose content. *Food Chemistry, 240*, 51–58.

Tang, M., Wang, L., Cheng, X., Wu, Y., & Ouyang, J. 2019. Non-starch constituents influence the in vitro digestibility of naked oat (Avena nuda L.) starch. *Food Chemistry, 297*, art. 124953.

Tao, K., Li, C., Yu, W., Gilbert, R. G., & Li, E. 2019. How amylose molecular fine structure of rice starch affects functional properties. *Carbohydrate Polymers, 204*, 24–31.

Tawheed, A., Naik, H. R., Zameer Hussain, S., Mir, M. A., & Abida, J. 2018. In vitro digestion, physicochemical and morphological properties of low glycemic index rice flour prepared through enzymatic hydrolysis. *International Journal of Food Properties, 21*(1), 2632–2645.

Teixeira, N. D. C., Queiroz, V. A. V., Rocha, M. C., Amorim, A. C. P., Soares, T. O., Monteiro, M. A. M., de Menezes, C. B., Schaffert, R. E., Garcia, M. A. V., & Junqueira, R. G. 2016. Resistant starch content among several sorghum (Sorghum bicolor) genotypes and the effect of heat treatment on resistant starch retention in two genotypes. *Food Chemistry, 197*, 291–296.

Tziotis, A., Seetharaman, K., Klucinec, J. D., Keeling, P., & White, P. J. 2005. Functional properties of starch from normal and mutant corn genotypes. *Carbohydrate Polymers, 61*(2), 238–247.

Vandeputte, G. E., Vermeylen, R., Geeroms, J., & Delcour, J. A. 2003. Rice starches. III. Structural aspects provide insight in amylopectin retrogradation properties and gel texture. *Journal of Cereal Science, 38*(1), 61–68.

Waigh, T. A., Gidley, M. J., Komanshek, B. U., & Donald, A. M. 2000. The phase transformations in starch during gelatinisation: A liquid crystalline approach. *Carbohydrate Research, 328*(2), 165–176.

Wang, L., Shi, D., Chen, J., Dong, H., & Chen, L. 2023a. Effects of Chinese chestnut powder on starch digestion, texture properties, and staling characteristics of bread. *Grain & Oil Science and Technology, 6*(2), 82–90.

Wang, L., Zhang, L., Wang, H., Ai, L., & Xiong, W. 2020. Insight into protein-starch ratio on the gelatinization and retrogradation characteristics of reconstituted rice flour. *International Journal of Biological Macromolecules, 146*, 524–529.

Wang, R., He, Z., Cao, Y., Wang, H., Luo, X., Feng, W., Chen, Z., Wang, T., & Zhang, H. 2023b. Impact of crystalline structure on the digestibility of amylopectin-based starch-lipid complexes. *International Journal of Biological Macromolecules, 242*, 125191.

Wang, S. J., Li, C. L., Copeland, L., Niu, Q., & Wang, S. 2015. Starch retrogradation: A comprehensive review. *Comprehensive Reviews in Food Science and Food Safety, 14*(5), 568–585.

Wang, W., Guan, L., Seib, P. A., & Shi, Y.-C. 2018. Settling volume and morphology changes in cross-linked and unmodified starches from wheat, waxy wheat, and waxy maize in relation to their pasting properties. *Carbohydrate Polymers, 196*, 18–26.

Wani, A. A., Singh, P., Shah, M. A., Wani, I. A., Götz, A., Schott, M., & Zacherl, C. 2013. Physico-chemical, thermal and rheological properties of starches isolated from newly released rice cultivars grown in Indian temperate climates. *LWT - Food Science and Technology, 53*(1), 176–183.

Woodbury, T. J., Grush, E., Allan, M. C., & Mauer, L. J. 2022. The effects of sugars and sugar alcohols on the pasting and granular swelling of wheat starch. *Food Hydrocolloids, 126*, art. 107433.

Wu, C., Wang, W., Jia, J., Guo, L., Zhang, C., & Qian, J.-Y. 2024. Effect of endogenous protein and lipid removal on the physicochemical and digestion properties of sand rice (Agriophyllum squarrosum) flour. *International Journal of Biological Macromolecules, 266*, art. 131269.

Wu, M., Wang, J., Ge, Q., Yu, H., & Xiong, Y. L. 2018. Rheology and microstructure of myofibrillar protein–starch composite gels: Comparison of native and modified starches. *International Journal of Biological Macromolecules, 118*, 988–996.

Xiao, Y., Liu, S., Shen, M., Jiang, L., Ren, Y., Luo, Y., & Xie, J. 2020. Effect of different Mesona chinensis polysaccharides on pasting, gelation, structural properties and *in vitro* digestibility of tapioca starch-Mesona chinensis polysaccharides gels. *Food Hydrocolloids, 99*, art. 105327.

Xiao, Y., Shen, M., Luo, Y., Ren, Y., Han, X., & Xie, J. 2020. Effect of Mesona chinensis polysaccharide on the pasting, rheological, and structural properties of tapioca starch varying in gelatinization temperatures. *International Journal of Biological Macromolecules, 156*, 137–143.

Xu, J., Fan, X., Ning, Y., Wang, P., Jin, Z., Lv, H., Xu, B., & Xu, X. 2013. Effect of spring dextrin on retrogradation of wheat and corn starch gels. *Food Hydrocolloids, 33*(2), 361–367.

Xu, J., Li, X., Chen, J., Dai, T., Liu, C., & Li, T. 2021. Effect of polymeric proanthocyanidin on the physicochemical and *in vitro* digestive properties of different starches. *LWT-Food Science & Technology, 148*, art. 111713.

Xu, N., Yu, P., Zhang, H., Ji, X., Wu, P., Zhang, L., & Wang, X. 2024. Effects of Laminaria japonica polysaccharide and coumaric acid on pasting, rheological, retrogradation and structural properties of corn starch. *International Journal of Biological Macromolecules, 263*, art. 130343.

Xu, X., Bean, S., Wu, X., & Shi, Y.-C. 2022. Effects of protein digestion on in vitro digestibility of starch in sorghum differing in endosperm hardness and flour particle size. *Food Chemistry, 383*, art. 132635.

Yang, Y., Jiao, A., Zhao, S., Liu, Q., Fu, X., & Jin, Z. 2021a. Effect of removal of endogenous non-starch components on the structural, physicochemical properties, and in vitro digestibility of highland barley starch. *Food Hydrocolloids, 117*, art. 106698.

Yang, Y., Xu, X., & Wang, Q. 2021b. Effects of potassium sulfate on swelling, gelatinizing and pasting properties of three rice starches from different sources. *Carbohydrate Polymers, 251*, art. 117057.

Yang, Z., Hao, H., Wu, Y., Liu, Y., & Ouyang, J. 2021c. Influence of moisture and amylose on the physicochemical properties of rice starch during heat treatment. *International Journal of Biological Macromolecules, 168*, 656–662.

Yang, Z., Zhang, Y., Wu, Y., & Ouyang, J. 2023. Factors influencing the starch digestibility of starchy foods: A review. *Food Chemistry, 406*, art. 135009.

Yu, L., & Christie, G. 2005. Microstructure and mechanical properties of orientated thermoplastic starches. *Journal of Materials Science, 40*(1), 111–116.

Zhang, B., Chen, L., Li, X., Li, L., & Zhang, H. 2015a. Understanding the multi-scale structure and functional properties of starch modulated by glow-plasma: A structure-functionality relationship. *Food Hydrocolloids, 50*, 228–236.

Zhang, C., Wang, M., Tan, Z., Ma, M., Sui, Z., & Corke, H. 2023a. Differential distribution of surface proteins/lipids between wheat A- and B-starch granule contributes to their difference in pasting and rheological properties. *International Journal of Biological Macromolecules, 240*, art. 124430.

Zhang, H., Sun, B., Zhang, S., Zhu, Y., & Tian, Y. 2015b. Inhibition of wheat starch retrogradation by tea derivatives. *Carbohydrate Polymers, 134*, 413–417.

Zhang, M., Golding, J. B., Pristijono, P., Yu, Y., Wang, P., Chen, G., Li, Y., Si, J., & Yang, H. 2024. Effects of Huangjing polysaccharides on the properties of sweet potato starch. *LWT, 204*, art. 116474.

Zhang, X., Shen, Y., Zhang, N., Bao, J., Wu, D., & Shu, X. 2019. The effects of internal endosperm lipids on starch properties: Evidence from rice mutant starches. *Journal of Cereal Science, 89*, art. 102804.

Zhang, Y., Li, B., Xu, F., He, S., Zhang, Y., Sun, L., Zhu, K., Li, S., Wu, G., & Tan, L. 2021. Jackfruit starch: Composition, structure, functional properties, modifications and applications. *Trends in Food Science & Technology, 107*, 268–283.

Zhang, Y., Li, D., Yang, N., Jin, Z., & Xu, X. 2018. Comparison of dextran molecular weight on wheat bread quality and their performance in dough rheology and starch retrogradation. *LWT, 98*, 39–45.

Zhang, Y., Li, M., You, X., Fang, F., & Li, B. 2020. Impacts of guar and xanthan gums on pasting and gel properties of high-amylose corn starches. *International Journal of Biological Macromolecules, 146*, 1060–1068.

Zhang, Y., Wang, Y., Yang, B., Han, X., He, Y., Wang, T., Sun, X., & Zhao, J. 2023b. Effects of zucchini polysaccharide on pasting, rheology, structural properties and in vitro digestibility of potato starch. *International Journal of Biological Macromolecules, 253*, art. 127077.

Zhang, Y., Xu, F., Wang, Q., Zhang, Y., Wu, G., Tan, L., & Zhang, Z. 2022. Effects of moisture content on digestible fragments and molecular structures of high amylose jackfruit starch prepared by improved extrusion cooking technology. *Food Hydrocolloids, 133*, art. 108023.

Zheng, J., Wang, N., Yang, J., You, Y., Zhang, F., Kan, J., & Wu, L. 2024. New insights into the interaction between bamboo shoot polysaccharides and lotus root starch during gelatinization, retrogradation, and digestion of starch. *International Journal of Biological Macromolecules, 254*, art. 127877.

Zhou, D.-N., Zhang, B., Chen, B., & Chen, H.-Q. 2017. Effects of oligosaccharides on pasting, thermal and rheological properties of sweet potato starch. *Food Chemistry, 230*, 516–523.

Zhou, H., Wang, C., Shi, L., Chang, T., Yang, H., & Cui, M. 2014. Effects of salts on physicochemical, microstructural and thermal properties of potato starch. *Food Chemistry, 156*, 137–143.

Zhu, B., Zhan, J., Chen, L., & Tian, Y. 2020. Amylose crystal seeds: Preparation and their effect on starch retrogradation. *Food Hydrocolloids, 105*, art. 105805.

Techniques for Measuring Starch

11

Yuanhui Chen, David Julian McClements, Hao Cheng, Zhengyu Jin, and Long Chen

11.1 INTRODUCTION

Starch is widely distributed in nature and is a kind of renewable natural resources. It occupies an important part of People's Daily diet, not only is a common staple food in life, providing carbohydrates for the body, but also can be considered to be used in food and other industries as thickeners, stabilizers, adhesives, emulsifiers, fat substitute, etc. (Mahmood et al., 2017). Furthermore, people have gradually realized the seriousness of energy consumption and environmental pollution, and they pay more and more attention to the deep development and application of renewable resources. Therefore, starch has attracted much attention because of its wide sources, green environmental protection, renewable nature, and other characteristics. With the development of additive manufacturing and bio-based materials, starch has also been widely used in many industries (Adewale, Yancheshmeh, & Lam, 2022; Amaraweera et al., 2021).

The structure and properties of starch impact its application as a functional ingredient in different industries. It is important to have suitable analytical methods to characterize the structure and properties of starch. The information provided by these methods can help to understand the molecular and physicochemical basis of starch functionality, which may contribute to the development of starch-based ingredients with enhanced performance. A variety of analytical instruments and experimental protocols have been developed to determine the composition, molecular structure, physical state, thermal properties, rheology, and digestibility of starch (Blazek & Gilbert, 2011; Teng et al., 2021). Understanding the basic principles, the information obtained, the range of application, and the advantages/disadvantages of different methods is important for those working with starch. Typically, it is important to employ an appropriate combination of different analytical methods to obtain a more complete understanding of starch properties.

It should be noted that natural starch has disadvantages such as poor stability and poor solubility in practical applications.

In order to overcome the deficiencies of natural starch, modified starch came into being (Compart et al., 2023). People have developed modified starch by using physical, chemical, or enzyme methods to treat the original starch, so that it can adapt to the application requirements of food, paper, plastic, textile, daily chemical, medicine, and other industries. Modified starches such as pregelatinized starch, cold-water-soluble starch, cross-linked starch, esterified starch, and debranched starch have been gradually developed and applied (Bangar et al., 2022; Wang et al., 2020). In order to better explore the structure–activity relationship between the properties and structure of modified starch, it is necessary to use tools and means to characterize its structure and properties. Therefore, the technologies of starch determination have been rapidly developed, the characterization methods are gradually diversified, and the data obtained are more scientific and accurate.

Currently, existing techniques can characterize starches from micro to macro, and on the basis of traditional technologies, more convenient and faster new technologies have been developed. The characterization of a certain index of starch is no longer limited to one technology, and the combination of multiple technologies can obtain data more accurately. In summary, this chapter will introduce the methods and techniques commonly used in starch measurement and characterization, hoping to provide more comprehensive references for starch and starch-based products in the future, as well as valuable help for future technological innovation.

11.2 COMPONENT ANALYSIS

Starches are macromolecular carbohydrates, which mainly consist of amylose and amylopectin (Pérez & Bertoft, 2010). Amylose is a linear polysaccharide consisting of glucose units held together by α-1,4-glycosidic bonds, whereas amylopectin is a branched polysaccharide consisting of linear chains with

DOI: 10.1201/9781003464396-11

glucose units held together by α-1,4-glycosidic bonds attached by α-1,6-glycosidic bonds at the branch points (Sárka & Dvorácek, 2017). The ratio and structure of these two starches affect the quality of the starch in many aspects, including viscosity, elasticity, swelling capacity, digestibility, gelatinization, and retrogradation (Liu et al., 2024; Stevnebo, Sahlström, & Svihus, 2006; Zavareze et al., 2012). Therefore, understanding the composition and proportion of starch is helpful to better understand the properties of starch and then achieve its further application.

The methods used to determine the amylose and amylopectin contents of starch can be classified into three categories: iodine colorimetry, spectral analysis, and chromatographic analysis (Sheng & Wei, 2022).

11.2.1 Iodine Colorimetry

The combination of amylose and iodine can produce a blue compound, which is usually measured at a specific wavelength using a spectrophotometer, called iodine colorimetry (Sheng & Wei, 2022).

Iodine colorimetry was first proposed in 1970 for the determination of amylose content (Williams, Kuzina, & Hlynka, 1970). Since then, it has experienced rapid development; on the basis of single wavelength method, dual wavelength method and multi-wavelength method have appeared.

Single wavelength colorimetry is an internationally recognized standard method. It is based on the fact that amylose forms helical inclusions with iodine, which appear blue and are characterized by an absorption maximum at 620 nm. Consequently, the amylose content can be quantified by measuring the absorption at 620 nm, provided that an appropriate calibration curve has been prepared. However, this relatively simple method is susceptible to interference from the presence of amylopectin, as the reddish amylopectin–iodine complex has a spectrum that overlaps with that of the bluish amylose–iodine complex. Therefore, the measurement results of single wavelength colorimetry are not always accurate.

This problem can be overcome using dual-wavelength iodine colorimetry, which involves measuring the absorption at both 618 and 550 nm. At these two wavelengths, the amylose–iodine and amylopectin–iodine complexes make different contributions to the overall absorption, which allows the amylopectin and amylose contents to be measured simultaneously using an appropriate mathematical model. The dual wavelength method has found increasing use because it can eliminate the interferences from amylopectin and is relatively simple to carry out. However, standard curves need to be prepared at both wavelengths using samples with known amylose and amylopectin content.

A more accurate multi-wavelength technique has been proposed to determine amylopectin, amylose, and starch simultaneously (Jarvis & Walker, 1993; Sene, Thevenot, & Prioul, 1997). The result of this method is more accurate, but the calculation formula is complicated. Because of the complexity of this method, it is not applicable.

11.2.2 Chromatographic Analysis

Chromatographic methods can also be used for the analysis of the composition of starches (Batey & Curtin, 1996; Lin et al., 2022):

High performance liquid chromatography (HPLC) is an efficient and rapid method for separating and then quantifying different substances in mixtures. It has been gradually used for the qualitative and quantitative analysis of starch and its components. The solvent and mobile phase must be carefully selected to ensure that the starch has good solubility and is not denatured prior to analysis (Batey & Curtin, 1996).

Gel permeation chromatography (GPC) has been developed to separate and quantify polymers in mixed solutions using columns packed with porous beads. It separates the polymers based on the differences in their hydration radius (effective size). GPC has been used to determine the amylose and amylopectin contents of starch, as these two molecules have different hydration radius. In recent years, GPC has been used more and more to identify the purity of amylose and amylopectin, and has become an important index to evaluate the purity of starch (Vilaplana, Hasjim, & Gilbert, 2012). However, its washing time is long and the requirement of chromatographic column is high.

It is worth noting that HPLC and GPC are often used in combination with starch debranching enzymes and hydrolyzing enzymes to obtain more detailed information about the molecular characteristics of the different components in starch (Shi et al., 2013).

11.2.3 Spectral Analysis

The near-infrared (NIR) region of the electromagnetic spectrum is particularly useful for the quantitative analysis of the composition of complex mixtures (Alamu et al., 2021). This technique has been employed to determine the amylose and amylopectin content of starches as these two types of starch absorb differently at different wavelengths in the NIR region. The amylose content in rice starch has been characterized by this method (Delwiche et al., 1995). By analyzing the spectral characteristics of starch in the NIR, it can quickly detect the content of starch components. When using NIR for the analysis of starch, it is important to first create correlation models by acquiring the spectra of numerous samples with known amylose and amylopectin contents. The main advantages of the NIR method are that it is rapid, simple, non-destructive, and does not require additional chemicals or reagents. However, the main disadvantage is that it must first be calibrated using a large number of samples with known compositions to build accurate predictive models.

When these technologies are used alone, the final results are often susceptible to error or provide limited information about the overall composition of starch (Vilaplana, Hasjim, & Gilbert, 2012). For this reason, two or more of these methods are often used in combination to improve the accuracy of the analysis. The principles, advantages, and disadvantages of each method are highlighted in Table 11.1.

TABLE 11.1 The techniques for starch component analysis and their principle, merit and demerit

TECHNIQUES	PRINCIPLE	MERIT AND DEMERIT	REFERENCES
Iodine colorimetry	The combination of amylopectin or amylose with iodine forms complexes with different colors. The light absorption is related to the concentration of amylopectin and amylose present	The operation is simple, and dual wavelength is a more accurate method. But the single wavelength method is easy to be interfered with and the accuracy is not high. The multi-wavelength rule is complex to calculate	Wang et al. (2010)
HPLC	Separates starch molecules based on their molecular weights, as this influences the time required to pass through the column. It is often used in conjunction with GPC	It has the advantages of high resolution and high sensitivity. HPLC system cannot clearly prove the accurate separation of amylose and amylopectin, which takes a long time and requires high requirements for experimental personnel and equipment	Flamme, Jurgens, and Jansen (1994)
GPC	It is based on the hydrodynamic radius of starch molecules to separate amylopectin and amylose. From the chromatogram of the starch molecules obtained, the amylose content was calculated by integrating the area	The sample pretreatment is complicated, but the molecular weight and distribution of starch can be obtained at the same time	Gérard et al. (2001)
NIR	According to Lambert–Beale absorption law, the composition content of the sample changes, and the corresponding NIR spectral shape characteristics also change. Therefore, starch can be qualitatively or quantitatively analyzed by using the physical information reflected in the near-infrared spectrum	It has the characteristics of rapid, micro and non-destructive testing. But it needs to spend more manpower and material resources	Alamu et al. (2021); Pandiselvam et al. (2023)
DSC	The single helical structure of amylose can interact with the polar groups of lysophosphatidylcholine to form complexes, and the enthalpy value is proportional to the content of amylose	This method is simple, fast, and accurate, but it requires special instruments and is expensive. The measurement results are easily affected by temperature, source, non-starch components, etc.	Mestres et al. (1996); Vilaplana, Hasjim, and Gilbert (2012)

11.3 MICROSCOPIC TECHNIQUES

Starch granules typically have dimensions that fall within the range of about 1–100 μm (Govindaraju et al., 2020). The size and shape of starch granules depend on their biological origin, maturity, and growing conditions. Commonly, the morphology of starch granules is characterized using microscopy methods (Horovitz et al., 2011), which can be classified according to their different operating principles. The optical microscopes are mainly used to observe the morphology of starch. Electron microscopes are widely used in the microstructure analysis of starch. The high resolution of electron microscope is obviously superior to that of optical microscope.

11.3.1 Polarizing Microscopes

Optical microscopies are such kinds of techniques that use the interaction of lens with light to magnify specimens (Xiao et al., 2020). Among them, polarized light microscope is often used to study starch particles. Due to different refractive indices of light in the crystalline and amorphous regions of starch, the polarization and interference of light are caused. This phenomenon causes starch particles to appear as black and white interference patterns under the polarizing microscope. The formation of polarizing cross is the direct evidence that starch particles have birefringence, and it is an important sign that starch particles have a spherulite structure (Sanders & Cohen, 2019).

11.3.2 Scanning Electron Microscopy (SEM)

In 1965, the first commercial scanning electron microscopy (SEM) appeared. After nearly 60 years of development, SEM has gradually developed in the direction of high resolution, large sample chamber, and miniaturization of electron microscopy.

The image-forming principle of SEM: In SEM instruments, an electron beam is emitted from an electron gun that then passes through a condenser lens, a deflection coil, and an objective lens before irradiating the sample. The secondary electrons generated at the surface of the sample are then recorded using a suitable detector, which results in a magnified image of the sample being created (Dery & Lou, 2023). SEM uses electron beams to provide information about the surface topology of samples. It has a higher magnification than optical microscopy and can therefore be used to provide information about smaller structural features in starch samples.

When using SEM to observe starch, the following points need to be paid special attention: (1) Because the SEM uses electron beam to scan the sample, the sample must be fixed first. (2) In order to avoid the electron beam colliding with the remaining gas molecules before it hits the sample surface, the SEM must be kept in a certain high vacuum environment. Therefore, the samples need to be pretreated by dehydration and critical point drying. (3) When using SEM to observe starch, it is generally necessary to gold-plate the surface of the sample. It is because starch is a kind of organic matter and does not have electrical conductivity. When the charge accumulates on the surface of the starch, it produces a repulsive force that interferes with the electron beam. As a result, the scan result is not accurate or even cannot be scanned. In addition, when the electron beam scans the sample, the incoming electrons will convert part of the incoming energy into heat energy, which increases the temperature of the sample's surface and subsurface layer. High temperatures tend to destroy the sample and increase the production of secondary electrons, thus affecting the clarity of the image. Therefore, it is necessary to gold-plate the starch surface when observing it by SEM.

In terms of starch particle morphology characterization, SEM is the most commonly used means, which is essential for the study of starch. SEM can produce images with higher magnification than optical microscopes. SEM can not only observe the shape and size of starch particles, but also clearly observe the cracks, pores, and equatorial depressions on the surface of starch particles, which is impossible to do with optical microscopy (Govindaraju et al., 2021).

11.3.3 Transmission Electron Microscopy (TEM)

Transmission electron microscopy (TEM) is mainly composed of five basic components: lighting system, imaging optical system, vacuum system, electrical system, and operation control system. TEM also uses an electron beam to provide information about the microstructure of starch samples, but it is based on passing the beam through the samples rather than generating secondary electrons at their surfaces (as is the case for SEM). The image-forming principle of TEM is that the electron beam passes through the sample, and then it is focused and amplified by the electromagnetic lens to produce the object image, which is projected on the fluorescent screen of the observed image (Nawrocka et al., 2021). The electron gun of TEM is at the top of the lens cone and mainly consists of a cathode and an anode. The cathode serves as the electron source and the anode is used to accelerate the electron beam. Electrons are emitted from the cathode, and the beam is focused through collecting lens. After the electron beam passes through the sample, it is imaged on the intermediate mirror through the objective lens, then amplified step by step through the intermediate lens and the projection lens, and finally imaged on the fluorescent screen.

TEM requires that the sample must be thin enough (<150 nm) for the electron beam to pass through (Pérez-Bermúdez et al., 2023). When starch was observed by TEM, a small amount of starch sample was first taken, dispersed it in ethanol, water, or other solvents, and ultrasounded for 3–15 minutes. The samples were dropped onto the carbon supporting film with a disposable dropper and then dried for TEM observation. TEM can obtain high-resolution, high-magnification images. When the cassava starch particles were observed by TEM, it was found that there were internal channels in the particles, which might be the form of chemical agents penetrating into the starch particles when the starch was chemically modified (Sívoli et al., 2009). TEM can visualize the internal structure of starch. However, the sample preparation procedures used are often time-consuming and laborious, and they may damage the structures being investigated.

11.3.4 Atomic Force Microscopy (AFM)

The image-forming principle of atomic force microscopy (AFM) is that it uses the interaction force between the probe and the sample to obtain the surface topography image of the sample (Morris, Woodward, & Gunning, 2011). Therefore, it does not require the electrical conductivity of the sample. In addition, AFM can characterize the surface morphology of samples at the nanoscale (Zhu, 2017).

During imaging, one end of the probe is fixed, while the other end is free to move (with a sharp tip). The laser is hit on the back of the tip of the cantilever of the probe, and the signal reflected by the laser is received by the four-quadrant of the photosensitive element. When the probe continues to approach the sample surface, due to different heights of the sample surface, the interaction force between the tip and the sample surface will change, thus deflecting the probe cantilever. The deflection of the cantilever will cause the reflection path of the illuminated laser to shift, so the position of the laser signal received on the four-quadrant will also change. The laser detector will convert the displacement signal of the reflected light spot for amplification, and then adjust the interaction force between the probe tip and the sample surface through the feedback system by adjusting the distance between the two. The signal acquisition system will record the movement of the scanner in real time, so as to obtain the information of the sample surface topography (Liu et al., 2008). In a nutshell, the tip of an AFM probe is moved in an x–y direction over the surface of a sample. The AFM probe may move upward or downward relative to the surface of the sample depending on its interaction with the sample. When the probe is near the sample surface, the interaction force between the tip and the sample is stronger, thereby deflecting the probe in the z-direction by a greater amount, which is quantified using a laser. The laser detector converts the displacement values measured at different x–y locations on the sample surface into a image of the sample's topography.

TABLE 11.2   Techniques, principles, and characteristics of starch micromorphology determination

MICROSCOPIC TECHNIQUE	PRINCIPLE	CHARACTERISTIC	REFERENCES
PLM	Polarization and birefringence of light	Dependent on starch crystals; the crystal of starch particles can be identified by the phenomenon of polarization cross	Xiao et al. (2020)
SEM	The physical information of the sample is excited by the interaction between the high-energy electron beam and the sample	The shape and surface morphology of starch can be observed, and the image can be three-dimensional and high resolution	Chen et al. (2021)
TEM	Similar to an optical microscope. The difference is that TEM uses a shorter wavelength electron beam instead of a light beam and an electromagnetic lens as the imaging lens	The sample is required to be thin, water-free, and tested under high vacuum conditions. The imaging has the advantages of high resolution and high magnification	Klang et al. (2012)
AFM	AFM generates images by "feeling" the surface with a sharp probe	The sample can be studied in liquid or gas environment, leaving the image of the sample in a "near native" state	Park, Xu, and Seetharaman (2011)
CLSM	Confocal imaging principle and laser scanning technology	CLSM has the characteristics of three-dimensional reconstruction and tomography in starch research, which can observe the external morphology and internal fine structure of starch particles	Chen et al. (2011)

AFM can be used to provide information about the fine structural details of starch granules. For instance, it has been used to detect hair-like structures on the surfaces of starch granules, which may reflect the organization of polymers within the granule depending on the shape and position of the hair-like extension (Park, Xu, & Seetharaman, 2011). Moreover, the sample preparation of AFM is simple, and the testing process has less destructive to the sample. In addition, AFM is not limited by environmental conditions when testing samples. The ultra-structure of the sample surface can be imaged in both atmospheric and solution environments (Zhu, 2017). The AFM takes the interaction force as the feedback signal, and its application field is no longer limited to whether the sample conducts electricity or whether the working environment is vacuum, which greatly broadens the research object. AFM has promoted the study of starch structure from micro to nano level.

11.3.5 Confocal Laser Scanning Microscopy (CLSM)

Confocal laser scanning microscopy (CLSM) is a new scanning imaging system based on optical microscope and various scanning microscopes.

The image-forming principle of CLSM: The laser beam is scanned on the surface of the sample, and the photoelectric detection device is used to receive the reflected light (or transmitted light) of the sample. The change in the structure of the sample changes the intensity of the reflected light (or transmitted light), so that the output current of the photoelectric detector changes, and the signal is processed and displayed on the computer screen simultaneously.

CLSM can be used as a tool for visualizing starch. This technique can not only observe the natural starch particles, but also image the gelatinization process and starch pastes. The technology does not require drying and gold spraying during the sample preparation process, and the starch can be stained with fluorescent dyeing solution and imaged in two or three dimensions by CLSM (van de Velde, van Riel, & Tromp, 2002). With the rapid development of optical, video, computer, and other technologies, laser confocal microscopy has gradually matured and become a new characterization method to reveal the internal structure of starch. It can continuously scan different layers of the observed sample layer by layer to obtain images of each layer.

In summary, there are a variety of different microscopic techniques available to provide information about the microstructure of starch, which differ in their operating principles and the dimensions of the structures they can detect (Table 11.2). Often, it is advantageous to combine two or more different microscopy methods to obtain more detailed information about the overall microstructure of starch.

11.4 THERMAL TECHNIQUES

Thermal techniques are useful for providing information about changes in the molecular structure and physicochemical properties of starch when it is heated or cooled. This information is useful because it influences the behavior of starch in many practical applications, such as thermal processing cooking, chilling, or freezing.

11.4.1 Differential Scanning Calorimeter (DSC)

Differential scanning calorimeter (DSC) provides valuable information about the thermal transitions that occur within

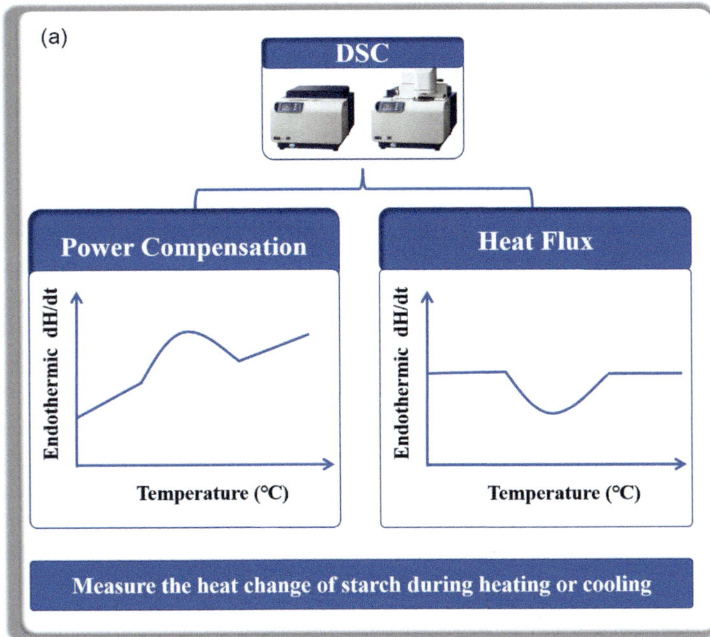

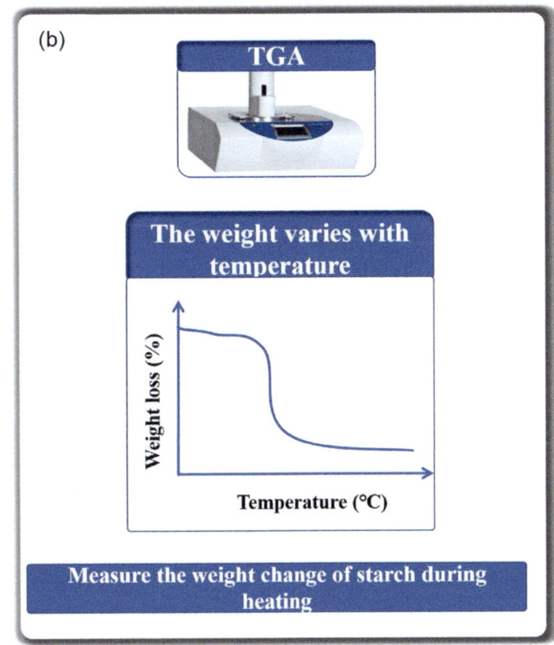

FIGURE 11.1 Schematic diagram of (a) DSC and (b) TGA curves of starch.

starch when it is heated or cooled under controlled conditions. Typically, the sample to be analyzed is placed in a small pan and an empty pan is used as a reference. The sample and reference pans are then heated or cooled at a controlled rate and any enthalpy changes are detected. DSC instruments usually detect these enthalpy changes using either power compensated or heat flux approaches (Figure 11.1a). Power compensated DSC: The sample and the reference are always maintained at the same temperature, the energy difference required to meet this condition is determined and is directly output as a signal (heat difference). Heat flux DSC: Give the same power to the sample and the reference, determine the temperature difference between the sample and the reference, and then convert the temperature difference into the heat difference as a signal output according to the heat flow equation (Danley, 2003). When the sample is analyzed by DSC, the amount of sample and the heating rate have great influence on the experimental results. For processes with small thermal effects or non-uniform samples, the sample amount can be appropriately increased, but the accuracy and resolution of the measured temperature will be reduced and the peaks will be wider. Therefore, a slower heating rate should be used.

With DSC, samples can be studied over a wide temperature range and with less sample usage. A variety of temperature programs can be used when measuring, and the instrument sensitivity is very high. Starch gelatinization can cause starch molecules to change from ordered states to disordered states. This process is accompanied by energy changes and can therefore be measured using DSC. The advantage is that DSC can study starch gelatinization in a wide range of starch/water ratios and can determine gelatinization temperatures above 100°C. The enthalpy of the phase transition can be estimated according to the DSC results (Sahoo et al., 2024). DSC can also be used to monitor the degree of retrogradation of starch, as the helical structures formed during aging are dissociated when the starch is heated, thereby leading to measurable enthalpy changes (Liu et al., 2010).

11.4.2 Thermogravimetric Analyzer (TGA)

Thermogravimetric analyzer (TGA) is a kind of thermal analysis technology for measuring the continuous changes of sample mass with temperature or time under controlled temperature and atmospheric conditions (Yashini et al., 2024). Commonly, a sample is placed in an appropriate sample holder and then its mass is recorded as it is heated at a fixed rate in a controlled atmosphere (such as nitrogen gas flushing). Alternatively, a sample can be heated to a particular temperature and then the mass recorded over time.

The TGA is mainly composed of host, auxiliary equipment, control system, data acquisition and processing system. The main instrument also includes temperature control system, furnace body, support component, atmosphere control system, sample temperature measurement system, and quality measurement system. The auxiliary equipment includes automatic sampler, pressure control device, illumination, and cooling device.

Thermogravimetric method can be used in the experiment of trace samples, and it has the advantages of simple operation, strong repeatability, high precision, sensitive and rapid response. However, in practical applications, the thermogravimetric method also has certain limitations, which are manifested as the single limitation of the sample mass change information to characterize its complex thermal behavior, and sample testing takes a long time.

The starch thermogravimetric curve can be obtained by heating the starch at a certain heating rate and measuring the relationship between the sample mass and temperature. The thermogravimetric curve can reflect the decomposition and water loss of starch during heating, thus revealing the thermal properties of starch (Sahoo et al., 2024). In the thermogravimetric curve of starch (Figure 11.1b), three main stages of weightlessness can usually be observed: (1) the moisture in the starch evaporates; (2) the starch is pyrolyzed into smaller organic molecules; (3) any remaining organic compounds are decomposed, leaving only low-volatility mineral-rich compounds (Zhang et al., 2024). Sometimes, the derivative of TGA curves is calculated to obtain differential thermogravimetric (DTG) curves, which reflect the rate of the mass change with temperature. These DTG curves can be used to identify the temperature ranges where the key thermal events occur. TGA and DTG curves of starch samples provide valuable information about their initial moisture and ash contents, as well as about the thermal stability of the structures formed by the starch molecules. They are particularly useful in applications where starches are subjected to high temperatures, such as the thermal decomposition of starch-based packaging materials during incineration.

In addition, TGA of starch is very important for the application of starch. In food industry, studying the thermogravimetric curve of starch can understand the gelatinization characteristics and stability of starch, which is helpful to optimize the quality of food processing. In the biomedicine industry, the release performance and stability of the drug can be evaluated by studying the thermal ablation of starch, and the drug formulation design and preparation process optimization can be guided. In materials science, thermogravimetric curves can be used to evaluate the thermal stability, thermal decomposition characteristics, and application temperature range of starch-based materials, providing guidance for material design and application. Therefore, studying the thermogravimetric curve of starch can help promote the wide application of starch in different fields and product optimization.

11.5 SPECTRAL TECHNIQUES

11.5.1 Fourier Transform Infrared Spectrometer (FTIR)

Fourier transform infrared spectrometer (FTIR) is composed of infrared optical station, computer, and printer. The infrared optical station is the most important part of FTIR and determines the performance indicators of FTIR, and the computer and printer are auxiliary equipment.

FTIR is a useful tool for the detection of functional groups, chemical bonds, and chemical compositions in starch, modified starch, and starch-based derivatives (Govindaraju et al., 2021). FTIR spectrometry is based on measuring the absorption of infrared light by a sample at different wavenumbers. Typically, a series of peaks is observed in the spectra, which are related to the presence of specific functional groups in a sample. Each functional group undergoes stretching and bending vibrations at a particular frequency, which leads to absorption of energy at a specific wavenumber. The intensity and location of these peaks depend on their interactions with other components in their environment. Consequently, FTIR spectroscopy can be used to provide information about the composition and interactions of starch samples.

FTIR is sensitive to changes in starch crystallization, chain conformation, and helical structure, and so it is commonly used to provide information about the short-range order of starch granules. After baseline correction, smoothing, and Fourier deconvolution of the infrared spectrum of starch samples, the ordered structure of starch can be evaluated according to the peak heights measured at specific wavenumbers (Warren, Gidley, & Flanagan, 2016). For instance, the peak intensity ratios of $1,045/1,022\,\text{cm}^{-1}$ and $1,022/995\,\text{cm}^{-1}$ of the FTIR spectrum can be used as indices of the ordered structures in starch. For instance, the higher the peak intensity ratio of $1,045/1,022\,\text{cm}^{-1}$, the higher the degree of short-range order (Pozo et al., 2018). The degradation behavior of starch can also be studied by infrared spectroscopy (Wu et al., 2006).

11.5.2 Raman Spectroscopy

Raman spectrum is a kind of scattering spectrum (Fan et al., 2012), which reflects the vibration of the molecule through the intensity and position of the Raman peak, and analyzes the different functional groups or chemical bonds in the sample to obtain the information of the molecular structure, so as to achieve the purpose of rapid analysis or identification. Raman spectroscopy has many advantages in the characterization of samples, such as non-destructive, no sample preparation is required, and multiple sample components can be measured simultaneously. Not only that, unlike FTIR spectroscopy, the sample is analyzed using Raman spectroscopy without interference from moisture (Liu et al., 2015).

According to the literature reports, the full width at half maximum (FWHM) at $480\,\text{cm}^{-1}$ of the Raman spectrum can be used to measure and compare the structural changes of the starch (Mutungi et al., 2012). The smaller the FWHM value at $480\,\text{cm}^{-1}$, the higher the short-range order degree of starch. Consequently, Raman spectroscopy can also be used to monitor the destruction of short-range ordered structures in starch during gelatinization (Schuster et al., 2000; Wang et al., 2016). After starch gelatinization, disordered molecular chains reassemble to form ordered structures, leading to aging. Raman spectroscopy has been used in the determination of starch aging (Fechner et al., 2005). In addition, studies have shown that Raman spectroscopy can also reflect the changes in amylose and amylopectin content in starch. However, when using this method, a calibration model needs to be established in advance, and it is best not to contain interfering substances.

The similarities and differences between FTIR and Raman spectroscopy techniques are summarized in Table 11.3. In practical applications, each technology has its own advantages and

TABLE 11.3 The similarities and differences between FTIR and Raman

SPECTRAL TECHNIQUES		FTIR	RAMAN	REFERENCES
Similarities	Molecular vibration analysis	Both provide information about molecular structure by analyzing molecular vibration		Nakajima, Kuroki, and Ikehata (2023)
	Qualitative analysis	Both can be used for qualitative analysis of compounds to identify different chemical bonds and functional groups by characteristic peaks		Rohman et al. (2019)
	Non-destructive	Both techniques are non-destructive and do not change the chemistry of the sample		Fechner et al. (2005)
Differences	Spectrum range	It is mainly concentrated in the mid-infrared region and is sensitive to vibrations of specific chemical bonds	Visible or near-infrared lasers are usually used, with a wide spectral range	Wu et al. (2023)
	Spectrum types	Absorption spectra	Scattered spectrum	Taylor and Donnelly (2020)
	Principle	When the frequency of infrared light matches the vibration frequency of molecules, the energy absorbed by molecules will undergo transitions of vibrational energy levels	Based on the Raman scattering effect, where energy is exchanged when photons interact with molecules, resulting in changes in the frequency of the scattered light	
	Sensibility	Sensitive to polar or asymmetric vibration patterns	Sensitive to non-polar or symmetrical vibration modes	
	Sample requirement	The testing process is disturbed by water molecules	The testing process is free from interference by water molecules	Liu et al. (2015)
	Technological difficulty	High requirements for sample preparation and environmental conditions, such as avoiding water interference	The Raman scattered signal is usually weak and requires a highly sensitive detector	Yuen et al. (2009)

disadvantages. Often, it is beneficial to combine the information obtained from both techniques to gain more detailed insights into the molecular structure of starch.

11.6 RHEOLOGICAL ANALYSIS

Rheological properties are an indispensable research direction in the research and production of starch. Because the rheological properties of starch can predict and explain the flow, deformation, and texture changes that occur when different starch-based foods are processed. Therefore, the rheological properties of starch and starch paste have become one of the hotspots in the process of structural modification of starch and the development and evaluation of starch-based foods.

11.6.1 Rheometer

The rheological properties of starch are closely related to starch gelatinization and retrogradation (Ai & Jane, 2015), which are often characterized by monitoring changes in the viscosity or elastic modulus of a starch suspension with temperature.

Dynamic shear rheometers are commonly used to measure the dynamic viscoelastic properties of starch samples (Lewandowicz, Le Thanh-Blicharz, & Szwengiel, 2024). An oscillating stress is typically applied to a starch suspension, and the resulting oscillating strain is measured (or vice versa). A mathematical analysis of the relationship between the amplitude and phase of the oscillating stress and strain is used to calculate the dynamic shear modulus (G) of the sample. For viscoelastic materials, the dynamic shear modulus can be separated into a storage modulus (G') and a loss modulus (G''). The storage modulus is related to the elastic properties of the material, whereas the loss modulus is related to the viscous properties. The dynamic shear modulus can also be represented as phase angle (δ): $\tan\delta = G''/G'$. For a predominantly viscous material, $\delta > 45°$ and $G'' > G'$, whereas for a predominantly elastic material, $\delta < 45°$ and $G'' < G'$. Measurements can be performed using small or large amplitude oscillating strains (SAOS or LAOS). SAOS measurements provide fundamental information about the rheological properties of starch-based materials because they are non-destructive and do not damage them. In contrast, LAOS measurements provide insights into the behavior of starch-based materials under the large stresses and strains that they may encounter during their practical application. It should be noted that SAOS measurements are carried out in the linear viscoelastic regime (LVR) of a material, where the stress is proportional to the strain. Information about the gelatinization and retrogradation of starch samples can be carried out by measuring the change in the dynamic shear modulus when they are heated and cooled under controlled conditions. Typically, these measurements are carried out under SAOS conditions so as not to damage the sample during the analysis.

The same instruments can be used to measure the change in the apparent viscosity of a starch-based material when the shear strain is increased at a constant temperature. Starch pastes are non-Newtonian fluids that exhibit strong shear thinning properties (Lagarrigue & Alvarez, 2001). The shear viscosity of starch pastes decreases as the shear rate is increased because of the breakdown of structures, such as dissociation and disentanglement of starch molecules, aggregates, and granules (Punia et al., 2020). Starch can be used as thickener. It is important to understand its static rheological properties for its application in starch food and improvement of production technology.

11.6.2 Rapid Visco Analysis (RVA)

A specialized instrument, known as a Rapid Visco Analyzer (RVA), has been developed that can be used to monitor changes in the rheological properties of starch suspensions when they are heated and cooled under controlled conditions. RVA instruments provide valuable information about the rheological changes that occur due to the gelatinization and retrogradation of starch. Different types of starch have different RVA profiles due to their different gelatinization/retrogradation properties (Blazek & Copeland, 2009; Md Zaidul et al., 2007). Because of the development and utilization of modified starch, the gelatinization and viscosity analysis of starch have become important bases for judging the quality and processing characteristics of starch.

The RVA is considered as a kind of modern rotary viscosity testers controlled by computer and equipped with special software. In the 1980s, Australian researchers developed RVA for the rapid detection of sprouted wheat, and after continuous improvement and development, RVA has been used in some cases as the standard method for detecting grain quality (Bartalné-Berceli et al., 2021). RVA is a flexible, automated instrument that incorporates both international standard methods and customizable test procedures (Jin, Kong, & Wang, 2019). During analysis, a uniform sample temperature is maintained using a computer-controlled stirrer, which also acts as a viscosity sensor. The resulting RVA profiles can be analyzed using dedicated software, which generates a number of useful parameters related to gelatinization and retrogradation. For instance, the peak viscosity (PV), final viscosity (FV), setback viscosity (SB), breakdown viscosity (BD), trough viscosity (TV), and pasting temperature (PT) of a starch suspension can be obtained when it is subjected to a standardized heating-holding-cooling procedure within an RVA instrument. Measurement of these parameters can be used to compare different starches and assess their suitability for different applications.

RVA is considered a sensitive rheological tool to assess the effect of chemical composition and structure on starch functional properties (Juhász & Salgó, 2008). The RVA profiles of starches have been shown to depend on granule size, amylose/amylopectin content, relative crystallinity, and crystal structure (Juhász & Salgó, 2008; Qian & Kuhn, 1999), that is, the rheological properties of starch were closely related

to the structure of starch itself. With the development of modified starch, RVA is often used in the characterization of cross-linked starch, esterified starch, oxidized starch, pregelatinized starch, irradiated starch, etc. (Fu et al., 2013; Nakorn, Tongdang, & Sirivongpaisal, 2009; Shih & Daigle, 2003).

11.7 X-RAY TECHNIQUES

Valuable information about the structural organization of the molecules within starch samples can be provided using X-ray techniques.

11.7.1 X-ray Diffraction (XRD)

X-ray diffraction (XRD) provides information about the crystalline structure of starches, as the diffraction angle depends on the spacing between the starch molecules in the crystals. The Bragg equation, $2d\,\sin\theta = n\lambda$, is typically used to analyze the XRD pattern obtained, where d is the spacing of the diffracting planes, θ is the half diffraction angle, n is the diffraction order, and λ is the wavelength of the X-rays used.

The proportion of crystalline and amorphous structures in starch granules varies from around 10%–50% depending on their source and play a critical role in determining their functionality (Lopez-Rubio et al., 2008). Consequently, it is important to be able to provide information about the degree of crystallinity of different kinds of starches. XRD can measure the long-range order of the crystalline regions in starch (Warren, Gidley, & Flanagan, 2016), which can be used to identify the types of crystals that are present. Figure 11.2a shows the XRD patterns of different kinds of crystals typically found in starch. XRD can also be used to determine the impact of starch modification methods on the type and amounts of crystals present in starch (Kalita, Kaushik, & Mahanta, 2014). The location, intensity, width, and area of the peaks in an XRD pattern can be used to characterize the nature of the crystals present in starch (Govindaraju et al., 2021).

11.7.2 Small Angle X-ray Scatterer (SAXS)

The small angle X-ray scatterer (SAXS) mainly consists of X-ray light source, collimation system, vacuum system, sample rack, and receiving system. Small angle X-ray scattering (SAXS) can also be used to provide valuable information about the crystalline structure of starches. When a very thin beam of X-rays passes through the sample, the X-rays are scattered in a very small angular domain near the original beam due to scattering of internal electrons, and the scattering intensity distribution is closely related to the difference of the electron density, the particle size, and distribution of the scatterers (Pikus, Olszewska, & Becal, 2006). X-ray scattering results from fluctuations in

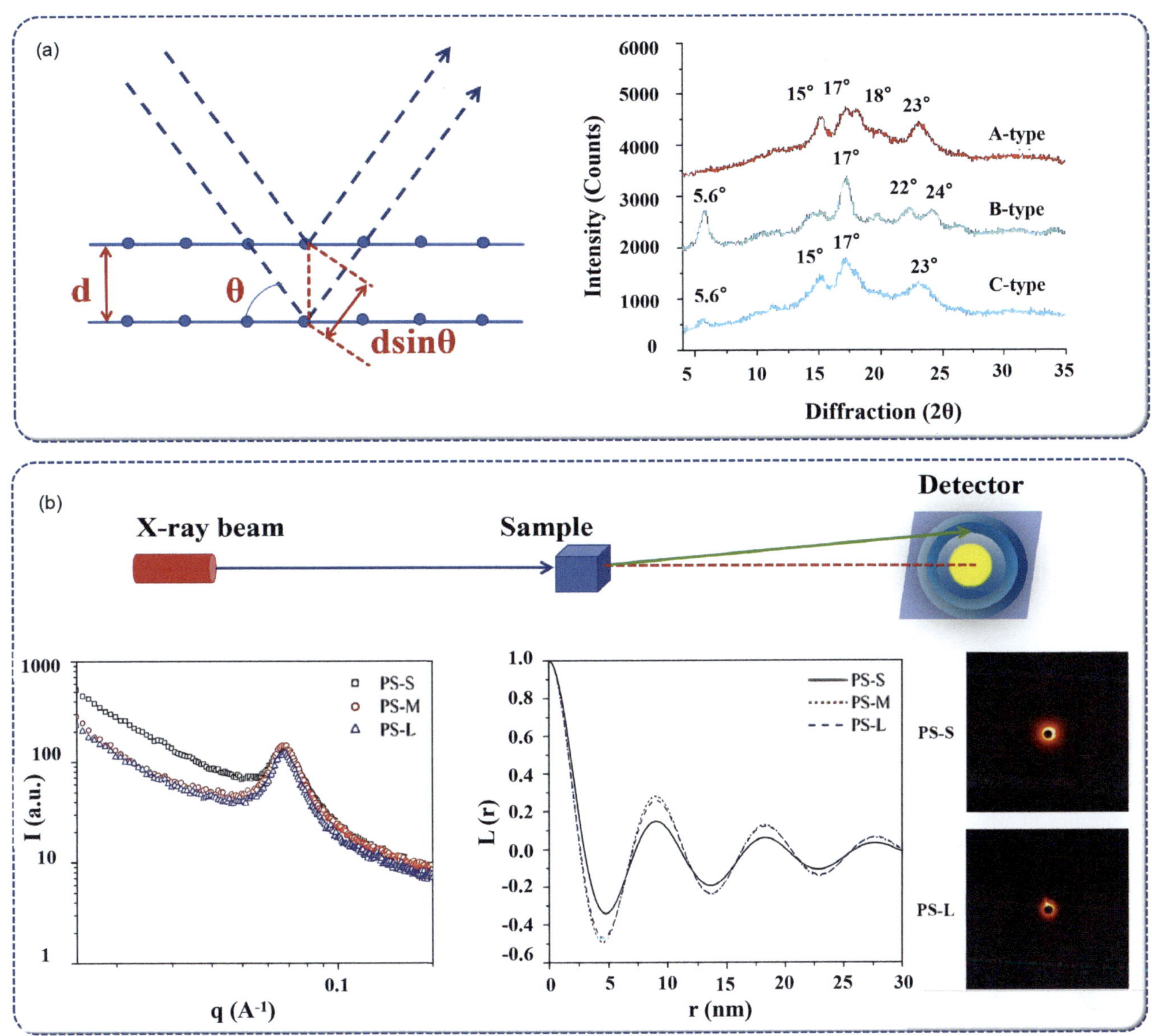

FIGURE 11.2   Examples of XRD (a) and SAXS (b) patterns (Chen et al., 2019; Wang et al., 2022).

electron density that occur over the range from 1 to 1,000 nm in materials. As a result, SAXS can provide information about the heterogeneity of materials on the nanoscale (Sujka, Jamroz, & Kwiatkowski, 2011). Electron density variations occur at these length scales within starch granules due to the presence of amorphous and crystalline regions (Bertoft, 2017). The most important application of SAXS in starch analysis is the quantitative determination of the lamellar structure of the semi-crystalline growth rings (Cardoso & Westfahl, 2010; Xu et al., 2020).

11.8 CONCLUSION

Knowledge of the composition and structure of different kinds of starches is essential for their successful application within the food and other industries. Based on its molecular, physicochemical, and functional properties, each type of starch has a range of applications where it is most suitable. A variety of analytical instruments and experimental protocols are available to provide information about the composition and structure of starch molecules and granules. It is important to understand the underlying principles behind each method so as to select the most appropriate one for providing the desired information, as well as to ensure that the measurements are carried out properly. Often, the results from a variety of complementary methods are combined to provide information at different length scales (from nanometer to micrometer) that influence the functionality of starch granules. For instance, one method may be used to provide information about starch composition, another about the internal nanostructure of starch granules, and yet another about the size and shape of the starch granules.

REFERENCES

Adewale, P, Yancheshmeh, MS, & Lam, E 2022. Starch modification for non-food, industrial applications: Market intelligence and critical review. *Carbohydrate Polymers*, 291, art. 119590.

Ai, Y, & Jane, JL 2015. Gelatinization and rheological properties of starch. *Starch-Starke*, 67, 213–224.

Alamu, EO, Nuwamanya, E, Cornet, D, Meghar, K, Adesokan, M, Tran, T, Belalcazar, J, Desfontaines, L, & Davrieux, F 2021. Near-infrared spectroscopy applications for high-throughput phenotyping for cassava and yam: A review. *International Journal of Food Science and Technology*, 56, 1491–1501.

Amaraweera, SM, Gunathilake, C, Gunawardene, OHP, Fernando, NML, Wanninayaka, DB, Dassanayake, RS, Rajapaksha, SM, Manamperi, A, Fernando, CAN, Kulatunga, AK, & Manipura, A 2021. Development of starch-based materials using current modification techniques and their applications: A review. *Molecules*, 26, 6880.

Bangar, SP, Ashogbon, AO, Singh, A, Chaudhary, V, & Whiteside, WS 2022. Enzymatic modification of starch: A green approach for starch applications. *Carbohydrate Polymers*, 287, art. 119265.

Bartalné-Berceli, M, Izsó, E, Gergely, S, & Salgó, A 2021. Effects of special additives in wheat dough system measured by Mixolab technique. *Czech Journal of Food Sciences*, 39, 460–468.

Batey, IL, & Curtin, BM 1996. Measurement of amylose/amylopectin ratio by high-performance liquid chromatography. *Starch-Starke*, 48, 338–344.

Bertoft, E 2017. Understanding starch structure: Recent progress. *Agronomy-Basel*, 7, art. 56.

Blazek, J, & Copeland, L 2009. Effect of monopalmitin on pasting properties of wheat starches with varying arnylose content. *Carbohydrate Polymers*, 78, 131–136.

Blazek, J, & Gilbert, EP 2011. Application of small-angle X-ray and neutron scattering techniques to the characterisation of starch structure: A review. *Carbohydrate Polymers*, 85, 281–293.

Cardoso, MB, & Westfahl, H 2010. On the lamellar width distributions of starch. *Carbohydrate Polymers*, 81, 21–28.

Chen, L, Ma, RR, McClements, DJ, Zhang, ZP, Jin, ZY, & Tian, YQ 2019. Impact of granule size on microstructural changes and oil absorption of potato starch during frying. *Food Hydrocolloids*, 94, 428–438.

Chen, L, McClements, DJ, Ma, Y, Yang, TY, Ren, F, Tian, YQ, & Jin, ZY 2021. Analysis of porous structure of potato starch granules by low-field NMR cryoporometry and AFM. *International Journal of Biological Macromolecules*, 173, 307–314.

Chen, P, Yu, L, Simon, GP, Liu, XX, Dean, K, & Chen, L 2011. Internal structures and phase-transitions of starch granules during gelatinization. *Carbohydrate Polymers*, 83, 1975–1983.

Compart, J, Singh, A, Fettke, J, & Apriyanto, A 2023. Customizing starch properties: A review of starch modifications and their applications. *Polymers*, 15, art. 3491.

Danley, RL 2003. New heat flux DSC measurement technique. *Thermochimica Acta*, 395, 201–208.

Delwiche, SR, Bean, MM, Miller, RE, Webb, BD, & Williams, PC 1995. Apparent amylose content of milled rice by near-infrared reflectance spectrophotometry. *Cereal Chemistry*, 72, 182–187.

Dery, B, & Lou, ZX 2023. Scanning electron microscopy (SEM) as an effective tool for determining the morphology and mechanism of action of functional ingredients. *Food Reviews International*, 39, 2007–2026.

Fan, DM, Ma, WR, Wang, LY, Huang, JL, Zhao, JX, Zhang, H, & Chen, W 2012. Determination of structural changes in microwaved rice starch using Fourier transform infrared and Raman spectroscopy. *Starch-Starke*, 64, 598–606.

Fechner, PM, Wartewig, S, Kleinebudde, P, & Neubert, RHH 2005. Studies of the retrogradation process for various starch gels using Raman spectroscopy. *Carbohydrate Research*, 340, 2563–2568.

Flamme, WT, Jurgens, HU, & Jansen, GG 1994. Quantitative-determination of amylose and amylopectin in starches by HPSEC with DMSO as solvent and eluant. *American Laboratory*, 26, 29-32.

Fu, JJ, Zhou, XY, Feng, DH, Wang, Q, & Deng, L 2013. Study on the rapid detection of irradiated rice based on RVA. *Nuclear Science and Techniques*, 24, art. S010310.

Gérard, C, Barron, C, Colonna, P, & Planchot, V 2001. Amylose determination in genetically modified starches. *Carbohydrate Polymers*, 44, 19–27.

Govindaraju, I, Chakraborty, I, Baruah, VJ, Sarmah, B, Mahato, KK, & Mazumder, N 2021. Structure and morphological properties of starch macromolecule using biophysical techniques. *Starch-Starke*, 73, art. 2000030.

Govindaraju, I, Pallen, S, Umashankar, S, Mal, SS, Melanthota, SK, Mahato, DR, Zhuo, GY, Mahato, KK, & Mazumder, N 2020. Microscopic and spectroscopic characterization of rice and corn starch. *Microscopy Research and Technique*, 83, 490–498.

Horovitz, O, Cioica, N, Jumate, N, Pojar-Fenesan, M, Balea, A, Liteanu, V, Mocanu, A, & Tomoaia-Cotisel, M 2011. SEM characterization of starch granules. *Studia Universitatis Babes-Bolyai Chemia*, 56, 211–219.

Jarvis, CE, & Walker, JRL 1993. Simultaneous, rapid, spectrophotometric determination of total starch, amylose and amylopectin. *Journal of the Science of Food and Agriculture*, 63, 53–57.

Jin, N, Kong, DD, & Wang, HY 2019. Effects of temperature and time on gelatinization of corn starch employing gradient isothermal heating program of rapid visco analyzer. *Journal of Food Process Engineering*, 42, art. E13264.

Juhász, R, & Salgó, A 2008. Pasting behavior of amylose amylopectin and, their mixtures as determined by RVA curves and first derivatives. *Starch-Starke*, 60, 70–78.

Kalita, D, Kaushik, N, & Mahanta, CL 2014. Physicochemical, morphological, thermal and IR spectral changes in the properties of waxy rice starch modified with vinyl acetate. *Journal of Food Science and Technology-Mysore*, 51, 2790–2796.

Klang, V, Matsko, NB, Valenta, C, & Hofer, F 2012. Electron microscopy of nanoemulsions: An essential tool for characterisation and stability assessment. *Micron*, 43, 85–103.

Lagarrigue, S, & Alvarez, G 2001. The rheology of starch dispersions at high temperatures and high shear rates: A review. *Journal of Food Engineering*, 50, 189–202.

Lewandowicz, J, Le Thanh-Blicharz, J, & Szwengiel, A 2024. Insight into rheological properties and structure of native waxy starches: Cluster analysis grouping. *Molecules*, 29, art. 2669.

Lin, LS, Zhao, SN, Li, EP, Guo, DW, & Wei, CX 2022. Structural properties of starch from single kernel of high-amylose maize. *Food Hydrocolloids*, 124,107349.

Liu, HS, Yu, L, Tong, Z, & Chen, L 2010. Retrogradation of waxy cornstarch studied by DSC. *Starch-Starke*, 62, 524–529.

Liu, P, Chen, L, Corrigan, PA, Yu, L, & Liu, ZD 2008. Application of atomic force microscopy on studying micro- and nano-structures of starch. *International Journal of Food Engineering*, 4, art. 8.

Liu, XC, Qiao, L, Kong, YX, Wang, HYT, & Yang, BJ 2024. Characterization of the starch molecular structure of wheat varying in the content of resistant starch. *Food Chemistry-X*, 21, 101103.

Liu, YQ, Xu, Y, Yan, YZ, Hu, DD, Yang, LZ, & Shen, RL 2015. Application of Raman spectroscopy in structure analysis and crystallinity calculation of corn starch. *Starch-Starke*, 67, 612–619.

Lopez-Rubio, A, Flanagan, BM, Gilbert, EP, & Gidley, MJ 2008. A novel approach for calculating starch crystallinity and its correlation with double helix content: A combined XRD and NMR study. *Biopolymers*, 89, 761–768.

Mahmood, K, Kamilah, H, Shang, PL, Sulaiman, S, Ariffin, F, & Alias, A 2017. A review: Interaction of starch/non-starch hydrocolloid blending and the recent food applications. *Food Bioscience*, 19, 110–120.

Md Zaidul, IS, Norulaini, N, Mohd Omar, AK, Yamauchi, H, & Noda, T 2007. Correlations of the composition minerals and RVA pasting properties of various potato starches. *Starch-Starke*, 59, 269–276.

Mestres, C, Matencio, F, Pons, B, Yajid, M, & Fliedel, G 1996. A rapid method for the determination of amylose content by using differential scanning calorimetry. *Starch-Starke*, 48, 2–6.

Morris, VJ, Woodward, NC, & Gunning, AP 2011. Atomic force microscopy as a nanoscience tool in rational food design. *Journal of the Science of Food and Agriculture*, 91, 2117–2125.

Mutungi, C, Passauer, L, Onyango, C, Jaros, D, & Rohm, H 2012. Debranched cassava starch crystallinity determination by Raman spectroscopy: Correlation of features in Raman spectra with X-ray diffraction and ^{13}C CP/MAS NMR spectroscopy. *Carbohydrate Polymers*, 87, 598–606.

Nakajima, S, Kuroki, S, & Ikehata, A 2023. Selective detection of starch in banana fruit with Raman spectroscopy. *Food Chemistry*, 401, art. 134166.

Nakorn, KN, Tongdang, T, & Sirivongpaisal, P 2009. Crystallinity and rheological properties of pregelatinized rice starches differing in amylose content. *Starch-Starke*, 61, 101–108.

Nawrocka, A, Piwonski, I, Sauro, S, Porcelli, A, Hardan, L, & Lukomska-Szymanska, M 2021. Traditional microscopic techniques employed in dental adhesion research-applications and protocols of specimen preparation. *Biosensors-Basel*, 11, art. 408.

Pandiselvam, R, Sruthi, NU, Kumar, A, Kothakota, A, Thirumdas, R, Ramesh, SV, & Cozzolino, D 2023. Recent applications of vibrational spectroscopic techniques in the grain industry. *Food Reviews International*, 39, 209–239.

Park, H, Xu, S, & Seetharaman, K 2011. A novel in situ atomic force microscopy imaging technique to probe surface morphological features of starch granules. *Carbohydrate Research*, 346, 847–853.

Pérez, S, & Bertoft, E 2010. The molecular structures of starch components and their contribution to the architecture of starch granules: A comprehensive review. *Starch-Starke*, 62, 389–420.

Pérez-Bermúdez, I, Castillo-Suero, A, Cortés-Inostroza, A, Jeldrez, C, Dantas, A, Hernández, E, Orellana-Palma, P, & Petzold, G 2023. Observation and measurement of ice morphology in foods: A review. *Foods*, 12, 3987.

Pikus, S, Olszewska, E, & Becal, M 2006. Investigation of the staling process of bread using the small-angle X-ray scattering (SAXS) method. *Flavour and Fragrance Journal*, 21, 37–41.

Pozo, C, Rodríguez-Llamazares, S, Bouza, R, Barral, L, Castaño, J, Müller, N, & Restrepo, I 2018. Study of the structural order of native starch granules using combined FTIR and XRD analysis. *Journal of Polymer Research*, 25, art. 266.

Punia, S, Sandhu, KS, Dhull, SB, Siroha, AK, Purewal, SS, Kaur, M, & Kidwai, MK 2020. Oat starch: Physico-chemical, morphological, rheological characteristics and its applications - A review. *International Journal of Biological Macromolecules*, 154, 493–498.

Qian, JY, & Kuhn, M 1999. Characterization of *Amaranthus cruentus* and *Chenopodium quinoa* starch. *Starch-Starke*, 51, 116–120.

Rohman, A, Windarsih, A, Lukitaningsih, E, Rafi, M, Betania, K, & Fadzillah, NA 2019. The use of FTIR and Raman spectroscopy in combination with chemometrics for analysis of biomolecules in biomedical fluids: A review. *Biomedical Spectroscopy and Imaging*, 8, 55–71.

Sahoo, B, Kumari, A, Sarkhel, S, Jha, S, Mukherjee, A, Jain, M, Mohan, A, & Roy, A 2024. Rice starch phase transition and detection during resistant starch formation. *Food Reviews International*, 40, 158–184.

Sanders, J, & Cohen, GM 2019. Observation of the optical & chemical properties of starch granules. *American Biology Teacher*, 81, 644–648.

Sárka, E, & Dvorácek, V 2017. New processing and applications of waxy starch (a review). *Journal of Food Engineering*, 206, 77–87.

Schuster, KC, Ehmoser, H, Gapes, JR, & Lendl, B 2000. On-line FT-Raman spectroscopic monitoring of starch gelatinisation and enzyme catalysed starch hydrolysis. *Vibrational Spectroscopy*, 22, 181–190.

Sene, M, Thevenot, C, & Prioul, JL 1997. Simultaneous spectrophotometric determination of amylose and amylopectin in starch from maize kernel by multi-wavelength analysis. *Journal of Cereal Science*, 26, 211–221.

Sheng, WJ, & Wei, CX 2022. Screening methods for cereal grains with different starch components: A mini review. *Journal of Cereal Science*, 108, 103557.

Shi, MM, Chen, Y, Yu, SJ, & Gao, QY 2013. Preparation and properties of RS III from waxy maize starch with pullulanase. *Food Hydrocolloids*, 33, 19–25.

Shih, FF, & Daigle, KW 2003. Gelatinization and pasting properties of rice starch modified with 2-octen-1-ylsuccinic anhydride. *Nahrung-Food*, 47, 64–67.

Sívoli, L, Pérez, E, Rodríguez, P, Raymúndez, MB, & Ayesta, C 2009. Microscopic techniques and of light dispersion used in the evaluation of the structure of the native starch of yuca (*Manihot esculenta* C). *Acta Microscopica*, 18, 195–203.

Stevnebo, A, Sahlström, S, & Svihus, B 2006. Starch structure and degree of starch hydrolysis of small and large starch granules from barley varieties with varying amylose content. *Animal Feed Science and Technology*, 130, 23–38.

Sujka, M, Jamroz, J, & Kwiatkowski, R 2011. Influence of α-amylolysis on the formation of electron density inhomogeneities on the surface of starch granules. *Starch-Starke*, 63, 17–23.

Taylor, EA, & Donnelly, E 2020. Raman and Fourier transform infrared imaging for characterization of bone material properties. *Bone*, 139, art. 115490.

Teng, C, Chen, D, Wu, GF, & Campanella, OH 2021. Non-invasive techniques to study starch structure and starchy products properties. *Current Opinion in Food Science*, 38, 196–202.

van de Velde, F, van Riel, J, & Tromp, RH 2002. Visualisation of starch granule morphologies using confocal scanning laser microscopy (CSLM). *Journal of the Science of Food and Agriculture*, 82, 1528–1536.

Vilaplana, F, Hasjim, J, & Gilbert, RG 2012. Amylose content in starches: Toward optimal definition and validating experimental methods. *Carbohydrate Polymers*, 88, 103–111.

Wang, JP, Li, Y, Tian, YQ, Xu, XM, Ji, XX, Cao, X, & Jin, ZY 2010. A novel triple-wavelength colorimetric method for measuring amylose and amylopectin contents. *Starch-Starke*, 62, 508–516.

Wang, SJ, Zhang, X, Wang, S, & Copeland, L 2016. Changes of multi-scale structure during mimicked DSC heating reveal the nature of starch gelatinization. *Scientific Reports*, 6, art. 28271.

Wang, X, Huang, LX, Zhang, CH, Deng, YJ, Xie, PJ, Liu, LJ, & Cheng, J 2020. Research advances in chemical modifications of starch for hydrophobicity and its applications: A review. *Carbohydrate Polymers*, 240, 116292.

Wang, Y, McClements, DJ, Long, J, Qiu, C, Sang, SY, Chen, L, Xu, ZL, & Jin, Z 2022. Structural transformation and oil absorption of starches with different crystal types during frying. *Food Chemistry*, 390, 133115.

Warren, FJ, Gidley, MJ, & Flanagan, BM 2016. Infrared spectroscopy as a tool to characterise starch ordered structure-A joint FTIR-ATR, NMR, XRD and DSC study. *Carbohydrate Polymers*, 139, 35–42.

Williams PC, Kuzina FD, & Hlynka I 1970. A rapid colorimetric-method for estimating the amylose content of starches and-flours. *Cereal Chemistry*, 47(4), 411–420.

Wu, H, Ran, XH, Zhang, KY, Zhuang, YG, & Dong, LS 2006. FTIR study on retrogradation behavior of cross-linked starch. *Chemical Journal of Chinese Universities-Chinese*, 27, 775–778.

Wu, ML, Li, YH, Yuan, Y, Li, S, Song, XX, & Yin, JY 2023. Comparison of NIR and Raman spectra combined with chemometrics for the classification and quantification of mung beans (*Vigna radiata* L.) of different origins. *Food Control*, 145, art. 109498.

Xiao, HX, Wang, SY, Xu, WZ, Yin, YQ, Xu, D, Zhang, L, Liu, GQ, Luo, FJ, Sun, SG, Lin, QL, & Xu, BC 2020. The study on starch granules by using darkfield and polarized light microscopy. *Journal of Food Composition and Analysis*, 92, art. 103576.

Xu, JC, Blennow, A, Li, XX, Chen, L, & Liu, XX 2020. Gelatinization dynamics of starch in dependence of its lamellar structure, crystalline polymorphs and amylose content. *Carbohydrate Polymers*, 229, 115481.

Yashini, M, Khushbu, S, Madhurima, N, Sunil, CK, Mahendran, R, & Venkatachalapathy, N 2024. Thermal properties of different types of starch: A review. *Critical Reviews in Food Science and Nutrition*, 64, 4373–4396.

Yuen, SN, Choi, SM, Phillips, DL, & Ma, CY 2009. Raman and FTIR spectroscopic study of carboxymethylated non-starch polysaccharides. *Food Chemistry*, 114, 1091–1098.

Zavareze, ED, El Halal, SLM, de los Santos, DG, Helbig, E, Pereira, JM, & Dias, ARG 2012. Resistant starch and thermal, morphological and textural properties of heat-moisture treated rice starches with high-, medium- and low-amylose content. *Starch-Starke*, 64, 45–54.

Zhang, Y, Xing, BF, Kong, DG, Gu, ZX, Yu, YJ, Zhang, YJ, & Li, DD 2024. Improvement of in vitro digestibility and thermostability of debranched waxy maize starch by sequential ethanol fractionation. *International Journal of Biological Macromolecules*, 254, 127895.

Zhu, F 2017. Atomic force microscopy of starch systems. *Critical Reviews in Food Science and Nutrition*, 57, 3127–3144.

Modification of Starch 12

James N. BeMiller

12.1 INTRODUCTION

Information about starch modifications is so huge that it is impossible to put it in a single chapter. Accordingly, this chapter gives some basic information about types of modification and the properties of modified starch products. Much of the material in this chapter is condensed from previous book chapters and review articles, which are cited in appropriate places and which should be consulted for specific information and references to original articles. Information not included in those articles is referenced herein, often to review articles. Research toward modification of additional types of starch and improvements in the production of and expansion of applications of modified starches continues.

12.1.1 Reasons for Modifying Starch

The attributes of starch (abundance, wide availability, in the form of cold-water-insoluble granules, relatively low cost, non-toxicity, source of nutritionally important D-glucose) make it an attractive material for food and other industrial applications, but native starches have shortcomings that limit their applications. They can, however, be transformed into versatile, widely used, and valuable commercial products via modifications that improve their properties and functionalities and thus expand their usefulness. The principal reasons why starches are modified for use in food products are given in Table 12.1.

Several chemical modifications that produce products called stabilized (or substituted) starches can be used to provide long-term product stability (Section 12.1.2.2). Several different chemical modifications that produce products called cross-linked starches can be used to provide tolerance to processing conditions and prevent overcooking (Section 12.1.2.2). Combinations of stabilization and cross-linking reactions are often used to produce starch products that provide stability and desirable texture. Other modifications can add or improve other functionalities. By proper modification, changes can be made in one or more of the characteristics (and others) of a starch listed in Table 12.2. The list in the table indicates that many modified starch products can be made from a single type of starch.

12.1.2 Types of Modification

12.1.2.1 Treatment types

A starch may be modified via plant breeding (Chapter 4) and by genetic engineering (Chapter 3). Then, that new starch (together with existing starches) may be modified by treatment with chemical reagents (Section 12.2), enzymes (Section 12.3), and/or physical methods (Section 12.4).

12.1.2.2 Functional types of chemical modifications

There are four general types of chemical modifications: (1) those that stabilize starch pastes and gels by means of retarding retrogradation – the products of which are known as stabilized or sometimes as substituted starches, (2) those that reduce susceptibility of a starch to processing conditions (often relatively high temperatures, high shear, and acidic or alkaline pII values), which are known as cross-linked starches, (3) those that reduce the viscosity of starch pastes and gels and allow higher contents of solids to be used without imparting excessive viscosity or gel strength, and (4) those that add a new functionality to the starch.

1. Solutions, pastes, and gels of most native starches are prone to retrogradation. Retrogradation is a process involving a series of molecular aggregations (largely of amylose molecules) via formation of intra- and intermolecular double helices, hydrogen-bonded associations, and molecular entanglements which insolubilize the starch as a hot paste is cooled and stored. The first indication of retrogradation is an increase in opacity, which is followed by precipitation or gelation, a hardening of semi-solid and solid food systems, and changes in the texture and water activity of the system. Retrogradation is related to the setback value calculated during determination of pasting and paste characteristics (Chapter 10). Gel strength/firmness, stability, texture, water holding and mobility, shelf life, and digestibility are also related to the extent of retrogradation.

DOI: 10.1201/9781003464396-12

TABLE 12.1   Principal reasons for modifying starch for use in food products (Mason 2009)

To provide long-term product stability

- To provide cold-storage stability
- To provide freeze–thaw stability
- To decrease syneresis
- To improve gel clarity

To provide tolerance to processing conditions and prevent overcooking

- To decrease or increase pasting temperature
- To decrease or increase peak viscosity
- To reduce breakdown due to acid, shear, and/or heat

To provide desirable texture

- To decrease or increase viscosity
- To decrease or increase gel formation
- To decrease or increase gel strength

TABLE 12.2   Some characteristics of a starch that can be altered by proper modification

Ability to bind or bind to other substances (Section 12.5.5)
Ability to act as an emulsifier (Section 12.5.3)
Ability to act as an emulsion stabilizer (Section 12.5.3)
Ability to encapsulate (Section 12.5.7)
Ability to coat/film formation (Section 12.5.8)
Adhesiveness (Section 12.5.6)
Cold-water swellability (Section 12.5.1.1)
Cooking/pasting characteristics (Section 12.5.1)

- Degree of breakdown
- Degree of setback
- Energy required to cook
- Gelatinization and pasting temperatures

Digestibility (Section 12.5.4)
Hydrophobicity/oil absorption capacity (Section 12.5.10)
Interactions with other substances/ingredients (Sections 12.5.3, 12.5.5, 12.5.6, 12.5.7, 12.5.8, 12.5.10, 12.5.12)
Oil absorption capacity/hydrophobicity (Section 12.5.10)
Paste and gel characteristics (Section 12.5.2)

- Clarity
- Freeze–thaw stability
- Gel strength
- Rheology
- Sheen
- Stability (of refrigerated and frozen foods, in high-salt environments)
- Syneresis
- Viscosity (hot, room-temperature, and cold paste or gel)

Process tolerance (Section 12.5.1.2)

- pH tolerance
- Shear tolerance
- Temperature tolerance

Solubility in room temperature and cold water (Section 12.5.1.1)
Thermoplasticity (Section 12.5.11)
Water-holding capacity (Section 12.5.9)
Water resistance (Section 12.5.10)

Non-cross-linked (see below) derivatized and oxidized starches are known as stabilized starches because they modify properties related to retrogradation and provide long-term product stability. Especially, they provide cold-storage and freeze–thaw stability, decrease syneresis, and improve gel clarity. Reactions that produce stabilized starches are esterification and etherification with monofunctional reagents and oxidation. Derivatization reactions that produce stabilized starches introduce groups that are bulkier than the reacted hydroxyl group, thereby sterically inhibiting close contact of solubilized starch molecules and reducing intermolecular associations. In some cases, the added functional group has an ionic charge, which increases the hydrophilic nature of the starch molecule and reduces intermolecular associations through electrostatic repulsion. Oxidation provides stabilization by introducing keto groups that alter the conformation of the polysaccharide chains and by introducing bulky, anionic carboxylate groups, both of which structural alterations reduce intermolecular associations.

Both the addition of substituent groups and the conversion of hydroxyl groups into carbonyl or carboxylate groups disrupt the crystalline structures of the alkali-swollen granules in the reaction media, even after they are neutralized, recovered, and dried, thereby reducing the relative crystallinity of the starch, and as a result, gelatinization and pasting temperatures and enthalpy values, which means that the starch cooks out more easily. The native starch used for modification makes a difference in product characteristics. For example, waxy maize starch inherently has a reduced tendency to make the opaque, cohesive, rubbery, and unstable/syneresing gels that amylose-containing starches make, and stabilization modifications of waxy starches makes their gels even more stable. The characteristics of gels made from amylose-containing starches can vary considerably depending on the starch source.

Stabilized starches are used in foods as thickeners and texturizers and to improve cold-storage and freeze–thaw stability. They are also employed in papermaking and other industries because of the stability of their pastes.

2. As stated in Sections 12.2.4.5, 12.2.5.3, and 12.2.5.8, modifications with di-functional reagents bridge/join adjacent starch molecules within granules and thus produce what are known as cross-linked starches. Cross-linking (via formation of intra- and intermolecular covalent bonds) creates a three-dimensional network structure that strengthens granules and reduces both the rate and extent of their swelling and subsequent disintegration during heating in water. Compared to native starch granules, cross-linked granules are more stable/less sensitive to processing conditions (high temperatures, extended cooking times, acidic environments, and high shear during mixing, pumping, and/or homogenization), thus reducing overcooking.

The use of a cross-linked starch product results in more swollen granules in the cooked product, thus increasing viscosity and improving texture (i.e., becoming heavier bodied and shorter textured), and hot paste and gel stability as compared to the native parent starch. However, there is not a linear correlation between the degree of cross-linking and viscosity as the temperatures of slurries of granules in water increase because, at low levels of cross-linking, granule strengthening allows them to swell to a greater extent (as compared to unmodified granules) before disintegrating so that viscosity of the resulting paste is increased. This effect will continue with increasing cross-linking to a point at which the cross-linking begins to restrict the degrees of swelling and the amounts of swollen granules at a given temperature begins to drop. This process will continue as the degree of cross-linking is raised to a point that granule swelling will be restricted/ inhibited as compared to unmodified granules so that higher temperatures must be used to cook the starch. Proper cross-linking also provides salt tolerance and reduces gel formation, gel cohesiveness, leaching of amylose molecules from cooked granules, and retrogradation/setback (improves storage stability). Starches from different sources reacted with the same amount of cross-linking reagent in the same way may have different degrees of changes in, or even opposing changes in, a specific property.

Cross-linked products are used in the preparation of processed foods. The contents of rapidly digestible starch (RDS) and slowly digestible starch (SDS) decrease, while the contents of resistant starch (RS, Chapter 20) and dietary fiber increase as a function of degree of cross-linking. Deep-fried battered food products made with cross-linked starch in the batter have greater crispness and oil uptake. Biodegradable films made from cross-linked starch products have improved mechanical properties. Cross-linking of starch provides an effective controlled release matrix for pharmaceutical tablets and a product that can replace some of the fat in cream.

3. A third common type of chemical modification involves reducing the size of the starch polysaccharide molecules. Native starches generally form high-viscosity pastes and strong gels at relatively low concentrations, but starches have functional properties other than simply thickening and gel formation, so preparation of high-solids solutions without generating an extremely high viscosity paste or gel is often desired. Products that allow formation of such solutions can be achieved by depolymerizing a native or modified starch to various degrees. Various terms are used to designate such products.

4. New attributes that may be added to a starch via chemical modification include (but are not limited to) an ability to act as an emulsifier, an ability to bind to other substances (including to flocculate), an ability to encapsulate, an ability to bind hydrophobic and ionic substances, and an ability to act as a thermoplastic material.

12.2 CHEMICAL MODIFICATIONS

Chemically modified starches are those whose structures have been changed by treatment with chemical reagents. Starches are chemically modified to enhance the positive attributes of either their granular or cooked states, to minimize the negative attributes of either their granular or cooked states, and/or to provide functionalities that native starches cannot provide, thereby expanding their applications. In the production of modified starches for food applications, treatment with small amounts of approved chemicals can dramatically improve certain properties of the native starch and/or introduce new functionalities. Changes to the properties of starches resulting from chemical modification include, but are not limited to, those listed in Table 12.2.

Chemical modifications of starch have been reported and reviewed many times. There are several sources that, although they were published years ago, are still useful because the basics of starch modification have changed little since their publication. They are Rutenberg and Solarek (1984) (a thorough review – primarily of reagents that had been used to modify starches up to 1983), Wurzburg (1986) (reviews of preparation and uses primarily of commercially available starches in the USA), and Chiu and Solarek (2009) (an update of Rutenberg and Solarek (1984)). The material presented in this section is an abbreviated version of the review by BeMiller and Fang (2025), which contains considerably more references. See also the reviews of Singh, Kaur, and McCarthy (2007), Chen et al. (2015), Haroon et al. (2016), Zia-ud-Din, Xiong, and Fei (2017), Masina et al. (2017), Chen, Kaur, and Singh (2018), Haq et al. (2019), and Galkowska, Kapuśniak, and Juszczak (2023).

12.2.1 Starch Molecules

There are two statements that should be made at the outset: (1) Starch is a generic term, and each starch from each plant source is unique and different from other starches, and (2) There are two definitions of the term 'modified starch'. One is legal and may differ from jurisdiction to jurisdiction. The second is chemical and is basically 'the addition of groups to or changes in the structures and properties of starch molecules and/or granules'. This chapter covers only the second definition.

A starch's properties, functionalities, and granular structures are determined by the structures of its component molecules,

and the chemical reactions used to modify starches are those that esterify, etherify, or oxidize the hydroxyl groups of the starch polysaccharide molecules or effect depolymerizations. Therefore, an understanding of the chemical structures of starch molecules is important. Starch molecules are polysaccharides. There are two primary starch polysaccharides: amylopectin (a constituent of all native starches) and amylose (a constituent of all native starches except the starch from waxy maize/corn a naturally occurring mutant of maize/corn) and the starches from a few mutagenized or genetically modified plants that also contain only amylopectin. The structures of amylose and amylopectin molecules and the nature and characteristics of native starch granules are presented in Chapters 6, 7, 9, and 10.

12.2.2 Major Types of Chemical Modification

All starches can be chemically modified by a variety of reagents. This chapter emphasizes products that have major commercial production and use and are, therefore, most often employed and established uses of the products. The chemical structures of the starch polysaccharide molecules dictate that there are two functional groups on the molecules that can be modified: (1) hydroxyl groups (on C2, C3, and C6 of the glucosyl units) which can be derivatized (i.e., etherified (Section 12.2.4) or esterified (Section 12.2.5)), oxidized (Section 12.2.7), and/or added to via graft polymerization (a special kind of derivatization (Section 12.2.10)) and (2) glycosidic bonds that can be cleaved by acid- or enzyme-catalyzed hydrolysis (Section 12.2.8; Chapters 13 and 14) and by alkalis after certain oxidations (Section 12.2.7.1).

Derivatization reactions are ordinary reactions of hydroxyl groups – differing only from other organic reactions of hydroxyl groups in that they are conducted in the presence of large amounts of water (because the great majority of all starch modifications are carried out in 30%–35% aqueous slurries of granules, i.e., in granule slurries coming directly from hydroclones during their isolation). Because most derivatization and other chemical modification reactions employ alkaline conditions, because starch granules swell in aqueous alkaline systems, and because many modifications (essentially all except cross-linking modifications) cause starch granules to swell (and disintegrate) more easily than native starch granules, slurries of granules being derivatized often contain a swelling inhibiting salt – the most effective for this purpose being sodium sulfate at a concentration of 7.5% (anhydrous) (17% decahydrate). The optimal reaction conditions in terms of pH and temperature are different for starches from different sources. After derivatization or other modification, the starch is recovered by centrifugation or filtration, washed, and dried. (Because the reagents used to esterify or etherify hydroxyl groups also react with water, byproducts are formed and are removed along with the salt used (or produced during neutralization) and any unreacted reagent by washing the product).

Most derivatization reactions (i.e., reactions that add ether (Section 12.2.4) or ester (Section 12.2.5) groups to starch polysaccharide molecules) are ordinary alkali-catalyzed, nucleophilic substitution reactions where a hydroxyl group ionized to an alkoxy anion is the nucleophile. In most alkali-catalyzed reactions of starch, the reactivity of the hydroxyl groups of the glucosyl units of the polysaccharide molecules is in the order of $O2 > O6 \gg O3$.

12.2.3 Amounts of Derivatization

Most derivatized starch products (i.e., starch ethers (Section 12.2.4) and esters (Section 12.2.5)) have only a small fraction of their hydroxyl groups modified. Amounts of derivatization that introduce ether and ester groups are determined and reported in research laboratories in DS (degree of substitution) or MS (moles of substitution or molar substitution) units. For most derivatization reactions, DS is used. DS = moles of substituent groups added to starch molecules ÷ moles of glucosyl units (1 mol of glucosyl unit = 162 g dry weight), which gives the average number of hydroxyl groups that have been modified. The average number of hydroxyl groups per glucosyl unit in both amylose and amylopectin molecules is 3, so the maximum DS value for starch is 3.0. For example, a DS of 0.2 (not unusual, but often at the high end, for a modified food starch) means that there are on average 2 substituted hydroxyl groups per 10 glucosyl units (in other words, 2 out of every 30 hydroxyl groups (or 6.7%) are derivatized). When propylene oxide (Section 12.2.4.2) or ethylene oxide (Section 12.2.4.1) are reacted with starch, hydroxypropyl ($-CH_2-CHOH-CH_3$) and hydroxyethyl ($-CH_2-CH_2OH$) ether groups, respectively, are formed. These new hydroxyl groups can also react with reagent molecules and form new hydroxyl groups, so more than one hydroxypropyl or hydroxyethyl group may be added in chains to a starch hydroxyl group. Therefore, the DS value may not indicate the average number of hydroxyl groups that have been derivatized. When such etherifications occur, MS values are used. The formula for calculating MS is the same as that for calculating DS, but because each time a hydroxyl group on a starch molecule is reacted, a hydroxyl group on the substituent group is formed, there is no upper limit to MS values, i.e., they can exceed 3. However, because the limits for the amounts of reagent molecules added to modified food starches are low, the values for DS and MS for food starches are believed to be essentially the same. Measures such as 'add on' are often used in industry.

Because any population of starch granules is heterogeneous, individual granules and individual starch polysaccharide molecules are likely modified to different degrees and have different substitution patterns, so determined DS and/or MS values are averages for the population of granules and their molecules used in the analysis. Methods for determination of DS and MS values are specific for the specific modification.

12.2.4 Ethers

Commercial starch ether (St-OR) products can be categorized based on the ionic nature of the product: (1) non-ionic ethers (hydroxyethyl and hydroxypropyl groups), (2) anionic ether (carboxymethyl groups), (3) cationic ethers (ammonium and amino groups), and (4) amphoteric ethers (both anionic and cationic groups). Because all etherification reactions are nucleophilic substitutions, all require a base as a catalyst. Reactions have been done in aqueous systems, in organic solvents, dry (Section 12.2.6), and via reactive extrusion (Section 12.2.6). The aqueous slurry method at a pH of ca. 11 is commonly used in the starch industry (Section 12.2.2). Salts are added to the reaction mixture to inhibit swelling, gelatinization, and pasting of the granules during modification.

12.2.4.1 Hydroxyethylstarch (hydroxyethyl starch)

Hydroxyethylstarch ($Starch-O-CH_2OH$) is produced via nucleophilic reaction of a starch with ethylene oxide in an alkaline medium containing a swelling-inhibiting salt to protect the derivatized starch from chemical gelatinization and pasting so that the modified starch granules can be recovered by centrifugation or filtration. Hydroxyethylstarches are used in coatings in the paper industry because they provide a uniform and stable viscosity, improve water retention, and increase coating capacity (Li et al. 2019). On a much smaller scale, they are used as plasma volume expanders, cryoprotectants, and carriers for drug delivery, but their overuse has raised health concerns.

12.2.4.2 Hydroxypropylstarch (hydroxypropyl starch)

Propylene oxide reacts with the hydroxyl groups of starch molecules and introduces hydroxypropyl groups ($Starch-O-CH_2-CHOH-CH_3$). As with hydroxyethylation, the reaction is effected in an alkaline medium with added sodium sulfate or sodium chloride to protect the derivatized starch granules from swelling and disintegrating, so that the modified starch granules can be recovered by centrifugation or filtration. Low-DS (Section 12.2.3) products are widely used in processed food products. Higher DS values result in further decreases in gelatinization enthalpy, retrogradation (Section 12.1.2.2), and syneresis during freeze–thaw cycles and increases in the degrees of granule disintegration and ambient-temperature water solubility, but there are limits to the degrees of substitution allowed for use as a food ingredient (Federal Regulations 1978; Galkowska, Kapuśniak & Juszczak 2023).

12.2.4.3 Cationic starches

Most commercial cationic starches usually contain quaternary ammonium groups ($Starch-O-CH_2-CHOH-CH_2\ NR_3^+$, where the R group is most often CH_3), but products containing tertiary amino groups are also cationic starches. The most widely used cationization reagents are 3-chloro-2-hydroxypropyltrimethylammonium chloride and 2,3-epoxypropyltrimethylammonium chloride (glycidyltrimethylammonium chloride (GTAC)). Tertiary amino etherifying agents, such as 2-chloroethyldiethylamine, are also used. Reaction of dry starch with hydroxymethyldimethylamine hydrochloride (via a Mannich reaction) has been studied.

Commercial cationization of starch commonly takes place in aqueous suspensions, but cationization can also be accomplished within a starch paste, in an organic solvent, and in a dry state (Section 12.2.6). During the cationization process in an aqueous suspension, a salt such as sodium sulfate or sodium chloride is present to prevent swelling and gelatinization. Some strategies for reducing the use of salts and their effect on wastewater, e.g., substitution of the salt with sucrose and addition of water miscible organic solvents to the aqueous alkaline media, have been tested. Benefits of application of a dry process (Section 12.2.6) and reactive extrusion (Section 12.2.6) include a lack of effluent, a decrease in starch loss caused by any washing process, and energy saving, but the cationic starch product so obtained contains the alkali used as a catalyst, any unreacted reagent, and byproducts. Another drawback of the dry process is that reaction in a dry state mainly occurs on granule surfaces, while reaction in aqueous suspensions occurs throughout granules so that all starch molecules are effectively derivatized and released as functional molecules upon pasting – a drawback that does not apply to reactive extrusion, which presumably results in rather uniform derivatization of the starch molecules owing to granule destruction. Granular starches after cationization commonly have decreased gelatinization and pasting temperatures and increased peak viscosities and breakdown values.

Because cationic starches are non-toxic, biodegradable, and relatively inexpensive, they are attractive substances for non-food industrial uses. Because of their ability to bind to cellulose fibers, cationic starches are used as wet-end additives and as sizing and coating agents in papermaking (Li et al. 2019; Sharma et al. 2020; Wei et al. 2021; Wilson 2005). When applied at the wet end, they increase the mechanical strength of the sheet and retention of fines and fillers, provide faster drainage, and reduce wastewater pollution.

Cationic starches are also used as textile sizing agents (Bismark, Zhu & Benjamin 2018). They have been found to be effective in flocculating both Gram positive and Gram negative bacteria in wastewater and microalgal biomass (so that it can be harvested). It has been claimed that cationic starches inhibit the growth of *Escherichia coli* and other fecal coliforms. Cationic high-amylose maize starch, in combination with sodium montmorillonites, has been reported to improve the mechanical properties and decrease the water-vapor permeability of films.

12.2.4.4 Carboxymethylstarch (carboxymethyl starch, sodium starch glycolate)

Carboxymethylstarch (an anionic starch; Starch-O-CH$_2$-CO$^-$) is produced by a nucleophilic reaction between a starch and monochloroacetate ions in an alkaline system. Upon carboxymethylation to DS values above 0.1 (Section 12.2.3), the starch becomes cold-water soluble so the alkali and the reagent are added to a suspension of starch in 2-propanol (isopropanol, isopropyl alcohol) or another alcohol containing a small amount of water to obtain granular products. The use of different alcohols and different starches has resulted in different maximum

DS values. Carboxymethylstarch with medium and high DS values can be made using a dry method (Section 12.2.6) in the presence of a small amount of water and/or an organic solvent.

Carboxymethylstarches are insoluble in acidic systems, have decreased gelatinization temperatures and paste and gel viscosities, and have increased water absorption, solubility, film-forming ability, and paste and gel clarity. They have had limited use as textile sizing agents. In the pharmaceutical industry, where they are widely used and alternatively known as sodium starch glycolates, carboxymethylstarches are used as tablet disintegrants/excipients (in which applications they are classified as superdisintegrants) and have been investigated for other applications, including controlled release complexes (Ab'lah et al. 2023; Berardi, Janssen & Dickhoff 2022; Garcia, Garcia & Faraco 2020; Hong, Liu & Gu 2016; Kumar et al. 2019; Lemos et al. 2021; Pooresmaeil & Namazi 2021; Sharma, Pahuja & Sharma 2019). Carboxymethylstarch films claimed to be cross-linked (Section 12.1.2.2) via citric acid esters and carboxymethylstarch-modified montmorillonite composite films have been prepared. Carboxymethylstarch cross-linked to starch acetate (Section 12.2.5.1) is claimed to function as an electrolyte in lithium-ion batteries. A water-soluble product consisting of carboxymethylstarch etherified with hexadecyl (cetyl) bromide via reactive extrusion (Section 12.2.6) was claimed to be a very good emulsifier.

Another anionic starch was made by reacting starch with 2-hydroxy-3-chloropropylcitric acid and evaluated for its effectiveness as a flocculant and its ability to bind cations.

12.2.4.5 Distarch ether

Modifications with difunctional reagents that bridge/join adjacent starch molecules within starch granules produce what are known as cross-linked starches (Section 12.1.2.2). Ether cross-links are introduced upon reaction with epichlorohydrin (ECH; 1-chloro-2,3-epoxypropane and other names). ECH cross-linked products (Starch chain-O-CH$_2$-CHOH-CH$_2$-O-Starch chain; a distarch glyceryl ether) are made and used much less than those with ester cross-links (Sections 12.2.5.3 and 12.2.5.7). The reaction conditions are similar to those used to make other starch ethers. As with other cross-linking reactions, cross-linking with ECH significantly modifies starch pasting and paste properties. As with other kinds of cross-linking, peak viscosity first increases, then decreases as higher levels of ECH cross-linking are achieved. Also, granule swelling and solubility decrease at higher levels of cross-linking.

ECH-cross-linked starch has been studied for applications as a binder/adhesive for particleboard, as a weakly basic ion-exchanger for treatment of wastewater when the starch was reacted with ECH in the presence of ammonium hydroxide and for sustained release of drugs in tablet form.

12.2.4.6 Benzyl ether

Many other ethers of starch have been made experimentally. Notable among them is the benzyl ether. Benzylation makes a starch more hydrophobic (more soluble in less polar solvents) and less hydrophilic (less soluble in water) (Bohrisch, Vorwerg & Radosta 2004; Cho & Lim 1998).

12.2.5 Esters

Although many organic and inorganic esters of starch have been prepared in the laboratory, only a few are produced commercially. Their primary use is in processed food products. Organic acid esters of starch (Starch-O-CO-R) are much less stable than starch ethers. (1) They will saponify in the alkaline conditions used to make them. To maximize DS values, esters of organic acids are usually prepared by reaction of a starch with an acid anhydride in an alkaline slurry with a lower pH value and temperature for shorter times than those that are used for preparation of starch ethers. Because the ester group is susceptible to saponification in a high-pH system, the reaction mixture is neutralized when the maximum DS has been obtained. Specific optimum conditions need to be determined for each starch, each reagent, and available equipment. (2) Organic esters of starch are also unstable at high temperatures, such as those that might be used in sterilization processes, which limit their use to certain food products. On the other hand, inorganic esters, i.e., phosphate (Sections 12.2.5.6 and 12.2.5.7) and sulfate esters (Section 12.2.5.8), of starch are quite stable.

12.2.5.1 Starch acetates (acetylated starch, acetylstarch)

Commercial starch acetate (Starch-O-CO-CH$_3$) products used as food ingredients are relatively low-DS products. Starch acetates can be prepared in various ways. Two processes are used commercially. One is by reaction of a starch with acetic anhydride. The pH used is usually ca. 8. Reaction times of 1–2 hours at ambient temperature are common. The other process involves a transesterification reaction using vinyl acetate as the acetyl donor and a pH of ca. 9.9. [The latter process is not used in the United States because starch acetates are used in food products and the byproduct of the reaction is acetaldehyde (a prohibited substance in food products) but is employed elsewhere.] There are mixed results as to the differences in DS values (Section 12.2.3) and characteristics of starches reacted with the same molar amounts of the two reagents. It has been reported that (1) when a legume starch was acetylated with the same molar amounts of acetic anhydride and vinyl acetate, the amylopectin of the sample modified by reaction with vinyl acetate had larger DS values than did the amylose and that the acetyl groups on the polysaccharide chains were more clustered in the sample reacted with vinyl acetate and (2) not only were reagent type and granule size important factors in determining the swelling and pasting behaviors of acetylated starches, but also the acetyl groups in the granules modified with the more rapidly reacting acetic anhydride were more concentrated near the granule surface, while the slower reacting vinyl acetate was able to diffuse throughout granules before reacting, giving a more uniform distribution of acetyl groups. Acetylation primarily takes place in amorphous regions and in outer crystalline lamellae of starch granules. Also, amylose becomes more acetylated than amylopectin, which is desirable because the primary purpose of acetylation is to reduce retrogradation of amylose (Section 12.1.2.2).

Relatively low-DS starch acetates are broadly used in food products. Like granules of other stabilized starches (Section 12.1.2.2), granules of starch acetates usually have greater water absorption capacity, swell more rapidly and to greater degrees, disintegrate more easily, have lower gelatinization and pasting temperatures, retrograde more slowly and to lesser degrees, and undergo reduced syneresis in various products under various conditions as compared to the native starch. However, the effect of acetylation on their pasting properties appears to vary with the type of starch and DS because contradictory results have been reported. Compared to gels of native starch controls, gels of low-DS acetylated starch have reduced strength, hardness, adhesiveness, and syneresis and increased stability and springiness. Both increases and decreases in gel cohesiveness have been reported. Acetylation increases the amount of RS (Chapter 20), but the resistance of acetylated starch to digestion as a function of the DS is unsettled.

Higher degrees of acetylation (and esterification with propionyl, butanoyl, and mixed ester groups) produce hydrophobic, thermoplastic products for potential use as food coating and biodegradable packaging materials. It has been known for almost 80 years that films with good tensile strength and pliability and low plasticizer requirements can be made from amylose triacetate, that amylose triacetate can be spun into fibers that have lower tensile strength but greater elongation at break compared to cellulose triacetate fibers, and that the tensile strength of amylose triacetate fibers can be greatly increased by orienting the fiber. Since then, it was established that cellulose diacetate is preferred for film and fiber manufacture and that amylopectin does not make suitable films and fibers, which prevents the use of whole starches (other than high-amylose starches) for manufacture of films and fibers. However, it has been found that filaments made from a blend of cellulose acetate and starch acetate have good tensile strength. It is also claimed that starch acetate fibers have potential tissue engineering applications and that starch acetate cross-linked to carboxymethylstarch (Section 12.2.4.4) functions as an electrolyte in lithium-ion batteries.

12.2.5.2 Starch 2-Octenylsuccinates [OS-Starch, OSA-Starch, Octenylsuccinylated Starch, (2-Octenyl)succinylstarch} and Succinates

Starch octenylsuccinates (one of two possible isomeric products being starch-O-CO-CH$_2$-CH(CO$_2^-$)-CH$_2$CH=CH-(CH$_2$)$_4$-CH$_3$) are prepared by reaction of a starch with *trans*-(2-octenyl)succinic anhydride (OSA). Commercial OS-starches (also referred to as lipophilic starches) are relatively low-DS products.

The OSA reagent is only slightly soluble in water. Its reaction with starch granules is believed to be somewhat different from reactions with most other reagents, but this is an unsettled matter. It was first concluded that the octenylsuccinyl ester groups were uniformly distributed across granule cross-sections (Shogren et al. 2000), but then Song et al. (2006) surmised that the reaction occurred on granule surfaces. Then, evidence was obtained that OS groups were on the outer surface of granules (Wetzel et al. 2010a, b), while Huang et al. (2010) concluded that the OS groups were distributed throughout the granules, but somewhat concentrated at granule surfaces. Following this work, it was suggested that, even though the reagent was present in the reaction medium as small droplets, reactions might not be limited to granule surfaces, but could take place throughout granules. Sui, Huber, and BeMiller (2013) proposed a process in which, because of the greater ordering of water molecules within hydrated granules than in the bulk water, ΔG for the transfer of the only slightly soluble, hydrophobic reagent from the bulk water to the ordered internal water would be negative, making transfer of the reagent from the bulk water to the internal water of hydrated granules favorable. The proposed mechanism would not only explain why reaction appears to take place throughout granules, but would also produce somewhat evenly derivatized starch polysaccharide molecules. To improve reaction efficiency and/or to obtain more uniform distributions of OS groups by opening up granule structures, various physical (Section 12.4) and starch hydrolase (Section 12.3) pretreatments have been employed (Altuna, Herrera & Foresti 2018).

It has been concluded that octenylsuccinylation of starch to a relatively high degree occurs in semi-crystalline granule regions and decreases the relative crystallinity of the starch. Results of a study of OSA-modified rice and tapioca starches suggested that OS groups were mainly located on amylose molecules, indicating that reaction occurred in amorphous regions. In an OSA-modified starch that did not contain amylose, viz., waxy maize starch, it was found that most OS groups were located at the O2 and O3 positions of the amylopectin molecules and that the OS groups were mainly located near the branching points of a low-DS modified product and near both the branching points and the non-reducing ends of a relatively high DS product.

Most applications of OS starches are in the production of processed food products. Adding hydrophobic octenylsuccinylate groups to the hydrophilic starch molecules turns starches into amphiphilic substances that can be employed as emulsifiers, emulsion stabilizers, encapsulating agents for protection of flavors and aromas, and fat replacers. Emulsions made with many OS-starch products can be spray-dried.

Not only do the hydrophobic octenyl groups concentrate at the interface of oil-in-water dispersions, but in addition, the bulky starch portion serves as an emulsion stabilizer because the large starch polysaccharide molecules to which the OS groups (which also contain negative charges) are attached function as protective colloids, providing emulsion stability. Emulsifying capacity and emulsion stability are influenced by the DS, the distribution of the OS groups, and the sizes and structures of the OS-starch molecules.

OSA-modification increases oil retention abilities and oil absorption capacities. OS-starches have been used to solubilize hydrophobic molecules, such as β-carotene, via self-assembled micelles and impart the mouthfeel of a fat when used to formulate reduced-calorie food products. He, Liu, and Zhang (2008) proposed that heat-moisture treated (Section 12.4.2.3) OS-starch is an uncompetitive inhibitor of the action of α-amylase.

Starch octenylsuccinates of higher DS values than are permitted for food use have improved film- and fiber-forming properties and show potential for application in edible films and coatings and biodegradable food packaging material. They have been investigated as wall materials of micro- and nano-encapsulants and tablets for controlled drug delivery and controlled release of food ingredients, such as flavors, pigments, and antioxidants.

Products made by reaction of unsubstituted succinic anhydride with a starch (up to the maximum allowed add-on) are approved for food use in the United States, but it is not known to the author that any such products are produced.

12.2.5.3 Acetylated distarch adipates

As mentioned in Section 12.2.4.5, modifications with difunctional reagents that bridge/join adjacent starch molecules within starch granules produce what are known as cross-linked starches (Section 12.1.2.2). Distarch adipates (Starch chain-O-CO-$(CH_2)_4$-CO-O-Starch chain) – made by reacting a starch with acetic acid-adipic acid dianhydride – are such products. Because starch hydroxyl groups can react with either the acetyl groups or the adipyl groups of the reagent and because there is excess acetic anhydride in the reagent, the starch molecules also become acetylated. Thus, the products of this reaction contain both adipic acid diester and acetyl groups. (Some adipic acid monoester groups may also be present.) Acetylated distarch adipates have also been prepared by first acetylating a starch and then reacting with the dianhydride reagent, as well as by first reacting a starch with the cross-linking reagent and then further acetylating it. Acetylated distarch adipates are employed as food ingredients.

12.2.5.4 Other means of cross-linking via organic acid diester and other linkages

Starches continue to be esterified with a variety of organic acids. One example is their treatment with citric acid, which is of particular interest because citric acid is a GRAS (generally regarded as safe) substance, but some of the claims are questionable. For example, claims that esterification occurred during stirring a suspension of starch in an aqueous solution of citric acid followed by recovery and drying of the starch at temperatures well below 100°C defy chemical logic. However, reacting a starch with an organic acid by first drying the starch impregnated with an aqueous or an alcohol-water solution of an organic acid and then heating the dry mixture at a temperature of 100°C or more apparently can be a successful, although inefficient, process. On the other hand, when a polycarboxylic acid (such as citric acid) is used, sufficient esterification to change the properties of the starch could occur because only a small amount of cross-linking can dramatically change the pasting and paste properties of the starch. Evidence for the occurrence of esterification is usually an increase in the content of RS (Chapter 20), decreases in granule swelling and solubility (all indirect evidence suggesting cross-linking), and indication of the formation of ester linkages via Fourier transform infrared (FTIR) spectroscopy. One problem is that treatment of starches with acids, whether in an aqueous system or in a dry state, removes some of the amorphous material between crystallites, and as a result, it not only increases the relative crystallinity of the starch, but also allows the crystallites to grow larger – processes that should increase the percentage of RS (Chapter 20). However, evidence that some esterification did occur during reactive extrusion (Section 12.2.6) of starches with citric acid has been reported, and it has been concluded that, after starch and citric acid were heated together at 100°C, diesters, i.e., cross-links, were formed, but they occurred intramolecularly within amylopectin molecules or intermolecularly between dextrin molecules (low-molecular weight fragments of starch molecules formed by acid-catalyzed hydrolysis (Section 12.2.8)) or between a dextrin molecule and a larger molecule; in other words, no cross-links between high-molecular-weight starch molecules were formed (Kapelko-Żeberska et al. 2016). Even though the exact nature of the products may not be established, they have been investigated for their ability to bind cations and to act as flocculants. Citric acid esters of starches have been shown to form pH-responsive nanoparticles for controlled release and to be solubility enhancers for pharmaceuticals.

Evidence has been obtained that impregnating starch granules with a solution (at the proper pH value) of an anionic non-starch polysaccharide and then drying and heating the granules resulted in some cross-linking. Starch xanthates are made by reaction of a starch with carbon disulfide in high pH systems. Starch xanthates are unstable, especially in the presence of more than 2% moisture, but become stable when the xanthate groups are converted to cross-links via oxidation to xanthides or through the use of divalent cations. Starch xanthides have been found to be effective as reinforcing agents in rubber, as slow-release encapsulating agents for agricultural pesticides, and in removing heavy metal ions from process water.

A thermoreversible thermoplastic starch product was prepared by introducing (via carbamoyl (not ester) linkages) relatively long furan/maleimide cross-linking chains formed by a Diels–Alder reaction.

12.2.5.5 Some other organic acid esters

Starches continue to be esterified experimentally with various organic acids, and advances in starch esterification have been reviewed (Otache et al. 2021). Many of these products are alkyl esters and biodegradable plastics (Rivard et al. 1995). Starch propionate has been compared to OS starch (Section 12.2.5.2) as an emulsion stabilizer (Hong et al. 2018). The preparation and functional and biological properties of starch butyrates have been reviewed (Zhang et al. 2024). Starch fatty acid esters, which can be thermoplastics that form flexible hydrophobic films (Sagar & Merrill 1995), have been prepared by transesterification from their corresponding vinyl esters.

Starch dodecenylsuccinate (related to OS-starch) has been prepared via reactive extrusion (Section 12.2.6). The maleic acid ester of starch (by reactive extrusion (Section 12.2.6)) is a thermoplastic material. Esterifying starch with betaine (N,N,N-trimethylglycine) results in a cationic product that has been evaluated as a coating material for improving the ink-jet printing quality of office papers (Auzely-Velty & Rinaudo 2003; Sharma et al. 2021a, b).

12.2.5.6 Recent investigations into techniques to improve reaction efficiency, to create new products, or to make the process greener

Approaches to improve esterification efficiency have been extensively and academically reviewed (Golachowski et al. 2015; Otache et al. 2021). New techniques include, but are not limited to, reaction in sub- and supercritical carbon dioxide; application before or during reaction of microwave irradiation, ultrasound (Section 12.4.3.1), pulsed electric fields (Section 12.4.3.3), and high-speed shear; pretreatments by grinding/ball-milling (Section 12.4.3.5) and heat-moisture conditions (Section 12.4.2.3); use of iodine as a catalyst; use of acetic anhydride in deep eutectic solvents without a catalyst (Portillo-Perez, Skov & Martinez 2024); and changes in the order of addition of the reagent and catalyst.

Starches have been esterified in the laboratory with fatty acids and other rather hydrophobic acids via lipase-catalyzed reactions with various donors in reactions that are usually quite rapid. However, all such lipase-catalyzed esterifications must be done in the virtual absence of water. Non-aqueous liquids that have been used include pyridine, dimethyl sulfoxide (DMSO), dimethylformamide (DMF), alcohols, and ionic liquids.

In addition, investigations continue in the use of different solvents and reaction promoters in efforts to make the processes greener. An example is the use of tartaric acid as the catalyst (Tupa et al. 2013).

12.2.5.7 Monostarch phosphates (starch phosphate monoesters)

Phosphate esters (Ramadan & Sitohy 2020) are considerably more stable than esters of organic acids and are prepared differently. Monostarch phosphates (Starch-O-PO_3^-) can be prepared in several different ways. The general basic way is as follows: A starch is impregnated with a mixture of monosodium dihydrogen phosphate and disodium monohydrogen phosphate by either adding the starch to a 40%–55% solution of the orthophosphates or adding the salts to a starch slurry. The pH of the 50°C–60°C slurry is adjusted (if necessary) to 5.5–6.0. After 10–30 minutes, the starch is collected by filtration (without washing), dried at a low temperature, and then heated at 120°C–170°C. Modifications of this process include dry-blending the starch and powdered orthophosphate salts, as well as and spraying solutions of the salts onto dry starch or a starch filter cake.

Because of their ionic nature, monostarch phosphate granules (even at low DS levels) may swell when placed in room-temperature water. They have reduced gelatinization and pasting temperatures as compared to those of their parent starch and their solutions and gels undergo little, if any, retrogradation (Section 12.5.2). Monostarch phosphates are used in processed foods. They generally produce high viscosities (which can be reduced by cooking in the presence of salt) and clear, stable pastes that have a long, cohesive texture and good cold-storage and freeze–thaw stability.

12.2.5.8 Distarch phosphates

As mentioned in Sections 12.2.4.5 and 12.2.5.3, modifications with di- or polyfunctional reagents that bridge/join adjacent starch molecules within starch granules produce what are known as cross-linked starches (Section 12.1.2.2). Distarch phosphates (Starch chain-O-PO$_2^-$-O-Starch chain) (Ramadan & Sitohy 2020) are such cross-linked starches. They can be and are made in different ways. The most used method is to react the starch with phosphoryl chloride (phosphorus oxychloride, POCl$_3$). An alternative approach is to react starch with sodium trimetaphosphate (STMP) or (perhaps more commonly), a mixture of sodium trimetaphosphate and sodium tripolyphosphate (STPP) (both sodium salts of anhydrides of phosphoric acid) using a variety of conditions. Starches cross-linked with the inorganic salts have about equal amounts of distarch and monostarch phosphate groups and, therefore, they are both cross-linked and stabilized (Section 12.1.2.2). Their properties differ somewhat from those cross-linked by reaction with POCl$_3$, probably because of differences in location of the cross-links, i.e., presumably throughout granules (via reaction with STPP and/or STMP) vs. only at granule surfaces (via reaction with POCl$_3$ (see the next paragraph)) and because they are dual-modified products. An amount of cross-linking (using either POCl$_3$ or STMP+STPP as the cross-linking reagent) too small to be measured leads to considerable changes in starch pasting properties. More highly derivatized products made using the salts have reduced digestibility and are a source of dietary fiber.

Reaction with phosphoryl chloride (POCl$_3$): POCl$_3$ (the acid chloride of phosphoric acid, commonly known as phosphorus oxychloride) reacts with both starch granules and water. It is so reactive that the reagent is used up within seconds. As a result, reactors are constructed with dispersers and baffles that ensure that the reagent is dispersed in small droplets, and the slurry of starch granules is mixed as rapidly as possible to ensure that as many granules as possible are exposed to the reagent and the cross-linking of granules approaches as much uniformity as possible. Even so, the use of this reagent is the most common way of cross-linking granular starch. Alkaline conditions are not necessary for the reaction to occur, but alkali is present to neutralize the hydrogen chloride (HCl) generated in the reaction. Because of the very rapid rate of reaction, at the degrees of cross-linking usually commercially employed, cross-linking with POCl$_3$ occurs only at granule surfaces.

Reaction with sodium trimetaphosphate (STMP) and sodium tripolyphosphate (STPP): Distarch phosphate can be produced in an aqueous slurry containing sodium sulfate or sodium chloride using STMP or preferably STMP/STPP in a ratio of 99:1 at a pH of 10–12. Reactive extrusion (Section 12.2.6) and microwave heating have been used to increase the reaction efficiency of STMP with starch in dry and low-moisture states. Cross-linking modifications using STMP at low levels mainly occur in amorphous regions, but some loss of the crystal structure may occur at high extents of cross-linking. As with cross-linking of starch with other reagents, cross-linking with STMP or STMP/STPP improves thermal and shear resistance; increases paste viscosity, yield stress, and gel strength; and decreases granule swelling and solubility, paste clarity, and gel cohesiveness. Cross-linking using STMP/STPP increased the percentage of RS (Chapter 20), and feeding of such products to laboratory animals reduced blood concentrations of total lipid, total triglyceride, and total cholesterol.

12.2.5.9 Starch sulfates (sodium starch sulfates)

Starch sulfates have been produced on small scales for many years. Reagents used (in non-aqueous solvents because of the reactivity of the reagents) have been various sulfur trioxide complexes in non-aqueous liquids, sulfuric acid in ethanol, and chlorosulfonic acid in pyridine. Starches can be sulfated in aqueous suspension via use of trisulfonated sodium amine (N(SO$_3$Na)$_3$) (Cui et al. 2007, 2011); however, as with other anionic starch derivatives (viz., monostarch phosphates (Section 12.2.5.7) and carboxymethylstarch (Section 12.2.4.4)), starch sulfates are very soluble in aqueous systems, and as a result, the products dissolve and must be precipitated for recovery. The use of starch sulfates for their anticoagulant and antiviral activities, to remove heavy metal ions from wastewater, and as paper and textile sizing agents has been explored.

12.2.6 Non-conventional Derivatization Conditions

Although an overwhelming amount of starch is chemically modified in aqueous slurry, other means of modification have been investigated, the principal one being reactive extrusion. Potential advantages of reactive extrusion include a lack of process effluents, a continuous process, a wide range of possible processing conditions, short reaction times, high throughput, enhanced contact between reactants, a 'homogeneous' reaction medium, a lack of the influence of granule structure on molecular reaction patterns, good heat transfer, a pregelatinized product (Section 12.4.2.1), and simultaneous partial depolymerization of starch molecules. Potential disadvantages include a need to dry and powder the product and the presence in the product of unreacted reagent, reagent byproducts, and catalyst (usually alkali), which rules out any use in foods without extensive washing (which creates effluents).

Dry processes have been used successfully to prepare novel modified starches and modified starches with higher DS levels and reaction efficiencies that can be obtained in an aqueous medium, but they suffer from the same problems as those made by reactive extrusion, viz., the products can only be used in industrial applications that can tolerate the presence of unreacted reagent, reagent byproducts, and alkali. The reaction of starch granules as a fluidized bed has been investigated. Note that, whenever the term dry is applied to starches, it refers to a starch in a powder form rather than being anhydrous. The moisture content of dry starch is the starch's normal moisture content (usually ca. 13% for cereal starches and ca. 18% for potato and cassava/tapioca starches).

The use of sub- and supercritical carbon dioxide, ionic liquids, and deep eutectic solvents as solvents has been explored (Fan & Picchioni 2020; Portillo-Perez, Skov & Martinez 2024).

12.2.7 Oxidations

Many oxidants will oxidize starch (Vanier et al. 2017). Oxidation of starch occurs at the hydroxyl groups on the C2, C3, and C6 carbon atoms of glucosyl units – sometimes with partial depolymerization of starch molecules. The keto groups that are formed by the oxidation of the secondary hydroxyl groups (on C2 and C3) form enediol structures – particularly under alkaline conditions – and these enediol structures can be oxidized to aldehyde groups, which can be further oxidized to carboxylate groups (BeMiller 2019, pp. 212–213). Oxidation of the primary hydroxyl groups (on C6) produces aldehyde groups, which then can be oxidized to carboxylate groups. Carboxylate groups make the products anionic. The contents of carbonyl and carboxylate groups are used to assess the degree of oxidation. The introduction of carbonyl (aldehyde and keto) groups simultaneously introduces a beta-alkoxy carbonyl system, and under alkaline conditions (especially as used in oxidations with hypochlorite ions (Section 12.2.7.1)), beta-elimination and depolymerization of the starch polysaccharide molecules occur (BeMiller2019, pp. 95–99). Factors affecting oxidation include the starch oxidized, the oxidant used, the oxidant concentration, the temperature employed, and any inorganic salts present. Oxidants usually favor the formation of one type of chemical group over the other. The carbonyl groups of an oxidized starch can subsequently participate in reactions of aldehydes and ketones – such as reacting with amino group-containing compounds to form imines (Karić et al. 2023).

There are often differences in the characteristics of oxidized starches. Such differences may originate from the type of oxidant, the degree of oxidation, or the starch. However, compared to the native starch, oxidized starches are generally whiter, require less energy to cook (because oxidations weaken granules), and form more stable gels and pastes (often with reduced viscosity). Oxidized starches commonly have lower values for peak viscosity, breakdown, setback, and final viscosity compared to the native starch. However, a higher peak viscosity may occur at a low level of hypochlorite oxidation (Section 12.2.7.1) as a result of the formation of hemiacetal cross-links between carbonyl and hydroxyl groups.

Oxidations usually alter the properties of starch gels, but the effects may be in opposite directions for different oxidants and starches. Oxidation usually decreases gel formation and retrogradation (Section 12.1.2.2). Many oxidized starches have improved film properties, viz., increased tensile strength and transparency and decreased elongation, solubility, surface roughness, and water vapor permeability.

In the food industry, oxidized starch has been used in coatings (due to the flowability and adhesiveness of its pastes), confections (as binders), dairy products (as texturizers), and many other applications. Oxidized starch has been used as a fat replacer and as a replacer of gum arabic for encapsulation of flavor.

Oxidized starches have been used for surface sizing (Wilson 2005) and in coatings for paper and textiles, in laundry finishing, as a binding material, and in biodegradable film formation (Cahyana et al. 2023). Nanocrystals of oxidized starch can be used for heavy metal ion removal. What is important for the papermaking and related industries is that hypochlorite-oxidized starches (as compared to an unmodified starch) have improved film-forming characteristics, have reduced gel-forming tendencies, are better dispersants and coating binders, and result in greater pigment retention.

12.2.7.1 Oxidation with hypochlorite ions

Sodium hypochlorite is the oxidant that is most widely used for preparation of oxidized starch because of its low cost and high reaction efficiency. In general, hypochlorite-oxidized starches have more carboxyl groups than carbonyl groups, with the difference becoming larger at greater extents of oxidation. Hypochlorite oxidation begins and primarily occurs in the amorphous regions of starch granules. At higher degrees of oxidation, the relative crystallinity of granules may be reduced. At low degrees of oxidation, oxidized starch may have some of the characteristics of a cross-linked starch (Section 12.1.2.2), as the carbonyl groups form hemiacetal cross-links. Although depolymerization of starch molecules following hypochlorite oxidation can always occur because of the alkaline conditions employed (BeMiller2019, pp. 95–99), depolymerization becomes more and more evident with increases in the degree of oxidation, and although oxidation and depolymerization occur in both amylose and amylopectin molecules, amylose molecules are more susceptible to depolymerization than amylopectin molecules. The degrees of oxidation and the resulting changes in physicochemical properties of oxidized starch differ between starches from different botanical sources.

As a result of depolymerization of the polysaccharide molecules and a weakening of granules, hypochlorite oxidation of starch decreases the hot paste viscosity and the energy required to gelatinize and paste/cook the starch. It also has a whiter color and gives rise to greater paste clarity than the non-oxidized native starch. It can be used as a dairy texturizer, a bread improver, a coating material, a binder, an adhesive, and in the preparation of spray-dried flavors. Adhesion of batters made from hypochlorite-oxidized starch is improved.

12.2.7.2 Oxidation with hydrogen peroxide

Hydrogen peroxide (H_2O_2) oxidation of starch is an environmentally friendly process, because H_2O_2 decomposes into oxygen and water with no harmful byproducts. However, compared to hypochlorite oxidation, peroxide oxidation (a reaction that can be achieved with and without a catalyst) is less efficient because, although hydrogen peroxide has a high oxidative potential, it is not very reactive toward organic functional groups. However, with the addition of transition metal ions, such as Cu(II) and Fe(II), its reaction efficiency can be improved considerably. Zhang et al. (2015) concluded that the oxidation of pasted starch with H_2O_2+Fe(II) at pH 6.2 gave a product in which 54.5% of the glucosyl units contained a carboxyl group and 56.2% of the glucosyl units contained a carbonyl group.

While carboxyl groups are the major functional groups in hypochlorite-oxidized starches, carbonyl groups are the major functional groups in peroxide-oxidized starches, and the two products have different properties.

12.2.7.3 Oxidation with other oxidants

Ozone is a highly oxidative, environmentally friendly, and easily generated oxidant (Raghunathan et al.2021). Ozone oxidation of starch involves three steps: (1) ozone is produced from oxygen in an ozone generator; (2) ozone-rich gas diffuses and passes through the starch (either in a dry powder form or in an aqueous suspension) in a reactor; and (3) unreacted ozone is decomposed to oxygen in an ozone destructor. In minutes, ozone oxidation generates amounts of carbonyl and carboxyl groups comparable to those produced by hydrogen peroxide and hypochlorite oxidations in hours. It does not show strong preferences between the formation of carbonyl and carboxyl groups. It has been reported that ozonation resulted in depolymerization of starch molecules without any visible changes in granule morphology.

Periodic acid and periodate and metaperiodate ions will oxidize starch. Periodate ions oxidize the hydroxyl groups on C2 and C3 of a glucosyl unit and cleave the C–C bond between C2 and C3, resulting in the formation of two aldehyde groups per glucosyl unit. Although the oxidized product (Yong & Liu 2024), which is called dialdehyde starch, has useful properties, particularly for production of wet-strength paper products, the reagents are too expensive for industrial production. Commercialization of dialdehyde starch was attempted in the 1960s by using an electrochemical process to regenerate reduced oxidant to periodic acid, but the process was still too expensive for industrial production. The properties of dialdehyde starch are unique among the oxidized starches. Periodate oxidation increases gelatinization and pasting temperatures, peak viscosity, and the breakdown and effects of depolymerization on both amylose and amylopectin molecules. Dialdehyde cassava/tapioca starch produced slowly biodegradable films with a brown color, low film moisture content, reduced water absorption, and improved mechanical properties, as compared to a control film made with native cassava starch. Nanoparticles of dialdehyde starch are claimed to be effective for delivery of anticancer drugs (Prasher & Sharma 2023). Dialdehyde starch has been investigated as a cross-linking agent for proteins, e.g., as required in the tanning of leather (Zhen & Ma 2002).

Chlorine dioxide and sodium chlorate (in a dry procedure) have been investigated as starch oxidants. Oxidation with sodium chlorate resulted in a high carboxyl group content. In part, in a search for biodegradable, non-polluting laundry detergent builders and because it specifically oxidizes primary hydroxyl groups to carboxylate groups, starch has been oxidized with TEMPO (2,2,6,6-tetramethylpiperidine-1-oxyl radical)+NaOCl+NaBr) to convert the polysaccharide molecules into polycarboxylates. Starches have also been oxidized with peroxydisulfate ions and other oxidants.

12.2.8 Reductions in Starch Polysaccharide Molecular Weights

Partial depolymerizations, i.e., reductions in the molecular weights of the starch polysaccharide molecules, of both native and derivatized starches are useful and established processes. Cleavage of starch molecules weakens granules and lowers the starch's pasting temperature, i.e., makes it easier to cook out. Various other changes in starch properties, such as reduced viscosity and shear resistance and increased solubility, may also result from partial depolymerizations (Karma et al. 2022). Partial depolymerization of starch polysaccharides can be achieved through physical (shear forces introduced by extrusion, sonication, (Section 12.4.3.1), or homogenization (Section 12.4.3.2)), certain types of radiation that produce radicals, and chemical (acid- and enzyme-catalyzed (e.g., α-amylase) treatments) (Section 12.3)). (Physical treatments are not used to prepare starch ingredients, except that extrusion is used to prepare pregelatinized starch products). Acid treatments can be done on starch in the granular form, while homogenization and enzymic treatments are effective with starch molecules in solution. Depolymerized starch products are often known as being 'thinned' and the processes of producing them are often called 'thinning'. The effects of acid-catalyzed hydrolysis on starch structure and functionality have been reviewed (Wang & Copeland 2015).

The primary reason for partially depolymerizing starches (including otherwise chemically modified starch products that are depolymerized either before or after other chemical modification) is to allow the preparation of high-solids solutions so as to make use of specific functionalities of the starch or starch product without generating a paste or gel of extremely high viscosity. Examples of the desirability of such products are in the preparation of adhesives and paper coating materials (unmodified or derivatized starches may be used in either) and in the preparation of products for food applications (for which modified/derivatized starches are usually used).

12.2.8.1 Acid-modified starches

Acid-catalyzed hydrolysis is the oldest and still the most used method for partial depolymerization, with hydrochloric acid being by far the most used acid. Starch molecules within granules are depolymerized by acids in reactions that become more rapid as the temperature increases. Acid-catalyzed hydrolysis occurs somewhat randomly, producing a mixture of molecules of different sizes. The least depolymerized starch products are known as acid-modified starches. Two processes are used to produce acid-modified starch products: either (1) hydrochloric acid is sprayed onto well-mixed starch or stirred moist starch is treated with hydrogen chloride gas, and the mixture is then heated until the desired degree of hydrolysis (which is very slight in any case) is obtained or (2) the pH of the slurry resulting after derivatization is lowered and the slurry is held at the modification temperature until the desired degree of hydrolysis is obtained. Then, the acid is neutralized, and the product is recovered by centrifugation or filtration, washed, and dried. Even though only a few glycosidic bonds are cleaved, the granules in the products (known as acid-modified, thin-boiling, fluid, or fluidity starches) disintegrate ('cook out' in industrial jargon) easily and largely dissolve when heated in water. In these processes used to make an acid-modified starch, depolymerization takes place to a relatively small degree, but only a small degree of average molecular weight reduction can result in a significant reduction in paste viscosity – meaning that higher solids solutions can be made. Attributes of acid-modified starches other than a reduction in viscosity include a reduction in the energy required to cook (because the granules are weakened); increases in solubility, the tendency to gel, the ability to form films, and adhesiveness and an improvement in texture. A wide range of viscosity grades of acid-modified starch products from 'thick-boiling' to 'thin-boiling' is available for papermaking and related industries, where they are used for surface sizing and as coating binders. In food and related industries, thin-boiling starches are used to form gels with improved clarity and increased gel strength, e.g., in the production of gum candies, such as jelly beans, jujubes, orange slices, and spearmint leaves and in processed cheese loaves. To prepare especially strong and fast-setting gels, high-amylose corn starch is used. Acid-modified starch is used as an excipient in pharmaceutical tablets.

12.2.8.2 Dextrins

A little more extensive depolymerization produces what are known as starch dextrins. Although the generic term dextrin means any fragment of a polysaccharide molecule, in the case of starch, it is most often applied to products obtained by heating dry acidified starch (pyroconversion). There are generally five steps in making such a dextrin: acidification, pre-drying at a low temperature, heating, cooling, and re-humidifying. Three types of reactions occur during dextrinization of a starch: hydrolysis of glycosidic bonds, transglycosylation([also known as transglycosidation]; which is the transfer of a glycosidic chain or a portion of a glycosidic chain from its original position to another position – usually forming glycosidic bonds other than the α-(1,4) and α-(1,6) bonds found in the structures of starch molecules) and reversion, which is the acid-catalyzed joining together of two carbohydrate molecules via creation of a new glycosidic bond. Hydrolysis depolymerizes the starch polymer molecules. Transglycosylation and reversion make more highly branched polymer structures containing linkages other than those in normal starch polysaccharide molecules. The relative amounts of these three reactions are a function of the moisture content of the starch and the temperature of the process. As an acidified starch (dry, but with some moisture present) is heated, glycosidic bonds undergo hydrolysis, some water is consumed in the reaction, and the temperature rises. After the temperature surpasses 100°C, dehydration occurs, and transglycosylation and reversion begin. The resulting dextrin is more highly branched than the native starch. Owing to the linkages other than those found in normal starch polysaccharide molecules, dextrins are less digestible than the parent starch. Treatment of them with an amylase to partially remove the digestible part makes them even less digestible.

Dextrins are classified by their cold-water solubility and color (white or yellow). Because they produce solutions of lower viscosity (as compared to the native parent starch), dextrins can be used at higher concentrations than the parent starches. Because they have film-forming and adhesive properties, they are used in coatings on products such as candies and pan-coated roasted nuts. They are also used as fillers, as encapsulating agents, and as carriers of flavors (especially spray-dried flavors). They increase crispness and browning in certain bakery products.

Further depolymerization via acids or enzymes produces even lower molecular weight products that no longer have the characteristics and properties of carbohydrate polymers. Such products are categorized as being converted or conversion products (Chapter 13).

12.2.8.3 Oxidized starches

As mentioned in Sections 12.2.7 and 12.2.7.1, oxidations (particularly under alkaline conditions and therefore particularly with hypochlorite ions) can result in the depolymerization of starch polysaccharide molecules.

12.2.9 Multiple Chemical Modifications

Industrially produced and used chemically modified starches, which in every case are tailor-made for specific applications, whether for food, papermaking, or other industrial uses, are often modified with more than one reagent in efforts to incorporate the benefits of each type of modification (Ashogbon 2021). Many modified food starches are made by a combination of cross-linking (Section 12.1.2.2) and stabilization (Section 12.1.2.2) of native starches so as to both reduce their tendency to produce opaque, cohesive, rubbery, syneresing, and otherwise unstable gels, as well as to make the products less sensitive to the high temperatures, extended cooking times, acidic environments, and high shear conditions of processing operations. (Acetylated adipic acid cross-linked starch has already been presented in Section 12.2.5.3.) Another common way to multiply modify a starch is to stabilize it (most often by etherification) and then lower the pH to depolymerize/thin the product so that it can be used at higher concentrations without producing excess viscosity (Sections 12.2.8 and 12.2.8.1). Such products also require less energy to cook. For use in the papermaking and related industries, stabilized and thinned starches are used in sizing operations, as coating binders, and as adhesives – all processes that require a high solids solution at a relatively low viscosity.

Multiple modifications of different starches with different allowable reagents to different allowable degrees offer endless product possibilities, as multiple modified starches can have a wide diversity of functional properties and attributes that allow them to be used in both specific processes and specific products and thus allow a starch producer to make scores of different products from a single type of starch. Characteristics that can be controlled/improved by multiple modifications of products that may be liquid, semi-solid, or solid include (but are not limited to) one or more of the following: adhesion, clarity of solutions/pastes, emulsion stabilization, film formation, flavor release, hydration rate, stability to acids, stability to heat, stability to shear, tackiness, temperature required to cook, and viscosity of hot paste and gel and moisture retention and control, mouthfeel, oil migration, sheen, shelf stability, and texture/consistency in the product in which it is used. Dual-modified food starch products other than those that have been derivatized and thinned or physically modified and derivatized are listed in Table 12.3.

12.2.10 Graft Polymerizations

Hydrophobic synthetic polymers have been attached to starch polysaccharide molecules by different means and for different purposes. Research in this area is quite active. The products of such attachments are known as starch-graft-copolymers (Jyothi 2010; Lele, Kumari & Niju 2018; Meimoun et al. 2018;

Zia et al. 2015). Synthetic polymers have been attached to starch polysaccharides via both ether and ester linkages. The ether type is produced by forming oxygen radicals on molecules of a starch or a modified starch and then introducing a polymerizable monomer(s). The radical initiator can be either chemical or physical. The large number of monomers that have been used are mostly vinyl or acrylol monomers, but grafted polymers have also been made via ring-opening reactions. Ester group attachments are made by ring-opening reactions of a starch with a lactone or lactide.

Polymers grafted to starch include, but are not limited to, polyacrylamide, poly(acrylamide-acrylic acid), poly((2-metha-cryloyloxyethyl)trimethylammonium chloride), poly(acrylamide/1-vinyl-2-pyrrolidone copolymer, starch-*g*-polyethylene, and poly(acrylamide-co-2-acrylamido-2-methylpropanesulfonic acid). Starches modified have included previously modified starches, such as a hydroxyethylstarch (Section 12.2.4.1), a cationic starch (Section 12.2.4.3), and an anionic starch (carboxymethylstarch (Section 12.2.4.4) or a starch sulfate (Section 12.2.5.8)). In such ways, amphoteric superabsorbent products have been made. Some products make partially biodegradable films that overcome the hydrophilic nature and brittleness of starch alone films. Others are claimed to be water superabsorbers, a filtration loss agent in drilling muds, flocculants for humic substances in textile dyeing wastewater and microalgal biomass, binders of cations, a compatibilizer for polymers such as polyalkanes, and an effective warp sizing agent for polyesters.

Another type of starch-graft-copolymer is produced via reaction of starch with a monomer that can be polymerized to make a polyester. The most prominent of such products is starch-*g*-poly(DL-lactic acid) (starch-*g*-polylactide, PLA), which is claimed to be useful as a thermoplastic material, a single-layer, flexible packaging material, a compatiblizer for starch-thermoplastic synthetic polymer blends, a drug carrier, and an encapsulator. Preparation and application of hydroxyethylstarch-*g*-polylactide has also been described.

Polyurethanes have also been grafted to starch. It should be mentioned that most starch-graft-copolymer products are undoubtably mixtures of the desired copolymer, homopolymer, unreacted starch, unreacted monomer, monomer byproduct(s), and catalyst (in various ratios depending on the process used) and that it is difficult to determine the actual amount of copolymer present in the mixture. Also, many of the starch-*g*-copolymers are claimed to be biodegradable, when only the starch part is biodegradable. Starch-*g*-poly(lactic acid) products should be completely biodegradable.

TABLE 12.3   Cross-linked and stabilized modified food starches

Acetylated distarch adipates
Acetylated distarch phosphates
Acetylated oxidized starch
Hydroxypropylated distarch phosphates
Phosphorylated distarch phosphates

TABLE 12.4   Major commercial chemical modifications and the type of product formed or reaction that occurs

1. Derivatized starches
    1.1. Stabilized/substituted products (Section 12.1.2.2)
        1.1.1. Hydroxyethylstarches (ether) (Section 12.2.4.1)
        1.1.2. Hydroxypropylstarches (ether) (Section 12.2.4.2)
        1.1.3. Cationic starches (ether) (Section 12.2,4.3)
        1.1.4. Starch acetates (organic acid ester) (Section 12.2.5.1)
        1.1.5. Starch octenylsuccinates (organic acid ester) (Section 12.2.5.2)
        1.1.6. Monostarch phosphates (inorganic acid ester) (Section 12.2.5.6)
    1.2. Cross-linked products (Section 12.1.2.2)
        1.2.1. Distarch phosphates (inorganic acid diester) (Section 12.2.5.7)
        1.2.2. Acetylated distarch adipates[a] (organic acid mono- and diester) (Section 12.2.5.3)
        1.2.3. Starch diglyceryl ethers (diether) (Section 12.2.4.5)
    1.3. Cross-linked and stabilized starches (Section 12.2.9)
        1.3.1. Hydroxypropylated distarch phosphates (ether and inorganic acid diester)
        1.3.2. Phosphorylated distarch phosphates (inorganic acid mono- and diester)
        1.3.3. Acetylated distarch adipates[a] (organic acid mono- and diester) (Section 12.2.5.3)
        1.3.4. Acetylated distarch phosphates (organic acid ester and inorganic acid diester)
2. Acid-modified starches
    2.1. Thinned products (lower-molecular weight products) (Section 12.2.7.1)
    2.2. Dextrins (depolymerization and rearrangement of glycosidic bonds) (Section 12.2.7.2)
3. Oxidized starches (transformation of hydroxyl groups into carbonyl and carboxyl groups) (Section 12.2.6)

[a] These two products are identical. Acetylated distarch adipates are listed twice because acetylation and cross-linking occur together.

12.2.11 Primary Commercial Chemical Modifications

Major commercial chemical modifications and the type of product formed or reaction that occurs are listed in Table 12.4.

12.2.12 The Relationship of the Natures of Starch Granules to Chemical Modification and Product Characteristics

Starch occurs in nature as insoluble granules (Chapter 9). Much of the usefulness of starch derives from the fact that these insoluble granules can be isolated, washed, modified, washed again, dried, and easily shipped and stored before use. Then, the dry powder can be dispersed in water without the use of special equipment and the resultant slurry cooked (i.e., the aqueous slurry can be heated to temperatures above what is known as the pasting temperature) to 'dissolve' the starch molecules and turn the slurry into a hot paste that has the desired useful properties. As already stated, each starch produced by a specific tissue of each type of plant differs from starches from other types of plants and their tissues. Among the differences in starches from different plant sources are different granule morphologies, pasting temperatures, and retrogradation propensities, and it is well established that each of these differences is related to the specific structures of its molecular components (Section 12.2.1). Granular structures and reactivities not only vary between starches of different botanical origin, but also from granule to granule in a population of granules from a single botanical source. In addition, growing conditions of the source plant may affect amylose content and amylopectin fine structure (Aboubacar et al. 2006), and small differences in amylopectin fine structure may affect the functional properties of the starch (Bertoft et al. 2016).

Chemical modification of starch occurs while the starch is in the form of granules, and the granular nature of the starch influences both granular and molecular reaction patterns. Starch granules contain what are called crystalline and amorphous regions, but the amorphous regions contain some crystallinity and the crystalline regions are not entirely crystalline. Some degree of granule swelling is generally necessary for efficient chemical derivatization to occur. Accumulated evidence indicates that chemical reactions probably occur via concerted processes that begin in amorphous granule regions and proceed progressively to crystalline regions as swelling of the weakened amorphous regions pulls apart crystallites. Therefore, both attributes of the starch itself (intrinsic factors) and reaction conditions (extrinsic factors) determine overall reactivity. Reaction conditions should be optimized so that swelling of granules (without incurring pasting) is promoted to maximize the accessibility of starch polysaccharide molecules within granules to catalyst (alkali) and reagent molecules. Factors that might affect reaction efficiency are listed in Table 12.5.

In any discussion of starch modification, it must be remembered that each starch (even starches from different cultivars of the same plant) is unique and that all populations of granules are heterogeneous, so not only does each type of starch react differently using identical conditions, each granule in a population of granules and each starch molecule within the granules become modified differently.

Only the granules of certain cereal starches have been found to contain channels and macropores (openings of the channels on the outside granule surface), but in these granules,

TABLE 12.5   Factors that might affect reaction efficiency (Han et al. 2006)

1. Intrinsic factors (Factors that are inherent to the specific starch granules being reacted)
   A. Natures of granules
      a. Shapes
      b. Sizes
      c. Granule organization, including the nature of the granule surface
      d. Presence of pores and channels
   B. Granular compositions
      a. Fine structures of the starch polysaccharide molecules
      b. Amylose–amylopectin ratio
      c. Average molecular weights of the starch polysaccharide molecules
      d. Non-starch components
2. Extrinsic factors (Reaction media and conditions)
   A. Reagent type and concentration
   B. Type and concentration of swelling-inhibiting salts
   C. pH
   D. Temperature
   E. Reaction time

the macropores and channels can have significant effects on where reaction occurs within granules and the nature of the products. In the cereal starches that contain macropores and channels, reagent solution fills the channels and the cavity at the hilum (roughly the center of granule), and penetration occurs mainly from the cavity outward. This means that, for rapidly reacting reagents, such as phosphorus oxychloride (Section 12.2.5.7), both internal surfaces (of the channels and cavity) and the outside surface are exposed to the reagent and that the reaction occurs on all these surfaces (and because of the very small amounts of reagent used, only there), with the reaction occurring primarily on inside surfaces. On the other hand, reactions with slowly reacting reagents, such as ethylene oxide and propylene oxide (Sections 12.2.4.1 and 12.2.4.2), have time to diffuse throughout the granule matrix before reacting, resulting in a more uniform reaction. In granules that do not contain macropores and channels (or in granules such as the A granules of wheat starch where the channels are plugged with protein), all reagent solution penetration occurs from the outside surface into the granule matrix.

12.2.13 Other Derivatizations

Modification of native starches and starches otherwise modified with other reagents and reaction types continues. Explored reagents include reactions with isocyanates, such as furfuryl isocyanate, to prepare carbamates (González et al. 2018). Explored derivatization reactions include click reactions (González et al. 2018; Tan et al. 2016; Uliniuc et al. 2013; Wei et al. 2020; Zhang et al. 2013).

12.3 ENZYMIC MODIFICATIONS

(The material presented in this section is a greatly abbreviated version of the review by Miao and BeMiller (2023)). Treating

granular starch with an amylolytic enzyme (such as an alpha-, beta- or glucoamylase) or a glycosyltransferase produces granules containing large pores and channels and a product known as porous starch (Cao et al. 2023; Chen et al. 2020; Latip et al. 2021; Sun et al. 2023). Porous starch granules are more efficient adsorbers than native starch granules and can be chemically modified to improve their usefulness as carriers. Porous starch can also be made by treatment with an acid or by certain physical treatments, but enzymic modification (perhaps in conjunction with an acid or physical treatment) is the most efficient means.

Enzymes have also been used to create less-digestible starch products, and treatment with specific enzymes is undoubtedly the frontier in the preparation of nutritionally beneficial products from starch (Banger et al. 2022; Miao & BeMiller 2023; Miao et al. 2018; Park et al. 2018; Sorndech, Tongta & Blennow 2018; Zhong et al. 2022). Among specific enzymes that have been used to produce functional, less-digestible products from starch are amylosucrases, α-transglycosylases, maltogenic amylases, and glucan branching enzymes (GBE). The action of each of these types of enzymes on a starch changes the structures of the starch polysaccharide molecules, and because human digestive enzymes are specific for the normal structures of these molecules, the modified molecules are less digestible. Although RS has received more attention than SDS (Chapter 20), the latter is also important, and the use of these enzymes has provided a route to SDS. The products obtained via treatment of a starch with these enzymes differ with the specific starch and specific enzyme used and treatment conditions. Some of the enzymes and some of the products have been commercialized.

Treatment of a starch with an enzyme almost always involves by heating a dilute aqueous slurry of a starch with stirring at a temperature above its pasting temperature to destroy the starch's granular structure and prepare a relatively dilute solution of the starch to which an enzyme is added. (In some cases, an enzyme is added to a food product system.)

Amylosucrase has several activities, the main one being the transfer of a glucosyl unit from a sucrose molecule to an α-1,4-linked glucosyl non-reducing end-unit of an acceptor molecule, such as an amylopectin molecule, with the concomitant release of fructose. Amylosucrase thereby elongates linear α-1,4-glucan chains. A commercial enzyme creates chain lengths ranging from 35 to 100 glucosyl units. The increased branch chain lengths result in more stable crystalline structures, and the products are classified as type 3 RS (Chapter 20). In a high-fat diet-induced mouse model, an amylosucrase-treated starch was indicated to have prebiotic characteristics and anti-obesity effects.

Starch glycosyl transferases such as α-transglycosylase, branching enzyme, and amylomaltase cleave an α-(1,4) bond of an α-(1,4)-linked glucan chain (a donor molecule) and transfer the cleaved part to a glycosyl acceptor (i.e., a glucosyl unit of another α-(1,4)-linked glucan chain) with the formation of a new α-(1,3), α-(1,4) or α-(1,6) linkage. Treatment of maltodextrin or amylose molecules with an α-transglycosylase produces a variety of α-glucans by introducing non-normal patterns of α-(1,3), α-(1,4), and α-(1,6) linkages. Three different families of α-transglycosylases have been identified: (1) one type produces linear oligo- and polysaccharides containing α-D-glucopyranosyl units connected by both (1,4) and (1,6) linkages, which are called isomalto/maltooligo- and polysaccharides; (2) another type produces similar polysaccharides with greater amounts of (1,6) linkages and lesser amounts of (1,4) linkages; and (3) a third type produces branched α(1,4:1,3)-glucans. Polysaccharide products from the first type of α-transglycosylases have been shown to function as SDS. The first type also produces oligosaccharides rich in α-(1,6) linkages, which function as prebiotics. Products from the second type of α-transglycosylases have been shown to have a relatively high percentage of RS and to produce high amounts of butyric acid when subjected to fermentation by fecal organisms, indicating a positive prebiotic effect on gastrointestinal health. The third type is also indicated to be slowly digestible. Thus, α-transglycosylases provide an approach for modification of starches into low-glycemic food ingredients with greater amounts of prebiotic and SDS or RS content (Chapter 20) than those found in other products (Gangoiti et al. 2020).

Maltogenic amylases catalyze hydrolysis of α-(1,4) bonds in glucan chains, primarily liberating maltose and oligosaccharides. The effect of this enzyme on amylopectin molecules is to shorten branch chain lengths. The result is a modest increase in the percentage of RS and a significant increase in the percentage of SDS. Some isomaltooligosaccharides and branched oligosaccharides with health benefits are also produced.

A family of slowly digestible products with high proportions of α-(1,6) linkages, known as highly branched maltodextrins (HBMD), can be made by the action of GBE or 4,6-α-glucanotransferase on starches. Treatment of amylopectin with 4,6-α-glucanotransferase produces high-molecular-weight HBMD (Ryu et al. 2022) and some cyclic products. Treatment of amylose with GBE produces medium-molecular-weight HBMD. Treatment of amylopectin with α-amylase produces low-molecular-weight HBMD called α-limit dextrin.

Because enzymes from different microbiological sources have different substrate specificities and produce different products using the same substrate, because microorganisms can be genetically modified, and because starch polysaccharides from different sources have different structures, the variety of products that can be made by enzymatic treatment of starches is almost limitless, and the use of combinations of enzymes extends the products that can be made even further.

12.4 PHYSICAL MODIFICATIONS

12.4.1 Introduction

Physical modifications of starches are those modifications that produce changes in starch properties through the action of physical treatments. In general, they do not introduce any chemical modification of the starch polysaccharide molecules (other than, in some cases, limited glyosidic bond cleavage (depolymerization) that results only in some decrease in average molecular weight). Physical treatments are generally divided into thermal and non-thermal treatments, although some treatments often categorized as non-thermal may have a thermal component.

The primary interest in physical modification seems to be that they are clean-label starches (sometimes referred to as label friendly or functional native starches), i.e., they can be classified as ingredients rather than additives and do not need to be identified as modified food starch, food starch modified, or similar designations on food labels, thus making them acceptable as natural products (Chapter 21). And some investigations of physical treatments of starches were done because the treatments had been investigated for non-thermal processing of foods, so it was of interest to determine what effect the treatment had on any starch or flour used as an ingredient. Physical modifications would be the preferred method of modification to improve the performance of poor-quality starches if proper functionalities can be realizedbecause physical modifications are usually easier to do and often less expensive than chemical modifications and produce no effluents containing salts, reagents, or reagent byproducts.

In general, physical treatments produce changes in the packing arrangements of the starch polysaccharide molecules within granules. Such structural changes can alter the properties and functionalities of the starch. Physical treatments can produce starches with properties somewhat like those obtained by chemical modifications (especially like those of lightly cross-linked starches, such as an increased tolerance to low pH, high temperatures, and high shear), but the changes are not as dramatic and as thermostable as those produced by chemical modification.

Different treatment conditions applied to the same starch and different starches subjected to the same treatment conditions give products with different characteristics, the latter because different starches are affected differently by physical treatments. For these reasons and because the conditions under which the starch is treated can vary considerably, contradictory results have often been reported. This section attempts to summarize the most frequently obtained results or the consensus on changes obtained. Much of the material presented in it is an abbreviated version of the book chapter by BeMiller (2018), which is an abbreviated version (with updates) of the review by BeMiller and Huber (2015). See also reviews by Zia-ud-Din, Xiong, and Fei (2017), Han, Shi, and Sun (2020), Maniglia et al. (2021a, b), Park and Kim (2021), Ye and Biak (2023), and Zailani et al. (2023).

12.4.2 Thermal Treatments

12.4.2.1 *Pregelatinized and destructured/ destructurized starches*

Commercial pregelatinized (instant) starches (Chapter 21) are manufactured using a process of cooking, drying, and milling of starch–water mixtures under conditions that allow little or no molecular reassociation. Pregelatinized products hydrate rapidly and are described as being cold-water soluble. Although they can be used without cooking, many commercial products are not completely soluble and generate additional viscosity when a dispersion of them is heated.

Pregelatinized products are used when heat cannot be applied to the product because of thermal lability of an ingredient, no processing step heats the ingredients to a temperature sufficient to paste the starch, no heat is available, or no heating equates to use convenience (especially in dry mixes to be used in the home). They are used for product thickening, moisture control, and texture provision and control. Both native and chemically modified starches can be pregelatinized. Pregelatinized modified starches retain much of the attributes contributed by the modification.

There are three basic types of equipment used to make pregelatinized starches. With either, the specific type of equipment used and operating parameters may vary from site to site. Those factors and different starch types used in the process means that different commercial products (even those produced from the same native starch) often have different characteristics. (1) In the original procedure, an aqueous slurry of a starch is applied to a drum heated with pressurized (high-temperature) steam or into the nip between counter-rotating drums heated with pressurized (high-temperature) steam. In both cases, the starch is rapidly gelatinized, pasted, and dried. The dry film is scraped from the drum and reduced to powders of various mesh sizes. Results of research on the effects of process variables using double-drum equipment verify that swollen particles contribute much more to the viscosity of resulting product dispersions than solubilized starch polysaccharides. (2) Another process employs extrusion. Because the variables include the type of starch, starch–water ratio, temperature, screw configuration and speed, and residence time, there is probably more variability in products made by extrusion. The extrudate must be dried and ground to a powder. (3) In yet another process, a starch slurry is gelatinized with superheated steam in a two-fluid nozzle, followed by a spray-drying chamber. Some depolymerization occurs, at least for those products

made using hot drums or extruders, with a greater amount being found in products exposed to the high shear produced in extruders.

A special type of pregelatinized starch that is thermoplastic and can be used in injection molding is called both destructured and destructurized starch (Butterfly Srl 1993; Warner-Lambert Co. 1988, 1989, 1990, 1991). The basic process used to make these products treats potato starch (which is naturally phosphorylated) or a monostarch phosphate (Section 12.2.5.7) in an extruder with a depolymerization catalyst (an acid) and other ingredients (such as an ethylene-acrylic acid copolymer, an ethylene-vinyl alcohol copolymer, a high-boiling plasticizer such as glycerol, a hydrogenated fat lubricant/release agent, lecithin, and reactive organosilanes). Claimed applications include manufacture of water-soluble pharmaceutical capsules, carriers for pharmaceutical and agrochemicals, and water-repellent films.

12.4.2.2 *Granular cold-water-swelling starches*

Another group of 'instant' starches consists of products which contain gelatinized, intact granules, i.e., amorphous granules, which swell extensively when placed in an aqueous system at room temperature (Majzoobi & Farahnaky 2021). These products are a special kind of pregelatinized starch and are often called cold-water-soluble starches. However, the classic pregelatinized starches are usually more soluble in room-temperature water (without application of shear) than these products, so the author prefers the term granular cold-water-swelling (GCWS) or simply cold-water-swelling (CWS) starches to describe them. GCWS starches produce viscosities and gel characteristics more like those of cook-up starches when added to room-temperature aqueous systems than classic pregelatinized starches. They can be made by four general methods: (1) heating an amylose-containing starch in an aqueous solution of an alcohol, (2) treating an amylose-containing starch with an alkaline, aqueous solution of an alcohol at room temperature, (3) rapidly heating an amylose-containing starch dispersion in a special spray-drying nozzle and drying the droplets in a spray drier, and (4) instantaneous controlled pressure drop (DIC). Mixtures of a normal (i.e., amylose-containing) starch and a waxy starch can also be used in these procedures. When waxy maize starch alone is heated in an aqueous alcohol solution and the product is added to hot water, the product dissolves and produces a hard-candy-like product upon cooling. Conversion of high-amylose maize starch into a GCWS product useful for the manufacture of confectionary, convenience, and other food products using a modified spray-drying system has been claimed.

Products made from a single starch using different methods, by varying the method conditions using a single starch, and by using a single method with different starches vary in characteristics, but they have the common characteristic that, when placed in an aqueous system at room temperature, their granules rapidly hydrate, swell, and lose their crystalline order without granule disintegration (even when little or no shear is applied). They produce the functionalities of the cook-up starches from which they are made without application of heat. A drawback is that, when GCWS starch granules of an amylose-containing starch are added to unheated water without shear, ensuing retrogradation can result in poor-quality dispersions. However, when those granules are dispersed in a sucrose or glucose syrup with rapid stirring, the resulting dispersion sets to a rigid gel that can be sliced, and the ability to swell in unheated aqueous systems can be useful for making desserts at home and in muffin batters containing fruit or nut pieces that would settle to the bottom when the batter thins as it is heated before the gelatinization temperature of the starch is reached and the batter thickens.

12.4.2.3 Heat-moisture treatments

Heat-moisture treatments (HMT) are hydrothermal processes that consist of heating starch granules at a temperature above the starch's glass transition temperature (at the moisture content employed) in a closed and sealed vessel (Chapter 21). Moisture contents employed are generally adjusted to 10%–40%. Processing temperatures generally range from 84°C to 140°C, with most investigations using temperatures above 94°C. Treatment times vary from 1 to >24 hours. Variables in the process include the type of starch, its moisture content, the temperature employed, and the time the starch is heated at that temperature. Because of the several variables, a large number of different HMT products are possible and, when a specific attribute of a treated starch is compared to that of the native starch, contradictory results are likely to have been reported. Jacobs and Delcour (1998), Hoover (2010), Zavareze and Dias (2011), BeMiller and Huber (2015), Iuga and Mironeasa (2020), Fonseca et al. (2021), Schafranski, Ito, and Lacerda (2021), and others (Chapter 21) have reviewed the large body of literature on heat-moisture treatments of starches and provided specific details on changes in starches under different conditions and should be consulted for those details.

Structural changes occur in both crystalline and amorphous regions of granules (Chapter 9). Changes that are generally agreed upon are that starches with A-type crystallinity undergo no change in the type of crystallinity upon HMT, starches with B-type crystallinity have their crystallinity changed to a C-type (partial conversion of B-type crystallinity to the A type) or completely to A-type crystallinity, and that starches with C-type crystallinity have their crystallinity changed to a greater proportion of the A-type crystallinity, with the B- to A-type crystallinity conversion being favored by higher temperatures and moisture contents. Often observed is an increase in the degree of crystallinity in A-type starches and a decrease in the degree of crystallinity in B-type starches. Possibly related to the increases in crystallinity, often, but not always, observed are increases in the onset (T_o), peak (T_p), and conclusion (T_c) gelatinization temperatures and a broadened phase-transition temperature range ($T_c - T_o$) (owing to a greater increase in T_c than in T_o) (Chapter 10).

Often found after HMT are decreases in swelling power and leaching of amylose from the swollen granules, with the reductions increasing with increasing moisture content and

temperature during treatment. HMT starches almost always have increased RVA pasting temperatures, decreased peak viscosities and breakdown, and increased hot-paste viscosities – giving the starch pasting characteristics like those of a lightly cross-linked starch (Section 12.1.2.2).

Both increases and decreases in SDS and RS (Chapter 20) have been reported. (Any assay method used to measure SDS and RS contents should have a heating step that employs the temperatures and times of processing and preparation, because the structures introduced by HMT, while being more stable than those in the native starch, are probably heat labile to some extent, and heating to temperatures that are likely to be encountered during food processing is not always employed in starch digestibility assays). HMT starches have found uses in food products (Wang, Li & Zheng 2021) and have been reported to make biodegradable films with improved characteristics as compared to the parent native starch (Cahyana et al. 2023).

Modifications to the standard technique have been investigated. Examples are (1) addition of lauric acid to maize starch prior to HMT, which results in the formation of amylose–lauric acid V-type complexes (Chapter 15) and a product that contains a maximum of 10% thermostable SDS and 18% thermostable RS when the moisture content of the treated starch was 50% and changes to the RVA pasting and paste attributes of the starch, and (2) pre-hydrolysis which also increases the thermostable RS content. Similarly, when HMT is conducted under slightly acidic conditions, thermostable RS contents increase.

Because the traditional method for HMT, which has been used for most laboratory investigations, is a procedure that would be expensive and difficult to do on a large scale, alternative methods have been investigated. These alternative methods include microwave heating, infrared heating, direct steam injection, heating in an aqueous alcohol, direct-vapor HMT, and reduced pressure-HMT. HMT has been conducted before, after, and simultaneously with chemical or other modification of starches.

12.4.2.4 Annealing

Annealing (another hydrothermal process) consists of holding starch granules in an excess of water (generally >39% w/w) at a temperature above the starch's glass transition temperature and below its gelatinization temperature for periods of time ranging from minutes to days (Chapter 21; BeMiller & Huber 2015; Fonseca et al. 2021; Iuga & Mironeasa 2020; Jacobs & Delcour 1998; Jayakody & Hoover 2008; Tester & Debon 2000; Zavareze & Dias 2011). Like HMT, the variables are the specific starch used, the temperature, and the duration of heating. Also, like HMT, when a specific attribute of a treated starch is compared to that of the native starch, increases, decreases, and no change are likely to have been reported because of the large numbers of starches and conditions that may be employed. Several of the observed property changes are in the same direction as those produced by HMT, while there are differences in others.

As compared with HMT, there is less consensus on the changes in starch properties imparted by annealing. One thing there seems to be general agreement about is that annealing increases the onset and peak gelatinization temperatures and decreases the phase-transition temperature range. Only slight to no increases in thermostable SDS and RS contents following annealing, then cooking, have been found.

12.4.2.5 Heating dry starch

Starch with <15% moisture heated to temperatures between 100°C and the temperature at which thermal degradation occurred was claimed to produce products with acid, shear, and temperature tolerances comparable to those of chemically cross-linked starches (Section 12.1.2.2). A low moisture content and alkalinity were found to facilitate the transformation.

Heating of dry starch can also be done with microwave irradiation. Effects produced by microwave irradiation are determined by the moisture content of the starch, the wattage and frequency of the microwave source, and the duration of the treatment, so many possible treatment conditions are possible. A major difference between microwave and conventional heating processes is that the latter are done in closed containers, while microwave heating is seldom done that way; as a result, the moisture contents of microwave-heated samples are reduced during the treatment.

Heating dry OS starches (Section 12.2.5.2) lowered gelatinization and pasting temperatures while increasing SDS contents (Han & BeMiller 2007).

12.4.3 'Non-Thermal' Treatments

Non-thermal treatments have been investigated as green, low-energy-requiring alternatives to chemical and enzymic modifications (Grgić et al. 2019; Raghunathan et al.2o21; Wang et al. 2023; Wu et al. 2022).

12.4.3.1 Treatments with ultrasound

Treatments with ultrasound are done in aqueous systems. Ultrasonic treatment (Chapter 21) generates areas of intense local heating and high shear stresses, so the treatment is not strictly non-thermal. Variables are the nature of the medium (specifically its vapor pressure and surface tension), natures and concentrations of dissolved gases (the atmosphere above the aqueous suspension being sonicated), temperature, treatment time, starch concentration, and ultrasound power, frequency, and amplitude, so treatment conditions can vary widely. Different treatment conditions applied to the same starch and different starches treated using the same conditions give different results, with the degree of change increasing with treatment duration. When conditions are such that changes occur, they often include damage to the granule surface that may include erosion, pitting, and cracking, increases in solubility, swelling power, and gel clarity, hardness, and adhesiveness, and decreases in paste and gel viscosities, as well as the consistency coefficient (k) of pastes – changes that are consistent with a general weakening of granule structure. Ultrasonic energy is generally insufficient to disrupt, or even to distort, granule crystallites, but it can effect changes in amorphous regions and perhaps disrupt and/or destroy double helical order

in both amorphous and crystalline regions. Depolymerization of starch polysaccharide molecules is also a possibility.

12.4.3.2 High-pressure treatments

There are two basic types of high-pressure treatments: (1) a static type (called ultrahigh pressure (UHP) and high hydrostatic pressure (HHP) treatments) as might be used in food processing, and (2) the use of food-processing homogenizers that produce turbulence, high shear, and cavitation as a result of forcing a starch slurry through an orifice under high pressure. The two types of treatment effect different changes in the starch.

UHP/HHP treatments consist of subjecting an aqueous slurry of starch granules to a pressure exceeding 400 MPa (Castro et al. 2020; Dominguez-Ayala et al. 2022). Variables are the type of starch, its concentration, and the pressure, temperature, and duration of treatment. Whatever the conditions used, partial or complete gelatinization of the starch with maintenance of the granular form occurs; i.e., cold-water-swelling starch is formed in various amounts. Gelatinization of granules of a starch affected by UHP occurs over a range of pressures because susceptibility to gelatinization varies from granule to granule. The higher the temperature, the lower is the pressure required for gelatinization of a granule and vice versa, and at any set of conditions, the degree of gelatinization is time dependent. Finally, the degree of gelatinization at any combination of pressure, temperature, and time decreases as the slurry concentration increases. Therefore, it is possible to obtain different degrees of gelatinization by manipulating the combination of slurry concentration, treatment duration, temperature, and pressure. Variables for treatments with high-pressure homogenizers and jets are the type of starch and the pressure and temperature employed. Changes in the starch have been reported include-granule deformation and fragmentation, partial gelatinization, increases in the onset, peak, and conclusion temperatures of gelatinization, and loss of crystallinity. All changes increase with increasing pressure. When a homogenizer is used, the number of passes/cycles is also a variable.

12.4.3.3 Treatments with Pulsed Electric Fields (PEF)

The effects of pulsed electric fields (PEF) on starch granules (Chapter 21; Zhu 2018) are a function of electric field strength and time (at a constant field strength), with the field strength predominating. PEF treatments result in damage to the crystalline structures of granules, making the starch more susceptible to enzyme-catalyzed hydrolysis. Some depolymerization of amylopectin molecules is indicated.

12.4.3.4 Treatments with cold plasma

Interest in treatment of starch with a cold plasma has grown in recent years (Leandro et al. 2024; Maniglia et al. 2021a, b; Okyere, Rajendran & Annor 2022; Zhu et al. 2023) – in part in efforts to make subsequent chemical modifications more efficient (Du et al. 2024; Taslikh et al. 2022), but also for its ability to modify starch granules. Reported changes in starch properties include granule damage (formation of cracks, pits, and holes), increased solubility (Liang et al. 2024), decreased paste viscosity (Liang et al. 2024), and improved film formation and properties (Chen et al. 2023; Guo et al. 2022). Also indicated are glycosidic bond cleavages (Chen et al. 2023; Du et al. 2024; Guo et al. 2022; Liang et al. 2024), introduction of carbonyl and carboxyl groups (Du et al. 2024), and cross-linking (Guo et al. 2022; Zhang et al. 2022), indicating that chemical changes occur.

12.4.3.5 Milling

The use of mechanical force is another method to change the characteristics of starches, but mechanical force can produce relatively high temperatures at the point of impact, so although milling is usually categorized as a non-thermal process, it is not strictly non-thermal and changes to the granular and/or molecular structures are likely caused by both thermal and mechanical energies. The temperature in ceramic, rolling ball mills usually used in laboratory research increase with milling time and the temperature at the point of impact can be quite high.

Starches have been milled with a variety of types of mills, including different types of ball mills. In recent years, milling of starch has often been referred to as micronization, probably because pharmaceutical micronizing mills have been used. However, micronization means reduction in particle size (comminution), and reduction in particle size may not occur when starch granules are milled. Commonly, cracks may appear on the surface of granules. Granules may become deformed, and fracture and fragments resulting from fracturing may agglomerate. Depending on the type of starch, the type and duration of milling, and the moisture content of the starch, various ratios of undamaged granules, damaged granules, and fragments of damaged granules and various extents of loss of crystallinity occur. When enough amorphous granules and granule fragments are produced, the starch has characteristics of a cold-water-soluble starch.

Changes, other than granule damage, that may occur in starch granules as a result of milling include increases in water vapor sorption, granule swelling, cold-water dissolution of starch polysaccharide molecules, and susceptibility to the action of amylases and reductions in double helix content, crystallinity, crystallite perfection, granule birefringence, paste viscosity, gel elasticity, and onset, peak, and conclusion gelatinization temperatures. Depolymerization, particularly of amylopectin, may occur. Fragments of amylopectin produced during milling have been related to stickiness in food products.

12.5 PRODUCTS WITH CHANGED PROPERTIES AND FUNCTIONALITIES AFFECTED BY MODIFICATIONS

Actual applications of modified starch are manifold. The properties and functionalities produced by a modification are a function of the source starch, so the effectiveness of a modified starch product in a specific application will likely differ from one type of starch to another. The listings below cover only a single modification, while many modified starch products are the result of two or more modifications (Section 12.2.9). Listed below in broad categories are general properties and functionalities affected by modifications, and some general applications that have been reported or suggested (not necessarily in use). More information about uses of modified starches in food products can be found in the review by Galkowska, Kapuśniak, and Juszczak (2023).

12.5.1 Products with Altered Gelatinization and Pasting Properties and Solubilization of Starch Molecules That Occur during Cooking of Starch

12.5.1.1 Products with reduced gelatinization and pasting temperatures that are easier to cook

Cooking starch with less energy input is a desirable characteristic in most commercial applications. All modifications that change the structures of the starch molecules (which are all chemical (Section 12.2) and enzymatic (Section 12.3) modifications except for cross-linking modifications (Section 1.2.2), which have an opposite effect) disrupt granular structures, thereby increasing the swelling and disintegration of granules and making them easier to gelatinize and paste (Chapter 10). Certain physical treatments (Section 12.4) also weaken the structure of granules and make them easier to gelatinize and paste.

Acid-modified starches (Section 12.2.8)
Carboxymethylstarches (Section 12.2.4.4) – (cold-water soluble) textile sizing
Granular cold-water-swelling starches (Section 12.4.2.2) – food products
UHP/HHP-treated starches (Section 12.4.3.2)
Hydroxyethylstarches (Section 12.2.4.1) – papermaking
Hydroxypropylstarches (Section 12.2.4.2) – food products
Milled starches (Section 12.4.3.5)
Monostarch phosphates (Section 12.2.5.7) – (cold-water swelling) food products
Oxidized starches (Section 12.2.7) – food products, papermaking
Pregelatinized starches (Section 12.4.2.1) – (partially cold-water soluble) food products
Starch acetates (Section 12.2.5.1) – food products
Starch dextrins (Section 12.2.8.2) – (cold-water soluble)
Starch octenylsuccinates (Section 12.2.5.2) – food products
Starch sulfates (Section 12.2.5.9) – (cold-water soluble)

12.5.1.2 Products with increase gelatinization and pasting temperatures that provide tolerance to processing conditions and prevent overcooking

As mentioned in Sections 12.1.2.2, 12.2.4.5, 12.2.5.3, and 12.2.5.8, cross-linking of starch granules is achieved by reacting starches with di- or poly-functional reagents to join starch polysaccharide chains lying close to each other within granules. Both inter- and intramolecular linkages may be formed. Intermolecular cross-linking is highly effective in delaying and restricting the changes to the internal structures of granules that result in granule swelling and disintegration during gelatinization and pasting. Industrially used cross-linking reagents are phosphoryl chloride ($POCl_3$) (Section 12.2.5.8), STMP (Section 12.2.5.8), mixtures of STMP and STPP (Section 12.2.5.8), acetic-adipic dianhydride (Section 12.2.5.1), and epichlorohydrin (Section 12.2.4.5). Certain physical treatments (Section 12.4) also strengthen the structures of granules and delay their gelatinization and pasting.

> Acetylated distarch adipates (Section 12.2.5.3) – food products
> Annealed starches (section 12.4.2.4) – food products
> Distarch phosphates (Section 12.2.5.8) – food products
> Epichlorohydrin cross-linked starch (Section 12.2.4.5) food products
> Heated dry starch (Section 12.4.2.5)
> Heat-moisture treated starches (Section 12.4.2.3) – food products

12.5.2 Products with Improved Paste and Gel Characteristics

12.5.2.1 Stabilized starches

All chemical modifications and most other modifications affect paste and gel characteristics. Changes in paste and gel characteristics, such as paste and gel viscosity and rheology and gel clarity, stability (during cold storage, during freezing, in high-salt environments, etc.), strength/firmness, texture, and water holding and mobility (syneresis) are related to retrogradation (Section 12.1.2.2; Chapter 10) – a process involving associations and crystallization of portions of starch polysaccharide molecules (primarily amylose) as a hot paste is cooled and stored. Non-cross-linked, derivatized, and oxidized starches are known as stabilized starches because they reduce the rate and extent of retrogradation and the negative attributes related to it, thus providing long-term product stability. Especially, they provide cold-storage and freeze–thaw stability, decrease syneresis, and improve gel clarity.

> Carboxymethylstarch (Section 12.2.4.4)
> Hydroxyethylstarch (Section 12.2.4.1) – paper production: stable viscosity

> Hydroxypropylstarch (Section 12.2.4.2) – food products: reduction in syneresis during refrigerated storage and freeze–thaw cycles
> Oxidized starches (Section 12.2.7) – food products: texturizer in dairy products; papermaking
> Starch acetates (Section 12.2.5.1) – food products

12.5.2.2 Cross-linked starches

The pastes of cross-linked starches are, in general, more viscous, heavier bodied, shorter textured, and more stable than the pastes of the native starch.

> Acetylated distarch adipates (Section 12.2.5.3) – food products
> Distarch phosphates (Section 12.2.5.7) – food products
> Epichlorohydrin cross-linked starch (Section 12.2.4.5) – food products

12.5.2.3 Other

> Acid-modified starches (Section 12.2.8.1) – confections with improved clarity and gel strength
> Granular cold-water-swelling starches (Section 12.4.2.2) – food products: fat mimetic, gum candies, instant mixes
> Heat-moisture treated starches (Section12. 4.2.3) – food products
> Milled starch (Section 12.4.3.5)
> Pregelatinized starches (Section 12.4.2.1) – food products: thickening and texture in cold water

12.5.3 Products That Act as Emulsifiers and Emulsion Stabilizers

> Starch octenylsuccinates (Section 12.2.5.2) – food and pharmaceutical products
> Starch propionates (Section 12.2.5.4)

12.5.4 Products with Reduced Digestibility

Modifications that reduce the percentage of rapidly digesting starch and increase the percentages of RS and SDS (Chapter 20) reduce starches' digestibility. Any chemical modification of starch (other than depolymerization) is likely to reduce its rate of digestion and resulting glycemic index value – probably owing to hinderance of binding of α-amylase to starch chains by the substituent groups of derivatized starches (Sections 12.2.4 and 12.2.5) or the change in chemical structures and conformations of starch chains in oxidized starches (Section 12.2.7). Owing to increased crystallinity, nongelatinized heat-moisture treated and annealed starches (Sections 12.4.2.3 and 12.4.2.4) also have reduced digestibility. The products listed below have been shown to be particularly resistant to digestion (Chapter 20):

Distarch phosphates (Section 12.2.5.8) – food products
Enzyme-modified dextrins (Section 12.2.8.2) – food products
Enzyme-modified starch (Section 12.3) – food products
Epichlorohydrin cross-linked starch (Section 12.2.4.5) – food products
Heated dry starch octenylsuccinates (Section 12.4.2.5)
Uncooked heat-moisture treated starches (Section 12.4.2.3) – food products
Starch octenylsuccinates (Section 12.2.5.2) – food products

12.5.5 Products with Increased Ability to Bind to Other Substances (Including to Flocculate)

Cationic starches (ethers) (Section 12.2.4.3) – paper production: drainage and retention aids, wet-end strength additives, sizing agents; textile sizing agents
Cationic starch (ester) (Section 12.2.5.4) – paper coating
Hydroxyethylstarches (Section 12.2.4.1) – paper production: coating binders, sizing agents
Citric acid esters of starches (Section 12.2.5.4) – cation binding
Oxidized starches (Section 12.2.7) – paper production: sizing agents and coating binders; food products; confection binders; laundry finishing agents; water treatment; heavy metal ion removal
Starch-graft-copolymers (Section 12.2.10) – flocculants
Starch sulfates (Section 12.2.5.9) – removal of heavy metal ions from wastewater; paper and textile sizing agents
Starch xanthides (Section 12.2.5.5) – removal of heavy metal ions from process water

12.6 PRODUCTS WITH GREATER ADHESIVENESS

Epichlorohydrin cross-linked starches (Section 12.2.4.5)
Milled starch (Section 12.4.3.5)
Starch acetates (Section 12.2.5.1) – food products
Starch dextrins (Section 12.2.8.2) – paper products

12.6.1 Products That Are Capable of Encapsulating

Citric acid esters (Section 12.2.5.4) – controlled/sustained release of pharmaceuticals; 'solubilization' of hydrophobic molecules
Epichlorohydrin cross-linked starch (Section 12.2.4.5) – controlled/sustained release of pharmaceutical and other products
Oxidized starches (Section 12.2.7) – food flavors
Porous starch (Section 12.3)
Starch octenylsuccinates (Section 12.2.5.2) – controlled/sustained release of flavors, pigments, and antioxidants in food and pharmaceutical products; 'solubilization' of hydrophobic molecules
Starch dextrins (Section 12.2.8.2) – encapsulation of food ingredients
Starch-graft-copolymers (Section 12.2.10)

12.6.2 Products with Greater Ability to Coat and Form Films or to Form Coatings or Films with Improved Properties

Acid-modified starches (Section 12.2.8.1) – paper products: sizing agents, coating binders
Carboxymethylstarch (Section 12.2.4.4)
Destructured/destructurized starches (Section 12.4.2.1) – water-repellent films
Fatty acid esters (Section 12.2.5.4) – biodegradable films
Oxidized starches (Section 12.2.7)
Starch acetates (high DS) (Section 12.2.5.1) – biodegradable films
Starch dextrins (Section 12.2.8.2) – food products: adhesives, films
Starch-graft-copolymers (Section 12.2.10) – paper products: sizing agents
Starch maleates (Section 12.2.5.5)
Starch octenylsuccinates (high DS) (Section 12.2.5.2)
Starch propionates and butyrates (Section 12.2.5.5) – biodegradable films

12.6.3 Products with Increased Solubility and Greater Water-Holding Capacity

Acid-modified starches (Section 12.2.8.1) – pharmaceutical tablet disintegrants/excipients
Carboxymethylstarch (Section 12.2.4.4) – pharmaceutical tablet disintegrants/excipients

Hydroxyethylstarch (Section 12.2.4.1) – paper production: water retention

Hydroxypropylstarch (Section 12.2.4.2) – food products: reduction in syneresis during refrigerated storage and freeze–thaw cycles

Monostarch phosphates (Section 12.2.5.7) – food products

Pregelatinized starches (Section 12.4.2.1) – food products: moisture control

Starch acetates (Section 12.2.5.1) – food products; pharmaceutical tablet disintegrants/excipients

Starch-graft-copolymers (Section 12.2.10) – superabsorbents; pharmaceutical tablet disintegrants/excipients

12.6.4 Products with Increased Hydrophobicity/Lipophilicity/Oil Absorption Capacity

Benzylstarch (Section 12.2.4.6)

Fatty acid esters (Section 12.2.5.4)

Starch esters (high DS) (Section 12.2.5.4)

Starch octenylsuccinates (Section 12.2.5.2) – food products

12.6.5 Products That Are Thermoplastic

Destructured/destructurized starches (Section 12.4.2.1)

Fatty acid esters of starches (Section 12.2.5.5)

Starch acetates (relatively high DS) (Section 12.2.5.1)

Starch carbamates (Section 12.2.5.4)

Starch-graft-copolymers (Section 12.2.10)

Starch maleates (Section 12.2.5.5)

Starch propionates and butyrates (Section 12.2.5.4)

12.6.6 Products with Other Attributes

Carboxymethylstarches (Section 12.2.4.4) – electrolytes in lithium-ion batteries

Destructured/destructurized starches (Section 12.4.2.1) – carriers for pharmaceuticals and agrochemicals

Dialdehyde starch (Section 12.2.7.3) – anticancer drug delivery, cross-linking of proteins

Hydroxyethylstarches (Section 12.2.4.1) – plasma volume expanders; cryoprotectants; carriers for drug delivery

Oxidized starches (Section 12.2.7) – food products: fat replacers; papermaking: pigment dispersers

Starch acetates (high DS) (Section 12.2.5.1) – filaments; lithium-ion batteries

Starch-graft-copolymers (Section 12.2.10) – compatibilizers for polymer blends

Starch dextrins (Section 12.2.8.2) – providers of crispness and browning in bakery products

Starch sulfates (Section 12.2.5.9) – anticoagulants; antiviral activities

Starch xanthides (Section 12.2.5.5) – reinforcing agents in rubber; slow-release agents for agricultural pesticides

REFERENCES

Ab'lah, N, Yusuf, CYL, Rojsitthisak, P & Wong, TW 2023. Reinvention of starch for oral drug delivery system design. *International Journal of Biological Macromolecules, 241,* 124506.

Aboubacar, A, Moldenhauer, KAK, McClung, AM, Beighley, DH & Hamaker, BR 2006. Effect of growth location in the United States on amylose content, amylopectin fine structure, and thermal properties of starches of long grain rice cultivars. *Cereal Chemistry, 83,* 93–98.

Altuna, L, Herrera, ML & Foresti, ML 2018. Synthesis and characterization of octenyl succinic anhydride modified starches for food applications: A review of recent literature. *Food Hydrocolloids, 80,* 97–110.

Ashogbon, AO 2021. Dual modification of various starches: Synthesis, properties and applications. *Food Chemistry, 342,* 128325.

Auzely-Velty, R & Rinaudo, M 2003. Synthesis of starch derivatives with labile cationic groups. *International Journal of Biological Macromolecules, 31,* 123–129.

Banger, SP, Ashogbon, AO, Singh, A, Chaudhary, V & Whiteside, WS 2022. Enzymatic modification of starch: A green approach for starch applications. *Carbohydrate Polymers, 287,* 119265.

BeMiller, JN 2018. Chapter 18: Physical modification of starch. In M Sjöö & L Nillson (eds.), *Starch in Food*, 2nd Ed, Elsevier: Amsterdam, 223–253,

BeMiller, JN 2019. *Carbohydrate Chemistry for Food Scientists*, 3rd Ed, Woodhead Publishing.

BeMiller, JN & Fang, F 2025. Chapter 5: Chemical modification of starch. In M Miao & O Campanella (eds.), *Starch-Based Materials: Principles and Applications*, Elsevier. (a review with 370 references; submitted 2022).

BeMiller, JN & Huber, KC 2015. Physical modification of food starch functionalities. *Annual Review of Food Science and Technology, 6,* 19–69.

Berardi, A, Janssen, PHM & Dickhoff, BHJ 2022. Technical insight into potential functional-related characteristics (FRCs) of sodium starch glycolate, croscarmellose sodium and crospovidone. *Journal of Drug Delivery Science and Technology, 70,* 103261.

Bertoft, E, Annor, GA, Shen, X, Rumpagaporn, P, Seetharaman, K & Hamaker, BR 2016. Small differences in amylopectin fine structure may explain large functional differences of starch. *Carbohydrate Polymers, 140,* 113–121.

Bismark, S, Zhu, Z & Benjamin, T 2018. Effects of differential degree of chemical modification on the properties of modified starches: Sizing. *Journal of Adhesion, 94,* 97–123.

Bohrisch, J, Vorwerg, W & Radosta, S 2004. Development of hydrophobic starch. *Starch/Stärke, 56,* 322–329.

Butterfly Srl 1990. Destructured-starch-based compositions for biodegradable plastic moldings. *European Patent Organization, EP 400531.*

Cahyana, Y, Verrell, C, Kriswanda, D, Aulia, GA, Yusra, NA, Marta, H, Sukri, N, Esirgapovich, SJ & Abduvakhitovna, SS 2023. Properties comparison of oxidized and heat moisture treated (HMT) starch-based biodegradable films. *Polymers, 15*, 2046.

Cao, F, Lu, S, Wang, L, Zheng, M & Quek, SY 2023. Modified porous starch for enhanced properties: Synthesis, characterization and applications. *Food Chemistry, 415*, 135765.

Castro, LMG, Azhulexandre, EMC, Saraiva, JA & Pintado, M 2020. Impact of high pressure on starch properties: A review. *Food Hydrocolloids, 106*, 105877.

Chen, G, Yang, B, Wang, J, Zhang, Y, Li, S & Chen, Y 2023. Facile production of low-viscosity mung bean starch through cold plasma treatment: Improvement of film formability. *Journal of Food Science, 88*, 2053–2063.

Chen, J, Wang, Y, Liu, Y & Xu, X 2020. Preparation, characterization, physicochemical property and potential application of porous starch: A review. *International Journal of Biological Macromolecules, 148*, 1169–1181.

Chen, Q, Yu, H, Wang, L, ul Abdin, Z, Chen, Y, Wang, J, Zhou, W, Yang, X, Kahn, RU, Zhang, H & Chen, C 2015. Recent progress in chemical modification of starch and its applications. *RSC Advances, 5*, 67459–67474.

Chen, Y-F, Kaur, L & Singh, J 2018. Chemical modification of starch. In M Sjöö & L Nilsson (eds.), Starch in Food, 2nd Ed, Woodhead Publishing, Duxford, pp. 283–321.

Chiu, C-W & Solarek, D 2009. Modification of starches. In J BeMiller & RL Whistler (eds.), *Starch: Chemistry and Technology*, 3rd Ed, Academic Press, San Diego, CA, pp. 629–655.

Cho, K-Y & Lim, S-T 1998. Preparation and properties of benzyl corn starches. *Starch/Stärke, 50*, 250–257.

Cui, D, Liu, M, Liang, R & Bi, Y 2007. Synthesis and optimization of the reaction conditions of starch sulfates in aqueous solutions. *Starch/Stärke, 59*, 91–98.

Cui, D, Liu, M, Zhang, B, Gong, H & Bi, Y 2011. Optimization of reaction conditions for potato starch sulphate and its chemical and structural characterization. *Starch/Stärke, 63*, 354–363.

Dominguez-Ayala, JE, Soler, A, Mendez-Montealvo, G & Valazquez, G 2022. Supramolecular structure and technofunctional properties of starch modified by high hydrostatic pressure (HHP): A review. *Carbohydrate Polymers, 291*, 119609.

Du, Z, Li, X, Zhao, X & Huang, Q 2024. Multi-scale structural disruption induced by radio-frequency air cold plasma accelerates enzymatic hydrolysis/hydroxypropylation of tapioca starch. *International Journal of Biological Macromolecules, 260*(Part 2), 129572.

Fan, Y & Picchioni, F 2020. Modification of starch: A review on the application of "green" solvents and controlled functionalization. *Carbohydrate Polymers, 241*, 116350.

Federal Regulations 1978. *Food starch, modified, Title 21, 172,892*, in *Federal Register, 43*, no. 11697.

Fonseca, LM, El Halal, SLM, Dias, ARG & Zavareze, EDR 2021. Physical modification of starch by heat-moisture treatment and annealing and their applications: A review. *Carbohydrate Polymers, 274*, 118665.

Galkowska, G, Kapuśniak, K & Juszczak, L 2023. Chemically modified starches as food additives. *Molecules, 28*, 7543.

Gangoiti, J, Corwin, SF, Lamothe, LM, Vafiadim, C, Hamaker, BR & Dijkhuizen, L 2020. Synthesis of novel α-glucans with potential health benefits through controlled glucose release in the human gastrointestinal tract. *Critical Reviews in Food Science and Nutrition, 60*, 123–146.

Garcia, MAVT, Garcia, CF & Faraco, AAG 2020. Pharmaceutical and biomedical applications of native and modified starch: A review. *Starch/Stärke, 72*, 190270.

Golachowski, A, Zięba, T, Kapelko-Żeberska, M, Drożdż, W, Gryszkin, A & Grzechac, M 2015. Current research addressing starch acetylation. *Food Chemistry, 176*, 350–356.

González, K, García-Astrain, C, Santamaria-Echart, A, Ugarte, L, Avérous, L, Eceiza, A & Gabilondo, N 2018. Starch/graphene hydrogels via click chemistry with relevant electrical and antibacterial properties. *Carbohydrate Polymers, 202*, 372–381.

Grgić, I, Ačkar, D, Barišić, V, Vlainić, M, Knežević, N & Knežević, ZM 2019. Nonthermal methods for starch modification – A review. *Journal of Food Processing and Preservation. 43*, e14242.

Guo, Z, Gou, Q, Yang, L, Yu, Q-L & Han, L 2022. Dielectric barrier discharge plasma: A green method to change structure of potato starch and improve physicochemical properties of potato starch films. *Food Chemistry, 370*, 130992.

Han, J-A & BeMiller, JN 2007. Preparation and physical properties of slowly digesting modified food starches. *Carbohydrate Polymers, 67*, 366–374.

Han, J-A, Gray, JA, Huber, KC & BeMiller, JN 2006. Derivatization of starch granules as influenced by the presence of channels and reaction conditions. In ML Fishman, PX Qi & L Wicker (eds.), *Advances in Biopolymers: Molecules, Clusters, Networks and Interactions*, American Chemical Society, Washington, DC, pp. 165–184.

Han, Z, Shi, R & Sun, D-W 2020. Effects of novel physical processing techniques on the multi-structures of starch. *Trends in Food Science & Technology, 97*, 126–135.

Haq, F, Yu, H, Wang, L, Teng, L, Haroon, M, Khan, RU, Mehmood, S, Bilal-Ul-Amin, Ullah, RS, Khan, A & Nazir, A 2019. Advances in chemical modifications of starches and their applications. *Carbohydrate Research, 476*, 12–35.

Haroon, M, Wang, L, Yu, H, Abbasi, NM, Zain-ul-Abdin, Saleem, M, Rizwan, UK, Ullah, RS, Chen, Q & Wu, J 2016. Chemical modification of starch and its use as an adsorbent material. *RSC Advances, 6*, 78264–78285.

He, J, Liu, J & Zhang, G 2008. Slowly digestible waxy maize starch prepared by octenylsuccinic anhydride esterification and heat-moisture treatment: Glycemic response and mechanism. *Biomacromolecules, 9*, 175–184.

Hong, L-F, Cheng, L-H, Gan, C-Y, Lee, CY & Peh, KK 2018. Evaluation of starch propionate as emulsion stabiliser in comparison with octenylsuccinate starch. *LWT – Food Science and Technology, 91*, 526–531.

Hong, Y, Liu, G & Gu, Z 2016. Recent advances of starch-based excipients used in extended release tablets: A review. *Drug Delivery, 23*, 12–20.

Hoover, R 2010. The impact of heat-moisture treatment on molecular structures and properties of starches isolated from different botanical sources. *Critical Reviews of Food Science, 50*, 835–847.

Huang, Q, Fu, X, He, X-W, Luo, F-X, Yu, S-J & Li, L 2010. The effect of enzymatic pretreatments on subsequent octenyl succinic anhydride modifications of cornstarch. *Food Hydrocolloids, 24*, 60–65.

Iuga, M & Mironeasa, S 2020. A review of the hydrothermal treatments impact on starch based systems. *Critical Reviews in Food Science and Nutrition, 60*, 3890–3915.

Jacobs, H & Delcour, JA 1998. Hydrothermal modifications of granular starch, with retention of the granular structure: A review. *Journal of Agricultural and Food Chemistry, 46*, 2895–2905.

Jayakody, L & Hoover, R 2008. Effect of annealing on the molecular structure and physicochemical properties of starches from different botanical origins – A review. *Carbohydrate Polymers, 74*, 691–703.

Jyothi, AN 2010. Starch graft copolymers: Novel applications in industry. *Composite Interfaces, 17*, 165–174.

Kapelko-Żeberska, M, Buksa, K, Szumny, A, Zięba, T & Gryszkin, A 2016. Analysis of molecular structure of starch citrate obtained by a well-established procedure. *LWT – Food Science and Technology, 69*, 334–341.

Karić, N, Vukčević, M, Maletić, M, Dimitrijević, S, Ristić, M, Grujić, AP & Trivunac, K 2023. Physico-chemical, structural, and adsorption properties of amino-modified starch derivatives for removal of (in)organic pollutants from aqueous solutions. *International Journal of Biological Macromolecules, 241,* 124527.

Karma, V, Gupta, AD, Yadav, DK, Singh, AA, Verma, M & Singh, H 2022. Recent developments in starch modification by organic acids: A review. *Starch/Stärke, 74,* 2200025.

Kumar, S, Sharma, B, Thakur, K, Bhardwaj, TR, Prasad, DN & Singh, RK 2019. Recent advances in the development of polymeric nanocarrier formulations for the treatment of colon cancer. *Drug Delivery Letters, 9,* 2–14.

Latip, DNH, Samsudin, H, Utra, U & Alias, AK 2021. Modification methods toward the production of porous starch. *Critical Reviews in Food Science and Nutrition, 61,* 2841–2862.

Leandro, GC, Laroque, DA, Monteiro, AR, Carciofi, BAM & Valencia, GA 2024. Current status and perspectives of starch powders: A review. *Journal of Polymers and the Environment, 32,* 510–523.

Lele, VV, Kumari, S & Niju, H 2018. Syntheses, characterization and applications of graft copolymers of sago starch – A review. *Starch/Stärke, 53,* 7–13.

Lemos, PVF, Marcelino, HR, Cardoso, LG, de Souza, CO & Druzian, JI 2021. Starch chemical modifications applied to drug delivery system; from fundamentals to FDA-approved raw materials. *International Journal of Biological Macromolecules, 184,* 218–234.

Li, H, Qi, Y, Zhao, Y, Chi, J & Cheng, S 2019. Starch and its derivatives for paper coatings: A review. *Progress in Organic Coatings, 135,* 213–227.

Liang, Y, Zheng, L, Yang, Y, Zheng, X, Xiao, D, Ai, B & Sheng, Z 2024. Dielectric barrier discharge cold plasma modifies the multiscale structure and properties of banana starch. *International Journal of Biological Macromolecules, 264*(Part 1), 130462.

Majzoobi, M & Farahnaky, A 2021. Granular cold-water swelling starch: Properties, preparation and applications, a review. *Food Hydrocolloids, 111,* 106393.

Maniglia, BC, Castanha, N, Le-Bail, P, Le-Bail, A & Augusto, PED 2021a. Starch modification through environmentally friendly alternatives: A review. *Critical Reviews in Food Science and Nutrition, 61,* 2482–2505.

Maniglia, BC, Castanha, N, Rojas, ML & Augusto, PED 2021b. Emerging technologies to enhance starch performance. *Current Opinion in Food Science, 37,* 26–36.

Masina, N, Choonara, YE, Kumar, P, du Toit, LC, Govender, M, Indermun, S & Pillay, V 2017. A review of chemical modification techniques of starch. *Carbohydrate Polymers, 157,* 1226–1236.

Mason, WR 2009. Starch use in food. In J BeMiller & RL Whistler (eds.), *Starch: Chemistry and Technology*, 3rd Ed, Academic Press, San Diego, CA, pp. 745–795.

Meimoun, J, Wiatz, V, Saint-Loup, R, Parcq, J, Favrelle, A, Bonnet, F & Zinck, P 2018. Modification of starch by graft copolymerization. *Starch/Stärke, 70,* 1600351.

Miao, M & BeMiller, JN 2023. Enzymatic approaches for structuring starch to improve functionality. *Annual Review of Food Science and Technology, 14,* 271–295.

Miao, M, Jiang, B, Jin, Z & BeMiller, JN 2018. Microbial starch-converting enzymes: Recent insights and perspectives. *Comprehensive Reviews in Food Science and Food Safety, 17,* 1238–1260.

Okyere, AY, Rajendran, S & Annor, GA 2022. Cold plasma technologies: Their effect on starch properties and industrial scale-up for starch modification. *Current Research in Food Science, 5,* 451–463.

Otache, MA, Duru, RU, Achugasim, O & Abayeh, OJ 2021. Advances in the modification of starch via esterification for enhanced properties. *Journal of Polymers and the Environment, 29,* 1365–1369.

Park, S & Kim, Y-R 2021. Clean label starch: Production, physico-chemical characteristics, and industrial applications. *Food Science and Biotechnology, 30,* 1–17.

Park, SH, Na, Y, Kim, J, Kang, SD & Park, K-H 2018. Properties and applications of starch modifying enzymes for use in the baking industry. *Food Science and Biotechnology, 27,* 299–312.

Pooresmaeil, M & Namazi, H 2021. Developments on carboxymethyl starch-based smart systems as promising drug carriers: A review. *Carbohydrate Polymers, 258,* 117654.

Portillo-Perez, GA, Skov, KB & Martinez, MM 2024. Starch esterification using deep eutectic solvents as chaotropic agents and reaction promoters. *Green Chemistry, 26,* 2225–2240.

Prasher, P & Sharma, M 2023. Dialdehyde starch nanoparticles: An emerging material for anticancer drug delivery. *Nanomedicine, 18,* 849–854.

Raghunathan, R, Pandiselvam, R, Kothakota, A & Khaneghah, AM 2021. The application of emerging non-thermal technologies for the modification of cereal starches. *LWT – Food Science and Technology, 138,* 110795.

Ramadan, MF & Sitohy, MZ 2020. Phosphorylated starches: Preparation, properties, functionality, and techno-applications. *Starch/Stärke, 72,* 1900302.

Rivard, C, Moens, L, Roberts, K, Brigham, J & Kelley, S 1995. Starch esters as biodegradable plastics: Effects of ester group chain length and degree of substitution on anaerobic biodegradation. *Enzyme and Microbial Technology, 17,* 848–852.

Rutenberg, MW & Solarek, D 1984. Starch derivatives: Production and uses. In RL Whistler, JN BeMiller & EF Paschall (eds.), *Starch: Chemistry and Technology*, 2nd Ed, Academic Press, Orlando, FL, pp. 311–388.

Ryu, J-J, Li, X, Lee, E-S, Li, D & Lee, B-H 2022. Slowly digestible property of highly branched α-limit dextrins produced by 4,6-α-glucanotransferase from *Streptococcus thermophilus* evaluated *in vitro* and *in vivo*. *Carbohydrate Polymers, 275,* 118685.

Sagar, AD & Merrill, EW 1995. Properties of fatty acid esters of starch. *Journal of Applied Polymer Science, 58,* 1647–1656.

Schafranski, K, Ito, VC & Lacerda, LG 2021. Impacts and potential applications: A review of the modification of starches by heat-moisture treatment (HMT). *Food Hydrocolloids, 117,* 106690.

Sharma, M, Aguado, R, Murtinho, D, Valente, AJM & Ferreira, PJT 2021a. Novel approach on the synthesis of starch betainate by transesterification. *International Journal of Biological Macromolecules, 182,* 1681–1689.

Sharma, M, Aguado, R, Murtinho, D, Valente, AJM & Ferreira, PJT 2021b. Synergetic effect of cationic starch (ether/ester) and Pluronics for improving inkjet printing quality of office papers. *Cellulose, 28,* 10609–10624.

Sharma, M, Aguado, R, Murtinho, D, Valente, AJM, Mendes de Sousa, AP & Ferreira, PJT 2020. A review on cationic starch and nanocellulose as paper coating components. *International Journal of Biological Macromolecules, 162,* 578–598.

Sharma, N, Pahuja, S & Sharma, N 2019. Immediate release tablets: A review. *International Journal of Pharmaceutical Sciences and Research, 10,* 3607–3618.

Shogren, RL, Viswanathan, A, Felker, F & Gross, RA 2000. Distribution of octenyl succinate groups in octenyl succinic anhydride modified waxy maize starch. *Starch/Stärke, 52,* 196–204.

Singh, J, Kaur, L & McCarthy, OJ 2007. Factors influencing the physico-chemical, morphological, thermal and rheological properties of some chemically modified starches for food applications – A review. *Food Hydrocolloids, 21,* 1–22.

Song, X, He, G, Ruan, H & Chen, Q 2006. Preparation and properties of octenyl succinic anhydride modified early *Indica* rice starch. *Starch/Stärke, 58,* 109–117.

Sorndech, W, Tongta, S & Blennow, A 2018. Slowly digestible- and non-digestible α-glucans: An enzymatic approach to starch modification and nutritional effects. *Starch/Stärke, 70,* 1700145.

Sui, Z, Huber, KC & BeMiller, JN 2013. Effects of the order of addition of reagents and catalyst on modification of maize starches. *Carbohydrate Polymers, 96,* 118–130.

Sun, C, Wei, Z, Xue, C & Yang, L 2023. Development, application and future trends of starch-based delivery systems for nutraceuticals: A review. *Carbohydrate Polymers, 308,* 120675.

Tan, W, Li, Q, Li, W, Dong, F & Guo, Z 2016. Synthesis and antioxidant property of novel 1,2,3-triazole-linked starch derivatives via 'click chemistry'. *International Journal of Biological Macromolecules, 82,* 404–410.

Taslikh, M, Abbasi, H, Mortazavian, AM, Ghasemi, JB, Naeimabadi, A & Nayebzadeh, K 2022. Effect of cold plasma treatment, cross-linking, and dual modification on corn starch. *Starch/Stärke, 74,* 2200008.

Tester, RF & Debon, SJJ 2000. Annealing of starch – A review. *International Journal of Biological Macromolecules, 27,* 1–12.

Tupa, M, Maldonado, L, Vazquez, A & Foresti, ML 2013. Simple organocatalytic route for the synthesis of starch esters. *Carbohydrate Polymers, 98,* 349–357.

Uliniuc, A, Popa, M, Drockenmuller, E, Boisson, F, Leonard, D & Hamaide, T 2013. Toward tunable amphiphilic copolymers via CuAAC click chemistry of oligocaprolactones onto starch backbone. *Carbohydrate Polymers, 96,* 259–261.

Vanier, NL, El Halal, SLM, Dias, ARG & Zavareze, EDR 2017. Molecular structure, functionality and applications of oxidized starches: A review. *Food Chemistry, 221,* 1546–1559.

Wang, N, Li, C, Miao, D, Hou, H, Dai, Y, Zhang, Y & Wang, B 2023. The effect of non-thermal physical modification on the structure, properties and chemical activity of starch: A review. *International Journal of Biological Macromolecules, 251,* 126200.

Wang, Q, Li, L & Zheng, X 2021. Recent advances in heat-moisture modified cereal starch: Structure, functionality and its application in starchy food systems. *Food Chemistry, 344,* 128700.

Wang, S & Copeland, L 2015. Effect of acid hydrolysis on starch structure and functionality: A review. *Critical Reviews in Food Science and Nutrition, 55,* 1081–1097.

Warner-Lambert Co. 1988. Destructurized starch and process for making same. *European Patent Organization,* EP 282451.

Warner-Lambert Co. 1989. Production of destructured modified starch. *United Kingdom Patent,* GB2214919.

Warner-Lambert Co. 1990. Destructurized starch manufacture. *European Patent Organization,* EP 391853.

Warner-Lambert Co. 1991. Polymer moldable blend compositions containing destructurized starch. *European Patent Organization,* EP 408503.

Wei, H, Li, W, Chen, H, Wen, X, He, J & Li, J 2020. Simultaneous Diels-Alder click reaction and starch hydrogel microsphere production via spray drying. *Carbohydrate Polymers, 241,* 116351.

Wei, Q, Zheng, H, Zhu, M, Han, X, Li, Y & Zhou, J 2021. Starch-based surface-sizing agents in paper industry: An overview. *Paper and Biomaterials, 6,* 54–61.

Wetzel, DL, Shi, Y-S & Reffner, JA 2010a. Synchrotron infrared confocal microspectroscopical detection of heterogeneity within chemically modified single starch granules. *Applied Spectroscopy, 64,* 282–285.

Wetzel, DL, Shi, Y-S & Schmidt, U 2010b. Confocal Raman and AFM imaging of individual granules of octenyl succinate modified and natural waxy maize starch. *Vibrational Spectroscopy, 53,* 173–177.

Wilson, P 2005. Surface sizing. In JM Gess & JM Rodriguez (eds.), *Sizing of Paper,* 3rd Ed, TAPPI Press, Atlanta, GA, pp. 211–236.

Wu, Z, Qiao, D, Zhao, S, Lin, Q, Zhang, B & Xie, F 2022. Nonthermal physical modification of starch: An overview of recent research into structure and property alterations. *International Journal of Biological Macromolecules, 203,* 153–175.

Wurzburg, OA (ed.) 1986. *Modified Starches: Properties and Uses.* CRC Press, Boca Raton, FL.

Ye, S-J & Biak, M-Y 2023. Characteristics of physically modified starches. *Food Science and Biotechnology, 32,* 875–883.

Yong, H & Liu, J 2024. Recent advances on the preparation conditions, structural characteristics, physicochemical properties, functional properties and potential applications of dialdehyde starch: A review. *International Journal of Biological Macromolecules, 259*(Part 1), 129261.

Zailani, MA, Kamilah, H, Husaini, A, Awang, S, Awang, ZR & Sarbini, SR 2023. Starch modifications via physical treatments and the potential in improving resistant starch content. *Starch/Stärke, 75,* 2200146.

Zavareze, EDR & Dias, ARG 2011. Impact of heat-moisture treatment and annealing in starches: A review. *Carbohydrate Polymers, 83,* 317–328.

Zhang, K, Zhang, Z, Zhao, M, Milosavljevic, V, Cullen, PJ, Scally, L, Sun, D-W & Tiwari, BK 2022. Low-pressure plasma modification of the rheological properties of tapioca starch. *Food Hydrocolloids, 125,* 107380.

Zhang, S, Liu, F, Peng, H, Peng, X, Jiang, S & and Wang, J 2015. Preparation of novel c-6 position carboxyl corn starch by a green method and its application in flame retardance of epoxy resin. *Industrial & Engineering Chemistry Research, 54,* 11944–11952.

Zhang, Y, Li, L, Sun, S, Cheng, L, Gu, Z & Hong, Y 2024. Structural characteristics, digestion properties, fermentation properties, and biological activities of butyrylated starch: A review. *Carbohydrate Polymers, 330,* 121825.

Zhang, Z, Shan, H, Chen, L, He, C, Zhuang, X & Chen, X 2013. Synthesis of pH-responsive starch nanoparticles grafted poly(L-glutamic acid) for insulin controlled release. *European Polymer Journal, 49,* 2082–2091.

Zhen, J & Ma, J 2002. Modification of starch and its application in leather making. *Journal of the Society of Leather Technologists and Chemists, 86,* 93–95.

Zhong, Y, Xu, J, Liu, X, Ding, L, Svensson, B, Herburger, K, Guo, K & Blennow, A 2022. Recent advances in enzyme biotechnology on modifying gelatinized and granular starch. *Trends in Food Science & Technology, 123,* 343–354.

Zhu, F 2018. Modifications of starch by electric field based techniques. *Trends in Food Science & Technology, 75,* 158–169.

Zhu, Q, Yao, S, Wu, Z, Li, D, Ding, T, Liu, D & Xu, E 2023. Hierarchical structural modification of starch via nonthermal plasma: A state-of-the-art review. *Carbohydrate Polymers, 311,* 120747.

Zia, F, Zia, KM, Zuber, M, Kamal, S & Aslam, N 2015. Starch based polyurethanes: A critical review updating recent literature. *Carbohydrate Polymers, 134,* 784–798.

Zia-ud-Din, Xiong, H & Fei, P 2017. Physical and chemical modification of starches: A review. *Critical Reviews in Food Science and Nutrition, 57,* 2691–2705.

Starch Bioconversion Products 13

Yong Wang and Sang-Ho Yoo

13.1 STARCH SYRUPS

13.1.1 Introduction

Starch syrup is a carbohydrate solution with multiple molecular weights produced by hydrolyzing starch to different degrees (Harni et al. 2021). Its main components are glucose, maltose, oligosaccharides, dextrin, etc. Starch syrups with different degrees of hydrolysis can be broadly categorized based on their dextrose equivalents (DE) (Chronakis 1998). The DE value of starch hydrolysate ranges from 0 to 100. As the degree of hydrolysis increases, the DE value rises, leading to a decrease in the degree of polymerization (DP) of the resulting starch oligomers (Biliaderis et al. 1999; Levine & Slade 1986). Starch solution has a value of 0, while dextrose solution has a DE value of 100. Its physical properties are manifested

as a colorless, transparent, and viscous liquid that is mainly used in candies, breads, canned goods, preserves, and medical syrups (Birch & Etheridge 1973; Hobbs 2009; Mohamed & Babucurr 2015). The main crops used for starch syrup production vary from place to place due to their geographical location. Currently, syrup production predominantly utilizes corn, wheat, and cassava starches. Nevertheless, alternative sources such as potato, barley, rice, and palm are also used in manufacturing (Eke-Ejiofor 2015; Lin et al. 2013; Linko et al. 1983; Nikvarz et al. 2021; Pontoh & Low 1995).

In the actual production process, various starch syrups are mainly prepared by enzymatic methods. Figure 13.1 provides a brief overview of the heat treatment, liquefaction, and saccharification of starch used to prepare starch syrup. Starch is hydrolyzed into small molecular maltodextrin and oligosaccharides after liquefaction, and then the targeted syrup is prepared due to the specificity of the relevant enzymes during the saccharification process (Li et al. 2023). Depending on the

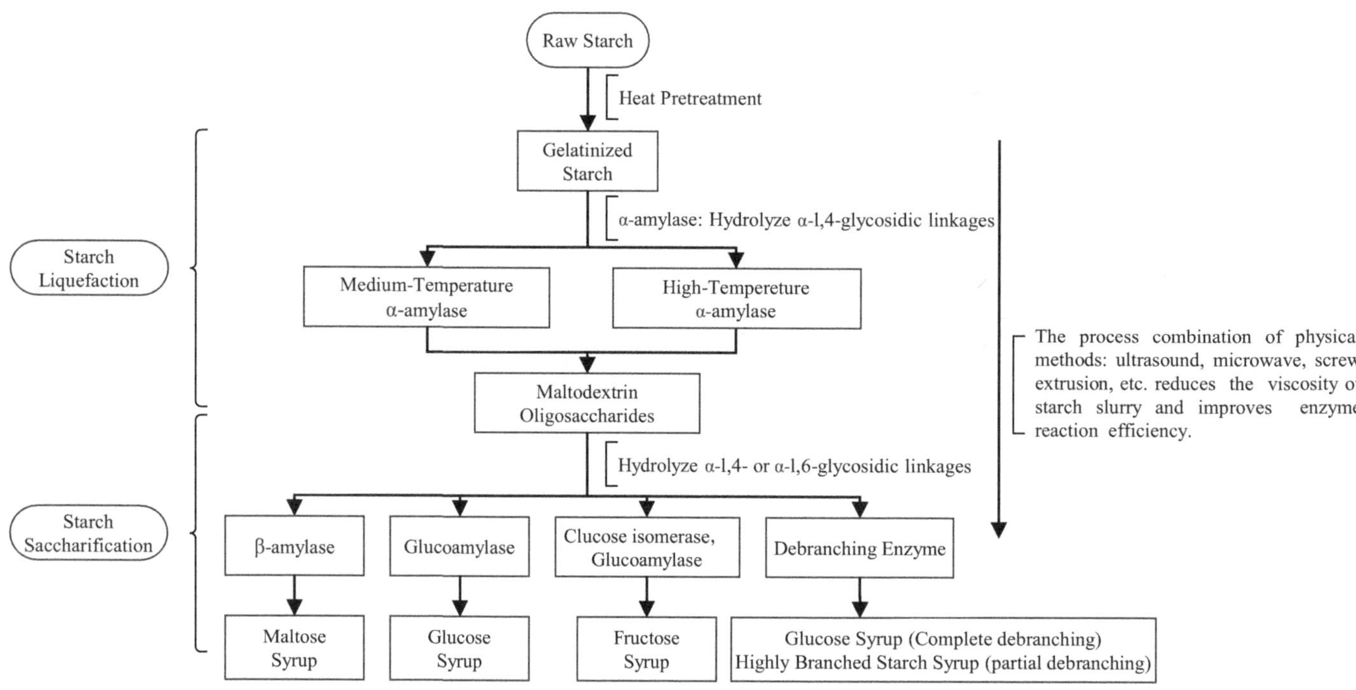

FIGURE 13.1 A brief process for the enzymatic preparation of starch syrup. (Made by the author.)

DOI: 10.1201/9781003464396-13

degree of hydrolysis, it is divided into low- (DP < 20), medium-, and high-conversion syrups. Low-conversion syrup, which can also be categorized as maltodextrin, consists mainly of dextrin and contains polysaccharides, oligosaccharides, and relatively small amounts of maltose and glucose. And high-conversion syrup as a dextrose, on particular saccharide in the syrup can be classified as high maltose and glucose. The physical and chemical characteristics of a syrup, including sweetness, browning, and viscosity, are typically influenced by specific groups of sugars or glycans (Lees 2012).

13.1.2 Dextrose

Dextrose, the most prevalent saccharide in nature, is abundant in honey, fruits, vegetables, and berries, either as a free monosaccharide or in a polymerized form of anhydrodextrose units. As a glucose polymer, glucose is found in starch, cellulose, and glycogen (Kim et al. 2004; Norman 1981). Glucose, which is colorless, highly soluble, and sweet, is a crucial and essential energy source for living organisms. In the food industry, it serves as a sweetener, preservative, and flavoring agent. Additionally, dextrose is extensively utilized in the pharmaceutical sector for intravenous nutrition, tableting, and various formulations. It is considered an important biological small molecule. Most of the commercially available dextrose is in the form of syrups, and some is sold in granular solid form (crystallized glucose) (Galant et al. 2015; Tirone & Brunicardi 2001). In solution, glucose occurs as two stereoisomers, specifically the α- and β-anomers. In aqueous solution, glucose exists in two forms that undergo reversible isomerization through open- and closed-loop reactions. These reactions play a crucial role in the metabolism of organisms, allowing glucose to adapt reversibly to the metabolic environment of the organism in different forms (Wasserman 2009). Glucose, as an important energy source in living organisms, is involved in numerous metabolic processes, including glycolysis, gluconeogenesis, glycogen synthesis and degradation, and tricarboxylic acid cycle reactions, by reacting with aldehyde, primary alcohol, secondary alcohol, and polyhydric alcohol, and it is an important source of energy and carbon for living organisms (Kearsley & Dziedzic 1995; Shendurse & Khedkar 2016).

By introducing heat-resistant bacterial α-amylase, based on the enzyme-enzyme process, acidic degradation products are effectively eliminated. This leads to decreased costs for chemical treatment and refining, a substantial boost in glucose production, and a reduction in refining requirements. In the enzymatic preparation of glucose, the starch slurry is first hydrolyzed (liquefied) by a thermostable α-amylase at pH 6–6.5 (Hua & Yang 2016). Subsequently, the substrate undergoes saccharification, where it is hydrolyzed into glucose by glucoamylase at an optimal pH.

13.1.3 Maltose Syrup

Maltose is a disaccharide formed by two glucose monomers joined through an α-1,4-glycosidic bond. It is a nutritional disaccharide produced from starch or amylose through processes such as liquefaction, saccharification, and refining. Its molecular formula is $C_{12}H_{22}O_{11} \cdot H_2O$, with a molecular weight of 360.31 (Li et al. 2022). Maltose appears as white crystals or crystalline powder, is highly soluble in water, and exhibits slight solubility in ethanol; sweetness is only about 40% of sucrose. Maltose, a product of deep starch hydrolysis, typically contains 40%–90% maltose on a dry basis. Maltose syrup possesses attributes including minimal viscosity, crystallization resistance, mild sweetness, and high stability. Due to these characteristics, maltose finds extensive applications in the pharmaceutical field (antibiotics, vaccines, and osmotic agents) and the food industry (food fermentation, sweeteners, and anti-crystallization agents) (Takusagawa & Jacobson 1978; Vaclavik et al. 2021).

α-Amylase from various sources is used during the liquefaction stage of the preparation of maltose syrup to form maltodextrins with varying molecular weights. Subsequently, β-amylases are used to separate maltose from maltodextrin molecules during the liquefaction process, and through precise control of reaction conditions, the desired maltose syrup is finally obtained (Mironescu et al. 2009). Based on the degree of hydrolysis, maltose syrup can be classified into normal maltose syrup (maltose dry base ≤ 60%), high maltose syrup (60%–80% of maltose dry base), and ultra-high maltose syrup (maltose dry base ≥ 80%) (Sun et al. 2014). Lin et al. (2013) pointed out that, based on previous research, the optimal DE range of maltose syrup starch during the liquefaction process is 8–11. When the DE value exceeds 11, maltose production declines significantly. On the other hand, if the DE value is under 8, viscosity rises sharply, adversely affecting the saccharification process. Li et al. (2021) utilized a substrate consisting of 50% maltodextrin and observed that as the concentration increased, maltose levels initially rose and then declined. This suggests a reduction in enzyme efficiency and the movement of water within the system. Maltose syrup types depend on their intended application and are mainly determined by the specific set of fermenters used in the process. High maltose and ultra-high maltose syrups are currently the most researched due to their higher sweetness, anti-crystallization properties, and lower viscosity (Doyle et al. 1989; Lin et al. 2013; Olsen 2002).

13.2 FRUCTOSE

13.2.1 Introduction

Fructose, as a monosaccharide, together with glucose, forms sucrose. The chemical formula is $C_6H_{12}O_6$, which is an isomer of hexanal. In contrast to glucose, the hexanone group is located on the second carbon atom in the molecular structure of fructose. The free form of fructose is found in large quantities in honey and a few other fruits such as figs, apples, pears, grapes, and other berries (Roeb & Weiskirchen 2021; Stricker et al. 2021). Fructose is stored in plants in the form of polysaccharides and serves as an energy source for overwintering in many plants (including chicory, Allium cepa, and asparagus). Fructose crystallizes as β-D-fructopyranose, possessing a molecular weight of 180 and a melting point ranging from 102°C to 104°C. At equilibrium at 25°C, fructose has a solubility of 80% in water (Hurtta et al. 2004).

Fructans are a class of oligosaccharides and polysaccharides where fructose units are polymerized via β-(2,1)- and/or β-(2,6)-glycosidic bonds to a sucrose molecule (Yoshida 2021). Fructans are mainly found in plants in the form of inulin, levan-type of fructan, and graminan.

Inulin, a functional sugar with known health benefits, is a linear fructan characterized by the polymerization of fructosyl units via β-(2,1) linkages, with the fructose chains attached to the fructose moiety of sucrose (Ni et al. 2024; Yoshida 2021). Unlike inulin, the new inulin variant has its fructose chain connected to the glucose in sucrose through β-(2,6) bonds (Anderson-Dekkers et al. 2021; Hughes et al. 2022). Levan is a type of fructan where fructose units are polymerized through β-(2,6) linkages and are attached to the fructose in sucrose. It is primarily derived from sources such as wheat and barley (Gallagher et al. 2007; Yoshida 2021). Graminan, another type of fructan with both β-(2,1) and β-(2,6) linkages to sucrose, accumulates in the vegetative tissues of plants like wheat and barley (Yoshida 2021). Fructans are complex in structure and diverse in form, and the large amount of fructans in plant tissues serves as a crucial energy source for plants during overwintering. The content and type of fructans in plants are also closely related to the freezing tolerance of plants (Bancal et al. 1991; Yoshida 2021).

13.2.2 Properties and Applications

Fructose participates in numerous metabolic reactions, predominantly occurring in the liver (Mayes 1993). When fructose is consumed by the body, it is primarily actively absorbed of the small intestine's cells, where it eventually finds its way into the liver. In the liver, fructose first undergoes a phosphorylation reaction catalyzed by fructokinase to produce fructose-1-phosphate. The resulting product undergoes a splitting reaction catalyzed by fructose-1-phosphate aldolase to produce dihydroxyacetone phosphate and glyceraldehyde-3-phosphate. The latter enters the glycolytic pathway, where it is converted to pyruvate, and can subsequently enters the citric acid cycle, ultimately contributing to the production of ATP, which provides energy for the body (Adeva-Andany et al. 2016).

Current research on fructose primarily focuses on its impact on various aspects of human health, including metabolism, obesity, weight control, cardiovascular conditions such as hypertension and hypercholesterolemia, gut microbiota, and inflammation. Table 13.1 succinctly compiles information regarding the metabolic effects of consuming a diet rich in fructose. While the existing body of research extensively explores fructose's physiological effects, there is limited investigation into its application as a substrate for the synthesis of functional carbohydrates. Park et al.'s study (2016), indicating a notable enhancement in turanose yield through the

incorporation of exogenous fructose during the synthesis process, underscores the potential for exploring fructose's role in producing functional carbohydrates. This aspect of research holds promise for further exploration and scientific inquiry.

Due to the widespread use of high fructose syrups in the food processing industries such as carbonated beverages, juice drinks, confectionery, and canned foods, increasing attention has been directed toward the link between high fructose syrups and various health issues, including obesity and metabolic syndrome. Recent epidemiologic and biochemical studies provide substantial evidence that excessive fructose consumption is a significant contributing factor in the development of metabolic syndrome. (Rutledge & Adeli 2007). Research on the human health effects of fructose has long been controversial, and in 2004, a study was published on the correlation between HFS and human obesity in the USA, based on the fact that the growing rate of obesity in the USA has been linked to the growing use of HFS in food and beverages (Bray et al. 2004). This correlation has not been confirmed by clear and strong data, but it has caused lasting damage to the reputation of sweeteners. Since then, scientists have been keen to study the health effects of fructose. Research has shown that consuming fructose and glucose results in similar metabolic responses, is deemed safe at normal intake levels, and does not negatively impact human health (Bray 2008; Forbes & Bowman 1993; White 2013). Recent research has also indicated that ingested fructose levels lead to changes in the metabolic capacity of the intestinal microbiota, profoundly influencing the intestinal microbiota by altering the composition and size of the intestinal flora and affecting the metabolic load of the intestinal flora (Drożdż et al. 2021; Montrose et al. 2021; Payne et al. 2012; Staudacher & Whelan 2017; Wilder-Smith et al. 2017; Zubiría et al. 2017). Furthermore, some studies indicate that a dose of fructose can reduce blood sugar levels in individuals with type 2 diabetes, thereby restoring the hyperglycemia-induced suppression of hepatic glucose production (Gupta et al. 2023; Hawkins et al. 2002; Moore et al. 2000; Moore et al. 2001). It has been observed that when actively absorbed in the intestine through the GLUT5 channel, fructose works synergistically with glucose, promoting fat storage, and subsequently regulates the body's carbohydrate metabolism (Du & Heaney 2012; Litherland et al. 2004).

13.2.3 Manufacture

The application of fructose syrup has a long history and is widely used globally. Today, the production process of crystallized fructose has gradually matured, representing an overall advanced level in the sugar industry. The enzymatic preparation of fructose includes three primary methods: hydrolyzing starch followed by isomerizing the resulting glucose residues, hydrolyzing sucrose, and hydrolyzing inulin (Lima et al. 2011). During starch hydrolysis, enzymes catalytically break down starch into glucose, and the resulting glucose residues may undergo an isomerization process. On the other hand, the enzymatic hydrolysis of sucrose breaks it down into glucose and fructose. Simultaneously, inulin, serving as a polysaccharide substrate,

TABLE 13.1   Metabolic effects of high dietary fructose sugars

HEALTH EFFECTS	CONCLUSION	REFERENCES
Metabolic diseases	High-fructose diet can induce hypertriglyceridemia	Macdonald (1966); Reaven et al. (1979)
	Hyperinsulinemia and hypertriglyceridemia are associated with hypertension in fructose-fed rats	Hwang et al. (1987); Sleder et al. (1980)
	Fructose-fed rats develop hypertriglyceridemia by developing insulin resistance and hyperinsulinemia	Zavaroni et al. (1982)
	Prolonged fructose administration to normal rats induces hyperinsulinemia and insulin resistance *in vivo*	Bezerra et al. (2000); Thorburn et al. (1989); Tobey et al. (1982)
	High-fructose feeding in rats results in altered lipid metabolism and reduced insulin sensitivity	Basciano et al. (2005); Kelley et al. (2004)
	Fructose consumption has been found to have moderate adverse effects on cardiometabolic risk factors and substantially elevate liver fat content in individuals with abdominal obesity	Taskinen et al. (2017)
	Dietary fructose specifically increases de novo lipogenesis, an atherogenic lipid profile, promotes dyslipidemia, decreases insulin sensitivity, and increases visceral adiposity in overweight/obese adults	Schwarz et al. (2017); Schwarz et al. (2015); Stanhope et al. (2009)
	When HFS is consumed, the interaction between fructose and glucose may lead to an increase in lipoprotein risk factors, thereby increasing the likelihood of developing cardiovascular disease and posing a potential threat to human health	Hieronimus et al. (2020); Stanhope et al. (2015)
Liver and intestinal fructose absorption	Glucose Transporter 5 (GLUT5) is essential for the intestinal absorption of dietary fructose	Barone et al. (2009); Douard and Ferraris (2008); Kim et al. (2017)
	Intestinal fructose catabolism attenuates fructose-induced hepatic lipogenesis	Jang et al. (2020)
	The synthesis and metabolic processing of endogenous fructose in the liver constitute a crucial pathway through which glucose plays a role in the onset of metabolic syndrome	Lanaspa et al. (2013)
	Fructose ingestion causes increased leaky gut, endotoxemia, and steatohepatitis with liver fibrosis	Cho et al. (2021)
	Excessive consumption of fructose, particularly in beverages, is a significant nutritional risk factor for metabolic-associated hepatic steatosis	Skenderian et al. (2020)

undergoes efficient hydrolysis into fructose through enzymatic catalysis. In this method, distinguished by its simplicity and forward-looking approach, the microbial expression of inulinase is enhanced by utilizing inulinase to catalyze the hydrolysis of inulin, with a specific emphasis on the cleavage of β-(2,1) linkages. This enzymatic strategy spans various microbial groups, including bacteria, yeast, and fungi. Nonetheless, strains associated with *Aspergilli* and *Kluyveromyces* sp. emerge as the most favored microbial sources for inulinase production, recognized for their proven efficiency and suitability in this enzymatic context (Singh et al. 2018). These three methods play a crucial role in fructose production, providing diversified production pathways to meet different raw materials and production needs (Amaral-Fonseca et al. 2021; Reilly & Antrim 2003; Ricca et al. 2007; Sánchez-Martínez et al. 2020; Walsh 2007).

Enzymatic preparation of fructose typically utilizes starch as a raw material, leveraging its abundant availability and cost-effectiveness. Initially, starch is converted into high fructose syrup through process steps such as liquefaction and saccharification. As high fructose syrup contains a significant amount of glucose, the conversion into fructose requires isomerization through the use of glucose isomerase (xylose isomerase) (Basso & Serban 2020; Buchholz & Seibel 2008; Schenck 2000; Sun et al. 2018; Swaisgood 2002). Alternatively, syrup obtained by hydrolysis of sucrose is also a source of fructose, and catalytic conversion using immobilized invertase can achieve this process more efficiently. The converted syrup produced is more easily utilized in commercial processes and offers greater economic benefits compared to sucrose (Chou & Jasovsky 1993).

13.2.4 Characterization

Fructose plays a vital role in determining the sweetness of fruits and the culinary quality of plant storage tissues such as potato tubers. Precisely tracking fructose levels is essential for predicting product quality. At present, the combination of high-performance liquid chromatography (HPLC) and high-sensitivity, high-resolution, and high-precision mass spectrometers, Evaporative Light Scattering Detection (ELSD) and refractive index detection (RID) are two widely used and highly regarded analytical techniques (Barreira et al. 2010; Duarte-Delgado et al. 2016; Soyseven et al. 2022; Yeganeh-Zare et al. 2022). This method allows for the rapid measurement of fructose content in plant tissues and has been validated for its high accuracy and robustness while requiring relatively low sample maintenance. In addition, HPLC-charged aerosol detection (CAD), high-performance anion exchange chromatography

(HPAEC), and pulsed amperometric detection (PAD) have been developed (EFSA Panel on Nutrition et al. 2024; Pitirollo et al. 2023; Sławińska et al. 2021).

Furthermore, mass spectrometry (MS) technology plays a pivotal role in fructose detection, providing high sensitivity and accuracy, making it an efficient analytical tool. Notably, liquid chromatography/mass spectrometry (LC/MS) coupling has higher selectivity and can effectively separate and detect fructose in plant tissues. This combined technology is able to cope with complex sample matrices, providing more detailed and reliable fructose determination results (Yan et al. 2021). However, it is important to mention that while LC/MS has distinct advantages in selectivity and sensitivity, in comparison, HPLC methods are faster and less labor-intensive in terms of analysis speed and operation. This makes HPLC still a viable option in certain situations, especially when rapid analysis of large batches of samples is required (Georgelis et al. 2018).

Moreover, UV detection and the combination of gas chromatography (GC) and MS have unique value for the measurement of sugars after derivatization. UV detection is commonly utilized to detect the light absorption properties of sugars, while fluorescence detection enhances detection sensitivity by exciting fluorescently labeled sugars (Buziau et al. 2020; Jalaludin & Kim 2021). The combination of GC and MS performs well in analyzing derivatization products of sugars, providing more information for specific studies. Consequently, selecting the most suitable detection method for fructose analysis should be guided by the specific requirements and objectives of the experiment. Depending on the depth of analysis, sample type, and availability of laboratory equipment, researchers have the flexibility to select the appropriate technology to ensure high-quality, comprehensive fructose measurement data.

13.3 TREHALOSE

13.3.1 Introduction

Trehalose, a non-reducing functional disaccharide, is composed of two glucose molecules connected by an α-(1,1)-glycosidic bond. It exhibits exceptional stability, resistant to high temperature, non-reducing properties, absence of Maillard reaction, and only undergoes decomposition under the influence of strong acids, strong alkali, and enzymes. Possessing a unique capability to protect bioactive substances, trehalose effectively safeguards cell structures and nutrients, enabling organisms to maintain vitality under extreme abnormal conditions, such as extreme temperatures, dryness, and high radiation (Jiang, Shen, et al. 2013). α-Trehalose has a symmetrical molecular structure with the molecular formula $C_{12}H_{22}O_{11}$, which was discovered by Wiggers in 1832 and isolated from ergot, and is 45% as sweet as sucrose (Nwaka & Holzer 1997). Bredereck elucidated its chemical structure in 1930 using NMR techniques (Bredereck 1930). Trehalose exists in three optical isomeric forms: α,α-trehalose, α,β-trehalose, and β,β-trehalose (Burek et al. 2015). Among them, α,α-trehalose is widely distributed in bacteria, fungi, plants, and animals, while α,β-trehalose and β,β-trehalose are uncommon in nature (Xin et al. 2017). It is recognized by the U.S. Food and Drug Administration in 2000 as a generally recognized safe compound.

13.3.2 Properties and Applications

Due to its distinctive physical and chemical characteristics, trehalose has been the focus of extensive biological research in recent years. A summary of the biochemical properties of trehalose is shown in Table 13.2. At present, trehalose is widely used in food, biomedicine, cosmetics and other fields. As people continue to develop trehalose, the application prospects will be broader.

Trehalose, owing to its non-reducing nature, presents a range of desirable properties, including moisturizing capabilities, anti-freezing attributes, high-temperature resistance, desiccation resistance, mild sweetness, and resistance to browning (Galmarini et al. 2011). In the formulation of functional foods, one key application is the use of trehalose, which plays a crucial role in preserving the natural structure of protein molecules in protein-rich foods, thus ensuring overall food quality. Moreover, its incorporation into food acts to maintain dryness, thereby reducing risks and costs associated with food storage (Nooshkam et al. 2019).

Trehalose, characterized by its low caloric value and mild sweetness, is increasingly regarded as a viable alternative to sucrose, particularly for individuals with obesity and diabetes. Its unique protective molecule function is of significant value in enhancing the stability of biological products during storage and transportation. In the realm of medicine, trehalose serves as a stabilizer for preserving vaccines, hormones, blood, and biological enzyme preparations, while also finding application as a sweetener for medications (Ohtake & Wang 2011; Thorat et al. 2022).

Beyond its applications in medicine, trehalose showcases its efficacy in agriculture by stabilizing biofilm structures and enhancing the resistance of biological tissues to various stress conditions, including low temperature, high temperature, saline-alkali, and drought. This versatility positions trehalose as a valuable component in agricultural breeding (Ohtake & Wang 2011).

13.3.3 Manufacture

Currently, the main techniques for commercially producing trehalose encompass microbial isolation, fermentation processes, and enzyme-catalyzed synthesis. The microbial extraction method, as a traditional preparation process, is now well-established, but its higher production cost and limited raw materials pose constraints on large-scale industrialized production of trehalose. In contrast, the enzyme synthesis method is simple, with low raw material cost and abundant enzyme source, making it highly favored by the trehalose industry (Su et al. 2014).

TABLE 13.2 A summary of the biochemical properties of trehalose

BIOCHEMICAL PROPERTIES	DETAILED INFORMATION ON THE PROPERTIES	REFERENCES
Good stability	Non-reducing sugar	Stick and Williams (2009)
	Lower glycoside oxygen bond energy (<1 kcal/mol)	Iturriaga et al. (2000)
	High glass transition temperature	Chen et al. (2022)
	Accumulation of trehalose increases plant cold tolerance	Romero et al. (1997)
	Trehalose effectively protects organisms from dehydration and freezing	Branca et al. (1999)
	Trehalose as a protein stabilizer	Lee et al. (2013); Olsson et al. (2016)
As an energy source	Trehalose plays a key role in prokaryotic and eukaryotic cells as an energy source during certain stages of cell development	Pohane et al. (2021); Sarkar and Sadhukhan (2022); Shukla et al. (2015)
As a structural component	As a key constituent of glycolipids in cellular membranes	Lederer (1976); Ohtake and Wang (2011)
Role in human health	Anti-inflammatory effect	Echigo et al. (2012); Minutoli et al. (2008)
	Antioxidant activity	Sadak et al. (2019)
	As a significant regulator of autophagy, inducing this process offers a therapeutic strategy for treating neurodegenerative and other diseases, such as Huntington's disease, prion disorders, amyotrophic lateral sclerosis, Parkinson's disease, and certain types of cancer	Aguib et al. (2009); Chaitanya et al. (2021); Kakoty et al. (2021); Khalifeh et al. (2019); Khalifeh et al. (2021)

The microbial extraction method typically employs yeast as the source of extraction. This method involves controlling the production conditions of various trehalose-containing organisms such as brewer's yeast and baker's yeast. Through this control, trehalose accumulates within the microbial strain, and it is then extracted using ethanol or another organic solvent (Ferreira et al. 1997; Li, Wang, et al. 2016).

Microbial fermentation, as the name suggests, involves the isolation and purification of trehalose from fermentation broth using microbial fermentation techniques. Advances in genetic technology have led to unprecedented developments in the preparation process for extracting trehalose through microbial fermentation. Highly productive strains were obtained by genetic engineering techniques, and the corresponding fermentation conditions were controlled, resulting in fermentation products with very high trehalose content. Although this method improved the yield of trehalose, the complexity of the fermentation metabolites and the low material conversion rate under industrialized conditions have posed significant challenges in the extraction and purification of trehalose (Du & Zhao 2012).

The enzyme synthesis method typically relies on glucose, maltose, or starch, which is converted to trehalose by the action of the corresponding trehalose (Liu, Yang, et al. 2018). In 1995, Japanese scholars obtained alginate synthase, which converts maltose to trehalose, for the first time in *Pimelobacter* sp. R48, *Thermus aquaticus*, and *Pseudomonas putida*, and since then, a new research direction on trehalase for the production of trehalose has been initiated (Nishimoto et al. 1996). By adjusting the reaction substrate, Bae et al. (2022) used amylosucrase derived from *Deinococcus deserti* to act on 2 M sucrose and 0.75 M fructose, and the yield of synthesized trehalose was increased to 246/L. The enzyme production of trehalose is not only cost-effective but also has a high enzyme conversion rate, making it an ideal way for industrialized production

(Liang et al. 2013). The five primary established pathways of trehalase are as follows (TPS: trehalose-6-phosphate synthase, TPP: trehalose-6-phosphate phosphatase, TreY: maltooligosyl trehalose synthase, TreZ: maltooligosyl trehalose trehalohydrolase, TreP: trehalose phosphorylase, TreT: trehalose glycosyltransferring synthase, and TreS: trehalose synthase): TPS-TPP, TreY-TreZ, TreP, TreT, and TreS pathways.

The pathway for trehalose synthesis is illustrated in Figure 13.2. The TPS-TPP pathway was first found in yeast (Magalhães et al. 2017), and in *Escherichia coli*, it is known as the OtsAB pathway (Li et al. 2012). Since this pathway requires the participation of uridine diphosphate (UDP), the cost of preparing trehalose is too high, so it is difficult to adopt this pathway to produce trehalose in industrialization. However, trehalose can be produced by malt starch in the TreY-TreZ pathway, with low production cost and high conversion rate, which is promising for industrialization (Paul et al. 2008). The TreS pathway, as investigated by Lin et al., unveiled a novel TreS variant sourced from *Thermomonospora curvata* DSM 43183; under optimized conditions at 35°C and pH 6.5, this enzyme demonstrated noteworthy efficacy in converting maltose into trehalose, achieving a maximum conversion rate exceeding 78% (Jiang, Lin, et al. 2013). In an independent study, Liu et al. employed a recombinant TreS expressed in *E. coli* BL21; the enzymatic conversion of maltose to trehalose transpired at 50°C and pH 8.0 over a 24-hour duration. Under these elevated temperature and alkaline conditions, 64.3% of maltose underwent effective transformation into trehalose. The stringency of these reaction conditions not only mitigated microbial contamination risks but also contributed to an augmented yield of trehalose (Liu et al. 2018). The TreS pathway is distinguished by its streamlined process, cost-effectiveness, high enzyme specificity, and despite a moderate conversion rate, it holds significant promise for industrial applications.

TPS-TPP pathway:

$$\text{UDP-glucose} + \text{Glucose-6-phosphate} \xrightarrow{\text{TPS}} \text{UDP} + \text{Trehalose-6-phosphate (1st step)}$$

$$\text{UDP} + \text{Trehalose-6-phosphate} \xrightarrow{\text{TPP}} \text{Trehalose} + \text{Pi (2nd step)}$$

TreY-TreZ pathway:

$$\text{Maltodextrin} \xrightarrow{\text{TreY}} \text{Malto-oligosaccharide trehalose} \xrightarrow{\text{TreZ}} \text{Trehalose} + \text{Glucose}$$

TreP pathway:

$$\text{Glucose-1-phosphate} + \text{Glucose} \xleftrightarrow{\text{TreP}} \text{Trehalose} + \text{Pi}$$

TreT pathway:

$$\text{ADP-glucose} + \text{glucose} \xrightarrow{\text{TreT}} \text{Trehalose} + \text{ADP}$$

TreS pathway:

$$\text{Maltose} \xrightarrow{\text{TreS}} \text{Trehalose}$$

FIGURE 13.2   The trehalose synthesis pathway. (Made by the author.)

13.3.4 Characterization

Trehalose, as a functional carbohydrate, is not limited to the aforementioned detection methods for fructose. Vanherp and other researchers (2021) employed magnetic resonance spectroscopy (MRS) to quantitatively analyze trehalose, using it as a biomarker for the diagnosis and monitoring of Cryptococcus *in vivo*. In addition, Mery et al. (2022) employed the mass spectrometry disaccharide method (MS-DS) to predict and diagnose invasive candidiasis by detecting trehalose in fungi.

These two methods have high resolution and provide powerful tools for a deep understanding of the physiological activity of trehalose in organisms. The application of MRS makes the quantitative determination of trehalose more intuitive and non-destructive, especially when analyzing biomarkers in *Cryptococcus*, providing the possibility of non-invasive *in vivo* monitoring. This holds significant potential for application prospects in follow-up treatment effects and early detection of disease changes. On the other hand, the high sensitivity and high resolution of mass spectrometric disaccharide assay (MS-DS) make it a powerful tool for the prediction and diagnosis of invasive candidiasis. By analyzing trehalose in the fungus, the method not only aids in early detection of disease but also provides a reliable means of assessing the invasiveness of Candida infections (Mery et al. 2016).

These advanced detection methods pave the way for a more comprehensive understanding of trehalose's function in various physiological and pathological conditions. The comprehensive utilization of these technologies can unveil the functions of trehalose in organisms more comprehensively and accurately, offering more precise information for the diagnosis and treatment of related diseases.

13.4 ISOMALTOOLI-GOSACCHARIDES

13.4.1 Introduction

Isomaltooligosaccharides (IMOs) represent a category of functional oligosaccharides characterized by low digestibility, and they are composed of glucose units connected by 2–10 α-(1,6) glycosidic linkages. A small number of glucose units are α-(1,4)-, α-(1,3)-, or α-(1,2)-glycosidic linkage (Sorndech et al. 2017). IMOs with DP≤2 are isomaltose, IMOs with DP≥3 include panose, isopantose, isomaltotriose, nigerose, nigerotriose, kojibiose, long-chain IMOs, and cyclic IMOs (Ketabi et al. 2011; Ruzanski et al. 2013; Shi et al. 2016; Sorndech, Nakorn, et al. 2018; Sorndech et al. 2017). According to different connection methods between glucose groups in IMOs molecules, IMOs can be divided into three categories: 1. 6-O-α-isomalt formed by connecting a linear chain to a glucose group through an α-1,4-glycosidic bond oligosaccharide-D-maltose. 2. Oligodextran composed of linear α-1,6 oligoglucan. 3. IMOs in which the above two types of isomaltose linear chains are linked to glucose via α-(1,2), α-(1,3), or α-(1,4) bonds (Goffin et al. 2010; Jacobs 2016; Sorndech, Nakorn, et al. 2018).

13.4.2 Properties and Applications

IMOs exhibit good sweetness, approximately 50%–60% of sucrose, with a soft sweetness. The sweetness of IMOs decreases as the DP increases. And it has good solubility and low viscosity (Jacobs 2016). The syrup of IMOs is colorless or light yellow, while the powdered sugar form is an amorphous white powder with good stability under acid and heat conditions. Processed products incorporating IMOs are moist, delicate, soft, and boast good palatability (Ketabi et al. 2011). IMOs are incorporated into beer, beverages, and dairy products to modify sweetness and enhance mouthfeel (Singla & Chakkaravarthi 2017).

IMOs are considered prebiotic substances, serving as a carbon source for intestinal flora, including *Bifidobacterium* and *Lactobacillus*. This application aids in maintaining intestinal pH homeostasis via fermentation, resulting in the production of metabolites like short-chain fatty acids. This process is essential for preserving a healthy balance of probiotic bacteria in the gut while inhibiting the growth of harmful microorganisms, thus promoting overall microbial equilibrium within the human digestive tract. (Candela et al. 2010; De Filippo et al. 2010; Rodriguez-Castaño et al. 2017; Zhang et al. 2010). Song et al. noted that it is slowly hydrolyzed by α-glucosidase in the human body as a low-digestible glycohydrate compound, and utilized this property as a slow-digesting material in the food industry to manage blood sugar levels and energy distribution in the mammalian gastrointestinal tract (Song et al. 2022). Additionally, IMOs possess the ability to stimulate bowel movements, relieve digestive disorders, and reduce intestinal inflammation (Hu et al. 2013; Wu et al. 2017; Yen et al. 2011). Song et al. increased the dietary fiber content of IMOs by modifying IMOs with dextransucrase derived from *Streptococcus mutans* UA159 using 10% sucrose and 20% IMOs (derived from corn starch) as substrates (Song, Kim, et al. 2020). IMOs also act as good bifidobacteria proliferation factors. Long-term use can promote the proliferation of bifidobacteria in the intestine, thereby enhancing human immunity (Kaulpiboon et al. 2015; Ketabi et al. 2011; Ojha et al. 2015). IMOs can be added as an additive to livestock feed. Since animals are difficult to digest and decompose IMOs, they are fermented and utilized by probiotics in the intestines, which can help animals improve their immunity and promote food digestion and absorption (Li et al. 2009; Thitaram et al. 2005). For more details on the health benefits of IMOs, see Table 13.3.

13.4.3 Manufacture

IMOs are found naturally in small amounts in honey, beer, and various fermented foods. Traditionally, the production of IMOs mainly relies on various hydrolases and transglucosidase. The production of IMOs mainly includes the following methods: Enzyme conversion method: (1) This method mainly uses dextrose sucrase; the glucose produced by the hydrolysis of sucrose under the action of dextran sucrase enzyme is connected with maltose through α-1,6-glucosidic bonds to generate IMOs (Bivolarski et al. 2013). (2) In industrial production, IMOs are primarily generated by enzymatically modifying starch using α-amylase, β-amylase, pullulanase, and glucosidase, see Figure 13.3 for details. In this method, starch is first liquefied and saccharified to produce products containing maltose-oligosaccharides, and finally, it is transglycosylated by α-glucosidase and separated and purified to obtain IMOs (Dobruchowska et al. 2012; Lee et al. 2002). However, this traditional production process has many

procedures, long time, complex control conditions, difficulty in continuous production, and low activity of α-glucosidase. Currently, growing research efforts focus on α-glucosidase to improve the conversion efficiency of IMOs.

At present, the goal of improving the yield of IMOs is primarily achieved by altering the reaction conditions (adjusting substrate concentration, temperature, pH, etc.) to modify the kinetic constants of the enzymatic reaction. In their comparative study, Mangas-Sanchez et al. examined the hydrolysis and transglycosylation activities of different glucosidases. They found that the β-transglucosidase from *Aspergillus niger* produced the highest yield (69%) when starting with an initial maltose concentration of 160 mM (Mangas-Sánchez & Adlercreutz 2015). Fourage et al., by raising the reaction temperature to 75°C, observed an increased transglycosylation rate of 42% with the β-glucosidase of *Thermus thermophilus* (Fourage et al. 2000). To enhance the production efficiency of IMOs, Niu et al. employed an enzyme mixture, comprising recombinant pullulanase, α-amylase, β-amylase, and α-transglucosidase, to simultaneously perform the liquefaction, saccharification, and transglycosylation of starch. This approach increased the yield of IMOs to 49.09%, while optimizing each individual step of the process (Niu et al. 2017). At present, in industrial production, the glucosidase derived from *A. niger* is often used but has poor heat resistance, with the optimal temperature being 50°C. This high-temperature intolerance limits the commercial application of α-glucosidase, making it challenging to operate the transglycosylation reaction at high temperatures. As a result, the reaction solution exhibits high viscosity and slow reaction rates, leading to increased production costs. Therefore, another major research direction is to develop efficient expression of thermostable α-glucosidase through genetic engineering, enabling the transglycosylation reaction in high-temperature environments. This includes methods such as mutating target genes and screening microorganisms with heat resistance. For instance, Park et al. expressed α-glucosidase derived from *hyperthermophilic Crenarchaea* in *E. coli*, and the resulting recombinant enzyme had a maximum activity at 95°C and pH 4.0 (Park et al. 2013). The T_{50} (the temperature at which the enzyme retains half of its original activity) of the mutant Q10Y constructed by Zhou et al. through rational design is 4°C higher than that of the wild type (Zhou et al. 2015). Ma et al. introduced mutations at amino acid position 694 in *A. niger* α-glucosidase, substituting it with Ala, Leu, Phe, and Trp. The resultant mutants (N694F and N694W) demonstrated enhanced yields compared to the wild-type *A. niger* enzyme. Notably, the mutants exhibited elevated concentrations of isomaltose and isomaltotrios, suggesting improved performance in transglycosylation reactions (Ma et al. 2017). Kumar et al. employed α-glucosidase derived from a novel strain of *A. niger* found in soil for an 18-hour reaction with maltose. This resulted in a noteworthy pantose yield of 63.2 g/L, and the accumulation of IMOs commenced at 12 hours, ultimately reaching a final yield of 20 g/L. Interestingly, a variation is observed at position 694 in the α-glucosidase from certain strains of *A. niger*, where threonine is substituted with asparagine, compared to the typical

TABLE 13.3   IMOs health benefits

NO.	IMOS HEALTH BENEFITS	REFERENCES
1	Low metabolic rate for digestion by human digestive enzymes	Ketabi et al. (2011); Sorndech (2022)
2	α-Glucosidase from *Bacillus subtilis* strain AP-1 generates long-chain IMOs with prebiotic properties, having a degree of polymerization (DP) ranging from 2 to 14. These IMOs are produced through the direct fermentation of maltose and are resistant to acidic conditions and gastrointestinal digestion. The enzyme promotes the growth of probiotic bacteria, leading to the production of short-chain fatty acids, while not enhancing the growth of pathogenic bacteria	Tiangpook et al. (2023)
3	In *in vitro* experiments, panose markedly boosted the population of *Bifidobacterium* and *Bifidobacterium lactis*, while reducing the number of *Bacteroidetes*, and significantly increased the production of butyrate and acetate, while reducing protein fermentation markers	Mäkeläinen et al. (2009)
4	In *in vitro* experiments, long-chain IMOs sourced from tapioca starch stimulate the growth of *Lactobacillus casei* (Probiotics)	Kaulpiboon et al. (2015)
5	IMOs stimulate the growth of probiotic *B. subtilis* CU1	Villéger et al. (2022)
6	IMOs can increase the growth of *Bifidobacteria* in the intestine and improve intestinal flora	Kohmoto et al. (1988)
7	IMOs can effectively improve defecation, stool volume, and colon microbial fermentation	Chen et al. (2001)
8	*In vitro* trial suggests that IMOs may benefit patients with type 2 diabetes	Bharti et al. (2015)
9	When IMOs are incorporated as dietary ingredients in the elderly, there is a reduction in plasma levels of total cholesterol and low-density lipoprotein (LDL) cholesterol, along with improvements in fecal microflora and intestinal function after 4 and 8 weeks	Yen et al. (2011)
10	IMOs are anticipated to serve as therapeutic ingredients for hemodialysis patients. Following a 4-week supplementation of IMOs, patients experienced a significant reduction in total cholesterol and triglycerides, along with a notable increase in high-density lipoprotein cholesterol	Wang et al. (2001)
11	Rat experiments show that cinnamaldehyde and IMOs are used as food ingredients in mice. After feeding for 12 weeks, IMOs can resist the antibacterial properties brought, by the anti-obesity agent cinnamaldehyde and may have adverse effects on the intestinal microbiota	Singh et al. (2017)
12	IMOs can reduce high-density lipoprotein in the serum of weaned piglets 25 days after birth and work together with Chinese herbal extracts to reduce the number of harmful intestinal flora	Pi et al. (2022)
13	A 4-week trial in 53 subjects aged 23–57 years showed that non-dairy creamer made from IMO significantly reduced cholesterol, triglycerides, and heart disease risk ratio scores	Mumpuni et al. (2022)
14	A symbiotic combination of *Bifidobacterium longum* Bif10 and *Bifidobacterium breve* Bif11, along with IMOs, can prevent dextran sodium sulfate-induced ulcerative colitis, improve intestinal flora imbalance in colitis mice, and increase the abundance of intestinal probiotic flora	Sharma et al. (2023)
15	By combining the use of *Lactobacillus helveticus* KM7 and IMOs, the quantity of intestinal flora of bees is adjusted to achieve synergistic regulation of the intestinal microbiota and immunity of adult worker bees of Apis cerana, thereby improving their survival rate	Fang et al. (2023)
16	IMOs synthesized by transglucosidase from *Thermoanaerobacter thermocopriae* achieves slow digestion properties by reducing its hydrolysis by human pancreatic α-amylase and significantly reduces postprandial blood glucose peaks in male mice	Um et al. (2023)

variant (Kumar et al. 2020). Punnatin et al. conducted a molecular dynamics analysis, revealing the significant influence of amino acids N226 and H227 in α-glucosidase from both the European honeybee (*Apis mellifera*) on the specific binding of the substrate maltose (Punnatin et al. 2020).

13.4.4 Characterization

Being a polysaccharide with a complex structure, IMOs face many challenges in its research and detection. Similar to common carbohydrates, commonly used analytical methods involve chromatography (GC or HPLC) and MS. In terms of chromatography, it includes gas chromatography (GC) and HPLC. GC is suitable for carbohydrate analysis because of its high resolution and sensitivity. However, its sample pretreatment and introduction of internal standard substances complicate the experiment. In contrast, the HPLC method is relatively simple to operate and does not require special gases, but the resolution is relatively low. When choosing a chromatography method, laboratory equipment, resolution requirements, and the characteristics of the IMOs need to be fully considered.

As an advanced detection method, MS boasts high sensitivity and resolution, enabling accurate identification and quantification. However, MS demands expensive instruments and operators with advanced technical skills, coupled

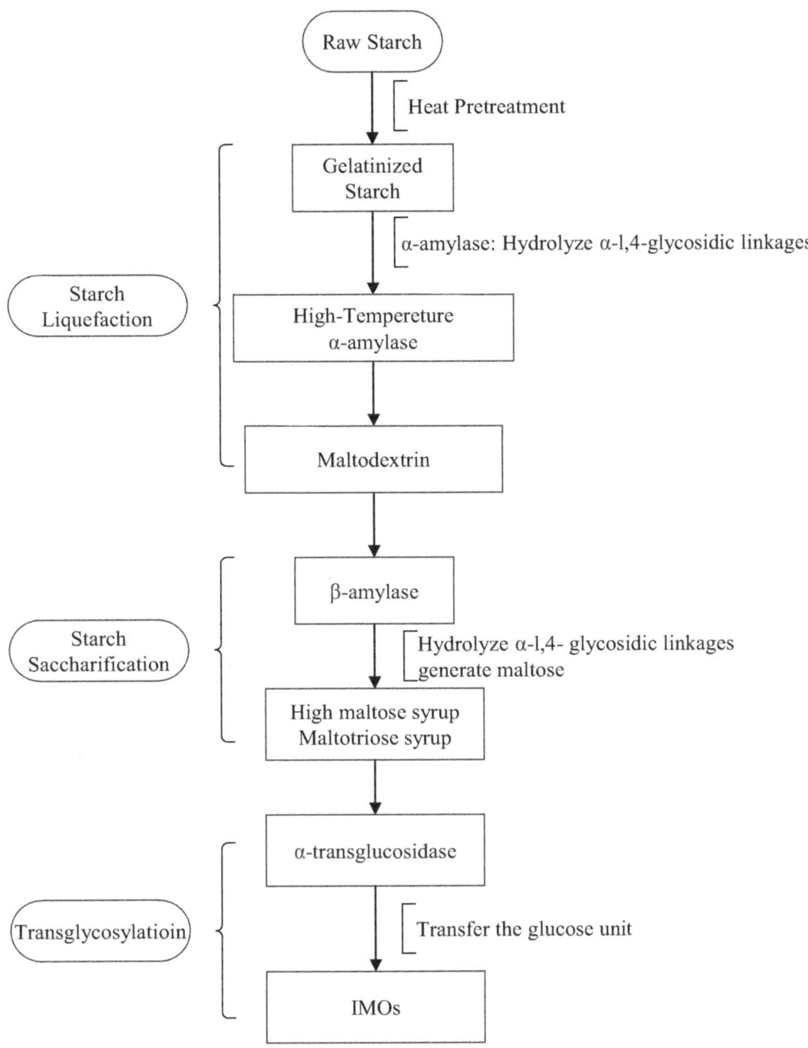

FIGURE 13.3   The main enzymatic processes for commercial production of IMOs. (Made by the author.)

with strict requirements for sample preparation. Taking into account the structural complexities of IMOs, the choice of detection method should align with the particular laboratory environment, analytical requirements, and available resources. In summary, chromatography is well-suited for general analysis, whereas MS is more appropriate for the precise identification of IMOs and the analysis of complex samples.

13.5 DIGESTION-RESISTANT GLUCANS

13.5.1 Introduction

"Digestion-resistant glucans" (DRGs) refers to those glucans that are not readily digestible in the human gastrointestinal tract and can resist breakdown by digestive enzymes (Sorndech, Tongta, et al. 2018). The classification of DRGs is challenging due to the complexity of dextran structures and the diversity of bonding patterns, there are no clear and detailed criteria, but generally, DRGs can include many types of starches and glucans, with their degree of resistance potentially affected by several factors, including their structure, origin, and processing method. Common examples encompass Digestion-resistant α-Glucan, β-Glucan, Arabinoxylan, and so on (Englyst et al. 1992; Englyst & Englyst 2005; Fadel et al. 2018; Romero Marcia et al. 2021; Ye et al. 2015). Owing to the specific structure of these glucans, their digestion and breakdown in the healthy human gut pose considerable challenges. By interacting with gut microbes through fermentation, these glucans play a pivotal role in preserving the equilibrium of the intestinal microecology and fostering the proliferation of beneficial flora. This, in turn, exerts a positive influence on the overall immune system and metabolic processes (Romero Marcia et al. 2021).

Digestion-resistant α-glucan is a subclass of dietary fibers, and most of the relevant studies have classified them according to the DP, with digestive-resistant starch being the most abundantly researched among the highly polymerized DRGs. Engylst et al.

simulated the human gastrointestinal environment in an *in vitro* environment, using amyloglucosidase and pancreatic α-amylase to hydrolyze and digest starch. The starch released in the form of glucose after 120 minutes was considered RS (Englyst et al. 1992). The most studied α-glucans are dextran (mainly α-(1,6) bonds), mutan (>50% α-(1,3) bonds), alternan (alternating α-(1,6) and α-(1,3)), reuteran (α-(1,4) linear segments interconnected by single α-(1,6) bridges), amylopectin (α-(1,4)-(1,6)), glycogen (α-(1,4)-(1,6)), pullulan (α-(1,4)-(1,6)), nigeran (α-(1,3)-(1,4)), elsinan (α-(1,3)-(1,4)), isolichenan (α-(1,3)-(1,4)), pseudonigeran (α-(1,3)), etc. (Gangoiti et al. 2018).

Resistant α-glucans with a low polymerization degree, such as resistant maltodextrin, have a special composition of α-glycosidic bonds, especially α-(1,2) and α-(1,3) linkages, that are completely resistant to amylolysis, or exhibit highly slowly digested α-(1,6) glycosidic linkages (Astina & Sapwarobol 2019; Englyst et al. 1992). β-Glucans exhibit considerable structural diversity and their sources can be classified according to the mode of glycosidic linkage. β-Glucans demonstrate significant structural diversity, with their sources categorized based on the glycosidic linkage patterns. Cereal β-glucans are predominantly characterized by β-(1,3)/(1,4)-glycosidic bonds, while brown algae primarily feature β-(1,3)-glycosidic bonds, and fungi exhibit a prevalent presence of β-(1,3)/(1,6)-glycosidic bonds. Furthermore, certain fungal-derived β-glucans also incorporate β-(1,4)-glycosidic linkages (Baky et al. 2022; Carvalho et al. 2021; Chen, Yang, et al. 2021; Mikkelsen et al. 2013). Arabinoxylan (AX) is polymerized from (1→4)-β-d-xylopyranose residue units into a linear main chain, through C(O)-2, C(O)-3 or C(O)-2,3 glycosidic bonds that connect α-L-arabinofuranosyl substituents (Carvalho et al. 2021). In AX, the xylan backbone typically features xylose residues that are mono- and disubstituted, with arabinosyl substitutions occurring in the form of monomers. The structure resembles a chain built from various sugar molecules and featuring some branch points. This unique structure allows AX to form a network-like structure in plant cell walls, providing increased strength and stability to the cell wall. This is essential for the support and structural stability of plant cell walls (Chen et al. 2019; Zhou et al. 2010).

In essence, DRGs represent a distinct class of glucans characterized by a unique molecular structure and biological activity. This intricate composition encompasses resistant starch, β-glucan, AX, and other variants, endowing DRGs with

TABLE 13.4 The classification summary of α-glucan, β-glucan, and AX

DRGS		MAIN LINKAGE TYPES	REFERENCES
α-Glucan	Dextran	Mainly α-(1,6) glucan chain segments and varying amounts of α-(1,2), α-(1,3), α-(1,4), and α-(1,6) linkages	Luanda and Badalamoole (2023)
	Mutan	With >50% α-(1,3) linkage and minor amounts of α-(1,6) linkage	Miyazaki (2023)
	Alternan	With alternating α-(1,6) and α-(1,3) linkages	Shetty et al. (2021)
	Reuteran	With α-(1,4) linear segments interconnected by a α-(1,6) bridges	Molina et al. (2021)
	Amylopectin	About 95% α-(1,4) and 5% α-(1,6) linkages	Kong (2020)
	Glycogen-like particle	α-(1,4) and α-(1,6) linkages, highly branched	Lee et al. (2022)
	Pullulan	The molecule is characterized by maltosyl units repeating throughout its structure, with two glucopyranose rings connected by α-(1,4)-glycosidic bonds and an additional glucopyranose ring linked through an α-(1,6)-glycosidic bond	Singh et al. (2021)
	Nigeran	Linear α-glucan with alternating α-(1,3) and α-(1,4) linkages	Togo et al. (2021)
	Elsinan	Maltotriose and maltotetraose units are interconnected via α-(1,3) glycosidic linkages	Wiater et al. (2020)
	Maltodextrin	The α-(1,4) bond connects adjacent glucose molecules on the main chain, while the α-(1,6) bond forms a branched structure	Brouns et al. (2007)
β-Glucans	β-(1,3)-Glucan	The fundamental β-glucans are structured through β-(1,3)-glycosidic bonds	Chioru and Chirsanova (2023)
	β-(1,6)-Glucan	A highly branched structure and a high proportion of β-(1,6) glucosidic linkages, minor amounts of β-(1,3) linkages	Manners et al. (1973)
	β-(1,3)-(1,6)-Glucan	The main chain is connected by β-(1,3), and the side chains are connected by β-(1,6) linkages	Lowman et al. (2011)
	β-(1,4)-(1,6)-Glucan	β-(1,4)-linked main chain and β-(1,6)-linked branch chain	Gonzaga et al. (2009)
	β-(1,3)-(1,2)-Glucan	β-(1,3) linkages and β-(1,2) linkages	Barsanti et al. (2011)
	β-(1,3)-(1,4)-Glucan	β-(1,3) linkages and β-(1,4) linkages	Bulmer et al. (2021)
Arabinoxylan		The principal chain consists of β-(1,4)-D-xylp units arranged in a linear sequence, with α-(1,2)- and/or α-(1,3)-glycosidic bonds linking additional units to this backbone	He et al. (2021)

distinctive physiological properties within the human body. For a comprehensive classification overview of α-glucan, β-glucan, and AX, refer to Table 13.4. The inherent resistant-digestion attributes of DRGs render them impervious to degradation by digestive enzymes in the human gastrointestinal system, facilitating enhanced fermentation by intestinal microorganisms. This exceptional property has spurred extensive research into these anti-digestive glucans.

13.5.2 Properties and Applications

Digestion-resistant α-glucans, such as SDS, RS, and resistant dextrins, are modified by physicochemical or enzymatic treatments that alter the structure of the glucan, in particular by adjusting the ratio of specific bonds, such as α-(1,6), α-(1,2), α-(1,3), and α-(1,4), in order to obtain a glycan with specific functional properties (Corwin 2020; Han, Lee, et al. 2021; Ma & Boye 2018). This structural adjustment allows these DRGs to exhibit different properties during digestion in the human body, such as a slower digestion rate, a lower glycemic response, while also fostering the formation of beneficial metabolites like short-chain fatty acids, under fermentation by intestinal microorganisms. Abundant evidence supports the beneficial impact of incorporating slowly digestible starch or resistant starch into functional foods on the regulation of postprandial blood sugar in individuals with hypertension and type 2 diabetes (Hanefeld & Schaper 2007; Martin & Montgomery 1996). Engineered to release sugar gradually during digestion, these specific starch types promote a steadier and more controlled rise in blood glucose levels, reducing the likelihood of significant fluctuations (Ceriello et al. 2004; Hanefeld & Schaper 2007). In summary, the varied enzymatic approaches for modifying DRGs, as elucidated in these studies, offer valuable insights into their potential applications, particularly in managing postprandial blood sugar levels in individuals with specific health conditions. Beyond their positive impact on blood sugar control, these functional starches may also influence satiety and contribute to weight management. Therefore, the incorporation of slowly digestible starch or resistant starch into foods not only enhances blood sugar levels but also holds promise for comprehensive health management. This provides instructive guidance for designing functional foods tailored to the needs of patients with specific chronic diseases (Hasek et al. 2018). Presently, glucansucrases are predominantly employed by researchers to act on sucrose, synthesizing α-glucans, or glucosyltransferase is used to act on starch or maltodextrin, yielding novel functional α-glucans (Gangoiti et al. 2018).

According to the CAZy (Carbohydrate-Active enZymes) classification system, which is based on amino acid sequences (http://www.cazy.org/), α-transglycosylases (EC 2.4.) primarily acting on sucrose and starch are mainly classified into GH13, GH31, members of the GH57 family, GH70, and GH77. These enzyme families encompass a diverse array of enzyme classes, including hydrolases such as amylase and α-glucosidase (EC 3.2.), as well as isomerases like isomaltulose synthase and trehalose synthase (EC 5.4.). GH13, GH31, GH57, and other family enzymes are crucial in the saccharification process, facilitating the breakdown and transformation of starch and sucrose. They contribute to the energy and carbon sources for organisms by hydrolyzing α-glycosidic bonds in starch or sucrose to release simple sugars such as glucose. Enzymes from the GH70 and GH77 families are also involved in this process. While these enzyme families exhibit distinct reaction and product specificities, both enzymes employ a comparable double-displacement α-retention mechanism. This process includes the creation of a covalent enzyme-glycosyl intermediate, which engages with an acceptor molecule—such as water or another carbohydrate—to generate either a hydrolysis product or a new α-glucan compound (Light et al. 2017). Members of these enzyme families display similar structures and activity mechanisms, sharing a catalytic (β/α) 8-barrel domain (Janeček et al. 2007; Meng et al. 2017). All of these enzymes possess four conserved amino acid sequence motifs, which include three catalytic residues—Asp (aspartic acid), Glu (glutamic acid), and another Asp (aspartic acid)—along with additional residues responsible for substrate binding (Leemhuis et al. 2013). These conserved structural features are crucial for achieving efficient catalytic activity while maintaining the high specificity in carrying out α-transglycosylation reactions. This shared catalytic mechanism serves as a foundation for comprehending the function of these enzymes and provides a basis for further research and industrial applications (Suzuki & Suzuki 2016).

β-Glucan is an abundant dietary fiber that can be obtained from several sources, including plants, fungi, and bacteria (abundant in mushrooms, algae, and cereals) (Rieder & Samuelsen 2012; Suzuki & Suzuki 2016). β-Glucan displays diverse biological effects, with various sources demonstrating distinct functional properties. Fungal-derived β-glucan has been shown to have a boosting effect on the immune system and also has anti-tumor potential (Zhu et al. 2015). The β-glucan present in cereal has been shown to aid in reducing cholesterol and blood sugar levels. Additionally, it is noted for its anti-aging, anti-inflammatory, and skin-health benefits (Du et al. 2014; Du et al. 2015; Du & Xu 2014; Rieder & Samuelsen 2012). Additionally, β-glucan exhibited high water absorption during dough preparation, hindering the hydration of gluten proteins and inhibiting the formation of reticulation, leading to a decrease in air-holding capacity and a decrease in dough ductility, which ultimately resulted in a decrease in bread volume (Brennan & Cleary 2007; Jacobs et al. 2008; Mann et al. 2005). Over the years, β-glucan has attracted much attention due to its rich source and biological activity, as well as its efficacy in immunomodulation and anti-tumor effects. Its physical and chemical properties have been intensively studied to elucidate its mechanism of action in the human body and to investigate its potential uses in medicine, nutrition, and healthcare.

AX exhibits various physical and chemical properties due to its diverse molecular structures and conformations. These properties, including solubility, viscosity, water absorption, and water retention, impact its functional applications in various fields, particularly the food industry (Malunga et al. 2017; Mann et al. 2005). The solubility of AX is affected by the interaction between molecular chains and between molecules and solvents, and it is also regulated by environmental factors (such as temperature, solvent pH, and ion concentration) (Ayala-Soto et al. 2015; Lazaridou et al. 2007). In bread making, AX shows good water holding capacity and plays a key role. Simultaneously, it serves as a dietary fiber with prebiotic properties that positively affect the homeostasis of the gut microbiota. Research has confirmed that AX intake helps to increase beneficial microorganisms, including *Bifidobacterium adolescentis, Bacteroides, Prevotella, Roseburia, and Streptococcaceae* (Geraylou et al. 2013; Saman et al. 2017; Walton et al. 2012; Zambrana et al. 2019). As a functional food ingredient, AX is widely used in bread, cereals, oatmeal, etc. to improve texture and water retention, and as a prebiotic, it is widely used in nutraceuticals and nutritional supplements to promote intestinal health and maintain overall health. Overall, AX has a wide range of applications in several fields, and its prebiotic nature and positive impact on intestinal health have made it a functional ingredient of interest in several industries.

13.5.3 Manufacture

Research on synthetic DRGs has witnessed substantial expansion. In particular, significant contributions have been made by various studies employing different enzymes to enzymatically modify starches. Park et al. utilized amylosucrase from *Neisseria subflava* (NsAS) to enzymatically modify gelatinized corn and rice starches, and this modification led to a significant increase in the starches' indigestibility rate, which reached up to 47.3% (Park et al. 2019). In another study, Ryu et al. employed recombinant amylosucrase (200 U/mL) from *Neisseria polysaccharea* to modify waxy and normal corn starches, demonstrating an increase in the branch chain length and a significant enhancement of insoluble resistant starch (Ryu et al. 2010). Additionally, Jung et al. utilized amylosucrase from *Deinococcus geothermalis* (DgAS) to modify Waxy rice starch, showing an increase in the chain lengths of DP 13–24 and DP 25–36 sections. When compared to starch treated with NpAS, enzymatic modification with DgAS resulted in a higher content of DP 13–24 in the modified starch (Jung et al. 2020). These studies highlight the diverse enzymatic approaches to enhance the properties of DRGs, providing valuable insights into their potential applications. Hong et al. utilized dual glycosyltransferases, namely *N. polysaccharea* amylosucrase and *Rhodothermus obamensis* glycogen branching enzyme, to synthesize highly branched α-glucans (with α-1,6 linkages constituting 7.5%–9.9% of the structure) from sucrose. This enzymatic approach contributes to the development of food materials with a low glycemic index (GI) by retarding digestive absorption in the gastrointestinal tract, showcasing significant potential (Hong et al. 2022).

In the industrial production of DRGs, the process is similar to the industrial preparation of IMOs mentioned in Section 13.4.3. First, starch granules are released by treatments such as crushing and wet leaching of starch raw materials (e.g., corn and potatoes) (Venkatachalam et al. 2021). Subsequently, after starch pasting, the starch is liquefied and saccharified by the synergistic action of various enzymes such as α-amylase, β-amylase, pentokinase, and glucosidase. This process gradually transforms the starch into a mixture containing glucan. The action of pentosanase further degrades the starch digest to form polysaccharides, which include chain-like molecules of different lengths. Next, unhydrolyzed starch and other impurities are removed through separation and refining steps. Finally, crystallization and drying operations are performed to obtain the dextran product for specific needs. The whole process requires careful control of the parameters (temperature, type and amounts of enzymes used in each step, reaction time, etc.) to achieve a final product that is a specific dextran that meets the purity requirements. This industrial production method is the basis for large-scale sugar production (Bivolarski et al. 2013).

Alternatively, microbial fermentation provides a viable approach for the production of β-glucan, offering the potential to generate this compound with specific structural attributes. Various microorganisms, including yeasts, fungi, and certain grains, can synthesize β-glucans, such as pullulan, gellan, curdlan, gum, dextran, laminarin, cellulose, lentinan, xanthan, and bacterial alginate (Venkatachalam et al. 2021). Current commercial production, however, predominantly relies on extraction from yeast, fungi, and cereal cell walls (Zhu et al. 2016). The selection of different microorganisms allows for targeted regulation of β-glucan structure, imparting diverse functions and application properties (Chioru & Chirsanova 2023). The field of enzymatic synthesis of β-glucans, employing glycosyl transferases, glycosynthases, and glycoside phosphorylases, is currently experiencing dynamic advancements. Him and colleagues utilized Glycosyltransferases from *Arabidopsis thaliana* to synthesize β-(1,3) glucan, employing UDP-glucose as a substrate. The significant obstacle in the feasibility process development of this method lies in the demanding requirement for expensive nucleotide sugar donors (Him et al. 2001). Takahashi et al. demonstrated that *Magnaporthe oryzae* Endotransglucosylase can act on β-(1,3)-(1,4)-glucan by cleaving β-(1,4)-glycosidic bonds, transferring glucan onto the receptor substrate. This enzyme exhibits both synthetic and hydrolytic activities in the context of β-glucans (Takahashi et al. 2013). The structural diversity and varying linkage proportions in β-glucans, coupled with their diverse sources, necessitate distinct enzymes for their synthesis. This dynamic interplay between enzyme specificity and the structural nuances of β-glucans represents an area of ongoing exploration and refinement in enzymatic production.

The enzymatic synthesis of AX has been challenging due to the complexity of its structure and solubility limitations. Typically, AX is extracted using a combination of enzymatic methods and other techniques. However, Senf et al. successfully addressed these challenges by employing glycosynthases along with chemical methods to achieve the synthesis of AX (Senf, Ruprecht et al. 2018). This innovative approach contributes to the advancement of AX synthesis in the field, offering a valuable alternative to traditional methods that often involve a combination of enzymatic and extraction strategies. Bhattacharya et al. utilized the specific action of arabinoxylanase (CtXyl5At) derived from *Clostridium thermocellum* to extract the water-soluble fraction of AX from wheat bran via an alkaline method.

They subsequently applied a dual-enzyme system that included GH43 α-L-arabinofuranosidase, designated as BaAXHd3, from *Bifidobacterium adolescentis*, along with CtXyl5At for hydrolysis. This enzymatic approach resulted in the synthesis and increased production of short-chain arabinoxylan oligosaccharides (DP 2–4) with prebiotic properties (Bhattacharya et al. 2020). Furthermore, Klangpeth et al.'s recent investigation into alkaline and enzymatic treatments, coupled with microwave pretreatment, significantly enhanced arabinoxylooligosaccharides (AXOs) production from rice husk. The noteworthy findings include achieving the highest AXOs yield of 9.01 g/100 through a 5-minute microwave pretreatment at 140°C (Klangpetch et al. 2022). These advancements contribute to the ongoing exploration of efficient AX synthesis methodologies, holding promise for diverse industrial applications.

13.5.4 Characterization

DRGs are polysaccharides widely distributed in nature. Depending on their origin, DRGs exhibit rich diversity in molecular weight and configuration. Since different glucans have different bond compositions and ratios, the detection and analysis of various glucans is a challenging task. A comprehensive understanding of the exact structure of purified polysaccharides is achieved through the combined application of spectroscopic, chemical, and separation methods.

The application of NMR spectroscopy for structural analysis is well-established, and it provides a robust method for investigating glucans in both solution and solid states. The combination of pertinent NMR experiments and methylation analysis (MA) methods is pivotal in determining the precise structure of tested polysaccharide (Synytsya & Novak 2014). NMR spectroscopy has proven effective in the structural analysis of various glucans. Synytsya et al. provided a comprehensive summary of the types and conformation of fungal glucans that can be detected using NMR-MA methods (Synytsya & Novák 2013). FTIR spectra are sensitive to the position and anomeric configuration of glycosidic bonds in dextran. By analyzing crude polymer fractions extracted from different raw materials, one can identify the types of glucans present (Šandula et al. 1999; Synytsya et al. 2009). In addition, size exclusion chromatography (SEC) and laser light scattering (LLS) were used to evaluate the range of molecular weights, uniformity, and branching degree of dextran (Synytsya & Novak 2014). You et al. utilized the Aqueous HPSEC-MALLS-RI System to analyze the molecular characteristics of amylose and amylopectin (such as molecular weight and uniformity) in corn starch with different solubility under different conditions (You & Lim 2000). For more information on carbohydrate detection, refer to the previous chapter.

The analysis of dextran preparations necessitates a synergistic application of methods to evaluate the purity, molecular size distribution, and structural arrangement of the dextran polymers. Each method has its advantages and limitations. When dealing with the structural analysis of intricate glucans, a comprehensive understanding is achieved through the combined use of spectroscopic, chemical, and separation techniques. This multi-layered analytical approach enhances the depth of structural information and offers valuable insights into the properties of glucans.

13.6 POROUS STARCH GRANULES

13.6.1 Introduction

Porous Starch Granules (PSGs) refer to the form of starch with a certain pore structure inside the starch granules (Zhang et al. 2012). The surface of these particles exhibits a honeycomb pore structure, and these pores extend into the interior of the particles, providing the particles a larger surface area and unique adsorption properties. Due to their good biocompatibility, non-toxicity, and low price, PSGs have been widely used in food, medicine, and health fields (Ogunsona et al. 2018; Oliyaei et al. 2019; Qian et al. 2020; Zhang et al. 2023). When natural PSGs are utilized as carriers for biologically active substances in the food industry and other sectors, they usually need to undergo appropriate modifications (Esterification, cross-linking, oxidation, etc.) through physical, chemical, or enzymatic methods to enhance their physical and chemical properties, thereby aligning with production requirements. These modification measures are designed to further improve the performance and applicability of PSGs to ensure that they can perform optimally in different application scenarios (Cao et al. 2023; Chen et al. 2020; Wang et al. 2022).

13.6.2 Properties and Applications

PSGs, a modified form of starch, retain the fundamental structure of starch while developing an intricate pore network on both the surface and interior (Sujka & Jamroz 2010; Zhao et al. 2018). The honeycomb structure with tiny pores offers advantages such as wide abundant raw materials and low production costs, a simple production process without the need for chemical reagents, safety, non-toxicity, and unrestricted dosage. It exhibits good mechanical properties in a dry state, maintains structural integrity in various solvents, is biodegradable, and showcases adaptability.

The preparation process allows control over pore characteristics and easy modification based on adsorbed substance properties, showcasing broad application potential across various fields (Zuo et al. 2012). At the same time, these pores extend into the interior of the particles, forming a complex pore network. According to the International Union of Pure and Applied Chemistry (IUPAC), pores are categorized into three main groups based on their diameters: micropores (less than 2 nm), mesopores (ranging from 2 to 50 nm), and macropores (greater than 50 nm) (Belingheri et al. 2015; Glenn et al. 2010; Hj. Latip et al. 2021). This imparts PSGs with an extremely large specific surface area, enabling outstanding performance in carrying and releasing small molecules. The PSGs form depressions inside, reinforcing the adsorption effect by securely capturing adsorbed substances and preventing easy detachment due to external gravity. This strong adsorption property helps stabilize volatile and unstable functional substances and achieve effective "hiding" (Wang et al. 2016; Wang et al. 2009). Porous starch adsorption is primarily physical and non-selective. It is suitable for adsorbing mixtures of substances with different polarities. The forms of adsorbed substances include powder, aqueous solution, oil solution, and organic solvent. The unique hollow structure of PSGs ensures its excellent adsorption performance. At present, porous starch particles serve as carriers in the microencapsulation process to adsorb functional substances and find extensive application across various sectors, including food, agriculture, medicine, and cosmetics. These research results are remarkable, not only improving the stability of active substances, but also successfully maintaining their activity (Drusch et al. 2006; Jiang et al. 2017; Li, Turner, et al. 2016; Zhang & Wang 2023).

13.6.3 Manufacture

At present, the preparation of PSGs mainly covers physical methods (Ultrasound, microwave assistance, extrusion, mechanical collision or freezing and thawing, etc.), chemical methods (Solvent exchange, acid hydrolysis, cross-linking agents, etc.), and biological enzymatic methods (Puncha-Arnon et al. 2020; Sun et al. 2023; Ubeyitogullari & Ciftci 2016). Among these methods, the enzymatic approach offers unique advantages. The preparation of PSGs by enzymatic methods has mild reaction conditions, which helps to maintain the natural structure and function of starch; simultaneously, biological enzymes act as catalysts in the reaction, enhancing both the efficiency and quality of the product. Moreover, compared to other methods, the enzymatic preparation process minimizes the use of chemical reagents, promoting environmental friendliness and reducing adverse effects on the environment (Chen et al. 2020; Liu, Wang, et al. 2018). Jung et al. employed α-amylase, glucoamylase, and β-amylase to modify PSGs derived from conventional corn starch, resulting in marked enhancements in both water retention and oil absorption capacities (Jung et al. 2017). Jung et al. initially utilized three amylolytic enzymes (glucoamylase from *A. niger*, β-amylase from barley, and α-amylase from *B. licheniformis*) to prepare PSGs. Subsequently, by employing amylosucrase from *Neisseria polysaccharea*, they modified PSGs generated by different enzymes, resulting in the development of an α-glucan layer on the surface of the granules. Through their research, the team demonstrated that this method enables the specific preparation of PSGs carriers with varying surface pore sizes. This specificity facilitates efficient loading of the active molecule crocin and delays the release of the active substance. The unique development of PSGs with different pore sizes, effectively loaded with crocin, significantly contributes to the design of biomaterials based on α-glucans. This advancement paves the way for the creation of biocompatible materials with tunable properties, offering versatile applications in various

fields (Jung et al. 2019). Therefore, the use of enzymatic methods to prepare PSGs is not only efficient, gentle, and environmentally friendly, but also has a simple process. The resulting PSGs have good adsorption properties and are more suitable for industrial production. With the deepening of research, many studies have begun to explore the collaborative application of enzymatic methods and other methods to further improve the performance of PSGs (Li et al. 2018; Majzoobi et al. 2015; Wu et al. 2011). Table 13.5 lists the research in recent years on the modification of various starches using single enzymes or their combinations to prepare PSGs (Figure 13.4).

This chapter focuses on introducing the application of bio-enzymatic methods in the preparation of PSGs. This method mainly uses exonuclease (amyloglucosidase (EC 3.2.1.3)) and endonuclease (α-amylase (EC 3.2.1.1)) to hydrolyze starch molecules to form a large number of deep pore structures (Fujii et al. 1988). At the same time, in order to further modify starch granules, other types of specific enzymes are increasingly used, such as glycogen branching enzyme (EC 2.4.1.18), and cyclodextrin glycosyltransferase (EC 2.4.1.19) (Das & Kayastha 2019; Gonzalez & Wang 2021; Guo et al. 2021; Zhang et al. 2019). By precisely controlling reaction conditions (such as pH, temperature, enzyme amount, and reaction time), PSGs can be prepared with different pore sizes and surface areas.

13.6.4 Characterization

During modification processes, including enzymatic treatments, the molecular structure of porous starch is altered. These structural changes lead to variations in the physical and chemical properties of the modified porous starch. To elucidate the modifications in structure and properties of PSGs and to further understand the link between structure and function, a comprehensive analysis utilizing FT-IR (Fourier transform infrared spectroscopy), SEM (scanning electron microscopy), XRD (X-ray diffraction), TEM (transmission electron microscopy), AFM (atomic force microscopy), NMR (nuclear magnetic resonance), and *in vitro* digestion experiments is employed. These methods can not only serve to characterize the structure of porous starch, but also offer detailed insights into various changes induced by chemical modification. The morphology and structure of PSGs are usually comprehensively characterized by microscopy techniques such as polarizing light microscopy (PLM), scanning electron microscopy (SEM), confocal laser scanning microscopy (CLSM), transmission electron microscopy (TEM), and atomic force microscopy (AFM) (Benavent-Gil & Rosell 2017 a; Cao et al. 2023; Chen et al. 2020).

PLM utilizes polarized light to illuminate the sample, revealing its optical properties and morphological information of the sample. SEM projects high-energy electron beams onto the sample surface to obtain high-resolution surface morphology information and reveals the microstructure and surface shape (Apinan et al. 2007). CLSM uses a laser light source to excite fluorescence and obtains three-dimensional images by controlling the laser focus, offering detailed insights into the internal organization and spatial distribution of particles

TABLE 13.5 Enzyme treatment methods, as well as morphology and characteristics of porous starch granules

SOURCES OF STARCH	USES OF ENZYME	MORPHOLOGY AND FUNCTIONALITY OF PSG OBTAINED BY ENZYMATIC MODIFICATION OF STARCH	REFERENCES
Corn	α-Amylase (*Bacillus licheniformis*), β-Amylase (Barley), Glucoamylase (*Aspergillus niger*)	1. The pore size of PSGs: α-Amylase > Glucoamylase > β-Amylase 2. Water or oil holding capacities: α-Amylase > Glucoamylase > β-Amylase	Jung et al. (2017)
Corn	Amyloglucosidase (EC 3.2.1.3), α-Amylase (EC 3.2.1.1), Cyclodextrin-glycosyltransferase (EC 2.4.1.19), Branching enzyme (EC 3.2.1.3)	1. The pore size of PSGs: Amyloglucosidase > α-Amylase > Cyclodextrin-glycosyltransferase 2. Water holding capacities: Amyloglucosidase > α-Amylase > Cyclodextrin-glycosyltransferase > Branching enzyme 3. The oil retention of PSGs: Treatment with α-Amylase significantly changes	Benavent-Gil and Rosell (2017b)
Normal dent corn and popcorn	α-Amylase (EC 3.2.1.1)	1. Crystallinity increased 2. The PSGs modified by popcorn starch enzyme have more dents on the surface, larger pores, and better water and oil absorption properties	Song, Zhong, et al. (2020)
Potato, corn, wheat, sweet potato	Amyloglucosidase (*Aspergillus niger*), α-Amylase (porcine pancreatin), Glycosyltransferase (*Aspergillus niger*), Branching enzyme (*Rhodothermus obamensis*)	Potato, corn, wheat, and sweet potato starches formed rich mesoporous structures (2–50 nm) after α-amylase and glucoamylase treatment, while glycosyltransferase and branching enzyme modification helped to form more macropores (>50 nm) Glycosyltransferase and branching enzyme modification resulted in larger particle size and specific surface area, as well as larger pore volume, and the generated PSGs have more significant adsorption capacity for dyes, oils, and heavy metal ions	Guo et al. (2020)
Rice	Amyloglucosidase (EC 3.2.1.3), Maltogenic α-amylase (EC 3.2.1.133)	The PSGs generated by Amyloglucosidase modification show large and shallow pore structures, while the PSGs generated by maltogenic α-amylase show small and deep pores. Both enzyme treatments resulted in an increase in the relative crystallinity of PSGs	Keeratiburana et al. (2020)
Tapioca	α-Amylase (*Bacillus licheniformis*, EC 3.2.1.1), Glucoamylase (*Aspergillus niger*, EC 3.2.1.3)	After 6 hours of hydrolysis, the PSGs produced by combining the two enzymes with modified starch exhibited larger single pores. Whether one enzyme is used alone or in combination, the pore size and specific surface area of the generated PSGs are increased	Prompiputtanapon et al. (2020)
Maize	Glucoamylase, α-Amylase	PSGs prepared by mixing the two enzymes exhibit a large and deep pore structure, a significantly increased particle size distribution and a significantly increased water and oil holding capacity	Han, Wen, et al. (2021)
Potato	α-Amylose	Forming nodules or pores with width of around 2–50 nm	Chen, McClements, et al. (2021)
Lotus seed	α-Amylase and Glucoamylase	After mixing α-amylase (50,000 U/g) and glucoamylase (100,000 U/g) to modify lotus seed starch, when the mass of the mixed enzyme added accounted for 1.5% of the dry weight of starch, the surface of PSGs was extremely rough, forming dense pores, and the water and oil absorption capacity reaches the maximum value. Compared with unmodified starch, the specific surface area and average pore diameter are significantly increased	Lin et al. (2022)
Cassava	Amyloglucosidase (*Aspergillus niger*), α-Amylase (*Bacillus licheniformis*)	Compared with natural starch, the pore size and surface area of PSGs prepared by mixing double enzymes increased	Figueroa-Flórez et al. (2023)
Edible canna	α-Amylase	Both the pore size and surface area increased (the control group had no pores on the surface of canna starch), as did the oil holding capacity, but the water holding capacity was not obvious	Purwitasari et al. (2023)

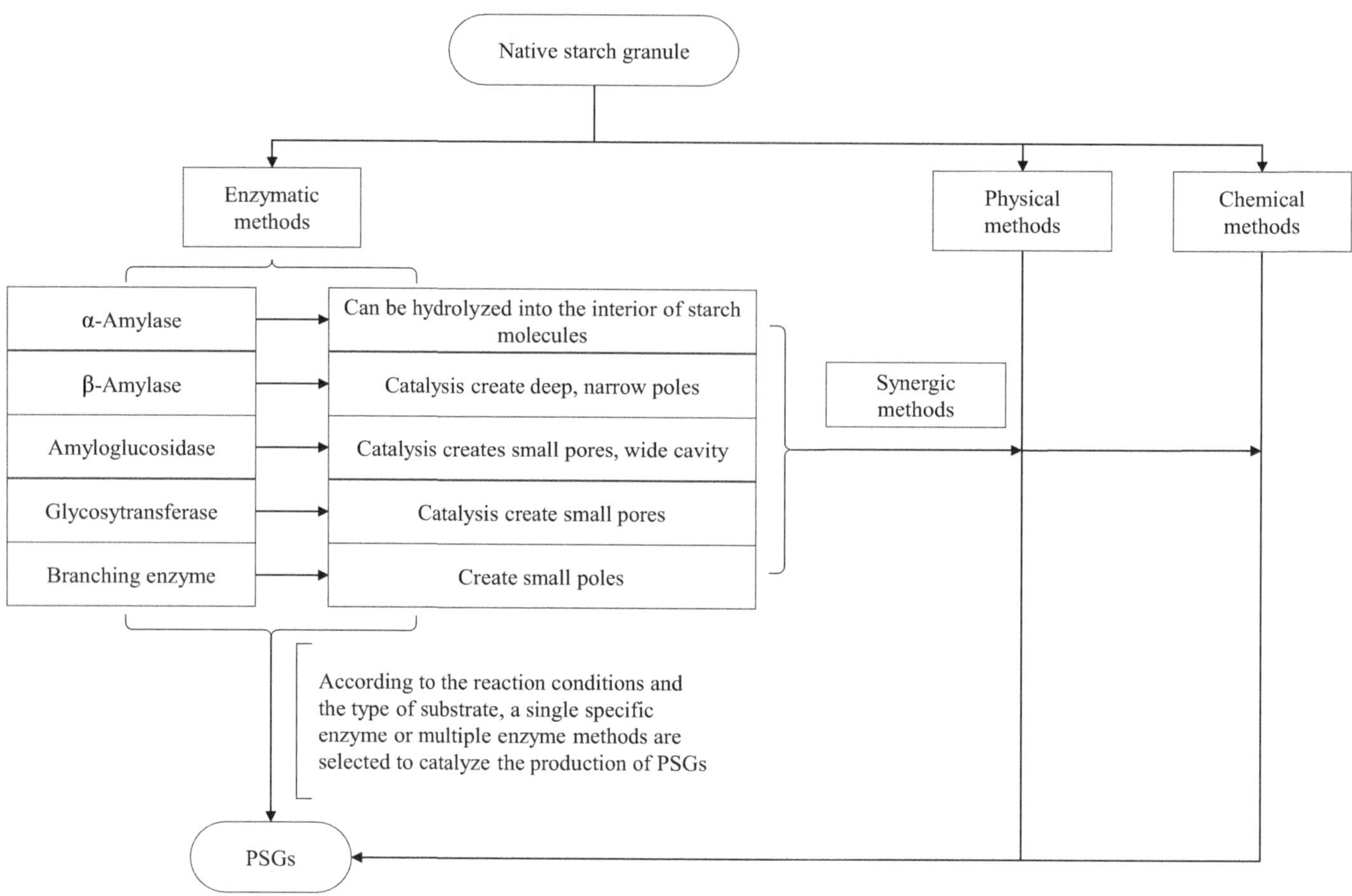

FIGURE 13.4   A summary for the enzymatic preparation of PSGs. (Made by the author.)

(Zhang et al. 2023). TEM and AFM can be used to gain in-depth understanding of the structure, pore size, distribution, and properties of molecular porous starch such as weight and pore depth (Chen et al. 2020).

FT-IR spectroscopy is highly responsive to structural modifications in PSGs (Chen et al. 2020). The distinctive band positions observed in the FT-IR spectrum of porous starch closely resemble those found in natural starch. However, with the enzymatic modification of starch granules, the starch granules form a pore structure, and the corresponding characteristic band intensity shows a downward trend, which is caused by the decrease in starch granule density (Chen et al. 2020). Ciardullo's FT-IR spectroscopy revealed the starch structure. The infrared absorption peak at $1,047\,cm^{-1}$ reflects the presence of crystalline regions, indicating the short-range ordered arrangement of starch macromolecules. In contrast, the absorption band at $1,022\,cm^{-1}$ corresponds to the structural characteristics of the amorphous region (Ciardullo et al. 2019). The absorbance ratios at $1,047\,cm^{-1}/1,022\,cm^{-1}$ serve as metrics for assessing the degree of short-term ordering in starch granules (Han, Wen, et al. 2021).

The structural attributes of porous starch can be elucidated through XRD analysis. Typically, natural starch reveals a semi-crystalline nature with distinct, sharp diffraction peaks in its XRD spectrum. Both native and porous starches display comparable crystalline patterns, with peaks appearing at the same positions. (Chen et al. 2020). However, there are differences regarding the change in relative crystallinity after enzyme preparation, which is affected by the type of enzyme, activity, dosage, and sources such as starch. In some instance, the relative crystallinity may increase, while in others, it may decrease (Jiang et al. 2017; Wang et al. 2016; Zhang et al. 2012).

13.7 CONCLUSION

Various types of starch syrups undergo different degrees of hydrolysis during the production process, resulting in syrups primarily composed of glucose and fructose, along with specific disaccharides or oligosaccharides. The proportions of these components significantly impact the physical and functional properties of starch syrups, thereby influencing their applications in the food and pharmaceutical industries.

The regular consumption of fructose from natural sources such as honey, fruits, and vegetables contribute to a healthy lifestyle. Despite studies suggesting a potential impact of a high fructose dietary profile on metabolism, normal fructose intake synergistically interacts with glucose through specialized metabolic pathways. In the liver, fructose catalyzes glucose absorption, while in the gut, glucose promotes fructose absorption, enhancing the body's ability to process dietary carbohydrates.

Trehalose, a functional disaccharide comprising two glucose molecules, exhibits unique properties extensively studied for applications in food, medicine, biology, and agriculture. As modern technology advances and research deepens on the catalytic pathway mechanism of trehalose synthase, its application fields are expected to broaden.

Driven by increasing consumer demand for functional foods, IMOs are widely utilized in the food industry as a low glycemic index, prebiotic ingredient, offering potential in the development of functional health foods. IMOs' low glycemic nature makes them suitable for glycemic control and weight management products. Additionally, their prebiotic properties support intestinal flora balance, making them essential in prebiotic-fortified foods and beverages, catering to health-conscious consumers. Despite challenges in efficient and economical industrial-scale production due to diverse IMOs, enzyme activities, and complex conditions, ongoing scientific research aims to streamline production methods for IMOs.

As a functional glucan, DRGs face challenges in research and industry due to their complex structure and composition. Despite being challenging to break down in the intestines, DRGs are fermented by microbial flora, promoting intestinal health. Industrially prepared using enzymes to modify glucans from various sources, DRGs possess a low glycemic index and exhibit various biological activities, making them valuable in food, medicine, and biotechnology applications.

The enzymatic method for PSGs (Porous Starch Granules) preparation offers specific modification advantages, with efficient, controllable, safe, non-toxic, and low-cost operations. Through comprehensive characterization using technologies like FT-IR, XRD, SEM, TEM, AFM, NMR, and CLSM, PSGs' unique properties are revealed, making them ideal carriers for functional substances. In particular, the synthesis of PSGs involves enzymatic modification of starch, yielding granules with abundant pores on the surface and extending to the center, contributing to their functionality. Widely used in food, medicine, and environmental protection, PSGs' future research focuses on understanding the mechanism of preparation, exploring diverse application fields, particularly in drug delivery, and establishing their safety as carriers. This pursuit aims to provide a more systematic and in-depth theoretical foundation for the enzymatic synthesis of PSGs, opening up possibilities for innovative applications.

Overall, starch bioconversion products, such as various types of syrups, fructose, trehalose, IMOs, DRGs and PSGs, play a key role in shaping the food industry. These products are derived through enzymatic processes and each has unique biochemical properties and applications. Starch syrup is widely used in various fields, while fructose aids in metabolism when consumed in moderate amounts. Trehalose is widely used in multiple industries due to its unique properties. As a low-glycemic prebiotic ingredient in functional foods, IMOs meets the needs of health-conscious consumers. DRGs have important application value in the fields of food, medicine and biotechnology due to their low glycemic index and various biological activities. Finally, the enzymatic preparation of PSGs has been proven to be a multifunctional carrier of functional substances, and its applications in food, medicine,

and environmental protection continue to expand. In this context, in-depth research on the preparation conditions of various enzyme conversion products and the development of enzymes is particularly important. Work in this area can not only optimize reaction conditions and improve yield and efficiency, but is also expected to innovate the types and properties of enzymes, providing a solid scientific foundation for wider and more efficient applications of starch bioconversion products.

REFERENCES

Adeva-Andany, M. M., N. Pérez-Felpete, C. Fernández-Fernández, C. Donapetry-García & C. Pazos-García (2016). Liver glucose metabolism in humans. *Bioscience Reports* **36**(6): e00416.

Aguib, Y., A. Heiseke, S. Gilch, C. Riemer, M. Baier, A. Ertmer & H. M. Schätzl (2009). Autophagy induction by trehalose counteracts cellular prion-infection. *Autophagy* **5**(3): 361–369.

Amaral-Fonseca, M., R. Morellon-Sterling, R. Fernandez-Lafuente & P. W. Tardioli (2021). Optimization of simultaneous saccharification and isomerization of dextrin to high fructose syrup using a mixture of immobilized amyloglucosidase and glucose isomerase. *Catalysis Today* **362**: 175–183.

Anderson-Dekkers, I., M. Nouwens-Roest, B. Peters & E. Vaughan (2021). Inulin. *Handbook of Hydrocolloids*, Elsevier: 537–562.

Apinan, S., I. Yujiro, Y. Hidefumi, F. Takeshi, P. Myllärinen, P. Forssell & K. Poutanen (2007). Visual observation of hydrolyzed potato starch granules by α-amylase with confocal laser scanning microscopy. *Starch-Stärke* **59**(11): 543–548.

Astina, J. & S. Sapwarobol (2019). Resistant maltodextrin and metabolic syndrome: a review. *Journal of the American College of Nutrition* **38**(4): 380–385.

Ayala-Soto, F. E., S. O. Serna-Saldívar, J. Welti-Chanes & J. A. Gutierrez-Uribe (2015). Phenolic compounds, antioxidant capacity and gelling properties of glucoarabinoxylans from three types of sorghum brans. *Journal of Cereal Science* **65**: 277–284.

Bae, J., S.-J. Jun, P.-S. Chang & S.-H. Yoo (2022). A unique biochemical reaction pathway towards trehalulose synthesis by an amylosucrase isolated from *Deinococcus deserti*. *New Biotechnology* **70**: 1–8.

Baky, M. H., M. Salah, N. Ezzelarab, P. Shao, M. S. Elshahed & M. A. Farag (2022). Insoluble dietary fibers: structure, metabolism, interactions with human microbiome, and role in gut homeostasis. *Critical Reviews in Food Science and Nutrition* **64**: 1–15.

Bancal, P., C. A. Henson, J. P. Gaudillère & N. C. Carpita (1991). Fructan chemical structure and sensitivity to an exohydrolase. *Carbohydrate Research* **217**: 137–151.

Barone, S., S. L. Fussell, A. K. Singh, F. Lucas, J. Xu, C. Kim, X. Wu, Y. Yu, H. Amlal & U. Seidler (2009). Slc2a5 (Glut5) is essential for the absorption of fructose in the intestine and generation of fructose-induced hypertension. *Journal of Biological Chemistry* **284**(8): 5056–5066.

Barreira, J. C., J. A. Pereira, M. B. P. Oliveira & I. C. Ferreira (2010). Sugars profiles of different chestnut (*Castanea sativa Mill.*) and almond (*Prunus dulcis*) cultivars by HPLC-RI. *Plant Foods for Human Nutrition* **65**: 38–43.

Barsanti, L., V. Passarelli, V. Evangelista, A. M. Frassanito & P. Gualtieri (2011). Chemistry, physico-chemistry and applications linked to biological activities of β-glucans. *Natural Product Reports* **28**(3): 457–466.

Basciano, H., L. Federico & K. Adeli (2005). Fructose, insulin resistance, and metabolic dyslipidemia. *Nutrition & Metabolism* **2**(1): 1–14.

Basso, A. & S. Serban (2020). Overview of immobilized enzymes' applications in pharmaceutical, chemical, and food industry. In J. M. Guisan, J. M. Bolivar, F. López-Gallego, & J. Rocha-Martín (eds.), *Immobilization of Enzymes and Cells: Methods and Protocols*, Humana Press: 27–63.

Belingheri, C., B. Giussani, M. T. Rodriguez-Estrada, A. Ferrillo & E. Vittadini (2015). Oxidative stability of high-oleic sunflower oil in a porous starch carrier. *Food Chemistry* **166**: 346–351.

Benavent-Gil, Y. & C. M. Rosell (2017a). Comparison of porous starches obtained from different enzyme types and levels. *Carbohydrate Polymers* **157**: 533–540.

Benavent-Gil, Y. & C. M. Rosell (2017b). Morphological and physicochemical characterization of porous starches obtained from different botanical sources and amylolytic enzymes. *International Journal of Biological Macromolecules* **103**: 587–595.

Bezerra, R. M., M. Ueno, M. S. Silva, D. Q. Tavares, C. R. Carvalho & M. J. Saad (2000). A high fructose diet affects the early steps of insulin action in muscle and liver of rats. *The Journal of Nutrition* **130**(6): 1531–1535.

Bharti, S. K., S. Krishnan, A. Kumar, A. K. Gupta, A. K. Ghosh & A. Kumar (2015). Mechanism-based antidiabetic activity of Fructo-and isomalto-oligosaccharides: validation by *in vivo*, *in silico* and *in vitro* interaction potential. *Process Biochemistry* **50**(2): 317–327.

Bhattacharya, A., A. Ruthes, F. Vilaplana, E. N. Karlsson, P. Adlecreutz & H. Stålbrand (2020). Enzyme synergy for the production of arabinoxylo-oligosaccharides from highly substituted arabinoxylan and evaluation of their prebiotic potential. *LWT* **131**: 109762.

Biliaderis, C. G., R. S. Swan & I. Arvanitoyannis (1999). Physicochemical properties of commercial starch hydrolyzates in the frozen state. *Food Chemistry* **64**(4): 537–546.

Birch, G. & I. Etheridge (1973). Chemical and physiological properties of glucose syrup components. *Starch-Stärke* **25**(7): 235–238.

Bivolarski, V., T. Vasileva, P. Bozov & I. Iliev (2013). Influence of different acceptors on synthesis of glucooligosaccharides by purified dextransucrase from Leuconostoc mesenteroides URE 13. *Comptes rendus de l'Académie bulgare des Sciences* **66**(10): 1405–1412.

Branca, C., S. Magazu, G. Maisano & P. Migliardo (1999). Anomalous cryoprotective effectiveness of trehalose: Raman scattering evidences. *The Journal of Chemical Physics* **111**(1): 281–287.

Bray, G. (2008). Fructose: should we worry? *International Journal of Obesity* **32**(7): S127–S131.

Bray, G. A., S. J. Nielsen & B. M. Popkin (2004). Consumption of high-fructose corn syrup in beverages may play a role in the epidemic of obesity. *The American Journal of Clinical Nutrition* **79**(4): 537–543.

Bredereck, H. (1930). Zur konstitution der trehalose. *Berichte der deutschen chemischen Gesellschaft (A and B Series)* **63**(4): 959–965.

Brennan, C. S. & L. J. Cleary (2007). Utilisation Glucagel® in the β-glucan enrichment of breads: a physicochemical and nutritional evaluation. *Food Research International* **40**(2): 291–296.

Brouns, F., E. Arrigoni, A. M. Langkilde, I. Verkooijen, C. Fässler, H. Andersson, B. Kettlitz, M. van Nieuwenhoven, H. Philipsson & R. Amadò (2007). Physiological and metabolic properties of a digestion-resistant maltodextrin, classified as type 3 retrograded resistant starch. *Journal of Agricultural and Food Chemistry* **55**(4): 1574–1581.

Buchholz, K. & J. Seibel (2008). Industrial carbohydrate biotransformations. *Carbohydrate Research* **343**(12): 1966–1979.

Bulmer, G. S., P. De Andrade, R. A. Field & J. M. van Munster (2021). Recent advances in enzymatic synthesis of β-glucan and cellulose. *Carbohydrate Research* **508**: 108411.

Burek, M., S. Waśkiewicz & I. Wandzik (2015). Trehalose–properties, biosynthesis and applications. *Methods* **3**: 9–10.

Buziau, A. M., J. L. Scheijen, C. D. Stehouwer, N. Simons, M. C. Brouwers & C. G. Schalkwijk (2020). Development and validation of a UPLC-MS/MS method to quantify fructose in serum and urine. *Journal of Chromatography B* **1155**: 122299.

Candela, M., S. Maccaferri, S. Turroni, P. Carnevali & P. Brigidi (2010). Functional intestinal microbiome, new frontiers in prebiotic design. *International Journal of Food Microbiology* **140**(2–3): 93–101.

Cao, F., S. Lu, L. Wang, M. Zheng & S. Y. Quek (2023). Modified porous starch for enhanced properties: synthesis, characterization and applications. *Food Chemistry* **415**: 135765.

Carvalho, V. S., L. Gómez-Delgado, M. Á. Curto, M. B. Moreno, P. Pérez, J. C. Ribas & J. C. G. Cortés (2021). Analysis and application of a suite of recombinant endo-β(1,3)-D-glucanases for studying fungal cell walls. *Microbial Cell Factories* **20**(1): 126.

Ceriello, A., M. Hanefeld, L. Leiter, L. Monnier, A. Moses, D. Owens, N. Tajima & J. Tuomilehto (2004). Postprandial glucose regulation and diabetic complications. *Archives of Internal Medicine* **164**(19): 2090–2095.

Chaitanya, N. S., A. Devi, S. Sahu & P. Alugoju (2021). Molecular mechanisms of action of Trehalose in cancer: a comprehensive review. *Life Sciences* **269**: 118968.

Chen, A., H. Tapia, J. M. Goddard & P. A. Gibney (2022). Trehalose and its applications in the food industry. *Comprehensive Reviews in Food Science and Food Safety* **21**(6): 5004–5037.

Chen, H.-L., Y.-H. Lu, J.-J. Lin & L.-Y. Ko (2001). Effects of isomalto-oligosaccharides on bowel functions and indicators of nutritional status in constipated elderly men. *Journal of the American College of Nutrition* **20**(1): 44–49.

Chen, J., Y. Wang, J. Liu & X. Xu (2020). Preparation, characterization, physicochemical property and potential application of porous starch: a review. *International Journal of Biological Macromolecules* **148**: 1169–1181.

Chen, J., J. Yang, H. Du, M. Aslam, W. Wang, W. Chen, T. Li, Z. Liu & X. Liu (2021). Laminarin, a major polysaccharide in stramenopiles. *Marine Drugs* **19**(10): 576.

Chen, L., D. J. McClements, Y. Ma, T. Yang, F. Ren, Y. Tian & Z. Jin (2021). Analysis of porous structure of potato starch granules by low-field NMR cryoporometry and AFM. *International Journal of Biological Macromolecules* **173**: 307–314.

Chen, Z., S. Li, Y. Fu, C. Li, D. Chen & H. Chen (2019). Arabinoxylan structural characteristics, interaction with gut microbiota and potential health functions. *Journal of Functional Foods* **54**: 536–551.

Chioru, A. & A. Chirsanova (2023). β-Glucans: characterization, extraction methods, and valorization. *Food and Nutrition Sciences* **14**(10): 963–983.

Cho, Y. E., D. K. Kim, W. Seo, B. Gao, S. H. Yoo & B. J. Song (2021). Fructose promotes leaky gut, endotoxemia, and liver fibrosis through ethanol-inducible cytochrome P450-2E1-mediated oxidative and nitrative stress. *Hepatology* **73**(6): 2180–2195.

Chou, C. C. & G. Jasovsky (1993). Advantages of Ecosorb™ precoats in liquid sugar production. *International Sugar Journal* **95**(1138): 425–430.

Chronakis, I. S. (1998). On the molecular characteristics, compositional properties, and structural-functional mechanisms of maltodextrins: a review. *Critical Reviews in Food Science and Nutrition* **38**(7): 599–637.

Ciardullo, K., E. Donner, M. R. Thompson & Q. Liu (2019). Influence of extrusion mixing on preparing lipid complexed pea starch for functional foods. *Starch-Stärke* **71**(7–8): 1800196.

Corwin, S. G. (2020). *Structural and Functional Properties of Enzymatically Modified Slow Digesting α-Glucans*, Purdue University.

Das, R. & A. M. Kayastha (2019). Enzymatic hydrolysis of native granular starches by a new β-amylase from peanut (Arachis hypogaea). *Food Chemistry* **276**: 583–590.

De Filippo, C., D. Cavalieri, M. Di Paola, M. Ramazzotti, J. B. Poullet, S. Massart, S. Collini, G. Pieraccini & P. Lionetti (2010). Impact of diet in shaping gut microbiota revealed by a comparative study in children from Europe and rural Africa. *Proceedings of the National Academy of Sciences* **107**(33): 14691–14696.

Dobruchowska, J. M., G. J. Gerwig, S. Kralj, P. Grijpstra, H. Leemhuis, L. Dijkhuizen & J. P. Kamerling (2012). Structural characterization of linear isomalto-/malto-oligomer products synthesized by the novel GTFB 4, 6-α-glucanotransferase enzyme from *Lactobacillus reuteri* 121. *Glycobiology* **22**(4): 517–528.

Douard, V. & R. P. Ferraris (2008). Regulation of the fructose transporter GLUT5 in health and disease. *American Journal of Physiology-Endocrinology and Metabolism* **295**(2): E227–E237.

Doyle, E. M., C. T. Kelly & W. M. Fogarty (1989). The high maltose-producing α-amylase of *Penicillium expansum*. *Applied Microbiology and Biotechnology* **30**: 492–496.

Drożdż, K., K. Nabrdalik, W. Hajzler, H. Kwiendacz, J. Gumprecht & G. Y. Lip (2021). Metabolic-associated fatty liver disease (MAFLD), diabetes, and cardiovascular disease: associations with fructose metabolism and gut microbiota. *Nutrients* **14**(1): 103.

Drusch, S., Y. Serfert, A. Van Den Heuvel & K. Schwarz (2006). Physicochemical characterization and oxidative stability of fish oil encapsulated in an amorphous matrix containing trehalose. *Food Research International* **39**(7): 807–815.

Du, B., Z. Bian & B. Xu (2014). Skin health promotion effects of natural beta-glucan derived from cereals and microorganisms: a review. *Phytotherapy Research* **28**(2): 159–166.

Du, B., C. Lin, Z. Bian & B. Xu (2015). An insight into anti-inflammatory effects of fungal beta-glucans. Trends in Food *Science & Technology* **41**(1): 49–59.

Du, B. & B. Xu (2014). Oxygen radical absorbance capacity (ORAC) and ferric reducing antioxidant power (FRAP) of β-glucans from different sources with various molecular weight. *Bioactive Carbohydrates and Dietary Fibre* **3**(1): 11–16.

Du, L. & A. P. Heaney (2012). Regulation of adipose differentiation by fructose and GluT5. *Molecular Endocrinology* **26**(10): 1773–1782.

Du, Y. & Y. Zhao (2012). Optimization of fermentation conditions for trehalose production by a marine yeast. *African Journal of Biotechnology* **11**(14): 3352–3357.

Duarte-Delgado, D., C. E. Ñústez-López, C. E. Narváez-Cuenca, L. P. Restrepo-Sánchez, S. E. Melo, F. Sarmiento, A. C. Kushalappa & T. Mosquera-Vásquez (2016). Natural variation of sucrose, glucose and fructose contents in Colombian genotypes of *Solanum tuberosum* Group Phureja at harvest. *Journal of the Science of Food and Agriculture* **96**(12): 4288–4294.

Echigo, R., N. Shimohata, K. Karatsu, F. Yano, Y. Kayasuga-Kariya, A. Fujisawa, T. Ohto, Y. Kita, M. Nakamura & S. Suzuki (2012). Trehalose treatment suppresses inflammation, oxidative stress, and vasospasm induced by experimental subarachnoid hemorrhage. *Journal of Translational Medicine* **10**(1): 1–13.

EFSA Panel on Nutrition, N. F., F. Allergens, D. Turck, T. Bohn, J. Castenmiller, S. De Henauw, K. I. Hirsch-Ernst, A. Maciuk, I. Mangelsdorf, H. J. McArdle & A. Naska (2024). Safety of isomaltulose syrup (dried) as a novel food pursuant to regulation (EU) 2015/2283. *EFSA Journal* **22**(1): e8491.

Eke-Ejiofor, J. (2015). Functional properties of starches, physico-chemical and rheological properties of glucose syrup made from cassava and different potato varieties. *International Journal of Recent Scientific Research* **6**(6): 4400–4406.

Englyst, H. N., S. Kingman & J. Cummings (1992). Classification and measurement of nutritionally important starch fractions. *European Journal of Clinical Nutrition* **46**: S33–S50.

Englyst, K. N. & H. N. Englyst (2005). Carbohydrate bioavailability. *British Journal of Nutrition* **94**(1): 1–11.

Fadel, A., A. M. Mahmoud, J. J. Ashworth, W. Li, Y. L. Ng & A. Plunkett (2018). Health-related effects and improving extractability of cereal arabinoxylans. *International Journal of Biological Macromolecules* **109**: 819–831.

Fang, P., Q. Lei, M. Lv, L. Xu, K. Dong, W. Zhao, D. Yue, Z. Cao & Q. Lin (2023). Effects of the combination of *Lactobacillus helveticus* and isomalto-oligosaccharide on survival, gut microbiota, and immune function in *Apis cerana* worker bees. *Letters in Applied Microbiology* **76**(12): ovad134.

Ferreira, J. C., V. M. Paschoalin, A. D. Panek & L. C. Trugo (1997). Comparison of three different methods for trehalose determination in yeast extracts. *Food Chemistry* **60**(2): 251–254.

Figueroa-Flórez, J. A., E. D. A. Dagobeth, E. Cadena-Chamorro, E. Rodríguez-Sandoval, J. G. Salcedo-Mendoza & H. J. Ciro-Velásquez (2023). Effect of physical and thermal pretreatments on enzymatic activity in the production of microporous cassava starch. *Agronomía Colombiana* **41**(1): e105089–e105089.

Forbes, A. L. & B. A. Bowman (1993). Health effects of dietary fructose. Am. J. Clin. Nutr 58 (5 Supplement): 721S–823S.

Fourage, L., M. Dion & B. Colas (2000). Kinetic study of a thermostable β-glycosidase of Thermus thermophilus. Effects of temperature and glucose on hydrolysis and transglycosylation reactions. *Glycoconjugate Journal* **17**: 377–383.

Fujii, M., T. Homma & M. Taniguchi (1988). Synergism of α-amylase and glucoamylase on hydrolysis of native starch granules. *Biotechnology and Bioengineering* **32**(7): 910–915.

Galant, A., R. Kaufman & J. Wilson (2015). Glucose: detection and analysis. *Food Chemistry* **188**: 149–160.

Gallagher, J. A., A. J. Cairns, L. B. Turner, S. Norio, B. Noureddine & O. Shuichi (2007). Fructan in temperate forage grasses; agronomy, physiology, and molecular biology. *Research Signpost*, **371661**(2): 15–-46.

Galmarini, M. V., C. van Baren, M. C. Zamora, J. Chirife, P. Di Leo Lira & A. Bandoni (2011). Impact of trehalose, sucrose and/or maltodextrin addition on aroma retention in freeze dried strawberry puree. *International Journal of Food Science & Technology* **46**(7): 1337–1345.

Gangoiti, J., T. Pijning & L. Dijkhuizen (2018). Biotechnological potential of novel glycoside hydrolase family 70 enzymes synthesizing α-glucans from starch and sucrose. *Biotechnology Advances* **36**(1): 196–207.

Georgelis, N., K. Fencil & C. M. Richael (2018). Validation of a rapid and sensitive HPLC/MS method for measuring sucrose, fructose and glucose in plant tissues. *Food Chemistry* **262**: 191–198.

Geraylou, Z., C. Souffreau, E. Rurangwa, G. E. Maes, K. I. Spanier, C. M. Courtin, J. A. Delcour, J. Buyse & F. Ollevier (2013). Prebiotic effects of arabinoxylan oligosaccharides on juvenile Siberian sturgeon (*Acipenser baerii*) with emphasis on the modulation of the gut microbiota using 454 pyrosequencing. *FEMS Microbiology Ecology* **86**(2): 357–371.

Glenn, G. M., A. P. Klamczynski, D. F. Woods, B. Chiou, W. J. Orts & S. H. Imam (2010). Encapsulation of plant oils in porous starch microspheres. *Journal of Agricultural and Food Chemistry* **58**(7): 4180–4184.

Goffin, D., B. Wathelet, C. Blecker, C. Deroanne, Y. Malmendier & M. Paquot (2010). Comparison of the glucooligosaccharide profiles produced from maltose by two different transglucosidases from *Aspergillus niger*. *Biotechnology, Agronomy and Society and Environment*, **14**(4): 607–616.

Gonzaga, M. L. C., D. P. Bezerra, A. P. N. N. Alves, N. M. N. de Alencar, R. de Oliveira Mesquita, M. W. Lima, S. de Aguiar Soares, C. Pessoa, M. O. de Moraes & L. V. Costa-Lotufo (2009). *In vivo* growth-inhibition of Sarcoma 180 by an α-(1→4)-glucan-β-(1→6)-glucan-protein complex polysaccharide obtained from *Agaricus blazei* Murill. *Journal of Natural Medicines* **63**: 32–40.

Gonzalez, A. & Y. J. Wang (2021). Surface removal enhances the formation of a porous structure in potato starch. *Starch-Stärke* **73**(7–8): 2000261.

Guo, L., J. Li, Y. Gui, Y. Zhu, B. Yu, C. Tan, Y. Fang & B. Cui (2020). Porous starches modified with double enzymes: structure and adsorption properties. *International Journal of Biological Macromolecules* **164**: 1758–1765.

Guo, L., Y. Yuan, J. Li, C. Tan, S. Janaswamy, L. Lu, Y. Fang & B. Cui (2021). Comparison of functional properties of porous starches produced with different enzyme combinations. *International Journal of Biological Macromolecules* **174**: 110–119.

Gupta, A., A. Jamal, D. A. Jamil & H. A. Al-Aubaidy (2023). A systematic review exploring the mechanisms by which citrus bioflavonoid supplementation benefits blood glucose levels and metabolic complications in type 2 diabetes mellitus. *Diabetes & Metabolic Syndrome: Clinical Research & Reviews* **17**: 102884.

Han, D.-J., B.-H. Lee & S.-H. Yoo (2021). Physicochemical properties of turanose and its potential applications as a sucrose substitute. *Food Science and Biotechnology* **30**: 433–441.

Han, X., H. Wen, Y. Luo, J. Yang, W. Xiao, X. Ji & J. Xie (2021). Effects of α-amylase and glucoamylase on the characterization and function of maize porous starches. *Food Hydrocolloids* **116**: 106661.

Hanefeld, M. & F. Schaper (2007). The role of alpha-glucosidase inhibitors (Acarbose). In C. E. Mogensen (ed.), *Pharmacotherapy of Diabetes: New Developments: Improving Life and Prognosis for Diabetic Patients*, Springer: 143–152.

Harni, M., S. Putri & T. Handayani (2021). Characteristics of glucose syrup from various sources of starch. *IOP Conference Series: Earth and Environmental Science* 757: 012064. IOP Publishing.

Hasek, L. Y., R. J. Phillips, G. Zhang, K. P. Kinzig, C. Y. Kim, T. L. Powley & B. R. Hamaker (2018). Dietary slowly digestible starch triggers the gut–brain axis in obese rats with accompanied reduced food intake. *Molecular Nutrition & Food Research* **62**(5): 1700117.

Hawkins, M., I. Gabriely, R. Wozniak, C. Vilcu, H. Shamoon & L. Rossetti (2002). Fructose improves the ability of hyperglycemia per se to regulate glucose production in type 2 diabetes. *Diabetes* **51**(3): 606–614.

He, H.-J., J. Qiao, Y. Liu, Q. Guo, X. Ou & X. Wang (2021). Isolation, structural, functional, and bioactive properties of cereal arabinoxylan—a critical review. *Journal of Agricultural and Food Chemistry* **69**(51): 15437–15457.

Hieronimus, B., V. Medici, A. A. Bremer, V. Lee, M. V. Nunez, D. M. Sigala, N. L. Keim, P. J. Havel & K. L. Stanhope (2020). Synergistic effects of fructose and glucose on lipoprotein risk factors for cardiovascular disease in young adults. *Metabolism* **112**: 154356.

Him, J. L., L. Pelosi, H. Chanzy, J. L. Putaux & V. Bulone (2001). Biosynthesis of (1→3)-β-d-glucan (callose) by detergent extracts of a microsomal fraction from *Arabidopsis thaliana*. *European Journal of Biochemistry* **268**(17): 4628–4638.

Hj. Latip, D. N., H. Samsudin, U. Utra & A. K. Alias (2021). Modification methods toward the production of porous starch: a review. *Critical Reviews in Food Science and Nutrition* **61**(17): 2841–2862.

Hobbs, L. (2009). Sweeteners from starch: production, properties and uses. In J. BeMiller & R. Whistler (eds.), *Starch*, Elsevier: 797–832.

Hong, M.-G., S.-H. Yoo & B.-H. Lee (2022). Effect of highly branched α-glucans synthesized by dual glycosyltransferases on the glucose release rate. *Carbohydrate Polymers* **278**: 119016.

Hu, Y., A. Ketabi, A. Buchko & M. Gänzle (2013). Metabolism of isomalto-oligosaccharides by Lactobacillus reuteri and bifidobacteria. *Letters in Applied Microbiology* **57**(2): 108–114.

Hua, X. & R. Yang (2016). Enzymes in starch processing. *Enzymes in Food and Beverage Processing*: 139–170.

Hughes, R. L., D. A. Alvarado, K. S. Swanson & H. D. Holscher (2022). The prebiotic potential of inulin-type fructans: a systematic review. *Advances in Nutrition* **13**(2): 492–529.

Hurtta, M., I. Pitkänen & J. Knuutinen (2004). Melting behaviour of D-sucrose, D-glucose and D-fructose. *Carbohydrate Research* **339**(13): 2267–2273.

Hwang, I. S., H. Ho, B. B. Hoffman & G. M. Reaven (1987). Fructose-induced insulin resistance and hypertension in rats. *Hypertension* **10**(5): 512–516.

Iturriaga, G., D. F. Gaff & R. Zentella (2000). New desiccation-tolerant plants, including a grass, in the central highlands of Mexico, accumulate trehalose. *Australian Journal of Botany* **48**(2): 153–158.

Jacobs, M. S., M. S. Izydorczyk, K. R. Preston & J. E. Dexter (2008). Evaluation of baking procedures for incorporation of barley roller milling fractions containing high levels of dietary fibre into bread. *Journal of the Science of Food and Agriculture* **88**(4): 558–568.

Jacobs, S. (2016). Membership categories and dues structures. In S. Jacobs (ed.), *Membership Essentials: Recruitment, Retention, Roles, Responsibilities, and Resources*, ASAE/Wiley: 125–137.

Jalaludin, I. & J. Kim (2021). Comparison of ultraviolet and refractive index detections in the HPLC analysis of sugars. *Food Chemistry* **365**: 130514.

Janeček, Š., B. Svensson & E. A. MacGregor (2007). A remote but significant sequence homology between glycoside hydrolase clan GH-H and family GH31. *FEBS Letters* **581**(7): 1261–1268.

Jang, C., S. Wada, S. Yang, B. Gosis, X. Zeng, Z. Zhang, Y. Shen, G. Lee, Z. Arany & J. D. Rabinowitz (2020). The small intestine shields the liver from fructose-induced steatosis. *Nature Metabolism* **2**(7): 586–593.

Jiang, L., M. Lin, Y. Zhang, Y. Li, X. Xu, S. Li & H. Huang (2013). Identification and characterization of a novel trehalose synthase gene derived from saline-alkali soil metagenomes. *PloS One* **8**(10): e77437.

Jiang, L., C. Shen, J. Dai & Q. Meng (2013). Di-rhamnolipids improve effect of trehalose on both hypothermic preservation and cryopreservation of rat hepatocytes. *Applied Microbiology and Biotechnology* **97**: 4553–4561.

Jiang, S., Z. Yu, H. Hu, J. Lv, H. Wang & S. Jiang (2017). Adsorption of procyanidins onto chitosan-modified porous rice starch. *LWT* **84**: 10–17.

Jung, H.-T., C.-S. Park, Y.-E. Shim, H. Shin, M.-Y. Baik, H.-S. Kim, S.-H. Yoo, D.-H. Seo & B.-H. Lee (2020). Enzymatically elongated rice starches by amylosucrase from *Deinococcus geothermalis* lead to slow down the glucose generation rate at the mammalian α-glucosidase level. *International Journal of Biological Macromolecules* **149**: 767–772.

Jung, Y.-S., M.-G. Hong, S.-H. Park, B.-H. Lee & S.-H. Yoo (2019). Biocatalytic fabrication of α-glucan-coated porous starch granules by amylolytic and glucan-synthesizing enzymes as a target-specific delivery carrier. *Biomacromolecules* **20**(11): 4143–4149.

Jung, Y.-S., B.-H. Lee & S.-H. Yoo (2017). Physical structure and absorption properties of tailor-made porous starch granules produced by selected amylolytic enzymes. *PloS One* **12**(7): e0181372.

Kakoty, V., S. KC, S. K. Dubey, C.-H. Yang & R. Taliyan (2021). Neuroprotective effects of trehalose and sodium butyrate on preformed fibrillar form of α-synuclein-induced rat model of Parkinson's disease. *ACS Chemical Neuroscience* **12**(14): 2643–2660.

Kaulpiboon, J., P. Rudeekulthamrong, S. Watanasatitarpa, K. Ito & P. Pongsawasdi (2015). Synthesis of long-chain isomaltooligosaccharides from tapioca starch and an in vitro investigation of their prebiotic properties. *Journal of Molecular Catalysis B: Enzymatic* **120**: 127–135.

Kearsley, M. & S. Dziedzic (1995). *Handbook of Starch Hydrolysis Products and their Derivatives,* Blackie Academic & Professional, Glasgow: 275.

Keeratiburana, T., A. R. Hansen, S. Soontaranon, A. Blennow & S. Tongta (2020). Porous high amylose rice starch modified by amyloglucosidase and maltogenic α-amylase. *Carbohydrate Polymers* **230**: 115611.

Kelley, G. L., G. Allan & S. Azhar (2004). High dietary fructose induces a hepatic stress response resulting in cholesterol and lipid dysregulation. *Endocrinology* **145**(2): 548–555.

Ketabi, A., L. Dieleman & M. Gänzle (2011). Influence of isomalto-oligosaccharides on intestinal microbiota in rats. *Journal of Applied Microbiology* **110**(5): 1297–1306.

Khalifeh, M., G. E. Barreto & A. Sahebkar (2019). Trehalose as a promising therapeutic candidate for the treatment of Parkinson's disease. *British Journal of Pharmacology* **176**(9): 1173–1189.

Khalifeh, M., G. E. Barreto & A. Sahebkar (2021). Therapeutic potential of trehalose in neurodegenerative diseases: the knowns and unknowns. *Neural Regeneration Research* **16**(10): 2026.

Kim, H.-J., H.-S. Chun & H.-Y. Kim (2004). Effects of corn syrup with different dextrose equivalent on quality attributes of black sesame Dasik, a Korean traditional snack. *Journal of the Korean Society of Food Science and Nutrition* **33**(8): 1414–1417.

Kim, M., I. I. Astapova, S. N. Flier, S. A. Hannou, L. Doridot, A. Sargsyan, H. H. Kou, A. J. Fowler, G. Liang & M. A. Herman (2017). Intestinal, but not hepatic, ChREBP is required for fructose tolerance. *JCI Insight* **2**(24): e96703.

Klangpetch, W., A. Pattarapisitporn, S. Phongthai, N. Utama-Ang, T. Laokuldilok, P. Tangjaidee, T. I. Wirjantoro & P. Jaichakan (2022). Microwave-assisted enzymatic hydrolysis to produce xylooligosaccharides from rice husk alkali-soluble arabinoxylan. *Scientific Reports* **12**(1): 11.

Kohmoto, T., F. Fukui, H. Takaku, Y. Machida, M. Arai & T. Mi Tsuoka (1988). Effect of isomalto-oligosaccharides on human fecal flora. *Bifidobacteria and Microflora* **7**(2): 61–69.

Kong, X. (2020). Fine structure of amylose and amylopectin. In S. Wang (ed.), *Starch Structure, Functionality and Application in Foods*, Springer: 29–39.

Kumar, S., A. Basu, K. Anu-Appaiah, B. Gnanesh Kumar & S. Mutturi (2020). Identification and characterization of novel transglycosylating α-glucosidase from Aspergillus neoniger. *Journal of Applied Microbiology* **129**(6): 1644–1656.

Lanaspa, M. A., T. Ishimoto, N. Li, C. Cicerchi, D. J. Orlicky, P. Ruzycki, C. Rivard, S. Inaba, C. A. Roncal-Jimenez & E. S. Bales (2013). Endogenous fructose production and metabolism in the liver contributes to the development of metabolic syndrome. *Nature Communications* **4**(1): 2434.

Lazaridou, A., C. G. Biliaderis & M. S. Izydorczyk (2007). Cereal beta-glucans: structures, physical properties, and physiological functions. In C. G. Biliaderis, & M. S. Izydorczyk (eds.), *Functional Food Carbohydrates*, CRC Press: 1–72.

Lederer, E. (1976). Cord factor and related trehalose esters. *Chemistry and Physics of Lipids* **16**(2): 91–106.

Lee, D., S.-D. Park, S.-J. Jun, J.-T. Park, P.-S. Chang & S.-H. Yoo (2022). Differentiated structure of synthetic glycogen-like particle by the combined action of glycogen branching enzymes and amylosucrase. *International Journal of Biological Macromolecules* **195**: 152–162.

Lee, H.-S., J.-H. Auh, H.-G. Yoon, M.-J. Kim, J.-H. Park, S.-S. Hong, M.-H. Kang, T.-J. Kim, T.-W. Moon & J.-W. Kim (2002). Cooperative action of α-glucanotransferase and maltogenic amylase for an improved process of isomaltooligosaccharide (IMO) production. *Journal of Agricultural and Food Chemistry* **50**(10): 2812–2817.

Lee, J., E.-W. Lin, U. Y. Lau, J. L. Hedrick, E. Bat & H. D. Maynard (2013). Trehalose glycopolymers as excipients for protein stabilization. *Biomacromolecules* **14**(8): 2561–2569.

Leemhuis, H., T. Pijning, J. M. Dobruchowska, S. S. van Leeuwen, S. Kralj, B. W. Dijkstra & L. Dijkhuizen (2013). Glucansucrases: three-dimensional structures, reactions, mechanism, α-glucan analysis and their implications in biotechnology and food applications. *Journal of Biotechnology* **163**(2): 250–272.

Lees, R. (2012). *Sugar Confectionery and Chocolate Manufacture*, Springer Science & Business Media.

Levine, H. & L. Slade (1986). A polymer physico-chemical approach to the study of commercial starch hydrolysis products (SHPs). *Carbohydrate Polymers* **6**(3): 213–244.

Li, C., Kong, H., Yang, Q., Gu, Z., Ban, X., Cheng, L., Hong, Y. & Li, Z. (2021). A temperature-mediated two-step saccharification process enhances maltose yield from high-concentration maltodextrin solutions. *Journal of the Science of Food and Agriculture* **101**: 3742–3748.

Li, G., X. Wei, R. Wu, W. Zhou, Y. Li, Z. Zhu & C. You (2022). Stoichiometric conversion of maltose for biomanufacturing by *in vitro* synthetic enzymatic biosystems. *BioDesign Research* 2022: 9806749.

Li, H., H. Su, S. B. Kim, Y. K. Chang, S.-K. Hong, Y.-G. Seo & C.-J. Kim (2012). Enhanced production of trehalose in *Escherichia coli* by homologous expression of *otsBA* in the presence of the trehalase inhibitor, validamycin A, at high osmolarity. *Journal of Bioscience and Bioengineering* **113**(2): 224–232.

Li, H., M. S. Turner & S. Dhital (2016). Encapsulation of *Lactobacillus plantarum* in porous maize starch. *LWT* **74**: 542–549.

Li, M., J. Li & C. Zhu (2018). Effect of ultrasound pretreatment on enzymolysis and physicochemical properties of corn starch. *International Journal of Biological Macromolecules* **111**: 848–856.

Li, N., H. Wang, L. Li, H. Cheng, D. Liu, H. Cheng & Z. Deng (2016). Integrated approach to producing high-purity trehalose from maltose by the yeast *Yarrowia lipolytica* displaying trehalose synthase (TreS) on the cell surface. *Journal of Agricultural and Food Chemistry* **64**(31): 6179–6187.

Li, Y.-J., G.-Y. Zhao, W. Du & T.-J. Zhang (2009). Effect of dietary isomaltooligosaccharides on nutrient digestibility and concentration of glucose, insulin, cholesterol and triglycerides in serum of growing pigs. *Animal Feed Science and Technology* **151**(3–4): 312–315.

Li, Z., H. Kong, Z. Li, Z. Gu, X. Ban, Y. Hong, L. Cheng & C. Li (2023). Designing liquefaction and saccharification processes of highly concentrated starch slurry: challenges and recent advances. *Comprehensive Reviews in Food Science and Food Safety* **22**(3): 1597–1612.

Liang, J., R. Huang, Y. Huang, X. Wang, L. Du & Y. Wei (2013). Cloning, expression, properties, and functional amino acid residues of new trehalose synthase from *Thermomonospora curvata* DSM 43183. *Journal of Molecular Catalysis B: Enzymatic* **90**: 26–32.

Light, S. H., L. A. Cahoon, K. V. Mahasenan, M. Lee, B. Boggess, A. S. Halavaty, S. Mobashery, N. E. Freitag & W. F. Anderson (2017). Transferase versus hydrolase: the role of conformational flexibility in reaction specificity. *Structure* **25**(2): 295–304.

Lima, D. M., P. Fernandes, D. S. Nascimento, L. Rita de Cássia & S. A. De Assis (2011). Fructose syrup: a biotechnology asset. *Food Technology and Biotechnology* **49**(4): 424.

Lin, Q., H. Xiao, G.-Q. Liu, Z. Liu, L. Li & F. Yu (2013). Production of maltose syrup by enzymatic conversion of rice starch. *Food and Bioprocess Technology* **6**: 242–248.

Lin, Y., L. Liu, L. Li, Y. Xu, Y. Zhang & H. Zeng (2022). Properties and digestibility of a novel porous starch from lotus seed prepared via synergistic enzymatic treatment. *International Journal of Biological Macromolecules* **194**: 144–152.

Linko, P., S. Hakulin & Y.-Y. Linko (1983). Extrusion cooking of barley starch for the production of glucose syrup and ethanol. *Journal of Cereal Science* **1**(4): 275–284.

Litherland, G. J., E. Hajduch, G. W. Gould & H. S. Hundal (2004). Fructose transport and metabolism in adipose tissue of Zucker rats: diminished GLUT5 activity during obesity and insulin resistance. *Molecular and Cellular Biochemistry* **261**: 23–33.

Liu, H., S. Yang, Q. Liu, R. Wang & T. Wang (2018). A process for production of trehalose by recombinant trehalose synthase and its purification. *Enzyme and Microbial Technology* **113**: 83–90.

Liu, J., X. Wang, H. Yong, J. Kan & C. Jin (2018). Recent advances in flavonoid-grafted polysaccharides: synthesis, structural characterization, bioactivities and potential applications. *International Journal of Biological Macromolecules* **116**: 1011–1025.

Lowman, D. W., L. J. West, D. W. Bearden, M. F. Wempe, T. D. Power, H. E. Ensley, K. Haynes, D. L. Williams & M. D. Kruppa (2011). New insights into the structure of (1→3, 1→6)-β-D-glucan side chains in the *Candida glabrata* cell wall. *PloS One* **6**(11): e27614.

Luanda, A. & V. Badalamoole (2023). Past, present and future of biomedical applications of dextran-based hydrogels: a review. *International Journal of Biological Macromolecules* **228**: 794–807.

Ma, M., M. Okuyama, M. Sato, T. Tagami, P. Klahan, Y. Kumagai, H. Mori & A. Kimura (2017). Effects of mutation of Asn694 in *Aspergillus niger* α-glucosidase on hydrolysis and transglucosylation. *Applied Microbiology and Biotechnology* **101**: 6399–6408.

Ma, Z. & J. I. Boye (2018). Research advances on structural characterization of resistant starch and its structure-physiological function relationship: a review. *Critical Reviews in Food Science and Nutrition* **58**(7): 1059–1083.

Macdonald, I. (1966). Influence of fructose and glucose on serum lipid levels in men and pre-and postmenopausal women. *The American Journal of Clinical Nutrition* **18**(5): 369–372.

Magalhães, R. S., K. C. De Lima, D. S. de Almeida, J. F. De Mesquita & E. C. Eleutherio (2017). Trehalose-6-phosphate as a potential lead candidate for the development of Tps1 inhibitors: insights from the trehalose biosynthesis pathway in diverse yeast species. *Applied Biochemistry and Biotechnology* **181**: 914–924.

Majzoobi, M., S. Hedayati & A. Farahnaky (2015). Functional properties of microporous wheat starch produced by α-amylase and sonication. *Food Bioscience* **11**: 79–84.

Mäkeläinen, H., O. Hasselwander, N. Rautonen & A. Ouwehand (2009). Panose, a new prebiotic candidate. *Letters in Applied Microbiology* **49**(6): 666–672.

Malunga, L. N., M. Izydorczyk & T. Beta (2017). Effect of water-extractable arabinoxylans from wheat aleurone and bran on lipid peroxidation and factors influencing their antioxidant capacity. *Bioactive Carbohydrates and Dietary Fibre* **10**: 20–26.

Mangas-Sánchez, J. & P. Adlercreutz (2015). Enzymatic preparation of oligosaccharides by transglycosylation: a comparative study of glucosidases. *Journal of Molecular Catalysis B: Enzymatic* **122**: 51–55.

Mann, G., E. Leyne, Z. Li & M. Morell (2005). Effects of a novel barley, Himalaya 292, on rheological and breadmaking properties of wheat and barley doughs. *Cereal Chemistry* **82**(6): 626–632.

Manners, D. J., A. J. Masson, J. C. Patterson, H. Björndal & B. Lindberg (1973). The structure of a β-(1→6)-D-glucan from yeast cell walls. *Biochemical Journal* **135**(1): 31–36.

Martin, A. E. & P. A. Montgomery (1996). Acarbose: an α-glucosidase inhibitor. *American Journal of Health-System Pharmacy* **53**(19): 2277–2290.

Mayes, P. A. (1993). Intermediary metabolism of fructose. *The American Journal of Clinical Nutrition* **58**(5): 754S–765S.

Meng, X., T. Pijning, M. Tietema, J. M. Dobruchowska, H. Yin, G. J. Gerwig, S. Kralj & L. Dijkhuizen (2017). Characterization of the glucansucrase GTF180 W1065 mutant enzymes producing polysaccharides and oligosaccharides with altered linkage composition. *Food Chemistry* **217**: 81–90.

Mery, A., S. Jawhara, N. François, M. Cornu, J. Poissy, M. Martinez-Esparza, D. Poulain, B. Sendid & Y. Guerardel (2022). Identification of fungal trehalose for the diagnosis of invasive candidiasis by mass spectrometry. *Biochimica et Biophysica Acta (BBA)-General Subjects* **1866**(4): 130083.

Mery, A., B. Sendid, N. François, M. Cornu, J. Poissy, Y. Guerardel & D. Poulain (2016). Application of mass spectrometry technology to early diagnosis of invasive fungal infections. *Journal of Clinical Microbiology* **54**(11): 2786–2797.

Mikkelsen, M. S., B. M. Jespersen, F. H. Larsen, A. Blennow & S. B. Engelsen (2013). Molecular structure of large-scale extracted β-glucan from barley and oat: identification of a significantly changed block structure in a high β-glucan barley mutant. *Food Chemistry* **136**(1): 130–138.

Minutoli, L., D. Altavilla, A. Bitto, F. Polito, E. Bellocco, G. Laganà, T. Fiumara, S. Magazù, F. Migliardo & F. S. Venuti (2008). Trehalose: a biophysics approach to modulate the inflammatory response during endotoxic shock. *European Journal of Pharmacology* **589**(1–3): 272–280.

Mironescu, M., I. D. Mironescu, A. Trifan & M. V. Ignatova Maya (2009). Influence of the liquefied starch composition and pH on the saccharification at the obtaining of maltose syrup. *Bulletin of the University of Agricultural and Veterinary Medicine* **66**(2): 364–369.

Miyazaki, T. (2023). Glycoside hydrolases active on microbial exopolysaccharide α-glucans: structures and function. *Essays in Biochemistry* **67**(3): 505–520.

Mohamed, I. O. & J. Babucurr (2015). Effect of date syrup on pasting, rheological, and retrogradation properties of corn starch gels. *Starch-Stärke* **67**(7–8): 709–715.

Molina, M., G. Cioci, C. Moulis, E. Séverac & M. Remaud-Siméon (2021). Bacterial α-glucan and branching sucrases from GH70 family: discovery, structure–function relationship studies and engineering. *Microorganisms* **9**(8): 1607.

Montrose, D. C., R. Nishiguchi, S. Basu, H. A. Staab, X. K. Zhou, H. Wang, L. Meng, M. Johncilla, J. R. Cubillos-Ruiz & D. K. Morales (2021). Dietary fructose alters the composition, localization, and metabolism of gut microbiota in association with worsening colitis. *Cellular and Molecular Gastroenterology and Hepatology* **11**(2): 525–550.

Moore, M. C., A. D. Cherrington, S. L. Mann & S. N. Davis (2000). Acute fructose administration decreases the glycemic response to an oral glucose tolerance test in normal adults. *The Journal of Clinical Endocrinology & Metabolism* **85**(12): 4515–4519.

Moore, M. C., S. N. Davis, S. L. Mann & A. D. Cherrington (2001). Acute fructose administration improves oral glucose tolerance in adults with type 2 diabetes. *Diabetes Care* **24**(11): 1882–1887.

Mumpuni, H., N. Yasmine, Y. Marsono, D. L. N. Fibri & A. Murdiati (2022). FiberCreme as a functional food ingredient reduces hyperlipidemia and risk of cardiovascular diseases in subjects with hyperlipidemia. *Preventive Nutrition and Food Science* **27**(2): 165.

Ni, D., S. Zhang, X. Liu, Y. Zhu, W. Xu, W. Zhang & W. Mu (2024). Production, effects, and applications of fructans with various molecular weights. *Food Chemistry* **437**: 137895.

Nikvarz, N., G. R. Khayati & S. Sharafi (2021). Bio-based ultraviolet protective packaging film preparation using starch with incorporated date palm syrup. *Materials Chemistry and Physics* **270**: 124794.

Nishimoto, T., T. Nakada, H. Chaen, S. Fukuda, T. Sugimoto, M. Kurimoto & Y. Tsujisaka (1996). Purification and characterization of a thermostable trehalose synthase from *Thermus aquaticus*. *Bioscience, Biotechnology, and Biochemistry* **60**(5): 835–839.

Niu, D., J. Qiao, P. Li, K. Tian, X. Liu, S. Singh & F. Lu (2017). Highly efficient enzymatic preparation of isomalto-oligosaccharides from starch using an enzyme cocktail. *Electronic Journal of Biotechnology* **26**: 46–51.

Nooshkam, M., M. Varidi & M. Bashash (2019). The Maillard reaction products as food-born antioxidant and antibrowning agents in model and real food systems. *Food Chemistry* **275**: 644–660.

Norman, B. E. (1981). New developments in starch syrup technology. *Enzymes and Food Processing* 1981: 15–50. Springer.

Nwaka, S. & H. Holzer (1997). Molecular biology of trehalose and the trehalases in the yeast *Saccharomyces cerevisiae*. *Progress in Nucleic Acid Research and Molecular Biology* **58**: 197–237.

Ogunsona, E., E. Ojogbo & T. Mekonnen (2018). Advanced material applications of starch and its derivatives. *European Polymer Journal* **108**: 570–581.

Ohtake, S. & Y. J. Wang (2011). Trehalose: current use and future applications. *Journal of Pharmaceutical Sciences* **100**(6): 2020–2053.

Ojha, S., S. Mishra & S. Chand (2015). Production of isomalto-oligosaccharides by cell bound α-glucosidase of Microbacterium sp. *LWT-Food Science and Technology* **60**(1): 486–494.

Oliyaei, N., M. Moosavi-Nasab, A. Tamaddon & M. Fazaeli (2019). Preparation and characterization of porous starch reinforced with halloysite nanotube by solvent exchange method. *International Journal of Biological Macromolecules* **123**: 682–690.

Olsen, H. S. (2002). Enzymes in starch modification. In R. J. Whitehurst & B. A. Law (eds.), *Enzymes in Food Technology*, CRC Press: 200–222.

Olsson, C., H. Jansson & J. Swenson (2016). The role of trehalose for the stabilization of proteins. *The Journal of Physical Chemistry B* **120**(20): 4723–4731.

Park, J.-E., S. H. Park, J. Y. Woo, H. S. Hwang, J. Cha & H. Lee (2013). Enzymatic properties of a thermostable α-glucosidase from acidothermophilic crenarchaeon *Sulfolobus tokodaii* strain 7. *Journal of Microbiology and Biotechnology* **23**(1): 56–63.

Park, M.-O., M. Chandrasekaran & S.-H. Yoo (2019). Production and characterization of low-calorie turanose and digestion-resistant starch by an amylosucrase from Neisseria subflava. *Food Chemistry* **300**: 125225

Park, M.-O., B.-H. Lee, E. Lim, J. Y. Lim, Y. Kim, C.-S. Park, H. G. Lee, H.-K. Kang & S.-H. Yoo (2016). Enzymatic process for high-yield turanose production and its potential property as an adipogenesis regulator. *Journal of Agricultural and Food Chemistry* **64**(23): 4758–4764.

Paul, M. J., L. F. Primavesi, D. Jhurreea & Y. Zhang (2008). Trehalose metabolism and signaling. *Annual Review of Plant Biology* **59**: 417–441.

Payne, A., C. Chassard & C. Lacroix (2012). Gut microbial adaptation to dietary consumption of fructose, artificial sweeteners and sugar alcohols: implications for host–microbe interactions contributing to obesity. *Obesity Reviews* **13**(9): 799–809.

Pi, G., J. Wang, W. Song, Y. Li & H. Yang (2022). Effects of isomalto-oligosaccharides and herbal extracts on growth performance, serum biochemical profiles and intestinal bacterial populations in early-weaned piglets. *Journal of Animal Physiology and Animal Nutrition* **106**(3): 671–681.

Pitirollo, O., M. Grimaldi, C. Corradini, S. Pironi & A. Cavazza (2023). HPAEC-PAD analytical evaluation of carbohydrates pattern for the study of technological parameters effects in low-FODMAP food production. *Molecules* **28**(8): 3564.

Pohane, A. A., C. R. Carr, J. Garhyan, B. M. Swarts & M. S. Siegrist (2021). Trehalose recycling promotes energy-efficient biosynthesis of the mycobacterial cell envelope. *mBio* **12**(1): 10.1128/mbio. 02801–02820.

Pontoh, J. & N. H. Low (1995). Glucose syrup production from Indonesian palm and cassava starch. *Food Research International* **28**(4): 379–385.

Prompiputtanapon, K., W. Sorndech & S. Tongta (2020). Surface modification of tapioca starch by using the chemical and enzymatic method. *Starch-Stärke* **72**(3–4): 1900133.

Puncha-Arnon, S., Y. Wandee, D. Uttapap, C. Puttanlek & V. Rungsardthong (2020). The effect of hydrolysis of cassava starch on the characteristics of microspheres prepared by an emulsification-crosslinking method. *International Journal of Biological Macromolecules* **161**: 939–946.

Punnatin, P., C. Chanchao & S. Chunsrivirot (2020). Molecular dynamics reveals insight into how N226P and H227Y mutations affect maltose binding in the active site of α-glucosidase II from European honeybee, *Apis mellifera*. *PloS One* **15**(3): e0229734.

Purwitasari, L., M. P. Wulanjati, Y. Pranoto & L. D. Witasari (2023). Characterization of porous starch from edible canna (*Canna edulis Kerr.*) produced by enzymatic hydrolysis using *thermostable* α-amylase. *Food Chemistry Advances* **2**: 100152.

Qian, J., Y. Chen, H. Yang, C. Zhao, X. Zhao & H. Guo (2020). Preparation and characterization of crosslinked porous starch hemostatic. *International Journal of Biological Macromolecules* **160**: 429–436.

Reaven, G. M., T. R. Risser, Y.-D. Chen & E. P. Reaven (1979). Characterization of a model of dietary-induced hypertriglyceridemia in young, nonobese rats. *Journal of Lipid Research* **20**(3): 371–378.

Reilly, P. & R. Antrim (2003). *Enzymes in Grain Wet Milling*, Ullmann's Encyclopedia of Industrial Chemistry.

Ricca, E., V. Calabrò, S. Curcio & G. Iorio (2007). The state of the art in the production of fructose from inulin enzymatic hydrolysis. *Critical Reviews in Biotechnology* **27**(3): 129–145.

Rieder, A. & A. B. Samuelsen (2012). Do cereal mixed-linked β-glucans possess immune-modulating activities? *Molecular Nutrition & Food Research* **56**(4): 536–547.

Rodriguez-Castaño, G. P., A. Caro-Quintero, A. Reyes & F. Lizcano (2017). Advances in gut microbiome research, opening new strategies to cope with a western lifestyle. *Frontiers in Genetics* **7**: 224.

Roeb, E. & R. Weiskirchen (2021). Fructose and non-alcoholic steatohepatitis. *Frontiers in Pharmacology* **12**: 634344.

Romero, C., J. M. Bellés, J. L. Vayá, R. Serrano & F. A. Culiáñez-Macià (1997). Expression of the yeast trehalose-6-phosphate synthase gene in transgenic tobacco plants: pleiotropic phenotypes include drought tolerance. *Planta* **201**: 293–297.

Romero Marcia, A. D., T. Yao, M.-H. Chen, R. E. Oles & S. R. Lindemann (2021). Fine carbohydrate structure of dietary resistant glucans governs the structure and function of human gut microbiota. *Nutrients* **13**(9): 2924.

Rutledge, A. C. & K. Adeli (2007). Fructose and the metabolic syndrome: pathophysiology and molecular mechanisms. *Nutrition Reviews* **65**(suppl_1): S13–S23.

Ruzanski, C., J. Smirnova, M. Rejzek, D. Cockburn, H. L. Pedersen, M. Pike, W. G. Willats, B. Svensson, M. Steup & O. Ebenhöh (2013). A bacterial glucanotransferase can replace the complex maltose metabolism required for starch to sucrose conversion in leaves at night. *Journal of Biological Chemistry* **288**(40): 28581–28598.

Ryu, J. H., B. H. Lee, D. H. Seo, M. Y. Baik, C. S. Park, R. Wang & S. H. Yoo (2010). Production and characterization of digestion-resistant starch by the reaction of *Neisseria polysaccharea* amylosucrase. *Starch-Stärke* **62**(5): 221–228.

Sadak, M. S., H. M. S. El-Bassiouny & M. G. Dawood (2019). Role of trehalose on antioxidant defense system and some osmolytes of quinoa plants under water deficit. *Bulletin of the National Research Centre* **43**(1): 1–11.

Saman, P., K. M. Tuohy, J. A. Vázquez, G. Gibson & S. S. Pandiella (2017). *In vitro* evaluation of prebiotic properties derived from rice bran obtained by debranning technology. *International Journal of Food Sciences and Nutrition* **68**(4): 421–428.

Sánchez-Martínez, M. J., S. Soto-Jover, V. Antolinos, G. B. Martínez-Hernández & A. López-Gómez (2020). Manufacturing of short-chain fructooligosaccharides: from laboratory to industrial scale. *Food Engineering Reviews* **12**: 149–172.

Šandula, J., G. Kogan, M. Kačuráková & E. Machová (1999). Microbial (1→3)-β-d-glucans, their preparation, physico-chemical characterization and immunomodulatory activity. *Carbohydrate Polymers* **38**(3): 247–253.

Sarkar, A. K. & S. Sadhukhan (2022). Imperative role of trehalose metabolism and trehalose-6-phosphate signaling on salt stress responses in plants. *Physiologia Plantarum* **174**(1): e13647.

Schenck, F. (2000). High fructose syrups-a review. *Indian Sugar* **50**(5): 281–287.

Schwarz, J.-M., S. M. Noworolski, A. Erkin-Cakmak, N. J. Korn, M. J. Wen, V. W. Tai, G. M. Jones, S. P. Palii, M. Velasco-Alin & K. Pan (2017). Effects of dietary fructose restriction on liver fat, de novo lipogenesis, and insulin kinetics in children with obesity. *Gastroenterology* **153**(3): 743–752.

Schwarz, J.-M., S. M. Noworolski, M. J. Wen, A. Dyachenko, J. L. Prior, M. E. Weinberg, L. A. Herraiz, V. W. Tai, N. Bergeron & T. P. Bersot (2015). Effect of a high-fructose weight-maintaining diet on lipogenesis and liver fat. *The Journal of Clinical Endocrinology & Metabolism* **100**(6): 2434–2442.

Senf, D., C. Ruprecht, S. Kishani, A. Matic, G. Toriz, P. Gatenholm, L. Wågberg & F. Pfrengle (2018). Tailormade polysaccharides with defined branching patterns: enzymatic polymerization of arabinoxylan oligosaccharides. *Angewandte Chemie International Edition* **57**: 11987–11992.

Sharma, S., R. Bhatia, K. Devi, A. Rawat, S. Singh, S. K. Bhadada, M. Bishnoi, S. S. Sharma & K. K. Kondepudi (2023). A synbiotic combination of *Bifidobacterium longum* Bif10 and *Bifidobacterium breve* Bif11, isomaltooligosaccharides and finger millet arabinoxylan prevents dextran sodium sulphate induced ulcerative colitis in mice. *International Journal of Biological Macromolecules* **231**: 123326.

Shendurse, A. & C. Khedkar (2016). Glucose: properties and analysis. *Encyclopedia of Food and Health* **3**: 239–247.

Shetty, P. R., U. R. Batchu, S. K. Buddana, K. S. Rao & S. Penna (2021). A comprehensive review on α-D-glucans: structural and functional diversity, derivatization and bioapplications. *Carbohydrate Research* **503**: 108297.

Shi, Q., Y. Hou, M. Juvonen, P. Tuomainen, I. Kajala, S. Shukla, A. Goyal, H. Maaheimo, K. Katina & M. Tenkanen (2016). Optimization of isomaltooligosaccharide size distribution by acceptor reaction of Weissella confusa dextransucrase and characterization of novel α-(1→2)-branched isomaltooligosaccharides. *Journal of Agricultural and Food Chemistry* **64**(16): 3276–3286.

Shukla, E., L. J. Thorat, B. B. Nath & S. M. Gaikwad (2015). Insect trehalase: physiological significance and potential applications. *Glycobiology* **25**(4): 357–367.

Singh, D. P., P. Khare, V. Bijalwan, R. K. Baboota, J. Singh, K. K. Kondepudi, K. Chopra & M. Bishnoi (2017). Coadministration of isomalto-oligosaccharides augments metabolic health benefits of cinnamaldehyde in high fat diet fed mice. *BioFactors* **43**(6): 821–835.

Singh, R., K. Chauhan, A. Pandey & C. Larroche (2018). Biocatalytic strategies for the production of high fructose syrup from inulin. *Bioresource Technology* **260**: 395–403.

Singh, R. S., N. Kaur, M. Hassan & J. F. Kennedy (2021). Pullulan in biomedical research and development-a review. *International Journal of Biological Macromolecules* **166**: 694–706.

Singla, V. & S. Chakkaravarthi (2017). Applications of prebiotics in food industry: a review. *Food Science and Technology International* **23**(8): 649–667.

Skenderian, S., G. Park & C. Jang (2020). Organismal fructose metabolism in health and non-alcoholic fatty liver disease. *Biology* **9**(11): 405.

Sławińska, A., E. Jabłońska-Ryś & A. Stachniuk (2021). High-performance liquid chromatography determination of free sugars and mannitol in mushrooms using corona charged aerosol detection. *Food Analytical Methods* **14**: 209–216.

Sleder, J., Y.-D. I. Chen, M. D. Cully & G. M. Reaven (1980). Hyperinsulinemia in fructose-induced hypertriglyceridemia in the rat. *Metabolism* **29**(4): 303–305.

Song, J. Y., Y.-M. Kim, B.-H. Lee & S.-H. Yoo (2020). Increasing the dietary fiber contents in isomaltooligosaccharides by dextransucrase reaction with sucrose as a glucosyl donor. *Carbohydrate Polymers* **230**: 115607.

Song, Y.-B., L. M. Lamothe, N. E. N. Rodriguez, D. R. Rose & B.-H. Lee (2022). New insights suggest isomaltooligosaccharides are slowly digestible carbohydrates, rather than dietary fibers, at constitutive mammalian α-glucosidase levels. *Food Chemistry* **383**: 132456.

Song, Z., Y. Zhong, W. Tian, C. Zhang, A. R. Hansen, A. Blennow, W. Liang & D. Guo (2020). Structural and functional characterizations of α-amylase-treated porous popcorn starch. *Food Hydrocolloids* **108**: 105606.

Sorndech, W. (2022). Isomaltooligosaccharides as Prebiotics and their Health Benefits. In P. S. Panesar & A. Anal (eds.), Probiotics, *Prebiotics and Synbiotics: Technological Advancements towards Safety and Industrial Applications*, John Wiley & Sons, Inc.: 361–377.

Sorndech, W., K. N. Nakorn, S. Tongta & A. Blennow (2018). Isomalto-oligosaccharides: recent insights in production technology and their use for food and medical applications. *LWT* **95**: 135–142.

Sorndech, W., D. Sagnelli, A. Blennow & S. Tongta (2017). Combination of amylase and transferase catalysis to improve IMO compositions and productivity. *LWT-Food Science and Technology* **79**: 479–486.

Sorndech, W., S. Tongta & A. Blennow (2018). Slowly digestible- and non-digestible α-glucans: an enzymatic approach to starch modification and nutritional effects. *Starch-Stärke* **70**(9–10): 1700145.

Soyseven, M., B. Sezgin & G. Arli (2022). A novel, rapid and robust HPLC-ELSD method for simultaneous determination of fructose, glucose and sucrose in various food samples: method development and validation. *Journal of Food Composition and Analysis* **107**: 104400.

Stanhope, K. L., V. Medici, A. A. Bremer, V. Lee, H. D. Lam, M. V. Nunez, G. X. Chen, N. L. Keim & P. J. Havel (2015). A dose-response study of consuming high-fructose corn syrup–sweetened beverages on lipid/lipoprotein risk factors for cardiovascular disease in young adults. *The American Journal of Clinical Nutrition* **101**(6): 1144–1154.

Stanhope, K. L., J. M. Schwarz, N. L. Keim, S. C. Griffen, A. A. Bremer, J. L. Graham, B. Hatcher, C. L. Cox, A. Dyachenko & W. Zhang (2009). Consuming fructose-sweetened, not glucose-sweetened, beverages increases visceral adiposity and lipids and decreases insulin sensitivity in overweight/ obese humans. *The Journal of Clinical Investigation* **119**(5): 1322–1334.

Staudacher, H. M. & K. Whelan (2017). The low FODMAP diet: recent advances in understanding its mechanisms and efficacy in IBS. *Gut* **66**(8): 1517–1527.

Stick, R. & S. Williams (2009). Disaccharides, oligosaccharides and polysaccharides. In R. V. Stick & S. J. Williams (eds.), *Carbohydrates: The Essential Molecules of Life*, Elsevier Science: 321–341.

Stricker, S., S. Rudloff, A. Geier, A. Steveling, E. Roeb & K.-P. Zimmer (2021). Fructose consumption—free sugars and their health effects. *Deutsches Ärzteblatt International* **118**(5): 71.

Su, J., T. Wang, C. Ma, Z. Li, Z. Li & R. Wang (2014). Homology modeling and function of trehalose synthase from *Pseudomonas putida* P06. *Biotechnology Letters* **36**: 1009–1013.

Sujka, M. & J. Jamroz (2010). Characteristics of pores in native and hydrolyzed starch granules. *Starch-Stärke* **62**(5): 229–235.

Sun, C., Z. Wei, C. Xuc & L. Yang (2023). Development, application and future trends of starch-based delivery systems for nutraceuticals: a review. *Carbohydrate Polymers* **308**: 120675.

Sun, J., H. Li, H. Huang, B. Wang, L. P. Xiao & G. Song (2018). Integration of enzymatic and heterogeneous catalysis for one-pot production of fructose from glucose. *ChemSusChem* **11**(7): 1157–1162.

Sun, Q., Y. Xing, C. Qiu & L. Xiong (2014). The pasting and gel textural properties of corn starch in glucose, fructose and maltose syrup. *PloS One* **9**(4): e95862.

Suzuki, E. & R. Suzuki (2016). Distribution of glucan-branching enzymes among prokaryotes. *Cellular and Molecular Life Sciences* **73**: 2643–2660.

Swaisgood, H. E., Whitaker, J. R., Voragen, A. G. J., & Wong, D. M. S. (2002). Use of immobilized enzymes in the food industry. In *Handbook of Food Enzymology*, CRC Press: 374–381.

Synytsya, A., K. Míčková, A. Synytsya, I. Jablonský, J. Spěváček, V. Erban, E. Kovaříková & J. Čopíková (2009). Glucans from fruit bodies of cultivated mushrooms *Pleurotus ostreatus* and *Pleurotus eryngii*: structure and potential prebiotic activity. *Carbohydrate Polymers* **76**(4): 548–556.

Synytsya, A. & M. Novák (2013). Structural diversity of fungal glucans. *Carbohydrate Polymers* **92**(1): 792–809.

Synytsya, A. & M. Novak (2014). Structural analysis of glucans. *Annals of Translational Medicine* **2**(2): 17.

Takahashi, M., K. Yoshioka, T. Imai, Y. Miyoshi, Y. Nakano, K. Yoshida, T. Yamashita, Y. Furuta, T. Watanabe & J. Sugiyama (2013). Degradation and synthesis of β-glucans by a *Magnaporthe oryzae* endotransglucosylase, a member of the glycoside hydrolase 7 family. *Journal of Biological Chemistry* **288**(19): 13821–13830.

Takusagawa, F. & R. A. Jacobson (1978). The crystal and molecular structure of α-maltose. *Acta Crystallographica Section B: Structural Crystallography and Crystal Chemistry* **34**(1): 213–218.

Taskinen, M. R., S. Söderlund, L. Bogl, A. Hakkarainen, N. Matikainen, K. Pietiläinen, S. Räsänen, N. Lundbom, E. Björnson & B. Eliasson (2017). Adverse effects of fructose on cardiometabolic risk factors and hepatic lipid metabolism in subjects with abdominal obesity. *Journal of Internal Medicine* **282**(2): 187–201.

Thitaram, S., C.-H. Chung, D. Day, A. Hinton Jr, J. Bailey & G. Siragusa (2005). Isomaltooligosaccharide increases cecal *Bifidobacterium* population in young broiler chickens. *Poultry Science* **84**(7): 998–1003.

Thorat, B. N., A. Sett & A. Mujumdar (2022). Drying of vaccines and biomolecules. *Drying Technology* **40**(3): 461–483.

Thorburn, A. W., L. H. Storlien, A. B. Jenkins, S. Khouri & E. Kraegen (1989). Fructose-induced *in vivo* insulin resistance and elevated plasma triglyceride levels in rats. *The American Journal of Clinical Nutrition* **49**(6): 1155–1163.

Tiangpook, S., S. Nhim, P. Prangthip, P. Pason, C. Tachaapaikoon, K. Ratanakhanokchai & R. Waeonukul (2023). Production of a series of long-chain isomaltooligosaccharides from maltose by *Bacillus subtilis* AP-1 and associated prebiotic properties. *Foods* **12**(7): 1499.

Tirone, T. A. & F. C. Brunicardi (2001). Overview of glucose regulation. *World Journal of Surgery* **25**(4): 461.

Tobey, T., C. Mondon, I. Zavaroni & G. Reaven (1982). Mechanism of insulin resistance in fructose-fed rats. *Metabolism* **31**(6): 608–612.

Togo, A., K. Uechi, O. Mizutani, S. Kimura & T. Iwata (2021). Synthesis and characterization of α-1, 3-alt-α-1, 4-glucan (nigeran) ester derivatives. *Polymer* **214**: 123343.

Ubeyitogullari, A. & O. N. Ciftci (2016). Phytosterol nanoparticles with reduced crystallinity generated using nanoporous starch aerogels. *RSC Advances* **6**(110): 108319–108327.

Um, H.-E., B.-R. Park, Y. M. Kim & B.-H. Lee (2023). Slow digestion properties of long-sized isomaltooligosaccharides synthesized by a transglucosidase from Thermoanaerobacter thermocopriae. *Food Chemistry* **417**: 135892.

Vaclavik, V. A., E. W. Christian & T. Campbell (eds.) (2021). Sugars, sweeteners, and confections. In *Essentials of Food Science*, Springer: 281–299.

Vanherp, L., J. Poelmans, A. Weerasekera, A. Hillen, A. R. Croitor-Sava, T. C. Sorrell, K. Lagrou, G. V. Velde & U. Himmelreich (2021). Trehalose as quantitative biomarker for in vivo diagnosis and treatment follow-up in cryptococcomas. *Translational Research* **230**: 111–122.

Venkatachalam, G., S. Arumugam & M. Doble (2021). Industrial production and applications of α/β linear and branched glucans. *Indian Chemical Engineer* **63**(5): 533–547.

Villéger, R., E. Pinault, K. Vuillier-Devillers, K. Grenier, C. Landolt, D. Ropartz, V. Sol, M. C. Urdaci, P. Bressollier & T.-S. Ouk (2022). Prebiotic isomaltooligosaccharide provides an advantageous fitness to the probiotic *Bacillus subtilis* CU1. *Applied Sciences* **12**(13): 6404.

Walsh, M. K. (2007). Immobilized enzyme technology for food applications. In R. Rastall (ed.), *Novel Enzyme Technology for Food Applications*, Elsevier: 60–84.

Walton, G. E., C. Lu, I. Trogh, F. Arnaut & G. R. Gibson (2012). A randomised, double-blind, placebo controlled cross-over study to determine the gastrointestinal effects of consumption of arabinoxylan-oligosaccharides enriched bread in healthy volunteers. *Nutrition Journal* **11**: 1–11.

Wang, H., J. Lv, S. Jiang, B. Niu, M. Pang & S. Jiang (2016). Preparation and characterization of porous corn starch and its adsorption toward grape seed proanthocyanidins. *Starch-Stärke* **68**(11–12): 1254–1263.

Wang, H., Y. Wang, K. Xu, Y. Zhang, M. Shi, X. Liu, C. Chi & H. Zhang (2022). Causal relations among starch hierarchical structure and physicochemical characteristics after repeated freezing-thawing. *Food Hydrocolloids* **122**: 107121.

Wang, H.-F., P.-S. Lim, M.-D. Kao, E.-C. Chan, L.-C. Lin & N.-P. Wang (2001). Use of isomalto-oligosaccharide in the treatment of lipid profiles and constipation in hemodialysis patients. *Journal of Renal Nutrition* **11**(2): 73–79.

Wang, Y., Z. Lu, H. Wu & F. Lv (2009). Study on the antibiotic activity of microcapsule curcumin against foodborne pathogens. *International Journal of Food Microbiology* **136**(1): 71–74.

Wasserman, D. H. (2009). Four grams of glucose. *American Journal of Physiology-Endocrinology and Metabolism* **296**(1): E11–E21.

White, J. S. (2013). Challenging the fructose hypothesis: new perspectives on fructose consumption and metabolism. *Advances in Nutrition* **4**(2): 246–256.

Wiater, A., A. Waśko, P. Adamczyk, K. Gustaw, M. Pleszczyńska, K. Wlizło, M. Skowronek, M. Tomczyk & J. Szczodrak (2020). Prebiotic potential of oligosaccharides obtained by acid hydrolysis of α-(1→3)-glucan from Laetiporus sulphureus: a pilot study. *Molecules* **25**(23): 5542.

Wilder-Smith, C., S. S. Olesen, A. Materna & A. Drewes (2017). Predictors of response to a low-FODMAP diet in patients with functional gastrointestinal disorders and lactose or fructose intolerance. *Alimentary Pharmacology & Therapeutics* **45**(8): 1094–1106.

Wu, Q., W. Liu, H. Chen, Y. Yin, D. Y. Hongwei, X. Wang & L. Zhu (2017). Fermentation properties of isomaltooligosaccharides are affected by human fecal enterotypes. *Anaerobe* **48**: 206–214.

Wu, Y., X. Du, H. Ge & Z. Lv (2011). Preparation of microporous starch by glucoamylase and ultrasound. *Starch-Stärke* **63**(4): 217–225.

Xin, R., S. Qi, C. Zeng, F. I. Khan, B. Yang & Y. Wang (2017). A functional natural deep eutectic solvent based on trehalose: structural and physicochemical properties. *Food Chemistry* **217**: 560–567.

Yan, S., M. Song, K. Wang, X. Fang, W. Peng, L. Wu & X. Xue (2021). Detection of acacia honey adulteration with high fructose corn syrup through determination of targeted α-dicarbonyl compound using ion mobility-mass spectrometry coupled with UHPLC-MS/MS. *Food Chemistry* **352**: 129312.

Ye, Z., V. Arumugam, E. Haugabrooks, P. Williamson & S. Hendrich (2015). Soluble dietary fiber (Fibersol-2) decreased hunger and increased satiety hormones in humans when ingested with a meal. *Nutrition Research* **35**(5): 393–400.

Yeganeh-Zare, S., K. Farhadi & S. Amiri (2022). Rapid detection of apple juice concentrate adulteration with date concentrate, fructose and glucose syrup using HPLC-RID incorporated with chemometric tools. *Food Chemistry* **370**: 131015.

Yen, C.-H., Y.-H. Tseng, Y.-W. Kuo, M.-C. Lee & H.-L. Chen (2011). Long-term supplementation of isomalto-oligosaccharides improved colonic microflora profile, bowel function, and blood cholesterol levels in constipated elderly people—a placebo-controlled, diet-controlled trial. *Nutrition* **27**(4): 445–450.

Yoshida, M. (2021). Fructan structure and metabolism in overwintering plants. *Plants* **10**(5): 933.

You, S. & S. T. Lim (2000). Molecular characterization of corn starch using an aqueous HPSEC-MALLS-RI system under various dissolution and analytical conditions. *Cereal Chemistry* **77**(3): 303–308.

Zambrana, L. E., S. McKeen, H. Ibrahim, I. Zarei, E. C. Borresen, L. Doumbia, A. Boré, A. Cissoko, S. Douyon & K. Koné (2019). Rice bran supplementation modulates growth, microbiota and metabolome in weaning infants: a clinical trial in Nicaragua and Mali. *Scientific Reports* **9**(1): 1–18.

Zavaroni, I., Y.-D. I. Chen & G. M. Reaven (1982). Studies of the mechanism of fructose-induced hypertriglyceridemia in the rat. *Metabolism* **31**(11): 1077–1083.

Zhang, B., D. Cui, M. Liu, H. Gong, Y. Huang & F. Han (2012). Corn porous starch: preparation, characterization and adsorption property. *International Journal of Biological Macromolecules* **50**(1): 250–256.

Zhang, C., S.-Y. Wang, C.-Y. Wu, J.-J. Li, L.-Z. Zhang, Z.-J. Wang, Q.-Q. Liu & J.-Y. Qian (2023). Effect of melting combined with ice recrystallization on porous starch preparation: pore-forming properties, granular morphology, functionality, and multi-scale structures. *Food Research International* **174**: 113463.

Zhang, H., R. Wang, Z. Chen & Q. Zhong (2019). Enzymatically modified starch with low digestibility produced from amylopectin by sequential amylosucrase and pullulanase treatments. *Food Hydrocolloids* **95**: 195–202.

Zhang, L., Y. Su, Y. Zheng, Z. Jiang, J. Shi, Y. Zhu & Y. Jiang (2010). Sandwich-structured enzyme membrane reactor for efficient conversion of maltose into isomaltooligosaccharides. *Bioresource Technology* **101**(23): 9144–9149.

Zhang, S. & T. Wang (2023). Preparation of enzymolysis porous corn starch composite microcapsules embedding organic sunscreen agents and its UV protection performance and stability. *Carbohydrate Polymers* **314**: 120903.

Zhao, A.-Q., L. Yu, M. Yang, C.-J. Wang, M.-M. Wang & X. Bai (2018). Effects of the combination of freeze-thawing and enzymatic hydrolysis on the microstructure and physicochemical properties of porous corn starch. *Food Hydrocolloids* **83**: 465–472.

Zhou, C., Y. Xue & Y. Ma (2015). Evaluation and directed evolution for thermostability improvement of a GH 13 thermostable α-glucosidase from *Thermus thermophilus* TC11. *BMC Biotechnology* **15**: 1–11.

Zhou, S., X. Liu, Y. Guo, Q. Wang, D. Peng & L. Cao (2010). Comparison of the immunological activities of arabinoxylans from wheat bran with alkali and xylanase-aided extraction. *Carbohydrate Polymers* **81**(4): 784–789.

Zhu, F., B. Du, Z. Bian & B. Xu (2015). Beta-glucans from edible and medicinal mushrooms: characteristics, physicochemical and biological activities. *Journal of Food Composition and Analysis* **41**: 165–173.

Zhu, F., B. Du & B. Xu (2016). A critical review on production and industrial applications of beta-glucans. *Food Hydrocolloids* **52**: 275–288.

Zubiría, M. G., S. E. Gambaro, M. A. Rey, P. Carasi, M. D. L. Á. Serradell & A. Giovambattista (2017). Deleterious metabolic effects of high fructose intake: the preventive effect of *Lactobacillus kefiri* administration. *Nutrients* **9**(5): 470.

Zuo, Y. Y. J., P. Hébraud, Y. Hemar & M. Ashokkumar (2012). Quantification of high-power ultrasound induced damage on potato starch granules using light microscopy. *Ultrasonics Sonochemistry* **19**(3): 421–426.

Starch Dextrins

14

Marie Sofie Møller and Birte Svensson

14.1 INTRODUCTION

To take full advantage of the large number of attractive attributes of starch, the raw starch has been subjected to degradation and modification (He et al. 2023; Miao et al. 2018; Punia Bangar et al. 2022; K Wu et al. 2024; W Yang et al. 2022; Zhong et al. 2022). The very large diversity of products encompasses α-glucan fragments or dextrins of different molecular sizes and structures, obtained from or related to the starch polysaccharides amylopectin (AP) and amylose (AM). AP consists of a core α-1,4-glucan chain carrying branches attached by α-1,6 linkages, and AM is an essentially linear α-1,4-glucan (Apriyanto, Compart & Fettke 2022; Bertoft 2017; Junejo et al. 2022; Nakamura & Kainuma 2022). In this chapter, starch-derived and starch-like enzyme products comprise (1) larger fragments, (2) medium-sized fragments, referred to as maltodextrins (MDs), and (3) maltooligosaccharides (MOSs) (Chen et al. 2021; X Li et al. 2024; Takata et al. 2010; Xue, Svensson & Bai 2022). The level of starch degradation is given by the dextrose equivalent (DE) value defined as the percentage of reducing end residues of the total glucose content (Hofman, van Buul & Brouns 2016; Pycia et al. 2017; Rong et al. 2009). DE values describe the degree of starch depolymerization and are regularly used in food science and biotechnology; the higher the DE, the more extensive the degradation and the shorter the average length of the dextrins (H Kong et al. 2018; Vargas-Campos et al. 2023; Xiao et al. 2022).

Starch can be modified using physical, chemical, and enzymatic methods (Boldrini 2023; X Chen et al. 2021; He et al. 2023; H-T Li et al. 2023; Miao & BeMiller 2023; Zhong et al. 2022). Notably, chemical modifications of starch, such as oxidation, cross-linking, acetylation, carboxymethylation, hydroxypropylation, and octenyl-succinylation, are well-established (Altuna, Herrera & Foresti 2018; Haq et al. 2019; H Jiang et al. 2024; Otache et al. 2021; Pycia et al. 2017; Y Wang et al. 2020). However, in particular, enzymatic modifications are increasingly employed to generate linear, branched, and cyclic large starch fragments, MDs, and MOSs for a wide range of applications (Y Chen et al. 2023a; F Li et al. 2023; X Li et al. 2024; Miao et al. 2018; Morin-Crini et al. 2021; H-J Ryu et al. 2023; W Yang et al. 2022). These reactions include enzyme-catalyzed synthesis of dextrins representing structural features, which can be even quite different from AP and AM (Bai et al. 2016; Gangoiti et al. 2016; X Li et al. 2023a; Sorndech, Tongta & Blennow 2018; Xue, Svensson & Bai 2022). The field of enzymatic modification and synthesis of starch and related α-glucans has a long history, but currently attracts much attention and undergoes substantial development to meet the need for green processes and sustainability.

14.1.1 Definition of Large Starch Fragments, Maltodextrins (MDs), and Maltooligosaccharides (MOSs)

Large starch fragments or dextrins are polysaccharides of high molecular weight (Mw) obtained from, albeit smaller than, AP (Mw 3×10^5–3×10^7 g/mol; DP 2,000–200,000) and AM (Mw 5×10^4–5×10^5 g/mol; DP 300–3,000). MDs are shorter, typically of DP 10–100 (Avaltroni, Bouquerand & Normand 2004; Chronakis 1998; Hofman, van Buul & Brouns 2016). Large starch fragments and MDs exist as polydisperse mixtures of different molecular size ranges, whereas MOSs, defined by DP 3–9, can be obtained as single molecules (Bláhová et al. 2023; Pan et al. 2017; Pullicin et al. 2018; Shad et al. 2023; Sorndech, Tongta & Blennow 2018). All three size-categories include linear, branched, and cyclic dextrins, which can contain α-1,4-glucan chains bifurcated at natural α-1,6 linkages or other branch points (Lee & Hamaker 2017; Takata et al. 2010; W Yang et al. 2022). Several modifying enzymes are used, either alone or together in sequential or one-pot reactions, to produce novel dextrins, MDs, MOSs, and glucoconjugates (Lang et al. 2014, 2023; X Li et al. 2020; Z Liu et al. 2022b; Sorndech, Nakorn, et al. 2018; Suksiri et al. 2021; W Wang et al. 2021). Dextrins include transglucosylated α-1,4-glucan products decorated in ways not seen in starch, for example, via α-1,3 or α-1,2 linkages (Brison et al. 2010; Gangoiti et al. 2017; Vuillemin et al. 2016; Wei et al. 2023b) or having linear α-1,3- or α-1,6-linked extensions of α-1,4-glucans (Dong et al. 2024; X Li et al. 2023a; Y Wu et al. 2023; Xue, Svensson & Bai 2022). MOSs, MDs, and larger dextrins also appear in cyclic forms, which are generated in intramolecular transglucosylation reactions. These are α-, β-, and γ-cyclodextrins (α-, β-, and γ-CDs) of DP 6, 7, and 8 (X Li et al. 2024; Morin-Crini et al. 2021),

"

large ring cyclodextrins (LR-CDs) of DP 9–100 (Krusong et al. 2022; F Li et al. 2023) as well as the larger cyclic cluster dextrins (CCDs) and highly branched cyclic dextrins (HBCDs) (X Li et al. 2023a; Takaha et al. 1996, 1998; Takata et al. 1996, 2003, 2010). Isomalto-/malto-polysaccharides (IMMPs) are synthesized from debranched starch by 4,6-α-glucanotransferase (Xue, Svensson & Bai 2022; Xue, Wang, et al. 2022). Isomaltomegalosaccharides (IMSs), composed of an α-1,4- and an α-1,6-linked segment, are obtained from MOSs of DP 6–7 by using dextran dextrinase, and recently also combined with CGTase to obtain products containing two α-1,4-linked segments (Lang et al. 2022, 2023). Although isomaltooligosaccharides (IMOSs) in the strict sense contain only α-1,6 linkages, commercial IMOSs by definition in addition have α-1,4 and sometimes α-1,2 and α-1,3 linkages (Goffin et al. 2011).

14.1.2 Sources of Starch for Production of Dextrins

Any type of starch can be used for the production of dextrins by enzyme-catalyzed modification. Popular sources are waxy and normal starches from corn, wheat, potato, cassava, and rice, but also high-amylose starches and modified starch products have been used (C Chen et al. 2020; Y Chen et al. 2023b; H Jiang et al. 2024; Krusong et al. 2022; Miao et al. 2018; Takata et al. 1996; D Wang et al. 2023; L Wang et al. 2020). Prior to enzymatic modification, customary pretreatment of starch includes heat-gelatinization or harsher conditions combining high temperatures and acids (Barczynska et al. 2012; Kapusniak et al. 2021; Z Liu et al. 2022b; Mao et al. 2021a; Wei et al. 2018).

In addition, AP, AM, and MDs isolated and purified from a large variety of starch sources are used to prepare different dextrins (Bai et al. 2015; Y Chen et al. 2023a; Wei et al. 2023a; Y Yang et al. 2023; Zhoukun et al. 2019). Initial treatment of starch and dextrins by the microbial debranching enzymes (DBEs) pullulanase or isoamylase after gelatinization or following gelatinization and enzymatic hydrolysis is recommended for production of linear dextrins (G Liu et al. 2018; J Park, Rho & Kim 2018; A-J Xie, Lee & Kim 2021; Xu et al. 2021; Xue, Svensson & Bai 2022). The molecular sizes of various obtained pretreated starches, debranched α-glucans, and dextrin products can be determined by analytical size exclusion chromatography (SEC) using molecular sieve resins covering suitable size ranges (Hu et al. 2017; Mao et al. 2021b; Ulbrich, Scholz & Flöter 2022; Wei et al. 2023a). While purification of single molecules in the form of monodisperse samples from such mixtures is practically unfeasible, fractionation of dextrins into narrower size ranges has been accomplished by solvent precipitation, membrane ultrafiltration, and preparative SEC (Chang et al. 2018; Hu et al. 2015; Ulbrich, Scholz & Flöter 2022; Wei et al. 2018; Zhen et al. 2021).

14.1.3 Enzymatic Production of Dextrins

A large number of glucoside hydrolases and transglucosylases of varying specificities and mechanisms have been applied to prepare useful starch-derived and related α-glucan dextrins (Figures 14.1 and 14.2). Different hydrolases (1) act in

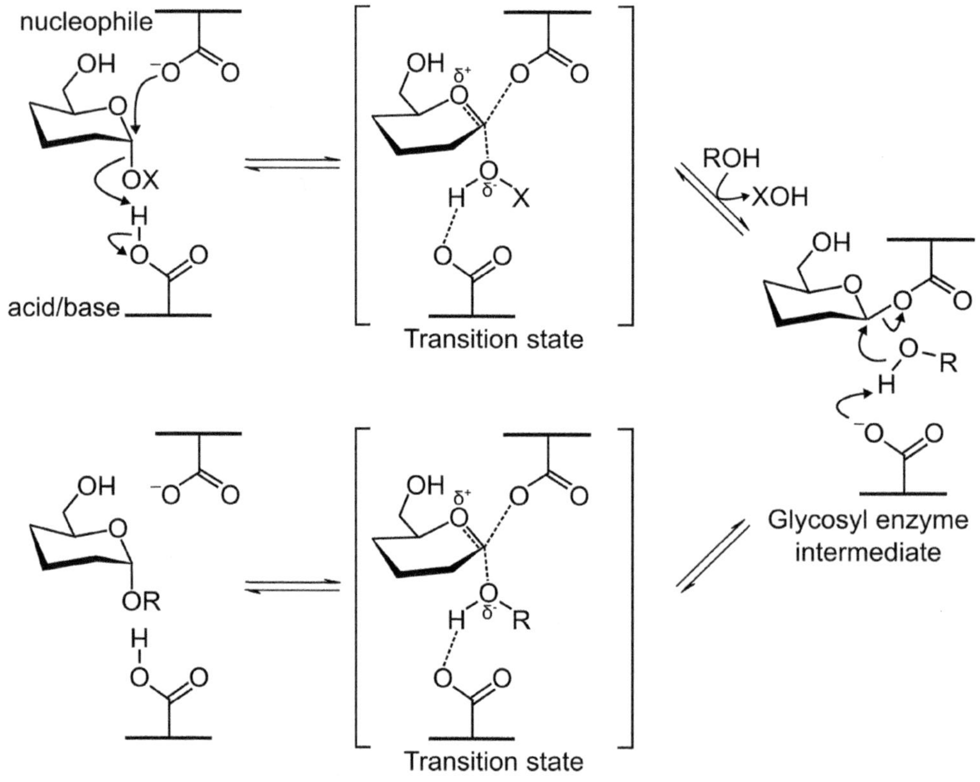

FIGURE 14.1   Schematic drawing of the retaining reaction mechanism for α-glucoside hydrolases. R=hydrogen in case of hydrolysis and a glucosyl containing unit in case of transglucosylation.

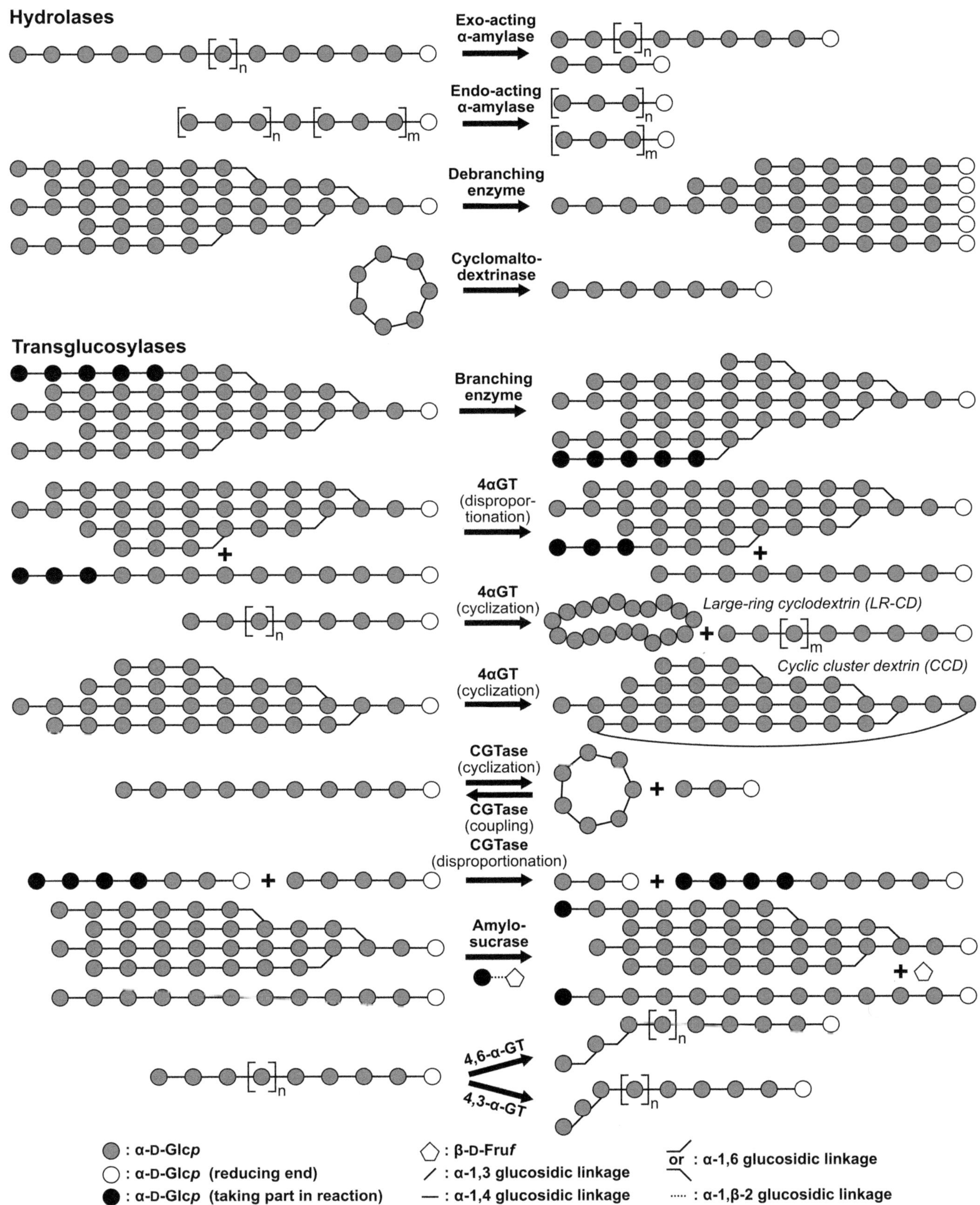

FIGURE 14.2 Schematic overview of enzymatic activities covered in the chapter. Cyclic cluster dextrin and highly branched cyclic dextrin (HBCD, not shown) are also made using branching enzyme.

endo-mode on α-1,4 linkages interior in α-glucan chains, (2) release MOSs in *exo*-mode reactions from non-reducing ends of α-1,4-glucans, and (3) hydrolyze α-1,6 branch points in AP and dextrins (H Kong et al. 2018; X Li et al. 2021; Møller, Henriksen & Svensson 2016; Pan et al. 2017; Paul et al. 2021). Transglucosylases introduce branches attached via α-1,6 linkages, as in AP, or non-natural α-1,2 or α-1,3 linkages, as well as α-1,3- and α-1,6-linked linear extensions of α-1,4-glucan chains, and are moreover generating cyclic dextrins (Y Chen et al. 2023a; Gangoiti et al. 2020; Gavgani et al. 2022; X. Li et al. 2023a; X Li et al. 2024; Takata et al. 2010; Wei et al. 2023b; Y Wu et al. 2023) (Figure 14.2). Glucoconjugates of non-carbohydrate aglycons, possessing improved water-solubility and bioavailability, are synthesized from donor and acceptor substrates by glucoside hydrolases in reverse hydrolysis reactions or by transglucosylases (Figure 14.1) (Han et al. 2020; Lambert et al. 2023; Marié et al. 2018; Vester-Christensen et al. 2023). There is a good choice of commercial enzymes available for different modifications (Lambert et al. 2023; X Li et al. 2024; Miao & BeMiller 2023; Pardhi et al. 2023; Paul et al. 2021; Sorndech, Tongta & Blennow 2018; Xu et al. 2021).

Enzymes acting on α-glucans are organized in glycoside hydrolase (GH) families in the CAZy database (www.cazy.org) of Carbohydrate-Active Enzymes based on protein sequence similarity (Drula et al. 2022). Typical starch-active hydrolases and transglucosylases belong to GH families 13, 14, 15, 31, 57, 70, and 77. The closely related GH families 13, 70, and 77 constitute clan GH-H and share the three invariant catalytic residues and the $(\beta/\alpha)_8$-barrel catalytic domain. The enormous GH13 family is currently divided into 49 subfamilies (www.cazy.org/), some of which include enzymes frequently used for starch and dextrin modification (Tables 14.1–14.5). Family GH70 contains transglucosylases that are able to transfer oligosaccharides or glucose from starch, MDs, MOSs, or sucrose resulting in branches on or linear extensions of 1,4-α-glucans (Dong et al. 2024; Gangoiti, Pijning & Dijkhuizen 2018, Gangoiti et al. 2020; Leemhuis et al. 2013; Y Wu et al. 2023) (Table 14.3). Very many starch-modifying enzymes are multidomain proteins and possess carbohydrate-binding modules (CBMs), which are organized in CBM families in the CAZy database (Drula et al. 2022). Currently, dedicated starch binding domains (SBDs) form 16 CBM families (Janeček et al. 2019).

Crystal structures are available for many starch-modifying enzymes. Here, examples are selected of the most important enzymes informing their ability to form linear (L), branched (B), cyclic (C), and glycoconjugate (G) products, mode of action and specificity, CAZy family (and where relevant CAZy subfamily), along with bound ligand(s) and corresponding subsites and binding sites occupied in structure-determined enzyme complexes (Table 14.1). The details listed are based on the atomic resolution of the molecular interactions between enzymes and MOS substrates or products, substrate analogs, and inhibitors at the active site crevice, pocket, or tunnel (Table 14.1). Structures of MOS-forming amylases, DBEs, branching enzymes, and cyclodextrinases, all belonging to GH13, as well as glucanotransferases of GH70 and α-1,6-glucosyltransferase of GH31, are shown in Figures 14.3–14.5. Known structures include a rare insight

into the accommodation adjacent to and at the catalytic site, respectively, of an α-1,6-branched substrate in the bacterial *Alicyclobacillus* sp. α-amylase AliC and in the plant DBE, barley limit dextrinase (Agirre et al. 2019; Møller et al. 2015).

Comparing enzyme structures helps us understand evolution, catalytic mechanisms, mode of actions, and substrate and product specificities. Thus, structural insights, typically along with multiple sequence alignment and phylogenetics data, can guide mutational analysis to identify functional residues and provide a basis for protein engineering and rational design of enzyme variants with desired properties (Buedenbender & Schulz, 2009; Hayashi et al. 2017; Møller, Henriksen & Svensson 2016; Vergès et al. 2017; Xie et al. 2020). Although enzyme tolerance to elevated temperatures, solvents, and detergents represents key characteristics required in industrial applications and is widely subjected to amelioration through protein engineering, this is not within the scope and gets little attention here. Conversely, the focus is on insights gained from the occupation of oligosaccharide ligands in starch hydrolase and transglucosylase complexes serving as starting points for comparison with predicted structures of related enzymes, as well as for ligand docking and computational analysis of details of the mechanism using, for example, Quantum Mechanics/Molecular Mechanics (QM/MM) (Morais, Nin-Hill & Rovira 2023; Neves, Fernandes & Ramos 2022). X-Ray crystallographic structures of enzymes from the different GH and GT families used for the production of dextrins (Table 14.1; Figures 14.3–14.5) are listed in the CAZy database (www.cazy.org/).

Structures have been determined of some of the maltotriose- (EC 3.2.1.116), maltotetraose- (EC 3.2.1.60), maltopentaose- (EC 3.2.1.-), and maltohexaose-forming amylases (EC 3.2.1.98) acting at non-reducing ends of α-1,4-glucans and referred to as maltooligosaccharide-forming amylases (MFAs) together with *endo*-acting α-amylases producing specific MOSs (Table 14.2; Figure 14.3). β-Amylase (EC 3.2.1.2) and maltogenic amylase (EC 3.2.1.133) releasing maltose, glucoamylase (EC 3.2.1.3) and α-glucosidase (EC 3.2.1.20) releasing glucose from non-reducing ends of dextrins, are not considered MFAs. α-1,6 branch point linkages in AP, MDs, and α- and β-limit dextrins are hydrolyzed by the DBEs, isoamylase (EC 3.2.1.68), pullulanase (EC 3.2.1.41), limit dextrinase (EC 3.2.1.142), and isopullulanase (EC 3.2.1.37) (Møller et al. 2016; Naik et al. 2023; Sim et al. 2014) (Table 14.1; Figure 14.4). Other enzymes, such as cyclomaltodextrinase (EC 3.2.1.54) and neopullulanase (EC 3.2.1.135), hydrolyze both α-1,4 and α-1,6 linkages, and some of these enzymes are able to catalyze ring-opening of CDs (Abdalla et al. 2021; Abe et al. 2005; Miao et al. 2018) (Table 14.1).

The transglucosylases found in GH families 13, 31, 57, 70, and 77 constitute a diverse group of retaining enzymes that apply a double displacement catalytic mechanism, similar to most starch hydrolases (Figures 14.1, 14.4, and 14.5) (Y Chen et al. 2023b; Gangoiti et al. 2020; X Li et al. 2023a, 2024) and transfer the reducing end of the donor segment in the covalent enzyme intermediate to an acceptor with the formation of a new glucosidic bond (X Li et al. 2023a, 2024; Tian et al. 2018; Xue, Svensson & Bai 2022)

TABLE 14.1 Examples of structure-determined ligand complexes of enzymes producing linear, branched, and cyclic large starch fragments, maltodextrins, maltooligosaccharides, and glucoconjugates

ENZYME[a]	ORGANISM	SPECIFICITY	CAZY[b]	PDB	LIGAND[c]	COMMENTS	LITERATURE
Hydrolases							
α-Amylase (L,B,G)	Bacillus subtilis	α-1,4 endo-hydrolase	GH13_5	1UA7	tg acarbose	Six subsites: −4 to +2[d]	Kagawa et al. (2003)
α-Amylase (L,B,G)	Bacillus subtilis	α-1,4 endo-hydrolase	GH13_5	1BAG	G5	E208Q[e] five subsites across the catalytic site	Fujimoto et al. (1998)
α-Amylase (L,B,G) (BA2)	Bacillus amyloliquefaciens/ Bacillus licheniformis (chimera)	α-1,4 endo hydrolase	GH13_5	1E3Z	tg acarbose	Ten subsites: −7 to +3	Brzozowski et al. (2000)
α-Amylase (L,B,G) (AliC)	Alicyclobacillus sp. 18711	α-1,4 endo-hydrolase	GH13_5	6GXV 6GYA	tg acarbose 6³-G-G3	Six subsites: −4 to +2 6³-G-G3 accommodated with an α-1,6-branch at +1(+1′) to +2	Agirre et al. (2019)
G4-Amylase (L)	Pseudomonas saccharophila STB07	α-1,4 exo-hydrolase	GH13	6IYG 6JQB	G4 tg acarbose	Four subsites: −4 to −1 Seven subsites: −4 to +3	Z Zhang et al. (2020)
G5-Amylase (L)	Bacillus stearothermophilus STB04	α-1,4 endo-hydrolase	GH13_5	6AG0	2×acarbose	Seven defined subsites: −6 to −3 and −1 to +2 Major products: G5 and G6	X Xie et al. (2019a)
G6-Amylase (L)	Alkalophilic Bacillus sp. 707	α-1,4 exo-hydrolase	GH13_5	1WPC	tg acarbose	Nine subsites: −6 to +3	Kanai et al. (2004)
α-Amylase (L) (MalS; G6-amylase)	Escherichia coli K12	α-1,4 endo-hydrolase	GH13_19	8IM8	G6, β-CD	Six subsites: −6 to −1 Primarily G6 production	An et al. (2023)
Limit dextrinase (L,B,G)	Hordeum vulgare (barley)	α-1,6 endo-hydrolase	GH13_13	4J3W 4J3X	α-limit dextrin pullulan fragment (G7)	E510A[e] three main chain subsites: 0′, +1, +2; three branch chain subsites: −3 to −1	Møller et al. (2015)
Pullulanase (L,B,G)	Klebsiella pneumonia	α-1,6 endo-hydrolase	GH13_13	2FHF	2×G4	Two parallel binding clefts. Four main chain subsites: 1′, 0′, +1, +2. Four branch chain subsites: −4 to −1	Mikami et al. (2006)
				5YNA/5YNE 5YN7/5YNC 5YNH/5YND	α-CD β-CD γ-CD	1 mM/10 mM 0.1 mM/1 mM 1 mM/10 mM	Saka et al. (2018)
Pullulanase (L,B,G)	Streptococcus pneumonia	α-1,6 endo-hydrolase	GH13_12	2YA0	2×G4	Two parallel binding clefts. Four main chain subsites: +1′, +1 to +3. Four branch chain subsites: −4 to −1	Lammerts Van Bueren et al. (2011)

(Continued)

TABLE 14.1 (Continued)   Examples of structure-determined ligand complexes of enzymes producing linear, branched, and cyclic large starch fragments, maltodextrins, maltooligosaccharides, and glucoconjugates

ENZYME[a]	ORGANISM	SPECIFICITY	CAZY[b]	PDB	LIGAND[c]	COMMENTS	LITERATURE
Isoamylase (L,B)	*Chlamydomonas reinhardtii*	α-1,6 *endo*-hydrolase	GH13_11	4OKD	G7	Seven subsites: -7 to -1 G at subsite $+2$	Sim et al. (2014)
Glycogen debranching enzyme (L,B) (TreX)	*Sulfolobus sulfataricus* P2	α-1,6 glucosidase α-1,4 transferase	GH13_11	2VR5 ,	acarbose G	Four subsites: -3 to -1. Covalent enzyme-acarbose intermediate. G at subsites $+2$ to $+3$	Woo et al. (2008)
Neopullulanase (L) (TVA I α-amylase)	*Thermoactinomyces vulgaris* R-47	α-1,4 (mainly) and α-1,6 hydrolase	GH13_21	2D0H 2D0G	P2 P5	D356N[e], D356N[e]/E396Q[e] Subsites: -5 to -1 (P2, pullulan fragment) -8 to $+2$ (P5, pullulan fragment)	Abe et al. (2005)
Neopullulanase (L,B,C) (TVA II α-amylase)	*Thermoactinomyces vulgaris* R-47	α-1,4 (pullulan, CD) α-1,6 (isopanose) hydrolase	GH13_20	1VB9	4^2-P2	Y374A near W356 (subsite $+2$) Subsites: -3, -2, -1, $+1(+2')$ $+2$ (made from 4^3-P2)	Mizuno et al. (2004)
Neopullulanase (L) (debranching enzyme)	*Nostoc punctiforme* PC73102	α-1,6>α-1,4 hydrolase	GH13_20	2WC7	tg acarbose	Five subsites: -4 to $+1$ Prefers α-1,6 over α-1,4 No CD hydrolysis No transglycosylation	Dumbrepatil et al. (2010)
Cyclomaltodextrinase (L,B)	*Thermus* sp.	α-1,4 hydrolase of CD>starch, pullulan	GH13_20	1GV1	β-CD	β-CD hosting F289 at active site. CDase is influenced by F47 from the N-terminal domain	HS Lee et al. (2002b)
Cyclomaltodextrinase (L, B)	*Flavobacterium* sp. 92	α-1,4 hydrolase of α-CDs>β-CDs>γ-CDs	GH13	3EDF, 3EDJ, 3EDK	α-CD, β-CD, γ-CD	T49P-(induces dimers of wild-type tetramer)-E340Q[e] shows productive binding	Buedenbender and Schulz (2009)
Glucoamylase (L)	*Aspergillus awamori* X100	α-1,4/(α-1,6) *exo*-glucohydrolase	GH15	1GAH 1GAI	acarbose,D-*gluco*-di-hydroacarbose	Four subsites: -1 to $+3$ Dual binding mode at -1 (K_i 10^{-12} M, 10^{-8} M)	Aleshin et al. (1996)
β-Amylase (L)	*Hordeum vulgare* (barley)	α-1,4 *exo*-maltohydrolase	GH14	2XFF 1XG9	acarbose 4-*O*-α-D-Glc-moranoline	Four subsites: $+1$ to $+4$ Two subsites: -2 to -1	Rejzek et al. (2011)
Transglucosylases							
Amylomaltase (L,C) (4αGT)	*Thermus aquaticus*	α-1,4 transglucosylase	GH77	5JIW	34-meric cycloamylose	Binding to an enzyme dimer, 16 residues bound to a monomer	Roth et al. (2017)
Amylomaltase (L,C) (MalQ; 4αGT)	*Escherichia coli*	α-1,4 transglucosylase	GH77	4S3R	tg: acarviosine-glucose-acarbose	Seven subsites: -4 to $+3$	Weiss, Skerra and Schiefner (2015)

(Continued)

TABLE 14.1 (Continued)   Examples of structure-determined ligand complexes of enzymes producing linear, branched, and cyclic large starch fragments, maltodextrins, maltooligosaccharides, and glucoconjugates

ENZYME[a]	ORGANISM	SPECIFICITY	CAZY[b]	PDB	LIGAND[c]	COMMENTS	LITERATURE
4αGT (L,C)	*Thermococcus litoralis*	α-1,4 transglucosylase	GH57	1K1Y	acarbose	Four subsites: −1 to +3	Imamura et al. (2003)
Branching enzyme (B,C)	*Escherichia coli*	α-1,4-glucan branching enzyme	GH13_9	4LQ1, 4LPC	G6, G7	Occupy six surface binding sites. Binding at donor site	Feng et al. (2015)
				8SDB	G8	Eight subsites: −8 to −1	Fawaz et al. (2023)
Branching enzyme (B,C) (BE1)	*Cyanobacterium* sp. NBRC 102756	α-1,4-glucan branching enzyme	GH13_9	5GQX	G7	W610N binding donor at seven subsites: −7 to −1	Hayashi et al. (2017)
Branching enzyme (B,C) (BEI)	*Oryzae sativa* (rice)	α-1,4-glucan branching enzyme	GH13_8	7ML5	maltododecaose	M12 bound at part of the acceptor binding site	Gavgani et al. (2022)
Amylosucrase (L)	*Neisseria polysaccharea*	sucrose α-1,4 Glc tgase	GH13_4	1JGI, 1MW0	sucrose, G7	E328Q[e] active site seven subsites: −1 to +6. Surface site binding of G7	Mirza et al. (2001) and Skov et al. (2002, 2006)
Amylosucrase (L)	*Neisseria polysaccharea*	sucrose α-1,4 Glc tgase	GH13_4	5N7J	docking sucrose and G7	Two subsites: −1 to +1. Surface binding site. Nine mutations at active site forms 2<DP<21	Vergès et al. (2017)
Amylosucrase (L)	*Deinococcus geothermalis*	sucrose α-1,4 Glc tgase	GH13_4	3UER	turanose	Two subsites: −1 to +1	Guérin et al. (2012)
Glucansucrase (L,B) (GTF180 ΔN)	*Lactobacillus reuteri* 180	sucrose α-1,4 to α-1,3/α-1,6 tgase	GH70	3KLK 3HZ3 3KLL	native sucrose maltose	Sucrose (donor) subsites −1 to +1. Maltose (acceptor) subsites +1 to +2, and +2 to +3	Vujičić-Žagar et al. (2010)
4,6-α-GT (L,B) (GtfB ΔV)	*Limosilactobacillus reuteri* N1	α-1,4 to α-1,6 tgase	GH70	8HW3	2×acarbose	Four subsites: −2 to +2 and an SBS	Dong et al. (2024)
				8HWK	G7	SBS	
4,6-α-GT (L,B) (GtfB ΔNΔV)	*Limosilactobacillus reuteri* NCC 2613	α-1,4 to α-1,6 tgase	GH70	7P39	acarbose	Four subsites: −1 to +3	Pijning et al. (2021)
4,6-α-GT (L,B) (GtfB ΔNΔV)	*Lactobacillus reuteri* 121	α-1,4 to α-1,6 tgase	GH70	5JBF	G5	D1015N[e] Five subsites: −5 to −1	Bai et al. (2017)
				5JBE	6⁴-G-G4	Wild-type subsites: −6 to −2	
6αGT (L,C) (Ps6GT31A)	*Paenibacillus* sp. 598K	α-glucosidase (broad) and α-1,6-tgase	GH31	5X7Q	G6	G at subsite −1 and G5 at the N-domain	Fujimoto et al. (2017b)
				5X7P	acarbose	Four subsites: −1 to +3	
				5X7R	IG6	G at subsite −1 and IG5 near subsites +1, +2 and acceptor subsites +1a to +4a	

(Continued)

TABLE 14.1 (*Continued*)   Examples of structure-determined ligand complexes of enzymes producing linear, branched, and cyclic large starch fragments, maltodextrins, maltooligosaccharides, and glucoconjugates

ENZYME[a]	ORGANISM	SPECIFICITY	CAZY[b]	PDB	LIGAND[c]	COMMENTS	LITERATURE
Branching sucrase (B) (ΔN_{123}-GBD-C2)	*Leuconostoc mesenteroides* NRRL B-1299	sucrose α-1,2 Glc tgase	GH70	4TVD 4TVC	G isomaltotriosyl-maltose	G at subsite −1 G, IG2, IG3 binding seen at an SBS in domain V important for branching	Brison et al. (2016)
Alternan-sucrase (L)	*Leuconostoc citreum* NRRL B-1395	sucrose α-1,3/α-1,6 Glc tgase	GH70	6T16 6T18	panose oligoalternan	Surface binding site identified in domain V with a role in elongation	Molina et al. (2020)
α-CGTase (C,L)	*Niallia circulans* 251	α-1,4 cyclo-tgase	GH13_2	1CXF	G4	D229N[e]/E257Q[e] Four subsites: −2 to +2 at active site	Knegtel et al. (1995)
				1CXE	α-CD	Two surface binding sites	
α-CGTase (C,L)	*Niallia circulans* 251	α-1,4 cyclo-tgase	GH13_2	1CXK, 1CXL	G9, 4-deoxy-G3	D229N[e]/E257Q[e] Nine subsites: −7 to +2	Uitdehaag et al. (1999b)
β-CGTase (C,L)	*Niallia circulans* 251	α-1,4 cyclo-tgase	GH13_2	3CGT	S-(α-D-Glc)-6-thio-β-CD	E257A[e]. Density for β-CD, but not for exocyclic G	Schmidt et al. (1998)
γ-CGTase (C,L)	*Niallia circulans* 251	α-1,4 cyclo-tgase	GH13_2	1D3C	γ-CD	E257Q/D229N[e]	Uitdehaag et al. (1999a)
CITase (C)	*Paenibacillus* sp. 598K	cycloisomaltooli-gosaccharide tgase	GH66	5X7H	IG7 (hydrolysis of CI7)	Seven subsites: −7 to −1	Fujimoto et al. (2017a)

α-Glucan phosphorylases

ENZYME[a]	ORGANISM	SPECIFICITY	CAZY[b]	PDB	LIGAND[c]	COMMENTS	LITERATURE
α-Glucan phosphorylase (L,G)	*Caldicellulosiruptor saccharolyticus*	kojibiose phosphorylase	GH65	3WIQ	kojibiose	2 G and 1 P_i bound at subsites −1 and +1	Okada et al. (2014)
α-1,4-Glucan phosphorylase (L) (MalP)	*Escherichia coli*	4-α-glucan phosphorylase	GT35	1L6I 1L5W	G1P/G5 G1P/G4	Contacts at subsites −1 to +4. Reaction in crystals shows water at subsite −1	Geremia et al. (2002)

[a]   L, linear; B, branched; C, cyclic; G, glycoconjugate.
[b]   CAZy family/subfamily number.
[c]   tg acarbose has undergone transglycosylation *in cristallo*.
[d]   Numbers indicate the occupied subsites.
[e]   Catalytic site mutant.

TABLE 14.2 Biochemical characterization and structure of maltooligosaccharide-forming amylases (MFAs)

ORGANISM	ACTION MODE	PRODUCT(S) (MINOR)	STRUCTURE LIGAND (PDB ID)	COMMENTS	LITERATURE
G3-Forming amylases					
Kitasatospora sp. MK-1785	exo	G3 (G1–G2)		Activity on soluble starch, AP, AM>γ-CDs (but hardly on α- and β-CDs). Hydrolyses G4–G7 to G3 and G1–G4 Forms glucoconjugates of phenols and alcohols with soluble starch as donor	Kamon et al. (2015)
Streptomyces griseus SGR5663	exo	G3 (G2–G6)		Optimal production from soluble starch of G3 at 20°C and 30°C and pH 3.5 (77%). TLC indicates *exo*-action	Kashiwagi et al. (2014)
Streptomyces avermitilis SAV6969	exo	G3 (G2–G6)		Optimal production from soluble starch of G3 at 20°C and 30°C and pH 3.5 (72%). TLC indicates *exo*-action	Kashiwagi et al. (2014)
G4-Forming amylases					
Pseudomonas stutzeri MO-19	exo	G4	native (2AMG)	Catalytic domain (the SBD was lost due to proteolysis)	Morishita et al. (1997)
Pseudomonas stutzeri MO-19	exo	G4	G4 (1JDC, 1JDD)	Co-crystallized G5 is cleaved to G4 by catalytic site mutant E219Q. Four subsites corresponding to −4 to −1	Yoshioka et al. (1997)
Pseudomonas stutzeri MO-19	exo	G4	G5 (1QI4, 1QI3, 1QPK, 1QI5)	Catalytic residues single mutants E219G, D193N, D193G, and D294G in complex with G5 cleaved to G4	Hasegawa, Kubota and Matsuura (1999)
Pseudomonas stutzeri MO-19	exo	G4	native (1GCY)	Intact with C-terminal SBD disordered in the crystal structure	Mezaki et al. (2001)
Pseudomonas stutzeri 7493	n.i.	G4 (G–G3)		Antarctic, psychrophilic, releasing 56% G4, 8% G3, 21% G2, and 14% G1 from soluble starch	Zhang and Zeng (2011)
Pseudomonas stutzeri AS22	exo	G4		Detergent tolerant, producing G4 from raw starch, starch, AP, AM, G5, G6, and G7	Maalej et al. (2013)
Pseudomonas saccharophila STB07	exo	G4 (G1–G3; G5–G7)	G4 (6ITG) tg acarbose (6JQB)	Four subsites: −4 to −1 Seven subsites: −4 to +3 (pseudomaltoheptaose) 65% G4 from corn starch	Z Zhang et al (2020)
Pseudomonas saccharophila STB07	exo	G4		Disulfide engineered A211C-S214C close to the active site has 2.6-fold increased half-life at 60°C	Y Wang et al. (2022)
Pseudomonas saccharophila STB07	exo	G4 (G1–G7)		Truncation of C-terminal linker-connected CBM20 increased G4 yield to 72% and reduced activity on starch	Duan et al. (2022)
G5-Forming amylases					
Saccharophagus degradans 2-40^T (SdMFA)	exo	G5 (G2–G4)		Truncation of C-terminal linker and CBM20 resulted in reduced activity and loss of cold adaptation	Ding et al. (2021)
Bacillus megaterium (BmMFA)	exo	G5 (G2–G4, G6–G7)		C-terminal SdMFA CBM20 fused to BmMFA increased G5 production	Ding et al. (2022)

(Continued)

TABLE 14.2 (*Continued*)   Biochemical characterization and structure of maltooligosaccharide-forming amylases (MFAs)

ORGANISM	ACTION MODE	PRODUCT(S) (MINOR)	STRUCTURE LIGAND (PDB ID)	COMMENTS	LITERATURE
Saccharophagus degradans 2-40[T]	exo	G5 (G2–G4, G6–G7)		Full length gives G5 as major product. C-terminal CBM20 truncation of SdMFA produces G2–G7	Ding et al. (2022)
Bacillus stearothermophilus STB04	endo	G5 (G1–G4; G6)		Produces G5, G6. Engineering for G5 enrichment by W139A/L/Y at subsites −5 and −6	X Xie et al. (2019b)
Bacillus stearothermophilus	endo	G5, G6 (G1–G4; G7)		*Endo*-acting initially producing most G7 from AP and AM, but G5 and G6 are major final products (G5>69%)	Y Wang et al. (2019)
Bacillus stearothermophilus STB04	endo	G5, G6	acarbose (6AG0)	Produces most G5 from G8, and most G6 from G9. Two acarbose molecules bound in the active site cleft show subsites −6 to −3 and −1 to +2	X Xie et al. (2019a)
Bacillus stearothermophilus STB04	endo	G6 (G1–G5; G7)		G5, G6. Engineered for G6 enrichment by mutation of Gly109 hydrogen bonding with substrate at subsite −5. G109F increased G6 to 40% and reduced G5 to 5%	Xie et al. (2020)
Bacillus cereus ATCC 14579	n.i.	G5 (G1–G4; G6)		G5 >65% from starch is >95% pure after AB-8 resin chromatography	Pan et al. (2023)

G6-Forming amylases

ORGANISM	ACTION MODE	PRODUCT(S) (MINOR)	STRUCTURE LIGAND (PDB ID)	COMMENTS	LITERATURE
Geobacillus stearothermophilus sp. US100	n.i.	G6>G5		Thermostable amylase producing G6 (G7) from starch and later G5. Mutants N315D_V450G produced G3/G4 and ΔI214G215 increased thermostability. M197A at subsite +3 is detergent tolerant	Ben Ali et al. (2006), Ben Ali, Mezghani and Bejar (1999), and Khemakhem et al. (2009)
Bacillus sp. 707	n.i.	G6>G5 (G2–G4)	native (1WP6) tg acarbose (1WPC)	Nine subsites: −6 to +3 (pseudomaltononaose)	Kanai et al. (2004)
Bacillus sp. 707 (mutant)	n.i.	G5>G6	G5 (2D3L) G6 (G1) (2D3N)	Five subsites: −6 to −2 / Six subsites: −7 to −2 and G at +2 Trp140 stacking at subsites −6, −5. W140Y/L produced mostly G5, less G6 and little G7 compared to wild type	Kanai et al. (2006)
Corallococcus sp. EGB	n.i.	G6 (G2–G7)		60% G6 formed from waxy corn starch / AmyM-Tr (C-terminal CBM20 truncated) improved bread quality more than AmyM	Z Li et al. (2015) and T Yang et al. (2022)
Escherichia coli K12	exo	G6/ G5>G4– G1	native (8IM8) docking of G6 and β-CD	Belongs to GH13_19; G6 produced initially is further hydrolyzed to MOSs (G5–G1). Low activity on β- and γ-CDs	An et al. (2023)

n.i., not informed.

TABLE 14.3 Enzymatic production of linear and branched dextrins using transglucosylases

ENZYME (CAZY FAMILY)[A]	ORGANISM	SUBSTRATES: DONOR (D), ACCEPTOR (A)	PRODUCT(S)	COMMENTS	LITERATURE
BE (GH13_9)	*Rhodothermus obamensis* STB05	Potato AM, AP, various starches (D&A)	Branching of AM and AP	Exceptionally high activity among BEs	Z Wang et al. (2019)
BE (GH13_9)	*Rhodothermus obamensis, Synechocystis* sp. PCC6803, or *Bifidobacterium thermophilum*	Waxy corn starch (D&A)	α-Amylase hydrolysis → highly branched α-limit dextrins	Removal of linear MOSs slowed digestion A chains increased by 15%–45% B1 chains decreased by 38%–44%	Shim et al. (2023)
BE (GH57)	*Thermococcus kodakarensis*	MD DE 6, 12, 19 (D&A)	Glucan dendrimers (EMDs)	MD12 was optimal substrate resulting in largest increase in DP<13 and in RS for the enzyme modified product (EMD)	Y Chen et al. (2023a)
BE (*Tf*BE; GH13_9) α-CGTase (*Tf*CGTase; GH13_2) α-Glucosidase (*An*GS; GH13)	*Thermobifida fusca* *Thermobifida fusca* *Aspergillus nidulans*	Pyrodextrin (D&A)	Increase in short branches	*Tf*BE reduced digestibility, which was further reduced by reacting with T*f*CGTase and *An*GS	Z Liu et al. (2022b)
Pullulanase (TVA II α-amylase; GH13_20)	*Thermoactinomyces vulgaris* R-47	Pullulan (D) and G (A)	6^3-α-G-G3 4^2-α-IG-IG2	TVA II α-amylase hydrolysis of α-1,4 linkages in pullulan and transfer to G forming α-1,4 or α-1,6 linkages	Tonozuka et al. (1994)
4αGT (GH77)	*Corallococcus* sp.	MOSs (DP 3–7), AM (D&A)	MD DP<12	G2 and AM showed disproportionation. No cycloamylose was produced	Zhoukun et al. (2019)
4αGT (PSGT; GH57)	*Pyrococcus* sp. ST04	pNPG6 (D&A)	pNPG10 (pNPG8, pNPG9)	MOSs (DP 3–7) disproportionation giving low amounts of G and G2. Mutants Y181A at subsite +2 and W219A at subsite +1 have reduced activity	Jung et al. (2022)
Pullulanase (GH13_14) 4,6-α-GT (GtfB ΔN; GH70)	*Bacillus acidopullulyticus* *Limosilactobacillus reuteri* 121	Pullulanase-debranched waxy starch (retrograded) reacted with 4,6-α-GT (D&A)	IMMPs: linear α-1,6 chains with short inlaid α-1,4-linked fragments at reducing end, non-reducing end and middle part	Detailed product structure identification using α-glucan specific enzymes IMMPs serve as novel dietary fiber. Highest yields after debranching	Xue, Wang, et al. (2022)

(Continued)

TABLE 14.3 (Continued)   Enzymatic production of linear and branched dextrins using transglucosylases

ENZYME (CAZY FAMILY)[a]	ORGANISM	SUBSTRATES: DONOR (D), ACCEPTOR (A)	PRODUCT(S)	COMMENTS	LITERATURE
4,6-α-GT (GH70)	Streptococcus thermophilus	Waxy corn starch, normal corn starch, amylomaize V and VII (D&A)	Increased content of short branches	Reduced digestibility and extended glycemic response in mice	Ryu et al. (2022)
4,3-α-GT (GtfB ΔN; GH70)	Lactobacillus fermentum NCC 2970	MOS DP 6,7; amylose V, AP, potato starch (D&A)	Products with α-1,3 and α-1,4 linkages	Disproportionation products of G7 and amylose V have α-1,4:α-1,3 ratio of 86:14 and 81:19 (4% disubstituted), respectively	Gangoiti et al. (2017)
4,6-α-GT (GtfB ΔN; GH70)	Limosilactobacillus fermentum NCC 3057	(1) MD DE 15–20 or 5–7 (D&A)	α-1,2 or α-1,3 and α-1,2/α-1,3-branched IMD	Branching increased solubility, resistance (dietary fiber), and reduced viscosity	Wei et al. (2023b)
α-1,2 branching sucrase (ΔN$_{123}$-GBD-C2; GH70)	Leuconostoc mesenteroides NRRL B-1299	(2) sucrose (D) linear IMD (A)	(MD 15–20 → DP 20; MD 5–7 → DP 30)		
α-1,3 branching sucrase (GH70)	Leuconostoc citreum NRRLB-724				
4,6-α-GT (GTF180 ΔN L940W; GH70)	Lactobacillus reuteri	Sucrose (D&A)	(1) IMOSs	Single-arm linear chimeric dextrin with α-1,4 non-reducing and α-1,6 reducing end segments possessing solubilization effect	Y Wu et al. (2023)
γ-CGTase (GH13_2)	Bacillus sp. FJAT-44876	(2) γ-CD (D), IMOSs (A)	(2) SLD (DP 21)		
4,6-α-GT (GtfB ΔN, GH70)	Lactobacillus reuteri 121	Amylose V, MDs (D&A)	IMMPs	α-1,6 linkage introduction was most efficient at high MD concentration	Bai et al. (2015)
4,6-α-GT (GTFB; GH70)	Lactobacillus reuteri 121	Starches, MDs (D&A)	IMMPs (linear and branched, up to DP 35)	Comparison of performance with different starches and MDs. High-amylose starch is the best substrate	Leemhuis et al. (2014)
GTase (GH70)	Azotobacter chroococcum	MD DE 6 (D&A)	Highly branched reuteran-like glucan	Short branch chains increased; digestibility decreased	Y Yang et al. (2023)
α-Glucosidase (AgdB; GH13)	Aspergillus nidulans	MD DE 15–20, 6–8, or 2 (D&A)	Average DP 7; methylation showed no 4,6-branch point residues	Slow digested dextrin (SDD) dietary fibers with highest prebiotic effect from high DE	Wei et al. (2023a)
CGTase (GH13_2)	Bacillus stearothermophilus NO$_2$				

(Continued)

TABLE 14.3 (*Continued*)   Enzymatic production of linear and branched dextrins using transglucosylases

ENZYME (CAZY FAMILY)[a]	ORGANISM	SUBSTRATES: DONOR (D), ACCEPTOR (A)	PRODUCT(S)	COMMENTS	LITERATURE
α-Amylase (1) α-Amylase (2) + β-amylase + pullulanase α-TGase (3)	*Bacillus licheniformis* (heat) *Aspergillus niger*	(1) Starch → MDs (2) MD → MOSs (3) MOSs → IMOs	IMOSs	α-Transglucosidase hydrolyzed α-1,4 linkages and formed α-1,6 linkages. Preparation steps for some routes used simultaneous BE and α-amylase treatment, others included the use of β-amylase	Goffin et al. (2011) and Sorndech et al. (2017)
Dextran dextrinase (GH15) α-amylase (GH13_24)	*Gluconobacter oxidans* Porcine pancreas	MOSs DP 6–7 (D&A)	L-IMS average DP 11	Linear type L-IMS (linear-isomaltomegalosaccharide) with α-1,4-linked reducing end segment shortened by α-amylase treatment	Lang et al. (2014)
Dextran dextrinase (GH15)	*Gluconobacter oxidans* ATCC 11894	MOS DP6/7 (20%) (D&A)	IMS average DP 14.5	α-1,6 non-reducing end and α-1,4 reducing end segments	Lang et al. (2022)
Dextran dextrinase (GH15)	*Gluconobacter oxidans* ATCC 11894	MOSs DP 3–7, short-chain AM, starch (D&A)	Dextran	Highest yield with short-chain AM	Naessens et al. (2005)
CGTase (Amano) (GH13_2)	*Bacillus macerans*	α-CD (D) S-IMS (single-segment-IMS) (A)	D-IMS (double-segment-IMS)	α-1,4 non-reducing end, α-1,6 middle, and α-1,4 reducing end segments	Lang et al. (2023)
Amylosucrase (GH13_4)	*Neisseria polysaccharea*	Sucrose (D&A)	AM DP 35–58	Larger size products with lower sucrose concentration	Potocki-Veronese et al. (2005)
Thermostable phosphorylase (GT35)	*Aquifex aeolicus* VF5	GlcA-1-P (D) G3 (A)/G4 (A)	GlcA-G3/ GlcA-G3-GlcA-G7	Synthesis by phosphorylase from G4 resulted in multiple products via disproportionation	Umegatani et al. (2012)

[a] Enzyme abbreviation (when relevant) and CAZy family (www.cazy.org) (Drula et al. 2022) in parenthesis.

TABLE 14.4 Enzymatic production of cyclic dextrins

ENZYME	ORGANISM	SUBSTRATE(S)	PRODUCT(S)	COMMENTS	LITERATURE
Cyclodextrins (CDs)					
α-CGTase	*Bacillus macerans* sp. 602-1	Soluble starch	α-CD/β-CD ratio improved from 6 to 12	Y167H (at subsite −6) with one (Y167HH) and two (Y167HHH) insertions improved α-CD purity	Yue et al. (2014)
Pullulanase β-CGTase (Toruzyme)	*Bacillus subtilis* 168 *Thermoanaerobacter* sp.	Potato starch	β-CD yield increased by 6%–7%. β-CD was 4–5-fold>α-CD and γ-CD	One-pot: Debranching using less-β-CD-inhibited pullulanase (F476Y/H/C near subsite +2) increased β-CD yield	X Li et al. (2020)
γ-CGTase	*Bacillus* sp. FJAT-44876	Cassava starch	γ-CD in 2.5–6-fold higher yield	High starch concentration and moderate heat was optimal	H Wu et al. (2022)
γ-CGTase	*Bacillus* sp. FJAT-44876	Cassava starch	γ-CD yield increased by 35%	Improved yield by thermostable mutant (G208S) in 10 mM Ca^{2+}	X Li et al. (2023b)
γ-CGTase	*Bacillus clarkii*	Potato starch	γ-CD 96% of a total CD yield 72% (favored by complexation)	Mutants F91N and F91L (subsite −3) and central binding Y186W doubled γ-CD forming activity	L Wang et al. (2020)
γ-CGTase Cyclomaltodextrinase (PpCDase)	*Bacillus* sp. *Palaeococcus pacificus*	Corn starch (CD mixtures)	γ-CD enriched from products α-CD, β-CD, and γ-CD (no complexant) by PpCDase	PpCDase hydrolyzed preferably α-CD and β-CD leaving γ-CD. Includes the analysis of simulated α-, β-, and γ-CD mixtures	Ji et al. (2020b)
γ-CGTase	*Bacillus clarkii* 7364	β-CD (D)+G2 (A)	γ-CD 86% of total CD yield 36%	Complexant increased γ-CD yield	Qiu et al. (2018)
β-CGTase (Toruzyme) Glucoamylase	*Thermoanaerobacter* sp. *Rhizopus* sp.	MD DE 4–7	G1-β-CD in 24% yield	Two steps: (i) β-CGTase produced β-CD used in coupling reaction to G_n-β-CD, (ii) trimmed by glucoamylase to G1-β-CD	L Xia et al. (2017)
Amylopullulanase (DAPu; GH57)	*Desulfurococcus amylolyticus* JCM9188	G2-β-CD (D&A)	$6^1,6^4$-di-G2-β-CD $6^1,6^3,6^5$-trio-G2-β-CD	G2 units transferred to G2-β-CD by transglucosylation	Y-U Park et al. (2018)
Limit dextrinase	*Hordeum vulgare* (barley)	G2- or G3-fluoride (D) α-CD, β-CD (A)	$G2_{1-4}$-α-CD $G2_{2-3}$-β-CD	Chemoenzymatic formation of CDs having from 1 to 4 α-1,6-G2 branches	Vester-Christensen et al. (2023)
6αGT (GH31) and CITase (GH66)	*Paenibacillus* sp. 598K	Starch → dextran by 6αGT, then dextran → CI by CITase	IMOSs of DP 4–10 CIs (cycloisomaltodextrins) of DP 4 and 5	6αGT transfers α-1,6 G to non-reducing end of G4. CITase from *Paenibacillus* required α-1,6 DP 4 for production of CI4 and CI5	Ichinose et al. (2017)
CITase (GH66)	*Bacillus circulans* T-3040	Soluble starch	CI7–CI12	CIT from *B. circulans* acted with IMOSs produced by a 135 kDa protein	Funane et al. (2014) and Suzuki et al. (2012)

(Continued)

TABLE 14.4 (Continued)   Enzymatic production of cyclic dextrins

ENZYME	ORGANISM	SUBSTRATE(S)	PRODUCT(S)	COMMENTS	LITERATURE
Large-Ring Cyclodextrins (LR-CDs), Cyclic Cluster Dextrins (CCDs), Highly Branched Cyclic Dextrins (HBCDs)					
Isoamylase (GH13_11) (1) (Megazyme)	*Pseudomonas* sp.	High-amylose starch debranched by isoamylase	LR-CD DP 5–16 & DP 23–36	Yield up to 76% depending on high AM content	J Park, Rho and Kim (2018)
4αGT (GH77) (2)	*Thermus aquaticus*				
4αGT (GH57)	*Archaeoglobus fulgidus* st. 7324/DSM8774	Amylose V	LR-CD DP 17–35	No cyclic compounds observed from potato starch as substrate	Paul et al. (2015)
Glycogen debranching enzyme (*Cg*GDE) (GH13_11) (1)	*Corynebacterium glutamicum*	Tapioca, corn, pea, potato starches	LR-CD DP 24–32	*Cg*GDE pretreatment gave much higher yield than isoamylase pretreatment	Suksiri et al. (2021)
4αGT (GH77) (2)	*Thermus filiformis*				
Amylomaltase (D-enzyme) (GH77)	*Solanum tuberosum* (potato)	AM	LR-CD DP 90 Minimum DP 17	Products containing cyclic cluster dextrins (CCDs) treated with glucoamylase for purification	Takaha et al. (1996)
Amylomaltase (4αGT) (GH77)	*Thermus aquaticus* 33923	AM	LR-CD DP>60 Minimum DP 22	Transglycosylation products were also obtained from G2–G7	Terada et al. (1999)
Branching enzyme (GH13_9)	*Bacillus stearothermophilus*	Waxy rice AP	HBCD DP 460–770	Highly branched cyclic dextrin	Takata et al. (1996)
Branching enzyme (GH13_9)	*Aquifex aeolicus*	Waxy corn starch	HBCD DP 1,200	Average DP 50 of ring carrying branch chains of average DP 16	Takata et al. (2003)
CGTase (GH13_2)	*Bacillus* sp. G-825-6	Soluble potato starch	DP 9–12	Semi-rational loop mutant library screen for high DP CDs (LR-CDs)	Sonnendecker and Zimmermann (2019)
			DP 8–12	Kinetics for CD/LR-CD DP 7–12 formation. Single mutants at subsites −2/−3. Suppression of β-CD and enriched for DP 11–12	Sonnendecker, Melzer and Zimmermann (2019)
Thermostable phosphorylase (GT35)	*Aquifex aeolicus* VF5	Highly branched cyclic dextrin (HCBD), 125 kDa, av. CL DP 15 (A) α-D-glucuronic acid 1-phosphate (D)	GlcA-ylation ratio 0.367–0.700/ non-reducing end	The GlcA-ylation ratio increased with increasing feed of GlcA-1-P	Takemoto et al. (2013)

D, donor; A, acceptor

TABLE 14.5 Examples of enzymatic production of glucoconjugates

ENZYME	ORGANISM	DONOR	ACCEPTOR	PRODUCT(S)	COMMENTS	LITERATURE
G3-forming amylase	*Kitasatospora* sp. MK-1785	G4 Soluble starch Soluble starch	G4 G Phenols, alcohols	G7 G4 G3-conjugates	α-1,4 *exo* hydrolase G3-Transfer to phenols and alcohols	Kamon et al. (2015)
Limit dextrinase	*Hordeum vulgare* (barley)	Pullulan G2-1-F, G2-1-F G1G3G3	pNPG1–pNPG4 α-CD, G2-β-CD Aromatic-β-glucosides	pNPG4–pNPG7 $G2_{1-4}$-α-CD, $G2_{2-3}$-β-CD G4-aromatic-β-glucosides	Pullulan fragment G1G3G3 is an excellent donor Aglycone binding at subsite +2 and β-G at subsite +1	Vester-Christensen et al. (2023)
Pullulanase (PulA)	*Thermotoga neapolitana*	Pullulan	Aesculin	G3-Aesculin	α-1,6 transfer of G3 from pullulan	Kang et al. (2011)
4αGT (GH57)	*Archaeoglobus fulgidus st. 7324/ DSM8774*	Soluble starch	β-D-dodecyl (C12)-maltoside	Up to G24/C12	An array of alkyl oligoglucosides was obtained	Paul et al. (2015)
CGTase	*Bacillus licheniformis*	Dextrin	Ginsenoside CK and ginsenoside F1	Ginsenoside F1 with 1–5 G added	Glycosylation of OH groups in ginsenoside F1s improved by acceptor accommodation design	Xiao et al. (2022)
CGTase	*Paenibacillus macerans*	MD	Sophoricoside (S)	G-S–G6-S	Engineered mutant affecting subsites −5 to −7 favored long chain-S synthesis	R Han et al. (2020)
CGtase	*Bacillus macerans*	β-CD	pNP-β-D-Glucopyranoside	α-1,4>α-1,3, α-1,6-glucosides	Regio-specificity influenced by α-(formation of α-1,4)/β-NP anomer	Strompen et al. (2015)
CGTase (CspCGT13)	*Carboxydocella* sp.	α-CD γ-CD	Dodecyl-β-G2 Dodecyl-β-G Dodecyl(C_{12})-β-G8	C_{12}G8, C_{12}G10	Mutants of F197, G263, E266 near acceptor subsites show increased coupling activity	Ara et al. (2021)
CGTase (Toruzyme)	*Thermoanaero-bacter* sp.	β-CD	Resveratrol	3-*O*-α-D-G/-G2-resveratrol 4′-*O*-α-D-G/-G2 resveratrol	α-, β-, and γ-CDs have essentially the same efficiency. Yield 35%	Marié et al. (2018)

(Continued)

TABLE 14.5 (*Continued*) Examples of enzymatic production of glucoconjugates

ENZYME	ORGANISM	DONOR	ACCEPTOR	PRODUCT(S)	COMMENTS	LITERATURE
CGTase (Toruzyme)	*Thermoanaerobacter* sp.	β-CD	Resveratrol	3-O-α-D-G/-G2-resveratrol 4'-O-α-D-G/-G2 resveratrol	Membrane reactor product separation to avoid enzymatic degradation. Yield 50%	Ioannou et al. (2021)
CGTase (Toruzyme)	*Thermoanaerobacter* sp.	β-CD	pNP-β-D-Glucopyranoside	α-1,6-Glucoside (α-1,3- and α-1,4-glucosides)	pNP-α-D-Glucopyranoside resulted in α-1,4-linked products	Strompen et al. (2015)
CGTase (Toruzyme)	*Thermoanaerobacter* sp.	α-CD	Baicalin (B)	BG1–BG9	The glucuronide residue in B is transglucosylated. BG1 and BG2 have increased antioxidant activity	Lambert et al. (2023)
CGTase (Amano)	*Bacillus macerans*					
CGTase (Toruzyme)	*Thermoanaerobacter* sp.	β-CD (α-CD, MD)	Rebaudioside A (RA)	Mono-α-1,4-G-RA	Glucoamylase trimming of products. RAG1 showed improved sensory quality	W Wang et al. (2021)
Glucoamylase	*Aspergillus oryzae*					
CGTase (Amano)	*Bacillus macerans*	α-CD, G6	Alkyl glucosides (*n*-cctyl-, *n*-decyl-, *n*-dodecyl-β-D-G)	C8G2–C8G7	Complexation of alkylglucosides by α-CD facilitated transglucosylation	Börner, Roger and Adlercreutz (2014)
CGTase (Amano)	*Bacillus macerans*	α-CD, soluble starch	2-(β-Glucosyloxy)-ethyl acrylate, 2-(β-Glucosyloxy)-ethyl methacrylate	(Gα)$_n$-G-β-EA (*n*=1–14)	Free radical polymerization of (Gα) $_n$-G-β-EA led to comb-shaped glycopolymers with pending G1–G15	Kloosterman, Spoelstra-van Dijk and Loos (2014)
Glucansucrase	*Leuconostoc mesenteroides* B-1299CB	Sucrose	Arbutin	G1-, G2-arbutin	Added 1 and 2 α-1,6-linked G extended β-linked glucose of arbutin, respectively	Moon et al. (2007)
Dextransucrase B-512F Alternansucrase B-23192	*Leuconostoc mesenteroides*	Sucrose	Flavonoids	Glycosylated luteolin, myricetin, quercetin	Both enzymes catalyzed transglycosylation. Yields 4%–49%	Bertrand et al. (2006)
Dextransucrase	*Lactobacillus reuteri* TMW 1.106	Sucrose	Gallic acid (GA) Caffeic acid (CA)	Mono-, oligo-, poly-glucosylated GA and CA	O3-mono-substituted, O4-oligo-, and poly-substituted	Klingel et al. (2019)

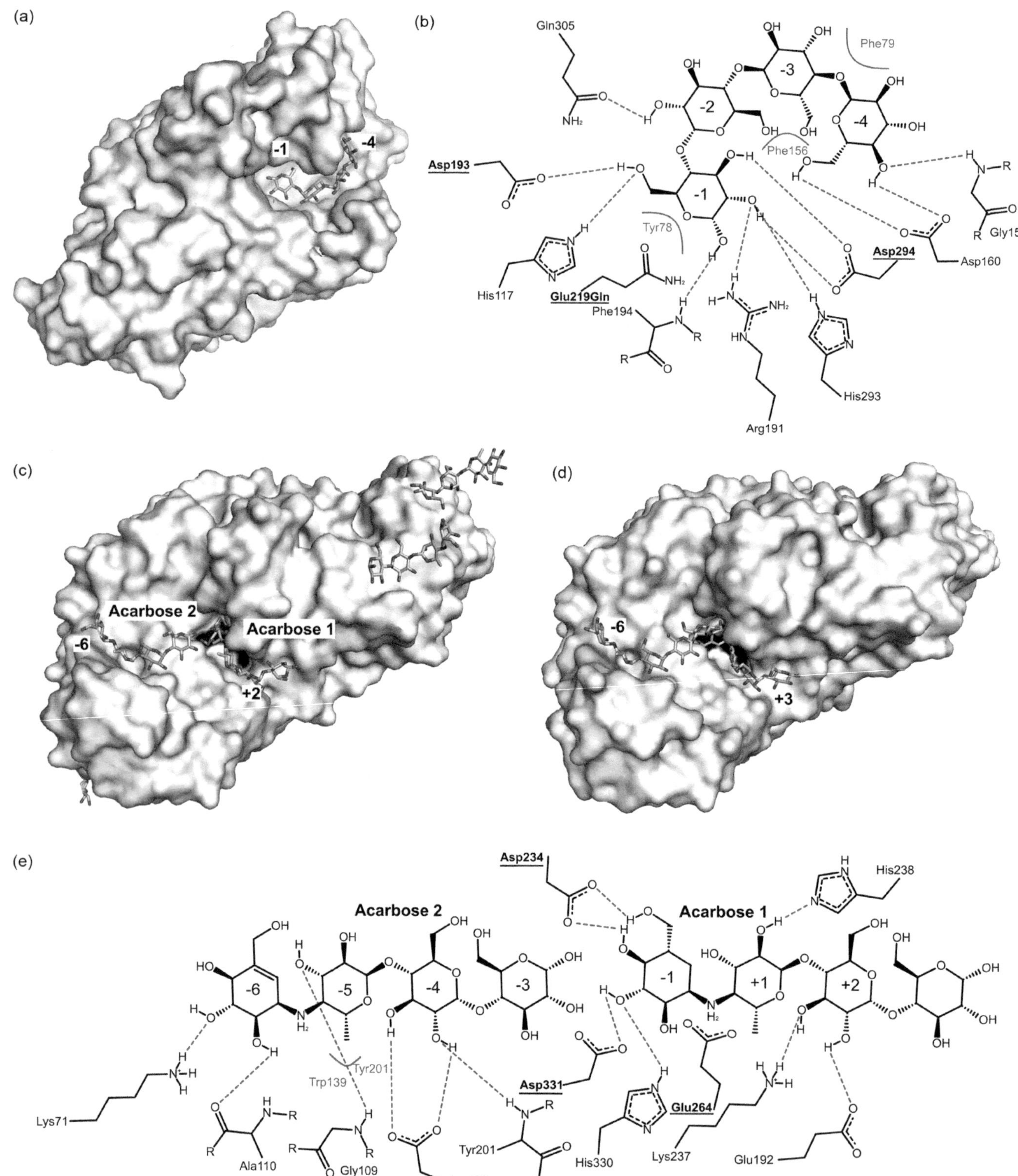

FIGURE 14.3   (a) G4-forming amylase from *Pseudomonas stutzeri* MO-19 in complex with G4 (PDB 1JDC) (Yoshioka et al. 1997). (b) Interactions at the active site of PDB 1JDC. Catalytic residues or mutants thereof are underlined. (c) G5-forming amylase from *Bacillus stearothermophilus* in complex with acarbose (PDB 6AG0) (X Xie et al. 2019a). (d) G6-forming amylase from *Bacillus* sp. 707 in complex with pseudomaltononaose (PDB 1WPC) (Kanai et al. 2004). The position of catalytic residues is highlighted in black in all cases. (e) Interactions at the active site of PDB 6AG0. Catalytic residues are underlined. The schematic drawings of the interactions between protein and ligands at the active site were generated using PoseEdit on the ProteinsPlus server (https://proteins.plus) (Diedrich et al. 2023; Schöning-Stierand et al. 2022).

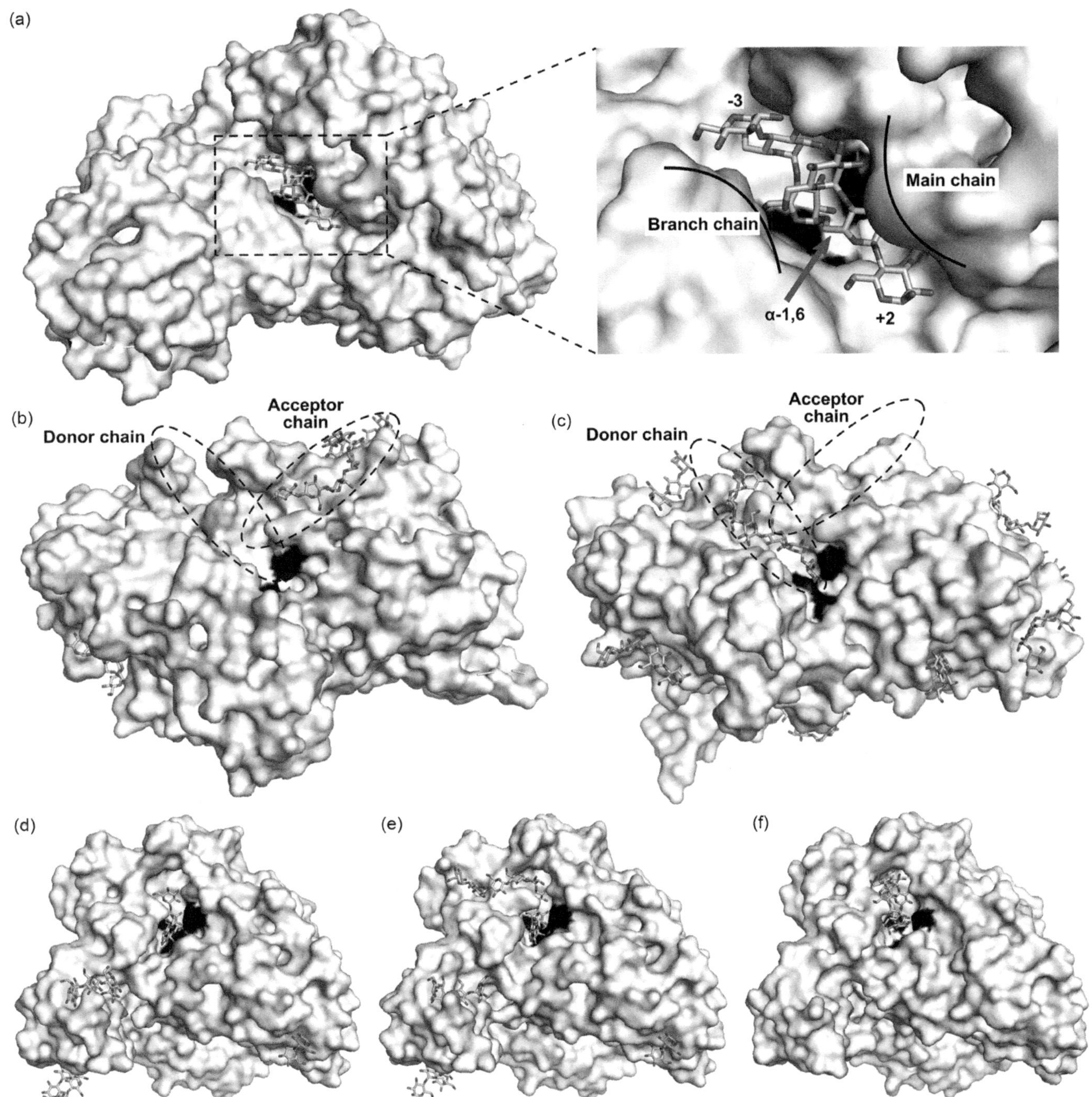

FIGURE 14.4 (a) Barley limit dextrinase in complex with a branched maltohexasaccharide (PDB 4J3W) (Møller et al. 2015). (b) Rice branching enzyme BEI in complex with maltododecaose at the acceptor binding site (PDB 7ML5) (Gavganı et al. 2022). (c) *Cyanobacterium* sp. NBRC 102756 branching enzyme BE1 in complex with maltoheptaose at the donor binding site (PDB 5GQX) (Hayashi et al. 2017). (d) CGTase from *Niallia circulans* 251 in complex with maltotetraose (PDB 1CXF) (Knegtel et al. 1995). (e) CGTase from *N. circulans* 251 in complex with maltononaose (PDB 1CXK) (Uitdehaag et al. 1999b). (f) CGTase from *N. circulans* 8 in complex with β-cyclodextrin (PDB 3CGT) (Schmidt et al. 1998). The position of catalytic residues is highlighted in black in all cases. Several surface binding sites are also occupied in PDB 1CXF, 1CXK, and 5GQX.

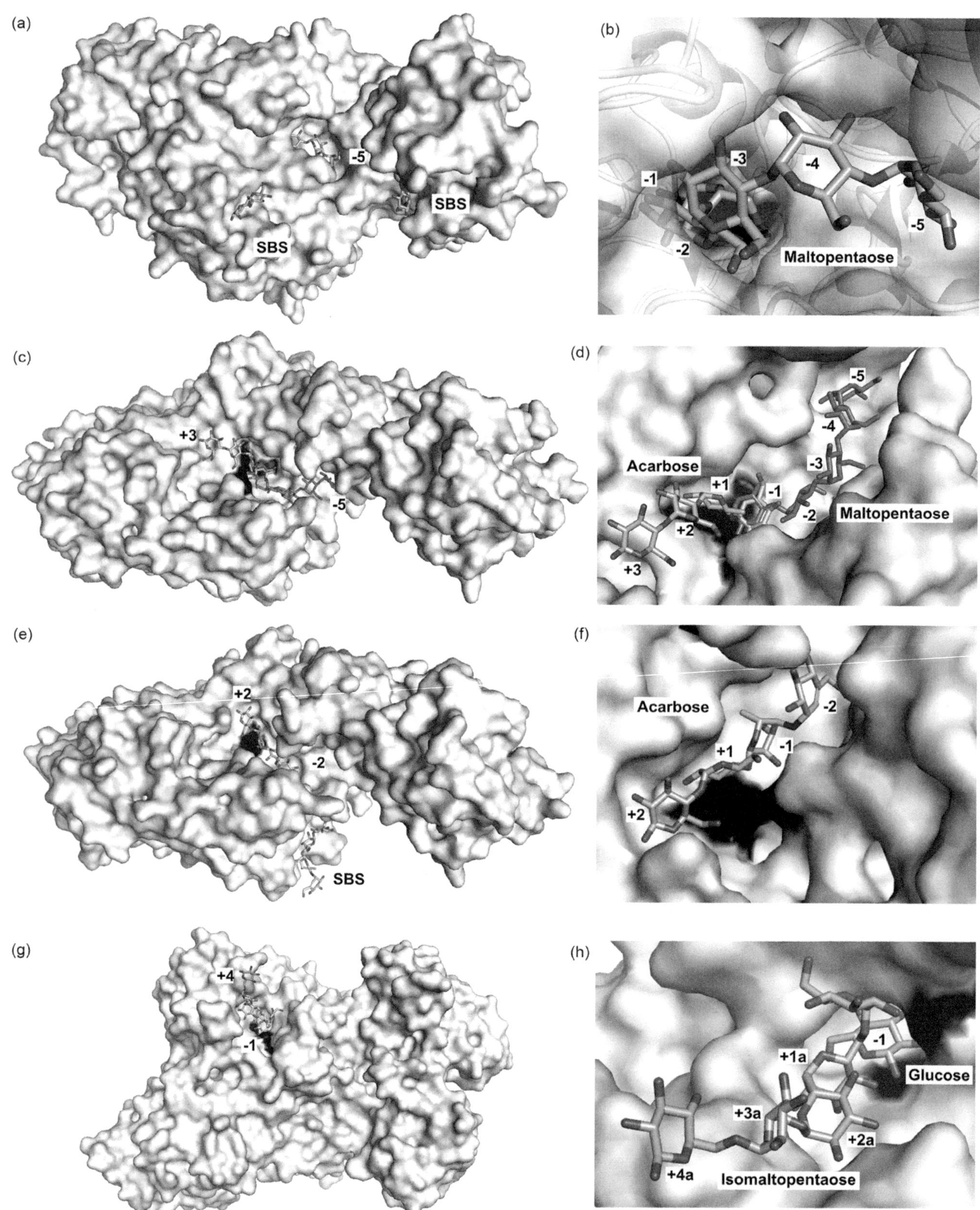

FIGURE 14.5 GH70 4,6-α-GT and GH31 6αGT structures. (a) *Limosilactobacillus reuteri* 121 GtfB ΔNΔV with maltopentaose bound in the active site (PDB 5JBF) (Bai et al. 2017). (b) Close-up of the active site of *L. reuteri* 121 GtfB ΔNΔV. (c) *L. reuteri* NCC2613 GtfB ΔNΔV in complex with acarbose (PDB 7P39) (Pijning et al. 2021) and with maltopentaose from PDB 5JBF superimposed. (d) Close-up of the active site of *L. reuteri* NCC2613 GtfB ΔNΔV with acarbose (light grey sticks), and the maltopentaose from PDB 5JBF (dark grey sticks). (e) *L. reuteri* N1 GtfB ΔV in complex with acarbose (PDB 8HW3) (Dong et al. 2024). (f) Close-up of the active site of the *L. reuteri* N1 GtfB ΔV. (g) *Paenibacillus* sp. 598K Ps6GT31A with isomaltopentaose and glucose bound in the active site (PDB 5X7R; molecule 2) (Fujimoto et al. 2017b). (h) Close-up of the active site of *Paenibacillus* sp. 598K Ps6GT31A. The position of catalytic residues is highlighted in black in all cases.

(Figure 14.1; Table 14.3). Transglucosylases include branching enzyme (BE; EC 2.4.1.18) of families GH13 and GH57, amylosucrase (AS; EC 2.4.1.4) of family GH13, 4-α-glucanotransferase (4αGT; EC 2.4.1.25) found in GH families 13, 57, and 77, α-6-glucosyltransferase (6αGT; EC 2.4.1.24) of family GH31, and 4,3-α- and 4,6-α-glucanotransferases (4,3-α-GT and 4,6-α-GT; EC 2.4.1.-) of family GH70 (Tables 14.1 and 14.3). Cyclodextrin glucanotransferases of family GH13 (CGTases; EC 2.4.1.19) form CDs of DP 6–8 by intramolecular transfer to $O4$ of the non-reducing end residue of the covalent enzyme intermediate (Uitdehaag et al. 1999a, b) (Tables 14.1 and 14.4; Figures 14.2 and 14.4). Amylomaltases (4αGTs) form cycloamyloses also called large-ring cyclodextrins (LR-CDs), while cyclic cluster dextrins (CCDs) and highly branched cyclic dextrins (HBCDs) are mostly formed using BEs (C Chen et al. 2020; Krusong et al. 2022; X Li et al. 2023a; J Park et al. 2018; Takaha et al. 1996). Intermolecular transglucosylation of acceptor molecules in disproportionation or coupling reactions (Figure 14.2), resulting in branching or linear extension of α-glucan chains, is catalyzed by enzymes from all three clan GH-H families 13, 70, and 77, making α-1,2, α-1,3, α-1,4, and α-1,6 linkages (Y Chen et al. 2023b; Gangoiti, Pijning & Dijkhuizen 2018; Wei et al. 2023b; Xue, Svensson & Bai 2022) (Table 14.3). 6αGT uses various disaccharides and MOSs in transglucosylations resulting in α-1,6-linked IMOSs, which in turn are cyclized by cycloisomaltooligosaccharide glucanotransferase (CITase) of GH66 to cycloisomaltooligosides (CIs) or mixed linked α-1,3/1,6 CI from α-glucan involving an α-1,3-isomaltosyltransferase (Aga et al. 2003; Ichinose et al. 2017; Kumar, Tissopi & Mutturi 2023; Nishimoto et al. 2002; Suzuki et al. 2012; W Yang et al. 2022). α-Glucosidases of family GH13, β-amylases of family GH14, and glucoamylases and dextranases of family GH15 are used more rarely (Lang et al. 2023; Z Liu et al. 2022b; Ulbrich, Scholz & Flöter 2022; W Wang et al. 2021). α-Glucosidases and glucoamylases can trim α-glucans by stepwise removal of α-1,4- and α-1,6-linked glucose from non-reducing ends (W Wang et al. 2021). Finally, dextrins are also synthesized by α-1,4-glucan phosphorylases of glycosyltransferase (GT) family 35 (Cifuente et al. 2024; Kadokawa, 2017, 2022; A Li et al. 2022; Ubiparip et al. 2018) and by inverting α-glucan phosphorylases of GH family 65, which form MOSs from disaccharides and α-G-1-P or as products in one-pot reactions with inorganic phosphate (De Beul et al. 2021; A Li et al. 2022; Nakai et al. 2013; Nihira, Nakai & Kitaoka 2012; Nihira et al. 2014; Takemoto et al. 2013; Umegatani et al. 2012) (Tables 14.1 and 14.3).

14.1.4 Applications of Dextrins

Dextrins have numerous applications spanning from food and nutrition over biomedicine and pharma to biomaterials, waste management and remediation. Many modified dextrins are intended as digestion resistant dextrins (RD) (Gangoiti et al. 2020; Kotatha et al. 2023; Trithavisup et al. 2019; Xie et al. 2021), which are not hydrolyzed by human enzymes of the gastrointestinal tract and can serve as prebiotics converted by gut bacteria to health beneficial metabolites such as short-chain fatty acids (Kumar et al. 2023; F Li et al. 2023; Z Liu et al. 2022b; Q Yang et al. 2023; T Yang et al. 2022). IMOSs, considered as dietary fibers, have been synthesized from corn starch using α-amylase and transglucosylase (Jeong et al. 2019; HS Lee et al. 2002a; Zeng, van Pijkeren & Pan 2023). Larger dextrins are important for emulsifying, texturizing, and gelation in food manufacture, as characterized by monitoring rheological behavior and microscopic structures (X Li et al. 2023a; Ma et al. 2024; Schädle, Bader-Mittermaier & Sanahuja 2022; Yu et al. 2022; Zhu et al. 2023). MDs and MOSs, important for texture, volume, and stabilization in baking and doughs, are supplemented as ingredients or generated *in situ* by enzymes added to the flour; DE and fine structure of MD ingredients are important for loaf volume (Huang et al. 2020; Y Li et al. 2022; T Yang et al. 2022; L Zhang et al. 2019). Dextrins can furthermore be part of encapsulations and delivery systems (Hoti et al. 2022; Periasamy 2020; Ribeiro & Veloso 2021), edible films, or anti-erosion coatings (Totosaus, Godoy & Ariza-Ortega 2020; Umoren & Eduok 2016). Dextrins also find various uses in both cosmetics and pharmaceuticals, where they serve as tablet filler excipients, and in medicine as glucoconjugates of drugs and traditional medicine products with improved solubility and bioavailability; glucoconjugates are also used as ingredients (Bertrand et al. 2006; Couto et al. 2019; Marié et al. 2018; Moon et al. 2018; Qi et al. 2013; Tse et al. 2018). Dextrins in biomaterials, besides their chemical and physicochemical attributes, offer the advantage of being biodegradable unlike synthetic polymers. This biodegradability is important for the environment and sustainable recycling of dextrins, eventually ending up as bioethanol (Gunawardene et al. 2021; Z Li, Wang & Shi 2019; Zaborowska & Bernat 2023). Mixtures of glucose, maltose, and short MOSs are used as syrups in foods and beverages (Nguyen et al. 2015; Pullicin et al. 2018). Composites containing dextrins are used for adsorption (Kamaraj et al. 2024; Y Li et al. 2024; Moradi & Panahandeh 2022; Ray et al. 2024). CDs and LR-CDs are used to host valuable guest molecules or for trapping unwanted contaminants or hazardous compounds in (1) controlled delivery, (2) improved reagent availability in chemical syntheses, and (3) remediation and waste management (Cao et al. 2020; Kfoury et al. 2019; X Li et al. 2024; Y Liu et al. 2023a; Morin-Crini et al. 2021).

14.2 LINEAR DEXTRINS

Starch-derived and starch-like linear dextrins are obtained by degradation of starch using hydrolases or synthesized in disproportionation and transglucosylation reactions from different donor and acceptor molecules (Chi et al. 2023; Lang et al. 2023; Leemhuis et al. 2013; te Poele et al. 2021; Xue, Wang, et al. 2022) (Figure 14.2).

14.2.1 Sources of Starch for Production of Linear Dextrins

Linear dextrins in principle can be produced from all types of starch often first subjected to DBE-catalyzed debranching to upgrade the starting materials used in the enzymatic modifications (Kotatha et al. 2023; Z Liu et al. 2022a; D Wang et al. 2023; Wangpaiboon et al. 2023; Xia et al. 2021; Xue, Wang, et al. 2022). Gradient non-solvent (e.g. alcohols) precipitation can be used to fractionate the α-glucan polysaccharides according to solubility (Chang et al. 2018; Hu & Goff, 2018; Hu et al. 2015; Pullicin et al. 2018). While retrogradation of linear α-glucan chains in processing often is problematic, it can also facilitate differential precipitation for purification (Z. Liu et al. 2022b; Xue, Wang, et al. 2022). Long linear dextrins were enriched from debranched waxy corn starch by heating-centrifugation (Chi et al. 2023).

14.2.2 Enzymatic Production of Linear Dextrins

Linear dextrins, representing large variation regarding molecular size and polydispersity, are prepared using enzymes individually or combined in sequential or one-pot type reactions (Abdalla et al. 2021; Miao et al. 2018; Xie, Lee & Sim 2021; Zheng et al. 2022). Enzymatic hydrolysis of gelatinized starch followed by separation, for example, by solvent precipitation, can lead to pure single MOSs (Ji et al. 2024). MDs and MOSs can be debranched by DBEs in starch hydrolysates (Xie, Lee & Sim 2021). MOSs of high purity are released by MFAs (Table 14.2) (Bláhová et al. 2023; Pan et al. 2017; Y. Wang et al. 2019). Long MOSs – or short MDs of > DP12 – have been produced by using disproportionating enzymes such as maltosyltransferase from *Thermotoga maritima* (Meissner & Liebl 1998). Longer linear α-1,4-glucans, including chimera containing α-1,3- and α-1,6-linked segments, can be synthesized using transglucosylases (Lang et al. 2023; X Li et al. 2023a; Y Wu et al. 2023; Xue, Wang, et al. 2022). Amylopullulanases, sometimes referred to as type II pullulanases, are found in GH13 and GH57 and form linear MDs from CDs or starch displaying also varying individual ability to act on α-1,4 and α-1,6 linkages and for hydrolyzing α-, β-, and γ-CDs (Li & Li, 2015; Li et al. 2021; Zhang et al. 2022). Thus, a versatile hyperthermophilic GH57 amylopullulanase from *Caldivirga maquinlingensis* IC-167 enabled production of amylodextrins of DP 6–96 from AM and of DP 1–76 from AP and potato starch, and it also catalyzed ring-opening of CDs, including G1-β-CD and G2-β-CD before debranching (Li & Li 2015). By contrast, G8-β-CD was debranched, followed by ring-opening of the β-CD, while the released G8 was a poor substrate (Li & Li 2015).

Pure MOSs of DP 6–8 can be obtained from starch using CGTases, which are forming CDs then hydrolyzed by cyclomaltodextrinases or by a ring-opening *Pyrococcus furiosus* α-amylase (Figure 14.2) (Cuong et al. 2016; Ji et al. 2019; X Li et al. 2024; Yang et al. 2006). CDs have been ring-opened by cyclomaltodextrinase or chemically by acetolysis or other acid treatments for use as synthons in organic chemical syntheses (Pélingre et al. 2021). Cyclomaltodextrinase from *Geobacillus thermopakistaniensis* preferably hydrolyzed α-CD to G6 (Aroob et al. 2019). G7 was made from starch in a one-pot reaction involving CGTase from *Gracilibacillus alcaliphilus* and hydrolysis of the formed β-CD by cyclomaltodextrinase from *Bacillus sphaericus* (Abdalla et al. 2021). The subsite +1 H233Y mutant of CGTase from alkalophilic *Bacillus* sp. 1.5 produced G7 from β-CD (Koo et al. 2015). MOSs have been derivatized at the reducing end, non-reducing end, or both ends as well as at a C6-position of interior residues (Bláhová et al. 2023). Non-reducing G7 was produced from β-CD in a cascade reaction by *Escherichia coli* containing cyclomaltodextrinase and maltooligosaccharide trehalose synthase and having two α-amylase genes deleted using CRISPR/Cas9 (Zheng et al. 2022). Linear MOSs were moreover prepared in disproportionation reactions catalyzed by *Bacillus stearothermophilus* NO$_2$ CGTase subsite +2 mutants (Kong et al. 2021). Long IMOSs of DP 8–14 were synthesized by α-1,6 glucose transfer to obtain panose-type α-1,6/α-1,4 containing products from G5, IG5, and MDs catalyzed by the C-terminal GH15 dextran dextrinase (DDase) domain, which follows a GH66 cycloisomaltooligosaccharide glucanotransferase (CITase) and two CBM35 domains in a multimodular, multifunctional enzyme from *Thermoanaerobacter thermocopriae* (Jeong et al. 2019). IMOSs are synthesized from MOSs by 6αGT of GH family 31 from *Paenibacillus* sp. 598K and can be cyclized by the GH66 CITase from the same organism to CIs (Fujimoto et al. 2017a; Ichinose et al. 2017). *Gluconobacter oxydans* DDase generates dextrans from MDs (Naessens et al. 2005).

14.2.2.1 Endo-Acting Glycoside Hydrolases

α-Amylases in nature degrade starch for energy and building blocks but are used industrially to produce different dextrins, and although larger fragments appear initially, MOSs are formed eventually (Ashok et al. 2024; Shad et al. 2023). *Endo*-acting α-amylases (EC 3.2.1.1) hydrolyzing α-1,4 linkages in AP and AM generating characteristic product profiles are prominent members of the large GH family 13 and moreover reported from GH families 57, 119, and 126 (Dhital et al. 2017; Janeček & Svensson, 2022; Z. Li, Wang & Shi 2019; Miao & BeMiller, 2023). Specific liquefying α-amylases form larger products, which in turn are degraded by saccharifying α-amylases to shorter MOSs (Kong et al. 2018; Z Li et al. 2023; Shinde & Vamkudoth, 2022). Notably, challenging substrates include retrograded linear α-1,4-glucan chains and raw starch granules. However, certain α-amylases, often equipped with SBDs or having surface binding sites (SBSs), can hydrolyze such substrates (Božić et al. 2020; Cockburn et al. 2015; Fu et al. 2024; Nielsen et al. 2009; Y Wang et al. 2023). This capability can also be gained by engineering α-amylases through SBD-fusion (Janeček et al. 2019; Juge et al. 2006; Peng et al. 2018; Y Wang et al. 2023). Raw starch granules are interesting for energy-saving processes avoiding heat-gelatinization, which calls for application of cold-adapted (psychrophilic) enzymes (Qian et al. 2023).

Recently, C-terminal fusion with one of two different SBDs of CBM20 to a psychrophilic α-amylase (AHA) from the Antarctic bacterium *Pseudoalteromonas haloplanktis* TAB23 improved the activity on raw starches by up to threefold. Interfacial kinetics analysis manifested that the AHA-SBD fusions empowered a higher density of enzyme attack sites on the granular surface than recognized by wild-type AHA, the data in fact also reflecting a significant difference in affinity between the two SBDs (Y Wang et al. 2023). Details of substrate interactions at the active site cleft of many α-amylases, including industrially important enzymes, are shown by crystal structures of complexes with the pseudomaltotetraoside inhibitor acarbose, *in cristallo* transglycosylated extensions of acarbose, or MOS substrates in the case of catalytic residue mutants (Table 14.1) (Brzozowski et al. 2000; Fujimoto et al. 1998; Kagawa et al. 2003).

DBEs are *endo*-hydrolases acting on α-1,6 linkages embodied by isoamylases, and pullulanases type I and type II of GH13 and GH57 (Cifuente et al. 2024; Møller et al. 2016; Naik et al. 2023; D Wang et al. 2023; W Xia et al. 2021). While the two former enzymes are strictly α-1,6 linkage specific, type II pullulanases hydrolyze both α-1,4 and α-1,6 linkages at either the same or two different catalytic domains, referred to as pullulanase types II_I and II_II, respectively (Lammerts Van Bueren et al. 2011; Møller et al. 2016; Saka et al. 2018; Sim et al. 2014; Xu et al. 2021) (Table 14.1; Figure 14.4). Pullulanases are highly active on pullulan – hence their name – which is a linear polysaccharide composed of α-1,6-linked maltotriose units (Møller et al. 2016; Xia et al. 2021; Xu et al. 2021). DBEs are useful for production of linear dextrins (Naik et al. 2023; Xu et al. 2021; Xue, Wang, et al. 2022). Waxy corn starch debranched by commercial pullulanase was crystallized as RD and subjected to acid treatment to remove amorphous regions, further reducing the digestibility (Xie et al. 2021). The GH13 DBE from *Nostoc punctiforme* (NPDE) displayed dual bond-type activity, preferring α-1,6 over α-1,4 linkages (Choi et al. 2009; Dumbrepatil et al. 2010). Toward G7–G12 and β-CDs with α-1,6-linked branches of DP 5–11, NPDE showed the highest activity on longer substrates; interestingly, NPDE could release glucose from reducing ends (Choi et al. 2009). The debranching enzyme TreX from *Sulfolobus sulfotaricus* hydrolyzed α-1,6 linkages in α-glucans and similarly released branches of DP 2–6 from G$_n$-β-CDs with the highest activity for the longest branch, G6 (Park et al. 2008).

Several DBEs have been structure-determined with MOSs or CD inhibitors at the active site (Y Li et al. 2024; Møller et al. 2015; Saka et al. 2018) (Table 14.1). The structure was obtained of an α-limit dextrin substrate bound in a barley limit dextrinase catalytic site mutant (Møller et al. 2015) (Figure 14.4a and b). Notably, other pullulanases accommodated two individual MOSs in parallel binding clefts for substrate main and branch chains (Lammerts Van Bueren et al. 2011; Mikami et al. 2006) (Table 14.1). Enzymes of subfamily GH13_20, including neopullulanases, maltogenic α-amylases, and cyclomaltodextrinases, have been structure-determined in complex with pullulan fragments of different lengths, acarbose, and β-CD (Abe et al. 2005; Dumbrepatil et al. 2010; Lee et al. 2002b; Mizuno et al. 2004) (Table 14.1).

14.2.2.2 Exo-Acting Amylases and MFAs

About 50 years ago, MFAs were discovered releasing MOSs of good purity from gelatinized starch and MDs (Pan et al. 2017), and structure, function, and application are reported for G3-, G4-, G5- and G6-forming amylases (An et al. 2023; Bláhová et al. 2023; Duan et al. 2022; Kamon et al. 2015; Kanai et al. 2006; Pan et al. 2017, 2023) (Table 14.2; Figure 14.3). MFAs with narrow specificity produce specific MOSs along with small amounts of other MOSs in the range DP 2–7, and some MFAs change the MOS profile with the progress of the reaction eventually leading to less product diversity (An et al. 2023; Ding et al. 2021; Duan et al. 2022; Z Li et al. 2015; Z Zhang et al. 2020).

A G3-forming amylase from *Bacillus koreensis* H12 released a mixture of G3 and G4 each in about 35% yield (Lekakarn et al. 2022), and a number of primarily G3-forming amylases showed good activity at 20°C and 30°C (Kashiwagi et al. 2014). There are no structures available of G3-forming amylases, whereas crystal structures of ligand complexes of G4-, G5- and G6-forming amylases provide a basis for rational engineering improving product specificity (X Xie et al. 2019b, 2020) (Tables 14.1 and 14.2). Native *Pseudomonas stutzeri* G4-forming amylase released only G4 from starch (Dellweg, John & Schmidt 1975). When crystallized with G5, the G4 product occupied four subsites from the site of catalysis in the pocket-shaped active site closed by Asp160 hydrogen bonding bidentately to *O*4 and *O*6 of the non-reducing end glucosyl residue at subsite −4 (Yoshioka et al. 1997) (Figure 14.3a and b). Pseudomaltoheptaose, a transglucosylation product of the inhibitor acarbose, was interacting with subsites −4 through +3 in *Pseudomonas saccharophila* STB07 G4-forming amylase and showed the same hydrogen bond pattern with the ligand at subsite −4 as observed for the *Pseudomonas stutzeri* enzyme in complex with G4 (Z Zhang et al. 2020). *Pseudomonas saccharophila* STB07 G4-forming amylase produced G4 in 65% yield from corn starch and showed increasing yields and product specificity with G5 through G8 as substrates (Z Zhang et al. 2020). SBDs of CBM20, 25, 26 and 74 are reported in MFAs (Janeček et al. 2019), and the *Pseudomonas stutzeri* G4-forming amylase contained a C-terminal SBD of CBM20. However, this domain was flexible and not ordered in the crystal structure (Mezaki et al. 2001). A related detergent tolerant enzyme was identified from *Pseudomonas stutzeri* AS22 (Maalej et al. 2013). Moreover, disulfide bond engineering of the enzyme from *Pseudomonas saccharophila* STB07 by the A211C/S214C double mutant close to the active site improved stability importantly, while S409C/Q412C introduced near the C-terminal linker-connected SBD of CBM20 was without effect (Y Wang et al. 2022). Truncation of the SBD reduced activity on gelatinized normal and waxy corn starch, but increased activity for shorter MDs. The G4 yield was highest (72%) using MDs of DE 7–9 and DE 16 as substrates, whereas the G4 purity was highest from polysaccharides (Duan et al. 2022). This indicated that the CBM20 domain played a role in facilitating interaction with substrate and in product specificity.

While there is no crystal structure of an *exo*-acting G5-forming amylase, the *endo*-acting BstMFA from *B.*

stearothermophilus STB04, releasing predominantly G5 and G6 (Y Wang et al. 2019), bound two acarbose molecules covering six glycone binding subsites at the active site cleft (X Xie et al. 2019a) (Table 14.2; Figure 14.3c). Notably, BstMFA also bound acarbose at four SBSs. Furthermore, the W139A/L/Y mutants of Trp139, that is stacking to the inhibitor at subsites −5 and −6 in the wild-type enzyme, dramatically increased G5 and reduced G6 yields (X Xie et al. 2019a) (Figure 14.3e). Conversely, favorable accommodation at subsite −6 was obtained for the G109D/N/F mutants of Gly109, which hydrogen bonds to substrate at subsite −5 (Figure 14.3e). This resulted in highest G6 and lowest G5 yields for G106D but was accompanied by some increase in G7 formation (X Xie et al. 2020). A C-terminal SBD of CBM20 in G5-forming amylase from *Saccharophagus degradans* conferred high yield of G5 from starch, and the same ability was achieved for *Bacillus megaterium* MFA by C-terminal fusion to a CBM20 (Ding et al. 2021, 2022) (Table 14.2).

In G6-forming amylase from alkalophilic *Bacillus* sp. 707, pseudomaltononaose bound at subsites −6 through +3 and stacked with Trp140 at subsite −6, which was proposed to define G6 specificity (Kanai et al. 2004) (Figure 14.3d). The W140L/Y mutants indeed gave reduced yields of G6 (Kanai et al. 2006). Mutations were also done to improve thermostability (Li, Duan & Wu 2016). The G6-forming MalS periplasmic α-amylase from *E. coli* has a binding pocket defined by Asp385 and Phe367 and a domain structure similar to the G6/G5-producing amylase from *Klebsiella pneumonia* as well as an N-terminal SBD of CBM69 (An et al. 2023) (Table 14.2). G6 was released initially followed by slower production of G5 and shorter MOSs. Recombinant G6-forming AmyM from *Corallococcus* sp. EBG, when supplied to wheat dough, worked as an anti-staling agent. It conferred increased viscous characteristics of the dough and slowed down the hardening of the bread due to reduced retrogradation (L. Zhang et al. 2019).

14.2.2.3 Transglucosylases

Transglucosylases catalyze the cleavage of α-glucans and transfer glycon moieties from the covalent enzyme intermediate to acceptor molecules with the formation of new glucosidic bonds (Figures 14.1 and 14.2). Transglucosylases used for the synthesis of linear dextrins include amylomaltase, also called 4-α-glucanotransferase (4αGT) of GH families 13, 57, and 77 (EC 2.4.1.25), 4,3-α-glucanotransferase (4,3-α-GT), and 4,6-α-glucanotransferase (4,6-α-GT) both from family GH70 (EC 2.4.1.-), and amylosucrase (AS) of GH family 13 (EC 2.4.1.4). The former enzymes use AM, MDs, and/or MOSs as donors in regiospecific transfer to non-reducing ends of α-glucans, yielding new types of linearly extended α-glucans, including bond-type chimera (Dong et al. 2024; Kralj et al. 2011; Pijning et al. 2022; te Poele et al. 2021; Y Wu et al. 2023). In contrast, ASs of GH13 and glucansucrases of GH70 use sucrose as the donor substrate (Table 14.3; Figures 14.2 and 14.5a–f) (Gangoiti et al. 2020; Meng et al. 2016; Moulis, André & Remaud-Simeon 2016; Potocki-Veronese et al. 2005). A subsite engineered AS with nine mutations at the active site synthesized linear α-1,4-glucans of 2 < DP < 21 from sucrose (Vergès et al. 2017). Crystal structures of MOSs bound to

the active site and to distant SBSs in AS mutants suggested a mechanism for the α-1,4-glucan chain elongation (Guérin et al. 2012; Mirza et al. 2001; Skov et al. 2002, 2006; Vergès et al. 2017). Notably, CGTases can catalyze disproportionation, coupling, cyclization, and hydrolytic reactions (Leemhuis et al. 2010; X Li et al. 2024) of which the first two are relevant for the synthesis of linear dextrins (Pardhi et al. 2023).

4αGT, also called amylomaltase in microorganisms and disproportionating enzyme (D-enzyme) in plants, similar to CGTase catalyzes cyclization, coupling, disproportionation, and hydrolysis, of which the three latter reactions lead to linear products (Y Chen et al. 2023b; Christensen et al. 2023, 2024; Jung et al. 2022; X Li et al. 2023a; Tian et al. 2018; Zhoukun et al. 2019) (Figure 14.2). MalQ from *E. coli* of GH77 accommodated transglycosylated acarbose at donor and acceptor binding subsites −4 through +3 (Weiss, Skerra & Schiefner, 2015) (Table 14.1). The structure of *Thermus aquaticus* amylomaltase (Przylas et al. 2000), when later obtained in complex with a 34-mer large-ring cyclodextrin (LR-CD), showed how the enzyme interacts with donor and acceptor molecules, highlighting residues shaping the product (Roth et al. 2017). 4αGT has been extensively used for starch modification in food science (Y Chen et al. 2023b). 4αGT from the hyperthermophilic *Thermococcus onnurineus* of GH57 catalyzed disproportionation transferring glucose (Kaila & Guptasarma 2019). Acarbose bound at subsites −1 through +3 in 4αGT of family GH57 from *Thermococcus litoralis* (Imamura et al. 2003) and bioinformatics analysis of loop lengths identified motifs controlling the levels of transglucosylation and hydrolysis (Xiang, Leemhuis & van der Maarel, 2022).

Linear IMOSs were produced from MOSs, MDs, AM, and starch by transglycosylation using 6αGT of GH family 31 or from sucrose/maltose mixtures by dextransucrase of GH70 (Ichinose et al. 2017; Shi et al. 2016). Different 4,6-α-GTs synthesized linear chimeric IMMPs from MD or starch, including single-arm linear dextrin having high flexibility and improved water solubility (Dong et al. 2024; Gangoiti et al. 2020; Y Wu et al. 2023; Xue, Wang, et al. 2022). β-CD was used as a starting material for cyclomaltodextrinase, which produced G7, which in turn was used by 4,6-α-GT to synthesize IMMPs (Y Liu et al. 2023b). 4,3-α-GT from *Limosilactobacillus fermentum* NCC synthesized an alternating α-1,3/α-1,4-glucan with α-3,4 branch points (Gangoiti et al. 2017; Pijning et al. 2021). 4,3-α-GT and 4,6-α-GT belong to family GH70 that furthermore contains reuteran-, alternan-, and dextransucrases, which together with α-1,2- and α-1,3-branching sucrases have been useful for the synthesis of novel α-glucans (Gangoiti et al. 2020; Wei et al. 2023b; Xue, Svensson & Bai 2022) (Table 14.3). Thus, a soluble linear MD of 2.3 kDa, obtained by low-temperature retrogradation of debranched waxy corn starch, was extended to IMMPs of 12 kDa by the N-domain truncated 4,6-α-GT (GtfB-ΔN) from *Limosilactobacillus reuteri* 121 (Xue, Svensson & Bai 2022; Xue, Wang, et al. 2022).

The large multidomain glucanotransferases of GH70 possess a circularly permuted catalytic domain and crystal structures are available for different specificities (Bai et al. 2017; Dong et al. 2024; Kralj et al. 2011; Meng et al. 2016; Pijning et al. 2021) (Table 14.1; Figure 14.5). In 4,6-α-GT

Lr121 ΔNΔV from *L. reuteri* a G5 molecule bound at subsites −5 through −1 of a catalytic site mutant and the product 6⁴-α-D-glucosyl-maltotetraose bound at subsites −6 through −2 in the wild-type enzyme. This illustrates the transglucosylation mechanism where the non-reducing end of the covalent intermediate will attach to the acceptor bound at subsites +1 and +2 (Bai et al. 2017) (Figure 14.5a and b). The full active site crevice can be seen for *L. reuteri* NCC 2613 GtfB ΔNΔV in complex with acarbose at subsites −1 to +3 and G5 bound to 4,6-α-GT Lr121 ΔNΔV superimposed at subsites −5 to −1 (Figure 14.4c and d) (Bai et al. 2017; Pijning et al. 2021). The active site of *L. reuteri* N1 GtfB ΔV is more open as observed for the complex with acarbose bound at subsites −2 to +2 and spanning the catalytic site (Figure 14.5e and f), illustrating the capacity to disproportionate linear MOSs and produce also α-1,6-branched oligosaccharides (Dong et al. 2024). Mutational analysis at acceptor binding subsites +1 and +2 in *L. reuteri* 121 GtfB ΔN gave further insights into the reaction and specificity as deduced from the variation of the α-1,6-/α-1,4-linkage ratio in the produced MDs (Jiang et al. 2022). The comparison of crystal structures of different MOS- and AM-active GH70 members from *L. reuteri* NCC 2613, *L. fermentum* NCC 2670, *L. reuteri* N1, and *L. reuteri* 121 explained their subtle functional differences (Bai et al. 2017; Dong et al. 2024; Pijning et al. 2021) (Figure 14.5a–f). Engineering active site loops in *L. reuteri* N1 GtfB provided an unprecedented toolbox for tailored synthesis of low-molecular-weight linear and branched dextrins containing α-1,6 linkages (Dong et al. 2024). A chimeric linear dextrin of average DP 21 was synthesized with α-1,6-glucan (dextran) reducing end and α-1,4-glucan non-reducing end segments in two steps. First, IMOSs were obtained from sucrose catalyzed by the 4,6-α-GT GTF180 ΔN L9940W mutant with suppressed α-4,3 specificity, and next using γ-CGTase, the IMOSs were extended by transfer of G8 from γ-CD to the non-reducing end (Y Wu et al. 2023).

Isomaltomegalosaccharides (IMMs), referring to isomaltodextrins of DP 10–100, can be prepared enzymatically from MOSs of DP 6–7 by transglucosylation catalyzed by dextran dextrinase of GH15 from *Gluconobacter oxidans* (Lang et al. 2014). The C-terminal GH15 domain of ClTase from *Thermoanaerobacter thermocopriae* was able to produce IMOSs of DP 2–14 (Jeong et al. 2019). At high G2 concentrations the GH31 α-glucosidase AgdB from *Aspergillus nidulans* formed IMOSs by transglucosylation (Kato et al. 2002). Additionally, when combined with a disproportionating CGTase, AgdB from MDs of different DE values produced soluble slow digestion dextrin (SDD) with α-1,6 linkages, functioning as a linear dietary fiber (Wei et al. 2023a).

14.2.2.4 *α-1,4-Glucan Phosphorylases*

α-Glucan phosphorylases of glycosyltransferase family 35 (GT35) (Drula et al. 2022) with covalently bound pyridoxal 5′-phosphate catalyze reversible phosphorolysis of α-1,4-glucans (Cifuente et al. 2024; Guo et al. 2024; Schinzel & Nidetzky 1999; Ubiparip et al. 2018). Using thermophilic isoamylase for debranching of corn starch combined with sequential addition of thermophilic 4αGT and α-glucan phosphorylase glucose-1-phosphate was obtained in high yields in a one-pot reaction, which can be used for production of artificial starch and MDs (Qi, You & Percival Zhang 2014; You et al. 2013; Zhou et al. 2016). By altering the enzyme addition procedure, a dual-enzyme synthesis using sucrose phosphorylase and α-glucan phosphorylase resulted in short and long linear α-glucans of DP<10 and DP>45, respectively (Guo et al. 2024). Inverting glucan phosphorylases of GH family 65 catalyze synthesis of disaccharides and oligosaccharides in one-pot reactions by phosphorolysis of maltose, or by using β-G-1-P as donor, and transfer of glucose to different acceptor carbohydrates accommodated due to enzyme engineering (Kadokawa 2022; A Li et al. 2022; Nakai et al. 2010a,b, 2013; Nihira, Nakai & Kitaoka 2012; Nihira et al. 2014).

14.2.3 Application of Linear Dextrins and Enzymes Producing Linear Dextrins

Linear dextrins, including resistant dextrins (RDs), are present in various food and healthcare products (X Li et al. 2021, 2023a; A-J Xie, Lee & Sim 2021). Dextrins are also used for coating of textile polymers (Aouay et al. 2022). Linear dextrins of V-type of varying DP, prepared by gradient ethanol precipitation from debranched normal starch, can be used for ethylene encapsulation (Z Liu et al. 2022a). Chimeric α-1,4/α-1,6 dextrins showed solubilizing effect on the hydrophobic stilbene resveratrol (Wu et al. 2023). MOSs have many different applications, in foods and nutrition as low-intensity sweeteners, anti-aging, aroma encapsulation, for texture, and, depending on the size, as prebiotics (Ji et al. 2024; Schädle, Bader-Mittermaier & Sanahuja 2022). In pharma MOSs are used in delivery systems, as biosensors, in syntheses of diagnostic enzyme substrates, and glycoconjugates can create porous or lamellar nanoparticles (Bláhová et al. 2023; Ji et al. 2024; Shinde & Vamkudoth 2022). Unlike MOSs, also shorter IMOSs are considered as prebiotics (Goffin et al. 2011).

Non-reducing G7 (N-G7), prepared by converting the α-1,4 linkage at the reducing end in G7, which is formed from β-CD by cyclomaltodextrinase, to an α-1,1 linkage as catalyzed by intramolecular transglucosylation by maltooligosyl trehalose synthase, can escape Maillard reactions (Zheng et al. 2022). *Exo-, endo-*, and debranching types of amylolytic enzymes and their products have played a role in breadmaking (Pan et al. 2017; Zhao et al. 2022). Moreover, 4,6-α-GT can produce alternating α-1,4-/α-1,6-linked linear glucans, which delay bread staling (te Poele et al. 2021). Anti-staling through suppression of retrogradation has also been achieved by adding G6-forming amylase from *Corallococcus* sp. EGB to wheat starch, which was used as a model system, and to flour for reaction *in situ* having positive impact on bread quality (T Yang et al. 2022; L Zhang et al. 2019). IMMPs synthesized from starch by 4,6-α-GT from *Lactobacillus reuteri* 121 are considered a novel type of dietary fiber (Leemhuis et al. 2014). Linear IMMs are used as amphiphilic carriers of dyes (Lang et al. 2014).

14.3 BRANCHED DEXTRINS

Branched dextrins are noncyclic α-1,4-glucans with α-1,6-branching as it occurs in AP or containing other types, namely α-1,2- and α-1,3-linked branches (Gangoiti et al. 2020; Junejo et al. 2022; X Li et al. 2023a; Meng et al. 2016; Wei et al. 2023b). Large starch fragments, MDs, and MOSs can all be branched as made from starch or AP hydrolysis by different α-amylases (Paul et al. 2021), eventually followed by fractionation using SEC, membrane filtration, or selective solvent precipitation (Hu et al. 2015; Ulbrich, Scholz & Flöter 2022; Zhen et al. 2021). Extensive starch hydrolysis by α-amylases generates branched MOSs, called α-limit dextrins (Jonathan et al. 2015; Lee & Hamaker 2017). The α-limit dextrin, $6^2,6^3$-di-α-D-glucosyl maltopentaose was indigestible to pancreatic α-amylase and glucoamylase (McCleary 2014). β-Amylases release maltose from non-reducing ends of starch and AP forming highly branched β-limit dextrins with good water-solubility, having potential as drug carrier targeting the mouth mucosa and as emulsifier after octenyl-succinylation that in addition reduces the *in vitro* digestibility (Tester & Qi 2011; Y Wang et al. 2020). Microbial BEs use AM, AP, and MDs as donors of α-1,4-glucan chains with distinct lengths to form branches attached by α-1,6-linkages on α-1,4-glucan acceptors (Ryu et al. 2022, 2023; Sun et al. 2023; Tetlow & Emes 2014; Y You et al. 2023). Branch points can be quantified by NMR spectroscopic analysis of α-1,4- and α-1,6-linked glucosyl residues (Bai et al. 2011; Han et al. 2018; Y Liu al. 2023b; Petersen et al. 2012). Well-established procedures exist to determine lengths and content of branches involving hydrolysis by α-1,6-specific DBEs, isoamylase and/or pullulanase, followed by analysis of chain-length distribution (CLD) using high-performance anion exchange chromatography with pulsed amperometric detection (HPAEC-PAD) (Blennow et al. 1998; Christensen et al. 2023; Kong et al. 2008; Song et al. 2020). Branch chains in AP and dextrins have been categorized as A chains of DP 6–12, B1 chains of DP 13–24, B2 chains of DP 25–36, and B3 chains of DP>37 (Hanashiro, Abe & Hizukuri 1996).

14.3.1 Sources of Starch for Production of Branched Dextrins

Branched dextrins can be obtained from AP originating from waxy and normal starches, typically containing <1% and 20%–30% AM, respectively (Shim et al. 2023; X Zhang, Leemhuis & van der Maarel 2020). Degradation by acid treatment of starch at high temperature leads to pyrodextrins having prebiotic potential (Barczynska et al. 2010, 2012; Kapusniak et al. 2021; F Li et al. 2023). Pyroconversion of cassava starch thus resulted in transglucosylated and repolymerized digestion resistant dextrins (RDs) considered as dietary fibers (Trithavisup, Krusong & Tananuwong 2019). RDs can be obtained from the pretreated dextrins using BE followed

by chain length trimming with γ-CGTase to increase resistance (X Han et al. 2018; Z Liu et al. 2022b). Isomaltodextrins (IMDs) have been made from MDs of DE 15–20 and 5–7 by 4,6-α-GT GtfB ΔN from *Limosilactobacillus fermentum* NCC 3057 and the resistance to digestion was enhanced by decorating with α-1,2- and α-1,3-linked branches as catalyzed by branching sucrases (Wei et al. 2023b).

14.3.2 Enzymes Producing Branched Dextrins

Branched dextrins of different molecular size and polydispersity are obtained by α-amylase hydrolysis of α-1,4 linkages typically followed by transglucosylation using enzymes of families GH13, 57, 70, and 77 (Gangoiti et al. 2020; X Li et al. 2023a; Miao & BeMiller 2023; Ryu et al. 2023; Sorndech, Tongta & Blennow 2018) (Table 14.3; Figure 14.2). Extra branches on dextrins can be introduced by BEs (Shim et al. 2023; Sun et al. 2023) or BEs in combination with other enzymes in sequential or one-pot reactions (Z Liu et al. 2022b). A variety of BEs exists having different chain length specificities (Fawaz et al. 2023; Gaenssle et al. 2021; Gavgani et al. 2022; X Zhang, Leemhuis & van der Maarel 2019). Branches can moreover be extended by 4αGTs, 4,3-α-GTs, 4,6-α-GTs, ASs and 4-α-glucan phosphorylases (Christensen et al. 2023; Gangoiti et al. 2020; X Li et al. 2023a; Ryu et al. 2023; Sorndech, Tongta & Blennow 2018), whereas trimming of branches is possible using glucoamylases, α-amylases, MFAs, or by CGTases and 4αGTs releasing CDs (Ji et al. 2020a; Y Wang et al. 2024). BEs are large multidomain proteins and crystal structures (Figure 14.4) and biochemical characterization have given insight into the reaction mechanism (Fawaz et al. 2023; Feng et al. 2015; Gavgani et al. 2022; Hayashi et al. 2017). Different glycogen branching enzymes (GBEs) cleaving α-1,4- and forming new α-1,6 linkages made branched MDs of varying size (X Zhang, Leemhuis & van der Maarel 2020). Transglucosylases of family GH70, forming dextrins with MOSs, MDs or sucrose as donor substrates have a sophisticated architecture, including a permuted catalytic domain, and their product specificities are understood based crystal structures (Bai et al. 2017; Brison et al. 2016; Dong et al. 2024; Gangoiti et al. 2018; Pijning et al. 2021; Vuillemin et al. 2016) (Tables 14.1, 14.3; Figure 14.5a–f). MD treated with *Azotobacter chroococcum* CGTase, catalyzing α-1,6 transfer, resulted in a reuteran-like glucan that contained more short branches and was more resistant to digestion (Y Yang et al. 2023). The branched 6^2-maltotriosyl-maltotriose was prepared by hydrolysis of pullulan and reverse hydrolysis reaction transglucosylation of the concentrated maltotriose syrup using *Bacillus acidopullulyticus* pullulanase (Yu et al. 2015). BEs are also used to synthesize highly branched cyclic dextrins (HBCDs) (Takata et al. 1996, 1997, 2010), including a version subsequently capped by anionic α-glucuronic acid residues at the non-reducing ends as catalyzed by phosphorylase from *Aquifex aeolicus* VF5, which can be used as drug carrier (Takemoto et al. 2013).

14.3.2.1 Hydrolytic Enzymes Producing Branched Dextrins

α-Amylases do not hydrolyze α-1,6 linkages and extensive degradation of starch leads to branched MDs and IMOSs along with linear MOSs (Jonathan et al. 2015; Lee & Hamaker 2017; Shad et al. 2023; Shim et al. 2023). IMOSs were prepared from starch using thermophilic liquefying α-amylase, fungal α-amylase, pullulanase, glucoamylase and transglucosidase (Goffin et al. 2011; Kaulpiboon et al. 2015; Xue, Svensson & Bai 2022). The products can be fractionated by solvent precipitation, membrane ultrafiltration, or SEC (Hu et al. 2015; Ulbrich, Scholz & Flöter 2022; Zhen et al. 2021). Remarkably, electron density appeared of an α-1,6-linked residue adjacent to subsite +1 in an α-limit dextrin bound to the AliC α-amylase from *Alicyclobacillus sp.* 18711 (Agirre et al. 2019). Using psychrophilic α-amylases allows degradation of raw starch granules without prior gelatinization, which can save energy (Y Liu et al. 2023c; Qian et al. 2023; Y Wang et al. 2023), and interfacial kinetics analysis showed highest densities of enzyme attack sites and activity for amylolytic enzymes, including DBE, acting on waxy compared to normal types of granular starches (Tian et al. 2023; Y Wang et al. 2023, 2024).

14.3.2.2 Branching Enzymes (BEs)

BEs from eukaryotes (plants and animals) and bacteria belong to subfamilies GH13_8 and GH13_9, respectively, in the CAZy database (www.cazy.org/), while GH family 57 contains BEs from archaea and bacteria (Y Chen et al. 2023a; Drula et al. 2022; Z Wang et al. 2019; Xiang et al. 2022). Other names are starch branching enzyme (SBE) from plants, Q-enzyme and GBE from animals and microorganisms. Most of the characterized BEs are from GH13 and have different specificities and applications. BEs cleave α-1,4 linkages in α-glucan donors and transfer MOS/MD segments from the covalent enzyme intermediate to α-glucan acceptor molecules with formation of α-1,6 linkages (Figure 14.2); substrates can be AM, AP, and MDs (El Mannai et al. 2021; X Zhang, Leemhuis & van der Maarel 2020). BEs are reported to transfer branches as short as DP 3 (Jo et al. 2015) and up to DP 14–60 (Palomo et al. 2009; Sawada et al. 2014), the specificity with regard to the length of transferred branches depending on organisms and BE isoforms (Y Chen et al. 2023a; Fawaz et al. 2023; Gavgani et al. 2022; I Park et al. 2019; Tetlow & Emes, 2014; K Wu et al. 2024; X Zhang, Leemhuis & van der Maarel 2019). BEs of GH13 were reported to form shorter branches than BEs of GH57; notably, BEs of GH57 from *Butyrivibrio fibrisolvens* also catalyzed α-1,4 transfer to linear MD of DP 18 thus elongating the α-1,4-glucan chain (Y Chen et al. 2023a; Gaenssle et al. 2021, 2022). The GH57 BE from *Thermus* transferred chains of DP 4–16, with preference for DP 6, while the GH57 BE from *Pyrococcus horikoshii* transferred branches of DP 10–12 (Gaenssle et al. 2021, 2022; Palomo et al. 2011; I Park et al. 2019). Bioinformatics analysis correlated details of loop structures in BEs of GH57 with levels of branching *versus* hydrolytic activity (Xiang, Leemhuis & van der Maarel 2022). Crystal structures of MOS complexes of BEs from prokaryotes and eukaryotes allowed identification of donor and acceptor binding sites (Tables 14.1 and 14.3; Figure 14.4b and c) (Fawaz et al. 2023; Feng et al. 2015; Gavgani et al. 2022; Hayashi et al. 2017). G7 bound at subsites −7 through −1 of the donor site to the W610N mutant of *Cyanothece* BE1 (Hayashi et al. 2017), while G12 was accommodated at the acceptor binding site of rice BEI (Gavgani et al. 2022). Notably, BEs contain up to 12 non-catalytic surface binding sites (SBSs) situated at a certain distance from the active site. Mutational analyses at SBSs in BEs from *Cyanothece* and *E. coli* showed that one SBS is clearly involved in activity. Additionally, two SBSs located on the C-terminal domain play roles in α-glucan binding, whereas two binding sites on the N-terminal SBD of CBM48 in *Cyanothece* BE1 seemed neither critical for activity nor for binding to AP, AM, or glycogen (El Mannai et al. 2021; Fawaz et al. 2023). Details on how BEs exploit the many SBSs during biosynthesis remain to be understood.

BE from *Thermobifida fusca* efficiently modified AP and increased branching in pyrodextrin, an RD, by cleaving chains of DP > 7 and creating branches of DP 2–6, leading to reduced α-1,4/α-1,6 ratios and conferring increased resistance from 44% to 70% by multienzyme treatment including *T. fusca* α-CGTase and *A. nidulans* α-glucosidase (Z Liu et al. 2022b). BEs and 4αGTs have also been used to synthesize cyclic cluster dextrins (CCDs) and highly branched cyclic dextrins (HBCDs), containing clusters from AP (Sorndech, Tongta & Blennow 2018; Takaha et al. 1996; Takata et al. 2003, 2010; Takii et al. 1999; Terada et al. 1999) (Table 14.4; Figure 14.2).

14.3.2.3 Other Branching Transglucosidases

Branched isomaltodextrins (IMDs) are potential dietary fibers synthesized from MDs of DE 15–20 and 5–7 by 4,6-α-GT of GH70 followed by adding α-1,2- and α-1,3-branches using two bond-type specific branching sucrases of GH70 (X Li et al. 2023a; Wei et al. 2023b). The crystal structure of α-1,2 branching sucrase ΔN_{123}-GBD-C2 from *Leuconostoc mesenteroides* NRRL B-1299 had glucose bound at subsite −1 at the catalytic site and isomaltotriosyl-maltose bound in a crevice of domain V assumed to be important for the transglucosylation (Brison et al. 2016). Different GH70 4,6-α-GTs using starch, AM, MD, and MOS as donor substrates are able to form α-1,3- and α-1,6-branched IMMPs and IMMs from starch and MD (Dong et al. 2024; Gangoiti et al. 2017, 2020; Pijning et al. 2021; Xue, Svensson & Bai 2022; Xue, Wang, et al. 2022) (Tables 14.1 and 14.3; Figure 14.5a–f). AS of GH13 is able to extend branches as shown for glycogen using sucrose as donor and explained based on crystal structures containing a number of SBSs (Albenne et al. 2007; Mirza et al. 2001; Skov et al. 2002, 2006). IMOSs have been obtained after starch degradation by using a combination of different hydrolases and treatment with α-glucosidases as well as by transglucosylation using 4αGT wild-type and mutant enzymes from *Thermus thermophilus* (Goffin et al. 2011; Kaulpiboon et al. 2015; Xue, Svensson & Bai 2022). Dextran with 6.2% α-1,4-branching points and 6.5% α-1,4 linkages within the main α-1,6 chain was synthesized from G4 by *Acetobacter capsulates* dextran dextrinase of GH15 (F Li et al. 2023; Yamamoto, Yoshikawa & Okada 1993).

14.3.3 Applications of Branched Dextrins

Branched dextrins serve in a broad range of applications as prebiotic candidates (Z Li et al. 2022; Włodarczyk et al. 2022), for food storage and freeze–thaw stability (Ma et al. 2024; Y You et al. 2023), pasting viscosity, inhibition of retrogradation, as well as for biomedical hydrogels, hydrogel–nanogel combinations, tissue matrices, encapsulation of cells, and in delivery systems (Liang et al. 2024; Miao et al. 2018; Silva et al. 2014; Wu et al. 2024). Highly branched dextrins produced by 4,6-α-GT were slowly digested *in vitro* and caused extended glycemic response in mice (Ryu et al. 2022). Excessive branching, reducing digestibility was achieved for cassava starch by using 4αGT and BE in combination (Sorndech et al. 2015). IMD prepared from MD of DE 3 is a highly branched dietary fiber of relatively low DP showing prebiotic properties (Nishimura, Tanabe & Yamamoto 2016; Takagaki et al. 2018; Tsusaki et al. 2009). Capping waxy corn starch high-molecular-weight dextrin (WS-1000) at non-reducing ends by α-1,6-linked residues catalyzed by 6αGT from *Paenibacillus alginolytic* resulted in a modified high-molecular-weight non-retrograding dextrin (MWS-1000), serving as a novel food additive affecting the viscoelastic behavior; MWS-1000, however, did not prevent retrogradation of the unmodified dextrin WS-1000 (Sumida et al. 2021). Branched dextrin produced from partially hydrolyzed AP has also been applied in composite polymers of bio-based polyurethane film (Konieczny & Loos 2019). Branched MOSs were used as substrates suitable for comprehensive enzyme specificity profiling (Jonathan et al. 2015).

14.4 CYCLIC DEXTRINS

Cyclic dextrins were discovered in the late 19th century and are subject to intensive research due to their unique interactions with other molecules (X Li et al. 2024; Morin-Crini et al. 2021). They are grouped into the cyclodextrins (CDs), consisting of 6, 7, or 8 α-1,4-linked glucosyl residues (α-, β- and γ-CDs) (Morin-Crini et al. 2021), and the large-ring cyclodextrins (LR-CDs) of DP from 9 to >100 (Krusong et al. 2022; Larsen 2002; Ngawiset et al. 2023; Sonnendecker & Zimmermann 2019). The even larger cyclic cluster dextrins (CCDs) and highly branched cyclic dextrins (HBCDs) are prepared using BE or 4αGT (Sorndech, Tongta & Blennow 2018; Takaha et al. 1996, 1998; Takaha & Smith 1999; Takata et al. 1997, 2003). CDs and LR-CDs are produced from starch by CGTases of family GH13 and by amylomaltases (4αGTs) found in families GH13, 57, and 77. GCTases form α-, β-, and γ-CDs, and rational protein engineering led to improved product specificity and promoted reactions using different acceptors (Leemhuis, Kelly & Dijkhuizen 2010; X Li et al. 2024). Production of CDs of DP 9–12 was favored by semi-rational engineering of a CGTase wild-type enzyme that formed β- and γ-CDs (Sonnendecker, Melzer & Zimmermann 2019). CDs with α-1,6-linked branches have been obtained by chemoenzymatic synthesis using donor fluorides, e.g., G-1-F and

G2-1-F and the DBE barley limit dextrinase in reverse hydrolysis reactions (Vester-Christensen et al. 2023), and from MDs catalyzed by β-CGTase and glucoamylase in a two-step process (L Xia et al. 2017). The multiply branched $6^1,6^4$-di-*O*-G2-β-CD and $6^1,6^3$, 6^5-tri-*O*-G2-β-CD have been produced in disproportionation reactions from 6-*O*-G2-β-CD by amylopullulanase from *Desulfurococcus amylolyticus* of family GH57 (Y-U Park et al. 2018). Notably, other CDs, including cycloisomaltooligosaccharides (CIs) and cycloalternan of DP 4 with alternating α-1,3 and α-1,6 linkages, were prepared from α-1,4-glucans using 6αGT followed by isomaltosyltransferase (Ichinose et al. 2017; Nishimoto et al. 2002; Xue, Svensson & Bai 2022) (Table 14.4). CIs of DP 7–17 were produced by *Bacillus circulans* T-3040 grown on soluble starch (Funane et al. 2014).

14.4.1 Sources of Starch for Production of Cyclic Dextrins

Starches of different origin such as corn, rice, potato, and cassava have been substrates for preparation of CDs using CGTases; moreover, MDs, short-chain AM, and AP have served as donors for 4αGT (Y Chen et al. 2023b; Pishtiyski & Zhekova 2006). CDs were obtained from cassava bagasse, a byproduct of starch production, by using the commercial CGTase Toruzyme (Novozymes) from *Thermoanaerobacter* sp. (Pinheiro et al. 2018). Yields of CDs and LR-CDs from starch were increased by DBE-catalyzed debranching of the substrate (Y Chen et al. 2023b; Y Li et al. 2024). CD inclusion complexes could favor CGTase-catalyzed production of specific CDs (Erichsen, Peters & Beeren 2023; T Liu et al. 2021; Sørensen et al. 2023). γ-CGTase from *Bacillus* sp. FJAT-44876 formed γ-CD in improved yields from cassava starch using moderate heat combined with high starch concentration (Wu et al. 2022).

14.4.2 Enzymes Producing Cyclic Dextrins

CDs of DP 6–8 are produced by CGTases catalyzing cyclization, coupling, disproportionation, and hydrolysis of α-1,4-linked glucans, and a large number of CGTases have been characterized with regard to specificity, product profile, and extent of hydrolysis (Benavent-Gil et al. 2021; Leemhuis, Kelly & Dijkhuizen 2010; X Li et al. 2023a, 2024; Lim et al. 2021; Morin-Crini et al. 2021; Pardhi et al. 2023) (Tables 14.1 and 14.4; Figures 14.2 and 14.4). CDs can regulate enzymatic processes as inhibitors and modulators of an array of starch active enzymes (X Li et al. 2024). Fewer CGTases, 4αGTs, and BEs generating LR-CDs have been characterized in detail (Y Chen et al. 2023b; Krusong et al. 2022; X Li et al. 2024; Sorndech, Tongta & Blennow 2018) (Table 14.4). Amylose is the best substrate for obtaining LR-CDs with higher DP. Yields of LR-CDs from tapioca and other starches likewise improved after enzymatic debranching (Y Chen et al. 2023a; Suksiri et al. 2021). CCD and HBCD are very large branched cyclic dextrins formed by BEs and 4αGT catalyzing intramolecular transglycosylation of AP (X Li et al. 2023a, 2024; Takaha et al. 1996, 1998; Takaha & Smith 1999; Takata et al. 2003; Takii et al. 1999; Terada et al. 1999). An average DP up to 1,200 is reported of HBCDs containing a cyclized ring of DP 50 that carries branches of average DP 16 originating from the AP clusters (Takata et al. 1996, 2003, 2010) (Table 14.4).

14.4.2.1 Cyclodextrin Glucanotransferases (CGTases)

CGTases are found in subfamily GH13_2 in the CAZy database (www.cazy.org/) and are rarely strictly CD specific but form a mixture of different CDs (X Li et al. 2024) (Table 14.4). Initially, CDs appear to have higher DP, but ultimately the final products of DP 6–8 accumulate (Endo et al. 1997a,b; Terada et al. 1997). CD specificity and yield can be manipulated by structure-guided CGTase protein engineering, which can also include targeted engineering of CGTases to be less susceptible to product inhibition (Leemhuis, Kelly & Dijkhuizen 2010; Zhekova et al. 2009), or by the addition of chelating agents referred to as guest molecules having preference for a specific CD (Erichsen, Peters & Beeren 2023; Sørensen et al. 2023). Thus, the costly δ-CD of DP 9 could be obtained in good yields from starch by combining an engineered CGTase with selectivity for LR-CD production and a recyclable bolaamphiphile. δ-CD has potential for the synthesis of mechanically interlocked molecules (Erichsen, Peters & Beeren 2023; Sonnendecker & Zimmermann 2019). γ-CD was produced from β-CD by the coupling activity of γ-CGTase from *Bacillus clarkii* 7364 in the presence of G2, and the yield was enhanced using chelation with 5-cyclohexadecen-1-one (Qiu et al. 2018).

Crystal structures of CGTases were solved early on (Table 14.1; Figure 14.4) (Schmidt et al. 1998) and allowed the design of enzyme mutants to enrich for specific CDs. γ-CD was the major product of the Y195R mutant of the central binding Tyr195 in an α-CD forming CGTase from *Paenibacillus* sp. 602-1, while Y195I gave higher yields of both β-CD and γ-CD (T Xie et al. 2014a,b). Analysis of the evolution of CGTases indicated product specificity to be associated with residues at subsites −3/−7, while substitutions at subsites −1, +1, and +2 changed the reaction mode, such as increasing hydrolysis (Kelly, Dijkhuizen & Leemhuis 2009). CIs can be obtained from starch using CI glucosyltransferase (CIT) (EC 2.4.2.248) of family GH66 and a GH31 6αGT (Funane et al. 2014; Ichinose et al. 2017; Suzuki et al. 2012; Xue, Svensson & Bai 2022). Crystal structures of complexes of 6αGT from *Paenibacillus* sp. 598K (Ps6GT31A) showed acarbose accommodated at the α-1,4-glucan binding site (subsites −1 to +3) and a single G at subsite −1, released from G6 or from IG6 hydrolyzed to IG5, which was seen in an acceptor binding site (subsites +1a to +4a) with the non-reducing end residue placed near subsites +1 and +2 (Figure 14.5g and h). The G5 product was bound to the N-terminal domain (Fujimoto et al. 2017b). Ps6GT31A synthesized dextrans from starch (Ichinose et al. 2017) and subsequent intramolecular transglycosylation of the dextran products catalyzed by GH66 from *Paenibacillus* sp. 598K gave CIs (Fujimoto et al. 2017a). Cycloisomaltotetraose (CI4) was created by CI4-forming glucanotransferases from *Agreia* sp. D1110 and *Microbacterium trichothecenolyticum*

D2006 using IMOSs as substrates, which were made from starch by 6αGT (Fujita et al. 2021).

14.4.2.2 Large-Ring Cyclodextrin (LR-CD) Producing Enzymes

Large-ring cyclodextrins (LR-CDs) are produced by 4αGTs (amylomaltases) belonging to GH13, 57, and 77 (Y Chen et al. 2023b; Krusong et al. 2022; X Li et al. 2024; Sorndech, Tongta & Blennow 2018). Most bacterial 4αGTs are found in GH family 77, and only a few are from GH13 and 57. The first structure was obtained of the 4αGT from *T. aquaticus* forming LR-CDs of DP >22 (Przylas et al. 2000; Terada et al. 1999). Later cocrystallization of this enzyme with an LR-CD, a 34-mer cycloamylose, showed interaction with the entire active site representing donor and acceptor binding regions (Roth et al. 2017) (Tables 14.1 and 14.4). The smallest reported 4αGT-produced LR-CD has DP 16 and the largest DP >100 (Bhuiyan et al. 2003; Takaha et al. 1996). A *Bacillus* sp. A2-5a β-CGTase initially produced CDs from DP 9 up to DP >60 from AM of 30 kDa (Terada et al. 1997). Structural comparisons with GH77 amylomaltase *Cg*AM from *Corynebacterium glutamicum* and other GH77 members led to suggestions of the basis for the unique ring-size specificity of *Cg*AM (Joo et al. 2016). This insight was exploited using structure-molecular dynamics-guided point mutation of two subdomains of the *Corynebacterium glutamicum* enzyme forming LR-CD of DP 36–40 compared to DP 29 obtained with the wild-type enzyme. Moreover, also the N-terminal domain was concluded to be important for transglycosylation (Ngawiset et al. 2023). LR-CD of DP 19–35 was prepared from AM using 4αGT from *T. aquaticus* fused to an SBD of CBM20 from a CGTase to improve the interaction with substrate (JH Park et al. 2007).

14.4.2.3 Enzymes Producing Branched Cyclic Dextrins

Branched β-CDs have higher solubility and bioavailability than β-CD. G-1,6-β-CD was synthesized in 24% yield from MD of DE 4–7 by the combined action of β-CGTase (Toruzyme, Novozymes) and glucoamylase that trimmed the branches originating from the MD substrate (L Xia et al. 2017). Disubstituted β-CDs, $6^1,6^3$ and $6^1,6^4$ di-G2-β-CD, were obtained from G2-1,6-β-CD by transglycosylation using the DBE TreX from *S. solfataricus* (Kang et al. 2008). Isoamylase catalyzed the formation of doubly branched di-G2-β-CD in reverse hydrolysis reactions (Abe et al. 1988). β-CDs with branch lengths up to DP 12 have been obtained in a disproportionation reaction from G2-1,6-β-CD and starch catalyzed by amylopullulanase from *D. amylolyticus* (HS Park et al. 2007; J-T Park et al. 2008; Y-U Park et al. 2018). *Klebsiella* pullulanase formed branched G2-α-, -β-, and -γ-CDs by reverse hydrolysis (Yim, Park & Park 1997). HBCDs and CCDs are produced by BE or 4αGT-catalyzed intramolecular cyclizing of AP (Sorndech, Tongta & Blennow 2018; Takaha et al. 1996; Takata et al. 1996, 2003, 2010; Takii et al. 1999). HBCD and CCD have very high DP ranging from 400 to 1,200, and with the cyclic part in an HBCD of average DP 50, carrying branches of average DP 16 originating from clusters in AP (Takata et al. 2003) (Table 14.4).

14.4.3 Application of Cyclic Dextrins

CDs are useful due to their unique ability to provide protection, solubilization, and delivery systems by hosting valuable guest molecules, such as pharmaceuticals and cosmetics, to contribute to biomaterials in chromatography, as adsorbents, and as enzyme inhibitors (Crini, Fenyvesi & Szente 2021; Y-R Lee et al. 2023; X Li et al. 2024; Morin-Crini et al. 2021; Periasamy 2020). CDs are thus used for decontamination by the removal of biopesticides and in materials for water treatment (Decool et al. 2024; Y Li et al. 2024). β-CD was moreover applied in metal corrosion inhibitors (Umoren & Eduok 2016). CD inclusion complexes were exploited during β-CD production from corn starch by *B. subtilis* sp. β-CGTase to catch valuable volatiles such as the antibacterial and anti-inflammatory carvacrol (T Liu et al. 2021).

LR-CDs (cycloamyloses) have a hydrophobic inner cavity convenient for encapsulations, as chiral selectors for enantiomeric pharmaceuticals, in drug delivery, and by increasing solubility (Y Chen et al. 2023b; Krusong et al. 2022; X Li et al. 2024; Sonnendecker et al. 2019; Takaha et al. 1996). LR-CD of DP 7–41 entrapped larger amounts of carnosic acid from rosemary, resulting in higher antioxidant and antimicrobial activity of the carnosic acid compared to complexes with branched dextrins of average Mw 112 kDa, MDs of DP 2–32, and β-CD (Park, Rho & Kim 2019). CCDs (Mw 462 kDa) were water soluble and applied in drug delivery (Kaneo et al. 2014). LR-CD prevented retrogradation and reduced viscosity during gelatinization for starches having large granules, but not for starches of small granules (W Li et al. 2024). HBCDs are used in composite particles as excipients for drug delivery in spray-dry powder inhalers (Kadota et al. 2015; Tse et al. 2018). HBCDs moreover improved mechanical properties and freeze–thaw stability of κ-carrageenan gels (Ma et al. 2024). CIs of DP 6–12 are flexible and water soluble; CI10 was best at forming inclusion complex with Victoria blue B (Funane et al. 2007). CI7 and CI8 are anti-plaque carbohydrates inhibiting Streptococcal glucansucrases found in the oral cavity (Kobayashi, Funane & Oguma 1995).

14.5 DEXTRIN CONJUGATES

Glucoconjugates of miscellaneous aglycons, such as aesculin, resveratrol, ginsenosides, and gallic acid, are synthesized enzymatically by glucoside hydrolases in reverse hydrolysis reactions or by transglucosidases (J Kang et al. 2011; Klingel et al. 2019; Marié et al. 2018; Z Xiao et al. 2022) (Figure 14.1 and Table 14.5). Glucoconjugates generally display enhanced accessibility/solubility conferred to the bioactive aglycon compounds having roles as drugs, ingredients, or amphipathic detergents (Ara et al. 2021; R Han et al. 2020; Lambert et al. 2023; W Wang et al. 2021; Z Xiao et al. 2022).

14.5.1 Sources of α-Glucans Used for Production of Dextrin Conjugates

Different donors including CDs, MDs, starch, pullulan, and sucrose have been applied in the synthesis of α-glucan conjugates (Table 14.5). Pullulan was thus used as donor for the synthesis of new α-1,6-linked glucon moieties in glucoconjugates using DBEs (Kang et al. 2011; Vester-Christensen et al. 2023). Notably, starch and CDs were used with *Bacillus macerans* CGTase to synthesize monoacrylate MOSs intended for polymerization resulting in a comb-shaped glycopolymer (Kloosterman, Spoelstra-van Dijk & Loos 2014).

14.5.2 Enzymes Used for Production of Dextrin Conjugates

Glucoside hydrolases, capable of catalyzing transglycosylation in reverse hydrolysis reactions, realized the formation of new α-glucan moieties or extension of existing ones on non-carbohydrate aglycons in accordance with the double displacement retaining mechanism (Figure 14.1). This activity has been exploited for several decades, and dextrin conjugates include products obtained using hydrolases or transglucosidases (Table 14.5) (Ara et al. 2021; Kang et al. 2011; Vester-Christensen et al. 2023). CGTases are useful for extending chains in glucosyl conjugates gaining water solubility (R Han et al. 2020; Ioannou et al. 2021; Leemhuis, Kelly & Dijkhuizen 2010; Marié et al. 2018; Strompen et al. 2015) (Table 14.5).

14.5.3 Application of Dextrin Conjugates

A wide range of glucoconjugates with glycons related to α-glucans has received attention for the increased water solubility, protection, and bioavailability of otherwise poorly soluble and vulnerable chemical compounds. It is beyond the scope of the present chapter to describe the numerous available glucoconjugates and their applications.

14.6 APPLICATION OF DEXTRINS IN FOOD/FEED, PHARMA/BIOMEDICINE, AND BIOMATERIALS/BIOREMEDIATION

While amylolytic enzymes are widely added for bioprocessing of starch in industrial settings, such as textile, paper, biofuel, detergent, and syrup production (Shad et al. 2023), MDs and MOSs can be used as more advanced additives and ingredients. MDs are thus used for electrospinning to provide fibers used in food applications for embedding of bioactive compounds (Vargas-Campos et al. 2023) and as a wall material for microcapsules (Xiao et al. 2022). Dextrins of DP 53 are used in nanocomposites for controlled delivery (Phillips et al. 2020). *In situ* action of *exo*-type and *endo*-type amylases, as well as debranching enzymes, are beneficial during bread making (Zhao et al. 2022). Good pasting and gelation properties are demonstrated for acid dextrinized potato starch with DE of 1.15–2.93 used in foods (Kapusniak et al. 2021). A large number of different α-glucans produced by enzymatic modifications are of interest as potential prebiotic dietary fibers (Gangoiti et al. 2020).

14.7 FUTURE PERSPECTIVES

In the future, the continuously improving possibility for rational design of enzymes acquiring structures having new functions will advance the modifications of starch and dextrins to be utilized in green processes. This can build on an enormous amount of experimental data spanning from structure-determined enzyme–ligand complexes to fast development of experimental and computational protein engineering techniques along with carbohydrate analytics, and chemical and physicochemical characterization of dextrin products. Application of such innovative approaches to custom-design novel dextrins for defined purposes is anticipated to be emerging, guided by existing data on characterization of numerous enzymes. To exploit starch and starch-like α-glucans by enzymatic tailor-making of useful dextrins is central for sustainable and environmentally friendly economically robust solutions for the future.

REFERENCES

Abdalla, M, Jiang, B, Hassanin, HAM, Zheng, L & Chen, J 2021, One-pot production of maltoheptaose (DP7) from starch by sequential addition of cyclodextrin glucotransferase and cyclomaltodextrinase. *Enzyme and Microbial Technology, 149*, 109847.

Abe, A, Yoshida, H, Tonozuka, T, Sakano, Y & Kamitori, S 2005, Complexes of *Thermoactinomyces vulgaris* R-47 α-amylase 1 and pullulan model oligossacharides provide new insight into the mechanism for recognizing substrates with α-(1,6) glycosidic linkages. *FEBS Journal, 272*, 6145–6153.

Abe, J, Hizukuri, S, Koizumi, K, Kubota, Y & Utamura, T 1988, Enzymic syntheses of doubly branched cyclomaltoheptaoses through the reverse action of *Pseudomonas* isoamylase. *Carbohydrate Research, 176*, 87–95.

Aga, H, Nishimoto, T, Kuniyoshi, M, Maruta, K, Yamashita, H, Higashiyama, T, Nakada, T, Kubota, M, et al. 2003, 6-α-Glucosyltransferase and 3-α-isomaltosyltransferase from *Bacillus globisporus* N75. *Journal of Bioscience and Bioengineering, 95*, 215–224.

Agirre, J, Moroz, O, Meier, S, Brask, J, Munch, A, Hoff, T, Andersen, C, Wilson, KS, et al. 2019, The structure of the AliC GH13 α-amylase from *Alicyclobacillus* sp. reveals the accommodation of starch branching points in the α-amylase family. *Acta Crystallographica Section D: Structural Biology, 75*, 1–7.

Albenne, C, Skov, LK, Tran, V, Gajhede, M, Monsan, P, Remaud-Siméon, M & André-Leroux, G 2007, Towards the molecular understanding of glycogen elongation by amylosucrase. *Proteins: Structure, Function and Genetics, 66*, 118–126.

Aleshin, AE, Stoffer, B, Firsov, LM, Svensson, B & Honzatko, RB 1996, Crystallographic complexes of glucoamylase with maltooligosaccharide analogs: Relationship of stereochemical distortions at the nonreducing end to the catalytic mechanism. *Biochemistry, 35*, 8319–8328.

Altuna, L, Herrera, ML & Foresti, ML 2018, Synthesis and characterization of octenyl succinic anhydride modified starches for food applications: A review of recent literature. *Food Hydrocolloids, 80*, 97–110.

An, Y, Tran, PL, Yoo, M-J, Song, H-N, Park, K-H, Kim, T-J, Park, J-T & Woo, E-J 2023, The distinctive permutated domain structure of periplasmic α-amylase (MalS) from glycoside hydrolase family 13 subfamily 19. *Molecules, 28*, 3972.

Aouay, M, Magnin, A, Putaux, J-L & Boufi, S 2022, Crosslinkable dextrin-coated latex via surfactant-free emulsion polymerization. *Colloids and Surfaces A: Physicochemical and Engineering Aspects, 632*, 127776.

Apriyanto, A, Compart, J & Fettke, J 2022, A review of starch, a unique biopolymer: Structure, metabolism and *in planta* modifications. *Plant Science, 318*, 111223.

Ara, KZG, Linares-Pastén, JA, Jönsson, J, Viloria-Cols, M, Ulvenlund, S, Adlercreutz, P & Karlsson, EN 2021, Engineering CGTase to improve synthesis of alkyl glycosides. *Glycobiology, 31*, 603–612.

Aroob, I, Ahmad, N, Aslam, M, Shaeer, A & Rashid, N 2019, A highly active α-cyclodextrin preferring cyclomaltodextrinase from *Geobacillus thermopakistaniensis*. *Carbohydrate Research, 481*, 1–8.

Ashok, PP, Dasgupta, D, Ray, A & Suman, SK 2024, Challenges and prospects of microbial α-amylases for industrial application: A review. *World Journal of Microbiology and Biotechnology, 40*, 44.

Avaltroni, F, Bouquerand, P & Normand, V 2004, Maltodextrin molecular weight distribution influence on the glass transition temperature and viscosity in aqueous solutions. *Carbohydrate Polymers, 58*, 323–334.

Bai, Y, Böger, M, van der Kaaij, RM, Woortman, AJJ, Pijning, T, van Leeuwen, SS, Lammerts van Bueren, A & Dijkhuizen, L 2016, *Lactobacillus reuteri* strains convert starch and maltodextrins into homoexopolysaccharides using an extracellular and cell-associated 4,6-α-glucanotransferase. *Journal of Agricultural and Food Chemistry, 64*, 2941–2952.

Bai, Y, Gangoiti, J, Dijkstra, BW, Dijkhuizen, L & Pijning, T 2017, Crystal structure of 4,6-α-glucanotransferase supports diet-driven evolution of GH70 enzymes from α-amylases in oral bacteria. *Structure, 25*, 231–242.

Bai, Y, Shi, YC, Herrera, A & Prakash, O 2011, Study of octenyl succinic anhydride-modified waxy maize starch by nuclear magnetic resonance spectroscopy. *Carbohydrate Polymers, 83*, 407–413.

Bai, Y, van der Kaaij, RM, Leemhuis, H, Pijning, T, van Leeuwen, SS, Jin, Z & Dijkhuizen, L 2015, Biochemical characterization of the *Lactobacillus reuteri* glycoside hydrolase family 70 GTFB type of 4,6-α-glucanotransferase enzymes that synthesize soluble dietary starch fibers. *Applied and Environmental Microbiology, 81*, 7223–7232.

Barczynska, R, Jochym, K, Slizewska, K, Kapusniak, J & Libudzisz, Z 2010, The effect of citric acid-modified enzyme-resistant dextrin on growth and metabolism of selected strains of probiotic and other intestinal bacteria. *Journal of Functional Foods, 2*, 126–133.

Barczynska, R, Slizewska, K, Jochym, K, Kapusniak, J & Libudzisz, Z 2012, The tartaric acid-modified enzyme-resistant dextrin from potato starch as potential prebiotic. *Journal of Functional Foods*, 4, 954–962.

Ben Ali, M, Khemakhem, B, Robert, X, Haser, R & Bejar, S 2006, Thermostability enhancement and change in starch hydrolysis profile of the maltohexaose-forming amylase of *Bacillus stearothermophilus* US100 strain. *Biochemical Journal*, 394, 51–56.

Ben Ali, M, Mezghani, M & Bejar, S 1999, A thermostable α-amylase producing maltohexaose from a new isolated *Bacillus* sp. US100: Study of activity and molecular cloning of the corresponding gene. *Enzyme and Microbial Technology*, 24, 584–589.

Benavent-Gil, Y, Rosell, CM & Gilbert, EP 2021, Understanding CGTase action through the relationship between starch structure and cyclodextrin formation. *Food Hydrocolloids*, 112, 106316.

Bertoft, E 2017, Understanding starch structure: recent progress. *Agronomy*, 7, 56.

Bertrand, A, Morel, S, Lefoulon, F, Rolland, Y, Monsan, P & Remaud-Simeon, M 2006, *Leuconostoc mesenteroides* glucansucrase synthesis of flavonoid glucosides by acceptor reactions in aqueous-organic solvents. *Carbohydrate Research*, 341, 855–863.

Bhuiyan, SH, Kitaoka, M & Hayashi, K 2003, A cycloamylose-forming hyperthermostable 4-α-glucanotransferase of *Aquifex aeolicus* expressed in *Escherichia coli*. *Journal of Molecular Catalysis B: Enzymatic*, 22, 45–53.

Bláhová, M, Štefuca, V, Hronská, H & Rosenberg, M 2023, Maltooligosaccharides: Properties, production and applications. *Molecules*, 28, 3281.

Blennow, A, Bay-Smidt, AM, Wischmann, B, Olsen, CE & Møller, BL 1998, The degree of starch phosphorylation is related to the chain length distribution of the neutral and the phosphorylated chains of amylopectin. *Carbohydrate Research*, 307, 45–54.

Boldrini, DE 2023, Starch-based materials for drug delivery in the gastrointestinal tract: A review. *Carbohydrate Polymers*, 320, 121258.

Börner, T, Roger, K & Adlercreutz, P 2014, Hydrophobic complexation promotes enzymatic surfactant synthesis from alkyl glucoside/cyclodextrin mixtures. *ACS Catalysis*, 4, 2623–2634.

Božić, N, Rozeboom, HJ, Lončar, N, Slavić, MŠ, Janssen, DB & Vujčić, Z 2020, Characterization of the starch surface binding site on *Bacillus paralicheniformis* α-amylase. *International Journal of Biological Macromolecules*, 165, 1529–1539.

Brison, Y, Fabre, E, Moulis, C, Portais, J-C, Monsan, P & Remaud-Siméon, M 2010, Synthesis of dextrans with controlled amounts of α-1,2 linkages using the transglucosidase GBD–CD2. *Applied Microbiology and Biotechnology*, 86, 545–554.

Brison, Y, Malbert, Y, Czaplicki, G, Mourey, L, Remaud-Simeon, M & Tranier, S 2016, Structural insights into the carbohydrate binding ability of an α-(1→2) branching sucrase from glycoside hydrolase family 70. *Journal of Biological Chemistry*, 291, 7527–7540.

Brzozowski, AM, Lawson, DM, Turkenburg, JP, Bisgaard-Frantzen, H, Svendsen, A, Borchert, TV, Dauter, Z, Wilson, KS, & Davies, GJ 2000, Structural analysis of a chimeric bacterial α-amylase. High-resolution analysis of native and ligand complexes. *Biochemistry*, 39, 9099–9107.

Buedenbender, S & Schulz, GE 2009, Structural base for enzymatic cyclodextrin hydrolysis. *Journal of Molecular Biology*, 385, 606–617.

Cao, C, Xu, L, Xie, P, Hu, J, Qi, J, Zhou, Y & Cao, L 2020, The characterization and evaluation of the synthesis of large-ring cyclodextrins (CD_9-CD_{22}) and α-tocopherol with enhanced thermal stability. *RSC Advances*, 10, 6584–6591.

Chang, R, Xiong, L, Li, M, Liu, J, Wang, Y, Chen, H & Sun, Q 2018, Fractionation of debranched starch with different molecular weights via edible alcohol precipitation. *Food Hydrocolloids*, 83, 430–437.

Chen, C, Lu, K, Hu, X, Liu, Y, Cui, SW & Miao, M 2020, Biofabrication, structure and characterization of an amylopectin-based cyclic glucan. *Food and Function*, 11, 2543–2554.

Chen, X, Liu, Y, Xu, Z, Zhang, C, Liu, X, Sui, Z & Corke, H 2021, Microwave irradiation alters the rheological properties and molecular structure of hull-less barley starch. *Food Hydrocolloids*, 120, 106821.

Chen, Y, Hu, X, Lu, K, Zhang, T & Miao, M 2023a, Biosynthesis of maltodextrin-derived glucan dendrimer using microbial branching enzyme. *Food Chemistry*, 424, 136373.

Chen, Y, McClements, DJ, Peng, X, Chen, L, Xu, Z, Meng, M, Ji, H, Long, J, et al. 2023b, Research progresses on enzymatic modification of starch with 4-α-glucanotransferase. *Trends in Food Science & Technology*, 131, 164–174.

Chi, C, Zhou, Y, Chen, B, He, Y & Zhao, Y 2023, A facile method for classifying starch fractions rich in long linear dextrin. *Food Hydrocolloids*, 135, 108182.

Choi, JH, Lee, H, Kim, YW, Park, JT, Woo, EJ, Kim, MJ, Lee, BH & Park, KH 2009, Characterization of a novel debranching enzyme from *Nostoc punctiforme* possessing a high specificity for long branched chains. *Biochemical and Biophysical Research Communications*, 378, 224–229.

Christensen, SJ, Madsen, MS, Zinck, SS, Hedberg, C, Sørensen, OB, Svensson, B & Meyer, AS 2023, Enzymatic potato starch modification and structure-function analysis of six diverse GH77 4-alpha-glucanotransferases. *International Journal of Biological Macromolecules*, 224, 105–114.

Christensen, SJ, Madsen, MS, Zinck, SS, Hedberg, C, Sørensen, OB, Svensson, B & Meyer, AS 2024, Bioinformatics and functional selection of GH77 4-α-glucanotransferases for potato starch modification. *New Biotechnology*, 79, 39–49.

Chronakis, IS 1998, On the molecular characteristics, compositional properties, and structural-functional mechanisms of maltodextrins: A review. *Critical Reviews in Food Science and Nutrition*, 38, 599–637.

Cifuente, JO, Colleoni, C, Kalscheuer, R & Guerin, ME 2024, Architecture, function, regulation, and evolution of α-glucans metabolic enzymes in prokaryotes. *Chemical Reviews*, 124, 4863–4934.

Cockburn, D, Nielsen, MM, Christiansen, C, Andersen, JM, Rannes, JB, Blennow, A & Svensson, B 2015, Surface binding sites in amylase have distinct roles in recognition of starch structure motifs and degradation. *International Journal of Biological Macromolecules*, 75, 338–345.

Couto, ARS, Aguiar, S, Ryzhakov, A, Larsen, KL & Loftsson, T 2019, Interaction of native cyclodextrins and their hydroxypropylated derivatives with parabens in aqueous solutions. Part 1: Evaluation of inclusion complexes. *Journal of Inclusion Phenomena and Macrocyclic Chemistry*, 93, 309–321.

Crini, G, Fenyvesi, É & Szente, L 2021, Outstanding contribution of Professor József Szejtli to cyclodextrin applications in foods, cosmetics, drugs, chromatography and biotechnology: A review. *Environmental Chemistry Letters*, 19, 2619–2641.

Cuong, NP, Lee, W-H, Oh, I-N, Thuy, NM, Kim, D-G, Park, J-T & Park, K-H 2016, Continuous production of pure maltodextrin from cyclodextrin using immobilized *Pyrococcus furiosus* thermostable amylase. *Process Biochemistry*, 51, 282–287.

De Beul, E, Jongbloet, A, Franceus, J & Desmet, T 2021, Discovery of a kojibiose hydrolase by analysis of specificity-determining correlated positions in glycoside hydrolase family 65. *Molecules*, 26, 6321.

Decool, G, Kfoury, M, Paitel, L, Sardo, A & Fourmentin, S 2024, Cyclodextrins as molecular carriers for biopesticides: A review. *Environmental Chemistry Letters*, *22*, 321–353.

Dellweg, H, John, M & Schmidt, J 1975, Amylases from *Pseudomonas stutzeri*. Affinity chromatography of exo-maltotetraohydrolase using Sephadex G-100 as an adsorbent. *European Journal of Applied Microbiology*, *1*, 191–198.

Dhital, S, Warren, FJ, Butterworth, PJ, Ellis, PR & Gidley, MJ 2017, Mechanisms of starch digestion by α-amylase: Structural basis for kinetic properties. *Critical Reviews in Food Science and Nutrition*, *57*, 875–892.

Diedrich, K, Krause, B, Berg, O & Rarey, M 2023, PoseEdit: Enhanced ligand binding mode communication by interactive 2D diagrams. *Journal of Computer-Aided Molecular Design*, *37*, 491–503.

Ding, N, Zhao, B, Ban, X, Li, C, Venkataram Prasad, BV, Gu, Z & Li, Z 2021, Carbohydrate-binding module and linker allow cold adaptation and salt tolerance of maltopentaose-forming amylase from marine bacterium *Saccharophagus degradans* 2–40T. *Frontiers in Microbiology*, *12*, 708480.

Ding, N, Zhao, B, Han, X, Li, C, Gu, Z & Li, Z 2022, Starch-binding domain modulates the specificity of maltopentaose production at moderate temperatures. *Journal of Agricultural and Food Chemistry*, *70*, 9057–9065.

Dong, J, Bai, Y, Wang, Q, Chen, Q, Li, X, Wang, Y, Ji, H, Meng, X, Pijning, T, Svensson, B, Dijkhuizen, L, Abou Hachem, M & Jin, Z 2024, Insights into the structure-function relationship of GH70 GtfB α-glucanotransferases from the crystal structure and molecular dynamic simulation of a newly characterized *Limosilactobacillus reuteri* N1 GtfB enzyme. *Journal of Agricultural and Food Chemistry*, *72*, 5391–5402.

Drula, E, Garron, ML, Dogan, S, Lombard, V, Henrissat, B & Terrapon, N 2022, The carbohydrate-active enzyme database: Functions and literature. *Nucleic Acids Research*, *50*, D571–D577.

Duan, K, Ban, X, Wang, Y, Li, C, Gu, Z & Li, Z 2022, Improving the product specificity of maltotetraose-forming amylase from *Pseudomonas saccharophila* STB07 by removing the carbohydrate-binding module. *Journal of Agricultural and Food Chemistry*, *70*, 13709–13718.

Dumbrepatil, AB, Choi, J, Park, JT, Kim, M, Kim, TJ, Woo, E & Park, KH 2010, Structural features of the *Nostoc punctiforme* debranching enzyme reveal the basis of its mechanism and substrate specificity. *Proteins: Structure, Function, and Bioinformatics*, *78*, 348–356.

El Mannai, Y, Deto, R, Kuroki, M, Suzuki, R & Suzuki, E 2021, Cyanobacterial branching enzymes bind to α-glucan via surface binding sites. *Archives of Biochemistry and Biophysics*, *702*, 108821.

Endo, T, Nagase, H, Ueda, H, Kobayashi, S & Nagai, T 1997a, Isolation, purification, and characterization of cyclomaltodecaose (ε-cyclodextrin), cyclomaltoundecaose (ζ-cyclodextrin) and cyclomaltotridecaose (θ-cyclodextrin). *Chemical and Pharmaceutical Bulletin*, *45*, 532–536.

Endo, T, Nagase, H, Ueda, H, Shigihara, A, Kobayashi, S & Nagai, T 1997b, Isolation, purification, and characterization of cyclomaltotetradecaose (ι-cyclodextrin), cyclomaltopentadecaose (κ-cyclodextrin) cyclomaltohexadecaose (λ-cyclodextrin), and cyclomaltoheptadecaose (μ-cyclodextrin). *Chemical and Pharmaceutical Bulletin*, *45*, 1856–1859.

Erichsen, A, Peters, GHJ & Beeren, SR 2023, Templated enzymatic synthesis of δ-cyclodextrin. *Journal of the American Chemical Society*, *145*, 4882–4891.

Fawaz, R, Bingham, C, Nayebi, H, Chiou, J, Gilbert, L, Park, SH & Geiger, JH 2023, The structure of maltooctaose-bound *Escherichia coli* branching enzyme suggests a mechanism for donor chain specificity. *Molecules*, *28*, 4377.

Feng, L, Fawaz, R, Hovde, S, Gilbert, L, Chiou, J & Geiger, JH 2015, Crystal structures of *Escherichia coli* branching enzyme in complex with linear oligosaccharides. *Biochemistry*, *54*, 6207–6218.

Fu, Z, Zhang, Z, Chu, M, Kan, N, Xiao, Y & Peng, H 2024, A starch-binding domain of α-amylase (AmyPG) disrupts the structure of raw starch. *International Journal of Biological Macromolecules*, *257*, 128673.

Fujimoto, Z, Kishine, N, Suzuki, N, Suzuki, R, Mizushima, D, Momma, M, Kimura, K & Funane, K 2017a, Isomaltooligosaccharide-binding structure of *Paenibacillus* sp. 598K cycloisomaltooligosaccharide glucanotransferase. *Bioscience Reports*. *37*, BSR20170253.

Fujimoto, Z, Suzuki, N, Kishine, N, Ichinose, H, Momma, M, Kimura, A & Funane, K 2017b, Carbohydrate-binding architecture of the multi-modular α-1,6-glucosyltransferase from *Paenibacillus* sp. 598K, which produces α-1,6-glucosyl-α-glucosaccharides from starch. *Biochemical Journal*, *474*, 2763–2778.

Fujimoto, Z, Takase, K, Doui, N, Momma, M, Matsumoto, T & Mizuno, H 1998, Crystal structure of a catalytic-site mutant α-amylase from *Bacillus subtilis* complexed with maltopentaose. *Journal of Molecular Biology*, *277*, 393–407.

Fujita, A, Kawashima, A, Mitsukawa, Y, Kitagawa, N, Watanabe, H, Mori, T, Nishimoto, T, Aga, H, & Ushio, S 2021, Purification and characterization of cycloisomaltotetraose-forming glucanotransferases from *Agreia* sp. D1110 and *Microbacterium trichothecenolyticum* D2006. *Bioscience, Biotechnology and Biochemistry*, *85*, 600–610.

Funane, K, Ichinose, H, Araki, M, Suzuki, R, Kimura, K, Fujimoto, Z, Kobayashi, M & Kimura, A 2014, Evidence for cycloisomaltooligosaccharide production from starch by *Bacillus circulans* T-3040. *Applied Microbiology and Biotechnology*, *98*, 3947–3954.

Funane, K, Terasawa, K, Mizuno, Y, Ono, H, Miyagi, T, Gibu, S, Tokashiki, T, Kawabata, Y, Kim, Y-M, Kimura, A & Kobayashi, M 2007, A novel cyclic isomaltooligosaccharide (cycloisomaltodecaose, CI-10) produced by *Bacillus circulans* T-3040 displays remarkable inclusion ability compared with cyclodextrins. *Journal of Biotechnology*, *130*, 188–192.

Gaenssle, ALO, Bax, HHM, van der Maarel, MJEC & Jurak, E 2021, GH13 Glycogen branching enzymes can adapt the substrate chain length towards their preferences via α-1,4-transglycosylation. *Enzyme and Microbial Technology*, *150*, 109882.

Gaenssle, ALO, van der Maarel, MJEC & Jurak, E 2022, The influence of amylose content on the modification of starches by glycogen branching enzymes. *Food Chemistry*, *393*, 133294.

Gangoiti, J, Corwin, SF, Lamothe, LM, Vafiadi, C, Hamaker, BR & Dijkhuizen, L 2020, Synthesis of novel α-glucans with potential health benefits through controlled glucose release in the human gastrointestinal tract. *Critical Reviews in Food Science and Nutrition*, *60*, 123–146.

Gangoiti, J, Pijning, T & Dijkhuizen, L 2018, Biotechnological potential of novel glycoside hydrolase family 70 enzymes synthesizing α-glucans from starch and sucrose. *Biotechnology Advances*, *36*, 196–207.

Gangoiti, J, van Leeuwen, SS, Gerwig, GJ, Duboux, S, Vafiadi, C, Pijning, T & Dijkhuizen, L 2017, 4,3-α-Glucanotransferase, a novel reaction specificity in glycoside hydrolase family 70 and clan GH-H. *Scientific Reports*, *7*, 39761.

Gangoiti, J, van Leeuwen, SS, Vafiadi, C & Dijkhuizen, L 2016, The Gram-negative bacterium *Azotobacter chroococcum* NCIMB 8003 employs a new glycoside hydrolase family 70 4,6-α-glucanotransferase enzyme (GtfD) to synthesize a reuteran like polymer from maltodextrins and starch. *Biochimica et Biophysica Acta (BBA) - General Subjects*, *1860*, 1224–1236.

Gavgani, HN, Fawaz, R, Ehyaei, N, Walls, D, Pawlowski, K, Fulgos, R, Park, S, Assar, Z, Ghanbarpour, A & Geiger, JH 2022, A structural explanation for the mechanism and specificity of plant branching enzymes I and IIb. *Journal of Biological Chemistry*, 298, 101395.

Geremia, S, Campagnolo, M, Schinzel, R & Johnson, LN 2002, Enzymatic catalysis in crystals of *Escherichia coli* maltodextrin phosphorylase. *Journal of Molecular Biology*, 322, 413–423.

Goffin, D, Delzenne, N, Blecker, C, Hanon, E, Deroanne, C & Paquot, M 2011, Will isomalto-oligosaccharides, a well-established functional food in Asia, break through the European and American market? The status of knowledge on these prebiotics. *Critical Reviews in Food Science and Nutrition*, 51, 394–409.

Guérin, F, Barbe, S, Pizzut-Serin, S, Potocki-Véronèse, G, Guieysse, D, Guillet, V, Monsan, P, Mourey, L, Remaud-Siméon, M, André, I & Trainer, S 2012, Structural investigation of the thermostability and product specificity of amylosucrase from the bacterium *Deinococcus geothermalis*. *Journal of Biological Chemistry*, 287, 6642–6654.

Gunawardene, OHP, Gunathilake, C, Amaraweera, SM, Fernando, NML, Wanninayaka, DB, Manamperi, A, Kulatunga, AK, Rajapaksha, SM, et al. 2021, Compatibilization of starch/synthetic biodegradable polymer blends for packaging applications: A review. *Journal of Composites Science*, 5, 300.

Guo, Y, Sun, S, Gu, M, Zhang, L, Cheng, L, Li, Z, Li, C, Ban, X, Hong, Y & Gu, Z 2024, Optimization and analysis of dual-enzymatic synthesis for the production of linear glucan. *International Journal of Biological Macromolecules*, 259, 129299.

Han, R, Ni, J, Zhou, J, Dong, J, Xu, G & Ni, Y 2020, Engineering of cyclodextrin glycosyltransferase reveals pH-regulated mechanism of enhanced long-chain glycosylated sophoricoside specificity. *Applied and Environmental Microbiology*, 86, e00004-20.

Han, X, Kang, J, Bai, Y, Xue, M & Shi, YC 2018, Structure of pyrodextrin in relation to its retrogradation properties. *Food Chemistry*, 242, 169–173.

Hanashiro, I, Abe, J & Hizukuri, S 1996, A periodic distribution of the chain length of amylopectin as revealed by high-performance anion-exchange chromatography. *Carbohydrate Research*, 283, 151–159.

Haq, F, Yu, H, Wang, L, Teng, L, Haroon, M, Khan, RU, Mehmood, S, Bilal-Ul-Amin, Ullah, RS, Khan A & Nazir, A 2019, Advances in chemical modifications of starches and their applications. *Carbohydrate Research*, 476, 12–35.

Hasegawa, K, Kubota, M & Matsuura, Y 1999, Roles of catalytic residues in α-amylases as evidenced by the structures of the product-complexed mutants of a maltotetraose-forming amylase. *Protein Engineering, Design and Selection*, 12, 819–824.

Hayashi, M, Suzuki, R, Colleoni, C, Ball, SG, Fujita, N & Suzuki, E 2017, Bound substrate in the structure of cyanobacterial branching enzyme supports a new mechanistic model. *Journal of Biological Chemistry*, 292, 5465–5475.

He, R, Li, S, Zhao, G, Zhai, L, Qin, P & Yang, L 2023, Starch modification with molecular transformation, physicochemical characteristics, and industrial usability: A state-of-the-art review. *Polymers*, 15, 2935.

Hofman, DL, van Buul, VJ & Brouns, FJPH 2016, Nutrition, health, and regulatory aspects of digestible maltodextrins. *Critical Reviews in Food Science and Nutrition*, 56, 2091–2100.

Hoti, G, Matencio, A, Rubin Pedrazzo, A, Cecone, C, Appleton, SL, Khazaei Monfared, Y, Caldera, F & Trotta, F 2022, Nutraceutical concepts and dextrin-based delivery systems. *International Journal of Molecular Sciences*, 23, 4102.

Hu, X & Goff, HD 2018, Fractionation of polysaccharides by gradient non-solvent precipitation: A review. *Trends in Food Science & Technology*, 81, 108–115.

Hu, X, Liu, C, Jin, Z & Tian, Y 2015, Fractionation of starch hydrolysate into dextrin fractions with low dispersity by gradient alcohol precipitation. *Separation and Purification Technology*, 151, 201–210.

Hu, X, Liu, C, Jin, Z & Tian, Y 2017, Preparative fractionation of dextrin by gradient alcohol precipitation. *Separation Science and Technology (Philadelphia)*, 52, 2704–2714.

Huang, Z, Wang, JJ, Chen, Y, Wei, N, Hou, Y, Bai, W & Hu, S-Q 2020, Effect of water-soluble dietary fiber resistant dextrin on flour and bread qualities. *Food Chemistry*, 317, 126452.

Ichinose, H, Suzuki, R, Miyazaki, T, Kimura, K, Momma, M, Suzuki, N, Fujimoto, Z, Kimura, A, & Funane, K 2017, *Paenibacillus* sp. 598K 6-α-glucosyltransferase is essential for cycloisomaltooligosaccharide synthesis from α-(1 → 4)-glucan. *Applied Microbiology and Biotechnology*, 101, 4115–4128.

Imamura, H, Fushinobu, S, Yamamoto, M, Kumasaka, T, Jeon, BS, Wakagi, T & Matsuzawa, H 2003, Crystal structures of 4-α-glucanotransferase from *Thermococcus litoralis* and its complex with an inhibitor. *Journal of Biological Chemistry*, 278, 19378–19386.

Ioannou, I, Barboza, E, Willig, G, Marié, T, Texeira, A, Darme, P, Renault, J-H & Allais, F 2021, Implementation of an enzyme membrane reactor to intensify the α-O-glycosylation of resveratrol using cyclodextrins. *Pharmaceuticals*, 14, 319.

Janeček, Š, Mareček, F, MacGregor, EA & Svensson, B 2019, Starch-binding domains as CBM families–history, occurrence, structure, function and evolution. *Biotechnology Advances*, 37, 107451.

Janeček, Š & Svensson, B 2022, How many α-amylase GH families are there in the CAZy database? *Amylase*, 6, 1–10.

Jeong, WS, Lee, Y-R, Hong, S-J, Choi, S-J, Choi, J-H, Park, S-Y, Woo, E-J, Kim, YM, & Park, B-R 2019, Carboxy-terminal region of a thermostable CITase from *Thermoanaerobacter thermocopriae* has the ability to produce long isomaltooligosaccharides. *Journal of Microbiology and Biotechnology*, 29, 1938–1946.

Ji, H, Bai, Y, Li, X, Wang, J, Xu, X & Jin, Z 2019, Preparation of malto-oligosaccharides with specific degree of polymerization by a novel cyclodextrinase from *Palaeococcus pacificus*. *Carbohydrate Polymers*, 210, 64–72.

Ji, H, Bai, Y, Li, X, Zheng, D, Shen, Y & Jin, Z 2020a, Structural and property characterization of corn starch modified by cyclodextrin glycosyltransferase and specific cyclodextrinase. *Carbohydrate Polymers*, 237, 116137.

Ji, H, Liu, J, McClements, DJ, Bai, Y, Li, Z, Chen, L, Qiu, C, Zhan, X, et al. 2024, Malto-oligosaccharides as critical functional ingredient: A review of their properties, preparation, and versatile applications. *Critical Reviews in Food Science and Nutrition*, 64, 3674–3686.

Ji, H, Wang, Y, Bai, Y, Li, X, Qiu, L & Jin, Z 2020b, Application of cyclodextrinase in non-complexant production of γ-cyclodextrin. *Biotechnology Progress*, 36, e2930.

Jiang, H, Feng, Y, Jane, J & Yang, Y 2024, Progress in understanding resistant-starch formation in hydroxypropyl starch: A minireview. *Food Hydrocolloids*, 149, 109628.

Jiang, Y, Li, X, Pijning, T, Bai, Y & Dijkhuizen, L 2022, Mutations in amino acid residues of *Limosilactobacillus reuteri* 121 GtfB 4,6-α-glucanotransferase that affect reaction and product specificity. *Journal of Agricultural and Food Chemistry*, 70, 1952–1961.

Jo, HJ, Park, S, Jeong, HG, Kim, JW & Park, JT 2015, *Vibrio vulnificus* glycogen branching enzyme preferentially transfers very short chains: N1 domain determines the chain length transferred. *FEBS Letters*, 589, 1089–1094.

Jonathan, MC, van Brussel, M, Scheffers, MS & Kabel, MA 2015, Characterisation of branched gluco-oligosaccharides to study the mode-of-action of a glucoamylase from *Hypocrea jecorina*. *Carbohydrate Polymers*, 132, 59–66.

Joo, S, Kim, S, Seo, H & Kim, KJ 2016, Crystal structure of amylomaltase from *Corynebacterium glutamicum*. *Journal of Agricultural and Food Chemistry*, *64*, 5662–5670.

Juge, N, Nøhr, J, Le Gal-Coëffet, MF, Kramhøft, B, Furniss, CSM, Planchot, V, Archer, DB, Williamson, G, et al. 2006, The activity of barley α-amylase on starch granules is enhanced by fusion of a starch binding domain from *Aspergillus niger* glucoamylase. *Biochimica et Biophysica Acta - Proteins and Proteomics*, 1764, 275–284.

Junejo, SA, Flanagan, BM, Zhang, B & Dhital, S 2022, Starch structure and nutritional functionality: Past revelations and future prospects. *Carbohydrate Polymers*, *277*, 118837.

Jung, J-H, Hong, S, Jeon, EJ, Kim, M-K, Seo, D-H, Woo, E-J, Holden, JF & Park, C-S 2022, Acceptor dependent catalytic properties of GH57 4-α-glucanotransferase from *Pyrococcus* sp. ST04. *Frontiers in Microbiology*, *13*, 1016675.

Kadokawa, J 2017, α-Glucan phosphorylase: A useful catalyst for precision enzymatic synthesis of oligo- and polysaccharides. *Current Organic Chemistry*, *21*, 1192–1204.

Kadokawa, J 2022, Glucan phosphorylase-catalyzed enzymatic synthesis of unnatural oligosaccharides and polysaccharides using nonnative substrates. *Polymer Journal*, *54*, 413–426.

Kadota, K, Nishimura, T, Hotta, D & Tozuka, Y 2015, Preparation of composite particles of hydrophilic or hydrophobic drugs with highly branched cyclic dextrin via spray drying for dry powder inhalers. *Powder Technology*, *283*, 16–23.

Kagawa, M, Fujimoto, Z, Momma, M, Takase, K & Mizuno, H 2003, Crystal structure of *Bacillus subtilis* α-amylase in complex with acarbose. *Journal of Bacteriology*, *185*, 6981–6984.

Kaila, P & Guptasarma, P 2019, An ultra-stable glucanotransferase-cum-exoamylase from the hyperthermophile archaeon *Thermococcus onnurineus*. *Archives of Biochemistry and Biophysics*, *665*, 114–121.

Kamaraj, M, Suresh Babu, P, Shyamalagowri, S, Pavithra, MKS, Aravind, J, Kim, W & Govarthanan, M 2024, β-cyclodextrin polymer composites for the removal of pharmaceutical substances, endocrine disruptor chemicals, and dyes from aqueous solution- A review of recent trends. *Journal of Environmental Management*, *351*, 119830.

Kamon, M, Sumitani, J, Tani, S, Kawaguchi, T, Kamon, M, Sumitani, J, Tani, S & Kawaguchi, T 2015, Characterization and gene cloning of a maltotriose-forming exo-amylase from *Kitasatospora* sp. MK-1785. *Applied Microbiology and Biotechnology*, *99*, 4743–4753.

Kanai, R, Haga, K, Akiba, T, Yamane, K & Harata, K 2004, Biochemical and crystallographic analyses of maltohexaose-producing amylase from alkalophilic *Bacillus* sp. 707. *Biochemistry*, *43*, 14047–14056.

Kanai, R, Haga, K, Akiba, T, Yamane, K & Harata, K 2006, Role of Trp140 at subsite −6 on the maltohexaose production of maltohexaose-producing amylase from alkalophilic *Bacillus* sp.707. *Protein Science*, *15*, 468–477.

Kaneo, Y, Taguchi, K, Tanaka, T & Yamamoto, S 2014, Nanoparticles of hydrophobized cluster dextrin as biodegradable drug carriers: Solubilization and encapsulation of amphotericin B. *Journal of Drug Delivery Science and Technology*, *24*, 344–351.

Kang, HK, Cha, H, Yang, TJ, Park, JT, Lee, S, Kim, YW, Auh, JH, Okada, Y, Kim, J-W, Cha, J, Kim, CH & Park, K-H 2008, Enzymatic synthesis of dimaltosyl-β-cyclodextrin via a transglycosylation reaction using TreX, a *Sulfolobus solfataricus* P2 debranching enzyme. *Biochemical and Biophysical Research Communications*, *366*, 98–103.

Kang, J, Park, KM, Choi, KH, Park, CS, Kim, GE, Kim, D & Cha, J 2011, Molecular cloning and biochemical characterization of a heat-stable type I pullulanase from *Thermotoga neapolitana*. *Enzyme and Microbial Technology*, *48*, 260–266.

Kapusniak, K, Wojcik, M, Wrobel, K, Rosicka-Kaczmarek, J & Kapusniak, J 2021, Assessment of physicochemical and thermal properties of soluble dextrin fiber from potato starch for use in fruit mousses. *Journal of the Science of Food and Agriculture*, *101*, 4125–4133.

Kashiwagi, N, Miyake, M, Hirose, S, Sota, M, Ogino, C & Kondo, A 2014, Cloning and starch degradation profile of maltotriose-producing amylases from *Streptomyces* species. *Biotechnology Letters*, *36*, 2311–2317.

Kato, N, Suyama, S, Shirokane, M, Kato, M, Kobayashi, T & Tsukagoshi, N 2002, Novel α-glucosidase from *Aspergillus nidulans* with strong transglycosylation activity. *Applied and Environmental Microbiology*, *68*, 1250–1256.

Kaulpiboon, J, Rudeekulthamrong, P, Watanasatitarpa, S, Ito, K & Pongsawasdi, P 2015, Synthesis of long-chain isomaltooligosaccharides from tapioca starch and an *in vitro* investigation of their prebiotic properties. *Journal of Molecular Catalysis B: Enzymatic*, *120*, 127–135.

Kelly, RM, Dijkhuizen, L & Leemhuis, H 2009, The evolution of cyclodextrin glucanotransferase product specificity. *Applied Microbiology and Biotechnology*, *84*, 119–133.

Kfoury, M, Auezova, L, Greige-Gerges, H & Fourmentin, S 2019, Encapsulation in cyclodextrins to widen the applications of essential oils. *Environmental Chemistry Letters*, *17*, 129–143.

Khemakhem, B, Ben Ali, M, Aghajari, N, Juy, M, Haser, R & Bejar, S 2009, Engineering of the α-amylase from *Geobacillus stearothermophilus* US100 for detergent incorporation. *Biotechnology and Bioengineering*, *102*, 380–389.

Klingel, T, Bindereif, B, Hadamjetz, M, Fischer, A, van der Schaaf, US & Wefers, D 2019, Enzymatic synthesis and characterization of mono-, oligo-, and polyglucosylated conjugates of caffeic acid and gallic acid. *Journal of Agricultural and Food Chemistry*, *67*, 13108–13118.

Kloosterman, WMJ, Spoelstra-van Dijk, G & Loos, K 2014, Biocatalytic synthesis of maltodextrin-based acrylates from starch and α-cyclodextrin. *Macromolecular Bioscience*, *14*, 1268–1279.

Knegtel, RMA, Strokopytov, B, Penninga, D, Faber, OG, Rozeboom, HJ, Kalk, KH, Dijkhuizen, L & Dijkstra, BW 1995, Crystallographic studies of the interaction of cyclodextrin glycosyltransferase from *Bacillus circulans* strain 251 with natural substrates and products. *Journal of Biological Chemistry*, *270*, 29256–29264.

Kobayashi, M, Funane, K & Oguma, T 1995, Inhibition of dextran and mutan synthesis by cycloisomaltooligosaccharides. *Bioscience, Biotechnology, and Biochemistry*, *59*, 1861–1865.

Kong, D, Wang, L, Su, L & Wu, J 2021, Effect of Leu[277] on disproportionation and hydrolysis activity in *Bacillus stearothermophilus* NO$_2$ cyclodextrin glucosyltransferase. *Applied and Environmental Microbiology*, *87*, e03151–20

Kong, H, Zou, Y, Gu, Z, Li, Z, Jiang, Z, Cheng, L, Hong, Y & Li, C 2018, Liquefaction concentration impacts the fine structure of maltodextrin. *Industrial Crops and Products*, *123*, 687–697.

Kong, X, Bertoft, E, Bao, J & Corke, H 2008, Molecular structure of amylopectin from amaranth starch and its effect on physicochemical properties. *International Journal of Biological Macromolecules*, *43*, 377–382.

Konieczny, J & Loos, K 2019, Bio-based polyurethane films using white dextrins. *Journal of Applied Polymer Science*, *136*, 47454.

Koo, YS, Lee, HW, Jeon, HY, Choi, HJ, Choung, WJ & Shim, JH 2015, Development and characterization of cyclodextrin glucanotransferase as a maltoheptaose-producing enzyme using site-directed mutagenesis. *Protein Engineering, Design and Selection*, *28*, 531–537.

Kotatha, D, Wandee, Y, Udchumpisai, W & Pattarapanawan, M 2023, Fortification of dietary fiber in cassava pulp by conversion of the remaining starch to resistant starch. *Future Foods*, *8*, 100265.

Kralj, S, Grijpstra, P, van Leeuwen, SS, Leemhuis, H, Dobruchowska, JM, van der Kaaij, RM, Malik, A, Oetari, A, Kamerling, JP & Dijkhuizen L 2011, 4,6-α-Glucanotransferase, a novel enzyme that structurally and functionally provides an evolutionary link between glycoside hydrolase enzyme families 13 and 70. *Applied and Environmental Microbiology, 77*, 8154–8163.

Krusong, K, Ismail, A, Wangpaiboon, K & Pongsawasdi, P 2022, Production of large-ring cyclodextrins by amylomaltases. *Molecules, 27*, 1446.

Kumar, S, Tissopi, T & Mutturi, S 2023, The successful synthesis of industrial isomaltooligosaccharides lies in the use of trans-glycosylating α-glucosidases: A review. *Carbohydrate Polymer Technologies and Applications, 5*, 100325.

Lambert, C, Lemagnen, P, Don Simoni, E, Hubert, J, Kotland, A, Paulus, C, De Bizemont, A, Bernard, S, Humeau, A & Auriol, D 2023, Enzymatic synthesis of α-glucosyl-baicalin through transglucosylation via cyclodextrin glucanotransferase in water. *Molecules, 28*, 3891.

Lammerts Van Bueren, A, Ficko-Blean, E, Pluvinage, B, Hehemann, JH, Higgins, MA, Deng, L, Ogunniyi, AD, Stroeher, UH, et al. 2011, The conformation and function of a multimodular glycogen-degrading pneumococcal virulence factor. *Structure, 19*, 640–651.

Lang, W, Kumagai, Y, Sadahiro, J, Maneesan, J, Okuyama, M, Mori, H, Sakairi, N & Kimura, A 2014, Different molecular complexity of linear-isomaltomegalosaccharides and β-cyclodextrin on enhancing solubility of azo dye ethyl red: Towards dye biodegradation. *Bioresource Technology, 169*, 518–524.

Lang, W, Kumagai, Y, Sadahiro, J, Saburi, W, Sarnthima, R, Tagami, T, Okuyama, M, Mori, H, Sakairi, N, Kim, D & Kimura, A 2022, A practical approach to producing isomaltomegalosaccharide using dextran dextrinase from *Gluconobacter oxydans* ATCC 11894. *Applied Microbiology and Biotechnology, 106*, 689–698.

Lang, W, Tagami, T, Kumagai, Y, Tanaka, S, Kang, H-J, Okuyama, M, Saburi, W, Mori, H, et al. 2023, Tunable structure of chimeric isomaltomegalosaccharides with double α-(1 → 4)-glucosyl chains enhances the solubility of water-insoluble bioactive compounds. *Carbohydrate Polymers, 319*, 121185.

Larsen, KL 2002, Large cyclodextrins. *Journal of Inclusion Phenomena and Macrocyclic Chemistry, 43*, 1–13.

Lee, BH & Hamaker, BR 2017, Number of branch points in α-limit dextrins impact glucose generation rates by mammalian mucosal α-glucosidases. *Carbohydrate Polymers, 157*, 207–213.

Lee, HS, Auh, JH, Yoon, HG, Kim, MJ, Park, JH, Hong, SS, Kang, MH, Kim, TJ, Moon, TW, Kim, JW & Park, KH 2002a, Cooperative action of α-glucanotransferase and maltogenic amylase for an improved process of isomaltooligosaccharide (IMO) production. *Journal of Agricultural and Food Chemistry, 50*, 2812–2817.

Lee, HS, Kim, MS, Cho, HS, Kim, JI, Kim, TJ, Choi, JH, Park, C, Lee, HS, Oh, B-H & Park, K-H 2002b, Cyclomaltodextrinase, neopullulanase, and maltogenic amylase are nearly indistinguishable from each other. *Journal of Biological Chemistry, 277*, 21891–21897.

Lee, Y-R, Jeong, H-M, Kim, J-S, Kim, E-A, Lee, E-H & Shim, J-H 2023, Enzymatic formation of cyclic maltooligosaccharides for the application of quercetin inclusion complex. *Carbohydrate Polymers, 310*, 120722.

Leemhuis, H, Dijkman, WP, Dobruchowska, JM, Pijning, T, Grijpstra, P, Kralj, S, Kamerling, JP & Dijkhuizen, L 2013, 4,6-α-Glucanotransferase activity occurs more widespread in *Lactobacillus strains* and constitutes a separate GH70 subfamily. *Applied Microbiology and Biotechnology, 97*, 181–193.

Leemhuis, H, Dobruchowska, JM, Ebbelaar, M, Faber, F, Buwalda, PL, van der Maarel, MJEC, Kamerling, JP & Dijkhuizen, L 2014, Isomalto/malto-polysaccharide, a novel soluble dietary fiber made via enzymatic conversion of starch. *Journal of Agricultural and Food Chemistry, 62*, 12034–12044.

Leemhuis, H, Kelly, RM & Dijkhuizen, L 2010, Engineering of cyclodextrin glucanotransferases and the impact for biotechnological applications. *Applied Microbiology and Biotechnology, 85*, 823–835.

Lekakarn, H, Bunterngsook, B, Pajongpakdeekul, N, Prongjit, D & Champreda, V 2022, A novel low temperature active maltooligosaccharides-forming amylase from *Bacillus koreensis* HL12 as biocatalyst for maltooligosaccharide production. *3 Biotech, 12*, 134.

Li, A, Benkoulouche, M, Ladeveze, S, Durand, J, Cioci, G, Laville, E & Potocki-Veronese, G 2022, Discovery and biotechnological exploitation of glycoside-phosphorylases. *International Journal of Molecular Sciences, 23*, 3043.

Li, F, Muhmood, A, Akhter, M, Gao, X, Sun, J, Du, Z, Wei, Y, Zhang, T, et al. 2023, Characterization, health benefits, and food applications of enzymatic digestion-resistant dextrin: A review. *International Journal of Biological Macromolecules, 253*, 126970.

Li, H-T, Zhang, W, Pan, W, Chen, Y, Bao, Y & Bui, AT 2023, Altered leaching composition of maize starch granules by irradiative depolymerization: The key role of degraded molecular structure. *International Journal of Biological Macromolecules, 253*, 126756.

Li, W, Xie, A, Li, X, Jin, Z, Svensson, B & Bai, Y 2024, Effects of large-ring cyclodextrin produced by 4-α-glucosyltransferase on gelatinization and retrogradation behavior of starches. *Food Hydrocolloids, 154*, 110090.

Li, X, Bai, Y, Ji, H, Wang, Y & Jin, Z 2020, Phenylalanine476 mutation of pullulanase from *Bacillus subtilis* str. 168 improves the starch substrate utilization by weakening the product β-cyclodextrin inhibition. *International Journal of Biological Macromolecules, 155*, 490–497.

Li, X, Ji, H, Zhai, Y, Bai, Y & Jin, Z 2021, Characterizing a thermostable amylopullulanase from *Caldisericum exile* with wide pH adaptation and broad substrate specificity. *Food Bioscience, 41*, 100952.

Li, X, Jin, Z, Bai, Y & Svensson, B 2024, Progress in cyclodextrins as important molecules regulating catalytic processes of glycoside hydrolases. *Biotechnology Advances, 72*, 108326.

Li, X & Li, D 2015, Preparation of linear maltodextrins using a hyperthermophilic amylopullulanase with cyclodextrin- and starch-hydrolysing activities. *Carbohydrate Polymers, 119*, 134–141.

Li, X, Wang, Y, Wu, J, Jin, Z, Dijkhuizen, L, Svensson, B & Bai, Y 2023a, Designing starch derivatives with desired structures and functional properties via rearrangements of glycosidic linkages by starch-active transglycosylases. *Critical Reviews in Food Science and Nutrition*, 1–14.

Li, X, Zheng, D, Wu, J, Jin, Z, Svensson, B & Bai, Y 2023b, Increasing γ-CD conversion rates by improving thermostability of *Bacillus* sp. FJAT-44876 γ-CGTase. *Food Bioscience, 51*, 102204.

Li, Y, Liu, F, Abdiryim, T & Liu, X 2024, Cyclodextrin-derived materials: From design to promising applications in water treatment. *Coordination Chemistry Reviews, 502*, 215613.

Li, Y, Zhao, F, Li, C, Ban, X, Gu, Z & Li, Z 2022, Fine structures of added maltodextrin impact stability of frozen bread dough system. *Carbohydrate Polymers, 298*, 120028.

Li, Z, Duan, X & Wu, J 2016, Improving the thermostability and enhancing the Ca²⁺ binding of the maltohexaose-forming α-amylase from *Bacillus stearothermophilus*. *Journal of Biotechnology, 222*, 65–72.

Li, Z, Kong, H, Li, Z, Gu, Z, Ban, X, Hong, Y, Cheng, L & Li, C 2023, Designing liquefaction and saccharification processes of highly concentrated starch slurry: Challenges and recent advances. *Comprehensive Reviews in Food Science and Food Safety*, 22, 1597–1612.

Li, Z, Liu, Y, Huang, Y, Tian, Y, Liu, J, Wang, S, Sun, P, Nie, Y, et al. 2022, Identification of the key structure, preparation conditions and properties of resistant dextrin for indigestibility based on simulated gastrointestinal conditions. *International Journal of Food Science and Technology*, 57, 7233–7244.

Li, Z, Wang, D & Shi, Y 2019, High-solids bio-conversion of maize starch to sugars and ethanol. *Starch-Stärke*, 71, 1800142.

Li, Z, Wu, J, Zhang, B, Wang, F, Ye, X, Huang, Y, Huang, Q & Cui, Z 2015, AmyM, a novel maltohexaose-forming α-amylase from *Corallococcus* sp. strain EGB. *Applied and Environmental Microbiology*, 81, 1977–1987.

Liang, X, Chen, L, McClements, DJ, Peng, X, Xu, Z, Meng, M & Jin, Z 2024, Bioactive delivery systems based on starch and its derivatives: Assembly and application at different structural levels. *Food Chemistry*, 432, 137184.

Lim, CH, Rasti, B, Sulistyo, J & Hamid, MA 2021, Comprehensive study on transglycosylation of CGTase from various sources. *Heliyon*, 7, e06305.

Liu, G, Ji, N, Gu, Z, Hong, Y, Cheng, L & Li, C 2018, Molecular interactions in debranched waxy starch and their effects on digestibility and hydrogel properties. *Food Hydrocolloids*, 84, 166–172.

Liu, T, Feng, C, Li, Z, Gu, Z, Ban, X, Hong, Y, Cheng, L & Li, C 2021, Efficient formation of carvacrol inclusion complexes during β-cyclodextrin glycosyltransferase-catalyzed cyclodextrin synthesis. *Food Control*, 130, 108296.

Liu, Y, Qiu, C, Li, X, McClements, DJ, Wang, C, Zhang, Z, Jiao, A, Long, J, Zhu, K, Wang, J & Jin, Z 2023a, Application of starch-based nanoparticles and cyclodextrin for prebiotics delivery and controlled glucose release in the human gut: A review. *Critical Reviews in Food Science and Nutrition*, 63, 6126–6137.

Liu, Y, Wu, Y, Ji, H, Li, X, Jin, Z, Svensson, B & Bai, Y 2023b, Cost-effective and controllable synthesis of isomalto/malto-polysaccharides from β-cyclodextrin by combined action of cyclodextrinase and 4,6-α-glucanotransferase GtfB. *Carbohydrate Polymers*, 310, 120716.

Liu, Y, Zhang, N, Ma, J, Zhou, Y, Wei, Q, Tian, C, Fang, Y, Zhong, R, Chen, G & Zhang, S 2023c, Advances in cold-adapted enzymes derived from microorganisms. *Frontiers in Microbiology*, 14, 1152847.

Liu, Z, Junejo, SA, Zhang, B, Fu, X & Huang, Q 2022a, Characteristics and ethylene encapsulation properties of V-type linear dextrin with different degrees of polymerisation. *Carbohydrate Polymers*, 277, 118814.

Liu, Z, Liu, J, Ren, L, Wu, J & Chen, S 2022b, Preparation of high-quality resistant dextrin through pyrodextrin by a multienzyme complex. *Food Bioscience*, 47, 101701.

Ma, W, Yuan, C, Cui, B, Gao, T, Guo, L, Yu, B, Zhao, M & Zou, F 2024, Highly-branched cyclic dextrin for improvement in mechanical properties and freeze-thaw stability of κ-carrageenan gels. *Food Hydrocolloids*, 148, 109497.

Maalej, H, Hmidet, N, Ghorbel-Bellaaj, O & Nasri, M 2013, Purification and biochemical characterization of a detergent stable α-amylase from *Pseudomonas stutzeri* AS22. *Biotechnology and Bioprocess Engineering*, 18, 878–887.

Mao, H, Chen, Z, Li, J, Zhai, X, Li, H, Wen, Y, Wang, J & Sun, B 2021a, Structural comparisons of pyrodextrins during thermal degradation process: The role of hydrochloric acid. *Food Chemistry*, 349, 129174.

Mao, H, Li, J, Chen, Z, Yan, S, Li, H, Wen, Y & Wang, J 2021b, Molecular structure of different prepared pyrodextrins and the inhibitory effects on starch retrogradation. *Food Research International*, 143, 110305.

Marié, T, Willig, G, Teixeira, ARS, Gazaneo Barboza, E, Kotland, A, Gratia, A, Courot, E, Hubert, J, Renault, JH & Allais, F 2018, Enzymatic synthesis of resveratrol α-glycosides from β-cyclodextrin-resveratrol complex in water. *ACS Sustainable Chemistry & Engineering*, 6, 5370–5380.

McCleary, BV 2014, Modification to AOAC official methods 2009.01 and 2011.25 to allow for minor overestimation of low molecular weight soluble dietary fiber in samples containing starch. *Journal of AOAC International*, 97, 896–901.

Meissner, H & Liebl, W 1998, *Thermotoga maritima* maltosyltransferase, a novel type of maltodextrin glycosyltransferase acting on starch and malto-oligosaccharides. *European Journal of Biochemistry*, 258, 1050–1058.

Meng, X, Gangoiti, J, Bai, Y, Pijning, T, Van Leeuwen, SS & Dijkhuizen, L 2016, Structure–function relationships of family GH70 glucansucrase and 4,6-α-glucanotransferase enzymes, and their evolutionary relationships with family GH13 enzymes. *Cellular and Molecular Life Sciences*, 73, 2681–2706.

Mezaki, Y, Katsuya, Y, Kubota, M & Matsuura, Y 2001, Crystallization and structural analysis of intact maltotetraose-forming exo-amylase from *Pseudomonas stutzeri*. *Bioscience, Biotechnology, and Biochemistry*, 65, 222–225.

Miao, M & BeMiller, JN 2023, Enzymatic approaches for structuring starch to improve functionality. *Annual Review of Food Science and Technology*, 14, 271–295.

Miao, M, Jiang, B, Jin, Z & BeMiller, JN 2018, Microbial starch-converting enzymes: Recent insights and perspectives. *Comprehensive Reviews in Food Science and Food Safety*, 17, 1238–1260.

Mikami, B, Iwamoto, H, Malle, D, Yoon, HJ, Demirkan-Sarikaya, E, Mezaki, Y & Katsuya, Y 2006, Crystal structure of pullulanase: Evidence for parallel binding of oligosaccharides in the active site. *Journal of Molecular Biology*, 359, 690–707.

Mirza, O, Skov, LK, Remaud-Simeon, M, Potocki de Montalk, G, Albenne, C, Monsan, P & Gajhede, M 2001, Crystal structures of amylosucrase from *Neisseria polysaccharea* in complex with D-glucose and the active site mutant Glu328Gln in complex with the natural substrate sucrose. *Biochemistry*, 40, 9032–9039.

Mizuno, M, Tonozuka, T, Uechi, A, Ohtaki, A, Ichikawa, K, Kamitori, S, Nishikawa, A & Sakano, Y 2004, The crystal structure of *Thermoactinomyces vulgaris* R-47 α-amylase II (TVA II) complexed with transglycosylated product. *European Journal of Biochemistry*, 271, 2530–2538.

Molina, M, Moulis, C, Monties, N, Guieysse, D, Morel, S, Cioci, G & Remaud-Siméon, M 2020, A specific oligosaccharide-binding site in the alternansucrase catalytic domain mediates alternan elongation. *Journal of Biological Chemistry*, 295, 9474–9489.

Møller, MS, Henriksen, A & Svensson, B 2016, Structure and function of α-glucan debranching enzymes. *Cellular and Molecular Life Sciences*, 73, 2619–2641.

Møller, MS, Windahl, MS, Sim, L, Bøjstrup, M, Abou Hachem, M, Hindsgaul, O, Palcic, M, Svensson, B, & Henriksen, A 2015, Oligosaccharide and substrate binding in the starch debranching enzyme barley limit dextrinase. *Journal of Molecular Biology*, 427, 1263–1277.

Moon, S, Lee, H, Mathiyalagan, R, Kim, Y, Yang, D, Lee, D, Min, J, Jimenez, Z, & Yang, DC 2018, Synthesis of a novel α-glucosyl ginsenoside F1 by cyclodextrin glucanotransferase and its *in vitro* cosmetic applications. *Biomolecules*, 8, 142.

Moon, YH, Nam, SH, Kang, J, Kim, Y-M, Lee, J-H, Kang, H-K, Breton, V, Jun, W-J, Park, K-D, Kimura, A & Kim, D 2007, Enzymatic synthesis and characterization of arbutin glucosides using glucansucrase from *Leuconostoc mesenteroides* B-1299CB. *Applied Microbiology and Biotechnology*, *77*, 559–567.

Moradi, O & Panahandeh, S 2022, Fabrication of different adsorbents based on zirconium oxide, graphene oxide, and dextrin for removal of green malachite dye from aqueous solutions. *Environmental Research*, *214*, 114042.

Morais, MAB, Nin-Hill, A & Rovira, C 2023, Glycosidase mechanisms: Sugar conformations and reactivity in endo- and exo-acting enzymes. *Current Opinion in Chemical Biology*, *74*, 102282.

Morin-Crini, N, Fourmentin, S, Fenyvesi, É, Lichtfouse, E, Torri, G, Fourmentin, M & Crini, G 2021, 130 years of cyclodextrin discovery for health, food, agriculture, and the industry: A review. *Environmental Chemistry Letters*, *19*, 2581–2617.

Morishita, Y, Hasegawa, K, Matsuura, Y, Katsube, Y, Kubota, M & Sakai, S 1997, Crystal structure of a maltotetraose-forming exo-amylase from *Pseudomonas stutzeri*. *Journal of Molecular Biology*, *267*, 661–672.

Moulis, C, André, I & Remaud-Simeon, M 2016, GH13 amylosucrases and GH70 branching sucrases, atypical enzymes in their respective families. *Cellular and Molecular Life Sciences*, *73*, 2661–2679.

Naessens, M, Cerdobbel, A, Soetaert, W & Vandamme, EJ 2005, Dextran dextrinase and dextran of *Gluconobacter oxydans*. *Journal of Industrial Microbiology & Biotechnology*, *32*, 323–334.

Naik, B, Kumar, V, Goyal, SK, Dutt Tripathi, A, Mishra, S, Joakim Saris, PE, Kumar, A, Rizwanuddin, S, Kumar, V & Rustagi, S 2023, Pullulanase: Unleashing the power of enzyme with a promising future in the food industry. *Frontiers in Bioengineering and Biotechnology*, *11*, 1139611.

Nakai, H, Dilokpimol, A, Abou Hachem, M & Svensson, B 2010b, Efficient one-pot enzymatic synthesis of α-(1→4)-glucosidic disaccharides through a coupled reaction catalysed by *Lactobacillus acidophilus* NCFM maltose phosphorylase. *Carbohydrate Research*, *345*, 1061–1064.

Nakai, H, Kitaoka, M, Svensson, B & Ohtsubo, K 2013, Recent development of phosphorylases possessing large potential for oligosaccharide synthesis. *Current Opinion in Chemical Biology*, *17*, 301–309.

Nakai, H, Petersen, BO, Westphal, Y, Dilokpimol, A, Abou Hachem, M, Duus, JØ, Schols, HA & Svensson, B 2010a, Rational engineering of *Lactobacillus acidophilus* NCFM maltose phosphorylase into either trehalose or kojibiose dual specificity phosphorylase. *Protein Engineering, Design and Selection*, *23*, 781–787.

Nakamura, Y & Kainuma, K 2022, On the cluster structure of amylopectin. *Plant Molecular Biology*, *108*, 291–306.

Neves, RPP, Fernandes, PA & Ramos, MJ 2022, Role of enzyme and active site conformational dynamics in the catalysis by α-amylase explored with QM/MM molecular dynamics. *Journal of Chemical Information and Modeling*, *62*, 3638–3650.

Ngawiset, S, Ismail, A, Murakami, S, Pongsawasdi, P, Rungrotmongkol, T & Krusong, K 2023, Identification of crucial amino acid residues involved in large ring cyclodextrin synthesis by amylomaltase from *Corynebacterium glutamicum*. *Computational and Structural Biotechnology Journal*, *21*, 899–909.

Nguyen, TTH, Cho, J-Y, Seo, Y-S, Woo, H-J, Kim, H-K, Kim, GJ, Jhon, D-Y & Kim, D 2015, Production of a low calorie mandarin juice by enzymatic conversion of constituent sugars to oligosaccharides and prevention of insoluble glucan formation. *Biotechnology Letters*, *37*, 711–716.

Nielsen, MM, Bozonnet, S, Seo, EN, Mótyán, JA, Andersen, JM, Dilokpimol, A, Abou Hachem, M, Gyémánt, G, Næsted, H, Kandra, L, Sigurskjold, BW & Svensson, B 2009, Two secondary carbohydrate binding sites on the surface of barley α-amylase 1 have distinct functions and display synergy in hydrolysis of starch granules. *Biochemistry*, *48*, 7686–7697.

Nihira, T, Nakai, H & Kitaoka, M 2012, 3-O-α-D-glucopyranosyl-L-rhamnose phosphorylase from *Clostridium phytofermentans*. *Carbohydrate Research*, *350*, 94–97.

Nihira, T, Nishimoto, M, Nakai, H, Ohtsubo, K & Kitaoka, M 2014, Characterization of two α-1,3-glucoside phosphorylases from *Clostridium phytofermentans*. *Journal of Applied Glycoscience*, *61*, 59–66.

Nishimoto, T, Aga, H, Mukai, K, Hashimoto, T, Watanabe, H, Kubota, M, Fukuda, S, Kurimoto, M, & Tsujisaka, Y 2002, Purification and characterization of glucosyltransferase and glucanotransferase involved in the production of cyclic tetrasaccharide in *Bacillus globisporus* C11. *Bioscience, Biotechnology, and Biochemistry*, *66*, 1806–1818.

Nishimura, N, Tanabe, H & Yamamoto, T 2016, Isomaltodextrin, a highly branched α-glucan, increases rat colonic H_2 production as well as indigestible dextrin. *Bioscience, Biotechnology, and Biochemistry*, *80*, 554–563.

Okada, S, Yamamoto, T, Watanabe, H, Nishimoto, T, Chaen, H, Fukuda, S, Wakagi, T & Fushinobu, S 2014, Structural and mutational analysis of substrate recognition in kojibiose phosphorylase. *The FEBS Journal*, *281*, 778–786.

Otache, MA, Duru, RU, Achugasim, O & Abayeh, OJ 2021, Advances in the modification of starch via esterification for enhanced properties. *Journal of Polymers and the Environment*, *29*, 1365–1379.

Palomo, M, Kralj, S, van der Maarel, MJEC & Dijkhuizen, L 2009, The unique branching patterns of *Deinococcus* glycogen branching enzymes are determined by their N-terminal domains. *Applied and Environmental Microbiology*, *75*, 1355–1362.

Palomo, M, Pijning, T, Booiman, T, Dobruchowska, JM, van der Vlist, J, Kralj, S, Planas, A, Loos, K, Kamerling, JP, Dijkstra, BW, van der Maarel, MJEC, Dijkhuizen, L & Leemhuis, H 2011, *Thermus thermophilus* glycoside hydrolase family 57 branching enzyme: Crystal structure, mechanism of action, and products formed. *Journal of Biological Chemistry*, *286*, 3520–3530.

Pan, S, Ding, N, Ren, J, Gu, Z, Li, C, Hong, Y, Cheng, L, Holler, TP, & Li, Z 2017, Maltooligosaccharide-forming amylase: Characteristics, preparation, and application. *Biotechnology Advances*, *35*, 619–632.

Pan, S, Wang, G, Sun, C, Du, L, Qi, X & Wei, Y 2023, A novel maltooligosaccharide-forming α-amylase from *Bacillus cereus* and its application in the preparation of maltopentaose product. *Process Biochemistry*, *128*, 68–75.

Pardhi, DS, Rabadiya, KJ, Panchal, RR, Raval, VH, Joshi, RG & Rajput, KN 2023, Cyclodextrin glucanotransferase: fundamentals and biotechnological implications. *Applied Microbiology and Biotechnology*, *107*, 5899–5907.

Park, HS, Park, JT, Kang, HK, Cha, H, Kim, DS, Kim, JW & Park, KH 2007, TreX from *Sulfolobus solfataricus* ATCC 35092 displays isoamylase and 4-α-glucanotransferase activities. *Bioscience, Biotechnology and Biochemistry*, *71*, 1348–1352.

Park, I, Park, M, Yoon, N & Cha, J 2019, Comparison of the structural properties and nutritional fraction of corn starch treated with thermophilic GH13 and GH57 α-glucan branching enzymes. *Foods*, *8*, 452.

Park, J, Rho, S & Kim, Y 2018, Feasibility and characterization of the cycloamylose production from high amylose corn starch. *Cereal Chemistry*, *95*, 838–848.

Park, J, Rho, S-J & Kim, Y-R 2019, Enhancing antioxidant and anti-microbial activity of carnosic acid in rosemary (*Rosmarinus officinalis* L.) extract by complexation with cyclic glucans. *Food Chemistry*, 299, 125119.

Park, J-H, Kim, H-J, Kim, Y-H, Cha, H, Kim, Y-W, Kim, T-J, Kim, Y-R & Park, K-H 2007, The action mode of *Thermus aquaticus* YT-1 4-α-glucanotransferase and its chimeric enzymes introduced with starch-binding domain on amylose and amylopectin. *Carbohydrate Polymers*, 67, 164–173.

Park, J-T, Park, H-S, Kang, H-K, Hong, J-S, Cha, H, Woo, E-J, Kim, J-W, Kim, M-J, Boos, W, Lee, S & Park, K-H 2008, Oligomeric and functional properties of a debranching enzyme (TreX) from the archaeon *Sulfolobus solfataricus* P2. *Biocatalysis and Biotransformation*, 26, 76–85.

Park, Y-U, Jung, J-H, Seo, D-H, Jung, D-H, Kim, J-H, Seo, E-J, Baek, N-I & Park, C-S 2018, GH57 amylopullulanase from *Desulfurococcus amylolyticus* JCM 9188 can make highly branched cyclodextrin via its transglycosylation activity. *Enzyme and Microbial Technology*, 114, 15–21.

Paul, CJ, Leemhuis, H, Dobruchowska, JM, Grey, C, Önnby, L, van Leeuwen, SS, Dijkhuizen, L & Karlsson, EN 2015, A GH57 4-α-glucanotransferase of hyperthermophilic origin with potential for alkyl glycoside production. *Applied Microbiology and Biotechnology*, 99, 7101–7113.

Paul, JS, Gupta, N, Beliya, E, Tiwari, S & Jadhav, SK 2021, Aspects and recent trends in microbial α-amylase: A review. *Applied Biochemistry and Biotechnology*, 193, 2649–2698.

Pélingre, M, Koffi Teki, DS-E, El-Abid, J, Chagnault, V, Kovensky, J & Bonnet, V 2021, Ring-opening of cyclodextrins: An efficient route to pure maltohexa-, hepta-, and octaoses. *Organics*, 2, 287–305.

Peng, H, Li, R, Li, F, Zhai, L, Zhang, X, Xiao, Y & Gao, Y 2018, Extensive hydrolysis of raw rice starch by a chimeric α-amylase engineered with α-amylase (AmyP) and a starch-binding domain from *Cryptococcus* sp. S-2. *Applied Microbiology and Biotechnology*, 102, 743–750.

Periasamy, R 2020, A systematic review on the significant roles of cyclodextrins in the construction of supramolecular systems and their potential usage in various fields. *Journal of Carbohydrate Chemistry*, 39, 189–216.

Petersen, BO, Meier, S & Duus, JØ 2012, NMR assignment of structural motifs in intact β-limit dextrin and its α-amylase degradation products in situ. *Carbohydrate Research*, 359, 76–80.

Phillips, J, Venter, J-L, Atanasova, M, Wesley-Smith, J, Oosthuizen, H, Emmambux, MN, Du Toit, EL & Focke, WW 2020, Dextrin nanocomposites as matrices for solid dosage forms. *ACS Applied Materials & Interfaces*, 12, 16969–16977.

Pijning, T, Gangoiti, J, te Poele, EM, Börner, T & Dijkhuizen, L 2021, Insights into broad-specificity starch modification from the crystal structure of *Limosilactobacillus reuteri* NCC 2613 4,6-α-glucanotransferase GtfB. *Journal of Agricultural and Food Chemistry*, 69, 13235–13245.

Pijning, T, te Poele, EM, de Leeuw, TC, Guskov, A & Dijkhuizen, L 2022, Crystal structure of 4,6-α-glucanotransferase GtfC-ΔC from thermophilic *Geobacillus* 12AMOR1: Starch transglycosylation in non-permuted GH70 enzymes. *Journal of Agricultural and Food Chemistry*, 70, 15283–15295.

Pinheiro, KH, Watanabe, LS, Nixdorf, SL, Barão, CE, Pimentel, TC, Matioli, G & de Moraes, FF 2018, Cassava bagasse as a substrate to produce cyclodextrins. *Starch-Stärke*, 70, 1800073.

Pishtiyski, I & Zhekova, B 2006, Effect of different substrates and their preliminary treatment on cyclodextrin production. *World Journal of Microbiology and Biotechnology*, 22, 109–114.

Potocki-Veronese, G, Putaux, J-L, Dupeyre, D, Albenne, C, Remaud-Siméon, M, Monsan, P & Buleon, A 2005, Amylose synthesized *in vitro* by amylosucrase: Morphology, structure, and properties. *Biomacromolecules*, 6, 1000–1011.

Przylas, I, Tomoo, K, Terada, Y, Takaha, T, Fujii, K, Saenger, W & Sträter, N 2000, Crystal structure of amylomaltase from *Thermus aquaticus*, a glycosyltransferase catalysing the production of large cyclic glucans. *Journal of Molecular Biology*, 296, 873–886.

Pullicin, AJ, Ferreira, AJ, Beaudry, CM, Lim, J & Penner, MH 2018, Preparation and characterization of isolated low degree of polymerization food-grade maltooligosaccharides. *Food Chemistry*, 246, 115–120.

Punia Bangar, S, Ashogbon, AO, Singh, A, Chaudhary, V & Whiteside, WS 2022, Enzymatic modification of starch: A green approach for starch applications. *Carbohydrate Polymers*, 287, 119265.

Pycia, K, Juszczak, L, Gałkowska, D, Socha, R & Jaworska, G 2017, Maltodextrins from chemically modified starches. Production and characteristics. *Starch-Stärke*, 69, 1600199.

Qi, P, You, C & Percival Zhang, Y-H 2014, One-pot enzymatic conversion of sucrose to synthetic amylose using enzyme cascades. *ACS Catalysis*, 4, 1311–1317.

Qi, X, Tester, R, Liu, Y & Mullin, M 2013, Applications of β-limit dextrin as a matrix forming excipient for fast disintegrating buccal dosage formats. *Journal of Pharmacy & Pharmaceutical Sciences*, 15, 669.

Qian, Y-F, Yu, J-Y, Xie, J & Yang, S-P 2023, A mini-review on cold-adapted enzymes from psychrotrophic microorganisms in foods: Benefits and challenges. *Current Research in Biotechnology*, 6, 100162.

Qiu, C, Wang, J, Fan, H, Bai, Y, Tian, Y, Xu, X & Jin, Z 2018, High-efficiency production of γ-cyclodextrin using β-cyclodextrin as the donor raw material by cyclodextrin opening reactions using recombinant cyclodextrin glycosyltransferase. *Carbohydrate Polymers*, 182, 75–80.

Ray, J, Sasmal, D, Jana, S & Tripathy, T 2024, Green synthesis of dextrin-graft-*poly*(allyl amine-*co*-methacrylic acid)/reduced graphene oxide composite and its application for dye and pesticide removal from the aqueous medium. *ChemistrySelect*, 9, e202302938.

Rejzek, M, Stevenson, CE, Southard, AM, Stanley, D, Denyer, K, Smith, AM, Naldrett, MJ, Lawson, DM, & Field, RA 2011, Chemical genetics and cereal starch metabolism: Structural basis of the non-covalent and covalent inhibition of barley β-amylase. *Molecular BioSystems*. 7, 718–730.

Ribeiro, JS & Veloso, CM 2021, Microencapsulation of natural dyes with biopolymers for application in food: A review. *Food Hydrocolloids*, 112, 106374.

Rong, Y, Sillick, M & Gregson, CM 2009, Determination of dextrose equivalent value and number average molecular weight of maltodextrin by osmometry. *Journal of Food Science*, 74, C34–C40.

Roth, C, Weizenmann, N, Bexten, N, Saenger, W, Zimmermann, W, Maier, T & Sträter, N 2017, Amylose recognition and ring-size determination of amylomaltase. *Science Advances*, 3, e1601386.

Ryu, H-J, Song, Y-B, Choi, W, Yoo, S-H & Lee, B-H 2023, Macromolecular α-glucans with α-1,3/α-1,4 branching structures produced using dual glycosyltransferases: Elucidation of physicochemical and slowly digestible properties. *International Journal of Biological Macromolecules*, 242, 124921.

Ryu, J-J, Li, X, Lee, E-S, Li, D & Lee, B-H 2022, Slowly digestible property of highly branched α-limit dextrins produced by 4,6-α-glucanotransferase from *Streptococcus thermophilus* evaluated *in vitro* and *in vivo*. *Carbohydrate Polymers*, 275, 118685.

Saka, N, Iwamoto, H, Malle, D, Takahashi, N, Mizutani, K & Mikami, B 2018, Elucidation of the mechanism of interaction between *Klebsiella pneumoniae* pullulanase and cyclodextrin. *Acta Crystallographica Section D: Structural Biology*, 74, 1115–1123.

Sawada, T, Nakamura, Y, Ohdan, T, Saitoh, A, Francisco, PB, Suzuki, E, Fujita, N, Shimonaga, T, Fujiwara, S, Tsuzuki, M, Colleoni, C & Ball, S 2014, Diversity of reaction characteristics of glucan branching enzymes and the fine structure of α-glucan from various sources. *Archives of Biochemistry and Biophysics*, *562*, 9–21.

Schädle, CN, Bader-Mittermaier, S & Sanahuja, S 2022, The effect of corn dextrin on the rheological, tribological, and aroma release properties of a reduced-fat model of processed cheese spread. *Molecules*, *27*, 1864.

Schinzel, R & Nidetzky, B 1999, Bacterial α-glucan phosphorylases. *FEMS Microbiology Letters*, *171*, 73–79.

Schmidt, AK, Cottaz, S, Driguez, H & Schulz, GE 1998, Structure of cyclodextrin glycosyltransferase complexed with a derivative of its main product β-cyclodextrin. *Biochemistry*, *37*, 5909–5915.

Schöning-Stierand, K, Diedrich, K, Ehrt, C, Flachsenberg, F, Graef, J, Sieg, J, Penner, P, Poppinga, M, Ungethüm, A & Rarey, M 2022, ProteinsPlus: A comprehensive collection of web-based molecular modeling tools. *Nucleic Acids Research*, *50*, W611–W615.

Shad, M, Hussain, N, Usman, M, Akhtar, MW & Sajjad, M 2023, Exploration of computational approaches to predict the structural features and recent trends in α-amylase production for industrial applications. *Biotechnology and Bioengineering*, *120*, 2092–2116.

Shi, Q, Hou, Y, Juvonen, M, Tuomainen, P, Kajala, I, Shukla, S, Goyal, A, Maaheimo, H, et al. 2016, Optimization of isomaltooligosaccharide size distribution by acceptor reaction of *Weissella confusa* dextransucrase and characterization of novel α-(1→2)-branched isomaltooligosaccharides. *Journal of Agricultural and Food Chemistry*, *64*, 3276–3286.

Shim, Y-E, Song, Y-B, Yoo, S-H & Lee, B-H 2023, Production of highly branched α-limit dextrins with enhanced slow digestibility by various glycogen-branching enzymes. *Carbohydrate Polymers*, *310*, 120730.

Shinde, VK & Vamkudoth, KR 2022, Maltooligosaccharide forming amylases and their applications in food and pharma industry. *Journal of Food Science and Technology*, *59*, 3733–3744.

Silva, DM, Nunes, C, Pereira, I, Moreira, ASP, Domingues, MRM, Coimbra, MA & Gama, FM 2014, Structural analysis of dextrins and characterization of dextrin-based biomedical hydrogels. *Carbohydrate Polymers*, *114*, 458–466.

Sim, L, Beeren, SR, Findinier, J, Dauville, D, Ball, SG, Henriksen, A & Palcic, MM 2014, Crystal structure of the *Chlamydomonas* starch debranching enzyme isoamylase ISA1 reveals insights into the mechanism of branch trimming and complex assembly. *Journal of Biological Chemistry*, *289*, 22991–23003.

Skov, LK, Mirza, O, Sprogøe, D, Dar, I, Remaud-Simeon, M, Albenne, C, Monsan, P & Gajhede, M 2002, Oligosaccharide and sucrose complexes of amylosucrase: Structural implications for the polymerase activity. *Journal of Biological Chemistry*, *277*, 47741–47747.

Skov, LK, Mirza, O, Sprogøe, D, van der Veen, B, Remaud-Simeon, M, Albenne, C, Monsan, P & Gajhede, M 2006, Crystal structure of the Glu328Gln mutant of *Neisseria polysaccharea* amylosucrase in complex with sucrose and maltoheptaose. *Biocatalysis and Biotransformation*, *24*, 99–105.

Song, Z, Zhong, Y, Tian, W, Zhang, C, Hansen, AR, Blennow, A, Liang, W & Guo, D 2020, Structural and functional characterizations of α-amylase-treated porous popcorn starch. *Food Hydrocolloids*, *108*, 105606.

Sonnendecker, C, Melzer, S & Zimmermann, W 2019, Engineered cyclodextrin glucanotransferases from *Bacillus* sp. G-825-6 produce large-ring cyclodextrins with high specificity. *MicrobiologyOpen*, *8*, e757.

Sonnendecker, C, Thürmann, S, Przybylski, C, Zitzmann, FD, Heinke, N, Krauke, Y, Monks, K, Robitzki, AA, Belder, D & Zimmermann, W 2019, Large-ring cyclodextrins as chiral selectors for enantiomeric pharmaceuticals. *Angewandte Chemie*, *131*, 6477–6480.

Sonnendecker, C & Zimmermann, W 2019, Change of the product specificity of a cyclodextrin glucanotransferase by semi-rational mutagenesis to synthesize large-ring cyclodextrins. *Catalysts*, *9*, 242.

Sørensen, J, Hansen, EL, Larsen, D, Elmquist, MA, Buchleithner, A, Florean, L & Beeren, SR 2023, Light-controlled enzymatic synthesis of γ-CD using a recyclable azobenzene template. *Chemical Science*, *14*, 7725–7732.

Sorndech, W, Meier, S, Jansson, AM, Sagnelli, D, Hindsgaul, O, Tongta, S & Blennow, A 2015, Synergistic amylomaltase and branching enzyme catalysis to suppress cassava starch digestibility. *Carbohydrate Polymers*, *132*, 409–418.

Sorndech, W, Nakorn, KN, Tongta, S & Blennow, A 2018, Isomalto-oligosaccharides: Recent insights in production technology and their use for food and medical applications. *LWT - Food Science and Technology 95*, 135–142.

Sorndech, W, Sagnelli, D, Blennow, A & Tongta, S 2017, Combination of amylase and transferase catalysis to improve IMO compositions and productivity. *LWT - Food Science and Technology*, *79*, 479–486.

Sorndech, W, Tongta, S & Blennow, A 2018, Slowly digestible- and non-digestible α-glucans: An enzymatic approach to starch modification and nutritional effects. *Starch-Stärke*, *70*, 1700145.

Strompen, S, Miranda-Molina, A, López-Munguía, A, Castillo, E & Saab-Rincón, G 2015, Acceptor-induced modification of regioselectivity in CGTase-catalyzed glycosylations of *p*-nitrophenyl-glucopyranosides. *Carbohydrate Research*, *404*, 46–54.

Suksiri, P, Ismail, A, Sirirattanachatchawan, C, Wangpaiboon, K, Muangsin, N, Tananuwong, K & Krusong, K 2021, Enhancement of large ring cyclodextrin production using pretreated starch by glycogen debranching enzyme from *Corynebacterium glutamicum*. *International Journal of Biological Macromolecules*, *193*, 81–87.

Sumida, R, Kishishita, S, Yasuda, A, Miyata, M, Mizote, A, Yamamoto, T, Mitsuzumi, H, Aga, H, Yamamoto, K & Kawai, K 2021, A novel dextrin produced by the enzymatic reaction of 6-α-glucosyltransferase. II. Practical advantages of the novel dextrin as a food modifier. *Bioscience, Biotechnology, and Biochemistry*, *85*, 1746–1752.

Sun, Y, Cheng, L, Hong, Y, Li, Z, Li, C, Ban, X & Gu, Z 2023, Preparation and characterization of cationic hyperbranched maltodextrins as potential carrier for siRNA encapsulation. *International Journal of Biological Macromolecules*, *225*, 786–794.

Suzuki, R, Terasawa, K, Kimura, K, Fujimoto, Z, Momma, M, Kobayashi, M, Kimura, A & Funane, K 2012, Biochemical characterization of a novel cycloisomaltooligosaccharide glucanotransferase from *Paenibacillus* sp. 598K. *Biochimica et Biophysica Acta (BBA) - Proteins and Proteomics*, *1824*, 919–924.

Takagaki, R, Ishida, Y, Sadakiyo, T, Taniguchi, Y, Sakurai, T, Mitsuzumi, H, Watanabe, H, Fukuda, S, & Ushio, S 2018, Effects of isomaltodextrin in postprandial lipid kinetics: Rat study and human randomized crossover study. *PLoS One*, *13*, e0196802.

Takaha, T & Smith, SM 1999, The functions of 4-α-glucanotransferases and their use for production of cyclic glucans. *Biotechnology and Genetic Engineering Reviews*, *16*, 257–280.

Takaha, T, Yanase, M, Takata, H, Okada, S & Smith, SM 1996, Potato D-enzyme catalyzes the cyclization of amylose to produce cycloamylose, a novel cyclic glucan. *Journal of Biological Chemistry*, *271*, 2902–2908.

Takaha, T, Yanase, M, Takata, H, Okada, S & Smith, SM 1998, Cyclic glucans produced by the intramolecular transglycosylation activity of potato D-enzyme on amylopectin. *Biochemical and Biophysical Research Communications*, *247*, 493–497.

Takata, H, Akiyama, T, Kajiura, H, Kakutani, R, Furuyashiki, T, Tomioka, E, Kojima, I & Kuriki, T 2010, Application of branching enzyme in starch processing. *Biocatalysis and Biotransformation*, *28*, 60–63.

Takata, H, Ohdan, K, Takaha, T, Kuriki, T & Okada, S 2003, Properties of branching enzyme from hyperthermophilic bacterium, *Aquifex aeolicus*, and its potential for production of highly-branched cyclic dextrin. *Journal of Applied Glycoscience*, *50*, 15–20.

Takata, H, Takaha, T, Nakamura, H, Fujii, K, Okada, S, Takagi, M & Imanaka, T 1997, Production and some properties of a dextrin with a narrow size distribution by the cyclization reaction of branching enzyme. *Journal of Fermentation and Bioengineering*, *84*, 119–123.

Takata, H, Takaha, T, Okada, S, Hizukuri, S, Takagi, M & Imanaka, T 1996, Structure of the cyclic glucan produced from amylopectin by *Bacillus stearothermophilus* branching enzyme. *Carbohydrate Research*, *295*, 91–101.

Takemoto, Y, Izawa, H, Umegatani, Y, Yamamoto, K, Kubo, A, Yanase, M, Takaha, T & Kadokawa, J 2013, Synthesis of highly branched anionic α-glucans by thermostable phosphorylase-catalyzed α-glucuronylation. *Carbohydrate Research*, *366*, 38–44.

Takii, H, Ishihara, K, Kometani, T, Okada, S & Fushiki, T 1999, Enhancement of swimming endurance in mice by highly branched cyclic dextrin. *Bioscience, Biotechnology, and Biochemistry*, *63*, 2045–2052.

te Poele, EM, van der Hoek, SE, Chatziioannou, AC, Gerwig, GJ, Duisterwinkel, WJ, Oudhuis, LAACM, Gangoiti, J, Dijkhuizen, L, & Leemhuis, II 2021, GtfC enzyme of *Geobacillus* sp. 12AMOR1 represents a novel thermostable type of GH70 4,6-α-glucanotransferase that synthesizes a linear alternating (α1 → 6)/(α1 → 4) α-glucan and delays bread staling. *Journal of Agricultural and Food Chemistry*, *69*, 9859–9868.

Terada, Y, Fujii, K, Takaha, T & Okada, S 1999, *Thermus aquaticus* ATCC 33923 amylomaltase gene cloning and expression and enzyme characterization: Production of cycloamylose. *Applied and Environmental Microbiology*, *65*, 910–915.

Terada, Y, Yanase, M, Takata, H, Takaha, T & Okada, S 1997, Cyclodextrins are not the major cyclic α-1,4-glucans produced by the initial action of cyclodextrin glucanotransferase on amylose. *Journal of Biological Chemistry*, *272*, 15729–15733.

Tester, RF & Qi, X 2011, β-limit dextrin: Properties and applications. *Food Hydrocolloids*, *25*, 1899–1903.

Tetlow, IJ & Emes, MJ 2014, A review of starch-branching enzymes and their role in amylopectin biosynthesis. *IUBMB Life*, *66*, 546–558.

Tian, Y, Wang, Y, Zhong, Y, Møller, MS, Westh, P, Svensson, B & Blennow, A 2023, Interfacial catalysis during amylolytic degradation of starch granules: Current understanding and kinetic approaches. *Molecules*, *28*, 3799.

Tian, Y, Xu, W, Zhang, W, Zhang, T, Guang, C & Mu, W 2018, Amylosucrase as a transglucosylation tool: From molecular features to bioengineering applications. *Biotechnology Advances*, *36*, 1540–1552.

Tonozuka, T, Sakai, H, Ohta, T & Sakano, Y 1994, A convenient enzymatic synthesis of 4²-α-isomaltosylisomaltose using *Thermoactinomyces vulgaris* R-47 alpha-amylase II (TVA II). *Carbohydrate Research*, *261*, 157–162.

Totosaus, A, Godoy, IA & Ariza-Ortega, TJ 2020, Structural and mechanical properties of edible films from composite mixtures of starch, dextrin and different types of chemically modified starch. *International Journal of Polymer Analysis and Characterization*, *25*, 517–528.

Trithavisup, K, Krusong, K & Tananuwong, K 2019, In-depth study of the changes in properties and molecular structure of cassava starch during resistant dextrin preparation. *Food Chemistry*, *297*, 124996.

Tse, JY, Kadota, K, Hirata, Y, Taniguchi, M, Uchiyama, H & Tozuka, Y 2018, Characterization of matrix embedded formulations for combination spray-dried particles comprising pyrazinamide and rifampicin. *Journal of Drug Delivery Science and Technology*, *48*, 137–144.

Tsusaki, K, Watanabe, H, Nishimoto, T, Yamamoto, T, Kubota, M, Chaen, H & Fukuda, S 2009, Structure of a novel highly branched α-glucan enzymatically produced from maltodextrin. *Carbohydrate Research*, *344*, 2151–2156.

Ubiparip, Z, Beerens, K, Franceus, J, Vercauteren, R & Desmet, T 2018, Thermostable alpha-glucan phosphorylases: Characteristics and industrial applications. *Applied Microbiology and Biotechnology*, *102*, 8187–8202.

Uitdehaag, JCM, Kalk, KH, van der Veen, BA, Dijkhuizen, L & Dijkstra, BW 1999a, The cyclization mechanism of cyclodextrin glycosyltransferase (CGTase) as revealed by a γ-cyclodextrin-CGTase complex at 1.8-Å resolution. *Journal of Biological Chemistry*, *274*, 34868–34876.

Uitdehaag, JCM, Mosi, R, Kalk, KH, van der Veen, BA, Dijkhuizen, L, Withers, SG & Dijkstra, BW 1999b, X-Ray structures along the reaction pathway of cyclodextrin glycosyltransferase elucidate catalysis in the α-amylase family. *Nature Structural Biology*, *6*, 432–436.

Ulbrich, M, Scholz, F & Flöter, E 2022, Chromatographic study of high amylose corn starch genotypes: Investigation of molecular properties after specific enzymatic digestion. *Starch-Stärke*, *74*, 2100303.

Umegatani, Y, Izawa, H, Nawaji, M, Yamamoto, K, Kubo, A, Yanase, M, Takaha, T & Kadokawa, JI 2012, Enzymatic α-glucuronylation of maltooligosaccharides using α-glucuronic acid 1-phosphate as glycosyl donor catalyzed by a thermostable phosphorylase from *Aquifex aeolicus* VF5. *Carbohydrate Research*, *350*, 81–85.

Umoren, SA & Eduok, UM 2016, Application of carbohydrate polymers as corrosion inhibitors for metal substrates in different media: A review. *Carbohydrate Polymers*, *140*, 314–341.

Vargas-Campos, L, Figueroa-Cárdenas, J de D, Tochihuitl-Vázquez, D, Ramírez-Bon, R, Yáñez-Limón, JM & Pérez-Robles, JF 2023, Study of the dextrose equivalent of maltodextrins in electrospinning using an ethanol/water mixture as the electrospinning solvent. *Food Hydrocolloids*, *139*, 108498.

Vergès, A, Barbe, S, Cambon, E, Moulis, C, Tranier, S, Remaud-Siméon, M & André, I 2017, Engineering of an efficient mutant of *Neisseria polysaccharea* amylosucrase for the synthesis of controlled size maltooligosaccharides. *Carbohydrate Polymers*, *173*, 403–411.

Vester-Christensen, MB, Holck, J, Rejzek, M, Perrin, L, Tovborg, M, Svensson, B, Field, RA & Møller, MS 2023, Exploration of the transglycosylation activity of barley limit dextrinase for production of novel glycoconjugates. *Molecules*, *28*, 4111.

Vuillemin, M, Claverie, M, Brison, Y, Séverac, E, Bondy, P, Morel, S, Monsan, P, Moulis, C, & Remaud-Simeon, M 2016, Characterization of the first α-(1→3) branching sucrases of the GH70 family. *Journal of Biological Chemistry*, *291*, 7687–7702.

Vujičić-Žagar, A, Pijning, T, Kralj, S, López, CA, Eeuwema, W, Dijkhuizen, L & Dijkstra, BW 2010, Crystal structure of a 117 kDa glucansucrase fragment provides insight into evolution and product specificity of GH70 enzymes. *Proceedings of the National Academy of Sciences*, *107*, 21406–21411.

Wang, D, Zhao, M, Wang, Y, Mu, H, Sun, C, Chen, H & Sun, Q 2023, Research progress on debranched starch: Preparation, characterization, and application. *Food Reviews International*, *39*, 6887–6907.

Wang, L, Xia, Y, Su, L & Wu, J 2020, Modification of *Bacillus clarkii* γ-cyclodextrin glycosyltransferase and addition of complexing agents to increase γ-cyclodextrin production. *Journal of Agricultural and Food Chemistry*, *68*, 12079–12085.

Wang, W, Sun, Y, Peng, P, Gu, G, Du, G, Xu, L & Xiao, M 2021, Two-step enzymatic conversion of rebaudioside A into a mono-α-1,4-glucosylated rebaudioside A derivative. *Journal of Agricultural and Food Chemistry*, *69*, 2522–2530.

Wang, Y, Huang, Z, Liu, Z, Luo, S, Liu, C & Hu, X 2020, Preparation and characterization of octenyl succinate β-limit dextrin. *Carbohydrate Polymers*, *229*, 115527.

Wang, Y, Li, C, Ban, X, Gu, Z, Hong, Y, Cheng, L & Li, Z 2022, Disulfide bond engineering for enhancing the thermostability of the maltotetraose-forming amylase from *Pseudomonas saccharophila* STB07. *Foods, 11*, 1207.

Wang, Y, Pan, S, Jiang, Z, Liu, S, Feng, Y, Gu, Z, Li, C & Li, Z 2019, A novel maltooligosaccharide-forming amylase from *Bacillus stearothermophilus*. *Food Bioscience*, *30*, 100415.

Wang, Y, Tian, Y, Christensen, SJ, Blennow, A, Svensson, B & Møller, MS 2024, An enzymatic approach to quantify branching on the surface of starch granules by interfacial catalysis. *Food Hydrocolloids*, *146*, 109162.

Wang, Y, Tian, Y, Zhong, Y, Suleiman, MA, Feller, G, Westh, P, Blennow, A, Møller, MS & Svensson, B 2023, Improved hydrolysis of granular starches by a psychrophilic α-amylase starch binding domain-fusion. *Journal of Agricultural and Food Chemistry*, *71*, 9040–9050.

Wang, Z, Xin, C, Li, C, Gu, Z, Cheng, L, Hong, Y, Ban, X & Li, Z 2019, Expression and characterization of an extremely thermophilic 1,4-α-glucan branching enzyme from *Rhodothermus obamensis* STB05. *Protein Expression and Purification*, *164*, 105478.

Wangpaiboon, K, Charoenwongpaiboon, T, Klaewkla, M, Field, RA & Panpetch, P 2023, Cassava pullulanase and its synergistic debranching action with isoamylase 3 in starch catabolism. *Frontiers in Plant Science*, *14*, 1114215.

Wei, B, Cai, C, Xu, B, Jin, Z & Tian, Y 2018, Disruption and molecule degradation of waxy maize starch granules during high pressure homogenization process. *Food Chemistry*, *240*, 165–173.

Wei, B, Wang, L, Chen, S, Su, L, Tao, X, Wu, J & Xia, W 2023b, Differentiated digestion resistance and physicochemical properties of linear and α-1,2/α-1,3 branched isomaltodextrins prepared by 4,6-α-glucanotransferase and branching sucrases. *Food Research International*, *171*, 113043.

Wei, B, Wang, L, Su, L, Tao, X, Chen, S, Wu, J & Xia, W 2023a, Structural characterization of slow digestion dextrin synthesized by a combination of α-glucosidase and cyclodextrin glucosyltransferase and its prebiotic potential on the gut microbiota *in vitro*. *Food Chemistry*, *426*, 136554.

Weiss, SC, Skerra, A & Schiefner, A 2015, Structural basis for the interconversion of maltodextrins by MalQ, the amylomaltase of *Escherichia coli*. *Journal of Biological Chemistry*, *290*, 21352–21364.

Włodarczyk, M, Śliżewska, K, Barczyńska, R & Kapuśniak, J 2022, Effects of resistant dextrin from potato starch on the growth dynamics of selected co-cultured strains of gastrointestinal bacteria and the activity of fecal enzymes. *Nutrients*, *14*, 2158.

Woo, EJ, Lee, S, Cha, H, Park, JT, Yoon, SM, Song, HN & Park, KH 2008, Structural insight into the bifunctional mechanism of the glycogen-debranching enzyme TreX from the archaeon *Sulfolobus solfataricus*. *Journal of Biological Chemistry*, *283*, 28641–28648.

Wu, H, Li, X, Ji, H, Svensson, B & Bai, Y 2022, Improved production of gamma-cyclodextrin from high-concentrated starch using enzyme pretreatment under swelling condition. *Carbohydrate Polymers*, *284*, 119124.

Wu, K, Li, C, Li, Z, Li, Z, Gu, Z, Ban, X, Hong, Y, Cheng, L & Kong, H 2024, Highly-branched modification of starch: An enzymatic approach to regulating its properties. *Food Hydrocolloids*, *147*, 109433.

Wu, Y, Li, X, Jin, Z, Svensson, B & Bai, Y 2023, A practical approach to producing the single-arm linear dextrin, a chimeric glucosaccharide containing an (α-1 → 4) linked portion at the nonreducing end of an (α-1 → 6) glucochain. *Carbohydrate Polymers*, *305*, 120520.

Xia, L, Bai, Y, Mu, W, Wang, J, Xu, X & Jin, Z 2017, Efficient synthesis of glucosyl-β-cyclodextrin from maltodextrins by combined action of cyclodextrin glucosyltransferase and amyloglucosidase. *Journal of Agricultural and Food Chemistry*, *65*, 6023–6029.

Xia, W, Zhang, K, Su, L & Wu, J 2021, Microbial starch debranching enzymes: Developments and applications. *Biotechnology Advances*, *50*, 107786.

Xiang, G, Leemhuis, H & van der Maarel, MJEC 2022, Structural elements determining the transglycosylating activity of glycoside hydrolase family 57 glycogen branching enzymes. *Proteins: Structure, Function and Bioinformatics*, *90*, 155–163.

Xiao, Z, Xia, J, Zhao, Q, Niu, Y & Zhao, D 2022, Maltodextrin as wall material for microcapsules: A review. *Carbohydrate Polymers*, *298*, 120113.

Xie, A-J, Lee, D-J & Lim, S-T 2021, Characterization of resistant waxy maize dextrins prepared by simultaneous debranching and crystallization followed by acidic or enzymatic hydrolysis. *Food Hydrocolloids*, *121*, 106942.

Xie, T, Hou, Y, Li, D, Yue, Y, Qian, S & Chao, Y 2014a, Structural basis of a mutant Y195I α-cyclodextrin glycosyltransferase with switched product specificity from α-cyclodextrin to β-/γ-cyclodextrin. *Journal of Biotechnology*, *182–183*, 92–96.

Xie, T, Song, B, Yue, Y, Chao, Y & Qian, S 2014b, Site-saturation mutagenesis of central tyrosine 195 leading to diverse product specificities of an α-cyclodextrin glycosyltransferase from *Paenibacillus* sp. 602-1. *Journal of Biotechnology*, *170*, 10–16.

Xie, X, Ban, X, Gu, Z, Li, C, Hong, Y, Cheng, L & Li, Z 2020, Structure-based engineering of a maltooligosaccharide-forming amylase to enhance product specificity. *Journal of Agricultural and Food Chemistry*, *68*, 838–844.

Xie, X, Li, Y, Ban, X, Zhang, Z, Gu, Z, Li, C, Hong, Y, Cheng, L, et al. 2019a, Crystal structure of a maltooligosaccharide-forming amylase from *Bacillus stearothermophilus* STB04. *International Journal of Biological Macromolecules*, *138*, 394–402.

Xie, X, Qiu, G, Zhang, Z, Ban, X, Gu, Z, Li, C, Hong, Y, Cheng, L & Li, Z 2019b, Importance of Trp139 in the product specificity of a maltooligosaccharide-forming amylase from *Bacillus stearothermophilus* STB04. *Applied Microbiology and Biotechnology*, *103*, 9433–9442.

Xu, P, Zhang, S-Y, Luo, Z-G, Zong, M-H, Li, X-X & Lou, W-Y 2021, Biotechnology and bioengineering of pullulanase: State of the art and perspectives. *World Journal of Microbiology and Biotechnology*, *37*, 43.

Xue, N, Svensson, B & Bai, Y 2022, Structure, function and enzymatic synthesis of glucosaccharides assembled mainly by α1 → 6 linkages: A review. *Carbohydrate Polymers*, *275*, 118705.

Xue, N, Wang, Y, Li, X & Bai, Y 2022, Enzymatic synthesis, structure of isomalto/malto-polysaccharides from linear dextrins prepared by retrogradation. *Carbohydrate Polymers*, *288*, 119350.

Yamamoto, K, Yoshikawa, K & Okada, S 1993, Structure of dextran synthesized by dextrin dextranase from *Acetobacter capsulatus* ATCC 11894. *Bioscience, Biotechnology, and Biochemistry*, *57*, 1450–1453.

Yang, Q, Guo, Y, Jiang, Y & Yang, B 2023, Structure identification of the oligosaccharides by UPLC-MS/MS. *Food Hydrocolloids*, *139*, 108558.

Yang, SJ, Lee, HS, Kim, JW, Lee, MH, Auh, JH, Lee, BH & Park, KH 2006, Enzymatic preparation of maltohexaose, maltoheptaose, and maltooctaose by the preferential cyclomaltooligosaccharide (cyclodextrin) ring-opening reaction of *Pyrococcus furiosus* thermostable amylase. *Carbohydrate Research*, *341*, 420–424.

Yang, T, Zhong, L, Jiang, G, Liu, L, Wang, P, Zhong, Y, Yue, Q, Ouyang, L, et al. 2022, Comparative study on bread quality and starch digestibility of normal and waxy wheat (*Triticum aestivum* L.) modified by maltohexaose producing α-amylases. *Food Research International*, *162*, 112034.

Yang, W, Su, L, Wang, L, Wu, J & Chen, S 2022, Alpha-glucanotransferase from the glycoside hydrolase family synthesizes α(1–6)-linked products from starch: Features and synthesis pathways of the products. *Trends in Food Science & Technology*, *128*, 160–172.

Yang, Y, Sun, Y, Zhang, T, Hamaker, BR & Miao, M 2023, Biofabrication, structure, and functional characteristics of a reuteran-like glucan with low digestibility. *Carbohydrate Polymers*, *305*, 120447.

Yim, DK, Park, YH & Park, YH 1997, Production of branched cyclodextrins by reverse reaction of microbial debranching enzymes. *Starch-Stärke*, *49*, 75–78.

Yoshioka, Y, Hasegawa, K, Matsuura, Y, Katsube, Y & Kubota, M 1997, Crystal structures of a mutant maltotetraose-forming exo-amylase cocrystallized with maltopentaose. *Journal of Molecular Biology*, *271*, 619–628.

You, C, Chen, H, Myung, S, Sathitsuksanoh, N, Ma, H, Zhang, X-Z, Li, J & Percival Zhang, Y-H 2013, Transformation of nonfood biomass to starch. *PNAS*, *110*, 7182–7187.

You, Y, Li, Y, Tao, J, Li, C, Gu, Z, Ban, X, Kong, H, Xia, H, Tong, Y & Li, Z 2023, Remarkable improvement in the storage stability of maltodextrin through 1,4-α-glucan branching enzyme modification. *Food Hydrocolloids*, *141*, 108696.

Yu, B, Bai, Y, Wu, Y & Jin, Z 2015, Preparation and identification of 6²–α-maltotriosyl-maltotriose using a commercial pullulanase. *International Journal of Food Properties*, *18*, 186–193.

Yu, S, Dong, K, Pora, BLR & Hasjim, J 2022, The roles of a native starch and a resistant dextrin in texture improvement and low glycemic index of biscuits. *Processes*, *10*, 2404.

Yue, Y, Song, B, Xie, T, Sun, Y, Chao, Y & Qian, S 2014, Enhancement of α-cyclodextrin product specificity by enriching histidines of α-cyclodextrin glucanotransferase at remote subsite −6. *Process Biochemistry*, *49*, 230–236.

Zaborowska, M & Bernat, K 2023, The development of recycling methods for bio-based materials: A challenge in the implementation of a circular economy: A review. *Waste Management & Research*, *41*, 68–80.

Zeng, M, van Pijkeren, JP & Pan, X 2023, Gluco-oligosaccharides as potential prebiotics: Synthesis, purification, structural characterization, and evaluation of prebiotic effect. *Comprehensive Reviews in Food Science and Food Safety*, *22*, 2611–2651.

Zhang, J, Li, L, Zhang, T & Zhong, J 2022, Characterization of a novel type of glycogen-degrading amylopullulanase from *Lactobacillus crispatus*. *Applied Microbiology and Biotechnology*, *106*, 4053–4064.

Zhang, J & Zeng, R 2011, Molecular cloning and expression of an extracellular α-amylase gene from an Antarctic deep sea psychrotolerant *Pseudomonas stutzeri* strain 7193. *World Journal of Microbiology and Biotechnology*, *27*, 841–850.

Zhang, L, Li, Z, Qiao, Y, Zhang, Y, Zheng, W, Zhao, Y, Huang, Y & Cui, Z 2019, Improvement of the quality and shelf life of wheat bread by a maltohexaose producing α-amylase. *Journal of Cereal Science*, *87*, 165–171.

Zhang, X, Leemhuis, H & van der Maarel, MJEC 2019, Synthesis of highly branched α-glucans with different structures using GH13 and GH57 glycogen branching enzymes. *Carbohydrate Polymers*, *216*, 231–237.

Zhang, X, Leemhuis, H & van der Maarel, MJEC 2020, Digestion kinetics of low, intermediate and highly branched maltodextrins produced from gelatinized starches with various microbial glycogen branching enzymes. *Carbohydrate Polymers*, *247*, 116729.

Zhang, Z, Jin, T, Xie, X, Ban, X, Li, C, Hong, Y, Cheng, L, Gu, Z, et al. 2020, Structure of maltotetraose-forming amylase from *Pseudomonas saccharophila* STB07 provides insights into its product specificity. *International Journal of Biological Macromolecules*, *154*, 1303–1313.

Zhao, F, Li, Y, Li, C, Ban, X, Gu, Z & Li, Z 2022, Exo-type, endo-type and debranching amylolytic enzymes regulate breadmaking and storage qualities of gluten-free bread. *Carbohydrate Polymers*, *298*, 120124.

Zhekova, B, Dobrev, G, Stanchev, V & Pishtiyski, I 2009, Approaches for yield increase of β-cyclodextrin formed by cyclodextrin glucanotransferase from *Bacillus megaterium*. *World Journal of Microbiology and Biotechnology*, *25*, 1043–1049.

Zhen, Y, Zhang, T, Jiang, B & Chen, J 2021, Purification and characterization of resistant dextrin. *Foods*, *10*, 185.

Zheng, L, Li, M, Jiang, B, Chen, J & Zhang, T 2022, Deletion of α-amylase genes via CRISPR/Cas9 decreases the side effects of hydrolysis towards nonreducing maltoheptaose preparation. *Food Bioscience*, *48*, 101801.

Zhong, Y, Xu, J, Liu, X, Ding, L, Svensson, B, Herburger, K, Guo, K, Pang, C, & Blennow, A 2022, Recent advances in enzyme biotechnology on modifying gelatinized and granular starch. *Trends in Food Science & Technology*, *123*, 343–354.

Zhou, W, You, C, Ma, H, Ma, Y & Zhang, Y-HP 2016, One-pot biosynthesis of high-concentration α-glucose 1-phosphate from starch by sequential addition of three hyperthermophilic enzymes. *Journal of Agricultural and Food Chemistry*, *64*, 1777–1783.

Zhoukun, L, Wenwen, Z, Lei, Z, Yanxin, W, Yajuan, Z, Yan, Q, Xue, L, Yan, H, & Zhongli, C 2019, Gene expression and biochemical characterization of a GH77 4-α-glucanotransferase CcGtase from *Corallococcus* sp. EGB. *Starch-Stärke*, *71*, 1800254.

Zhu, Y, Yuan, C, Cui, B, Guo, L & Zhao, M 2023, Pickering emulsion stabilized by linear dextrins: Effect of the chain length. *Food Hydrocolloids*, *136*, 108298.

Starch Inclusion Complexes

15

Yongfeng Ai

15.1 INTRODUCTION

Starch is a biopolymer found in abundance in nature. Almost all the starch used in the industry is extracted from higher plants, which synthesize starch in a semi-crystalline granular form mainly for energy reserve. Starch is composed of two glycans of α-D-glucopyranose: essentially linear amylose and highly branched amylopectin (Chapters 6 and 7). Granule size, organization of amylose and amylopectin within granules, amylose content, and branch chain-length distribution of amylopectin are important structural features that determine the functional properties and digestibility of starch (Chapters 9, 10 and 20) (Ai & Jane 2024; Junejo et al. 2022; Ren et al. 2021). In addition to amylose and amylopectin, minor components – either endogenous (e.g., lipids and proteins) or exogenous (e.g., iodine, alcohols, emulsifiers, and phenolic compounds) – can interact with amylose and/or amylopectin to notably influence the functional attributes and digestibility of starch. The most extensively studied interaction is starch-guest inclusion complexation, in which amylose and long branch chains of amylopectin undergo a conformational change to develop a single, left-handed helix, as shown in Figure 15.1. The complexing agent is situated in the hydrophobic central cavity of the helix. The single-helical complex is primarily stabilized by hydrophobic interactions and van der Waals forces (Putseys, Lamberts, & Delcour 2010; Shi et al. 2021). The structure, physicochemical properties, and digestibility of such inclusion complexes are determined by the structures of both starch chain and complexing agent as well as the applied complexation conditions (Di Marco, Ixtaina, & Tom 2022; Putseys, Lamberts, & Delcour 2010; Tan & Kong 2020).

Starch inclusion complexes can be found in some native starches (e.g., normal and high-amylose maize, rice, and oat) in the form of amylose-lipid complexes (ALC) (Morrison, Law, & Snape 1993; Tan et al. 2007), or they can be generated during processing (Di Marco, Ixtaina, & Tom 2022; Hasjim et al. 2010; Tan & Kong 2020). Complexation between starch and polyiodide ions is utilized to: (1) measure amylose content of starch, either as isolated starch or in flour (Bates, French, & Rundle 1943; Chrastil 1987; Jane et al. 1999; Li, Li et al. 2021); and (2) determine the extent of starch complexation with other ligands (sometimes referred to as "complexing index" in the

literature) by quantitating the iodine binding capacity of the sample (Sun et al. 2021; Tang & Copeland 2007). The interactions between starch and complexing ligands can be used to modify the physicochemical properties of starch (Ai, Hasjim, & Jane 2013; Chen et al. 2023; Gelders, Goesaert, & Delcour 2006). Nutritionally, the complex formation with lipids or phenolic compounds is an effective approach to reducing starch digestibility, and the developed ALC is defined as type 5 resistant starch (RS5) (Ai, Hasjim, & Jane 2013; Hasjim et al. 2010; Tan & Kong 2020). Moreover, the inclusion complex formation with starch is applied to encapsulate target complexing compounds, which can increase the solubility and dissolution rate, enhance the stability, control the release profile, and/or improve the bioavailability of the compounds (Cohen et al. 2011; Di Marco, Ixtaina, & Tom 2022; Lesmes, Barchechath, & Shimoni 2008; Li et al. 2019). This book chapter focuses on the structural characteristics, physicochemical properties, digestibility, and applications of starch inclusion complexed and discusses their interrelationships.

15.2 STRUCTURE OF STARCH INCLUSION COMPLEXES

Because of the more linear structure and longer chain, amylose has stronger complexing ability than amylopectin. The size of amylose inclusion helix is determined by the cross-section size of the complexing agent: amylose develops an inclusion complex of 6 glucose units per turn with linear-chain ligands, such as free fatty acids (FFAs), polyiodide ions, and *n*-butanol (Figure 15.1) (Rundle & Edwards 1943; Shi et al. 2021; Takeo, Tokumura, & Kuge 1973); amylose develops an inclusion complex of 7 glucose units per turn with branched-chain or cyclic compounds having larger cross sections, such as *iso*-butanol, *tert*-butanol, menthone, thymol, and 1,1,2,2-tetrachloroethane (Shi et al. 2021; Yamashita & Hirai 1966; Zaslow 1963); and amylose develops an inclusion complex of 8 glucose units per turn with cyclic or bicyclic compounds having even larger cross sections, such as α-naphthol and salicylic acid (Shi et al. 2019; Shi et al. 2021; Uchino et al. 2002; Yamashita & Monobe 1971). The amylose helices of six, seven, and eightfold show different

DOI: 10.1201/9781003464396-15

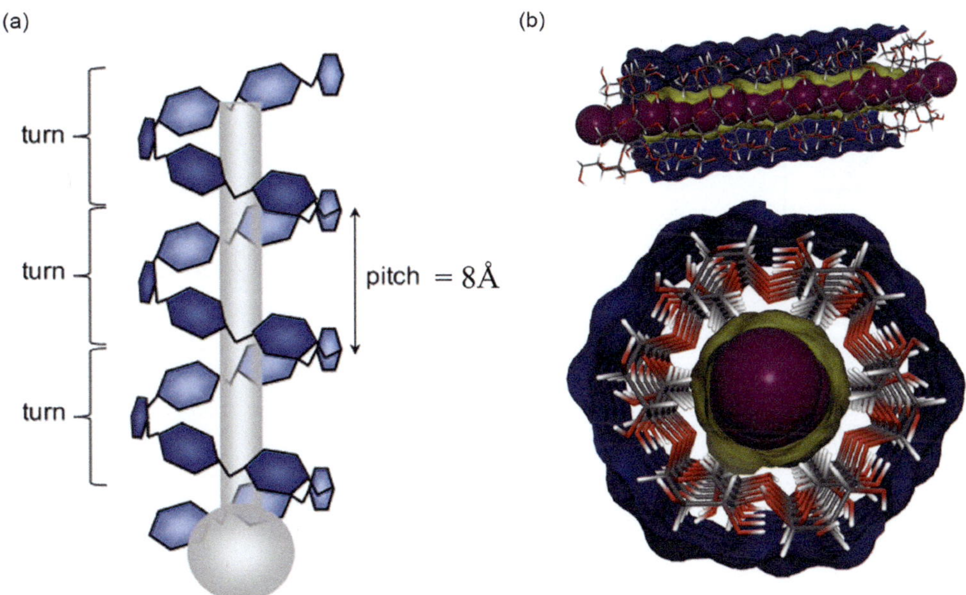

FIGURE 15.1   (a) Schematic illustration of left-handed single-helical complex between amylose and lipid. The lipid molecule has the whole hydrocarbon chain situated in the hydrophobic central cavity of the helix and the polar head located outside the helix. Adapted from Carlson et al. (1979) and Putseys, Lamberts, and Delcour (2010). (b) Schematic illustration of left-handed single-helical complex between amylose and polyiodide ion (shown as purple space-filling spheres). Atoms are depicted with CPK coloring, where blue and yellow represent hydrophilic and hydrophobic surfaces, respectively. Images were adapted from Mottiar and Altosaar (2011) and built based on structural models from Immel and Lichtenthaler (2000).

V-type crystalline structures (known as V_6, V_7, and V_8) (Shi et al. 2021). In addition, the presence of water (i.e., "hydrate form") or complexing ligands between helices affects the cell size and packing pattern of amylose single-helical crystallites to give different X-ray diffractograms (Helbert & Chanzy 1994; Le Bail, Rondeau, & Buléon 2005; Shi et al. 2021).

After complexing with long-chain FFAs, aldehydes, and alcohols (i.e., aliphatic compounds containing different functional groups on the terminal carbon) as well as their derivatives (e.g., monoglycerides), the resultant amylose inclusion complexes are present in two separate thermodynamic states, namely Form I and II, mainly depending on the crystallization temperature: 60°C–70°C and below typically leads to Form I, whereas 80°C and above typically leads to Form II (Biliaderis 1992; Seneviratne & Biliaderis 1991; Tufvesson & Eliasson 2000; Tufvesson, Wahlgren, & Eliasson 2003a). Previous research has described the differences between these two forms: Form I is a random distribution of helical complexes developed by rapid nucleation at a low crystallization temperature, while Form II has a more ordered/crystalline structure derived from a higher crystallization temperature or from partial melting and rearrangement of Form I during hydrothermal treatment (Biliaderis, Page, & Maurice 1986; Kong, Perez-Santos, & Ziegler 2019; Tufvesson et al. 2001; Tufvesson, Wahlgren, & Eliasson 2003a). According to the literature, Form II has a higher melting temperature (roughly 15°C–25°C higher in peak temperature) and greater relative crystallinity than Form I, as demonstrated in Figure 15.2, indicating that the former is thermodynamically more stable (Tufvesson & Eliasson 2000; Tufvesson et al. 2001). Moreover, after complete melting of amylose inclusion complexes in a differential scanning calorimeter (DSC),

immediate cooling and rescan reveal the re-formation of Form I or both Form I and II complexes, depending on the complexing agents and the heating and cooling conditions (Tufvesson & Eliasson 2000; Tufvesson, Wahlgren, & Eliasson 2003a, 2003b). The findings indicate that the formation of amylose inclusion complexes is an instant and reversible process.

Nevertheless, not all complexing ligands can develop Form II complexes based on prior research. For example, FFAs with a chain length ≤ 10 carbons are unable to generate Form II complexes with potato amylose even after prolonged heat treatment at 100°C for 24 hours, suggesting that Form I complexes with short-chain FFAs are unfavorable for lamellar stacking into Form II structure (Tufvesson, Wahlgren, & Eliasson 2003b). In another study, the strong complexing agent, cetyltrimethylammonium bromide, only develops Form I complexes with amylose in potato starch even after extended heat treatment at 100°C for 24 hours (Tufvesson & Eliasson 2000). Because cetyltrimethylammonium bromide carries one negative charge, the resulting Form I complexes electronically repel each other, which is detrimental for further development of Form II structure.

15.2.1 Influence of Complexing Agent Structure on Starch Inclusion Complexes

When amylose complexes with FFAs, aldehydes, alcohols, and monoglycerides, the melting temperatures of the products increase with a longer hydrocarbon chain of the complexing ligands, indicating enhanced thermal stability (Ai, Hasjim, &

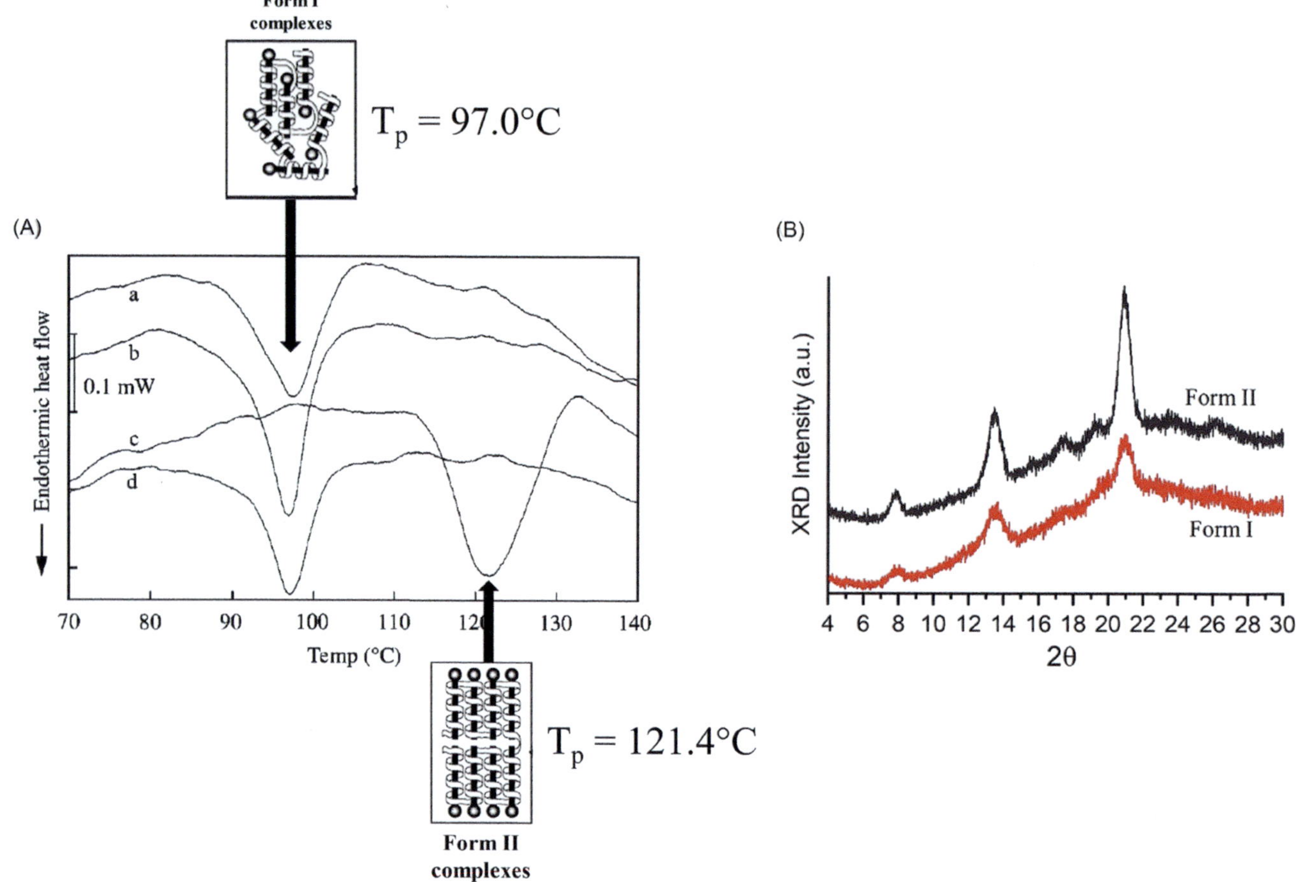

FIGURE 15.2 (A) Differential scanning calorimetry (DSC) thermograms of potato starch-glycerol monopalmitin complexes prepared after gelatinization and subsequent hydrothermal treatments. (a) Form I complexes with melting peak temperature (T_p) of 97.0°C; (b) rescan of sample a; (c) Form II complexes with T_p of 121.4°C; and (d) rescan of sample c. Adapted from Tufvesson et al. (2001). (B) X-ray diffractograms of amylose-decanol complexes in Form I and II. Adapted from Kong, Perez-Santos, and Ziegler (2019).

Jane 2013; Tufvesson, Wahlgren, & Eliasson 2003a, 2003b). It is important to note that complexing ligands with a long hydrocarbon chain tend to self-assemble into an ordered array (i.e., crystallites), rather than complexing with starch chains. Thus, proper dispersion of complexing ligands to disrupt the ordered array and separate them into individual molecules is critical for their complexation with starch (Kong, Perez-Santos, & Ziegler 2019; Tapanapunnitkul et al. 2008; Tufvesson, Wahlgren, & Eliasson 2003b). By contrast, the presence of double bond in the hydrocarbon chain of the abovementioned complexing ligands noticeably lowers the melting temperatures of the obtained complexes because the bent-chain structure from the double-bond kink is less desirable for the single-helical complex formation than a straight-chain structure without double bond (Ai, Hasjim, & Jane 2013; Tufvesson, Wahlgren, & Eliasson 2003a, 2003b). The observation of the thermal stability of starch inclusion complexes increasing with the chain length but decreasing with the degree of unsaturation of the complexing agent is valid for both Form I and II (Tufvesson, Wahlgren, & Eliasson 2003b).

Phenolic compounds, a group of substances sharing the base structure of a six-carbon aromatic ring with one or more covalently bound hydroxyl groups, have also received

extensive research attention as a distinct group of complexing agents with starch. There is limited evidence of V-type complexation with genistein, naringenin, gallic acid, and caffeic acid under the applied test conditions in different studies (Cohen et al. 2008; Cohen et al. 2011; He et al. 2024; Li, Tian et al. 2021; Liu, Chen et al. 2019). However, other studies do not provide evidence for such V-type complexation with ferulic acid (Karunaratne & Zhu 2016), anthocyanins (Miao et al. 2021), or proanthocyanidins (Xiao et al. 2021). Main factors restricting the development of V-type inclusion complexes with phenolic compounds include their bulky size and lack of hydrophobicity (Zhu 2015). Thus, non-inclusion complexes between starch and phenolic compounds have been proposed, which are driven and stabilized by hydrogen bonds and van der Waals forces (Bordenave, Hamaker, & Ferruzzi 2014; Chai, Wang, & Zhang 2013; Zhao et al. 2019). Other complexation patterns between starch and phenolic compounds have also been suggested based on the data from specific complexing and characterization conditions (Amoako & Awika 2019; Li, Ndiaye et al. 2020). Further research is needed to elucidate the structural characteristics of such complexes under a broad range of processing conditions.

15.2.2 Influence of Amylose Structure on Starch Inclusion Complexes

Godet, Bizot, and Buleon (1995) have prepared highly crystalline amylose-FFA complexes from five amylose fractions, which have number-average degrees of polymerization (DP) of 20, 30, 40, 80, and 900, respectively, with three FFAs: caprylic, lauric, and palmitic acids. Complexation and crystallization are carried out in water-dimethyl sulfoxide mixture at 90°C. Yields and melting temperatures of the derived amylose-FFA complexes largely increase with longer amylose chains, indicating the importance of amylose chain length for the crystallization of single-helical complexes. Moreover, short amylose of DP 20 remains in the solution and fails to form precipitate with any of the three FFAs. Similar findings have been reported in another study (Gelders et al. 2004). Amylose fractions of DP 20, DP 60, DP 400, and DP 950 (peak reading) are subjected to complexation with docosanoic acid or glycerol monostearate in dimethyl sulfoxide-water mixture (1:15) at 60°C or 90°C. Yields, relative crystallinity, and melting temperatures and enthalpy changes of the resulting ALC generally increase with amylose chain lengths from DP 20 to DP 400. However, relative crystallinity and thermal stability of ALC with DP 950 amylose fraction are marginally lower than those with DP 400, which is attributed to the longest amylose fraction increasing conformational disorders and causing crystal defaults in ALC. Furthermore, molecular-weight distributions of the prepared ALC show that the minimum DP for complexation and precipitation with docosanoic acid and glycerol monostearate are 40 and 35, respectively, regardless of the complexation temperatures (Gelders et al. 2004).

15.2.3 Influence of Amylopectin Structure on Starch Inclusion Complexes

Compared with essentially linear amylose, highly branched amylopectin has substantially poorer complexing ability with different compounds. The main limiting factors include the steric hindrance from the highly branched structure and the short branch chains of amylopectin. It is reasonable to infer that outer long branch chains of amylopectin can develop inclusion complexes with hydrophobic agents, resembling the structure of amylose inclusion complexes (Gelders, Goesaert, & Delcour 2006). Prior research has provided evidence that sodium dodecyl sulfate, cetyltrimethylammonium bromide, and retinyl palmitate can form V-type complexes with amylopectin under the applied conditions according to the characterization results using DSC and/or wide-angle X-ray diffraction (WAXD) (Gudmundsson 1992; Gudmundsson & Eliasson 1990; Ma, Floros, & Ziegler 2011). On the contrary, other studies have reported that amylopectin cannot develop V-type complexes with FFAs, monoglycerides, or lecithin (Ai, Hasjim, & Jane 2013; Chen et al. 2023; Liu, Kang et al. 2019).

Enzymatic debranching of amylopectin using either pullulanase or isoamylase releases an increased proportion of linear α-1,4-glucan chains to enhance the complexing ability (Liu, Kang et al. 2019). Interestingly, amylosucrase has been employed to extend the branch chain length of waxy maize starch (>99% amylopectin) dispersed in water, which can then effectively form single-helical complexes with free lauric, myristic, and palmitic acids, as confirmed by the measurement of complexing index, DSC, and/or WAXD (Kim et al. 2017; Lim et al. 2019; Zhang et al. 2023). In contrast, the native counterpart cannot generate such complexes with the same FFAs. The data from the three studies also support that outer long branch chains of native amylopectin can develop V-type complexes with ligands. The absence of evidence for the formation of V-type complexes in some studies can be explained by the low concentration, which is below the detection limit of DSC and WAXD (Garcia et al. 2016; Liu, Kang et al. 2019).

15.2.4 Influence of Starch Modifications on Starch Inclusion Complexes

Previous research has investigated the effects of combined chemical and enzymatic modifications of starch on the complexation with different FFAs (Arijaje & Wang 2017, 2015; Arijaje et al. 2014). After chemical modification (e.g., acetylation and hydroxypropylation), enzymatic debranching by isoamylase and pullulanase, and optional β-amylase treatment, the starch is complexed with FFAs under desired conditions. In the subsequent step, the authors separate the complexes into insoluble and soluble fractions through centrifugation at 7,000 g for 10 minutes: the insoluble complexes are collected as the precipitate, and the soluble complexes in the supernatant are recovered by precipitating with a fourfold volume of anhydrous ethanol. The influence of the sequential modifications on the complex formation is mainly dependent on the added FFA type. For example, Arijaje and Wang (2017) have reported that acetylation of potato starch – imparting a degree of substitution (DS) of 0.045 and 0.091 for low- and high-acetylated starches, respectively – improves the complexation with linoleic acid in the insoluble complexes but shows little improvement in the formation of the soluble complexes; in another study, Arijaje and Wang (2015) have demonstrated that acetylation of potato starch [DS = 0.041 (low) and 0.091 (high)] enhances the formation of both soluble and insoluble starch-oleic acid complexes as compared to its unacetylated counterpart; and in a third study, Arijaje et al. (2014) have illustrated that low acetylation of potato starch (DS = 0.041) promotes the complexation with stearic acid in insoluble complexes, whereas high acetylation (DS = 0.078) increases the formation of soluble complexes. The chemical modifications tend to reduce the melting temperatures of both soluble and insoluble complexes. Furthermore, controlled β-amylolysis to manipulate the starch chain length is able to further enhance the complex formation. Consequently, the authors have concluded that a combination of optimum DS of acetylation or hydroxy-propylation and a favored DP range of starch is important for maximizing its complexation with FFAs. In two separate studies, octenyl succinic anhydride modification alone improves the emulsifying effect of starch, which favors the dispersion of α-lipoic acid and lipids in water to promote the complexation with starch (Li, Park, & Lim 2018; Wang et al. 2021). The utilization of chemical and/or enzymatic modifications of starch to manipulate its complexation with various bioactive compounds can be used to encapsulate them for enhanced stability and bioavailability (Gonzalez et al. 2018; Li, Park, & Lim 2018; Wang et al. 2021).

15.3 STARCH INCLUSION COMPLEXES ON STARCH FUNCTIONAL PROPERTIES

15.3.1 Gelatinization Properties

According to the literature, simple addition of common food lipids, including soy lecithin, myristic acid, palmitic acid, stearic acid, oleic acid, and linoleic acid, does not obviously alter the gelatinization properties of starch (Ai, Hasjim, &

Jane 2013; Chen et al. 2023; Zhou et al. 2007). When sodium dodecyl sulfate – a strong complexing agent – is added to native tapioca, wheat, and potato starches at concentrations ranging from 0.6% to 17.3% (w/w), the complex formation decreases the starch gelatinization enthalpy changes, which is attributed to its exothermic effect (Radhika & Moorthy 2008; Svensson, Autio, & Eliasson 1998). For native high-amylose maize and barley starches, the DSC thermograms show a broad endothermic transition: Peak I mainly corresponds to the melting of amylopectin crystallites, and Peak II represents two endothermic events – melting of long-chain double-helical crystallites of amylose/intermediate component (IC) and melting of ALC (Figure 15.3a) (Jiang, Lio, et al. 2010; Liu, Reimer, & Ai 2020; Ren et al. 2023). After gelatinization and cooling, immediate rescan of high-amylose maize and barley starches shows a single endothermic peak, which corresponds to the melting of ALC and possibly the melting of some retrograded amylose chains (Figure 15.3b) (Liu, Reimer, & Ai 2020; Ren et al. 2023). Jiang, Lio, et al. (2010) have also reported that the area of ALC peak increases with kernel development of high-amylose maize, consistent with increased concentrations of amylose/IC and endogenous lipids of isolated high-amylose maize starch samples from a later stage of kernel development (Figure 15.3c). Defatting of high-amylose maize starches using methanol reduces their RS contents from 10.6%–43.4% to 9.0%–28.9%, suggesting that ALC contributes to the high RS contents of native high-amylose maize starches (more discussion in Section 15.4) (Jiang, Campbell, et al. 2010).

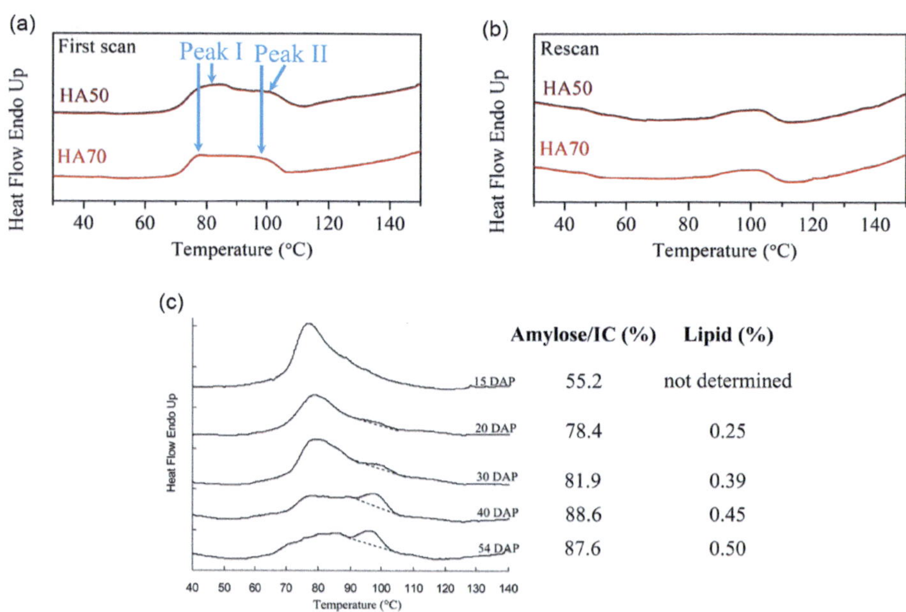

FIGURE 15.3   (a) Differential scanning calorimetry (DSC) thermograms of high-amylose maize starches: HA50 and HA70 consisting of 59.1% and 68.2% apparent amylose, respectively. Peak I mainly corresponds to the melting of amylopectin crystallites, and Peak II represents two endothermic transitions: melting of long-chain double-helical crystallites of amylose/intermediate component (IC) and melting of amylose-lipid complexes (ALC). (b) Rescan of HA50 and HA70. Sub-figures a and b were adapted from Liu, Reimer, and Ai (2020). (c) DSC thermograms of high-amylose maize starch samples from GEMS-0067 line at different developmental stages in the 2007 crop year. The peak area above the dashed line indicates the melting of ALC. DAP, days after pollination. The contents of amylose and intermediate component (amylose/IC) and lipids of the starch samples are listed in the two columns on the right-hand side. Sub-figure c was adapted from Jiang, Lio, et al. (2010).

15.3.2 Pasting Properties

After amylose and long branch chains of amylopectin complex with different agents (e.g., lipids, alcohols, and polyiodide ions), they promote physical entanglement between amylose and amylopectin molecules, thereby restricting the swelling of starch granules during heating in water (Chen & Jane 1994; Patel, Saibene, & Seetharaman 2006; Tester & Morrison 1990). The impacts of ALC on the pasting properties of starch have been most extensively investigated. Overall, ALC increases the pasting temperature, decreases the peak viscosity, and results in an opaque appearance of the obtained starch paste (Ai, Hasjim, & Jane 2013; Craig et al. 1989; Debet & Gidley 2006). Wheat starch is known to have a high concentration of endogenous phospholipids, which can easily develop inclusion complexes with amylose during pasting. Thus, wheat starch has the highest pasting temperature and the lowest peak viscosity among all the normal cereal starches (Figure 15.4a) (Debet & Gidley 2006). Upon the removal of endogenous lipids using sodium dodecyl sulfate as an effective detergent, pasting temperature and peak viscosity of wheat starch become comparable to those of tapioca starch (Figure 15.4b and c).

In a previous study, exogenous corn oil (exclusively consisting of triglycerides), soy lecithin (exclusively consisting of phospholipids), and four FFAs varying in hydrocarbon chain length and degree of unsaturation, including palmitic acid, stearic acid, oleic acid, and linoleic acid, are added to normal maize, tapioca, and waxy maize starches at a mass ratio of lipid:starch = 1:10 (dry starch basis, dsb) to examine their impacts on starch pasting properties (Figure 15.5) (Ai, Hasjim, & Jane 2013). After the addition of soy lecithin, the pasting temperature of normal maize starch decreases from 80.1°C to 72.3°C and the peak viscosity increases from 185.7 to 231.0 RVU (Figure 15.5a), which is attributable to soy lecithin removing endogenous lipids from normal maize starch, similar to the influence of washing with sodium dodecyl sulfate on the pasting properties of wheat starch, as illustrated in Figure 15.4. With the presence of soy lecithin, normal maize starch exhibits a pasting temperature and peak viscosity close to those of tapioca starch (Figure 15.5a and b).

By contrast, the addition of FFAs elevates the pasting temperature but reduces the peak viscosity (except for linoleic acid) of normal maize starch (Ai, Hasjim, & Jane 2013). Palmitic acid, linoleic acid, soy lecithin, and oleic acid increase the final viscosity of the starch to different extents. The triggering of a considerably greater final viscosity of normal maize starch at approximately 65°C by the addition of palmitic acid can be associated with the melting temperature of this FFA (62.1°C). Amylose-palmitic acid complexes in the swollen granules and free palmitic acid can solidify and become rigid after cooling to 65°C, which thus remarkably increases the final viscosity of normal maize starch-palmitic acid blend.

Different from the influence on normal maize starch, the presence of various lipids does not effectively alter the pasting temperature of tapioca starch (Figure 15.5b) (Ai, Hasjim, & Jane 2013). The noted difference is attributed to the fact that tapioca starch granules possess little endogenous lipids and swell rapidly to develop pasting viscosity. When the heating temperature reaches the pasting temperature (69.1°C) and above, dispersed amylose molecules complex with the added lipids in the aqueous medium, which reduces the peak viscosity of tapioca starch. Overall, the addition of lipids exhibits substantially less influence on the pasting properties of waxy maize starch due to the lack of amylose (Figure 15.5c).

Chen et al. (2023) have examined the influence of adding 8.0% (w/w, dsb) FFAs, including myristic acid, palmitic acid, stearic acid, oleic acid, and linoleic acid, on the functional attributes of waxy, normal, and high-amylose maize starches at 95°C–140°C heating in Rapid Visco Analyser. The majority of the FFAs markedly increase the final viscosity of normal maize starch at 95°C and 120°C heating and that of high-amylose maize starch at 120°C heating, but the improving effect disappears at 140°C heating. At 95°C heating, the presence of the lipids does not alter the pasting profile of waxy maize starch, which is consistent with the results reported in Figure 15.5c; at 120°C and 140°C heating, waxy maize starch molecules also show physical interactions with the exogenous FFAs to lower the holding strength and final viscosity, despite the absence of amylose and ALC in this waxy starch and its mixtures with different FFAs.

Because of the restrictive effect of lipids on starch viscosity development, adding FFAs (myristic, stearic, and docosanoic acids) and monoglyceride at 4% level (w/w, dsb) to maize starches with 25%, 50%, and 70% amylose has been shown to lower the expansion ratios, percentages of water-soluble carbohydrates, and water-solubility indices, but increase the bulk densities of the starch extrudates (Bhatnagar & Hanna 1994). In contrast, the addition of tristearin at the same percentage does not significantly alter those quality attributes because this triglyceride does not appear to complex with amylose according to the measurements of iodine-binding capacity and iodine spectra of the starch-tristearin extrudates. In other studies, the complexation with polyiodide ions (commonly added as I_2-KI solution) is used to stain starch granules during heating to understand their swelling behavior (Ji et al. 2022; Lin et al. 2013). The complex formation with short-chain alcohols (e.g., ethanol) inhibits starch granule swelling during gelatinization, which has been utilized to prepare granular cold-water-soluble starch (Chen & Jane 1994; Jane et al. 1986; Singh & Singh 2003). The generated pregelatinized starch provides instant and higher viscosity and has a smoother texture and greater processing tolerance than conventional drum-dried pregelatinized starch.

15.3.3 Gelling Ability

Starch hydrogel is considered as a hydrophilic polymeric network with the ability to hold a considerable amount of water or biological fluids in their porous structure. Starch hydrogel differs from starch paste in that the former has a self-standing, viscoelastic structure with no fluidity (Wang, Liu, & Ai 2022). Gelation of starch plays a vital role in the quality attributes of numerous starchy food products, such as glass noodles, puddings, and confections (Kasemsuwan, Bailey, & Jane 1998; Marfil, Anhê, & Telis 2012; Ren et al. 2021). Different from other food hydrocolloids (e.g., pectin, agar, κ-carrageenan, and

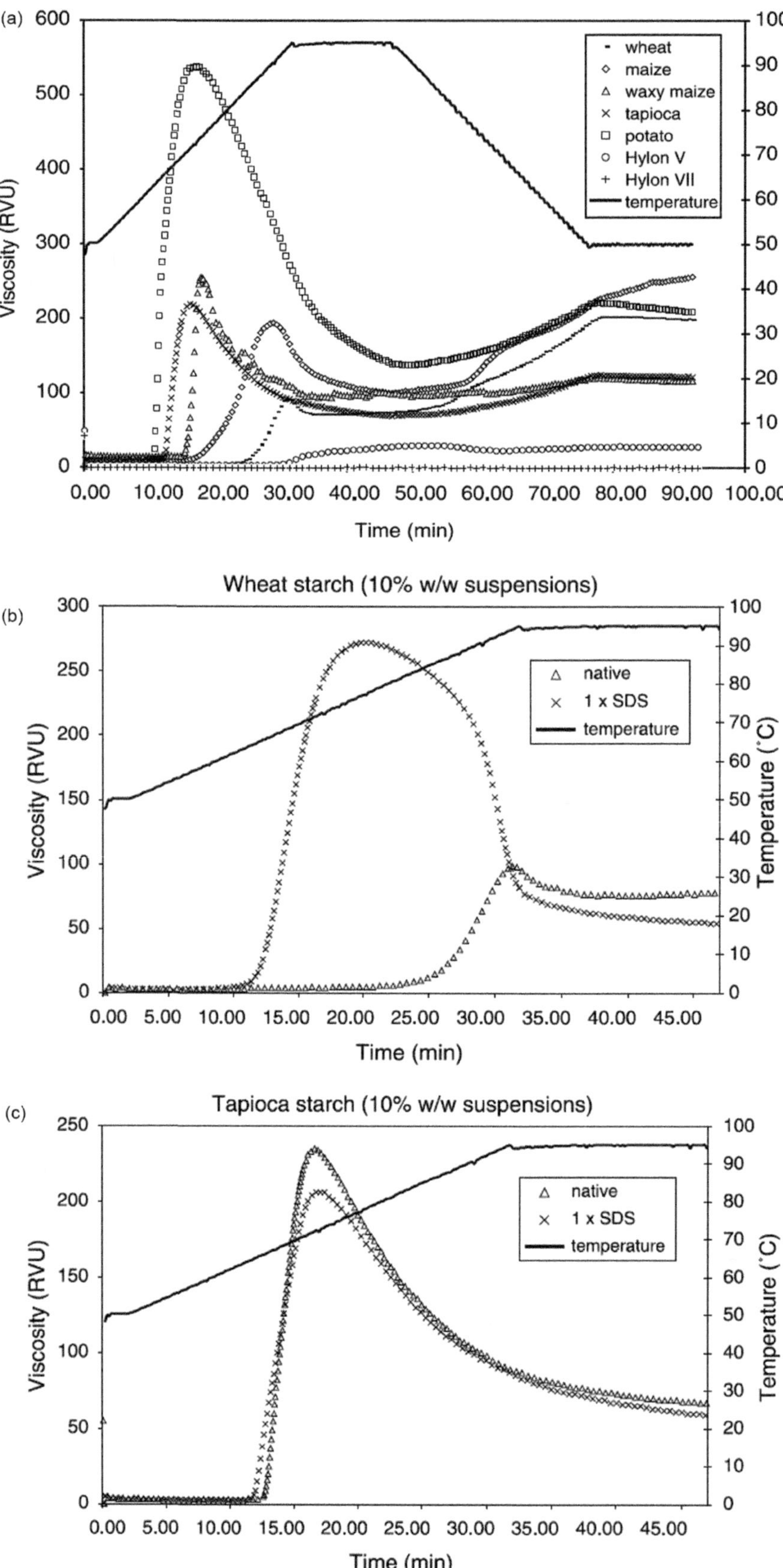

FIGURE 15.4 (a) Pasting profiles of wheat, maize, waxy maize, tapioca, potato, and high-amylose maize (Hylon V and VII) starches. Pasting profiles of wheat starch (b) and tapioca starch (c) before (native) and after washing ("1×SDS") once with 2% (w/v) sodium dodecyl sulfate (SDS) at room temperature followed by extensive water washing. The pasting profiles of different starch samples were measured by Rapid Visco Analyser at 10% (w/w) starch concentration. Adapted from Debet and Gidley (2006).

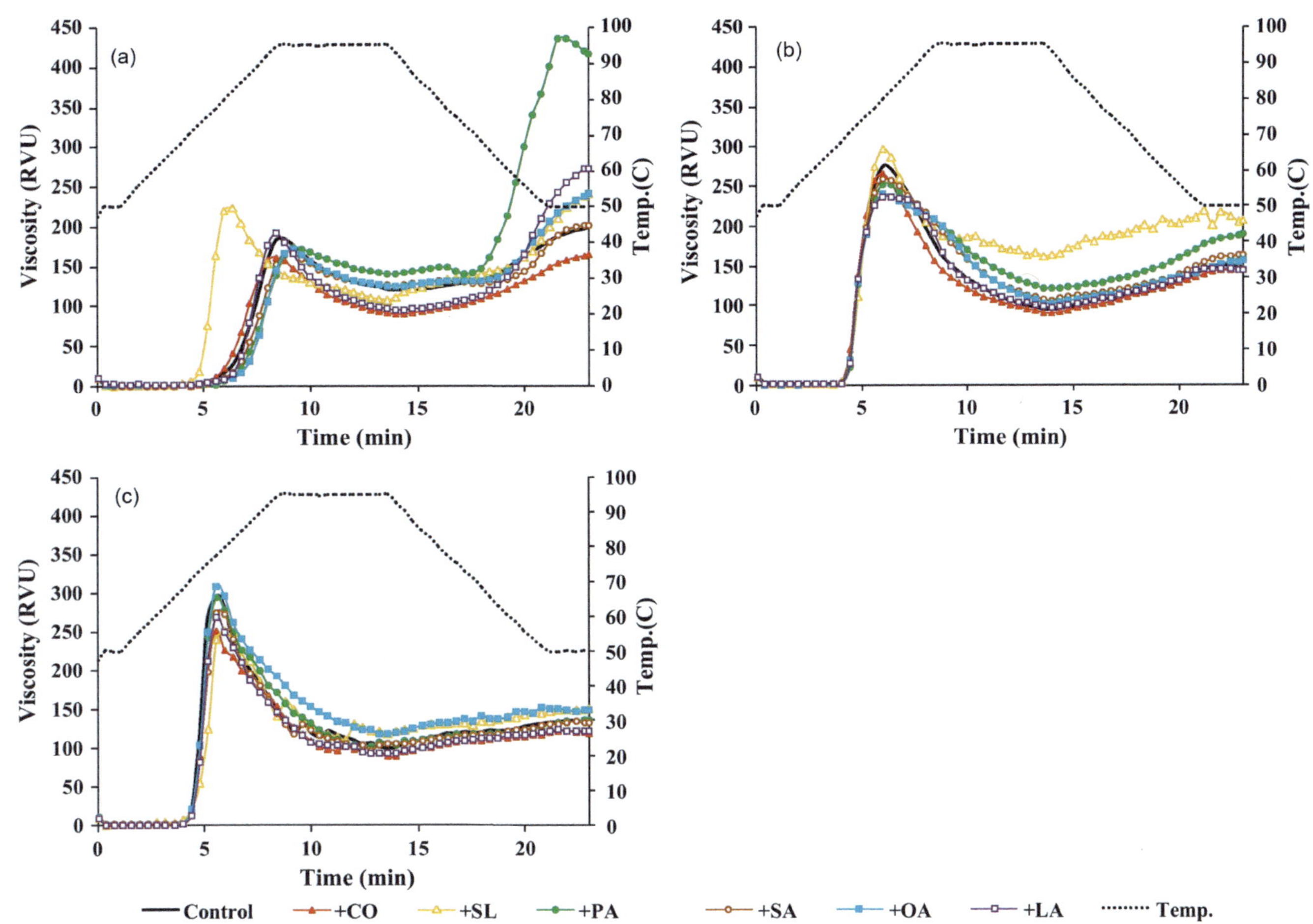

FIGURE 15.5 Pasting profiles of starches and starch-lipid blends (starch:lipid = 10:1, w/w, dsb) measured using a Rapid Visco Analyser. Starch suspensions (28.0 g total weight) with 8% starch (w/w, db) were used for the analysis. (a) Normal maize starch; (b) tapioca starch; and (c) waxy maize starch. CO, corn oil; SL, soy lecithin; PA, palmitic acid; SA, stearic acid; OA, oleic acid; LA, linoleic acid. Adapted from Ai, Hasjim, and Jane (2013).

alginate), native granular starch requires gelatinization and pasting prior to gelation. Upon cooking, pasting, and cooling, swollen starch granules/remnants with well-maintained integrity (also referred to as "starch ghost" in previous literature) are essential for the formation of a firm gel (Liu, Yuan, et al. 2019; Ring 1985; Wang, Liu, & Ai 2022). For certain cereal starches, such as normal maize and wheat, ALC plays a critical role in maintaining the integrity of such swollen starch granules for strong gel formation at a starch concentration of 5%–10% (w/w) (Liu, Yuan, et al. 2019; Takahashi & Seib 1988). On the contrary, some other native starches without ALC, such as waxy maize, waxy potato, and tapioca, swell extensively during heating (Figure 15.4a), lose the granular structure, and eventually become well dispersed in the aqueous medium, which thus fail to develop a self-standing, viscoelastic hydrogel at a concentration of 5%–10% (w/w) (Liu, Yuan, et al. 2019). Nevertheless, when exogenous complexing lipids are added to normal and high-amylose maize starches at 95°C–140°C cooking to develop extra amounts of ALC, the resulting ALC aggregate and precipitate from the aqueous medium, which tends to reduce the gel strength of these two starches (Ai, Hasjim, & Jane 2013; Byars et al. 2012; Chen et al. 2023).

Recently, normal and high-amylose maize starches are subjected to steam jet-cooking, and the obtained dispersions are blended with an aqueous solution of sodium palmitate to develop amylose inclusion complexes, which possess polyelectrolyte properties due to the anionic charge of sodium palmitate and thus prevent amylose from retrogradation in the dispersions (Byars, Fanta, & Kenar 2013; Kenar et al. 2014). Stable aqueous dispersions of these complexes are transformed into hydrogels when acid (e.g., acetic and hydrochloric) is added to partly convert the complexed sodium palmitate to palmitic acid. The linear viscoelastic moduli of the derived hydrogels increase with a higher amylose level (Byars, Fanta, & Kenar 2013). This type of hydrogel from high-amylose maize starch is further transformed to xerogel, cryogel, and aerogel through air drying, freeze drying, and solvent exchange with ethanol and subsequent supercritical carbon dioxide drying, respectively. Based on the characterization of the gels of different physical forms using Fourier transform infrared spectroscopy, WAXD, and scanning electron microscopy, these gels have molecular and microscopic structures distinctly different from those of gels prepared from native granular starches (Kenar et al. 2014; Liu, Yuan, et al. 2019). The gels generated from amylose-sodium palmitate V-type complexes are employed to stabilize silver nanoparticles (Fanta et al. 2013).

15.3.4 Retrogradation Properties

Retrogradation refers to the return of a solvated, dispersed, and amorphous state of starch to an insoluble, aggregated, or crystalline condition upon cold storage. At the molecular level, two adjacent starch chains tend to recrystallize to form double helices when gelatinized starch is cooled (Abd Karim, Norziah, & Seow 2000; Ren et al. 2021). Both amylopectin and amylose molecules can undergo retrogradation. Retrogradation is responsible for syneresis (expulsion of liquid from gel) of a starch gel during storage, staling of bread, and the loss of viscosity and precipitation formation in soups and sauces containing starch as a thickener.

When amylose and long branch chains of amylopectin develop single-helical complexes with other compounds, such as FFAs, monoglycerides, phospholipids, and surfactants, they will have a notably lower tendency to re-form double helices and thus a lesser extent of retrogradation (Eliasson & Ljunger 1988; Gudmundsson 1992; Gudmundsson & Eliasson 1990; Huang & White 1993). One good example is that monoglycerides are commonly included in the formula of bread to reduce starch retrogradation, which can improve the softness and extend the shelf life of bread (Gray & Bemiller 2003; Purhagen, Sjöö, & Eliasson 2011). Similar concepts have been applied to inhibit the staling and improve the textural properties of steamed bread (Kang et al. 2021), tortillas (Mariscal-Moreno et al. 2019), and instant rice noodles (Chen et al. 2022).

15.4 DIGESTIBILITY OF STARCH INCLUSION COMPLEXES

Starch is a main energy source in both human foods and animal feeds. After consumption, starch is hydrolyzed by amylolytic enzymes into glucose for energy supply. Long-term intake of high-glycemic foods (i.e., giving high postprandial glycemic response after consumption) has been reported to be linked to higher risks of diabetes, obesity, cardiovascular diseases, cancers, and other metabolic syndrome (Schwingshackl & Hoffmann 2013; Turati et al. 2015; Wang et al. 2015; Willett, Manson, & Liu 2002). The prevalence of these diseases has fostered intensive research interest in developing new methods to reduce starch digestibility in human foods. The inclusion complex formation with food-grade ligands, particularly lipids, has been employed for this purpose. ALC is accepted as type 5 resistant starch (RS5) according to the literature (Ai, Hasjim, & Jane 2013; Hasjim et al. 2010; Tan & Kong 2020). Single-helical complexation with lipids (Form I) and further growth of lamellar crystallites (Form II) protect amylose from enzymatic hydrolysis (Jane & Robyt 1984; Seneviratne & Biliaderis 1991). Additionally, the formed ALC inhibits

the swelling of starch granules (Section 15.3.2), thus further decreasing the susceptibility of starch molecules to amylolytic enzymes (Ai, Hasjim, & Jane 2013; Ai et al. 2014; Lauro, Poutanen, & Forssell 2000; Tester & Morrison 1990).

Cooking with soy lecithin, palmitic acid, stearic acid, oleic acid, and linoleic acid at 10% level (w/w, dsb) effectively decreases *in vitro* digestibility of normal maize, tapioca, and high-amylose maize starches due to the formation of ALC as confirmed by DSC scanning (Figure 15.6a-d) (Ai, Hasjim, & Jane 2013). It is noteworthy that the decreases in percentages of enzymatic hydrolysis of the three starches at 120 minutes after cooking with soy lecithin and FFAs are positively correlated with the onset melting temperatures of the formed ALC, suggesting that ALC with better thermal stability is more resistant to amylolysis (Ai, Hasjim, & Jane 2013; Eliasson & Krog 1985; Tan & Kong 2020). However, the lipid addition does not significantly change the digestion of waxy maize starch due to the lack of amylose (Figure 15.6 c), consistent with the absence of influence on the pasting profile of this starch, as illustrated in Figure 15.5c. DSC scanning and WAXD show no complex formation between amylose and corn oil that exclusively contains triglycerides. However, cooking of normal maize, tapioca, and high-amylose maize starches with corn oil under the same conditions also substantially reduces the enzymatic hydrolysis rates of the three starches, but not that of waxy maize starch (Figure 15.6a-d). ^{13}C-nuclear magnetic resonance spectra reveal the inclusion complex formation between amylose and corn oil, which thereby diminishes the digestibility of the above three starches after cooking (Ai, Hasjim, & Jane 2013).

Isothermal crystallization of amylose and glycerol monostearate (mass ratio=5:1) at 60°C and 90°C for 24 hour results in Form I and II complexes, respectively (Seneviratne & Biliaderis 1991). The hydrolysis using *Bacillus subtilis* α-amylase or porcine pancreatic α-amylase reveals that Form II complexes possess stronger enzymatic resistance than Form I, suggesting that the α-amylolysis rate and extent of ALC are inversely correlated with the degree of organization of helices into larger domains of ordered chains in the lamellar structure. In a different study of complexing glycerol monopalmitin with potato or high-amylose maize starch, isothermal crystallization at 80°C for 4 hour yields Form I complexes and that at 100°C for 24 hour yields Form II complexes. *In vitro* starch hydrolysis assay reveals no significant difference in the digestibility of Form I and II complexes (Tufvesson et al. 2001). The discrepancy in the *in vitro* digestibility of Form I and II complexes between these two studies could be linked to: (1) the different raw materials and conditions used to prepare ALC of the two physical forms; and (2) the two different *in vitro* starch hydrolysis assays applied. In the same study of Tufvesson et al. (2001), autoclaving and cooling of high-amylose maize starch without glycerol monopalmitin addition shows an even slower starch hydrolysis rate, suggesting that retrograded amylose (type 3 resistant starch, RS3) possesses greater enzymatic resistance than RS5. Another study reports that double helices formed between elongated branch chains of amylopectin (RS3) have higher enzymatic resistance than single-helical complexes between elongated branch chains of amylopectin

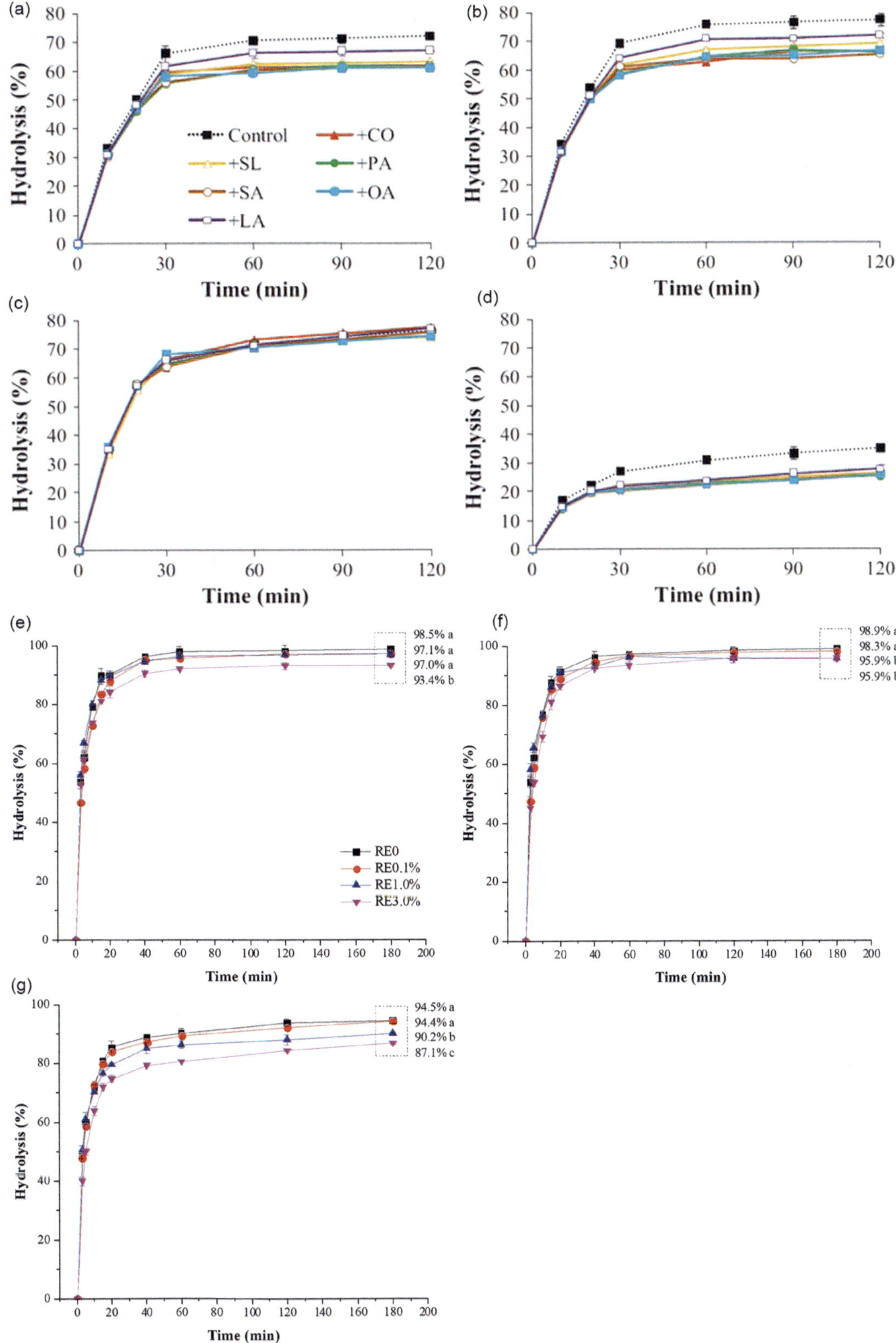

FIGURE 15.6 Enzymatic hydrolysis rates of cooked native starches and those cooked with 10% (w/w, dsb) different lipids: (a) normal maize starch; (b) tapioca starch; (c) waxy maize starch; and (d) high-amylose maize starch. CO, corn oil; SL, soy lecithin; PA, palmitic acid; SA, stearic acid; OA, oleic acid; and LA, linoleic acid. Porcine pancreatic α-amylase was used for the hydrolysis at 37°C and pH 6.9. Adapted from Ai, Hasjim, and Jane (2013). Enzymatic hydrolysis rates of starch in milled pet food extrudates with and without the addition of rosemary extract (RE): (e) rice diets; (f) round pea diets; and (g) high-amylose wrinkled pea diets. Values at the end of digestion curves in each sub-figure represent starch hydrolysis at 180 minute of RE0 (without RE addition), RE0.1% (with 0.1% RE addition), RE1.0% (with 1.0% RE addition), and RE3.0% (with 3.0% RE addition) samples from top to bottom. Values followed by the same letter in each sub-figure are not significantly different at $p < 0.05$. Modified Englyst Method was used for the *in vitro* starch hydrolysis. Adapted from Ren et al. (2024).

and added FFAs (RS5) (Lim et al. 2019). These studies tend to suggest that different physical forms of amylose exhibit enzymatic resistance in a descending order of retrograded amylose (RS3)>amylose-lipid complexes (RS5)>amorphous amylose.

The inhibitory effect of ALC formation on starch digestion has also been demonstrated *in vivo*. A new type of RS5 is developed by debranching preheated high-amylose maize starch with isoamylase, followed by complexation with palmitic acid (Hasjim et al. 2010). The modification enhances the RS content of high-amylose maize starch from 41.5% to 52.7% after cooking. RS bread is baked by substituting 60% of wheat flour with RS5. After feeding the RS bread to 20 male human subjects, postprandial glycemic response is reduced to 55% and insulinemic response to 43% as compared to the control white bread (as 100%). In a separate study, high-amylose maize starch-ascorbyl palmitate inclusion complexes are generated using two different methods, followed by annealing in 40% (v/v) aqueous ethanol at 75°C for 10 minutes and subsequent acid hydrolysis in 1 N HCl with 2% (w/v) solids at 40°C for 2 days (Guo et al. 2023). The prepared RS5 samples are fed to C57BL/6J mice to examine their glycemic responses. High-amylose maize starch-ascorbyl palmitate inclusion complexes that are prepared using the "empty" V-type method and subsequently subjected to annealing and acid hydrolysis show the lowest glycemic response, even significantly lower than that of raw high-amylose maize starch. The data from the two studies suggest that the new RS5 ingredients can be utilized for reducing postprandial glycemic and insulinemic responses, which can potentially lower the risks of type 2 diabetes and other associated chronic diseases.

Despite the potential health benefits, the use of ALC (RS5) is limited in the food industry. One major concern is the considerably higher caloric content of lipid than starch (9 vs. 4 kcal/g in theory), which impairs the practical value of RS5 in food applications as low-calorie foods are more desirable to consumers nowadays. In one previous study, when stearic acid is used to complex with debranched high-amylose maize starch to prepare RS5, the majority of the added stearic acid (68.2%–99.0%) in the RS5 diet is not absorbed and thus excreted in the feces of Fisher-344 rats during the 9-week feeding period (Ai et al. 2014). The low absorption of stearic acid is attributed to the complexation with amylose and debranched amylopectin, as well as the presence of uncomplexed stearic acid in a crystalline form (Berry & Sanders 2005; Kritchevsky 1994). In this regard, saturated fatty acids and their derivatives (e.g., monoglycerides) with a melting temperature well above the human body temperature may be nutritionally more favorable for preparing RS5 in comparison with unsaturated fatty acids and their derivatives, as the former will likely have less bioavailability and thus a lower caloric content. Additional research is required to understand how the lipids used to develop RS5 will affect the bioavailability of both starch and lipids in the complexes, as well as the total caloric content of the new RS ingredients.

Phenolic compounds are another group of complexing ligands that have received much research attention for decreasing starch digestibility. Prior studies have demonstrated the reducing effect of phenolic compounds on starch digestion (Figure 15.6e-g) (Barros, Awika, & Rooney 2012; Li, Griffin, et al. 2020; Ren et al. 2024; Sun & Miao 2020). The inhibitory influence on starch digestibility can be partly explained by physical interactions between starch molecules and phenolic compounds, which develop either single-helical complexes stabilized by hydrophobic interactions and van der Waals forces or weaker complexes stabilized by hydrogen bonding and van der Waals forces (Barros, Awika, & Rooney 2012; Chai, Wang, & Zhang 2013; Hernández, Gutiérrez, & Bello-Pérez 2022). Amylose or long branch chains of amylopectin tend to have stronger interactions with phenolic compounds, which is therefore more favorable for RS formation (Figure 15.6g) (Barros, Awika, & Rooney 2012; Ren et al. 2024). Furthermore, phenolic compounds can inhibit the activities of starch-hydrolyzing enzymes, such as α-amylase and α-glucosidase, thereby reducing starch digestibility (Rasouli et al. 2017; Sun, Wang, & Miao 2020; Wang et al. 2019).

Rosemary extract, which contains carnosic acid, carnosol, and rosmarinic acid as the three major phenolic compounds, is added at 0, 0.1%, 1.0%, and 3.0% (w/w) levels to extrude dry pet foods with rice, round pea, or wrinkled pea flour, aiming to decrease the starch digestibility (Figure 15.6e-g) (Ren et al. 2024). Adding rosemary extract at 1.0% (except for rice) and 3.0% significantly reduces the starch digestibility of extruded pet foods *in vitro*, and the greatest reduction is found with the high-amylose wrinkled pea sample (Figure 15.6g), suggesting that phenolic compounds in rosemary extract have stronger interactions with amylose to decrease starch digestibility. Although 0.1% addition of rosemary extract does not notably lower the *in vitro* and *in vivo* starch digestibility (Figure 15.6e-g), the treatment consistently delays the peak times of postprandial blood glucose responses to all the three extruded diets in beagles (Ren et al. 2024).

Despite the great potential to reduce starch digestibility and enhance antioxidant properties of foods, the inclusion levels of phenolic compounds in previous studies are very high, ranging from 5% to 30% (w/w) of starch (Cohen et al. 2011; Liu, Chen, et al. 2019), which can impart undesirable taste and color in the final products. For example, beagles – a dog breed known for the good appetite for extruded dry pet food – are observed to have a strong distaste for extruded dry pet food with the addition of rosemary extract at a low level of 1% (w/w) (Ren et al. 2024). Understanding how the incorporation of phenolic compounds influences the taste, aroma, and color of final products will be critical for their food and feed uses.

Because the inclusion complex formation enhances the enzymatic resistance of starch *in vitro* and *in vivo* as indicated above, a notable portion of RS can be passed into the large intestine for microbial fermentation. Previous research has examined the relationships among the structural features, physicochemical properties, digestibility, and fermentability of various RS ingredients from starch inclusion complexes (Li et al. 2023; Yang et al. 2024; Zhou, Fu, et al. 2021). Both the starch structure and lipid type can be used to control the digestibility and fermentability of V-type inclusion complexes. There is evidence that the fermentation of RS from V-type inclusion complexes promotes the growth of beneficial bacteria and induces the production of short-chain fatty acids (e.g., acetic, propionic, and butyric), which can potentially enhance the colon health of humans (Li et al. 2023; Yang et al. 2024; Zhou, Fu, et al. 2021).

15.5 APPLICATIONS OF STARCH INCLUSION COMPLEXES

As described in the previous sections, starch molecules – particularly amylose – can develop inclusion complexes with a variety of compounds, which resemble the structure of cyclodextrin-guest inclusion complexes to some extent. The unique structure of such single-helical complexes can be utilized for a wide range of applications. Good examples of using starch inclusion complexation for different purposes are provided in this part.

15.5.1 Determination of Starch Amylose Content

Amylose-polyiodide helical complexes exhibit a typical dark blue color with the absorption λ_{max} close to 640 nm. Amylose with DP of 40–50 and above can form blue complexes with polyiodide ions, while the color of the complexes changes to purple, red, and eventually orange as the DP of amylose gradually decreases to 19–25 (Mould 1954; Mould & Synge 1954; Yu, Houtman, & Atalla 1996). The color differences result from different numbers of iodine molecules that align to develop a polyiodide ion in the central cavity of the inclusion helix (Figure 15.1b): the longer the polyiodide ion forms in the cavity to provide a longer electron relay, the darker the color is. The inclusion complex formation between amylose and polyiodide ions has been employed to determine the amylose content of starch (Bates, French, & Rundle 1943; Chrastil 1987; Jane et al. 1999; Li, Li, et al. 2021). Iodine potentiometric titration method measures the iodine affinity of an isolated starch, which can be converted to "apparent amylose content" after dividing by the iodine affinity of pure amylose of 0.20 (Table 15.1) (Bates, French, & Rundle 1943; Jane et al. 1999). However, the long branch chains of amylopectin/intermediate component (IC) can also complex with polyiodide ions to give additional iodine affinity to inflate the amylose content. Subtracting the iodine affinity of amylopectin/IC from the whole starch can be applied to determine "absolute amylose content" of starch, as presented in Table 15.1. The differences between "apparent amylose content" and "absolute amylose content" ("A – B" in Table 15.1) show a clear order of B-type > C-type > A-type starches, consistent with the rank order of their amylopectin branch chain lengths (Jane et al. 1999). Iodine calorimetric method has also been developed to determine "apparent amylose content" of starch (Chrastil 1987; Li, Li, et al. 2021). One major advantage of this method lies in that it can quantify the apparent amylose content of a flour sample directly (i.e., no requirement of isolating starch), which can then be converted to "apparent amylose content of starch" after dividing by the total starch content of the flour (Sun et al. 2023; Yuan & Ai 2022).

15.5.2 Encapsulation of Flavor/Aroma and Bioactive Compounds

A diverse group of flavor/aroma compounds, such as alcohols, aldehydes, lactones, and terpenes, are strong complexing agents with amylose as characterized using different techniques according to the literature (Arvisenet, Voilley, & Cayot 2002; Gao et al. 2020; Heinemann et al. 2001; Itthisoponkul et al. 2007; Jouquand, Ducruet, & Le Bail 2006). In accordance with the understanding as presented in Section 15.2, the cross-section sizes of the flavor compounds determine V_6, V_7, or V_8 crystalline structure after the complex formation, and the thermal stability of the complexes increases with a longer hydrocarbon chain (Gao et al. 2020; Heinemann et al. 2001; Itthisoponkul et al. 2007). The inclusion complexation affects the binding capacity of starch matrix and the release profiles of the target compounds (Conde-Petit, Escher, & Nuessli 2006; Jouquand, Ducruet, & Le Bail 2006; Ma et al. 2019; Rutschmann & Solms 1990). Nonetheless, the impacts of another two factors on the release profiles of flavor compounds cannot be ignored: (1) nonspecific physical interactions between flavor compounds and amylopectin; and (2) the adsorption of flavor compounds on the starch matrix (Arvisenet, Voilley, & Cayot 2002; Boutboul et al. 2002). Recently, pre-formed "empty" V-type starches varying in inner helical cavity sizes, namely V_{6h} (hydrate form of V_6), V_7, and V_8, are prepared from high-amylose maize starch to encapsulate nine beany off-flavor compounds (Zhou, Hopfer, & Kong 2024). Upon complexation with these off-flavor compounds, most of the V-type inclusion complexes retain their crystalline structures. The study also reveals that both V_{6h} and V_8 starches largely exhibit more effective encapsulation efficiency than V_7 for the studied beany off-flavor compounds. The generated "empty" V-type starches will have great potential to scavenge beany off-flavor in soy and other legume products.

Amylose inclusion complexes have also been utilized to encapsulate various bioactive compounds, such as α-lipoic acid (Li et al. 2019; Li, Park, & Lim 2018), bioactive fatty acids (e.g., conjugated linoleic and α-linolenic) (Lalush et al. 2005; Lesmes et al. 2009), fatty acid esters of vitamins and phenolic compounds (Kenar et al. 2016; Kong & Ziegler 2014; Ma, Floros, & Ziegler 2011; Zhou, Liu, et al. 2021), and genistein (Cohen et al. 2008; Cohen et al. 2011). The benefits of encapsulating those bioactive compounds in such inclusion complexes include, but not limited to, increased solubility and dissolution rate, enhanced thermal, photochemical, and oxidative stability, controlled release, and improved bioavailability (Cohen et al. 2011; Di Marco, Ixtaina, & Tom 2022; Lesmes, Barchechath, & Shimoni 2008; Li et al. 2019). Different methods can be applied to prepare amylose inclusion complexes for encapsulation purposes according to a recent review article (Di Marco, Ixtaina, & Tom 2022). In addition, the molecular structure of starch can be tailored through chemical and/or enzymatic modifications to improve the performance in encapsulation, as highlighted in Section 15.2.4.

TABLE 15.1 Iodine affinities, and apparent and absolute amylose contents of isolated starches

STARCH	IODINE AFFINITY[b]		AMYLOSE CONTENT (%)[c]		
	STARCH	AMYLOPECTIN	APPARENT (A)[d]	ABSOLUTE (B)[e]	A − B
A-type[a]					
Normal maize	5.88±0.14	1.78±0.03	29.4	22.5	6.9
Rice	5.00±0.02	1.11±0.00	25.0	20.5	4.5
Wheat	5.75±0.15	0.8±0.2	28.8	25.8	3.0
Barley	5.1±0.3	0.50±0.04	25.5	23.6	1.9
Cattail millet	0.50±0.04	1.08±0.03	19.8	15.3	4.5
Mung bean	7.58±0.06	2.07±0.03	37.9	30.7	7.2
Chinese taro	2.75±0.06	0.00±0.00	13.8	13.8	0.0
Tapioca	4.7±0.2	1.39±0.07	23.5	17.8	5.7
B-type[a]					
Amylomaize V	10.4±0.1	6.79±0.04	52.0	27.3	24.7
Amylomaize VII	13.6±0.2	9.3±0.1	68.0	40.2	27.8
Potato	7.20±0.01	4.6±0.1	36.0	16.9	19.1
Green leaf canna	8.64±0.00	5.31±0.03	43.2	22.7	20.5
C-type[a]					
Water chestnut	5.79±0.19	3.08±0.02	29.0	16.0	13.0

Source: Adapted from Jane et al. (1999).

[a] Determined using wide-angle X-ray diffraction (WAXD).

[b] Average from at least three replicates±standard deviation of each sample, except for defatted green leaf canna starch, of which the iodine affinity was average from two replicates.

[c] Iodine affinity for pure amylose was assigned as 0.20.

[d] Calculated as: $C = 100 \times IA_S/0.20$, where C is the percentage of apparent amylose content and IA_S is the iodine affinity of the whole defatted starch.

[e] Calculated as: $C = (IA_S - IA_{AP+IC})/[0.20 - (IA_{AP+IC}/100)]$, where C is the percentage of absolute amylose content, IA_S is the iodine affinity of whole defatted starch, and IA_{AP+IC} is the iodine affinity of the amylopectin and the intermediate component (IC) mixture.

15.5.3 Development of Fat Replacers

ALC is prepared by stirring 5% (w/w) gelatinized maize starch dispersion with stearic acid (5% of starch) at 75°C for 60 minutes, which is used to replace shortening at 50% and 100% (w/w) levels in white pan bread formulation (Lee et al. 2020). The fat replacement with ALC at the two levels decreases the specific volume of bread from 6.66 to 6.39 and 5.16 mL/g, respectively, and reduces the caloric content from 4,515 to 4,338 and 4,186 Cal/g, respectively. The use of ALC also retards starch retrogradation, which is associated with the formation of larger amounts of ALC during the baking and cooling of bread (Section 15.3.4). In another study, stearic acid is also used to complex with isolated teff starch or commercial white maize starch to develop ALC through cooking in Rapid Visco Analyser with extended holding time of 1 hour 51 minutes at 91°C. ALC is utilized to replace sunflower oil at 50% and 80% (w/w) levels to prepare mayonnaise-type emulsions (Teklehaimanot, Duodu, & Emmambux 2013). Overall, ALC outperforms the corresponding native starches without modification in this application due to the weaker gelation ability and better stability of the former. The fat replacement considerably reduces the caloric content of the mayonnaise too. Future research can focus on controlling starch molecular structure and utilizing food-grade complexing agents to create starch inclusion complexes desired for replacing fat in diverse food products.

15.6 CONCLUSIONS

In the past decades, research has greatly expanded our knowledge on the structure, physicochemical properties, and digestibility of starch inclusion complexes with diverse compounds, and new technologies have also been developed to prepare such complexes for various final applications. The single-helical complex formation can be employed to modify starch functionality, produce RS-rich ingredients, measure starch amylose content, encapsulate flavor/aroma and bioactive compounds, prepare fat replacers, and achieve other goals. Future research efforts can focus on: (1) applying new starch modification methods, novel processing techniques, and suitable complexing agents to design starch inclusion complexes with desired functional attributes and nutritional profiles; (2) examining potential physiological benefits of inclusion complexes with lipids, phenolic compounds, and other bioactive substances in human and animal models; and (3) exploring new applications of single-helical complexes with starch as an abundant, affordable, biorenewable, and versatile polysaccharide.

REFERENCES

Abd Karim, A, Norziah, MH, & Seow, CC. 2000. Methods for the study of starch retrogradation. *Food Chemistry, 71*(1), 9–36.

Ai, Y, Hasjim, J, & Jane, J. 2013. Effects of lipids on enzymatic hydrolysis and physical properties of starch. *Carbohydrate Polymers, 92*(1), 120–127.

Ai, Y, & Jane, J. 2024. Understanding starch structure and functionality. In L Nilsson (Ed.), *Starch in Food: Structure, Function and Applications*, 3rd ed., (pp. 55–77). Cambridge, MA: Woodhead Publishing.

Ai, Y, Zhao, Y, Nelson, B, Birt, DF, Wang, T, & Jane, J. 2014. Characterization and *in vivo* hydrolysis of amylose–stearic acid complex. *Cereal Chemistry, 91*(5), 466–472.

Amoako, DB, & Awika, JM. 2019. Resistant starch formation through intrahelical V-complexes between polymeric proanthocyanidins and amylose. *Food Chemistry, 285*, 326–333.

Arijaje, EO, & Wang, YJ. 2015. Effects of chemical and enzymatic modifications on starch-oleic acid complex formation. *Journal of Agricultural and Food Chemistry, 63*(16), 4202–4210.

Arijaje, EO, & Wang, YJ. 2017. Effects of chemical and enzymatic modifications on starch-linoleic acid complex formation. *Food Chemistry, 217*, 9–17.

Arijaje, EO, Wang, YJ, Shinn, S, Shah, U, & Proctor, A. 2014. Effects of chemical and enzymatic modifications on starch-stearic acid complex formation. *Journal of Agricultural and Food Chemistry, 62*(13), 2963–2972.

Arvisenet, G, Voilley, A, & Cayot, N. 2002. Retention of aroma compounds in starch matrices: Competitions between aroma compounds toward amylose and amylopectin. *Journal of Agricultural and Food Chemistry, 50*(25), 7345–7349.

Barros, F, Awika, JM, & Rooney, LW. 2012. Interaction of tannins and other sorghum phenolic compounds with starch and effects on in vitro starch digestibility. *Journal of Agricultural and Food Chemistry, 60*(46), 11609 11617.

Bates, FL, French, D, & Rundle, RE. 1943. Amylose and amylopectin content of starches determined by their iodine complex formation. *Journal of the American Chemical Society, 65*, 142–148.

Berry, SE, & Sanders, TA. 2005. Influence of triacylglycerol structure of stearic acid-rich fats on postprandial lipaemia. *Proceedings of the Nutrition Society, 64*(2), 205–212.

Bhatnagar, S, & Hanna, MA. 1994. Amylose-lipid complex formation during single-screw extrusion of various corn starches. *Cereal Chemistry, 71*(6), 582–587.

Biliaderis, CG. 1992. Structures and phase transitions of starch in food systems. *Food Technology, 46*, 98–109.

Biliaderis, CG, Page, CM, & Maurice, TJ. 1986. On the multiple melting transitions of starch/monoglyceride systems. *Food Chemistry, 22*(4), 279–295.

Bordenave, N, Hamaker, BR, & Ferruzzi, MG. 2014. Nature and consequences of non-covalent interactions between flavonoids and macronutrients in foods. *Food & Function, 5*(1), 18–34.

Boutboul, A, Giampaoli, P, Feigenbaum, A, & Ducruet, V. 2002. Influence of the nature and treatment of starch on aroma retention. *Carbohydrate Polymers, 47*(1), 73–82.

Byars, JA, Fanta, GF, & Kenar, JA. 2013. Effect of amylopectin on the rheological properties of aqueous dispersions of starch-sodium palmitate complexes. *Carbohydrate Polymers, 95*(1), 171–176.

Byars, JA, Fanta, GF, Kenar, JA, & Felker, FC. 2012. Influence of pH and temperature on the rheological properties of aqueous dispersions of starch-sodium palmitate complexes. *Carbohydrate Polymers, 88*(1), 91–95.

Carlson, TL-G, Larsson, K, Dinh-Nguyen, N, & Krog, N. 1979. A study of the amylose-monoglyceride complex by Raman spectroscopy. *Starch-Stärke, 31*(7), 222–224.

Chai, Y, Wang, M, & Zhang, G. 2013. Interaction between amylose and tea polyphenols modulates the postprandial glycemic response to high-amylose maize starch. *Journal of Agricultural and Food Chemistry, 61*(36), 8608–8615.

Chen, J, Cai, H, Yang, S, Zhang, M, Wang, J, & Chen, Z. 2022. The formation of starch-lipid complexes in instant rice noodles incorporated with different fatty acids: Effect on the structure, *in vitro* enzymatic digestibility and retrogradation properties during storage. *Food Research International, 162*, 111933.

Chen, J, & Jane, J. 1994. Properties of granular cold-water-soluble starches prepared by alcoholic-alkaline treatments. *Cereal Chemistry, 71*(6), 623–626.

Chen, X, Ren, Y, Cai, Y, Huang, X, Zhou, L, Ai, Y, & Jiang, B. 2023. Interactions between exogenous free fatty acids and maize starches varying in amylose content at high heating temperatures. *Food Hydrocolloids, 143*, 108855.

Chrastil, J. 1987. Improved colorimetric determination of amylose in starches or flours. *Carbohydrate Research, 159*(1), 154–158.

Cohen, R, Orlova, Y, Kovalev, M, Ungar, Y, & Shimoni, E. 2008. Structural and functional properties of amylose complexes with genistein. *Journal of Agricultural and Food Chemistry, 56*(11), 4212–4218.

Cohen, R, Schwartz, B, Peri, I, & Shimoni, E. 2011. Improving bioavailability and stability of genistein by complexation with high-amylose corn starch. *Journal of Agricultural and Food Chemistry, 59*(14), 7932–7938.

Conde-Petit, B, Escher, F, & Nuessli, J. 2006. Structural features of starch-flavor complexation in food model systems. *Trends in Food Science & Technology, 17*(5), 227–235.

Craig, SAS, Maningat, CC, Seib, PA, & Hoseney, RC. 1989. Starch paste clarity. *Cereal Chemistry, 66*(3), 173–182.

Debet, MR, & Gidley, MJ. 2006. Three classes of starch granule swelling: Influence of surface proteins and lipids. *Carbohydrate Polymers, 64*(3), 452–465.

Di Marco, AE, Ixtaina, VY, & Tom, MC. 2022. Analytical and technological aspects of amylose inclusion complexes for potential applications in functional foods. *Food Bioscience, 47*, 101625.

Eliasson, AC, & Krog, N. 1985. Physical properties of amylose-monoglyceride complexes. *Journal of Cereal Science, 3*(3), 239–248.

Eliasson, AC, & Ljunger, G. 1988. Interactions between amylopectin and lipid additives during retrogradation in a model system. *Journal of the Science of Food and Agriculture, 44*(4), 353–361.

Fanta, GF, Kenar, JA, Felker, FC, & Byars, JA. 2013. Preparation of starch-stabilized silver nanoparticles from amylose-sodium palmitate inclusion complexes. *Carbohydrate Polymers, 92*(1), 260–268.

Gao, Q, Zhang, B, Qiu, L, Fu, X, & Huang, Q. 2020. Ordered structure of starch inclusion complex with C10 aroma molecules. *Food Hydrocolloids, 108*, 105969.

Garcia, MC, Pereira-da-Silva, MA, Taboga, S, & Franco, CML. 2016. Structural characterization of complexes prepared with glycerol monoestearate and maize starches with different amylose contents. *Carbohydrate Polymers, 148*, 371–379.

Gelders, GG, Goesaert, H, & Delcour, JA. 2006. Amylose-lipid complexes as controlled lipid release agents during starch gelatinization and pasting. *Journal of Agricultural and Food Chemistry, 54*(4), 1493–1499.

Gelders, GG, Vanderstukken, TC, Goesaert, H, & Delcour, JA. 2004. Amylose-lipid complexation: A new fractionation method. *Carbohydrate Polymers, 56*(4), 447–458.

Godet, MC, Bizot, H, & Buleon, A. 1995. Crystallization of amylose-fatty acid complexes prepared with different amylose chain lengths. *Carbohydrate Polymers, 27*(1), 47–52.

Gonzalez, A, Wang, YJ, Staroszczyk, H, Brownmiller, C, & Lee, SO. 2018. Effect of hydroxypropylation and beta-amylase treatment on complexation of debranched starch with naringenin. *Starch-Stärke, 70*(5–6), 1700263.

Gray, J, & Bemiller, J. 2003. Bread staling: Molecular basis and control. *Comprehensive Reviews in Food Science and Food Safety, 2*(1), 1–21.

Gudmundsson, M. 1992. Effects of an added inclusion-amylose complex on the retrogradation of some starches and amylopectin. *Carbohydrate Polymers, 17*(4), 299–304.

Gudmundsson, M, & Eliasson, AC. 1990. Retrogradation of amylopectin and the effects of amylose and added surfactants/emulsifiers. *Carbohydrate Polymers, 13*(3), 295–315.

Guo, J, Ellis, A, Zhang, Y, Kong, L, & Tan, L. 2023. Starch-ascorbyl palmitate inclusion complex, a type 5 resistant starch, reduced in vitro digestibility and improved in vivo glycemic response in mice. *Carbohydrate Polymers, 321*, 121289.

Hasjim, J, Lee, SO, Hendrich, S, Setiawan, S, Ai, Y, & Jane, J. 2010. Characterization of a novel resistant-starch and its effects on postprandial plasma-glucose and insulin responses. *Cereal Chemistry, 87*(4), 257–262.

He, T, Zhao, L, Wang, L, Liu, L, Liu, XW, Dhital, S, Hu, ZY, & Wang, K. 2024. Gallic acid forms V-amylose complex structure with starch through hydrophobic interaction. *International Journal of Biological Macromolecules, 260*, 129408.

Heinemann, C, Conde-Petit, B, Nuessli, J, & Escher, F. 2001. Evidence of starch inclusion complexation with lactones. *Journal of Agricultural and Food Chemistry, 49*(3), 1370–1376.

Helbert, W, & Chanzy, H. 1994. Single crystals of V amylose complexed with n-butanol or n-pentanol: Structural features and properties. *International Journal of Biological Macromolecules, 16*(4), 207–213.

Hernández, HAR, Gutiérrez, TJ, & Bello-Pérez, LA. 2022. Can starch-polyphenol V-type complexes be considered as resistant starch? *Food Hydrocolloids, 124*, 107226.

Huang, J, & White, PJ. 1993. Waxy corn starch: Monoglyceride interaction in a model system. *Cereal Chemistry, 70*(1), 42–47.

Immel, S, & Lichtenthaler, FW. 2000. The hydrophobic topographies of amylose and its blue iodine complex. *Starch-Stärke, 52*(1), 1–8.

Itthisoponkul, T, Mitchell, JR, Taylor, AJ, & Farhat, IA. 2007. Inclusion complexes of tapioca starch with flavour compounds. *Carbohydrate Polymers, 69*(1), 106–115.

Jane, J, Chen, YY, Lee, LF, McPherson, AE, Wong, KS, Radosavljevic, M, & Kasemsuwan, T. 1999. Effects of amylopectin branch chain length and amylose content on the gelatinization and pasting properties of starch. *Cereal Chemistry, 76*(5), 629–637.

Jane, J, Craig, SAS, Seib, PA, & Hoseney, RC. 1986. Characterization of granular cold water-soluble starch. *Starch-Stärke, 38*(8), 258–263.

Jane, J, & Robyt, JF. 1984. Structure studies of amylose-V complexes and retro-graded amylose by action of alpha amylases, and a new method for preparing amylodextrins. *Carbohydrate Research, 132*(1), 105–118.

Ji, L, Zhang, H, Cornacchia, L, Sala, G, & Scholten, E. 2022. Effect of gelatinization and swelling degree on the lubrication behavior of starch suspensions. *Carbohydrate Polymers, 291*, 119523.

Jiang, H, Campbell, M, Blanco, M, & Jane, J. 2010. Characterization of maize amylose-extender (*ae*) mutant starches: Part II. Structures and properties of starch residues remaining after enzymatic hydrolysis at boiling-water temperature. *Carbohydrate Polymers, 80*(1), 1–12.

Jiang, H, Lio, JY, Blanco, M, Campbell, M, & Jane, J. 2010. Resistant-starch formation in high-amylose maize starch during kernel development. *Journal of Agricultural and Food Chemistry, 58*(13), 8043–8047.

Jouquand, C, Ducruet, V, & Le Bail, P. 2006. Formation of amylose complexes with C6-aroma compounds in starch dispersions and its impact on retention. *Food Chemistry, 96*(3), 461–470.

Junejo, SA, Flanagan, BM, Zhang, B, & Dhital, S. 2022. Starch structure and nutritional functionality-Past revelations and future prospects. *Carbohydrate Polymers, 277*, 118837.

Kang, X, Yu, B, Zhang, H, Sui, J, Guo, L, Abd El-Aty, A, & Cui, B. 2021. The formation and in vitro enzymatic digestibility of starch-lipid complexes in steamed bread free from and supplemented with different fatty acids: Effect on textural and retrogradation properties during storage. *International Journal of Biological Macromolecules, 166*, 1210–1219.

Karunaratne, R, & Zhu, F. 2016. Physicochemical interactions of maize starch with ferulic acid. *Food Chemistry, 199*, 372–379.

Kasemsuwan, T, Bailey, T, & Jane, J. 1998. Preparation of clear noodles with mixtures of tapioca and high-amylose starches. *Carbohydrate Polymers, 36*(4), 301–312.

Kenar, JA, Compton, DL, Little, JA, & Peterson, SC. 2016. Formation of inclusion complexes between high amylose starch and octadecyl ferulate via steam jet cooking. *Carbohydrate Polymers, 140*, 246–252.

Kenar, JA, Eller, FJ, Felker, FC, Jackson, MA, & Fanta, GF. 2014. Starch aerogel beads obtained from inclusion complexes prepared from high amylose starch and sodium palmitate. *Green Chemistry, 16*(4), 1921–1930.

Kim, HI, Kim, HR, Choi, SJ, Park, CS, & Moon, TW. 2017. Preparation and characterization of the inclusion complexes between amylosucrase-treated waxy starch and palmitic acid. *Food Science and Biotechnology, 26*(2), 323–329.

Kong, L, Perez-Santos, DM, & Ziegler, GR. 2019. Effect of guest structure on amylose-guest inclusion complexation. *Food Hydrocolloids, 97*, 105188.

Kong, L, & Ziegler, GR. 2014. Molecular encapsulation of ascorbyl palmitate in preformed V-type starch and amylose. *Carbohydrate Polymers, 111*, 256–263.

Kritchevsky, D. 1994. Stearic acid metabolism and atherogenesis: History. *The American Journal of Clinical Nutrition, 60*(6), 997S–1001S.

Lalush, I, Bar, H, Zakaria, I, Eichler, S, & Shimoni, E. 2005. Utilization of amylose-lipid complexes as molecular nanocapsules for conjugated linoleic acid. *Biomacromolecules, 6*(1), 121–130.

Lauro, M, Poutanen, K, & Forssell, P. 2000. Effect of partial gelatinization and lipid addition on α-amylolysis of barley starch granules. *Cereal Chemistry, 77*(5), 595–601.

Le Bail, P, Rondeau, C, & Buléon, A. 2005. Structural investigation of amylose complexes with small ligands: Helical conformation, crystalline structure and thermostability. *International Journal of Biological Macromolecules, 35*(1–2), 1–7.

Lee, HS, Kim, KH, Park, SH, Hur, SW, & Auh, JH. 2020. Amylose-lipid complex as a fat replacement in the preparation of low-fat white pan bread. *Foods, 9*, 194.

Lesmes, U, Barchechath, J, & Shimoni, E. 2008. Continuous dual feed homogenization for the production of starch inclusion complexes for controlled release of nutrients. *Innovative Food Science & Emerging Technologies, 9*(4), 507–515.

Lesmes, U, Cohen, SH, Shener, Y, & Shimoni, E. 2009. Effects of long chain fatty acid unsaturation on the structure and controlled release properties of amylose complexes. *Food Hydrocolloids, 23*(3), 667–675.

Li, D, Wang, X, Wang, J, Wang, M, Zhou, J, Liu, S, Zhao, J, Li, J, & Wang, H. 2023. Structural characterization of different starch–fatty acid complexes and their effects on human intestinal microflora. *Journal of Food Science, 88*(8), 3562–3576.

Li, J, Li, L, Zhu, J, & Ai, Y. 2021. Utilization of maltogenic α-amylase treatment to enhance the functional properties and reduce the digestibility of pulse starches. *Food Hydrocolloids, 120*, 106932.

Li, J, Tian, L, Fang, Y, Chen, W, & Huang, G. 2021. Ultrasonic-assisted preparation of maize starch-caffeic acid complex: Physicochemical and digestion properties. *Starch-Stärke, 73*(1–2), 2000084.

Li, M, Griffin, LE, Corbin, S, Neilson, AP, & Ferruzzi, MG. 2020. Modulating phenolic bioaccessibility and glycemic response of starch-based foods in wistar rats by physical complexation between starch and phenolic acid. *Journal of Agricultural and Food Chemistry, 68*(46), 13257–13266.

Li, M, Ndiaye, C, Corbin, S, Foegeding, EA, & Ferruzzi, MG. 2020. Starch-phenolic complexes are built on physical CH-π interactions and can persist after hydrothermal treatments altering hydrodynamic radius and digestibility of model starch-based foods. *Food Chemistry, 308*, 125577.

Li, Y, Kim, YJ, Reddy, CK, Lee, SJ, & Lim, ST. 2019. Enhanced bioavailability of alpha-lipoic acid by complex formation with octenylsuccinylated high-amylose starch. *Carbohydrate Polymers, 219*, 39–45.

Li, Y, Park, EY, & Lim, S-T. 2018. Stabilization of alpha-lipoic acid by complex formation with octenylsuccinylated high amylose starch. *Food Chemistry, 242*, 389–394.

Lim, JH, Kim, HR, Choi, SJ, Park, CS, & Moon, TW. 2019. Complexation of amylosucrase-modified waxy corn starch with fatty acids: Determination of their physicochemical properties and digestibilities. *Journal of Food Science, 84*(6), 1362–1370.

Lin, JH, Kao, WT, Tsai, YC, & Chang, YH. 2013. Effect of granular characteristics on pasting properties of starch blends. *Carbohydrate Polymers, 98*(2), 1553–1560.

Liu, P, Kang, X, Cui, B, Gao, W, Wu, Z, & Yu, B. 2019. Effects of amylose content and enzymatic debranching on the properties of maize starch-glycerol monolaurate complexes. *Carbohydrate Polymers, 222*, 115000.

Liu, S, Reimer, M, & Ai, Y. 2020. In vitro digestibility of different types of resistant starches under high-temperature cooking conditions. *Food Hydrocolloids, 107*, 105927.

Liu, S, Yuan, TZ, Wang, X, Reimer, M, Isaak, C, & Ai, Y. 2019. Behaviors of starches evaluated at high heating temperatures using a new model of Rapid Visco Analyzer - RVA 4800. *Food Hydrocolloids, 94*, 217–228.

Liu, Y, Chen, L, Xu, H, Liang, Y, & Zheng, B. 2019. Understanding the digestibility of rice starch-gallic acid complexes formed by high pressure homogenization. *International Journal of Biological Macromolecules, 134*, 856–863.

Ma, R, Tian, Y, Zhang, H, Cai, C, Chen, L, & Jin, Z. 2019. Interactions between rice amylose and aroma compounds and their effect on rice fragrance release. *Food Chemistry, 289*, 603–608.

Ma, UVL, Floros, JD, & Ziegler, GR. 2011. Formation of inclusion complexes of starch with fatty acid esters of bioactive compounds. *Carbohydrate Polymers, 83*(4), 1869–1878.

Marfil, PHM, Anhê, ACBM, & Telis, VRN. 2012. Texture and microstructure of gelatin/corn starch-based gummy confections. *Food Biophysics, 7*(3), 236–243.

Mariscal-Moreno, RM, Figueroa-Cárdenas, JDD, Santiago-Ramos, D, & Rayas-Duarte, P. 2019. Amylose lipid complexes formation as an alternative to reduce amylopectin retrogradation and staling of stored tortillas. *International Journal of Food Science & Technology, 54*(5), 1651–1657.

Miao, LG, Xu, Y, Jia, CH, Zhang, BJ, Niu, M, & Zhao, SM. 2021. Structural changes of rice starch and activity inhibition of starch digestive enzymes by anthocyanins retarded starch digestibility. *Carbohydrate Polymers, 261*, 117841.

Morrison, WR, Law, RV, & Snape, CE. 1993. Evidence for inclusion complexes of lipids with V-amylose in maize, rice and oat starches. *Journal of Cereal Science, 18*(2), 107–109.

Mottiar, Y, & Altosaar, I. 2011. Iodine sequestration by amylose to combat iodine deficiency disorders. *Trends in Food Science & Technology, 22*(6), 335–340.

Mould, D. 1954. Potentiometric and spectrophotometric studies of complexes of hydrolysis products of amylose with iodine and potassium iodide. *Biochemical Journal, 58*(4), 593–600.

Mould, D, & Synge, R. 1954. The electrophoretic mobility and fractionation of complexes of hydrolysis products of amylose with iodine and potassium iodide. *Biochemical Journal, 58*(4), 585–593.

Patel, BK, Saibene, D, & Seetharaman, K. 2006. Restriction of starch granule swelling by iodine during heating. *Cereal Chemistry, 83*(2), 173–178.

Purhagen, JK, Sjöö, ME, & Eliasson, AC. 2011. Starch affecting anti-staling agents and their function in freestanding and pan-baked bread. *Food Hydrocolloids, 25*(7), 1656–1666.

Putseys, JA, Lamberts, L, & Delcour, JA. 2010. Amylose-inclusion complexes: Formation, identity and physico-chemical properties. *Journal of Cereal Science, 51*(3), 238–247.

Radhika, GS, & Moorthy, SN. 2008. Effect of sodium dodecyl sulphate on the physicochemical, thermal and pasting properties of cassava starch. *Starch-Stärke, 60*(2), 87–96.

Rasouli, H, Hosseini-Ghazvini, SM-B, Adibi, H, & Khodarahmi, R. 2017. Differential α-amylase/α-glucosidase inhibitory activities of plant-derived phenolic compounds: A virtual screening perspective for the treatment of obesity and diabetes. *Food & Function, 8*(5), 1942–1954.

Ren, Y, Bokshowan, E, Warkentin, TD, Weber, LP, & Ai, Y. 2024. Effects of rosemary extract addition on starch bioavailability and antioxidant properties of extruded pet foods. *Animal Feed Science and Technology, 313*, 116005.

Ren, Y, Liang, W, Zhong, Y, Hebelstrup, KH, Blennow, A, & Ai, Y. 2023. Impact of a full range of amylose level on pasting and gelling properties of barley starches at high-temperature heating. *Starch-Stärke, 75*(1–2), 2200167.

Ren, Y, Yuan, TZ, Chigwedere, CM, & Ai, Y. 2021. A current review of structure, functional properties, and industrial applications of pulse starches for value-added utilization. *Comprehensive Reviews in Food Science and Food Safety, 20*(3), 3061–3092.

Ring, SG. 1985. Some studies on starch gelation. *Starch-Stärke, 37*(3), 80–83.

Rundle, RE, & Edwards, FC. 1943. The configuration of starch in the starch-iodine complex. IV. An X-ray diffraction investigation of butanol-precipitated amylose. *Journal of the American Chemical Society, 65*, 2200–2203.

Rutschmann, M, & Solms, J. 1990. Formation of inclusion complexes of starch with different organic compounds. II. Study of ligand binding in binary model systems with decanal, 1-naphthol, monostearate and monopalmitate. *Lebensmittel- Wissenschaft und -Technologie, 23*, 70–79.

Schwingshackl, L, & Hoffmann, G. 2013. Long-term effects of low glycemic index/load vs. high glycemic index/load diets on parameters of obesity and obesity-associated risks: A systematic review and meta-analysis. *Nutrition, Metabolism and Cardiovascular Diseases, 23*(8), 699–706.

Seneviratne, HD, & Biliaderis, CG. 1991. Action of α-amylases on amylose-lipid complex superstructures. *Journal of Cereal Science, 13*(2), 129–143.

Shi, L, Hopfer, H, Ziegler, GR, & Kong, L. 2019. Starch-menthol inclusion complex: Structure and release kinetics. *Food Hydrocolloids, 97*, 105183.

Shi, L, Zhou, J, Guo, J, Gladden, I, & Kong, L. 2021. Starch inclusion complex for the encapsulation and controlled release of bioactive guest compounds. *Carbohydrate Polymers, 274*, 118596.

Singh, J, & Singh, N. 2003. Studies on the morphological and rheological properties of granular cold water soluble corn and potato starches. *Food Hydrocolloids, 17*(1), 63–72.

Sun, G, Ni, P, Lam, E, Hrapovic, S, Bing, D, Yu, B, & Ai, Y. 2023. Exploring the functional attributes and *in vitro* starch and protein digestibility of pea flours having a wide range of amylose content. *Food Chemistry, 405*, 134938.

Sun, L, & Miao, M. 2020. Dietary polyphenols modulate starch digestion and glycaemic level: A review. *Critical Reviews in Food Science and Nutrition, 60*(4), 541–555.

Sun, L, Wang, Y, & Miao, M. 2020. Inhibition of α-amylase by polyphenolic compounds: Substrate digestion, binding interactions and nutritional intervention. *Trends in Food Science & Technology, 104*, 190–207.

Sun, S, Jin, Y, Hong, Y, Gu, Z, Cheng, L, Li, Z, & Li, C. 2021. Effects of fatty acids with various chain lengths and degrees of unsaturation on the structure, physicochemical properties and digestibility of maize starch-fatty acid complexes. *Food Hydrocolloids, 110*, 106224.

Svensson, E, Autio, K, & Eliasson, AC. 1998. The effect of sodium dodecylsulfate on gelatinization and gelation properties of wheat and potato starches. *Food Hydrocolloids, 12*(2), 151–158.

Takahashi, S, & Seib, PA. 1988. Paste and gel properties of prime corn and wheat starches with and without native lipids. *Cereal Chemistry, 65*(6), 474–483.

Takeo, K, Tokumura, A, & Kuge, T. 1973. Complexes of starch and its related materials with organic compounds. Part X. X-ray diffraction of amylose–fatty acid complexes. *Starch-Stärke, 25*, 357–362.

Tan, I, Flanagan, BM, Halley, PJ, Whittaker, AK, & Gidley, MJ. 2007. A method for estimating the nature and relative proportions of amorphous, single, and double-helical components in starch granules by C CP/MAS NMR. *Biomacromolecules, 8*(3), 885–891.

Tan, L, & Kong, L. 2020. Starch-guest inclusion complexes: Formation, structure, and enzymatic digestion. *Critical Reviews in Food Science and Nutrition, 60*(5), 780–790.

Tang, MC, & Copeland, L. 2007. Analysis of complexes between lipids and wheat starch. *Carbohydrate Polymers, 67*(1), 80–85.

Tapanapunnitkul, O, Caiseri, S, Peterson, DG, & Thompson, DB. 2008. Water solubility of flavor compounds influences formation of flavor inclusion complexes from dispersed high-amylose maize starch. *Journal of Agricultural and Food Chemistry, 56*(1), 220–226.

Teklehaimanot, WH, Duodu, KG, & Emmambux, MN. 2013. Maize and teff starches modified with stearic acid as potential fat replacer in low calorie mayonnaise-type emulsions. *Starch-Stärke, 65*(9–10), 773–781.

Tester, RF, & Morrison, WR. 1990. Swelling and gelatinization of cereal starches. I. Effects of amylopectin, amylose, and lipids. *Cereal Chemistry, 67*(6), 551–557.

Tufvesson, F, & Eliasson, AC. 2000. Formation and crystallization of amylose-monoglyceride complex in a starch matrix. *Carbohydrate Polymers, 43*(4), 359–365.

Tufvesson, F, Skrabanja, V, Björck, I, Elmstahl, HL, & Eliasson, AC. 2001. Digestibility of starch systems containing amylose-glycerol monopalmitin complexes. *LWT - Food Science and Technology, 34*(3), 131–139.

Tufvesson, F, Wahlgren, M, & Eliasson, AC. 2003a. Formation of amylose-lipid complexes and effects of temperature treatment. Part 1. Monoglycerides. *Starch-Stärke, 55*(2), 61–71.

Tufvesson, F, Wahlgren, M, & Eliasson, AC. 2003b. Formation of amylose-lipid complexes and effects of temperature treatment. Part 2. Fatty acids. *Starch-Stärke, 55*(3–4), 138–149.

Turati, F, Galeone, C, Gandini, S, Augustin, LS, Jenkins, DJ, Pelucchi, C, & La Vecchia, C. 2015. High glycemic index and glycemic load are associated with moderately increased cancer risk. *Molecular Nutrition & Food Research, 59*(7), 1384–1394.

Uchino, T, Tozuka, Y, Oguchi, T, & Yamamoto, K. 2002. Inclusion compound formation of amylose by sealed-heating with salicylic acid analogues. *Journal of Inclusion Phenomena and Macrocyclic Chemistry, 43*(1–2), 31–36.

Wang, H, Wang, J, Liu, Y, Ji, Y, Guo, Y, & Zhao, J. 2019. Interaction mechanism of carnosic acid against glycosidase (α-amylase and α-glucosidase). *International Journal of Biological Macromolecules, 138*, 846–853.

Wang, J, Ren, F, Yu, J, Copeland, L, & Wang, S. 2021. Octenyl succinate modification of starch enhances the formation of starch–lipid complexes. *Journal of Agricultural and Food Chemistry, 69*(49), 14938–14950.

Wang, Q, Xia, W, Zhao, Z, & Zhang, H. 2015. Effects comparison between low glycemic index diets and high glycemic index diets on HbA1c and fructosamine for patients with diabetes: A systematic review and meta-analysis. *Primary Care Diabetes, 9*(5), 362–369.

Wang, X, Liu, S, & Ai, Y. 2022. Gelation mechanisms of granular and non-granular starches with variations in molecular structures. *Food Hydrocolloids, 129*, 107658.

Willett, W, Manson, J, & Liu, S. 2002. Glycemic index, glycemic load, and risk of type 2 diabetes. *The American Journal of Clinical Nutrition, 76*(1), 274S–280S.

Xiao, Y, Zheng, MZ, Yang, S, Li, ZF, Liu, MH, Yang, XB, Lin, N, & Liu, JS. 2021. Physicochemical properties and *in vitro* digestibility of proso millet starch after addition of proanthocyanidins. *International Journal of Biological Macromolecules, 168*, 784–791.

Yamashita, Y, & Hirai, N. 1966. Single crystals of amylose V complexes. II. Crystals with 7_1 helical configuration. *Journal of Polymer Science Part A-2: Polymer Physics, 4*, 161–171.

Yamashita, Y, & Monobe, K. 1971. Single crystals of amylose V complexes. III. Crystals with 8_1 helical configuration. *Journal of Polymer Science Part A-2: Polymer Physics, 9*, 1471–1481.

Yang, D, Guo, Q, Li, R, Chen, L, & Zheng, B. 2024. Amylose content controls the V-type structural formation and in vitro digestibility of maize starch-resveratrol complexes and their effect on human gut microbiota. *Carbohydrate Polymers, 327*, 121702.

Yu, X, Houtman, C, & Atalla, RH. 1996. The complex of amylose and iodine. *Carbohydrate Research, 292*, 129–141.

Yuan, TZ, & Ai, Y. 2022. Pasting and gelation behaviors and in vitro digestibility of high-amylose maize starch blended with wheat or potato starch evaluated at different heating temperatures. *Food Hydrocolloids, 131*, 107783.

Zaslow, B. 1963. Characterization of a second helical amylose modification. *Biopolymers, 1*, 165–169.

Zhang, H, Wang, HY, Zhang, QC, Wang, T, Feng, W, Chen, ZX, Luo, XH, & Wang, R. 2023. Fabrication and characterization of starch-lipid complexes using chain-elongated waxy corn starches as substrates. *Food Chemistry, 398*, 133847.

Zhao, B, Sun, S, Lin, H, Chen, L, Qin, S, Wu, W, Zheng, B, & Guo, Z. 2019. Physicochemical properties and digestion of the lotus seed starch-green tea polyphenol complex under ultrasound-microwave synergistic interaction. *Ultrasonics Sonochemistry, 52*, 50–61.

Zhou, DP, Liu, SW, Song, HY, Liu, X, & Tang, XZ. 2021. Preparation and characterization of chemically modified high amylose maize starch-ascorbyl palmitate inclusion complexes in mild reaction condition. *LWT - Food Science and Technology, 142*, 110983.

Zhou, J, Hopfer, H, & Kong, L. 2024. Odor-scavenging capabilities of pre-formed "empty" V-type starches for beany off-flavor compounds. *Food Hydrocolloids, 147*, 109315.

Zhou, Q, Fu, X, Dhital, S, Zhai, H, Huang, Q, & Zhang, B. 2021. In vitro fecal fermentation outcomes of starch-lipid complexes depend on starch assembles more than lipid type. *Food Hydrocolloids, 120*, 106941.

Zhou, Z, Robards, K, Helliwell, S, & Blanchard, C. 2007. Effect of the addition of fatty acids on rice starch properties. *Food Research International, 40*(2), 209–214.

Zhu, F. 2015. Interactions between starch and phenolic compound. *Trends in Food Science & Technology, 43*(2), 129–143.

Starch Plastification

16

Gregory M. Glenn, Randal L. Shogren, and Xing Jin

16.1 INTRODUCTION

Starch is an abundant, biodegradable, biocompatible, renewable and low-cost feedstock for both food and industrial uses. Many companies are exploring ways to use starch to replace plastics or other non-renewable products that negatively impact the environment. The global commercial supply of starch comes primarily from corn (75%), but other important sources include cassava root (14%), wheat (7%) and potato tubers (4%) (Vilpoux & Santos Silveira Junior 2023). Estimates of the global production of starches exceed 134 Mmt (ResearchandMarkets 2023). Important industrial uses for starch include biofuel production, where in the United States, more than 35% of corn production is used in ethanol production (Voegele 2023). Other important non-food markets for starch include the paper and chemical industries and biopolymers including starch-based bioplastics (Greencellfoam 2023; Novamont 2023; Plantic 2023) and poly(lactic acid) resins made by starch fermentation (Prismane 2021; Vilpoux & Santos Silveira Junior 2023). The plastification or plasticization of starch is important since it determines how starch can be processed and the applications to which it is suited. This chapter addresses starch plastification, different methods of processing starches and their application in making various films and foam products.

16.2 PLASTICIZERS

Most applications for starch, including paper coatings, food additives, adhesives and bioplastics, involve its use in a plasticized form. Even fermentation products of starch such as ethanol and lactic acid typically require starch to be in a plasticized form for reasonable enzyme conversion rates. Native, crystalline starch granules are melted usually in the presence of a suitable solvent such as water, allowing the starch molecules to freely move, intermix and entangle, thus forming a continuous material with the desired physical properties. The melting behavior of starch has been described in several reviews (Ai & Jane 2015; Ratnayake & Jackson 2008) and is briefly discussed in Section 16.4.1.

For amorphous and semi-crystalline polymers, an important determinant of physical properties is the glass transition temperature or T_g. Due to the abundance of starch hydroxyl groups and strong hydrogen bonding between them, T_g for dry starch (~230°C) (Orford et al. 1989) is near to or above the decomposition temperature (>200°C). Therefore, plastification of starch is required in order to process it into the desired shape.

Plasticizers for starch can be either external or internal. External plasticizers are typically small, hydrophilic molecules which can strongly interact with starch hydroxyls and create free volume (Montilla-Buitrago et al. 2021; Zhang, Rempel & Liu 2014). The most common are water and glycols such as glycerol or sorbitol, although many others have been studied in detail as we shall explore below. Starch that has added plasticizers of low vapor pressure and has been processed with heat and shear to a mostly amorphous form or "destructurized" is commonly termed thermoplastic starch or TPS. Internal plasticizers refer to the chemical modification of starch hydroxyls to give side chains that separate starch chains, increase free volume and inhibit recrystallization. Common modifications include esterification, etherification and graft copolymerization.

Water is by far the most common plasticizer for starch. Starch readily absorbs moisture from the air so that it always contains some water depending on the ambient relative humidity (Myllärinen et al. 2002). The water content of starch increases with humidity reaching about 12 wt.% at 50% humidity and 17% at 80% humidity (at room temperature or 23°C). T_g for starch decreases monotonically with increasing water content: 230°C at 0%, 110°C at 10% and 23°C at 22% (Orford et al. 1989; Shogren 1992). Although such data are usually obtained by DSC, computational approaches can provide reasonable estimates. For example, the Couchman–Karasz equation is often used to model the change in T_g with diluent addition (Couchman & Karasz 1978; Lourdin et al. 1997). It is simply the average of T_g of the components weighted by mole fraction and delta C_p. It does not, however, take into account starch-diluent interactions. Advances in molecular dynamics simulations have allowed the prediction of T_g for starch and added plasticizers by computer calculations at an atomistic level (Momany, Willett & Schnupf 2009; Özeren et al. 2020a; 2020b; Sanders et al. 2021).

DOI: 10.1201/9781003464396-16

Since most starches are mixtures of mostly linear amylose and highly branched amylopectin, the properties of both polymers as well as their blend morphologies need to be considered. Bizot et al. (1997) found that pure pea amylose had a slightly higher T_g (127°C) than waxy maize starch (111°C), both at 10% moisture. They ascribed this to the greater number of chain ends and branching points in amylopectin. Amylose can phase separate from amylopectin depending on starch processing conditions and composition (Rindlav-Westling, Stading & Gatenholm 2002).

Therefore, starch processed with water alone is a rigid, somewhat brittle polymer with mechanical properties similar to polystyrene under typical ambient room conditions. Brittleness can be reduced by selecting a starch with higher amylose content (Shogren & Jasberg 1994), by orientation (Shogren 2007) and by selective chemical modification (Kim, Jane & Lamsal 2017). Physical aging, the gradual embrittlement due to sub-T_g volume relaxations, can be reduced by increasing amylose content (Shogren & Jasberg 1994). Common commercial applications involve replacements for polystyrene where water resistance is not critical such as foam peanuts, foam sheets and clear, dry food containers as well as grease-resistant coatings (see Sections 16.5 and 16.6). When the latter are composed of high amylose starches, crystallization can occur rapidly leading to coatings that maintain their integrity sufficiently for wet food contact.

Many other plasticizers have been studied such as polyhydric alcohols, amides, amino acids, carboxylic acids, IL and deep eutectic solvents. Besides water, glycerol is the most commonly added plasticizer since it is an inexpensive waste product of biodiesel production, compatible and biodegradable (Ben, Samsudin & Yhaya 2022; Lourdin, Bizot & Colonna 1997; Lourdin et al. 1997; Van Soest et al. 1996). Other polyols have also been studied including sorbitol (Esmaeili, Pircheraghi & Bagheri 2017; Lim et al. 2020; Lourdin et al. 1997), xylitol (Muscat et al. 2012), various sugars (Teixeira et al. 2007; Zdanowicz, Staciwa & Spychaj 2019) and oligomers of ethylene glycol (Lourdin et al. 1997; Shogren 1992). Amide and amine-based plasticizers have included urea (de Souza Gamarano et al. 2022; Lourdin et al. 1997), formamide (Ma & Yu 2004), ethanolamine (Özeren et al. 2020b) and amino acids (Niazi, Zijlstra & Broekhuis 2015; Stein & Greene 1997). Polycarboxylic acids have been used both as plasticizers and cross-linking agents although some loss of starch molecular weight occurs on thermal processing. Those studied have included citric, oxalic and tartaric acids (Gebresas, Szabó & Marossy 2023; Li et al. 2022), as well as malic acid (Niazi, Zijlstra & Broekhuis 2015).

Combinations of two or more of the above plasticizers have sometimes yielded enhanced properties. For example, glycerol and citric can react to form oligomeric polyesters, which confer reduced plasticizer migration and higher tensile strength (Castro et al. 2023). Mixing sorbitol and glycerol enhanced mechanical properties and reduced retrogradation (Esmaeili, Pircheraghi & Bagheri 2017).

Unlike water, most of these plasticizers have a low vapor pressure so that they remain in the starch during processing and afterward. Usually, sufficient plasticizer is added to starch to lower T_g to below room temperature, giving the TPS much increased flexibility but lower tensile strength (typically < 6 MPa)

(Zhang, Rempel & Liu 2014). Besides low strength, another major downside of TPS is enhanced absorption of water at high humidities due to osmotic pressure and exposure to liquid water resulting in a soft material. Thus, TPS is seldom used alone but usually blended with a hydrophobic polymer to protect against excess water absorption. An example of a commercial product is compostable garbage bags made from blends of TPS and PBAT.

Antiplasticization can occur at lower levels of certain plasticizers such as glycerol (Lourdin, Bizot & Colonna 1997; Myllärinen et al. 2002; Shogren, Swanson & Thompson 1992). This is manifested as a minimum in elongation to break at around 12%–15% glycerol. At these lower levels, glycerol may fill in the spaces between starch chains effectively increasing the rigidity of the system. As glycerol is increased above 20%, T_g drops below room temperature and elongation to break increases as the material becomes rubbery.

At higher levels of certain plasticizers, amylose and amylopectin may crystallize, leading to increases in strength and decreases in elongation. For example, Van Soest et al. (1996) found that tensile strength of extruded potato starch-glycerol increased from 3 to 7 MPa and elongation decreased from 105% to 55% after B-type crystallinity increased to 30% after high humidity storage. At high levels, some plasticizers such as glycerol will partially phase separate.

In contrast, starch recrystallization is inhibited with a mixed glycerol/urea plasticizer and elongation to break is enhanced and stabilized (Paluch et al. 2022; Shogren, Swanson & Thompson 1992; Zdanowicz 2021). This is an example of a deep eutectic solvent (DES). The weakly acidic glycerol interacts with weakly basic urea to inhibit crystallization of the latter. Urea also has a favorable electrostatic attraction with weakly acidic starch. Zdanowicz, Wilpiszewska and Spychaj (2018) have reviewed DES as solvent and plasticizers for starch including choline chloride/urea, choline chloride/glycerol, imidazole/citric acid and various other combinations.

In addition to their activity as solvents for cellulose, starch and other polysaccharides, ionic liquids (IL) have been studied recently as plasticizers for starch (Ptak, Zarski & Kapusniak 2020; Ren et al. 2020; Yu et al. 2023). These consist of a large, organic cation in which the charge is delocalized through conjugation and a smaller inorganic or organic anion. The cation interacts with the hydroxyl oxygen of starch, which has a partial negative charge, while the anion interacts with the hydroxyl hydrogen. These interactions effectively disrupt the hydrogen bonding network of starch, allowing the chains to move more freely. TPS plasticized with 30% BMIMCl, for example, has a very high elongation to break (~400%), low tensile strength (0.6 MPa) and absorbs less moisture at 50% r.h. (13%) compared to TPS with 30% glycerol (20%) (Sankri et al. 2010). Recrystallization of starch also is generally inhibited by IL plasticizers although many increase water absorption relative to glycerol at high humidities (Yu et al. 2023). Disadvantages of many IL may include relatively high cost, lack of biodegradability and leaching. However, IL made with cholinium salts of carboxylic acids have relatively low cost and are biodegradable (Colomines et al. 2016). Cholinium citrate acted as a simultaneous plasticizing and cross-linking agent leading to an increase in strength and decrease in water sorption (Zdanowicz, Wilpiszewska & Spychaj 2018).

Many recent studies have been devoted to improving the mechanical properties of TPS by adding various biodegradable polyesters, natural fibers and nanofillers (Dufresne & Castaño 2017; Surendren et al. 2022; Tan et al. 2022). Polyesters studied have included PLA, PBAT and PHAs. The advantages of adding TPS to these polymers are to lower cost and accelerate biodegradation. Fiber and fillers usually increase modulus and tensile strength but decrease elongation at break. Such aspects will be discussed in more detail in Chapter 17.

Chemical modifications designed to internally plasticize starch have been studied but less extensively than for external plasticizers. Most commercially available chemically modified starches are low DS (<0.05) products, which are used for food and paper applications. At these low DS values, the granular structure of starch is preserved, allowing easy and inexpensive separation of reaction byproducts by washing with water. There have been few published reports on T_g of such modified starches, but it is likely that values are similar to unmodified starches. A major effect of the modifications is to make interactions with water and other plasticizers more energetically favorable and inhibit starch recrystallization (Gutierrez-Montiel et al. 2023).

At intermediate to high DS (1.5–3.0), some modified starches become truly thermoplastic, that is melt processable without added external plasticizer. For example, high amylose starch hexanoates of DS 1.7 and 2.6 have T_g of 114°C and 71°C, tensile strengths of 43 and 9 MPa and elongations to break of 30% and 110% (Winkler, Vorwerg & Rihm 2014). Thermal and mechanical properties varied with fatty acid chain length: T_g remained about the same up to C18; T_m for the fatty acid side chains were observed for C12 and higher; tensile strengths declined as side chain length increased, while elongation at break was maximal for C12. Thus, starch fatty esters of intermediate DS are hydrophobic (Ojogbo, Blanchard & Mekonnen 2018) and biodegradable (Zarski et al. 2020), and they could replace petroleum-based plastics. There has been little commercial development, however, due to the high cost of the reactants and solvents (Xu, Andrews & Shi 2020). More efficient synthetic methods are needed to achieve economic viability.

Starch grafted with various polymers can behave similar to thermoplastics. Bagley et al. (1977) found that starch grafted with around 50% polystyrene or methyl acrylate could be extruded into sheets with no added water or other plasticizer. Recent work has focused on *in situ* grafting of starch with polyesters in twin-screw extruders (TSE), which will be a topic of Section 16.3.

16.3 REACTIVE EXTRUSION

Reactive extrusion (REX) is the process whereby chemical reactions are conducted in an extruder (Zhuang et al. 2023). TSE are usually chosen for REX due to their flexibility in terms of screw design, feeding options and residence time. Co-rotating TSE of fully intermeshing, self-wiping design have excellent pumping action and thus can accept powders, pellets, or liquid injection at multiple points along the barrel. A wide variety of conveying elements both forward and reverse and kneading blocks are available for different mixing intensities. Residence time can be controlled by adjusting screw speed, feed rate and screw selection. TSE have several independent heating zones so that temperature may be varied as reactants travel down the barrel.

There are several advantages of REX for the preparation of internally plasticized starch over conventional stirred batch reactor type processes (Giri, Tambe & Narayan 2018). Extruders can handle highly viscous mixtures so little or even no solvent may be needed. This minimizes the use of toxic solvents and the need to provide for their recovery and reuse. Where water is the solvent, minimization of water can improve the reaction efficiency since many reagents that react with starch hydroxyls also undergo side reactions with water. TSE have excellent mixing capabilities so this increases interfacial area and reaction rate when starch and reactants occupy separate phases. Finally, extruders typically operate at high temperatures and pressures so that reactions often are completed in several minutes.

Like batch reactions, the modified starches made by REX may need to be purified by extraction of any unreacted starting materials and byproducts. If these happen to be volatile, they may be removed by vacuum stripping through an open port near the end of the extruder. In some cases such as grafting of existing polymers onto starch, there may not be byproducts, and the small amount of catalyst used may be safely left in the product.

Early as well as more recent research efforts on REX of starch have focused on preparation of low DS products for the paper and food industries (Moad 2011; Xie et al. 2006). These included cationic starches (Berzin & Vergnes 2018; Carr 1994), oxidized starches (Gutierrez-Montiel et al. 2023), hydroxypropylated starches (de Graaf & Janssen 2003), carboxymethyl starches (Milotskyi & Bliard 2018), hexadecyl carboxymethyl starch (Xie et al. 2017), phosphorylated starches (Landerito & Wang 2005), starch acetates (Aguilar-Palazuelos et al. 2023), starch succinates (Wang, Shogren & Willett 1997), starch alkenylsuccinates (Fonseca-Florido et al. 2019) and starch fatty acid esters (Miladinov & Hanna 2000). These mostly low (<0.2) DS modifications help to improve processability of TPS formulations and reduce tendency of starch to recrystallize. There have been few reports of the preparation of high DS starches by REX, so this may represent an opportunity for future research.

Starch graft copolymers have received much attention recently, especially starches grafted with biodegradable polymers (Formela et al. 2018). Several studies have examined REX of starch with anhydrides or polycarboxylic acids, followed by transesterification with polyesters. Narayan and coworkers (Raquez et al. 2008b; Stagner et al. 2011) extruded starch, glycerol plasticizer and maleic anhydride (2%–8%) to give maleated thermoplastic starches (MTPS). Glycerol was also found to react with starch via esterification to free carboxyl groups and by hemiacetal formation with aldehyde groups of starch reducing chain ends. MTPS and PBAT were then melt blended by REX to give MTPS-PBAT graft copolymers through transesterification, as evidenced by Soxhlet extraction, reduced phase sizes and improved mechanical properties (Hablot et al. 2013;

Raquez et al. 2008a). It should be noted that the MTPS-PBAT graft copolymer represented a small portion of the blend and served mainly to reduce the interfacial tension between the starch and polyester phases.

REX has also been used to prepare citrate esters (Ye et al. 2019). REX of starch, glycerol, citric acid and PBAT gave starch-PBAT graft copolymers due to citric esterification/transesterification reactions (Olivato et al. 2012). Similarly, blown films made with starch, glycerol, PBS and PBAT had improved mechanical properties when citric acid was included as a compatibilizer (Beluci et al. 2023).

Numerous types of starch graft copolymers have also been prepared by first grafting maleic anhydride onto polymers in a free radical catalyzed reaction, followed by REX of starch with the maleated polymer to form ester-linked grafts (Surendren et al. 2022). Polymers studied frequently have included PLA (Chauhan, Raghu & Raj 2021; Huneault & Li 2007; Wang, Yu & Ma 2007) and PBAT (Nabar et al. 2005). Starch-PLA grafts were also prepared by REX of TPS with PLA, glycidyl methacrylate and benzoyl peroxide (Palai et al. 2019).

Another approach to preparing starch polyester graft copolymers by REX has been to generate free radicals by peroxide decomposition followed by starch and polyester radical combination. Examples of systems studied include waxy maize starch, PBS and benzoyl peroxide (Yan & Dou 2021), starch, plasticizer, PCL and hydrogen peroxide/ferrous sulfate (Kalambur & Rizvi 2006), starch, PHB and dicumyl peroxide (Xu et al. 2017).

In summary, REX is an efficient, fast and versatile technique for preparing chemically modified plasticized starches. Relatively small amounts of starch-graft-polyester (<5%) can give major improvements in mechanical properties of TPS/polyester blends. Chemically modified starches of low DS such as hydroxypropyl can be easier to melt process, exhibit less retrogradation and show improved compatibility with polyesters (Wadaugsorn et al. 2022). In the next section, processing methods and properties of plasticized starch will be discussed in more detail.

16.4 PROCESSING PROPERTIES OF STARCH

Thermal processing of starches in water or other plasticizers can be much more complicated than with conventional thermoplastic resins (Donmez et al. 2021; Xie et al. 2007). As mentioned earlier, the most common plasticizer for starch is water, and there are several types of equipment commonly used to thermally process water-plasticized starches including steam jacketed kettles, jet cookers, extruders, and to a lesser extent, microwave heating (Bowser 2017; Palav & Seetharaman 2006). The processing properties of water plasticized starches vary depending on the rate of heating, water content, shear and atmospheric pressure (Donmez et al. 2021). Some of the common processing conditions are addressed in the following sections.

16.4.1 Low-Shear Thermal Processing at Atmospheric Pressure

When heating starch suspensions in excess water (>63%) under low-shear conditions, the suspensions increase in viscosity and a paste forms as the granules go through an important thermal transition known as gelatinization (Donmez et al. 2021; Wang et al. 1991). The phase changes leading up to gelatinization include swelling, loss of birefringence, melting and eventual solubilization (Altay & Gunasekaran 2006; Balet et al. 2019). Under low-shear conditions, gelatinization is primarily dependent on temperature and water content (Liu et al. 2009).

From an analytic view, a single peak temperature from DSC endotherms is observed that corresponds to the gelatinization temperature (Wang et al. 1991). From a mechanistic view, as heat is applied, water begins to infiltrate the amorphous regions of the granules causing them to swell. The initial interaction of water with the granule is mostly reversible, but eventually the water molecules begin to disrupt hydrogen bonding between adjacent polymers in the amorphous regions of the granules leading to further swelling (Balet et al. 2019). The physical changes of the granules are accompanied by the leaching of amylose polymers primarily into the aqueous phase (Balet et al. 2019; Han et al. 2019). The complete gelatinization of starch granules may happen gradually and over a temperature range that varies depending on the starch source (Chen et al. 2017; Donmez et al. 2021). The process involves swelling, the loss of granule birefringence, the melting of less stable starch crystallites, granule rupture, and the eventual complete dissolution of the granule itself resulting in an amorphous paste (Balet et al. 2019; Gudmundsson 1994; Liu et al. 2020; Wang et al. 1991).

From a rheology perspective, the viscosity of a starch suspension increases as it is heated and transitions into forming a paste. After reaching a peak maximum in viscosity, the viscosity decreases which is attributable to the solubilization of the crystalline regions of the granule remnants (Balet et al. 2019). The pasting behavior of starches is important in many food and industrial products. Various batch cooking processes form starch pastes to enhance and stabilize the viscosity of products. Starch from different plant sources can vary considerably in their response to heating. For example, potato starch granules heated to 60°C swell rapidly and leach more than 50% of their amylose content. In contrast, cassava starch granules heated to 60°C swell slightly and only leach out 8.5% of their amylose content (Han et al. 2019). High amylose corn starch granules swell only slightly at 100°C, but swell rapidly at 120°C (Chen et al. 2017).

A gelatinized starch paste is thermodynamically unstable, and as it cools, it transitions from an amorphous to a more ordered or crystalline state in a process known as retrogradation (Gudmundsson 1994). There are many changes that occur in a starch paste during retrogradation including an increase in paste viscosity, the formation of a gel network, recrystallization and exudation of water (syneresis) (Liu et al. 2021). Amylose and amylopectin behave differently during retrogradation. Amylopectin is responsible for long-term changes such as crystallization and densification leading to syneresis. Amylose is primarily responsible for short-term changes such as the formation of a rigid gel network (McGrane et al. 2004). It is during cooling of a starch paste that a gel network forms from intermolecular hydrogen bonding between water molecules and the mostly amylose molecules that leach from starch granules during gelatinization (McGrane et al. 2004).

Low shear thermal processing may also include high pressure processing (HPP) (Balakrishna, Wazed & Farid 2020; Castro et al. 2020; Liu et al. 2020). In HPP, the starch is subjected to hydrostatic pressures typically ranging from 100 to 600 MPa, but can include pressures as high as 1,000 MPa (Balakrishna, Wazed & Farid 2020; Leite et al. 2017). HPP has minimal effect on unplasticized, dry starch powder (Leite et al. 2017). However, HPP of starches plasticized with water can result in gelatinization at low temperatures (Leite et al. 2017). HPP-induced starch gelatinization is often referred to as "cold gelatinization", but temperature treatment with HPP ranging from 25°C to 80°C has been reported (Balakrishna, Wazed & Farid 2020). HPP induces starch gelatinization but in a way different from temperature-induced gelatinization at atmosphere and results in different gel properties. HPP alters non-covalent chemical linkages in and between starch molecules. It also disrupts starch crystallites to a greater degree than with thermally induced gelatinization (Liu et al. 2020). There is variation among different starches in the way they behave to HPP. Under some conditions, the starch granule morphology is somewhat preserved, while under other conditions, the granules swell, distort, lose birefringence and largely solubilize (Balakrishna, Wazed & Farid 2020; Liu et al. 2020). HPP provides a non-chemical alternative to starch modification that can significantly alter the thermal properties of starch, as well as its pasting properties and digestibility (Castro et al. 2020).

16.4.2 High-Shear Thermal Processing at Elevated Pressure

Extrusion has become an important thermal processing technology for starch and starch-rich compositions (Choton et al. 2020; Ek & Ganjyal 2020; Montilla-Buitrago et al. 2021). Extrusion systems use a heated barrel and typically one or two screws with various configurations that heat, mix and convey a plasticized starch feed material through a die. While being conveyed through the extruder, the plasticized starch is subject to intense shear fields and elevated temperatures and pressures (Montilla-Buitrago et al. 2021). Extrusion processing of water-plasticized starches can be done at low moisture levels (12%–16%) and much higher temperatures (100°C–160°C) to produce expanded products (Chinnaswamy & Hanna 1988; Ek & Ganjyal 2020; Lai & Kokini 1991). Starches behave differently to temperature under low moisture conditions. Rather than a single DSC peak corresponding to the gelatinization temperature, a second DSC peak appears and shifts to higher temperatures as the moisture content decreases (Wang et al. 1991). Under such conditions, the starch tends to melt rather than gelatinize (Donmez et al. 2021). In extrusion processing, the loss of crystallinity and melting of starch granules occurs due to a combination of high temperature exposure, mechanical shear and the physical destructurization of the starch granules.

The intense shear conditions produced in extrusion processing unavoidably lead to starch fragmentation and random chain splitting of amylose and amylopectin (Colonna et al. 1984). Amylose is less susceptible to macromolecular degradation during extrusion than amylopectin (Lai & Kokini 1991). In one study, Colonna et al. (1984) reported that the average molecular weight of amylopectin decreased 15 folds due to extrusion, while the molecular weight of amylose only decreased by a factor of 1.5. Macromolecular degradation can be reduced in some cases by increasing the processing temperature and decreasing the screw speed (Liu et al. 2009).

16.4.3 Hydrothermal Processing

Hydrothermal processing of starch refers to a heat ($>T_g$) and moisture/plasticizer treatment that forms irreversible modifications without destroying the granular structure (Dutta & Sit 2022; Iuga & Mironeasa 2020; Zavareze & Dias 2011). In hydrothermal processes, the critical parameters of moisture, temperature and heating time need to be controlled. It is a simple, low cost and environmentally friendly process for modifying the properties of starches without the use of chemical treatments and the associated waste and/or safety concerns (Dutta & Sit 2022; Iuga & Mironeasa 2020). Hydrothermal processing can be subcategorized into two separate processes, annealing (ANN) and heat-moisture treatment (HMT). ANN treatment, which can be lengthy, uses relatively high (<76% but preferably 40%–55%, w/w) water contents and temperatures above T_g but below the gelatinization temperature (Dutta & Sit 2022; Zavareze & Dias 2011). In contrast, HMT uses temperatures above the gelatinization temperature but with insufficient moisture levels (commonly <35% w/w) to gelatinize and melt the starch granules (Dutta & Sit 2022; Zavareze & Dias 2011).

Dutta and Sit (2022) found that ANN and HMT treatment decreased the solubility and water vapor permeation and increased tensile strength of potato starch films compared to films made of native starches. Majzoobi et al. (2015) reported that biodegradable films from HMT treated rice starch had lower water solubility, rigidity and extensibility, but higher water vapor permeability compared to native rice starch. Xu et al. (2018) found significant differences in whether ANN treatments of potato starch were applied as a continuous vs. a repeated process. Güllich et al. (2023) compared ANN-treated and chemically oxidized wheat starches. Starch films made of ANN-treated starches had less cracking and better mechanical properties compared to chemically oxidized wheat starch. Malumba et al. (2022) found that coupling the ANN process with incubation in dilute acid or base did not change the granular structure of wheat starch, but did affect the gelatinization temperature. Hoyos-Leyva et al. (2015) reported that HMT treatments of banana starch using 32.2% moisture and temperatures of 100°C maximized the amount of resistant and slowly digestible starch content for food applications.

16.5 STARCH SHEETS/FILMS

In recent years, starch-based sheets and films have been widely studied because of the natural abundance and sustainability of starch and the biodegradability, high barrier properties to O_2 and CO_2, low odor, low toxicity and relatively high transparency of some starch-based films (Lauer & Smith 2020; Su et al. 2023; Thakur et al. 2019). Moreover, starch sheets and films are widely studied as a replacement for non-biodegradable, traditional plastic sheets and films, especially for packaging applications. At present, the scale of related real applications for starch-based sheets and films is still very limited due to issues of water sensitivity and inferior mechanical properties. Much of the current research addresses these problems, while new or expanded uses of starch sheets and films are being explored (Diyana et al. 2021).

16.5.1 Applications

The use of starch-based coatings and films has become established in the food and pharmaceutical sectors (García-Guzmán et al. 2022; Jimenez et al. 2012; Kramer 2009; Shah et al. 2016; Singh et al. 2022). Examples of the use of starch-based edible films and coatings include batter coating for fried foods (Kurek & Ščetar 2017), coatings on vegetables and fruits to extend shelf life, reduce moisture loss and improve surface gloss (Sapper & Chiralt 2018), and coatings for pharmaceutical tablets (Labelle, Ispas-Szabo & Mateescu 2020; Thakur et al. 2019). The early Apollo astronauts ate foods coated with films of high-amylose starch to minimize the formation of crumbs that could disperse in the weightless environment of space (Jiang et al. 2020; Kramer 2009). Examples of edible films are glutinous rice paper wrapping for candies or dry seasonings, oral-dissolving breath strips, drug delivery systems and pharma capsules (Thakur et al. 2019). In addition, there are many studies on starch sheets and films as active food packaging with antimicrobial and antioxidant functionalities (Lauer & Smith 2020; Priyadarshi et al. 2022; Shah et al. 2015).

16.5.2 Processing Methods

Traditional edible films, such as Vietnamese rice papers, are prepared in a way similar to the making of thin pancakes. Drum drying is typically used in modern commercial production of wafer paper and spring-roll wraps, etc. (Anko 2023; Primus 2020). There are many studies of starch sheets and films in the literature made by casting a solution containing approximately 5% starch (Cui et al. 2021).

Sheets and films have also been prepared by sheet extrusion or film blowing using conventional equipment designed for plastic melt processing (Dang & Yoksan 2021). These technologies have certainly enabled much higher production efficiency than solution casting methods, while developments on formulations and technology details are normally needed. As introduced in Section 16.2, the required thermoplasticity for melt processing can be obtained with either external plasticizers or internal plastification using chemical modification. Plantic Technologies (Plantic 2023), an Australian company that was recently acquired (Kuraray 2023), extrudes sheets of hydroxypropylated high-amylose starch, which can be thermoformed into pouches, trays, lids, etc. with good mechanical properties.

Extrusion film blowing has been used for many years to make biodegradable trash bags from composites of starch blended with poly(butylene adipate-co-terephthalate) (World-Centric 2023). Starch-based Mater-Bi resin of Novamont (Novara, Italy) is widely used to produce extrusion blown agricultural mulch films (Novamont 2023). Thunwall et al. (2008) observed that extrusion blowing films using native potato starch and glycerol produced poor films that had a sticky surface, areas of imperfections and poor mechanical strength. In contrast, a satisfactory result was obtained when an oxidized and hydroxypropylated starch was used. The average number of carboxylic acid groups per anhydroglucose unit was 0.04 and the degree of substitution of hydroxypropyl groups was 0.11. Thunwall et al. (2008) determined that the lower hydrophilicity of the modified starch reduced the stickiness of the film surface and the relatively lower melting temperature led to better homogeneity.

Other less common technologies have been reported for making starch films and fibrous membranes, such as electrospinning or solution blow spinning, but these methods have less production efficiency and are primarily of interest in medical or other niche applications (Estevez-Areco et al. 2020; Kong & Ziegler 2014; Li et al. 2021; Temesgen et al. 2021).

16.5.3 Improving Film Properties

Nowadays, food-grade modified starches are routinely used to improve paste viscosity, freeze–thaw stability, film forming and to reduce retrogradation during processing, as well as to improve texture of foods (Kramer 2009; Shah et al. 2016). Physical modifications include pregelatinizing, annealing, dry heating or treating with osmotic pressure, pulsed electric field, ultrasound, microwave and gamma irradiation (Shah et al. 2016). Chemical modifications include a low degree of substitution of starch hydroxyl groups via hydroxypropylation, carboxymethylation, acetylation, propylation or creating new carboxyl groups via oxidation (Shah et al. 2016). Modification can also use enzymes of hydrolases or glucanohydrolases (Abbas, Khalil & Hussin 2010).

Of the two major components of starch (amylose and amylopectin), amylose has proven to have good film forming properties and produce sheets or films with good mechanical properties (Faisal et al. 2022). Minor differences in the amylose content of starches may not result in significant improvements in film properties (Basiak, Lenart & Debeaufort 2017). However, high amylose starches (>50%) have been commercially available for many years and provide an important feedstock for excellent starch films (Faisal et al. 2022). Treating starch granules with a mineral acid to partially depolymerize amylopectin produces compositions that behave more like amylose-rich mixtures (Kramer 2009).

Often combined with substitution modification in commercial production, cross-linking with phosphorous oxychloride, sodium trimetaphosphate or adipic anhydride is done to reduce fragmentation of starch granules, stabilize viscosity, improve textural properties and reduce retrogradation in cooked foods (Korma et al. 2016; Kramer 2009; Singh, Kaur & McCarthy 2007). In academic studies, solution-cast starch-based sheets and films are often cross-linked with multifunctional carboxylic acids (Azeredo & Waldron 2016), especially citric acid (Cheng et al. 2021; Garavand et al. 2017) to improve water stability.

Starch films have inherently high water-vapor permeation rates and moisture sorption, which limit many applications (Long et al. 2023). As introduced in Section 16.2, plasticizers are often needed to process starch into useful products. Interestingly, below a certain threshold of plasticizer content, water vapor permeability is lower than in unplasticized sample, and the opposite is observed above the threshold (Talja et al. 2007). Similarly, below a threshold of relative humidity, absorbed moisture content is lower in plasticized sample than in unplasticized sample, and the opposite is observed above the threshold. The humidity threshold decreases with increasing plasticizer content (Talja et al. 2007). Additionally, the effect was observed with starches of both high and low amylose content and occurred with glycerol or xylitol or mixtures of both polyols (Muscat et al. 2012; Myllärinen et al. 2002).

The evidence from these studies are indicative of an anti-plasticization effect. The results may be systematically associated with the antiplasticization effect where a polymer mixed with a low content of plasticizer behaves even more brittle or rigid than the neat polymer (Mascia et al. 2020). The more commonly recognized plasticization effect with more flexible plasticized samples occurs when plasticizer content is above a threshold, which usually is about 10–30 wt.% depending on the exact system. At a level below the threshold, small plasticizer molecules are locally bound to polymer chains and reduce the free volume of the system. Above the threshold, plasticizer small molecules begin forming clusters and increase the free volume of the system. The antiplasticization mechanism was used in studies of polysaccharide–polyol mixtures (Van der Sman 2019), foods (Ubbink 2018), starch–urea (Wang, Cheng & Zhu 2014) and starch–glycerol–nanoclay (Yu et al. 2008) systems.

Bangar et al. (2021) listed typical values for water vapor permeability, oxygen permeability and tensile properties of starch-based films. Sheets and films of starch-based blends and composites have much improved properties compared to that of neat thermoplastic starch (Cheng et al. 2021). The most convenient way to effectively improve starch-based sheet or film properties is by blending the starch with biodegradable plastics, which includes polyesters such as poly(lactic acid), polycaprolactone, poly(butylene succinate) and polyhydroxyalkanoates (Cheng et al. 2021). Examples of commercial products are the aforementioned biodegradable composting trash bags and mulching films. In addition to physical blends of starch and other biopolymers, useful composite films can be made via grafting, coating, multi-layer casting and extrusion (Carneiro da Silva, Rios & Campomanes Santana 2023). Poly(vinyl alcohol) is often used to toughen starch-based sheets and films, due to its own excellent

mechanical properties and its excellent compatibility with starch (Cheng et al. 2021; Estevez-Areco et al. 2020). Unfortunately, biodegradation of poly(vinyl alcohol) only happens in limited environments (Alonso-López, López-Ibáñez & Beiras 2021). Polydopamine (Xu et al. 2023) and gelatin (Zhang et al. 2023) are also used to make starch-based composite films.

Nanofillers have been added to starch mixtures for film forming (Shah et al. 2015). Nanofillers can be effective at about 5 wt.% of a composite when they are uniformly dispersed in a starch solution via methods that may include sonication, ultra-sonication and homogenization (Shah et al. 2015). Clay minerals comprised of silicate sheets are naturally abundant and of low cost and can be used in starch films (Madhumitha et al. 2018). De Carvalho, Curvelo and Agnelli (2001) were the first to melt intercalate clay in a starch matrix. Park et al. (2002) used transmission electron microscopy to show the dispersion of clay nanoplatelets in starch films. In both studies, compounding was done in a batch melt mixer. Park et al. (2002) used native potato starch, glycerol, water, an unmodified Na$^+$ montmorillonite or one of three grades modified with different organic ammonium cations to make films. Blends of 5 wt.% of unmodified montmorillonite in starch films produced good dispersion and intercalation, resulting in higher tensile strength and lower water vapor permeation. Later, Qiao, Jiang and Sun (2005) compounded acetylated starch, glycerol with montmorillonite by batch melt mixing, followed by a hot press step to make composite sheets. Organically modified montmorillonite at 5 wt.% based on starch content showed a greater positive effect than unmodified montmorillonite on melt viscosity, equilibrium torque, tensile strength and storage modulus.

Gao et al. (2012) used a twin-screw extruder to compound and pelletize composite materials that were subsequently extrusion blown into film. The composite materials consisted of hydroxypropyl distarch phosphate with glycerol and chemically modified montmorillonites at 35 and 6 wt.% of the starch content, respectively. The results of X-ray diffraction and transmission electron microscopy showed the starch was intercalated the best into clay having medium hydrophilicity. Furthermore, the composites had a much lower T_g and peak temperature than other samples. Tang, Alavi and Herald (2008) compounded native cornstarch with glycerol, water and montmorillonite nanoclay using twin-screw extrusion. The extrudate was dried, ground into powder and mixed with water to form a solution that was cast into a film. Starch composites containing 6 and 5 wt.% of montmorillonite and glycerol, respectively, showed the greatest amount of clay exfoliation, highest tensile strength and the lowest water vapor permeability.

Silver nanoparticles at 10 wt.% based on starch content have been dispersed in solution-cast films of starch (Božanić et al. 2011) or starch/nanoclay composites (Abreu et al. 2015) to provide antimicrobial activity. Antimicrobial films are of particular interest in food packaging applications. Antimicrobial properties were also obtained in starch films by adding ZnO nanorods (Estevez-Areco et al. 2020), TiO$_2$ or CuO nanoparticles (Muñoz-Gimena et al. 2023), chitosan, nanoclay and conventional antimicrobial agents such as potassium sorbate and grapefruit seed extract (Jha 2020). For extrusion blown films of hydroxypropyl starch with glycerol (30 wt.% of starch), Zhu et al. (2021) observed that nano-ZnO and/or nano-SiO$_2$ (3 wt.% of starch) significantly increased tensile strength and decreased elongation to break. The combination of nano-ZnO and nano-SiO$_2$ greatly improved the smoothness of the film surface. Other nano-sized materials used in starch composites include cellulose nanofibers and nanocrystals (Cataño et al. 2023; Muñoz-Gimena et al. 2023). These composites have greater mechanical strength and reduced water vapor permeance.

Estevez-Areco et al. (2020) built a transparent composite bilayer film for active food packaging. The outer layer was prepared first by sheet-extruding thermoplastic starch containing ZnO nanorods which provided antimicrobial functionality. The composite also contained an inner layer formed by electrospinning an aqueous solution of poly(vinyl alcohol) containing rosemary extract which provided antioxidant properties. The composite bilayer film had a 42% lower water vapor permeation rate and 50%–100% higher Young's modulus, stress and strain at break, and tensile toughness compared to a neat starch film.

16.6 STARCH FOAMS

There are several reviews devoted solely to starch foams that provide extensive details of starch foam processes and products (Soykeabkaew, Thanomsilp & Suwantong 2015). Some aspects pertaining to the food industry will be addressed here, but the primary focus will be on non-food products and processes.

16.6.1 Food and Snack Processes

The food industry has long exploited processing conditions that convert starch-rich grains or milled flours into a cellular foam structure with unique and desirable textural and digestibility qualities (Swarnakar, Mohapatra & Das 2022). Popcorn is the earliest example of a foam made from grain. It has been used as a food in different parts of the world for more than 6,500 years (Grobman et al. 2012). The mechanisms involved in popping popcorn have been well documented (Hoseney, Zeleznak & Abdelrahman 1983; Shimoni, Dirks & Labuza 2002; Sweley, Rose & Jackson 2013). Water (10%–15%) plasticized starchy endosperm is encased in a pericarp that surrounds each kernel and acts as a miniature pressure vessel. Pressure inside the pericarp builds to about 135 psi as kernel temperature reaches 177°C (Hoseney, Zeleznak & Abdelrahman 1983). Popping occurs when the pericarp suddenly ruptures allowing superheated moisture inside the starch component to act as a blowing agent driving the expansion of the starchy endosperm into porous, thin-walled foam with low bulk density (Da Silva et al. 1993; Gulati & Datta 2016; Hoseney, Zeleznak & Abdelrahman 1983; Sweley, Rose & Jackson 2013).

Other food processes have subsequently been developed that mimic the conditions created in popcorn popping including the use a puffing gun or an oven puffing process (Mariotti et al. 2006).

A puffing gun typically involves a pressure vessel that is loaded with grain and tightly closed before pressurizing with superheated steam (Mariotti et al. 2006; Mishra, Joshi & Panda 2014). Once the contents are heated, the vessel is instantaneously opened to release the pressure and eject the puffed product (Mariotti et al. 2006). The rice cake popper is a simple variation of a puffing gun (Orts et al. 2000).

Oven puffing is a process conducted at atmospheric pressure (Mariotti et al. 2006). With oven puffing, high temperatures (175°C–220°C) are required to ensure that the water-plasticized starch can rapidly transform to a soft rubbery state, and the moisture component can convert to steam and act as a blowing agent before too much water is lost to evaporation. Variations of heating processes for puffing at atmospheric pressures include the use of hot (250°C) sand, hot oil (200°C–220°C), a high-temperature, short-time (HTST) fluidized bed (240°C–270°C), microwaving, infrared heating and hot air (Castro-Campos et al. 2021; Chinnaswamy & Bhattacharya 1983; Mishra, Joshi & Panda 2014; Shavandi, Javanmard & Basiri 2023).

16.6.2 Extruded Foams

Of all the processes developed for puffing starch-based foods, extrusion has proven to be the most impactful. Extrusion is described as a HTST cooking process. Temperatures typically range from 100°C to 180°C, and residence time may range from 30 to 120 seconds (Maskan & Altan 2011). Single- and twin-screw extruders as described earlier can be found that are configured in different ways to optimize the processing parameters of interest. By rotating, the screws convey, mix, heat, pressurize and force the cooked/melted ingredients and superheated moisture out of a die, where the sudden drop in pressure causes the moisture to instantaneously form steam. The steam acts as a blowing agent that forms bubbles in molten extrudate (Gray & Chinnaswamy 1995; Gu, Kowalski & Ganjyal 2017; Maskan & Altan 2011). The global market for extruded starch-based flours and other food ingredients is projected to reach nearly $100 billion by 2026 (MarketsandMarkets 2023).

Much of the technology developed for food extrusion has provided a foundation for the extrusion of starches (Chinnaswamy & Hanna 1988; Hutchinson, Mantle & Smith 1989; Mercier et al. 1979; Shogren, Fanta & Doane 1993). In the 1930s, General mills used single-screw extrusion to make expanded cereal products (Riaz 2000). Single screw extrusion was primarily used to make expanded snack foods and pet foods until twin-screw extrusion was developed (Anderson 1969; Mercier et al. 1979; Riaz 2000). A twin-screw extruder designed for plastics was used to make puffed corn and sorghum grits using moisture levels as low as 15% (Anderson 1969; Anderson, Conway & Peplinski 1970; Anderson et al. 1969). Later, Mercier and colleagues used twin-screw extrusion to report the processing of cereal flours and various starches including potato and manioc (Mercier 1977; Mercier et al. 1975, 1979). Chinnaswamy and Hanna (1988) used extrusion to optimize the expansion and foaming characteristics of corn starch. Since these early studies, there have been many reports where extrusion was used to make foams using various sources of starch. Among other properties, these studies characterize the mechanical properties, bulk density and microstructure (Chinnaswamy & Hanna 1988; Hutchinson, Mantle & Smith 1989; Lin et al. 1995; Tatarka & Cunningham 1998; Warburton, Donald & Smith 1990).

Lacourse and Altieri (1989) were among the first to use an extrusion process for creating packaging foam sheets and loose-fill (Lacourse & Altieri 1989, 1991). Commercial starches from maize, waxy maize and potato with relatively low amylose content (<45%) all foamed well, but had a relatively open-cell structure and were brittle (Lacourse & Altieri 1989). Foams made of high amylose corn starch (HACS, 70% amylose) had lower bulk densities and higher resiliency compared to starches with lower amylose content (Lacourse & Altieri 1989). Their patent included chemically modified starches including starch esters and starch ethers such as hydroxypropyl starch, which lowers the gelatinization temperature of HACS and enhances expansion, uniformity, processability and resiliency (Lacourse & Altieri 1989; Woggum, Sirivongpaisal & Wittaya 2015). Additives such as polyvinyl alcohol (PVA) were also disclosed as a material that improved the resiliency of the starch foam (Lacourse & Altieri 1989, 1991). Extruded HACS foam had a bulk density and resiliency similar to polystyrene (PS) foam used in packaging applications although the compressive strength was lower.

Starch foam is inherently susceptible to moisture, and there are many studies that disclose the use of different additives to improve moisture resistance including PVA (Combrzyński et al. 2020; Roesser et al. 2000), poly-ε-caprolactone (PCL) (Bastioli et al. 1998), cellulose acetate (Bastioli et al. 1994), poly(ethylene vinyl alcohol) (Bastioli et al. 1994), ethylene-acrylic acid (Bastioli et al. 1994), polyalkylene glycols (Neumann & Seib 1993), poly(lactic acid) (PLA) (Fang & Hanna 2000; Guan, Eskridge & Hanna 2005; Willett & Shogren 2002), poly(hydroxyester ether) (Willett & Shogren 2002), poly(hydroxybutyrate-co-valerate) (Willett & Shogren 2002) and poly(hydroxyamino ether) (PHAE) (Nabar, Narayan & Schindler 2006b). Researchers have also reported improved moisture resistance in foams made with starches chemically modified by etherification with alkylene oxide (Roesser et al. 2000) or by using a starch grafted with plastic moieties such as poly(methyl acrylate) (Bagley et al. 1977; Chen, Gordon & Imam 2004).

Narayan and colleagues have described in detail the extrusion process used for making commercial foam packaging sheets from hydroxypropyl HACS (Arif et al. 2007; Nabar, Narayan & Schindler 2006a; Yang, Graiver & Narayan 2013). High-amylose starch is especially suitable for producing thermoplastic materials because amylose can easily form crystallites and entanglements (Chen et al. 2017), and amylose is less prone to embrittlement caused by chain scission under high shear processing conditions (Lai & Kokini 1991; Shogren & Jasberg 1994).

The screw configuration, screw speed, temperature of heating zones and other extrusion parameters were described for making commercial foam sheets (Nabar, Narayan & Schindler 2006a). A hydroxypropylated HACS with a moisture content of 7%–10% was used including a polymer additive, poly(hydroxyamino ether) (PHAE), at preferred levels of 7% based on starch content. PHAE is a thermoplastic that helps provide moisture resistance,

increases toughness and reduces brittleness (Nabar, Narayan & Schindler 2006a; White et al. 2000). In spite of the use of additives, the extruded foam was not dimensionally stable when stored in different humidities (Arif et al. 2007) and required a protective film for some applications (Greencellfoam 2023). Research has continued for additives that can confer moisture and humidity stability to these packaging foams (Yang, Graiver & Narayan 2013). Nevertheless, the manufacturer touts the benefit of a packaging foam that can easily dissolve in water and be disposed of in a home drain (Greencellfoam 2023). The product marks a trend toward maintaining the inherent moisture sensitivity of starch to facilitate easy and safe disposal, as well as using an external moisture barrier film if moisture resistance is needed (Greencellfoam 2023; Storopak 2023).

16.6.3 Aerogel/Microcellular Starch Foam

Organic aerogels have been made from various aqueous gels including gelatin, agar, cellulose, and egg albumin (Kistler 1931). Semi-rigid starch aquagels or hydrogels are made from gelatinized starch pastes and form solid gels by starch retrogradation (Gudmundsson 1994). Starch aquagels form a three-dimensional structure with very small pores that can be used for separating proteins in gel electrophoresis (Maity et al. 2022). The small pore structure can also be very useful in making starch foams with unique properties. Starch aerogels are formed by replacing the water within the three-dimensional aquagel structure with air. Simply removing water from aquagels by evaporation results in the collapse of the gel structure due to surface tension (Glenn & Irving 1995; Kistler 1931).

The aquagel structure can be better preserved by using a solvent exchange process to reduce surface tension. For example, the surface tension created at the solvent/air interface within a pore is 73.05, 24.05 and 1.16 mN/m for water, ethanol and liquid CO_2, respectively (De Marco & Reverchon 2017; Wexler & Lide 1992). Aquagels can be dried with no surface tension by critical point drying (CPD).

Starch aerogels prepared through solvent exchange and CPD form a porous, open-cell starch matrix with pore sizes ranging from 2 μm to sub-micrometer size (Glenn & Irving 1995). Mean densities reported for starch aerogels are 0.15, 0.26 and 0.29 g/cm³ for HACS, wheat and corn starch aerogels, respectively. Thermal conductivity of HACS aerogels was in the range of super-insulating aerogels (0.024 W/mK) (Glenn & Irving 1995). More recent studies have explored aerogels from amylose–sodium palmitate complexes (Kenar et al. 2014). The gels were dried by CPD and formed aerogels with densities similar to that of HACS aerogels (0.12–0.18 g/cm³) and BET surface areas ranging from 313 to 362 m²/g. Druel et al. (2017) reported pea starch aerogels with densities and insulative values similar to HACS aerogels.

The process for making starch aerogels from aquagels by solvent exchange and CPD can be a lengthy and costly process. Eliminating unnecessary drying steps can help reduce costs. With some starches, aquagels can be dehydrated in ethanol and oven-dried without the need of CPD. In such cases, the starch foam is considered a xerogel instead of an aerogel. Xerogel panels made from wheat and corn starch have properties similar to aerogel panels made by CPD (Glenn & Irving 1995). In contrast, HACS aquagels require dehydration and CPD to minimize surface tension and produce panels of foam aerogels.

Another way to improve the efficiency of making starch aerogels is to minimize the sample size. The solvent exchange step is a diffusion rate dependent process requiring lengthy equilibration periods depending on the thickness of the samples. Small aquagel beads and particles require much shorter dehydration times, and when thoroughly dehydrated in anhydrous ethanol, they can be simply dried in an oven to produce xerogel beads with unique properties (Glenn & Klamczynski 2012; Glenn et al. 2010). Beads 0.25–1 mm in diameter were made by injecting a hot starch paste through nozzles into a stream of chilled vegetable oil (Buttery, Glenn & Stern 1999). The beads were collected, dehydrated in anhydrous ethanol and oven-dried. The beads had many pores in the submicron range and a significant fraction in the range of 5–14 Å, as determined by gas adsorption porosimetry (Buttery, Glenn & Stern 1999). Buttery, Glenn and Stern (1999) found that the beads were effective in sorbing volatile compounds, especially polar compounds in a manner similar to other sorbents. They suggested that the beads could be useful in dried food products as a flavor carrier.

Nanoporous microbeads were formed by collecting small droplets of atomized hot starch paste in an ethanol bath (Glenn et al. 2010). The microbeads were in the size range of 1–5 μm and were dehydrated in anhydrous ethanol and either oven-dried or CPD to obtain a porous foam aerogel microbead (Figure 16.1). Microparticles were also formed by shearing a starch aquagel in a solution of ethanol using a high shear mixer (Glenn & Klamczynski 2012). Emulsion-gelation techniques have also been used to form starch gel microbeads with controllable size ranges (García-González, Camino-Rey et al. 2012; García-González, Uy et al. 2012).

The inherent biocompatibility and biodegradability of starch aerogels make them ideal materials for agricultural applications. Starch aerogels have been used to encapsulate active ingredients for agricultural and pharmaceutical products (García-González, Uy et al. 2012; Glenn & Klamczynski 2012; Glenn et al. 2007, 2010; Yu et al. 2023). Active ingredients can be introduced by dissolution in the dehydrating solvent during the preparation of the aerogel. This allows the active ingredient to penetrate the pores and internal surface area of the starch aerogel. The active ingredient remains behind once the dehydrating solvent (e.g., ethanol or liquid CO_2) is evaporated (De Marco & Reverchon 2017; Glenn et al. 2010). Starch aerogels have been used as carriers for pharmaceuticals including lovastatin (Wu et al. 2011), eprosartan mesylate (Kothawade & Chaudhari 2021), α-tocopherol (De Marco & Reverchon 2017), menadione (De Marco & Reverchon 2017), ibuprofen (Mehling et al. 2009), paracetamol (Mehling et al. 2009) and model compounds including ketoprofen (García-González, Uy et al. 2012).

Starch aerogels are also an attractive material for bone scaffolds due to the biocompatibility of starch, its ability to biodegrade within the body and its ability to promote cell adhesion and proliferation using human osteoblasts (Santos-Rosales et al. 2020; Silva et al. 2007). The fiber-like network that comprises the matrix of starch aerogels closely mimics the extracellular matrix (Santos-Rosales et al. 2020). Composite materials of PCL and starch aerogel microspheres were successfully used in studies on bone regeneration (Goimil et al. 2017).

Another potential application of aerogel particles is as a replacement for mineral fillers used in the paper industry. Aerogel microparticles have an ability to effectively scatter light and have strong opacifying properties due to their high surface area, porosity and brightness (Saari et al. 2005). However, the moisture sensitivity of the aerogels limits their effectiveness as a paper filler or coating. Cross-linking chemicals including glutaraldehyde, epichlorohydrin and alkyl ketene dimer (AKD) were reported to improve water resistance of starch aerogel powders (Bolivar et al. 2007; El-Tahlawy, Venditti & Pawlak 2007, 2008; Patel et al. 2009). New developments in the technology of starch aerogels are frequently reported and will no doubt lead to new applications (Zheng et al. 2020; Zhu 2019).

16.6.4 Baked Starch Foam

Although extrusion has become the most impactful process for making starch foam, it doesn't lend itself to making molded products. A baking process that has been used for over a century in making molded foam food products such as cake ice cream cones and wafer cookies has been developed for making starch-based foam plates, bowls, clamshells and other products. The process involves baking a starch-based dough

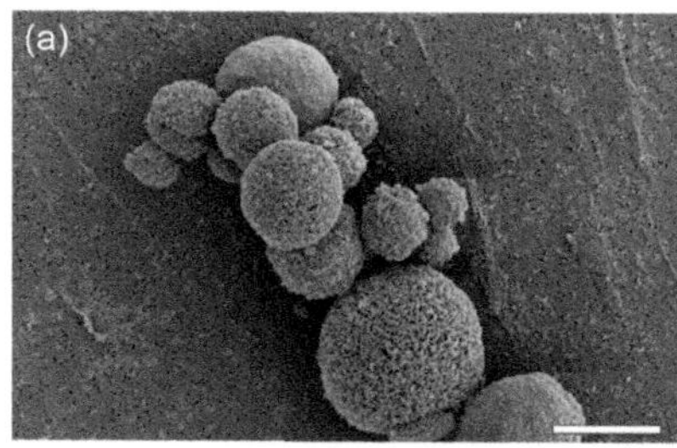
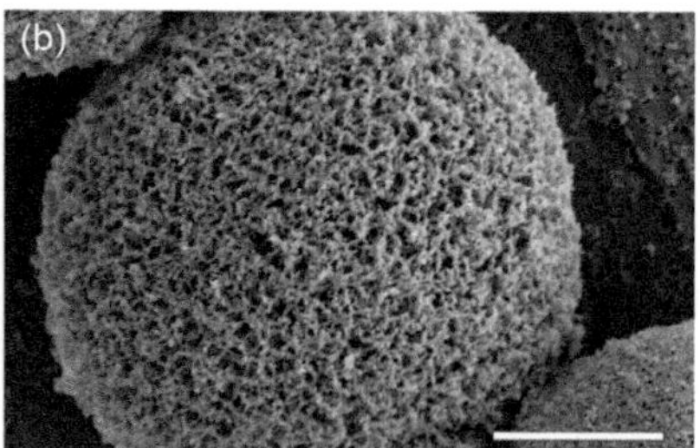

FIGURE 16.1   SEM micrographs of nanoporous microspheres of high-amylose corn starch (HACS) aerogels. Scale bars = 10 and 2 μm for (a) and (b), respectively (Glenn et al., 2010).

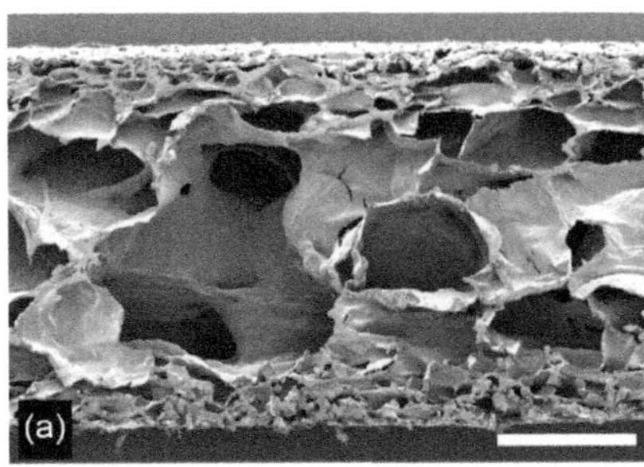

FIGURE 16.2   SEM of foam cross-sectional view (a, Scale bar = 0.5 mm) and photograph of baked foam plate (b).

in a hot, vented mold. The heat causes steam to form and the dough to expand to fill the mold cavity and dry. Tiefenbacher (1993) showed that dough formulations made of a composite of water, starch, plant fibers and other food additives could be baked into trays that were lightweight with good thermal insulation. The baked foam consisted of a porous interior and a denser outer skin where the foam contacted the mold surfaces (Figure 16.2) (Glenn et al. 2001; Shogren et al. 1998a). Baked starch-based formulations have been developed that produce foam products similar to molded polystyrene foam containers in appearance (Figure 16.2).

Numerous studies have investigated the effect of fiber (Chiellini et al. 2009; Glenn, Orts & Nobes 2001; Moo-Tun, Iñiguez-Covarrubias & Valadez-Gonzalez 2020; Nugroho et al. 2022; Tapia-Blácido et al. 2022; Thinyom & Pattavarakorn 2021), starch type (Chiellini et al. 2009; Kaewtatip et al. 2014; Lawton, Shogren & Tiefenbacher 1999; Shogren et al. 1998a) and fillers (Glenn, Orts & Nobes 2001; Moo-Tun, Iñiguez-Covarrubias & Valadez-Gonzalez 2020) on the properties of baked starch-based foams. The major functional limitation of baked starch foams is their moisture sensitivity. Multiple studies have investigated ways to improve moisture resistance including the use of rubber latex (Shey et al. 2006), PVA (Nugroho et al. 2022; Shogren, Lawton & Tiefenbacher 2002; Shogren et al. 1998b), monostearyl citrate (Shogren, Lawton & Tiefenbacher 2002), chemical cross-linking (Uslu & Polat 2012) and laminating films (Andersen & Hodson 1996; Glenn et al. 2001; Tapia-Blácido et al. 2022).

Baked starch-based formulations were developed that produced foam plates and bowls that looked very similar to extruded polystyrene foam containers in appearance (Andersen & Hodson 1996; The-compete-package 2023). The containers were laminated with a degradable film coating that provided moisture resistance and improved the mechanical properties. Commercial production of plates and bowls began in the United States in 2009 with limited success, although some product continues to be marketed (The-compete-package 2023). In addition, an injection molding process was developed for making customized molded baked foam containers (PaperFoam 2023).

A relatively new baking process has been developed for making much thicker molded foam panels for cushioning applications (Alimadadi & Uesaka 2016; Glenn et al. 2023; Lohtander et al. 2022). The primary foam ingredients include

FIGURE 16.3   Starch/fiber foam packaging cushion made by baking process.

water, a foaming surfactant and fiber. The foam is made by rapidly mixing the ingredients to form a wet, stiff foam. Starch can be added as a binder, and the foam can be compression molded into panels or other shapes before baking to remove the water and create a molded foam with excellent cushioning properties (Glenn et al. 2023). The addition of starch as well as reinforcing paperboard elements increased the compressive strength of the foam. These foam products are being developed as a replacement for polystyrene foam packaging materials (Figure 16.3).

16.7 CONCLUSION

Starch is a versatile, renewable and compostable natural polymer that is inexpensive and available in abundant supply. Starch plastification using water or other known plasticizers allows starch to be processed as a continuous matrix material with a wide range of physical properties suitable for various films and foams that have many applications in the food and industrial sectors. The biocompatibility and biodegradability of starch make it ideal for medical applications. Although starch doesn't have the moisture resistance required to replace many plastic products, starch products are slowly gaining market share. This is especially true in single-use packaging products, where the environmental benefits of starch-based products can sometimes outweigh the drawbacks of moisture sensitivity. Further research into starch plastification could lead to better compatibility with other polymers and the development of starch composites with better moisture resistance or other physical properties.

REFERENCES

Abbas, K, Khalil, SK, & Hussin, ASM 2010. Modified starches and their usages in selected food products: A review study. *Journal of Agricultural Science, 2*(2), 90.

Abreu, AS, Oliveira, M, de Sá, A, Rodrigues, RM, Cerqueira, MA, Vicente, AA, & Machado, A 2015. Antimicrobial nanostructured starch based films for packaging. *Carbohydrate Polymers, 129*, 127–134.

Aguilar-Palazuelos, E, Fitch-Vargas, PR, Pérez-Vega, LF, Camacho-Hernández, IL, de Jesús Zazueta-Morales, J, & Calderón-Castro, A 2023. Functional characterization of edible films based on reactive extrusion acetylated corn starch. *Journal of Food Measurement and Characterization, 17*(3), 2363–2373.

Ai, Y, & Jane, Jl 2015. Gelatinization and rheological properties of starch. *Starch/Stärke, 67*(3–4), 213–224.

Alimadadi, M, & Uesaka, T 2016. 3D-oriented fiber networks made by foam forming. *Cellulose, 23*(1), 661–671.

Alonso-López, O, López-Ibáñez, S, & Beiras, R 2021. Assessment of toxicity and biodegradability of poly (vinyl alcohol)-based materials in marine water. *Polymers, 13*(21), 3742.

Altay, F, & Gunasekaran, S 2006. Influence of drying temperature, water content, and heating rate on gelatinization of corn starches. *Journal of Agricultural and Food Chemistry, 54*(12), 4235–4245.

Andersen, PJ, & Hodson, S 1996. *Molded articles having an inorganically filled organic polymer matrix.* (United States Patent No. 5,545,450). USPTO.

Anderson, R 1969. Gelatinization of corn grits by roll-and extrusion-cooking. *Journal of Cereal Science, 14*, 4–7.

Anderson, R, Conway, H, & Peplinski, A 1970. Gelatinization of corn grits by roll cooking, extrusion cooking and steaming. *Starch/Stärke, 22*, 130–134.

Anderson, R, Conway, H, Pfeifer, V, & Griffin, E 1969. Roll and extrusion-cooking of grain sorghum grits. *Cereal Science Today, 14*(11), 372–376.

Anko. 2023. *Spring roll equipment food production equipment.* Retrieved 01-02-2024 from https://www.anko.com.tw/en/category/D11.html

Arif, S, Burgess, G, Narayan, R, & Harte, B 2007. Evaluation of a biodegradable foam for protective packaging applications. *Packaging Technology and Science: An International Journal, 20*(6), 413–419.

Azeredo, HM, & Waldron, KW 2016. Crosslinking in polysaccharide and protein films and coatings for food contact–A review. *Trends in Food Science & Technology, 52*, 109–122.

Bagley, E, Fanta, G, Burr, R, Doane, W, & Russell, C 1977. Graft copolymers of polysaccharides with thermoplastic polymers. A new type of filled plastic. *Polymer Engineering & Science, 17*(5), 311–316.

Balakrishna, AK, Wazed, MA, & Farid, M 2020. A review on the effect of high pressure processing (HPP) on gelatinization and infusion of nutrients. *Molecules, 25*(10), 2369.

Balet, S, Guelpa, A, Fox, G, & Manley, M 2019. Rapid visco analyser (RVA) as a tool for measuring starch-related physiochemical properties in cereals: A review. *Food Analytical Methods, 12*, 2344–2360.

Bangar, SP, Purewal, SS, Trif, M, Maqsood, S, Kumar, M, Manjunatha, V, & Rusu, AV 2021. Functionality and applicability of starch-based films: An eco-friendly approach. *Foods, 10*(9), art. 2181.

Basiak, E, Lenart, A, & Debeaufort, F 2017. Effect of starch type on the physico-chemical properties of edible films. *International Journal of Biological Macromolecules, 98*, 348–356.

Bastioli, C, Bellotti, V, Del Giudice, L, Lombi, R, & Rallis, A 1994. *Expanded articles of biodegradable plastic materials.* (United States Patent No. 5,360,830). USPTO.

Bastioli, C, Bellotti, V, Del Tredici, G, Montino, A, & Ponti, R 1998. *Biodegradable foamed plastic materials.* (United States Patent No. 5,736,586). USPTO.

Beluci, NCL, dos Santos, J, de Carvalho, FA, & Yamashita, F 2023. Reactive biodegradable extruded blends of thermoplastic starch and polyesters. *Carbohydrate Polymer Technologies and Applications, 5*, art. 100274.

Ben, ZY, Samsudin, H, & Yhaya, MF 2022. Glycerol: Its properties, polymer synthesis, and applications in starch based films. *European Polymer Journal, 175*, art. 111377.

Berzin, F, & Vergnes, B 2018. Preparation of cationic starches by reactive extrusion: Experiments and modelling. In *Biomass extrusion and reaction technologies: Principles to practices and future potential*, eds A Ayoub, & L Lucia, ACS Publications, Washington DC, pp. 67–88.

Bizot, H, Le Bail, P, Leroux, B, Davy, J, Roger, P, & Buleon, A 1997. Calorimetric evaluation of the glass transition in hydrated, linear and branched polyanhydroglucose compounds. *Carbohydrate Polymers, 32*(1), 33–50.

Bolivar, AI, Venditti, RA, Pawlak, JJ, & El-Tahlawy, K 2007. Development and characterization of novel starch and alkyl ketene dimer microcellular foam particles. *Carbohydrate Polymers, 69*(2), 262–271.

Bowser, TJ 2017. *Steam basics for food processors.* Oklahoma Cooperative Extension Service, Issue. FAPC-142. https://extension.okstate.edu/fact-sheets/print-publications/fapc-food-and-agricultural-products-center/steam-basics-for-food-processors-fapc-142.pdf

Božanić, DK, Djoković, V, Dimitrijević-Branković, S, Krsmanović, R, McPherson, M, Nair, PS, Georges, MK, & Radhakrishnan, T 2011. Inhibition of microbial growth by silver–starch nanocomposite thin films. *Journal of Biomaterials Science, Polymer Edition, 22*(17), 2343–2355.

Buttery, RG, Glenn, GM, & Stern, DJ 1999. Sorption of volatile flavor compounds by microcellular cereal starch. *Journal of Agricultural and Food Chemistry, 47*(12), 5206–5208.

Carneiro da Silva, LR, Rios, Ad O, & Campomanes Santana, RM 2023. Polymer blends of poly (lactic acid) and starch for the production of films applied in food packaging: A brief review. *Polymers from Renewable Resources, 14*(2), 108–153.

Carr, M 1994. Preparation of cationic starch containing quaternary ammonium substituents by reactive twin-screw extrusion processing. *Journal of Applied Polymer Science, 54*(12), 1855–1861.

Castro, JM, Montalbán, MG, Martínez-Pérez, N, Domene-López, D, Pérez, JM, Arrabal-Campos, FM, Fernández, I, Martín-Gullón, I, & García-Quesada, JC 2023. Thermoplastic starch/polyvinyl alcohol blends modification by citric acid–glycerol polyesters. *International Journal of Biological Macromolecules, 244,* art. 125478.

Castro, LMG, Alexandre, EMC, Saraiva, JA, & Pintado, M 2020. Impact of high pressure on starch properties: A review. *Food Hydrocolloids, 106,* art. 105877.

Castro-Campos, FG, Cabrera-Ramírez, AH, Morales-Sánchez, E, Rodríguez-García, ME, Villamiel, M, Ramos-López, M, & Gaytán-Martínez, M 2021. Impact of the popping process on the structural and thermal properties of sorghum grains (sorghum bicolor l. Moench). *Food Chemistry, 348,* art. 129092.

Cataño, FA, Moreno-Serna, V, Cament, A, Loyo, C, Yáñez-S, M, Ortiz, JA, & Zapata, PA 2023. Green composites based on thermoplastic starch reinforced with micro-and nano-cellulose by melt blending-A review. *International Journal of Biological Macromolecules,* 248, art. 125939.

Chauhan, S, Raghu, N, & Raj, A 2021. Effect of maleic anhydride grafted polylactic acid concentration on mechanical and thermal properties of thermoplasticized starch filled polylactic acid blends. *Polymers and Polymer Composites, 29*(9_suppl), S400–S410.

Chen, L, Gordon, SH, & Imam, SH 2004. Starch graft poly(methyl acrylate) loose-fill foam: Preparation, properties and degradation. *Biomacromolecules, 5*(1), 238–244.

Chen, X, Du, X, Chen, P, Guo, L, Xu, Y, & Zhou, X 2017. Morphologies and gelatinization behaviours of high-amylose maize starches during heat treatment. *Carbohydrate Polymers, 157,* 637–642.

Cheng, H, Chen, L, McClements, DJ, Yang, T, Zhang, Z, Ren, F, Miao, M, Tian, Y, & Jin, Z 2021. Starch-based biodegradable packaging materials: A review of their preparation, characterization and diverse applications in the food industry. *Trends in Food Science & Technology, 114,* 70–82.

Chiellini, E, Cinelli, P, Ilieva, V, Imam, S, & Lawton, J 2009. Environmentally compatible foamed articles based on potato starch, corn fiber, and poly (vinyl alcohol). *Journal of Cellular Plastics, 45*(1), 17–32.

Chinnaswamy, R, & Bhattacharya, K 1983. Studies on expanded rice. Optimum processing conditions. *Journal of Food Science, 48*(6), 1604–1608.

Chinnaswamy, R, & Hanna, MA 1988. Optimum extrusion-cooking conditions for maximum expansion of corn starch. *Journal of Food Science, 53*(3), 834–836.

Choton, S, Gupta, N, Bandral, JD, Anjum, N, & Choudary, A 2020. Extrusion technology and its application in food processing: A review. *The Pharma Innovation Journal, 9*(2), 162–168.

Colomines, G, Decaen, P, Lourdin, D, & Leroy, E 2016. Biofriendly ionic liquids for starch plasticization: A screening approach. *RSC Advances, 6*(93), 90331–90337.

Colonna, P, Doublier, J, Melcion, J, De Monredon, F, & Mercier, C 1984. Extrusion cooking and drum drying of wheat starch. *Cereal Chemistry, 61*(6), 538–554.

Combrzyński, M, Matwijczuk, A, Wójtowicz, A, Oniszczuk, T, Karcz, D, Szponar, J, Niemczynowicz, A, Bober, D, Mitrus, M, & Kupryaniuk, K 2020. Potato starch utilization in ecological loose-fill packaging materials–Sustainability and characterization. *Materials, 13*(6), art. 1390.

Couchman, P, & Karasz, F 1978. A classical thermodynamic discussion of the effect of composition on glass-transition temperatures. *Macromolecules, 11*(1), 117–119.

Cui, C, Ji, N, Wang, Y, Xiong, L, & Sun, Q 2021. Bioactive and intelligent starch-based films: A review. *Trends in Food Science & Technology, 116,* 854–869.

Da Silva, W, Vidal, B, Martins, M, Vargas, H, Pereira, C, Zerbetto, M, & Miranda, LC 1993. What makes popcorn pop. *Nature, 362*(6419), 417–417.

Dang, KM, & Yoksan, R 2021. Thermoplastic starch blown films with improved mechanical and barrier properties. *International Journal of Biological Macromolecules, 188,* 290–299.

De Carvalho, A, Curvelo, AAS, & Agnelli, J 2001. A first insight on composites of thermoplastic starch and kaolin. *Carbohydrate Polymers, 45*(2), 189–194.

de Graaf, RA, & Janssen, LP 2003. The hydroxypropylation of starch in a self-wiping twin screw extruder. *Advances in Polymer Technology: Journal of the Polymer Processing Institute, 22*(1), 56–68.

De Marco, I, & Reverchon, E 2017. Starch aerogel loaded with poorly water-soluble vitamins through supercritical CO_2 adsorption. *Chemical Engineering Research and Design, 119,* 221–230.

de Souza Gamarano, D, Pereira, IM, Mottin, AC, & Ayres, E 2022. Thermoplastic starch–urea, a feasible alternative to release nitrogen as fertilizer. *Macromolecular Symposia, 406*(1), art. 2200043.

Diyana, Z, Jumaidin, R, Selamat, MZ, Ghazali, I, Julmohammad, N, Huda, N, & Ilyas, R 2021. Physical properties of thermoplastic starch derived from natural resources and its blends: A review. *Polymers, 13*(9), art. 1396.

Donmez, D, Pinho, L, Patel, B, Desam, P, & Campanella, OH 2021. Characterization of starch–water interactions and their effects on two key functional properties: Starch gelatinization and retrogradation. *Current Opinion in Food Science, 39,* 103–109.

Druel, L, Bardl, R, Vorwerg, W, & Budtova, T 2017. Starch aerogels: A member of the family of thermal superinsulating materials. *Biomacromolecules, 18*(12), 4232–4239.

Dufresne, A, & Castaño, J 2017. Polysaccharide nanomaterial reinforced starch nanocomposites: A review. *Starch/Stärke, 69*(1–2), art. 1500307.

Dutta, D, & Sit, N 2022. Comparison of properties of films prepared from potato starch modified by annealing and heat–moisture treatment. *Starch/Stärke, 74*(11–12), art. 2200110.

Ek, P, & Ganjyal, GM 2020. Basics of extrusion processing. In *Extrusion cooking*, 2nd Ed, ed Girish M. Ganjyal, UK: Woodhead Publishing, pp. 1–28.

El-Tahlawy, K, Venditti, R, & Pawlak, J 2008. Effect of alkyl ketene dimer reacted starch on the properties of starch microcellular foam using a solvent exchange technique. *Carbohydrate Polymers, 73*(1), 133–142.

El-Tahlawy, K, Venditti, RA, & Pawlak, JJ 2007. Aspects of the preparation of starch microcellular foam particles crosslinked with glutaraldehyde using a solvent exchange technique. *Carbohydrate Polymers, 67*(3), 319–331.

Esmaeili, M, Pircheraghi, G, & Bagheri, R 2017. Optimizing the mechanical and physical properties of thermoplastic starch via tuning the molecular microstructure through co-plasticization by sorbitol and glycerol. *Polymer International, 66*(6), 809–819.

Estevez-Areco, S, Guz, L, Candal, R, & Goyanes, S 2020. Active bilayer films based on cassava starch incorporating ZnO nanorods and PVA electrospun mats containing rosemary extract. *Food Hydrocolloids, 108,* art. 106054.

Faisal, M, Kou, T, Zhong, Y, & Blennow, A 2022. High amylose-based bio composites: Structures, functions and applications. *Polymers, 14*(6), art. 1235.

Fang, Q, & Hanna, MA 2000. Functional properties of polylactic acid starch-based loose-fill packaging foams. *Cereal Chemistry, 77*(6), 779–783.

Fonseca-Florido, H, Soriano-Corral, F, Yañez-Macías, R, González-Morones, P, Hernández-Rodríguez, F, Aguirre-Zurita, J, Ávila-Orta, C, & Rodríguez-Velázquez, J 2019. Effects of multiphase transitions and reactive extrusion on in situ thermoplasticization/succination of cassava starch. *Carbohydrate Polymers, 225,* art. 115250.

Formela, K, Zedler, Ł, Hejna, A, & Tercjak, A 2018. Reactive extrusion of bio-based polymer blends and composites-current trends and future developments. *Express Polymer Letters, 12*(1), 24–57.

Gao, W, Dong, H, Hou, H, & Zhang, H 2012. Effects of clays with various hydrophilicities on properties of starch–clay nanocomposites by film blowing. *Carbohydrate Polymers, 88*(1), 321–328.

Garavand, F, Rouhi, M, Razavi, SH, Cacciotti, I, & Mohammadi, R 2017. Improving the integrity of natural biopolymer films used in food packaging by crosslinking approach: A review. *International Journal of Biological Macromolecules, 104*, 687–707.

García-González, CA, Camino-Rey, M, Alnaief, M, Zetzl, C, & Smirnova, I 2012. Supercritical drying of aerogels using CO_2: Effect of extraction time on the end material textural properties. *The Journal of Supercritical Fluids, 66*, 297–306.

García-González, CA, Uy, J, Alnaief, M, & Smirnova, I 2012. Preparation of tailor-made starch-based aerogel microspheres by the emulsion-gelation method. *Carbohydrate Polymers, 88*(4), 1378–1386.

García-Guzmán, L, Cabrera-Barjas, G, Soria-Hernández, CG, Castaño, J, Guadarrama-Lezama, AY, & Rodríguez Llamazares, S 2022. Progress in starch-based materials for food packaging applications. *Polysaccharides, 3*(1), 136–177.

Gebresas, GA, Szabó, T, & Marossy, K 2023. A comparative study of carboxylic acids on the cross-linking potential of corn starch films. *Journal of Molecular Structure, 1277*, art. 134886.

Giri, P, Tambe, C, & Narayan, R 2018. Using reactive extrusion to manufacture greener products: From laboratory fundamentals to commercial scale. In *Biomass extrusion and reaction technologies: Principles to practices and future potential*, eds A Ayoub & L Lucia, ACS Publications, Washington, DC, pp. 1–23.

Glenn, G, Orts, W, Klamczynski, A, Shogren, R, Hart-Cooper, W, Wood, D, Lee, C, & Chiou, B-S 2023. Compression molded cellulose fiber foams. *Cellulose, 30*(6), 3489–3503.

Glenn, GM, & Irving, DW 1995. Starch-based microcellular foams. *Cereal Chemistry, 72*(2), 155–161.

Glenn, GM, & Klamczynski, A 2012. *Starch foam microparticles.* (United States Patent No. 8163309 B2). USPTO.

Glenn, GM, Klamczynski, AP, Shey, J, Chiou, BS, Holtman, KM, Wood, DF, Ludvik, C, Hoffman, GD, Orts, WJ, & Imam, S 2007. Controlled release of 2-heptanone using starch gel and polycaprolactone matrices and polymeric films. *Polymers for Advanced Technologies, 18*(8), 636–642.

Glenn, GM, Klamczynski, AP, Woods, DF, Chiou, B, Orts, WJ, & Imam, SH 2010. Encapsulation of plant oils in porous starch microspheres. *Journal of Agricultural and Food Chemistry, 58*(7), 4180–4184.

Glenn, GM, Orts, WJ, & Nobes, GAR 2001. Starch, fiber and $CaCO_3$ effects on the physical properties of foams made by a baking process. *Industrial Crops and Products, 14*(3), 201–212.

Glenn, GM, Orts, WJ, Nobes, GAR, & Gray, GM 2001. In situ laminating process for baked starch-based foams. *Industrial Crops and Products, 14*(2), 125–134.

Goimil, L, Braga, MEM, Dias, AMA, Gómez-Amoza, JL, Concheiro, A, Alvarez-Lorenzo, C, de Sousa, HC, & García-González, CA 2017. Supercritical processing of starch aerogels and aerogel-loaded poly(ε-caprolactone) scaffolds for sustained release of ketoprofen for bone regeneration. *Journal of CO_2 Utilization, 18*, 237–249.

Gray, DR, & Chinnaswamy, R 1995. Role of extrusion in food processing. In *Food processing*, ed AG Gaonkar, Amsterdam: Elsevier Science B.V. pp. 241–268.

Greencellfoam. 2023. *Green cell foam.* Retrieved 10/27 from https://greencellfoam.com/green-cell-foam

Grobman, A, Bonavia, D, Dillehay, TD, Piperno, DR, Iriarte, J, & Holst, I 2012. Preceramic maize from Paredones and Huaca Prieta, Peru. *Proceedings of the National Academy of Sciences, 109*(5), 1755–1759.

Gu, B-J, Kowalski, RJ, & Ganjyal, GM. 2017. *Food extrusion processing: An overview.* Fact sheet-Washington State University Extension; 264E, Issue. https://hdl.handle.net/2376/12875

Guan, J, Eskridge, KM, & Hanna, MA 2005. Acetylated starch-polylactic acid loose-fill packaging materials. *Industrial Crops and Products, 22*(2), 109–123.

Gudmundsson, M 1994. Retrogradation of starch and the role of its components. *Thermochimica Acta, 246*(2), 329–341.

Gulati, T, & Datta, AK 2016. Coupled multiphase transport, large deformation and phase transition during rice puffing. *Chemical Engineering Science, 139*, 75–98.

Güllich, LMD, Rosseto, M, Rigueto, CVT, Biduski, B, Gutkoski, LC, & Dettmer, A 2023. Film properties of wheat starch modified by annealing and oxidation. *Polymer Bulletin, 80*(12), 12881–12893.

Gutierrez-Montiel, E, Ávila-Orta, CA, Cabrera-Canales, ZE, Covarrubias-Gordillo, CA, Reyes-Rodríguez, PY, Rodríguez-Velázquez, JG, Cabello-Romero, JN, & Fonseca-Florido, HA 2023. Effect of oxidized, maleate and dual chemical modification through extrusion on thermoplastic starch properties. *Polymer Bulletin, 81*(3), 2525–2544.

Hablot, E, Dewasthale, S, Zhao, Y, Zhiguan, Y, Shi, X, Graiver, D, & Narayan, R 2013. Reactive extrusion of glycerylated starch and starch–polyester graft copolymers. *European Polymer Journal, 49*(4), 873–881.

Han, H, Hou, J, Yang, N, Zhang, Y, Chen, H, Zhang, Z, Shen, Y, Huang, S, & Guo, S 2019. Insight on the changes of cassava and potato starch granules during gelatinization. *International Journal of Biological Macromolecules, 126*, 37–43.

Hoseney, R, Zeleznak, K, & Abdelrahman, A 1983. Mechanism of popcorn popping. *Journal of Cereal Science, 1*(1), 43–52.

Hoyos-Leyva, JD, Bello-Pérez, LA, Agama-Acevedo, E, & Alvarez-Ramirez, J 2015. Optimising the heat moisture treatment of Morado banana starch by response surface analysis. *Starch-Stärke, 67*(11–12), 1026–1034.

Huneault, MA, & Li, H 2007. Morphology and properties of compatibilized polylactide/thermoplastic starch blends. *Polymer, 48*(1), 270–280.

Hutchinson, R, Mantle, S, & Smith, A 1989. The effect of moisture content on the mechanical properties of extruded food foams. *Journal of Materials Science, 24*, 3249–3253.

Iuga, M, & Mironeasa, S 2020. A review of the hydrothermal treatments impact on starch based systems properties. *Critical Reviews in Food Science and Nutrition, 60*(22), 3890–3915.

Jha, P 2020. Effect of plasticizer and antimicrobial agents on functional properties of bionanocomposite films based on corn starch-chitosan for food packaging applications. *International Journal of Biological Macromolecules, 160*, 571–582.

Jiang, J, Zhang, M, Bhandari, B, & Cao, P 2020. Current processing and packing technology for space foods: A review. *Critical Reviews in Food Science and Nutrition, 60*(21), 3573–3588.

Jimenez, A, Fabra, MJ, Talens, P, & Chiralt, A 2012. Edible and biodegradable starch films: A review. *Food and Bioprocess Technology, 5*, 2058–2076.

Kaewtatip, K, Poungroi, M, Holló, B, & Mészáros Szécsényi, K 2014. Effects of starch types on the properties of baked starch foams. *Journal of Thermal Analysis and Calorimetry, 115*, 833–840.

Kalambur, S, & Rizvi, SS 2006. An overview of starch-based plastic blends from reactive extrusion. *Journal of Plastic Film & Sheeting, 22*(1), 39–58.

Kenar, JA, Eller, FJ, Felker, FC, Jackson, MA, & Fanta, GF 2014. Starch aerogel beads obtained from inclusion complexes prepared from high amylose starch and sodium palmitate. *Green Chemistry, 16*(4), 1921–1930.

Kim, H-Y, Jane, J-L, & Lamsal, B 2017. Hydroxypropylation improves film properties of high amylose corn starch. *Industrial Crops and Products, 95*, 175–183.

Kistler, SS 1931. Coherent expanded aerogels and jellies. *Nature, 127*(3211), 741.

Kong, L, & Ziegler, GR 2014. Fabrication of pure starch fibers by electrospinning. *Food Hydrocolloids, 36*, 20–25.

Korma, SA, Kamal-Alahmad, SN, Ammar, A-F, Zaaboul, F, & Zhang, T 2016. Chemically modified starch and utilization in food stuffs. *International Journal of Nutrition and Food Sciences, 5*(4), 264–272.

Kothawade, SN, & Chaudhari, PD 2021. Development of biodegradable porous starch foam for improving oral delivery of eprosartan mesylate. *Journal of Advanced Scientific Research, 12*(03 Suppl 1), 120–126.

Kramer, ME 2009. Structure and function of starch-based edible films and coatings. In *Edible films and coatings for food applications*, eds ME Embuscado & KC Huber, Springer, Dordrecht, pp. 113–134.

Kuraray. 2023. *News and events*. Retrieved December 18, 2023 from https://kuraray.us.com/2015/08/10/kuraray-is-pleased-to-announce-completion-of-the-acquisition-of-all-of-the-shares-in-plantic-technologies-limited/

Kurek, M, & Ščetar, M 2017. Edible coatings minimize fat uptake in deep fat fried products: A review. *Food Hydrocolloids, 71*, 225–235.

Labelle, MA, Ispas-Szabo, P, & Mateescu, MA 2020. Structure-functions relationship of modified starches for pharmaceutical and biomedical applications. *Starch/Stärke, 72*(7–8), art. 2000002.

Lacourse, L, & Altieri, P 1989. *Biodegradable packaging material and the method of preparation thereof.* (United States Patent No. 4,863,655A). USPTO.

Lacourse, L, & Altieri, P 1991. *Biodegradable shaped products and the method of preparation thereof.* (United States Patent No. 5,043,196). USPTO.

Lai, L, & Kokini, J 1991. Physicochemical changes and rheological properties of starch during extrusion (a review). *Biotechnology Progress, 7*(3), 251–266.

Landerito, NA, & Wang, YJ 2005. Preparation and properties of starch phosphates using waxy, common, and high-amylose corn starches. II. Reactive extrusion method. *Cereal Chemistry, 82*(3), 271–276.

Lauer, MK, & Smith, RC 2020. Recent advances in starch-based films toward food packaging applications: Physicochemical, mechanical, and functional properties. *Comprehensive Reviews in Food Science and Food Safety, 19*(6), 3031–3083.

Lawton, J, Shogren, R, & Tiefenbacher, K 1999. Effect of batter solids and starch type on the structure of baked starch foams. *Cereal Chemistry, 76*(5), 682–687.

Leite, TS, de Jesus, ALT, Schmiele, M, Tribst, AA, & Cristianini, M 2017. High pressure processing (HPP) of pea starch: Effect on the gelatinization properties. *LWT-Food Science and Technology, 76*, 361–369.

Li, J, He, H, Zhang, H, Xu, M, Gu, Q, & Zhu, Z 2022. Preparation of thermoplastic starch with comprehensive performance plasticized by citric acid. *Journal of Applied Polymer Science, 139*(25), art. e52401.

Li, X, Liu, J, Lu, Y, Hou, T, Zhou, J, Wang, A, Zhang, X, & Yang, B 2021. Centrifugally spun starch/polyvinyl alcohol ultrafine fibrous membrane as environmentally-friendly disposable nonwoven. *Journal of Applied Polymer Science, 138*(40), art. 51169.

Lim, WS, Ock, SY, Park, GD, Lee, IW, Lee, MH, & Park, HJ 2020. Heat-sealing property of cassava starch film plasticized with glycerol and sorbitol. *Food Packaging and Shelf Life, 26*, art. 100556.

Lin, Y, Huff, HE, Parsons, MH, Iannotti, E, & Hsieh, F 1995. Mechanical properties of extruded high amylose starch for loose-fill packaging material. *LWT-Food Science and Technology, 28*(2), 163–168.

Liu, H, Xie, F, Yu, L, Chen, L, & Li, L 2009. Thermal processing of starch-based polymers. *Progress in Polymer Science, 34*(12), 1348–1368.

Liu, X, Chao, C, Yu, J, Copeland, L, & Wang, S 2021. Mechanistic studies of starch retrogradation and its effects on starch gel properties. *Food Hydrocolloids, 120*, art. 106914.

Liu, Y, Chao, C, Yu, J, Wang, S, Wang, S, & Copeland, L 2020. New insights into starch gelatinization by high pressure: Comparison with heat-gelatinization. *Food Chemistry, 318*, art. 126493.

Lohtander, T, Herrala, R, Laaksonen, P, Franssila, S, & Österberg, M 2022. Lightweight lignocellulosic foams for thermal insulation. *Cellulose, 29*(3), 1855–1871.

Long, J, Zhang, W, Zhao, M, & Ruan, C-Q 2023. The reduce of water vapor permeability of polysaccharide-based films in food packaging: A comprehensive review. *Carbohydrate Polymers, 321*, art. 121267.

Lourdin, D, Bizot, H, & Colonna, P 1997. "Antiplasticization" in starch-glycerol films? *Journal of Applied Polymer Science, 63*(8), 1047–1053.

Lourdin, D, Coignard, L, Bizot, H, & Colonna, P 1997. Influence of equilibrium relative humidity and plasticizer concentration on the water content and glass transition of starch materials. *Polymer, 38*(21), 5401–5406.

Ma, X, & Yu, J 2004. The plasticizers containing amide groups for thermoplastic starch. *Carbohydrate Polymers, 57*(2), 197–203.

Madhumitha, G, Fowsiya, J, Mohana Roopan, S, & Thakur, VK 2018. Recent advances in starch–clay nanocomposites. *International Journal of Polymer Analysis and Characterization, 23*(4), 331–345.

Maity, A, Kesh, SS, Palai, S, & Egbuna, C 2022. Electrophoretic techniques. In *Analytical techniques in biosciences-from basics to applications*, eds C Egbuna, K Patrick-Iwuanyanwu, MA Shah, JC Ifemeje, & A Rasul, New York: Academic Press, pp. 59–72.

Majzoobi, M, Pesaran, Y, Mesbahi, G, Golmakani, MT, & Farahnaky, A 2015. Physical properties of biodegradable films from heat-moisture-treated rice flour and rice starch. *Starch/Stärke, 67*(11–12), 1053–1060.

Malumba, P, Delatte, S, Doran, L, & Blecker, C 2022. The effect of annealing under acid or alkaline environment on the physicochemical and functional properties of wheat starch. *Food Hydrocolloids, 125*, art. 107452.

Mariotti, M, Alamprese, C, Pagani, M, & Lucisano, M 2006. Effect of puffing on ultrastructure and physical characteristics of cereal grains and flours. *Journal of Cereal Science, 43*(1), 47–56.

MarketsandMarkets. 2023. *Global food extrusion market.* Retrieved October 21, 2023 from https://www.marketsandmarkets. com/Market-Reports/food-extrusion-market-221423108. html#:~:text=2023%20and%20beyond%3F-,The%20 Food%20Extrusion%20Market%20is%20projected%20to%20 reach%20USD%2099.7,USD%2073.1%20billion%20in%20 2021

Mascia, L, Kouparitsas, Y, Nocita, D, & Bao, X 2020. Antiplasticization of polymer materials: Structural aspects and effects on mechanical and diffusion-controlled properties. *Polymers, 12*(4), art. 769.

Maskan, M, & Altan, A. (eds) 2011. *Advances in food extrusion technology.* Taylor and Francis Group, Florida, USA: CRC Press.

McGrane, SJ, Mainwaring, DE, Cornell, HJ, & Rix, CJ 2004. The role of hydrogen bonding in amylose gelation. *Starch/Stärke, 56*(3–4), 122–131.

Mehling, T, Smirnova, I, Guenther, U, & Neubert, RH 2009. Polysaccharide-based aerogels as drug carriers. *Journal of Non-Crystalline Solids, 355*(50–51), 2472–2479.

Mercier, C 1977. Effect of extrusion-cooking on potato starch using a twin screw french extruder. *Starch/Stärke, 29*(2), 48–52.

Mercier, C, Charbonniere, R, Gallant, D, & Guilbot, A 1979. Structural modification of various starches by extrusion-cooking with a twin screw french extruder. In *Polysaccharides in food*, eds JMV Blanchard & JR Mitchell, Butterworth, London, pp. 153–170.

Mercier, C, Charbonniere, R, Grebaut, J, & De la Gueriviere, J 1975. Formation of amylose-lipid complexes by twin-screw extrusion cooking of manioc starch. *Cereal Chemistry, 57*(1), 4–9.

Miladinov, V, & Hanna, M 2000. Starch esterification by reactive extrusion. *Industrial Crops and Products, 11*(1), 51–57.

Milotskyi, R, & Bliard, C 2018. Carboxymethylation of plasticized starch by reactive extrusion (REX) with high reaction efficiency. *Starch/Stärke, 70*(11–12), art. 1700275.

Mishra, G, Joshi, DC, & Panda, BK 2014. Popping and puffing of cereal grains: A review. *Journal of Grain Processing and Storage, 1*(2), 34–46.

Moad, G 2011. Chemical modification of starch by reactive extrusion. *Progress in Polymer Science, 36*(2), 218–237.

Momany, FA, Willett, J, & Schnupf, U 2009. Molecular dynamics simulations of a cyclic-DP-240 amylose fragment in a periodic cell: Glass transition temperature and water diffusion. *Carbohydrate Polymers, 78*(4), 978–986.

Montilla-Buitrago, CE, Gómez-López, RA, Solanilla-Duque, JF, Serna-Cock, L, & Villada-Castillo, HS 2021. Effect of plasticizers on properties, retrogradation, and processing of extrusion-obtained thermoplastic starch: A review. *Starch/Stärke, 73*(9–10), art. 2100060.

Moo-Tun, NM, Iñiguez-Covarrubias, G, & Valadez-Gonzalez, A 2020. Assessing the effect of PLA, cellulose microfibers and $CaCO_3$ on the properties of starch-based foams using a factorial design. *Polymer Testing, 86*, art. 106482.

Muñoz-Gimena, PF, Oliver-Cuenca, V, Peponi, L, & López, D 2023. A review on reinforcements and additives in starch-based composites for food packaging. *Polymers, 15*(13), art. 2972.

Muscat, D, Adhikari, B, Adhikari, R, & Chaudhary, DS 2012. Comparative study of film forming behaviour of low and high amylose starches using glycerol and xylitol as plasticizers. *Journal of Food Engineering, 109*(2), 189–201.

Myllärinen, P, Partanen, R, Seppälä, J, & Forssell, P 2002. Effect of glycerol on behaviour of amylose and amylopectin films. *Carbohydrate Polymers, 50*(4), 355–361.

Nabar, Y, Narayan, R, & Schindler, M 2006a. Twin-screw extrusion production and characterization of starch foam products for use in cushioning and insulation applications. *Polymer Engineering & Science, 46*(4), 438–451.

Nabar, Y, Narayan, R, & Schindler, M 2006b. Twin-screw extrusion production and characterization of starch foam products for use in cushioning and insulation applications. *Polymer Engineering & Science, 46*(4), 438–451.

Nabar, Y, Raquez, JM, Dubois, P, & Narayan, R 2005. Production of starch foams by twin-screw extrusion: Effect of maleated poly(butylene adipate-co-terephthalate) as a compatibilizer. *Biomacromolecules, 6*(2), 807–817.

Neumann, P, & Seib, P 1993. *Starch-based. Biodegradable packing filler and method of preparing same.* (United States Patent No. 5,208,267). USPTO.

Niazi, MB, Zijlstra, M, & Broekhuis, AA 2015. Influence of plasticizer with different functional groups on thermoplastic starch. *Journal of Applied Polymer Science, 132*(22) art. 42012

Novamont. 2023. Mater bi mulching film. Retrieved December 18, 2023 from https://materbi.com/en/solutions/agriculture/ mulching-film/

Nugroho, A, Maharani, DM, Legowo, AC, Hadi, S, & Purba, F 2022. Enhanced mechanical and physical properties of starch foam from the combination of water hyacinth fiber (eichhornia crassipes) and polyvinyl alcohol. *Industrial Crops and Products, 183*, art. 114936.

Ojogbo, E, Blanchard, R, & Mekonnen, T 2018. Hydrophobic and melt processable starch-laurate esters: Synthesis, structure–property correlations. *Journal of Polymer Science Part A: Polymer Chemistry, 56*(23), 2611–2622.

Olivato, J, Grossmann, M, Yamashita, F, Eiras, D, & Pessan, L 2012. Citric acid and maleic anhydride as compatibilizers in starch/ poly (butylene adipate-co-terephthalate) blends by one-step reactive extrusion. *Carbohydrate Polymers, 87*(4), 2614–2618.

Orford, P, Parker, R, Ring, S, & Smith, A 1989. Effect of water as a diluent on the glass transition behaviour of malto-oligosaccharides, amylose and amylopectin. *International Journal of Biological Macromolecules, 11*(2), 91–96.

Orts, W, Glenn, G, Nobes, G, & Wood, D 2000. Wheat starch effects on the textural characteristics of puffed brown rice cakes. *Cereal Chemistry, 77*(1), 18–23.

Özeren, HD, Guivier, M, Olsson, RT, Nilsson, F, & Hedenqvist, MS 2020a. Ranking plasticizers for polymers with atomistic simulations: PVT, mechanical properties, and the role of hydrogen bonding in thermoplastic starch. *ACS Applied Polymer Materials, 2*(5), 2016–2026.

Özeren, HD, Olsson, RT, Nilsson, F, & Hedenqvist, MS 2020b. Prediction of plasticization in a real biopolymer system (starch) using molecular dynamics simulations. *Materials & Design, 187*, art. 108387.

Palai, B, Biswal, M, Mohanty, S, & Nayak, SK 2019. In situ reactive compatibilization of polylactic acid (PLA) and thermoplastic starch (TPS) blends; synthesis and evaluation of extrusion blown films thereof. *Industrial Crops and Products, 141*, art. 111748.

Palav, T, & Seetharaman, K 2006. Mechanism of starch gelatinization and polymer leaching during microwave heating. *Carbohydrate Polymers, 65*(3), 364–370.

Paluch, M, Ostrowska, J, Tyński, P, Sadurski, W, & Konkol, M 2022. Structural and thermal properties of starch plasticized with glycerol/urea mixture. *Journal of Polymers and the Environment, 30*, 1–13.

PaperFoam. 2023. *Eco friendly packaging.* Retrieved November 08, 2023 from https://paperfoam.com/

Park, HM, Li, X, Jin, CZ, Park, CY, Cho, WJ, & Ha, CS 2002. Preparation and properties of biodegradable thermoplastic starch/clay hybrids. *Macromolecular Materials and Engineering, 287*(8), 553–558.

Patel, S, Venditti, RA, Pawlak, JJ, Ayoub, A, & Rizvi, SS 2009. Development of cross-linked starch microcellular foam by solvent exchange and reactive supercritical fluid extrusion. *Journal of Applied Polymer Science, 111*(6), 2917–2929.

Plantic. 2023. *Planntic technologies limited*. Retrieved December 18, 2023 from https://plantic.com.au/company/about.html

Primus. 2020. *Improving the production process*. Retrieved January 02, 2024 from https://www.waferpaper.com/blog/production-process/improving-the-production-process

Prismane. 2021. *Polylactic acid market overview-2021*. Retrieved August 09, 2024 from https://prismaneconsulting.com/report-details/asia-pacific-polylactic-acid-pla-market-study

Priyadarshi, R, Roy, S, Ghosh, T, Biswas, D, & Rhim, J-W 2022. Antimicrobial nanofillers reinforced biopolymer composite films for active food packaging applications-A review. *Sustainable Materials and Technologies, 32*, art. e00353.

Ptak, S, Zarski, A, & Kapusniak, J 2020. The importance of ionic liquids in the modification of starch and processing of starch-based materials. *Materials, 13*(20), art. 4479.

Qiao, X, Jiang, W, & Sun, K 2005. Reinforced thermoplastic acetylated starch with layered silicates. *Starch/Stärke, 57*(12), 581–586.

Raquez, JM, Nabar, Y, Narayan, R, & Dubois, P 2008a. In situ compatibilization of maleated thermoplastic starch/polyester melt-blends by reactive extrusion. *Polymer Engineering & Science, 48*(9), 1747–1754.

Raquez, JM, Nabar, Y, Srinivasan, M, Shin, B-Y, Narayan, R, & wDubois, P 2008b. Maleated thermoplastic starch by reactive extrusion. *Carbohydrate Polymers, 74*(2), 159–169.

Ratnayake, WS, & Jackson, DS 2008. Starch gelatinization. *Advances in Food and Nutrition Research, 55*, 221–268.

Ren, F, Wang, J, Xie, F, Zan, K, Wang, S, & Wang, S 2020. Applications of ionic liquids in starch chemistry: A review. *Green Chemistry, 22*(7), 2162–2183.

ResearchandMarkets. 2023. *Starch - global strategic business report*. #2832332. https://www.globenewswire.com/en/news-rclcase/2023/02/28/2617539/28124/en/Starch-Global-Market-to-Reach-199-8-Million-Metric-Tons-by-2030-Usc-of-Starch-as-a-Fat-Replacer-Drives-Growth.html#:~:text=The%20global%20market%20for%20Starch,the%20analysis%20period%202022%2D2030

Riaz, MN (ed) 2000. *Extruders in food applications*. CRC Press, London.

Rindlav-Westling, Å, Stading, M, & Gatenholm, P 2002. Crystallinity and morphology in films of starch, amylose and amylopectin blends. *Biomacromolecules, 3*(1), 84–91.

Roesser, D, Nevling, J, Rawlins, D, & Billmers, R 2000. *Biodegradable expanded starch products and the method of preparation*. (United States Patent No. 6,107,371). USPTO.

Saari, J, Kataja, K, Qvintus-Leino, P, Kuutti, L, Peltonen, S, Mikkonen, H, & Joyce, M 2005. Development of non-mineral starch based pigments 6 fillers: Trials with developed non-mineral starch based pigments and fillers. In *5th International Scientific and Technical Advances in Fillers and Pigments for Papermakers*, Pira International Ltd., Bonn, GmbH.

Sanders, JM, Misra, M, Mustard, TJ, Giesen, DJ, Zhang, T, Shelley, J, & Halls, MD 2021. Characterizing moisture uptake and plasticization effects of water on amorphous amylose starch models using molecular dynamics methods. *Carbohydrate Polymers, 252*, art. 117161.

Sankri, A, Arhaliass, A, Dez, I, Gaumont, AC, Grohens, Y, Lourdin, D, Pillin, I, Rolland-Sabaté, A, & Leroy, E 2010. Thermoplastic starch plasticized by an ionic liquid. *Carbohydrate Polymers, 82*(2), 256–263.

Santos-Rosales, V, Alvarez-Rivera, G, Hillgärtner, M, Cifuentes, A, Itskov, M, García-González, CA, & Rege, A 2020. Stability studies of starch aerogel formulations for biomedical applications. *Biomacromolecules, 21*(12), 5336–5344.

Sapper, M, & Chiralt, A 2018. Starch-based coatings for preservation of fruits and vegetables. *Coatings, 8*(5), art. 152.

Shah, U, Gani, A, Ashwar, BA, Shah, A, Ahmad, M, Gani, A, Wani, IA, & Masoodi, F 2015. A review of the recent advances in starch as active and nanocomposite packaging films. *Cogent Food & Agriculture, 1*(1), art. 1115640.

Shah, U, Naqash, F, Gani, A, & Masoodi, F 2016. Art and science behind modified starch edible films and coatings: A review. *Comprehensive Reviews in Food Science and Food Safety, 15*(3), 568–580.

Shavandi, M, Javanmard, M, & Basiri, A 2023. Novel infrared puffing: Effect on physicochemical attributes of puffed rice (Oryza sativa L.). *Food Science & Nutrition, 11*(5), 2141–2151.

Shey, J, Imam, S, Glenn, G, & Orts, W 2006. Properties of baked starch foam with natural rubber latex. *Industrial Crops and Products, 24*(1), 34–40.

Shimoni, E, Dirks, E, & Labuza, T 2002. The relation between final popped volume of popcorn and thermal–physical parameters. *LWT-Food Science and Technology, 35*(1), 93–98.

Shogren, R 2007. Effect of orientation on the physical properties of potato amylose and high-amylose corn starch films. *Biomacromolecules, 8*(11), 3641–3645.

Shogren, RL 1992. Effect of moisture content on the melting and subsequent physical aging of cornstarch. *Carbohydrate Polymers, 19*(2), 83–90.

Shogren, RL, Fanta, GF, & Doane, WM 1993. Development of starch based plastics-A reexamination of selected polymer systems in historical perspective. *Starch/Stärke, 45*(8), 276–280.

Shogren, RL, & Jasberg, BK 1994. Aging properties of extruded high-amylose starch. *Journal of Environmental Polymer Degradation, 2*, 99–109.

Shogren, RL, Lawton, JW, Doane, WM, & Tiefenbacher, KF 1998a. Structure and morphology of baked starch foams. *Polymer, 39*(25), 6649–6655.

Shogren, RL, Lawton, JW, & Tiefenbacher, KF 2002. Baked starch foams: Starch modifications and additives improve process parameters, structure and properties. *Industrial Crops and Products, 16*(1), 69–79.

Shogren, RL, Lawton, JW, Tiefenbacher, KF, & Chen, L 1998b. Starch–poly (vinyl alcohol) foamed articles prepared by a baking process. *Journal of Applied Polymer Science, 68*(13), 2129–2140.

Shogren, RL, Swanson, CL, & Thompson, AR 1992. Extrudates of cornstarch with urea and glycols: Structure/mechanical property relations. *Starch/Stärke, 44*(9), 335–338.

Silva, GA, Coutinho, OP, Ducheyne, P, Shapiro, IM, & Reis, RL 2007. The effect of starch and starch-bioactive glass composite microparticles on the adhesion and expression of the osteoblastic phenotype of a bone cell line. *Biomaterials, 28*(2), 326–334.

Singh, GP, Bangar, SP, Yang, T, Trif, M, Kumar, V, & Kumar, D 2022. Effect on the properties of edible starch-based films by the incorporation of additives: A review. *Polymers, 14*(10), art. 1987.

Singh, J, Kaur, L, & McCarthy, O 2007. Factors influencing the physico-chemical, morphological, thermal and rheological properties of some chemically modified starches for food applications—A review. *Food Hydrocolloids, 21*(1), 1–22.

Soykeabkaew, N, Thanomsilp, C, & Suwantong, O 2015. A review: Starch-based composite foams. *Composites Part A: Applied Science and Manufacturing, 78*, 246–263.

Stagner, J, Dias Alves, V, Narayan, R, & Beleia, A 2011. Thermoplasticization of high amylose starch by chemical modification using reactive extrusion. *Journal of Polymers and the Environment, 19*, 589–597.

Stein, TM, & Greene, RV 1997. Amino acids as plasticizers for starch-based plastics. *Starch/Stärke, 49*(6), 245–249.

Storopak. 2023. *Renature®*. Retrieved October 27, 2023 from https://www.storopack.us/products/flexible-protective-packaging/loose-fill/renaturer/

Su, C-Y, Li, D, Wang, L-J, & Wang, Y 2023. Biodegradation behavior and digestive properties of starch-based film for food packaging–A review. *Critical Reviews in Food Science and Nutrition, 63*(24), 6923–6945.

Surendren, A, Mohanty, AK, Liu, Q, & Misra, M 2022. A review of biodegradable thermoplastic starches, their blends and composites: Recent developments and opportunities for single-use plastic packaging alternatives. *Green Chemistry, 24*(22), 8606–8636.

Swarnakar, AK, Mohapatra, M, & Das, SK 2022. A review on processes, mechanisms, and quality influencing parameters for puffing and popping of grains. *Journal of Food Processing and Preservation, 46*(10), art. e16891.

Sweley, JC, Rose, DJ, & Jackson, DS 2013. Quality traits and popping performance considerations for popcorn (zea mays everta). *Food Reviews International, 29*(2), 157–177.

Talja, RA, Helén, H, Roos, YH, & Jouppila, K 2007. Effect of various polyols and polyol contents on physical and mechanical properties of potato starch-based films. *Carbohydrate Polymers, 67*(3), 288–295.

Tan, SX, Andriyana, A, Ong, HC, Lim, S, Pang, YL, & Ngoh, GC 2022. A comprehensive review on the emerging roles of nanofillers and plasticizers towards sustainable starch-based bioplastic fabrication. *Polymers, 14*(4), art. 664.

Tang, X, Alavi, S, & Herald, TJ 2008. Effects of plasticizers on the structure and properties of starch–clay nanocomposite films. *Carbohydrate Polymers, 74*(3), 552–558.

Tapia-Blácido, DR, Aguilar, GJ, de Andrade, MT, Rodrigues-Júnior, MF, & Guareschi-Martins, FC 2022. Trends and challenges of starch-based foams for use as food packaging and food container. *Trends in Food Science & Technology, 119*, 257–271.

Tatarka, P, & Cunningham, R 1998. Properties of protective loose-fill foams. *Journal of Applied Polymer Science, 67*(7), 1157–1176.

Teixeira, EM, Da Róz, A, Carvalho, A, & Curvelo, A 2007. The effect of glycerol/sugar/water and sugar/water mixtures on the plasticization of thermoplastic cassava starch. *Carbohydrate Polymers, 69*(4), 619–624.

Temesgen, S, Rennert, M, Tesfaye, T, & Nase, M 2021. Review on spinning of biopolymer fibers from starch. *Polymers, 13*(7), art. 1121.

Thakur, R, Pristijono, P, Scarlett, CJ, Bowyer, M, Singh, S, & Vuong, QV 2019. Starch-based films: Major factors affecting their properties. *International Journal of Biological Macromolecules, 132*, 1079–1089.

The-compete-package. 2023. *Earthshell biodegradable plates and bowls*. Retrieved November 08, 2023 from https://www.the-complete-package.com/View/EarthShell-Biodegradable-Plates-And-Bowls

Thinyom, K, & Pattavarakorn, D 2021. Baked starch-based bio-composite foam filled with cassava wastes. *Proceeding National & International Conference, 1*(14), art. 235.

Thunwall, M, Kuthanová, V, Boldizar, A, & Rigdahl, M 2008. Film blowing of thermoplastic starch. *Carbohydrate Polymers, 71*(4), 583–590.

Tiefenbacher, KF 1993. Starch-based foamed materials—use and degradation properties. *Journal of Macromolecular Science, Part A: Pure and Applied Chemistry, 30*(9–10), 727–731.

Ubbink, J 2018. Plasticization and antiplasticization in amorphous food systems. *Current Opinion in Food Science, 21*, 72–78.

Uslu, M-K, & Polat, S 2012. Effects of glyoxal cross-linking on baked starch foam. *Carbohydrate Polymers, 87*(3), 1994–1999.

Van der Sman, R 2019. Phase separation, antiplasticization and moisture sorption in ternary systems containing polysaccharides and polyols. *Food Hydrocolloids, 87*, 360–370.

Van Soest, JJ, Hulleman, S, De Wit, D, & Vliegenthart, J 1996. Changes in the mechanical properties of thermoplastic potato starch in relation with changes in b-type crystallinity. *Carbohydrate Polymers, 29*(3), 225–232.

Vilpoux, OF, & Santos Silveira Junior, JF 2023. Global production and use of starch. In *Starchy crops morphology, extraction, properties and applications*, eds M Pascoli Cereda & O François Vilpoux, New York: Academic Press, pp. 43–66.

Voegele, E 2023. Usda maintains 2023-'24 forecast for corn use in ethanol. *Ethanol Producer Magazine*, (September 12, 2023). https://ethanolproducer.com/articles/usda-maintains-forecast-for-2023-24-corn-use-in-ethanol

Wadaugsorn, K, Panrong, T, Wongphan, P, & Harnkarnsujarit, N 2022. Plasticized hydroxypropyl cassava starch blended PBAT for improved clarity blown films: Morphology and properties. *Industrial Crops and Products, 176*, art. 114311.

Wang, J-L, Cheng, F, & Zhu, P-X 2014. Structure and properties of urea-plasticized starch films with different urea contents. *Carbohydrate Polymers, 101*, 1109–1115.

Wang, L, Shogren, RL, & Willett, JL 1997. Preparation of starch succinates by reactive extrusion. *Starch/Stärke, 49*(3), 116–120.

Wang, N, Yu, J, & Ma, X 2007. Preparation and characterization of thermoplastic starch/PLA blends by one-step reactive extrusion. *Polymer International, 56*(11), 1440–1447.

Wang, SS, Chiang, W, Zhao, B, Zheng, X, & Kim, I 1991. Experimental analysis and computer simulation of starch-water interactions during phase transition. *Journal of Food Science, 56*(1), 121–124.

Warburton, SC, Donald, AM, & Smith, AC 1990. The deformation of brittle starch foams. *Journal of Materials Science, 25*(9), 4001–4007.

Wexler, A, & Lide, D. 1992. *Handbook of chemistry and physics* (Vol. 73). CRC Press, London.

White, J, Winkler, M, Glass, T, Reddy, H, & Everett, J 2000. Development of new family of thermoplastics employing poly (hydroxyamino ether) chemistry. *Plastics, Rubber and Composites, 29*(8), 395–400.

Willett, JL, & Shogren, RL 2002. Processing and properties of extruded starch/polymer foams. *Polymer, 43*(22), 5935–5947.

Winkler, H, Vorwerg, W, & Rihm, R 2014. Thermal and mechanical properties of fatty acid starch esters. *Carbohydrate Polymers, 102*, 941–949.

Woggum, T, Sirivongpaisal, P, & Wittaya, T 2015. Characteristics and properties of hydroxypropylated rice starch based biodegradable films. *Food Hydrocolloids, 50*, 54–64.

World-Centric. 2023. *Compostable food scraps, kitchen and yard waste bags*. Retrieved December 18, 2023 from https://store.worldcentric.com/store/bags

Wu, C, Wang, Z, Zhi, Z, Jiang, T, Zhang, J, & Wang, S 2011. Development of biodegradable porous starch foam for improving oral delivery of poorly water soluble drugs. *International Journal of Pharmaceutics, 403*(1), 162–169.

Xie, F, Liu, H, Chen, P, Xue, T, Chen, L, Yu, L, & Corrigan, P 2007. Starch gelatinization under shearless and shear conditions. *International Journal of Food Engineering, 2*(5), art. 6.

Xie, F, Yu, L, Liu, H, & Chen, L 2006. Starch modification using reactive extrusion. *Starch/Stärke, 58*(3–4), 131–139.

Xie, M, Duan, Y, Li, F, Wang, X, Cui, X, Bacha, U, Zhu, M, Xiao, Z, & Zhao, Z 2017. Preparation and characterization of modified and functional starch (hexadecyl corboxymethyl starch) ether using reactive extrusion. *Starch/Stärke, 69*(5–6), art. 1600061.

Xu, H, Chen, L, Xu, Z, McClements, DJ, Cheng, H, Qiu, C, Long, J, Ji, H, Meng, M, & Jin, Z 2023. Structure and properties of flexible starch-based double network composite films induced by dopamine self-polymerization. *Carbohydrate Polymers, 299*, art. 120106.

Xu, J, Andrews, TD, & Shi, YC 2020. Recent advances in the preparation and characterization of intermediately to highly esterified and etherified starches: A review. *Starch/Stärke, 72*(3–4), art. 1900238.

Xu, M, Saleh, ASM, Gong, B, Li, B, Jing, L, Gou, M, Jiang, H, & Li, W 2018. The effect of repeated versus continuous annealing on structural, physicochemical, and digestive properties of potato starch. *Food Research International, 111*, 324–333.

Xu, P, Zeng, Q, Cao, Y, Ma, P, Dong, W, & Chen, M 2017. Interfacial modification on polyhydroxyalkanoates/starch blend by grafting in-situ. *Carbohydrate Polymers, 174*, 716–722.

Yan, Y, & Dou, Q 2021. Effect of peroxide on compatibility, microstructure, rheology, crystallization, and mechanical properties of PBS/waxy starch composites. *Starch/Stärke, 73*(3–4), art. 2000184.

Yang, Z, Graiver, D, & Narayan, R 2013. Extrusion of humidity-resistant starch foam sheets. *Polymer Engineering & Science, 53*(4), 857–867.

Ye, J, Luo, S, Huang, A, Chen, J, Liu, C, & McClements, DJ 2019. Synthesis and characterization of citric acid esterified rice starch by reactive extrusion: A new method of producing resistant starch. *Food Hydrocolloids, 92*, 135–142.

Yu, J, Liu, X, Xu, S, Shao, P, Li, J, Chen, Z, Wang, X, Lin, Y, & Renard, CM 2023. Advances in green solvents for production of polysaccharide-based packaging films: Insights of ionic liquids and deep eutectic solvents. *Comprehensive Reviews in Food Science and Food Safety, 22*(2), 1030–1057.

Yu, JH, Wang, JL, Wu, X, & Zhu, PX 2008. Effect of glycerol on water vapor sorption and mechanical properties of starch/clay composite films. *Starch/Stärke, 60*(5), 257–262.

Zarski, A, Bajer, K, Raszkowska-Kaczor, A, Rogacz, D, Zarska, S, & Kapusniak, J 2020. From high oleic vegetable oils to hydrophobic starch derivatives: II. Physicochemical, processing and environmental properties. *Carbohydrate Polymers, 243*, art. 116499.

Zavareze, ER, & Dias, ARG 2011. Impact of heat-moisture treatment and annealing in starches: A review. *Carbohydrate Polymers, 83*(2), 317–328.

Zdanowicz, M 2021. Deep eutectic solvents based on urea, polyols and sugars for starch treatment. *International Journal of Biological Macromolecules, 176*, 387–393.

Zdanowicz, M, Staciwa, P, & Spychaj, T 2019. Low transition temperature mixtures (LTTM) containing sugars as potato starch plasticizers. *Starch/Stärke, 71*(9–10), art. 1900004.

Zdanowicz, M, Wilpiszewska, K, & Spychaj, T 2018. Deep eutectic solvents for polysaccharides processing. A review. *Carbohydrate Polymers, 200*, 361–380.

Zhang, W, Azizi-Lalabadi, M, Jafarzadeh, S, & Jafari, SM 2023. Starch-gelatin blend films: A promising approach for high-performance degradable food packaging. *Carbohydrate Polymers*, 320, art. 121266.

Zhang, Y, Rempel, C, & Liu, Q 2014. Thermoplastic starch processing and characteristics—A review. *Critical Reviews in Food Science and Nutrition, 54*(10), 1353–1370.

Zheng, Q, Tian, Y, Ye, F, Zhou, Y, & Zhao, G 2020. Fabrication and application of starch-based aerogel: Technical strategies. *Trends in Food Science & Technology, 99*, 608–620.

Zhu, F 2019. Starch based aerogels: Production, properties and applications. *Trends in Food Science & Technology, 89*, 1–10.

Zhu, J, Gao, W, Wang, B, Kang, X, Liu, P, Cui, B, & Abd El-Aty, A 2021. Preparation and evaluation of starch-based extrusion-blown nanocomposite films incorporated with nano-ZnO and nano-SiO$_2$. *International Journal of Biological Macromolecules, 183*, 1371–1378.

Zhuang, Y, Saadatkhah, N, Morgani, MS, Xu, T, Martin, C, Patience, GS, & Ajji, A 2023. Experimental methods in chemical engineering: Reactive extrusion. *The Canadian Journal of Chemical Engineering, 101*(1), 59–77.

Starch Blends and Composites

17

Marianne Ayumi Shirai and Juliano Zanela

17.1 STARCH BLENDS

Native starch is semi-crystalline, and due to its chemical structure, its materials are hydrophilic and brittle, limiting its application in the packaging sector. Therefore, different strategies have been studied to improve starch material properties and maintain biodegradability. Blending starch with different biopolymers is an interesting alternative for improving the properties of starch-based materials, as starch is widely available, comes from a renewable source, and helps to reduce costs. Furthermore, starch has adequate thermal stability to be melted with synthetic plastics, making it possible to produce new biodegradable materials (Reis et al. 2014, 2018; Shirai et al. 2015, 2016; Zanela et al. 2019).

Thermochemical processing techniques like extrusion, casting, melt blending, and compression molding have been employed in producing starch-based blends. The processing technology, the proportion of biopolymers in the blend, and the use of additives determine the physical and chemical interactions. These are important aspects in achieving the starch blend properties. In the literature, several types of starch blends can be found. This chapter will focus on the starch blend with agropolymers and biodegradable polyesters obtained by different processing technologies.

17.1.1 Starch and Agropolymers Blends

Agropolymers are polymers obtained from products or byproducts of agricultural origin, such as polysaccharides and proteins. Blended films have shown unique properties as a result of a change in the network structure of the film matrix, which can significantly improve their mechanical and barrier properties more than any pure film (Gontard et al. 1993; Wu et al. 2023).

Starch and protein are hydrocolloids with good film-forming ability that have long been investigated as combined formulations for edible film and coating because they exhibit good miscibility and can be cross-linked through intermolecular interactions. Some examples of starch–protein blends are starch–gelatin (Zhang et al. 2023), Starch–pea protein isolate (Huntrakul et al. 2020), starch–soy protein, starch–whey protein, starch–albumin (Hosseini et al. 2021), starch–zein (Masanabo et al. 2022; Pérez et al. 2021), and starch–wheat gluten (Rivera Leiva et al. 2022). Also, alternative proteins, such as sesame protein, are investigated (Ertan et al. 2024).

Different polysaccharides are employed in starch-based blend production, such as mucilage, gums, and pectin. Mucilage is part of the coating of some seeds (chia, linseed, yellow mustard, etc.), and it has strong interactions with polysaccharides and proteins (Lai & Liang 2012; Zeng & Lai 2016), resulting in a reinforced network with improvement in the mechanical and morphological properties. Mucilage from different sources (chia, *Dioscorea opposita* Thunb., shoeblack leaves, okra, basil, fenugreek, flaxseed, etc.) has been extracted and used in the starch-based blend (Aaliya et al. 2024; Mohite & Chandel 2020). In some cases, bioactive properties were observed, such as in the case of the ora-pro-nobis mucilage-starch blend, which presented antioxidant capacity varying from 3.9% to 23.3% (Oliveira et al. 2022).

Citrus pectin and starch present high compatibility for film-making due to their strong adhesion (Fishman et al. 1996). For example, plasticized blends of citrus pectin-pea starch and sugar beet pectin-pea starch resulted in strong films with an increase in the tensile strengths and a decrease in the WVP, without compromising the elongation at break (Bai et al. 2023). The effect of addition of different concentrations of high methoxylation pectin and 4% of lignin microparticles as a reinforcing and active agent in the cassava starch films properties were evaluated by Oliveira Begali et al. (2021). Pectin and lignin microparticles improved the mechanical strength, thermal stability, UV protection, oxygen permeability and provided antioxidant capacity to the films.

In the starch–gum blends, low concentration of acacia gum resulted in substantial improvement in corn starch films' physical and chemical properties. In this context, ethylene scavenging film based on corn starch, gum acacia, and sepiolite

DOI: 10.1201/9781003464396-17

was produced by casting method. The addition of sepiolite in the corn starch–acacia gum films showed improved tensile strength, WVP, crystallinity, and thermal stability (Upadhyay et al. 2022). Potato starch–guar gum films were chemically modified with succinic, malic, and tartaric acids and observed increased stretchability, transparency, solubility, and water absorption capacity of the films, while tensile strength and WVP of the films diminished significantly. The authors concluded that although the organic acids were added with the aim to act as a cross-linker, they acted as a plasticizer in the films due to the concentration incorporated (Nandi & Guha 2021).

A starch–chitosan blend is another way to improve starch-based material properties. The literature stated that incorporation of chitosan into corn starch films improved the mechanical properties of these films. Tensile strength and elongation at break of the corn starch/chitosan films were raised with the higher different concentrations of chitosan and showed a maximum elongation at break (Ren et al. 2017).

17.1.2 Starch and Biodegradable Polyesters Blends

The biodegradable polyesters such as polylactic acid (PLA), poly(butylene co-terephthalate adipate) (PBAT), polyvinyl alcohol (PVA), and poly(ε-caprolactone) (PCL) show mechanical and barrier properties compared to synthetic non-biodegradable polymers; however, their cost is higher, and the production of the blend with starch is an interesting alternative to improve the functional properties of the starch-based material, reduce cost, and maintain the biodegradability.

PLA and thermoplastic starch (TPS) are biodegradable polymers from renewable source, and the mixture of these polymers has been extensively studied because PLA shows properties similar to PET and PP. However, TPS and PLA are thermodynamically immiscible due to the poor interfacial interaction between the hydrophilic starch granules and the hydrophobic PLA. To overcome these limitations, several researches have been conducted on compatibilization strategy and processing technology to increase the compatibility between phases (Villadiego et al. 2022). In this sense, Shirai et al. (2016) observed significant improvement in the mechanical and barrier properties of the TPS/PLA sheets produced by extrusion and calendaring due to the citric acid addition. In the same way, Reis et al. (2018), producing trays of TPS/PLA (30%) covered with a beeswax emulsion by extrusion, calendaring, and thermopressing, observed improved barrier properties and adequate processability.

PBAT is a petroleum-based aliphatic aromatic co-polyester polymer with full biodegradability under composting conditions (Shahlari & Lee 2012), and it has similar tensile properties to low-density polyethylene. Biodegradable films based on TPS and PBAT blends are already being produced on a pilot scale by blown extrusion. The obtained materials have interesting mechanical and barrier properties that can be used as packaging for different products. Also, TPS/PBAT blend is a promising matrix for incorporating bioactive compounds aiming to obtain active biodegradable packaging for foods, as observed by different researchers (Balan et al. 2021; Silva et al. 2019; Campos et al. 2019; Mücke et al. 2021; Zanela et al. 2021).

PVA is commercially available in various grades defined by molecular weight and hydrolysis degree (HD), which determine polymer water solubility and the final properties when blended with starch. In this sense, Zanela et al. (2019) produced PVA/starch sheets by extrusion-calendering and observed good compatibility between PVA and starch; however, the molecular weights (MW) and HD of PVA influenced in the mechanical properties. In the PVA/starch foam, the viscoelastic properties, thermal stabilities, and mechanical performances were significantly improved with increasing HD of PVA. These enhancements were attributed to hydrogen bonding interactions and intermolecular entanglements between starch and PVA chains. Additionally, the incorporation of PVA resulted in a remarkable reduction in the water absorption capacity of the foams (Liu et al. 2023).

PCL is a linear polyester manufactured by ring-opening polymerization of a seven-membered lactone with low processing temperature ($T_m \sim 60°C$), good compatibility with many polymeric materials, and high flexibility. The addition of even low concentrations of PCL can improve some properties of TPS, such as mechanical resistance, moisture sensitivity, and shrinkage (Correa et al. 2017; Corrêa et al. 2022; Kaseem et al. 2012).

17.2 STARCH COMPOSITES

The composite system generally is formed with two phases, a continuous phase denominated matrix and other dispersed phase called reinforcement or filler (organic or inorganic materials with specific shapes) (Niranjana Prabhu & Prashantha 2018). For the starch composites production, different inorganic and organic fillers (minerals, metals, ceramics, natural fibers, or even other types of polymers) are employed. The use of hybrid fillers, which consist of the combination of both inorganic and organic fillers properties, has been extensively reported with promising results (Osman et al. 2020).

Aiming to improve mechanical properties and decrease hydrophilic properties of starch-based materials, it was proposed to introduce nanofillers into the polymer matrix instead of macrofillers such as cellulose fibers (Balakrishnan, Gopi & Geethamma, et al. 2018; Gopi et al. 2019), cotton, jute, and bamboo (John & Thomas 2008). Macro and micro fillers usually have poor distribution in the polymer matrix, resulting in local stress; the film morphology is not homogeneous, and the boundary between the polymer and the fillers is visible. Therefore, the incorporation of nanofillers is a widely recognized technique because for an equal amount of filler, if the dispersion exhibits homogeneity, the influence on the macroscopic characteristics will be more pronounced due to the smaller size, attributable to the augmented polymer/filler interface (Cho et al. 2006; Kwaśniewska et al. 2020).

Nanoclay, such as montmorillonite, talc, sepiolite, and kaolin, has been incorporated in the starch-based material.

Significant improvement in the barrier properties is observed, which is commonly associated with the layered arrangement of additive particles in the polymer matrix. Considering the use of metallic and metal oxide nanofillers, a variety of metal nanoparticles produced from gold, copper, silver, palladium, and cobalt have been incorporated in starch polymeric matrix for a wide range of applications. These nanoparticles show unique physicochemical properties that guarantee active effects, such as antimicrobial (e.g., Ag and Cu), scavenging of gas molecules (e.g., Fe or Pd), and antioxidant (e.g., Se), but also enhanced polymer intrinsic properties (e.g., thermal, mechanical, optical, and rheological properties), leading to polymer-metallic composites with potential use as active food packaging (Santos et al. 2020; He et al. 2019; Videira-Quintela et al. 2021).

Polysaccharides as nanofillers have been studied in different forms, including cellulose nanocrystals (CNC), cellulose nanofibrils (CNF), starch nanoparticles, and chitosan nanoparticles. CNC, also known as nanowhiskers, are a material with high strength, low density, large specific surface area, high aspect ratio, and hydrophilicity (Muñoz-Gimena et al. 2023). The addition of 1.5%–2.5% of wood CNC to extruded native potato starch material formed strong hydrogen-hydroxyl bonds, which improved the mechanical properties and reduced swelling and enzymatic degradation (Nessi et al. 2019). In contrast, when CNC content is excessive, it aggregated due to the intermolecular hydrogen bonding interactions that occur between the hydroxyl groups present on the surface of the nanocrystals, resulting in a negative impact on material properties (Azeredo et al. 2017; Nessi et al. 2019; Sucinda et al. 2021). Differently, CNF possesses both crystalline and amorphous regions. Potato starch and pineapple leaf CNF nanocomposite films showed UV resistant and increased water barrier properties (Balakrishnan, Gopi, & Thomas 2018).

Chitosan offers multiple valuable properties for nanobiocomposite development due to its non-toxic, biodegradable, biofunctional, and biocompatible features, coupled with antimicrobial properties (Dutta et al. 2009). Chitosan nanoparticles can be obtained through various methods, including ionic gelation, reverse emulsion, precipitation, and polyelectrolyte complexation (Shapi'i et al. 2020). The chemical similarity between the starch matrix and chitosan nanoparticles creates a strong interfacial interaction, resulting in a reinforcement effect, increasing tensile strength, thermal stability, and creating a more tortuous path that lowers WVP (Chang et al. 2010). Additionally, starch films added of chitosan nanoparticles were more efficient at inhibiting the growth of Gram-positive (*Bacillus. cereus* and *Staphylococcus aureus*) bacteria compared to Gram-negative (*Escherichia coli* and *Salmonella typhi*) bacteria (Shapi'i et al. 2020).

17.3 BIODEGRADABLE PACKAGING MATERIALS

According to the Plastics Europe report, in 2021, the packaging sector was the most significant world market, accounting for 44% of the total plastics used. However, just 1.5% of total plastics production was bio-based/bio-attributed (Plastics Europe 2022). Packaging is characterized by commonly single-use materials with a short period of use before disposal, causing severe environmental concerns due to their non-biodegradability.

After use, the packaging is mainly landfilled, and in the case of non-biodegradable materials, it may cause ecological problems. However, if substituted for biodegradable and bio-based polymers, when these packages are landfilled, they result in carbon dioxide and water (Blanco et al. 2020)—being a more environmentally friendly option.

Many biodegradable polymers have been developed in the past years. However, despite these new polymers' presenting adequate properties to substitute the non-biodegradable ones, they offer a drawback limiting their use: their high costs compared with conventional polymers such as PET or LDPE.

A viable way to overcome the high-cost issues of biodegradable materials is to blend them with another polymer at a reduced cost. Starch is a promising option because it is a fully renewable and low-cost polymer (Nascimento et al. 2024; Zanela et al. 2019).

Despite starch's economic and environmental advantages and its ability to produce films, starch presents some drawbacks for pure use in packaging applications, mainly related to poor mechanical and barrier properties and high sensitivity to water. The literature shows the use of different starch and biodegradable polymers aimed at obtaining packaging with adequate properties, as well as the use of different packaging production methods, such as extrusion, thermopressing, and casting.

Extrusion is the packaging production method most used by the industry for thin film production, and starch can be processed in this way, facilitating the scale-up and stimulating the commercial production of these biodegradable materials.

The botanical source can differ in the film properties due to the inherent characteristics of the starch originating from different sources, mainly the amylose/amylopectin ratio. Yang et al. (2023) produced films with starches from five botanical sources (cassava, corn, pea, rice, and sweet potato). The tensile strength of the obtained films was ordered as follows: pea> corn> sweet potato> rice> cassava; the elongation at break presented the opposite order, with cassava starch film presenting the higher elongation, and pea starch film, the lower. Pea starch presented higher amylose content; amylose is a linear chain instead of the highly branched amylopectin chain, so their arrangement is more similar to conventional polymers, which possess long and linear chains and can help explain their better mechanical properties.

Starch naturally occurs in granules due to branched amylopectin chains that can form crystalline zones. In its native form, starch's melting point is higher than the thermal decomposition. For their use as packaging, they need to be plasticized and melted, making them suitable for the thermoplastic extrusion process (Kaseem et al. 2012).

This fact commonly leads to using plasticizers to improve the processability and properties of starch-based materials. Adequate plasticizers for starch materials consist of hydrophilic and low-molecular-mass molecules, such as water, polyols (glycerol, sorbitol, xylitol), organic acids (citric acid, maleic acid), and urea, are most used in the literature. Still, sugars, fatty acids, and amino acids can plastify the starch films. These compounds are able to interact with the polymeric starch chains, changing the starch from a crystalline to an amorphous structure, reducing their intermolecular forces, and allowing greater mobility between the starch chains. This results in a decrease in brittleness, a reduction of the glass transition temperature, and an increase in the material flexibility and workability, making the starch to be processed in conventional plastics production machinery, such as thermoplastic extruders (Galdeano et al. 2009; Niazi et al. 2015; Tang & Alavi 2011; Zullo & Iannace 2009).

Research into the development of starch-based materials has long been in vogue, leading to the maturation and consolidation of knowledge about processing, compatibility with plasticizers and other polymers, and ways to overcome some of these difficulties, not that this area of knowledge has ended. Still, the solid base of knowledge established allows research to advance in practical applications, leading to the development of packaging films and the study of their application simulating real-use situations. A strong focus is on developing active or intelligent packaging based mainly on using natural compounds, such as essential oils, plant extracts, or industry co-products.

Starch was used in films with different polymers, and one of the most used and researched is the PBAT. The PBAT is a petroleum-based biodegradable polymer that is flexible and presents mechanical properties similar to that found in the low-density polyethylene (LDPE) (Muthuraj et al. 2018).

Olivato et al. (2013) produced plastic bags of cassava starch and PBAT (55:45% w/w ratio, respectively), using glycerol as a plasticizer and tartaric acid as a compatibilizer for the blend—the film with more adequate properties (88.2% of polymers, 11.0% glycerol, and 0.8% of tartaric acid) is in accordance of the Brazilian standard for plastic bags for transport of products sold at retail, indicating that these materials reach the requirements for the proposed use, demonstrating that the starch:PBAT blend can be an alternative.

A critical drawback of starch-based films is their inherent sensitivity to water, as demonstrated in cassava starch materials blended with different PVA grades and obtained by extrusion. The mechanical properties are directly influenced by the relative humidity of the environment, with depletion of mechanical properties according to increased humidity occurring due to the plastificant effect that water exerts on starch for conventional plastics (Zanela et al. 2016).

A way to overcome these drawbacks is incorporating fatty acids or vegetable oils in the blends, making water vapor transmission in the films difficult due to their hydrophobic character and by forming complexes with the fatty acids. The amylose molecules can form single helical complexes with fatty acids entrapped in the helical structure's interior, named V-type amylose complexes. The V-type complexes can modify the functional properties of the starches, and one way is retarding the starch recrystallization (Kang et al. 2020; Wang et al. 2019). The recently produced starch films are almost completely amorphous. However, the aging of the film promotes the retrogradation of the starch (formation of crystalline zones), leading to changes in material properties (Jiménez

et al. 2012), which is not desirable when thinking in commercial packaging production.

Different starches films, pure or in blends with another polymer, are produced and evaluated by incorporating fatty acids or vegetable oils. Cassava starch and carboxymethyl cellulose films with the addition of buriti oil and natural surfactant extracted from *Yucca schidigera* plant; the presence of oil and surfactant increased tensile strength (1,100%) and opacity by 375%, while the Water Vapor Permeability (WVP) reduces by 47% and the water solubility by 70% (Santos et al. 2023). According to the authors, the presence of the surfactant can easily interact with the buriti oil and with the amorphous starch and sodium groups in the CMC chain, reinforcing the inter- and intra-molecular interactions between the film components.

Similar behavior was observed in apple pectin/cassava starch films incorporated with *Laurus nobilis* L. oil and oleic acid (Taqi et al. 2014). Tensile strength and elongation at break reduce according to oil or oleic acid increase. The oxygen permeability and WVP show a reduction related to the increase of oil and oleic acid. Still, the *Laurus nobilis* L. oil presents lower values at the same oleic acid concentration in films.

Evaluating the incorporation of palmitic (C16:0), stearic (C18:0), and oleic acid (C18:1) in corn starch films and evaluating their properties after 5 weeks of aging, Jiménez et al. (2012) observed that mechanical properties and WVP of the films reduce with the addition of fatty acids in the films; however, after the 5 weeks aging time, the WVP values increase. The film became stiffer and had a reduced elongation, with the saturated fatty acid showing better WVP results. The increase in the crystalline zones with the aging of the films was noticeable.

Liu et al. (2016) observed a similarity between sweet potato starch and sweet potato starch-based films added with stearic (C18:0), oleic (C18:1), and linoleic acid (C18:2). The saturated fatty acid tested (stearic acid) strongly bonding with amylose than the unsaturated fatty acids, indicating better mechanical and barrier properties.

These are some of the strategies used to improve and make the starch-based packages more competitive compared to those made with conventional petrochemical polymers.

17.4 BIOMEDICAL MATERIALS

In addition to the characteristics that make starch a good option in packaging production, such as availability and low cost, starch can also be advantageous for biomedical applications. Besides low cost and availability, their potential for biomedical and pharmaceutical applications is because starch is non-toxic and non-immunogenic, biocompatible, and biodegradable; they can be used in bone tissue applications as scaffolds, hydrogels, or drug delivery. However, starch presents poor mechanical properties for some specific uses in biomedical applications (Lee & Hwang 2023; Zakaria et al. 2017). Due to that, blending with other polymers is a viable way to develop materials with adequate properties.

To be apt to act as a replacer for natural tissues, the biomaterials need to meet a large number of requirements to achieve biocompatibility, mainly conditioned mostly by its surficial features, such as surface composition and degradation, hydrophilic–hydrophobic character, wettability, roughness, stiffness, among others. A biodegradable material needs to exhibit properties to replace the tissue and be completely resorbed in a velocity compatible with cellular growth (Jurak et al. 2021).

The starch shows the capacity to meet many biomedical application demands, pure or in blends with biocompatible polymers. One of these potential applications is in aerogel production. Aerogels are nanostructured ultralight materials in which the gel's liquid is replaced with a gas. They present a high specific superficial area, usually several hundred square meters per gram. Aerogels can also be made from inorganic matrices such as silica or metal oxides, polysaccharides, or proteins, such as albumin, gelatin, cellulose, agar, and starch (Zhu 2019; Zou & Budtova 2021).

Milovanovic et al. (2024) have produced aerogels of corn starch that incorporated four natural bioactive agents: thymol, citronellol, carvacrol, and eugenol. The aerogels presented a specific area of 225 m²/g and maintained the controlled release of bioactive compounds for 3 days. Thymol showed higher antimicrobial activity. Instead, eugenol possessed a higher antioxidant capacity. The author claims that due to the high absorption, the produced aerogels present potential for biomedical application because they can maintain the sorption of exudates in skin wound healing. Due to the large surface area and porosity characteristics, the starch-based aerogel presents a higher capacity for aqueous fluid absorption, which helps the healing process, maintaining a pH and moisture balance (Zheng et al. 2020).

Bone grafts and implants are traditional therapies used for bone problems. However, there is an interest in developing scaffolds absorbable by the body, with starch being a viable option. An important requirement for a bone tissue scaffold is to present a 3D porous structure that permits cell and tissue adhesion and growth in the desired shape (Wu et al. 2017).

Scaffolds for bone tissue regeneration based on a blend of corn starch, ethylene vinyl alcohol, and cellulose acetate were produced by different methods. For extrusion, three different blowing agents were tested, aiming to release CO_2, causing the expansion of extruded material, and consequently promoting a porose material; compression molding–particle leaching and solvent casting–particulate leaching were also tested. The porosity ranged from 40% to 70%. Porosity is an essential factor for scaffolds, being responsible for the cell and nutrient ingress and cell fixation, enabling cell multiplication. Other properties, such as mechanical and degradation properties, are adequate for the aim (Gomes et al. 2002), demonstrating the viability of starch as scaffolds for bone tissue engineering.

The same research group tested the cytotoxicity of similar materials previously described (corn starch: ethylene vinyl alcohol and corn starch: cellulose acetate) with hydroxyapatite. The authors observed that none of the materials showed complete cytotoxicity for the L929 fibroblasts (connective mouse tissue) and did not promote significant alterations in cell morphology besides adequate cellular adhesion and proliferation

(Marques et al. 2002); this indicates that this starch-based material is promissory for tissue engineering.

17.5 BIOACTIVE STARCH-BASED CARRIERS

Starch has been studied to act as a carrier of interest molecules mainly in two ways: entrapping the active molecule to protect them from the external environment as microcapsules, for example, or dispersing the active molecules directly in the polymer, which can be an essential oil, a plant extract, or a drug, among others.

Protecting these functional molecules of the environment is an essential issue because these molecules are usually sensitive to heat, light, and oxygen. Many of them are volatile substances that can decay their concentration quickly and not exercise the desirable role, so these methodologies can promote a controlled release of the substance, guaranteeing its effect for a more extended time.

There are many encapsulation processes, such as spray drying, spray cooling, coacervation, and fluidized bed. All the processes aimed at protecting the core material from adverse environmental conditions are based on the creation of an external coating, usually composed of polysaccharides, gums, proteins, lipids, and others, that are namely wall material, and the selection of the wall material is related to the nature of the core material (the bioactive substance that wishes to protect), because the wall and core material need to be compatible between them. Maltodextrin (a starch derivative) is commonly used as wall material due to its good retention capacity; Arabic gum is a polysaccharide often used in wall materials due to its good emulsifier and film-forming capacities too (Ballesteros et al. 2017; Fang & Bhandari 2010; Marcillo-Parra et al. 2021).

Starch has been researched for its use as wall material, in native or modified form, aiming to improve the emulsifier capacities of starch. Arshad et al. (2018) evaluated substituting Arabic gum with native and octenyl succinic anhydride and succinylated sorghum starches for microencapsulation by spray-drying nutmeg oleoresin. The author observed better results with native and octenyl succinic anhydride sorghum starches.

Microencapsulation by spray drying of rosemary essential oil with the partial or total substitution of Arabic gum for a commercially modified waxy maize starch is especially suited for encapsulation, maltodextrin, and inulin. The authors did not observe a significant difference in encapsulation efficiency by substituting Arabic gum for modified starch or a blend of modified starch:maltodextrin (1:1 m/m). The authors claim that using modified starch and maltodextrin can be profitable because they have a high volatile retention capacity and the advantage of being a relatively inexpensive wall material (Fernandes et al. 2014). These findings show the potential use of starch as a wall material for protecting sensible molecules from unfavorable environmental conditions.

As discussed above, one way to incorporate the bioactive agents in the film is their direct incorporation in the polymer blends or even the microcapsules incorporation in these blends. Adding a compound with some specific and desirable activity leads to what is known as active packaging. Biodegradable films are a field that has received much attention nowadays, too, so a logical step is the union of both research fields, leading to the development of active biodegradable packaging, and based on all the advantages and properties of starch-based packaging materials, it can be an excellent carrier for active molecules.

Active packaging can be described as a package that intends to extend the shelf life or enhance the safety or sensory qualities of the packaged food, and this is achieved due to the interaction of the product with the package and the environment. The active packaging can present antioxidants, antimicrobials, oxygen scavenging, ethylene or moisture absorbers, or other valuable properties for shelf-life extension or the maintenance of the safety and quality levels of the packaged food (Realini & Marcos 2014; Said et al. 2023).

The active biodegradable packaging films can contain synthetic active agents, as observed in films obtained by blown extrusion of cassava starch and PBAT with potassium sorbate as an antimicrobial agent (Andrade-Molina et al. 2013).

Despite the effective results, the tendency is to use natural agents, such as essential oils or plant extracts, due to the deleterious effects of some synthetic food additives used in foods as carcinogenic or teratogenic effects, in addition to a growing demand consumers for more health and natural foods (Naveena et al. 2008), reducing or eliminating the synthetic additives if possible.

Many works evaluated the presence of active molecules in starch-based films. Some authors tested the direct incorporation of milled plants with well-known properties in the films (Kechichian et al. 2010), incorporated cinnamon and clove powder into cassava starch films, and presented equal or higher counts of viable mold and yeast count when compared to control film for bread slice packaging. Zanela et al. (2021) incorporated milled *Baccharis dracunculifolia* leaves (a Brazilian native plant with excellent antioxidant properties) in blow-extruded cassava starch/PBAT films. When tested in food simulants, the films presented antioxidant activity, indicating that the active molecules resisted the extrusion process. In common, both works report a reduction of mechanical properties with the increase of active material concentration and the reduction of transparency of the films, which were expected due to the nature of the material incorporated.

Natural extracts obtained from plants or coproducts are another way to produce active packaging films, and Luchese et al. 2018) produced cassava starch films added with blueberry pomace. The films are darker according to increased blueberry pomace content, and the release was correlated with the initial concentration. Casagrande et al. (2021) produced a lyophilized extract of *B. dracunculifolia* and incorporated it in the cassava starch/PVA film; the authors observed a darkening of the film and a decay in mechanical properties according to the increase of lyophilized extract added. However, the obtained films show antioxidant properties and antimicrobial

effects against foodborne pathogens, indicating the potential of the natural plant extracts for active packaging.

Balan et al. (2021) and Medeiros et al. (2019) tested active biodegradable extruded blown films with oregano essential oil neat or microencapsulated (maltodextrin and Arabic gum as wall material) and incorporated in wheat flour:PBAT and cassava starch:PBAT films, respectively. In both works, despite the starch type used, the authors observed that microencapsulation maintained a high amount of essential oil in the film compared to the essential oil freely added, which was mainly lost probably during thermoplastic extrusion process. The films show antioxidant and antimicrobial properties, and the microencapsulated essential oil reduces the film's mechanical properties; instead, in free essential oil incorporation, the essential oil has a plasticizing effect in starch:PBAT films, with increased elongation of films.

In banana starch casting films added with a nanoemulsion of lemongrass and rosemary essential oil, the essential oil promoted a plasticizer effect, too. Still, the nanoemulsion instead of microencapsulation did not cause significant alteration in the color of the films (Restrepo et al. 2018).

The results showed that starch has more significant potential for use in many areas, such as biomedical or packaging, with the advantage of being widely available, biocompatible, relatively low cost, and fully renewable and biodegradable. Starch research is in vogue, with a wide field of possibilities for new uses or applications.

REFERENCES

Aaliya, B, Sunooj, KV, Vijayakumar, A, Krina, P, Navaf, M, Parambil Akhila, P, Raviteja, P, Mounir, S, Lackner, M, George, J & Nemţanu, MR 2024. Fabrication and characterization of talipot starch-based biocomposite film using mucilages from different plant sources: A comparative study. *Food Chemistry*, 438, art. 138011.

Andrade-Molina, TPC, Shirai, MA, Grossmann, MVE & Yamashita, F 2013. Active biodegradable packaging for fresh pasta. *LWT - Food Science and Technology*, 54, 25–29.

Arshad, H, Ali, TM & Hasnain, A 2018. Native and modified sorghum starches as wall materials in microencapsulation of nutmeg oleoresin. *International Journal of Biological Macromolecules*, 114, 700–709.

Azeredo, HMC, Rosa, MF & Mattoso, LHC 2017. Nanocellulose in bio-based food packaging applications. *Industrial Crops and Products*, 97, 664–671.

Bai, W, Vidal, NP, Roman, L, Portillo-Perez, G & Martinez, MM 2023. Preparation and characterization of self-standing biofilms from compatible pectin/starch blends: Effect of pectin structure. *International Journal of Biological Macromolecules*, 251, art. 126383.

Balakrishnan, P, Gopi, S, Geethamma, VG, Kalarikkal, N & Thomas, S 2018. Cellulose nanofiber vs nanocrystals from pineapple leaf fiber: A comparative studies on reinforcing efficiency on starch nanocomposites. *Macromolecular Symposia*, 380, art. 1800102.

Balakrishnan, P, Gopi, SMSS & Thomas, S 2018. UV resistant transparent bionanocomposite films based on potato starch/cellulose for sustainable packaging. *Starch-Stärke*, 70, art. 1700139.

Balan, G, Paulo, AFS, Correa, LG, Alvim, ID, Ueno, CT, Coelho, AR, Ströher, GR, Yamashita, F., Sakanaka, LS & Shirai, MA 2021. Production of wheat flour/PBAT active films incorporated with oregano oil microparticles and its application in fresh pastry conservation. *Food and Bioprocess Technology*, 14, 1587–1599.

Ballesteros, LF, Ramirez, MJ, Orrego, CE, Teixeira, JA & Mussatto, SI 2017. Encapsulation of antioxidant phenolic compounds extracted from spent coffee grounds by freeze-drying and spray-drying using different coating materials. *Food Chemistry*, 237, 623–631.

Blanco, I, Ingrao, C & Siracusa, V 2020. Life-cycle assessment in the polymeric sector: A comprehensive review of application experiences on the Italian scale. *Polymers*, 12, art. 1212.

Campos, SS, Oliveira, A, Moreira, TFM, Silva, TBV, Silva, MV, Pinto, JA, Bilck, AP, Gonçalves, OH, Fernandes, IP, Barreiro, MF, Yamashita, F, Valderrama, P, Shirai, MA & Leimann, FV 2019. TPCS/PBAT blown extruded films added with curcumin as a technological approach for active packaging materials. *Food Packaging and Shelf Life*, 22, art. 100424.

Casagrande, M, Zanela, J, Wagner Júnior, A, Yamashita, F, Busso, C, Wouk, J, Radaelli, JC & Malfatti, CRM 2021. Optical, mechanical, antioxidant and antimicrobial properties of starch/polyvinyl alcohol biodegradable film incorporated with baccharis dracunculifolia lyophilized extract. *Waste and Biomass Valorization*, 12, 3829–3848.

Chang, PR, Jian, R, Yu, J, & Ma, X 2010. Fabrication and characterisation of chitosan nanoparticles/plasticised-starch composites. *Food Chemistry*, 120, 736–740.

Cho, J, Joshi, MS & Sun, CT 2006. Effect of inclusion size on mechanical properties of polymeric composites with micro and nano particles. *Composites Science and Technology*, 66, 1941–1952.

Corrêa, AC, Campos, A, Claro, PIC, Guimarães, GGF, Mattoso, LHC & Marconcini, JM 2022. Biodegradability and nutrients release of thermoplastic starch and poly (ε-caprolactone) blends for agricultural uses. *Carbohydrate Polymers*, 282, art. 119058.

Correa, AC, Carmona, VB, Simão, JA, Mattoso, LHC & Marconcini, JM 2017. Biodegradable blends of urea plasticized thermoplastic starch (UTPS) and poly(ε-caprolactone) (PCL): Morphological, rheological, thermal and mechanical properties. *Carbohydrate Polymers*, 167, 177–184.

Dutta, PK, Tripathi, S, Mehrotra, GK & Dutta, J 2009. Perspectives for chitosan based antimicrobial films in food applications. *Food Chemistry*, 114, 1173–1182.

Ertan, K, Sahin, S & Sumnu, G 2024. Effects of alkaline pH and gallic acid enrichment on the physicochemical properties of sesame protein and common vetch starch-based composite films. *International Journal of Biological Macromolecules*, 257, art. 128743.

Fang, Z & Bhandari, B 2010. Encapsulation of polyphenols – A review. *Trends in Food Science & Technology*, 21, 510–523.

Fernandes, RVDB, Borges, SV & Botrel, DA 2014. Gum arabic/starch/maltodextrin/inulin as wall materials on the microencapsulation of rosemary essential oil. *Carbohydrate Polymers*, 101, 524–532.

Fishman, ML, Coffin, DR, Unruh, JJ & Ly, T 1996. Pectin/starch/glycerol films: Blends or composites? *Journal of Macromolecular Science*, 33, 639–654.

Galdeano, MC, Mali, S, Grossmann, MVE, Yamashita, F & García, MA 2009. Effects of plasticizers on the properties of oat starch films. *Materials Science and Engineering: C,* 29, 532–538.

Gomes, ME, Godinho, JS, Tchalamov, D, Cunha, AM & Reis, RL 2002. Alternative tissue engineering scaffolds based on starch: Processing methodologies, morphology, degradation and mechanical properties. *Materials Science and Engineering: C*, 20, 19–26.

Gontard, N, Guilbert, S & Cuq, JL 1993. Water and glycerol as plasticizers affect mechanical and water vapor barrier properties of an edible wheat gluten film. *Journal of Food Science*, 58, 206–211.

Gopi, S, Amalraj, A, Jude, S, Thomas, S & Guo, Q 2019. Bionanocomposite films based on potato, tapioca starch and chitosan reinforced with cellulose nanofiber isolated from turmeric spent. *Journal of the Taiwan Institute of Chemical Engineers*, 96, 664–671.

He, X, Deng, H & Hwang, H 2019. The current application of nanotechnology in food and agriculture. *Journal of Food and Drug Analysis*, 27, 1–21.

Hosseini, SN, Pirsa, S & Farzi, J 2021. Biodegradable nano composite film based on modified starch-albumin/MgO; antibacterial, antioxidant and structural properties. *Polymer Testing*, 97, art. 107182.

Huntrakul, K, Yoksan, R, Sane, A & Harnkarnsujarit, N 2020. Effects of pea protein on properties of cassava starch edible films produced by blown-film extrusion for oil packaging. *Food Packaging and Shelf Life*, 24, art. 100480.

Jiménez, A, Fabra, MJ, Talens, P & Chiralt, A 2012. Effect of re-crystallization on tensile, optical and water vapour barrier properties of corn starch films containing fatty acids. *Food Hydrocolloids*, 26, 302–310.

John, M & Thomas, S 2008. Biofibres and biocomposites. *Carbohydrate Polymers*, 71, 343–364.

Jurak, M, Wiącek, AE, Ładniak, A, Przykaza, K & Szafran, K 2021. What affects the biocompatibility of polymers? *Advances in Colloid and Interface Science*, 294, art. 102451.

Kang, X, Liu, P, Gao, W, Wu, Z, Yu, B, Wang, R, Cui, B, Qiu, L & Sun, C 2020. Preparation of starch-lipid complex by ultrasonication and its film forming capacity. *Food Hydrocolloids*, 99, art. 105340.

Kaseem, M, Hamad, K & Deri, F 2012. Thermoplastic starch blends: A review of recent works. *Polymer Science Series A*, 54, 165–176.

Kechichian, V, Ditchfield, C, Veiga-Santos, P & Tadini, CC 2010. Natural antimicrobial ingredients incorporated in biodegradable films based on cassava starch. *LWT - Food Science and Technology*, 43, 1088–1094.

Kwaśniewska, A, Chocyk, D, Gładyszewski, G, Borc, J, Świetlicki, M & Gładyszewska, B 2020. The influence of kaolin clay on the mechanical properties and structure of thermoplastic starch films. *Polymers,* 12, art. 73.

Lai, LS, & Liang, HY 2012. Chemical compositions and some physical properties of the water and alkali-extracted mucilage from the young fronds of *Asplenium australasicum* (J. Sm.) Hook. *Food Hydrocolloids*, 26(2), 344–349.

Lee, CS & Hwang, HS 2023. Starch-based hydrogels as a drug delivery system in biomedical applications. *Gels*, 9, art. 951.

Liu, F, Zhang, Y, Xiao, X, Cao, Y, Jiao, W, Bai, H, Yu, L & Duan, Q 2023. Effects of polyvinyl alcohol content and hydrolysis degree on the structure and properties of extruded starch-based foams. *Chemical Engineering Journal*, 472, art. 144959.

Liu, P, Sun, S, Hou, H & Dong, H 2016. Effects of fatty acids with different degree of unsaturation on properties of sweet potato starch-based films. *Food Hydrocolloids*, 61, 351–357.

Luchese, CL, Garrido, T, Spada, JC, Tessaro, IC & Caba, K 2018. Development and characterization of cassava starch films incorporated with blueberry pomace. *International Journal of Biological Macromolecules*, 106, 834–839.

Marcillo-Parra, V, Tupuna-Yerovi, DS, González, Z & Ruales, J 2021. Encapsulation of bioactive compounds from fruit and vegetable by-products for food application – A review. *Trends in Food Science & Technology*, 116, 11–23.

Marques, AP, Reis, RL & Hunt, JA 2002. The biocompatibility of novel starch-based polymers and composites: *In vitro* studies. *Biomaterials*, 23, 1471–1478.

Masanabo, MA, Ray, SS & Emmambux, MN 2022. Properties of thermoplastic maize starch-zein composite films prepared by extrusion process under alkaline conditions. *International Journal of Biological Macromolecules*, 208, 443–452.

Medeiros, JAS, Blick, AP, Galindo, MV, Alvim, ID, Yamashita, F, Ueno, CT, Shirai, MA, Grosso, CRF, Corradini, E & Sakanaka, LS 2019. Incorporation of oregano essential oil microcapsules in starch-poly(butylene adipate co-terephthalate) (PBAT) films. *Macromolecular Symposia*, 383, 1–7.

Milovanovic, S, Markovic, D, Jankovic - Castvan, I & Lukic, I 2024. Cornstarch aerogels with thymol, citronellol, carvacrol, and eugenol prepared by supercritical CO_2-assisted techniques for potential biomedical applications. *Carbohydrate Polymers*, 331, art. 121874.

Mohite, AM & Chandel, D 2020. Formulation of edible films from fenugreek mucilage and taro starch. *SN Applied Sciences*, 2, art. 1900.

Mücke, N, Silva, TBV, Oliveira, A, Moreira, TFM, Venancio, CDS, Marques, LLM, Valderrama, P, Gonçalves, OH, Silva-Buzanello, RA, Yamashita, F, Shirai, MA, Genena, AK & Leimann, FV 2021. Use of water-soluble curcumin in TPS/PBAT packaging material: Interference on reactive extrusion and oxidative stability of chia oil. *Food and Bioprocess Technology*, 14, 471–482.

Muñoz-Gimena, PF, Oliver-Cuenca, V, Peponi, L & López, D 2023. A review on reinforcements and additives in starch-based composites for food packaging. *Polymers*, 15, art. 2972.

Muthuraj, R, Misra, M & Mohanty, AK 2018. Biodegradable compatibilized polymer blends for packaging applications: A literature review. *Journal of Applied Polymer Science*, 135, art. 45726.

Nandi, S & Guha, P 2021. Organic acid-compatibilized potato starch/guar gum blend films. *Materials Chemistry and Physics*, 268, art. 124714.

Nascimento, JV, Silva, KA, Giuliangeli, VC, Mendes, ALD, Piai, LP, Michels, RN, Dal Bosco, TC, Ströher, GR & Shirai, MA 2024. Starch-PVA based films with Clitoria ternatea flower extract: Characterization, phenolic compounds release and compostability. *International Journal of Biological Macromolecules*, 255, art. 128232.

Naveena, BM, Sen, AR, Vaithiyanathan, S, Babji, Y & Kondaiah, N 2008. Comparative efficacy of pomegranate juice, pomegranate rind powder extract and BHT as antioxidants in cooked chicken patties. *Meat Science*, 80, 1304–1308.

Nessi, V, Falourd, X, Maigret, JE, Cahier, K, D'Orlando, A, Descamps, N, Gaucher, V, Chevigny, C & Lourdin, D 2019. Cellulose nanocrystals-starch nanocomposites produced by extrusion: Structure and behavior in physiological conditions. *Carbohydrate Polymers*, 225, art. 115123.

Niazi, MBK, Zijlstra, M & Broekhuis, AA 2015. Influence of plasticizer with different functional groups on thermoplastic starch. *Journal of Applied Polymer Science*, 132, art. 42012.

Niranjana Prabhu, T & Prashantha, K 2018. A review on present status and future challenges of starch based polymer films and their composites in food packaging applications. *Polymer Composites*, 39, 2499–2522.

Olivato, JB, Grossmann, MVE, Bilck, AP, Yamashita, F & Oliveira, LM 2013. Starch/polyester films: simultaneous optimisation of the properties for the production of biodegradable plastic bags. *Polímeros*, 23, 32–36.

Oliveira, NL, Oliveira, ACS, Silva, SH, Rodrigues, AA, Borges, SV, Oliveira, JE & Resende, JV 2022. Development and characterization of starch-based films added ora-pro-nobis mucilage and study of biodegradation and photodegradation. *Journal of Applied Polymer Science*, 139, art. 52108.

Oliveira Begali, D, Ferreira, LF, Oliveira, ACS, Borges, SV, Sena Neto, AR, Oliveira, CR, Yoshida, MI & Sarantopoulos, CIGL 2021. Effect of the incorporation of lignin microparticles on the properties of the thermoplastic starch/pectin blend obtained by extrusion. *International Journal of Biological Macromolecules*, 180, 262–271.

Osman, AF, Ashafee, A, Moh, TL, Adnan, SA & Alakrach, A 2020. Influence of hybrid cellulose/bentonite fillers on structure, ambient, and low temperature tensile properties of thermoplastic starch composites. *Polymer Engineering & Science*, 60, 810–822.

Pérez, PF, Ollé Resa, C, Gerschenson, LN & Jagus, RJ 2021. Addition of zein for the improvement of physicochemical properties of antimicrobial tapioca starch edible film. *Food and Bioprocess Technology*, 14, 262–271.

Plastics Europe. (2022). Plastics – The Facts 2022. *Plastic Europe*, 81 p. Available on: https://plasticseurope.org/knowledge-hub/plastics-the-facts-2022/

Realini, CE & Marcos, B 2014. Active and intelligent packaging systems for a modern society. *Meat Science*, 98, 404–419.

Reis, MO, Olivato, JB, Bilck, AP, Zanela, J, Grossmann, MVE & Yamashita, F 2018. Biodegradable trays of thermoplastic starch/poly (lactic acid) coated with beeswax. *Industrial Crops and Products*, 112, 481–487.

Reis, MO, Zanela, J, Olivato, JB, Garcia, PS, Yamashita, F & Grossmann, MVE 2014. Microcrystalline cellulose as reinforcement in thermoplastic starch/poly(butylene adipate-co-terephthalate) films. *Journal of Polymers and the Environment*, 22, 545–552.

Ren, L, Yan, X, Zhou, J, Tong, J & Su, X 2017. Influence of chitosan concentration on mechanical and barrier properties of corn starch/chitosan films. *International Journal of Biological Macromolecules*, 105, 1636–1643.

Restrepo, AE, Rojas, JD, García, OR, Sánchez, LT, Pinzón, MI & Villa, CC 2018. Mechanical, barrier, and color properties of banana starch edible films incorporated with nanoemulsions of lemongrass (*Cymbopogon citratus*) and rosemary (*Rosmarinus officinalis*) essential oils. *Food Science and Technology International*, 24, 705–712.

Rivera Leiva, AF, Hernández-Fernández, J & Ortega Toro, R 2022. Active films based on starch and wheat gluten (Triticum vulgare) for shelf-life extension of carrots. *Polymers*, 14, art. 5077.

Said, NS, Howell, NK & Sarbon, NM 2023. A review on potential use of gelatin-based film as active and smart biodegradable films for food packaging application. *Food Reviews International*, 39, 1063–1085.

Santos, CA, Ingle, AP & Rai, M 2020. The emerging role of metallic nanoparticles in food. *Applied Microbiology and Biotechnology*, 104, 2373–2383.

Santos, RWS, Santana Neto, DC, Leite, MS, Jorge, RMM & Dantas, TLP 2023. Buriti oil and natural surfactant improve the properties of cassava starch and carboxymethylcellulose films. *Packaging Technology and Science*, 36, 281–292.

Shahlari, M & Lee, S 2012. Mechanical and morphological properties of poly(butylene adipate- co -terephthalate) and poly(lactic acid) blended with organically modified silicate layers. *Polymer Engineering & Science*, 52, 1420–1428.

Shapi'i, RA, Othman, SH, Nordin, N, Kadir Basha, R & Nazli Naim, M 2020. Antimicrobial properties of starch films incorporated with chitosan nanoparticles: *In vitro* and *in vivo* evaluation. *Carbohydrate Polymers*, 230, art. 115602.

Shirai, MA, Müller, CMO, Grossmann, MVE & Yamashita, F 2015. Adipate and citrate esters as plasticizers for poly(lactic acid)/thermoplastic starch sheets. *Journal of Polymers and the Environment*, 23(1). https://doi.org/10.1007/s10924-014-0680-9

Shirai, MA, Zanela, J, Kunita, MH, Pereira, GM, Rubira, AF, Müller, CMO, Grossmann, MVE & Yamashita, F 2016. Influence of carboxylic acids on poly(lactic acid)/thermoplastic starch biodegradable sheets produced by galendering-extrusion. *Advances in Polymer Technology*, 37, art. 21671.

Silva, TBV, Moreira, TFM, Oliveira, A, Bilck, AP, Gonçalves, OH, Ferreira, ICFR, Barros, L, Barreiro, M, Yamashita, F, Shirai, MA & Leimann, FV 2019. *Araucaria angustifolia* (Bertol.) Kuntze extract as a source of phenolic compounds in TPS/PBAT active films. *Food & Function*, 10, 7697–7706.

Sucinda, EF, Abdul Majid, MS, Ridzuan, MJM, Cheng, EM, Alshahrani, HA & Mamat, N 2021. Development and characterisation of packaging film from Napier cellulose nanowhisker reinforced polylactic acid (PLA) bionanocomposites. *International Journal of Biological Macromolecules*, 187, 43–53.

Tang, X & Alavi, S 2011. Recent advances in starch, polyvinyl alcohol based polymer blends, nanocomposites and their biodegradability. *Carbohydrate Polymers*, 85, 7–16.

Taqi, A, Mutihac, L & Stamatin, I 2014. Physical and barrier properties of apple pectin/cassava starch composite films incorporating *Laurus nobilis* L. oil and oleic acid. *Journal of Food Processing and Preservation*, 38, 1982–1993.

Upadhyay, A, Kumar, P, Kardam, SK & Gaikwad, KK 2022. Ethylene scavenging film based on corn starch-gum acacia impregnated with sepiolite clay and its effect on quality of fresh broccoli florets. *Food Bioscience*, 46, art. 101556.

Videira-Quintela, D, Martin, O & Montalvo, G 2021. Recent advances in polymer-metallic composites for food packaging applications. *Trends in Food Science and Technology*, 109, 230–244.

Villadiego, KM, Tapia, MJA, Useche, J, Macías, DE 2022. Thermoplastic starch (TPS)/polylactic acid (PLA) blending methodologies: A review. *Journal of Polymer and Environment*, 30, 75–91.

Wang, R, Liu, P, Cui, B, Kang, X & Yu, B 2019. Effects of different treatment methods on properties of potato starch-lauric acid complex and potato starch-based films. *International Journal of Biological Macromolecules*, 124, 34–40.

Wu, D, Bäckström, E & Hakkarainen, M 2017. Starch derived nano-sized graphene oxide functionalized bioactive porous starch scaffolds. *Macromolecular Bioscience*, 17, art. 1600397.

Wu, H, Li, T, Peng, L, Wang, J, Lei, Y, Li, S, Li, Q, Yuan, X, Zhou, M & Zhang, Z 2023. Development and characterization of antioxidant composite films based on starch and gelatin incorporating resveratrol fabricated by extrusion compression moulding. *Food Hydrocolloids*, 139, art. 108509.

Yang, N, Gao, W, Zou, F, Tao, H, Guo, L, Cui, B, Lu, L, Fang, Y, Liu, P & Wu, Z 2023. The relationship between molecular structure and film-forming properties of thermoplastic starches from different botanical sources. *International Journal of Biological Macromolecules*, 230, art. 123114.

Zakaria, NH, Muhammad, N & Abdullah, MMAB 2017. Potential of starch nanocomposites for biomedical applications. *IOP Conference Series: Materials Science and Engineering*, 209, art. 012087.

Zanela, J, Casagrande, M, Radaelli, JC, Dias, AP, Wagner Júnior, A, Malfatti, CRM & Yamashita, F 2021. Active biodegradable packaging for foods containing baccharis dracunculifolia leaf as natural antioxidant. *Food and Bioprocess Technology*, 14, 1301–1310.

Zanela, J, Casagrande, M, Reis, MO, Grossmann, MVE & Yamashita, F 2019. Biodegradable sheets of starch/polyvinyl aAlcohol (PVA): Effects of PVA molecular weight and hydrolysis degree. *Waste and Biomass Valorization*, 10, 319–326.

Zanela, J, Casagrande, M, Shirai, MA, Lima, VA & Yamashita, F 2016. Biodegradable blends of starch/polyvinyl alcohol/glycerol: Multivariate analysis of the mechanical properties. *Polímeros*, 26, 193–196.

Zeng, WW, Lai, LS 2016. Characterization of the mucilage extracted from the edible fronds of bird's nest fern (*Asplenium australasicum*) with enzymatic modifications, *Food Hydrocolloids*, 53, 84–92.

Zhang, W, Azizi-Lalabadi, M, Jafarzadeh, S & Jafari, SM 2023. Starch-gelatin blend films: A promising approach for high-performance degradable food packaging. *Carbohydrate Polymers*, 320, art. 121266.

Zheng, L, Zhang, S, Ying, Z, Liu, J, Zhou, Y & Chen, F 2020. Engineering of aerogel-based biomaterials for biomedical applications. *International Journal of Nanomedicine*, 15, 2363–2378.

Zhu, F 2019. Starch based aerogels: Production, properties and applications. *Trends in Food Science & Technology*, 89, 1–10.

Zou, F & Budtova, T 2021. Tailoring the morphology and properties of starch aerogels and cryogels via starch source and process parameter. *Carbohydrate Polymers*, 255, art. 117344.

Zullo, R & Iannace, S 2009. The effects of different starch sources and plasticizers on film blowing of thermoplastic starch: Correlation among process, elongational properties and macromolecular structure. *Carbohydrate Polymers*, 77, 376–383.

Starch Electrospun Fiber

18

Razia Rehman, Anjum Hamid Rather, Rumysa Saleem Khan,
Muheeb Rafiq, Syed Naeim Raza, Touseef Amna,
M. Shamshi Hassan, Nisar A. Khan, and Faheem A. Sheikh

18.1 INTRODUCTION

Starch, a polymeric carbohydrate frequently known as amylum, comprises abundant glucose entities linked by glycosidic bonds. Green plants synthesize this polysaccharide to store energy (Doi et al. 2002). It is insoluble in alcohol and cold water; pure starch is a colorless and flavorless powder. It comprises two distinct varieties of molecules: branched amylopectin and straight and spiral amylose. This polysaccharide is composed of various D-glucose molecules called amylose. Through 1,4-glycosidic linkages, they are interlinked together. On the one hand, amylose is considered water soluble. It gives a vivid blue color in the presence of iodine, which gets entrapped in the amylose helices. This indicates that some of amylose's molecules have a branched nature, although it was previously believed to be entirely unbranched (Lineback 1986). Conversely, amylopectin is a polymer that is water insoluble and is constructed from numerous D-glucose units that are connected at the branch via α-1,4-glycosidic and α-1,6-glycosidic bonds. The weight ratio of starch is usually 75%–80% amylopectin to 20%–25% amylose, depending on the type of plant. Plants use light energy during photosynthesis to convert carbon dioxide into glucose. This glucose is either converted into the chemical energy needed for general metabolism or stored in amyloplasts as starch granules to be converted into organic substances, including nucleic acids, fats, proteins, and polysaccharides. Plants synthesize glucose first and later convert it to a complex substance, i.e., starch. This substance is granulated and semi-crystalline, and it is the primary form of energy storage in green plants (Bailey & Long 1916; Zobel 1988). The structure of amylose and amylopectin is provided in Figure 18.1.

One of the methods to utilize this material's property in the field of research and development is transforming it into nanofibers. There are several methods for creating nanofibers, including electrospinning (Bhardwaj & Kundu 2010), phase separation (Alghoraibi & Alomari 2018), physical drawing (Zhang et al. 2020), and wet chemical reduction process (Kong et al. 2015). However, electrospinning is the most promising technique for fabricating nanofibers, among other methods, since it is comparatively quick, simple to use, affordable, and valuable (Hu et al. 2014). Several investigations focus on the final product of starch-based electrospun nanofibers, which can be altered by varying the electrospinning parameters (Deitzel et al. 2001; Rutledge & Fridrikh 2007; Kong et al. 2015). These attributes encompass high surface area-to-volume ratios (Jaiturong et al. 2018), high porosity (Liu et al. 2017), good adherence, excellent elasticity, and exceptional mechanical strength, which a pristine starch lacks. Electrospun fibers have demonstrated versatility and numerous uses in tissue engineering (Sundaramurthi et al. 2014), medication delivery (Williams et al. 2012), wound dressing (Sylvester et al. 2020), wastewater treatment (Cui et al. 2020), and filtration (Bhardwaj & Kundu 2010; Liu et al. 2017). Owing to their excellent physicochemical, biocompatibility, and non-toxicity properties, several researchers are directing their efforts toward making starch-based nanofibers from different sources, including tapioca starch (Sutjarittangtham et al. 2014), potato starch (Ainurofiq & Choiri 2015; Cárdenas et al. 2016), and corn starch (Wang et al. 2011; Kong & Ziegler 2014a, 2014b), utilizing electrospinning process for several biomedical applications, including wound dressing, tissue engineering, and delivery of drugs (Wang et al. 2019). This chapter has covered the electrospinning procedure and various parameters influencing it. Furthermore, an extensive analysis of starch-based electrospun nanofibers and their possible uses in tissue engineering and drug administration is presented.

18.2 PRINCIPLE OF ELECTROSPINNING

The electrokinetics of electrospinning reveals the electrification of a liquid droplet to create a jet (Taylor cone), which is then ejected out and extended to form fibers. The fundamental electrospinning apparatus is so simplistic that it may be utilized in any laboratory, as Figure 18.2 illustrates its basic setup (Li & Xia 2004). The major parts are a conductive collector, a

DOI: 10.1201/9781003464396-18

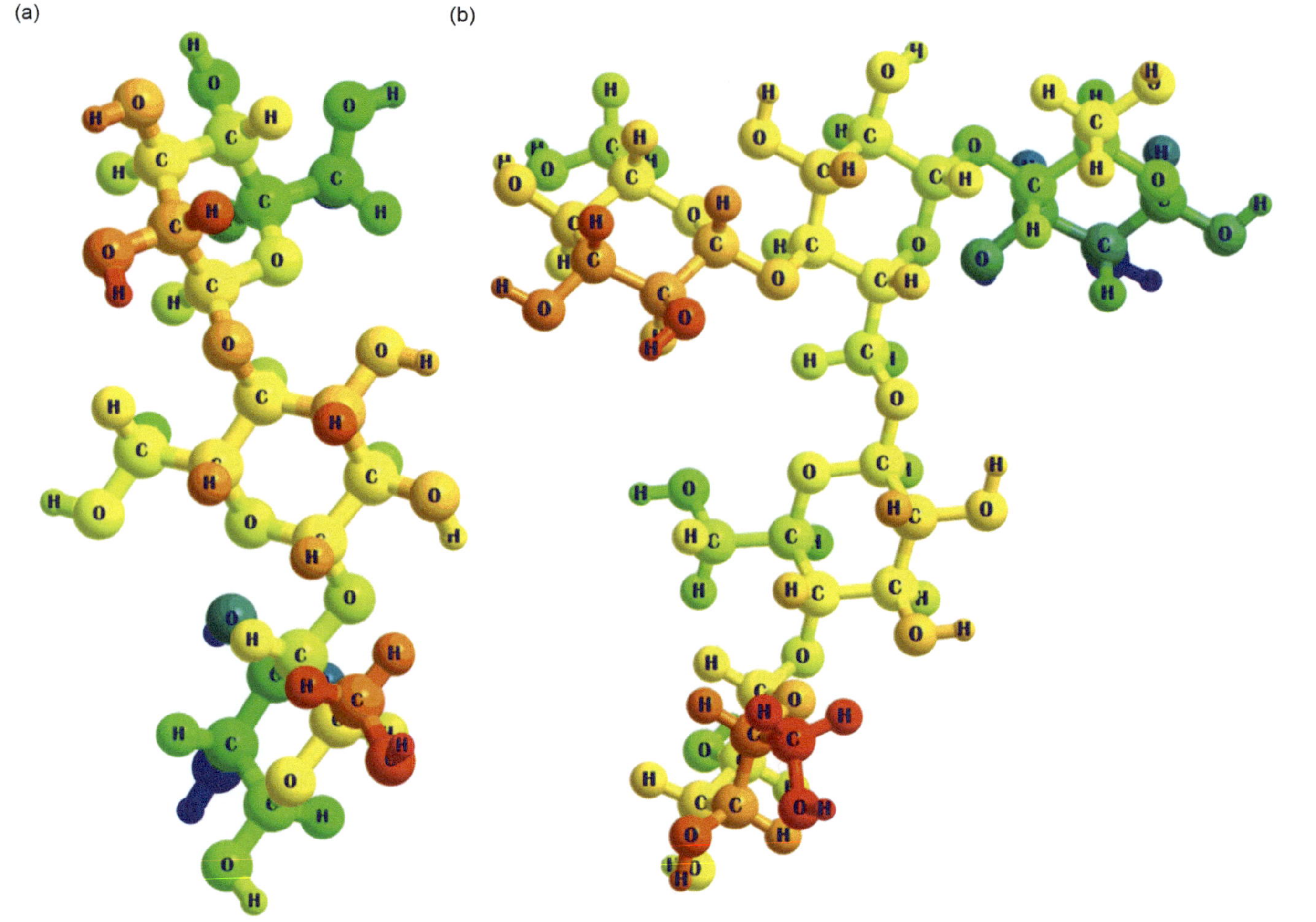

(a)   (b)

FIGURE 18.1   Structure of amylose (a) and amylopectin (b); two primary constituents of starch.

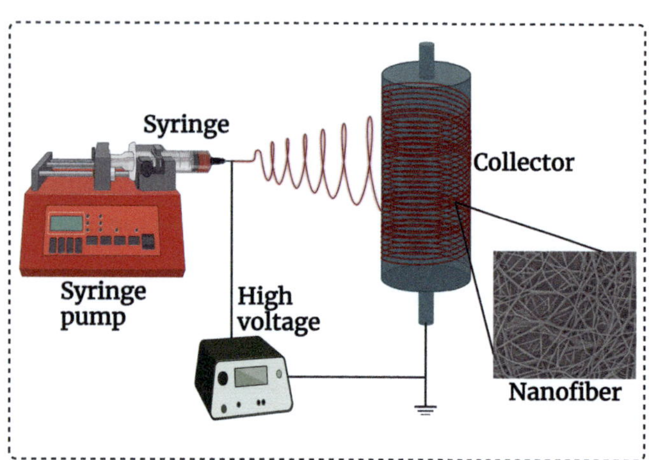

FIGURE 18.2   Electrospinning process: Voltage is applied to a polymer dispersion, which develops a Taylor cone and eventually collects on a mandrel as a fiber mat in a 2D sheet form.

droplet that has been electrified is distorted to create a Taylor cone because of the surface charges on them. Later, a jet of these droplets' experiences severe whipping motions after initially extending in a straight line owing to bending instabilities. Upon compression into smaller dimensions, the jet solidifies and adheres solid fibers to the grounded collector (Liu et al. 2017; Ashraf et al. 2019). The electrospinning process usually consists of four consecutive steps:

i. Liquid droplet charging and formation of a Taylor cone or cone-shaped projectile.
ii. The linear extension of the charged projectile.
iii. In the presence of an electric field, electrical bending instability, also called whipped instability, intensifies as the jet compresses.
iv. On a grounded collector, the jet solidifies and gathers into solid fibers (Li & Xia 2004; Sun et al. 2014; Liao et al. 2018).

spinneret, a syringe pump, and a power supply with high voltage. One of two possible power supplies is alternating or direct current. A suspended droplet is produced during electrospinning when surface tension forces the liquid out of the spinneret. This

The diameter, morphology, porosity, surface area, and mechanical strength of nanofibers are the key features that are essential for fabricating an ideal material for a particular application. Several factors can influence these properties during electrospinning (Beachley & Wen 2010). It is critical to carefully examine the electrospinning

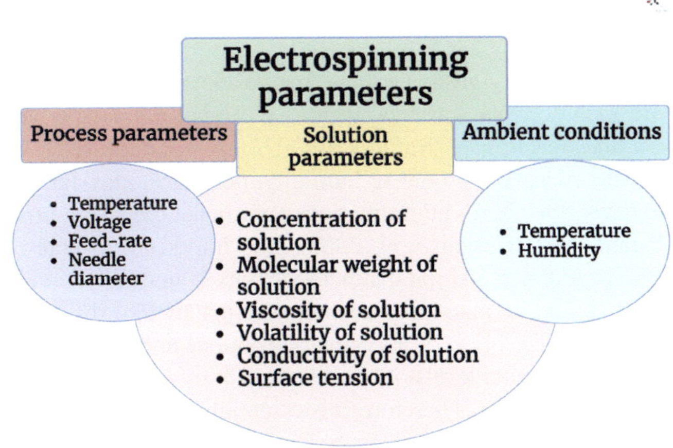

FIGURE 18.3 The parameters necessary to consider during electrospinning processes that can directly influence the formation and properties of fibers for various applications. These parameters will differ when changing the polymer.

parameters to generate nanofibers that deliver drugs and other uses better. These comprise process parameters (such as voltage, flow rate, and tip-to-mandrel distance), solution parameters (such as viscosity, polymer molecular weight and concentration, solvent volatility, and conductivity), and ambient conditions (such as temperature and humidity). These processing parameters are shown in Figure 18.3 and are discussed in forthcoming sections.

18.2.1 Role of Voltage

The voltage that acts on the solution jet has a direct influence on electrospinning. High voltage enables the creation of a significant proportion of narrow-diameter fibers while reducing the applied voltage, producing fibers with large diameters (Huan et al. 2015). However, the fiber diameter remains constant once the voltage exceeds a threshold. The fiber's diameter is affected by the viscoelastic characteristics, coulombic forces, and surface tension. Moreover, the columbic forces are significantly influenced by the applied voltage. In addition, the three forces are equilibrated at a low voltage, which leads to a limited size range in the fibers (Huan et al. 2015).

18.2.2 Role of Feed Rate

When the polymeric solution is fed at a faster rate during electrospinning, it forms beaded fibers with bigger diameters. Conversely, a lower feed rate rewards the fibers with smaller diameters. To properly drive superfine threads toward the collector and produce a Taylor cone, the feed rate needs to be at its optimum level. Instead of the polymeric threads being driven in the direction of the collector, the stability of the Taylor cone is impacted once the flow rate surpasses the optimal value, causing either beaded fibers (i.e., defected fibers) or polymeric solution to drip (leading to wastage of polymeric content). Furthermore, if too much solution drips from the needle tip during electrospinning, a large piece of the

polymer melts and lands on the collector. This causes the unevaporated solution to distort nanofibers that have already been deposited. For this reason, optimizing this parameter is crucial to getting the desired outcomes (Haghi & Akbari 2007).

18.2.3 Needle Diameter

Generally, it is believed that the sizeable internal diameter of the needle will result in a larger diameter in fiber and vice versa (Tan et al. 2005). This phenomenon is exceptionally applicable to most polymeric polymers. Regarding this, numerous investigations have examined the effects of needle gauges (18, 22, and 26 gauges) on electrospun poly(methyl methacrylate) nanofibers. These studies indicate a statistically insignificant correspondence between the average diameter of nanofiber and the needle. Overall, outcomes were unaffected by the needle diameter used in the fabrication procedure (Macossay et al. 2007; Hekmati et al. 2013).

18.2.4 Concentration of Solution

The optimal solution concentration is the determining factor to consider for the fabrication of fibers. The increase in concentration directly impacts the viscosity of the solution's viscosity, which leads to an elevation in the diameter of nanofibers. Moreover, as the concentration of the solution increases from low to high, the spherical shape transforms into a spindle-like structure (Haghi & Akbari 2007). At an optimal concentration, the solution forms fibers that have a uniform structure, unlike the lower concentration, which tends to produce fibers with irregular shapes (i.e., a beaded structure) (Kong & Ziegler 2013).

18.2.5 Solution's Molecular Weight

Concerning molecular weight, the utilization of low molecular weight polymer results in the formation of beads in nanofibers, while larger diameter fibers are created with an increase in solution's molecular weight. Thus, the molecular weight must be optimal for favorable outcomes (Casper et al. 2003). It considerably impacts how a solution behaves, including its concentration and viscosity. As a result, this needs to be considered during the electrospinning process as it significantly affects the determination of fiber morphology.

18.2.6 Viscosity of Solution

The primary factor determining the shape of electrospun fibers is the viscosity of the solution. As a consequence of the increased viscosity of the solution, the diameter of the nanofibers expands. Solutions characterized by low viscosity yield nanofibers with the smallest diameter (Liu et al. 2017).

18.2.7 Volatility of Solvent

When electrospun fibers are utilized for targeted drug delivery, the solvent's volatility becomes crucial. The enhanced volatility of the solvent used to solubilize the polymer affects the generation of fibers that are more porous and consist of a larger surface area (Balaji et al. 2015). Nevertheless, fibers with increased pore size and decreased surface area result from using solvents with lower volatility (Megelski et al. 2002). Hence, solvent volatility is an essential consideration for producing highly porous fibers. When less volatile solvents are used to create the polymeric dispersion, any remaining solvent residue is trapped in the nanofiber matrix and may impact the structure of fibrous mats.

18.2.8 Conductivity of Solution

The jet's electrostatic interaction is enhanced by the solution's conductivity, producing fibers with a minor diameter through concurrent jet elongation. In addition, the formation of fibers with an increased diameter occurs in the solution characterized by low conductivity (Huan et al. 2015). The utilization of high-conductive solutions in electrospinning may result in the generation of tiny strands that resemble fibers found in spider webs (Uyar & Besenbacher 2008; Drosou et al. 2018). Since conductivity is a solvent-specific property, the choice of solvent affects the conductivity of the entire spinning solution when creating a polymeric dispersion. The conductivity and solvent properties can effectively affect the desired diameter and structure of fibers, leading to the anticipated outcomes.

18.2.9 Surface Tension

The type of solvent utilized for polymer dissolution influences the solution's surface tension. Because of lowered surface tension, it is possible to create fibers without irregularity and carry out electrospinning at a lower electric field (Liu et al. 2008). Solutions with increased surface tension result in fibers with distorted morphology.

18.2.10 Temperature

Temperature significantly influences fiber morphology since large diameters of nanofibers are obtained at higher temperatures (Pillay et al. 2013). In this connection, the viscosity of the polymeric solution is significantly influenced by temperature. Elevated temperatures can speed up the drying process of the fibers, resulting in a beneficial outcome. Conversely, colder temperatures may lead to decreased fiber accumulation on the collector and blockage of the needle during the fabrication process. Thus, closely monitoring the temperature during the fabrication process can significantly enhance the quality of the nanofiber.

18.2.11 Humidity

Greater humidity levels facilitate the development of nanofibers with high porosity, a critical factor for promoting cellular adhesion and penetration (Rafiq et al. 2023). A direct correlation is established between relative humidity, pore size, and diameter. However, the rise in humidity does not impact the shape or size of nanofibers (Casper et al. 2003). It is important to note that various polymers exhibit distinct reactions to humidity, and this factor should be taken into account. An instance of this can be observed in the fibers of polyvinylpyrrolidone and cellulose acetate, where higher levels of humidity result in an enlargement of the mean diameter. Therefore, it is essential to adjust this parameter based on the type and composition of the polymer utilized in nanofiber production (De Vrieze et al. 2009).

18.3 ELECTROSPINNING OF STARCH

Amylose possesses a linear arrangement of molecules and exhibits a remarkable ability to aggregate. Consequently, previous studies on starch nanofibers focused on producing amylose fibers. However, amylose is just a small part of ordinary starches and is expensive to purify; therefore, it is not advisable to perform electrospinning on it alone. These issues thus provide significant challenges to the large-scale synthesis of amylose nanofibers (Kong & Ziegler 2012a). Since amylose is a linear polymer, its molecular mobility is sufficiently unrestricted to allow it to preferentially align into parallel alignments, creating hydrogen bonds. However, amylopectin is highly branched, which prevents it from moving quickly or forming alignments and linkages (Kong & Ziegler 2012b). Amylopectin should theoretically have a detrimental effect on the spinning process and the fiber's strength (Kong & Ziegler 2012b). However, as amylopectin makes up the majority of natural starch, methods for creating fibers by electrospinning natural starch, even though it contains a lot of amylopectin, ensure the commercial viability of starch nanofiber production. Accordingly, several strategies have been investigated to achieve this purpose (Liu et al. 2017). As an illustration, in a study, researchers successfully electrospun starch fibers with a diameter in the micron range out of 15 wt.% maize starch (gelose 80). Wet electrospinning was employed to re-spin the starch, which was dissolved in a 95% dimethyl sulfoxide (DMSO) solution and entailed submerging the grounded collector in an ethanol bath. In addition, nanofibers were subjected to heating in the presence of ethanol (50% v/v) and glutaraldehyde (25% v/v) to improve the crystallinity and cross-linking, resulting in fibrous mat's enhanced water stability (Kong & Ziegler 2014a, 2014b). Figure 18.4 shows schematics for the strategy and the optical micrographs of the prepared fibers.

By electrospinning Hylon VII corn starch, researchers could fabricate starch fibers. Here, formic acid was utilized as a

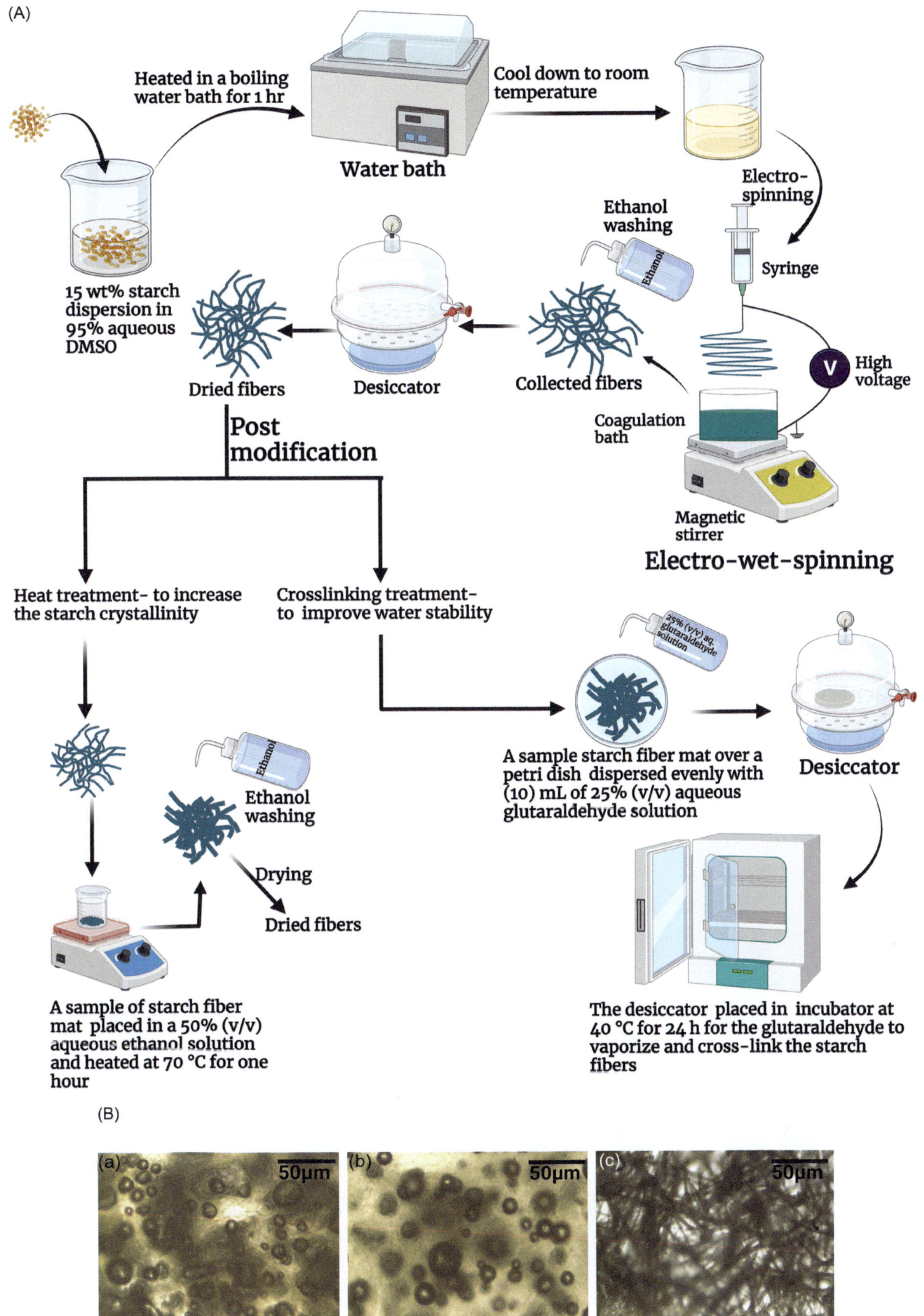

FIGURE 18.4   (A) Schematics for the preparation and post-modification of maize starch nanofibers. (B) Electrospun starch fiber optical micrographs: (a) as-spun fibers, (b) heat-treated fibers, and (c) cross-linked fibers by vapor phase glutaraldehyde. The scale bar in all figures is 50 μm (Kong & Ziegler 2014a, 2014b).

solvent at different concentrations to electrospun 17 wt.% starch effectively. Upon analyzing the synthesized fiber mats, it was revealed that dispersion containing formic acid concentrations of 100%, 90%, and 80% produced fibers with sizes ranging from 80 to 300 nm (Lancuski et al. 2015). Another study examined tapioca and glutinous rice starches for electrospinning with water as the sole solvent. Before electrospinning, the polymeric dispersions were stirred for 15 minutes at a temperature of 80°C. According to the results, glutamic rice starch could not produce nanofibers via electrospinning. However, tapioca starch exhibited fiber diameters between 1.3 and 14.5 µm (Jaiturong et al. 2012). It was due to the presence of a significant amount of amylopectin that hindered the efficient production of nanofibers from starch. Further exploration and inquiry must be conducted with unconventional methodologies. Researchers have made an effort to investigate alternative methods in this area. For instance, the fabrication of fibers from potato starch (73.35% amylopectin) and maize starch (68.89% amylopectin) using centrifugal or force spinning method (Amalorpava Mary et al. 2013; Padron et al. 2013) produced fibers with a diameter within sub-micron range, as given in Figure 18.5. This illustrates how the force or centrifugal spinning may be utilized as a substitute to create nanofibers from starch with a significant amylopectin content (Li et al. 2016).

Furthermore, multiple research investigations are currently conducted to effectively electrospun starch, considering the diverse and extensive applications of these nanofibers (Hermansson & Svegmark 1996; Wang & Copeland 2015). As was already indicated, the main challenge to the successful electrospinning of these nanofibers from native starch is the starch's high amylopectin content. Moreover, it impacts the

mechanical and aqueous stability of starch nanofibers (Lancuski et al. 2015). Keeping this in consideration, numerous researchers tried electrospinning starch with changes like physical, chemical, and enzymatic interventions (Xu et al. 2009; Wang et al. 2012). Successful electrospinning of starch acetate was accomplished by the researchers in another work. Here, formic acid, ethanol, and water were used as solvents in different proportions to produce nanofibers utilizing starch acetate with varying acetylation levels. By altering other factors, such as the solvent ratios and starch concentrations, degree of substitution, etc., the research explored their effect on the microarchitecture and nano-morphology of manufactured nanofibers, as shown in Figure 18.6. The diagram also illustrates the synthesis of starch acetate and its subsequent electrospinning process. The findings indicated that using a solvent solution consisting of 90% (v/v) formic acid and water produced uniform fibers and devoid of bead-like structures (Xu, Yang & Yang 2009).

In an additional study, researchers successfully electrospun the oxidized starch treated with HCl. In this work, only DMSO was exploited as a solvent system to perform electrospinning. The successful production of fibers occurred when modified starch and DMSO were dispersed at a concentration of 19 wt.%. The results were reproduced using a 5 wt.% native starch/DMSO solution. However, unlike the original starch, the modification resulted in a decrease in the viscosity of the solution (Wang et al. 2012). Figure 18.7 demonstrates schematically the preparation of modified potato starch and its electrospinning.

Another study used modified starch to create poly(ethylene-alt-maleic anhydride) and starch blends. Here, DMSO was employed as a solvent to generate nanofibers. These

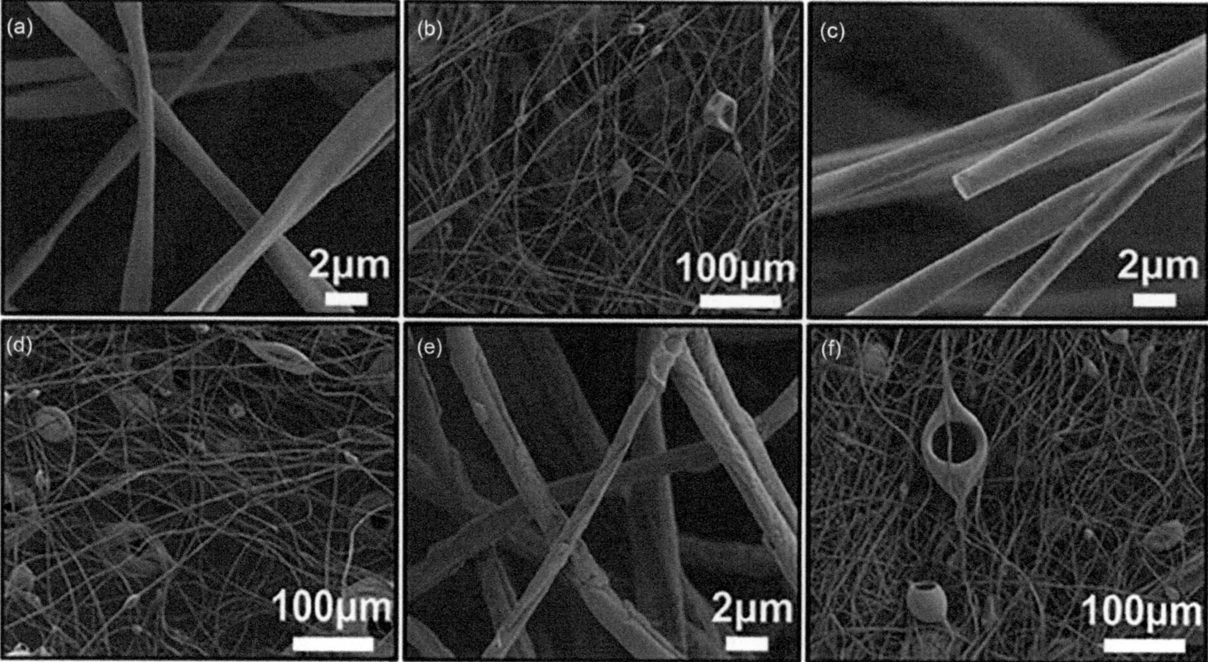

FIGURE 18.5   SEM micrographs of starch-based fibers generated by centrifugal spinning: (a, b) amylose-rich corn starch, (c, d) native corn starch, and (e, f) potato starch. (Reprinted with copyright permission from Li et al. 2016.)

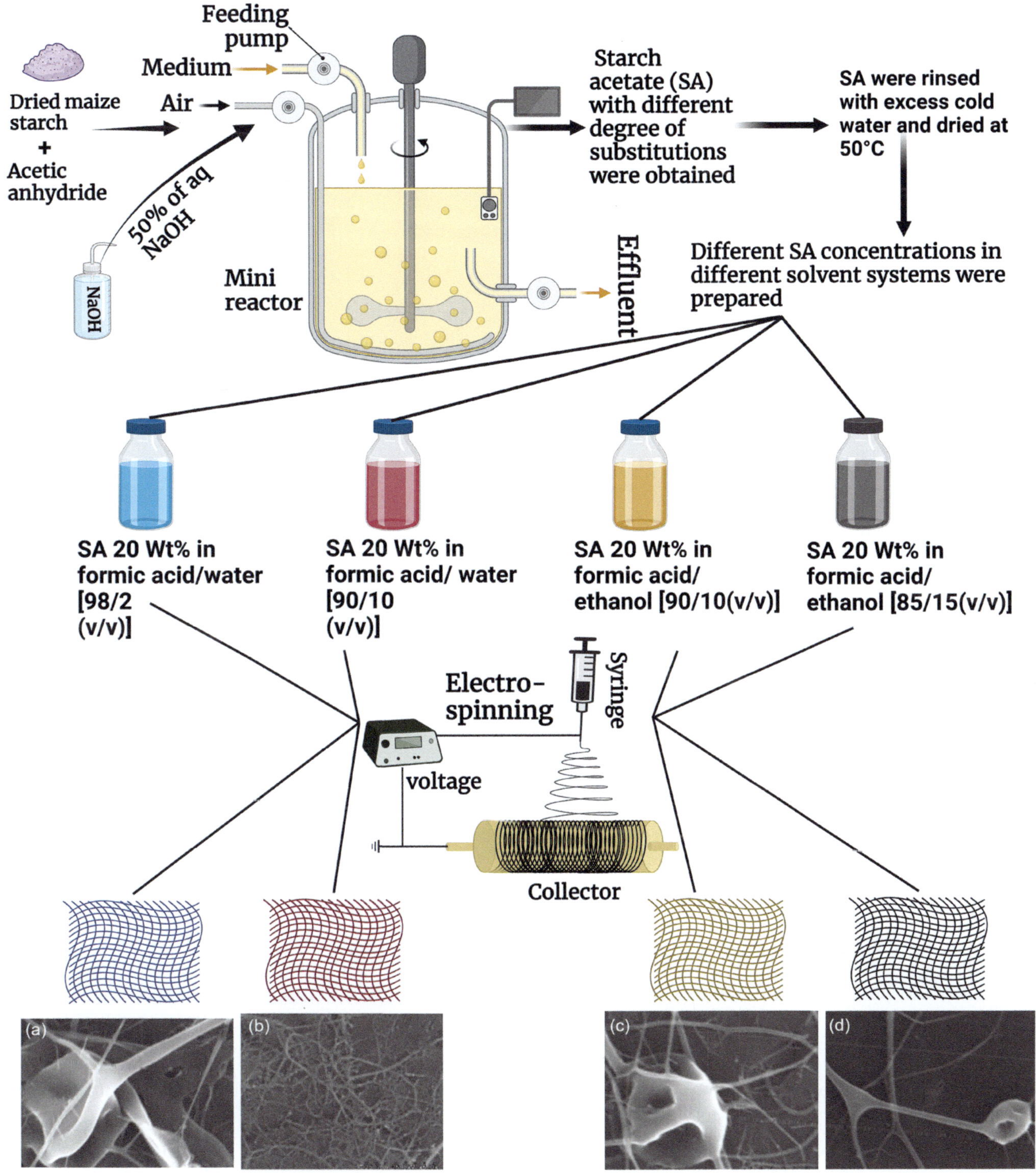

FIGURE 18.6   Diagrammatic depiction of the preparation of starch acetate and the production of various starch acetate nanofibrous mats in different solvent systems. The SEM morphologies of these mats are also provided in (a)–(d). SEM micrograph of electrospun starch acetate nanofiber at (a, b) 98% v/v, and 90% v/v formic acid:water solvent system. (c, d) 90% v/v and 85% v/v formic acid:ethanol solvent system. (Reprinted with permission from Xu, Yang & Yang 2009.)

nanofibrous meshes were characterized, and the alteration was discovered to improve the intermolecular bond formation, enhancing the electrospun nanofiber's thermal stability (Oktay et al. 2013). The hydrophilic property of starch allows for complete absorption by the human body with no unfavorable or allergic responses (Zhang et al. 2011; Liu et al. 2017; Krishnakumar et al. 2019). Therefore, starch has been investigated as a scaffold material in tissue engineering, wound dressings, and drugs delivery (Liu et al. 2017). However, starch's hydrophilic nature prevents it from being utilized in producing

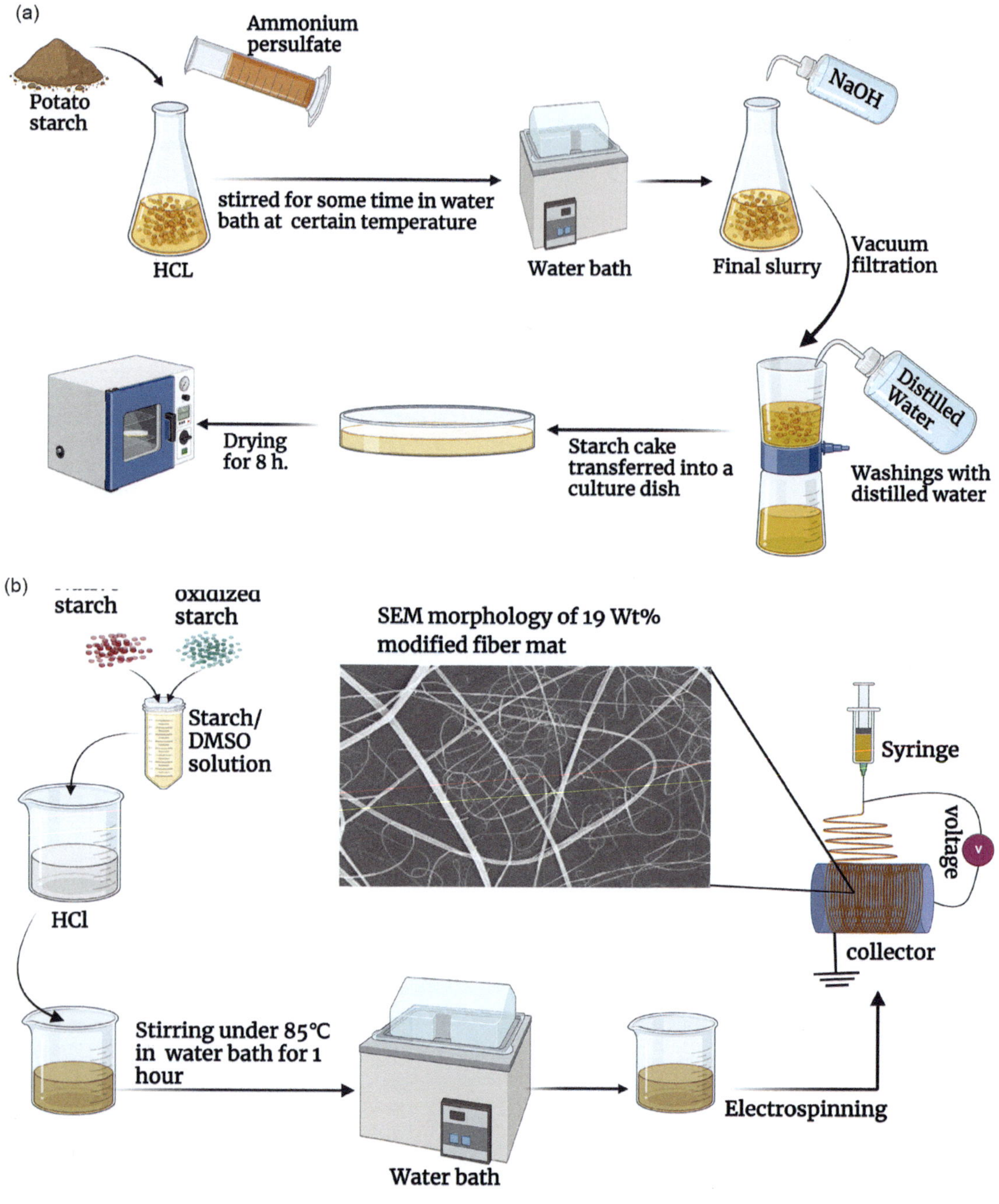

FIGURE 18.7   Schematics for the preparation of (a) acidified-oxidized potato starch and (b) starch fibers using electrospinning.

electrospun nanofibers for biomedical use. The aqueous physiological environment causes the starch nanofibers to swell or disintegrate instantly. In order to enhance the hydrophobic properties of starch nanofibers, several strategies have opted to blend starch with hydrophobic synthetic polymers (e.g., polylactide, poly(lactide-co-glycolide), poly(vinylidene fluoride), and poly(ε-caprolactone)). However, these approaches may compromise the advantages offered by starch (Wang et al. 2011; Komur et al. 2017; Amini et al. 2018). Numerous research groups have tried to enhance the water resistance of starch nanofibers through glutaraldehyde cross-linking. But,

glutaraldehyde is a very poisonous chemical (Draye et al. 1998; Gough et al. 2002; Fürst & Banerjee 2005). Following the process of cross-linking, there is a possibility that unreacted glutaraldehyde may remain in the nanofibers. The presence of unreacted glutaraldehyde might potentially result in additional harmful consequences, such as the cross-linking of proteins within a living system. In Table 18.1, we have provided the publication details of the work that has been conducted on starch nanofibers for applications in diverse fields since last decade.

TABLE 18.1 Work done on starch electrospun nanofibers from 2015 onward

PUBLICATION TITLE	NAME OF JOURNAL	REFERENCE
Crosslinked starch nanofibers with high mechanical strength and excellent water resistance for biomedical applications	*Biomedical Materials*	Lv et al. (2020)
Development of antimicrobial and antioxidant electrospun soluble potato starch nanofibers loaded with carvacrol	*International Journal of Biological Macromolecules*	Fonseca et al. (2019)
Electrospun potato starch nanofibers for thyme essential oil encapsulation: Antioxidant activity and thermal resistance	*Journal of the Science of Food and Agriculture*	Fonseca et al. (2020)
Electrospun starch nanofibers as a delivery carrier for carvacrol as anti-glioma agent	*Starch*	Fonseca et al. (2022)
Starch nanofibers as vehicles for folic acid supplementation: Thermal treatment, UVA irradiation and in vitro simulation of digestion	*Journal of the Science of Food and Agriculture*	Fonseca et al. (2021a)
Investigation of the effects of starch on the physical and biological properties of polyacrylamide (PAAm)/starch nanofibers	*Progress in Biomaterials*	Taghavian et al. (2017)
Development and evaluation of novel nanofibers based on mango kernel starch obtained by electrospinning	*Polymer Testing*	Gomez-Caturla et al. (2022)
Preparation and characterization of PVDF/Starch nanocomposite nanofibers using electrospinning method	*Materials Today: Proceedings*	Amini et al. (2018)
Antibacterial and physical effects of cationic starch nanofibers containing carvacrol@casein nanoparticles against Bacillus cereus in soy products	*International Journal of Food Microbiology*	Cui et al. (2022)
Fabrication of drug-loaded starch-based nanofibers via electrospinning technique	*Biointerface Research in Applied Chemistry*	Tuah et al. (2021)
Effects of hydrogen bonding on starch granule dissolution, spinnability of starch solution, and properties of electrospun starch fibers	*Polymer*	Wang et al. (2018)
Effect of carvacrol encapsulation in starch-based nanofibers: Thermal resistance and antioxidant and antimicrobial properties	*Food Processing and Preservation*	Fonseca et al. (2021b)
Starch based nanofibrous scaffolds for wound healing applications	*Bioactive Materials*	Waghmare et al. (2018)
Organic solvent-free starch-based green electrospun nanofiber mats for curcumin encapsulation and delivery	*International Journal of Biological Macromolecules*	Du et al. (2023)
Fabrication and characterization of starch-TPU based nanofibers for wound healing applications	*Materials Science and Engineering: C*	Mistry et al. (2021)
Porous electrospun starch rich polycaprolactone blend nanofibers for severe haemorrhage	*International Journal of Biological Macromolecules*	Dev & Hemamalini (2018)
Preparation of hydroxypropyl starch/polyvinyl alcohol composite nanofibers films and improvement of hydrophobic properties	*International Journal of Biological Macromolecules*	Pan et al. (2022)
Optimizing the activity of immobilized phytase on starch blended polyacrylamide nanofibers-nanomembranes by response surface methodology	*Fibers and Polymers*	Taghavian et al. (2015)
Potential of novel electrospun core-shell structured polyurethane/starch (hyaluronic acid) nanofibers for skin tissue engineering: In vitro and in vivo evaluation	*International Journal of Biological Macromolecules*	Movahedi et al. (2020)
Crosslinked carboxymethyl starch nanofiber mats: Preparation, water resistance and exudates control ability	*European Polymer Journal*	Luan et al. (2021)
Suitability of starch/carvacrol nanofibers as biopreservatives for minimizing the fungal spoilage of bread	*Carbohydrate Polymers*	Fonseca et al. (2021c)
Starch-derived nanographene oxide paves the way for electrospinnable and bioactive starch scaffolds for bone tissue engineering	*Biomacromolecules*	Wu et al. (2017)
AgNPs-incorporated nanofiber mats: Relationship between AgNPs size/content, silver release, cytotoxicity, and antibacterial activity	*Materials Science and Engineering: C*	Lv et al. (2021)
Improving the hydrophobicity and mechanical properties of starch nanofibrous films by electrospinning and cross-linking for food packaging applications	*LWT - Food Science and Technology*	Zhu et al. (2022)

(Continued)

TABLE 18.1 (*Continued*)   Work done on starch electrospun nanofibers from 2015 onward

PUBLICATION TITLE	NAME OF JOURNAL	REFERENCE
Hydrophobic interface starch nanofibrous film for food packaging: From bioinspired design to self-cleaning action	*Journal of Agricultural and Food Chemistry*	Cai et al. (2021)
Aging time of soluble potato starch solutions for ultrafine fibers formation by electrospinning	*Starch*	Fonseca et al. (2019)
Pore-structure-optimized CNT-carbon nanofibers from starch for rechargeable lithium batteries	*Materials*	Jeong et al. (2016)
Preparation of glutinous rice starch/polyvinyl alcohol copolymer electrospun fibers for using as a drug delivery carrier	*Asian Journal of Pharmaceutical Sciences*	Jaiturong et al. (2018)
Effect of amylose content on the preparation for carboxymethyl starch/pullulan electrospun nanofibers and their properties as encapsulants of thymol	*Food Hydrocolloids*	Liang & Gao (2023)
The antibacterial and anti-inflammatory investigation of *Lawsonia inermis*-gelatin-starch nano-fibrous dressing in burn wound	*International Journal of Biological Macromolecules*	Hadisi et al. (2018)
Fabrication of nanofibers based on hydroxypropyl starch/polyurethane loaded with the biosynthesized silver nanoparticles for the treatment of pathogenic microbes in wounds	*Polymers*	El-Hefnawy et al. (2022)
Novel polyvinyl alcohol/starch electrospun fibers as a strategy to disperse cellulose nanocrystals into poly(lactic acid)	*Polymers*	López de Dicastillo et al. (2017)
Starch/tea polyphenols nanofibrous films for food packaging application: From facile construction to enhance mechanical, antioxidant and hydrophobic properties	*Food Chemistry*	Zhang et al. (2021)
Evaluation of the effects of starch on polyhydroxybutyrate electrospun scaffolds for bone tissue engineering applications	*International Journal of Biological Macromolecules*	Asl et al. (2021)
Silver nanoparticles impregnated in tapioca starch biofilm made using the electrospinning technique: A cutting-edge material for food packaging	*Biomass Conversion and Biorefinery*	Alqahtani (2022)
PLA/starch biodegradable fibers obtained by the electrospinning method for micronutrient mineral release	*AIMS Materials Science*	Donizette Malafatti et al. (2023)
Starch-based electrospun nanofibrous scaffold incorporated with bioactive compound from *Cayratia trifolia* promotes wound healing by activating VEGF and PDGF signaling pathways	*Applied Biological Research*	Meganathan & Panagal (2023)
Effect of vapor-phase glutaraldehyde crosslinking on electrospun starch fibers	*Carbohydrate Polymers*	Wang et al. (2016)
Electrospun halloysite nanotube loaded polyhydroxybutyrate-starch fibers for cartilage tissue engineering	*International Journal of Biological Macromolecules*	Movahedi & Karbasi (2022)
An antimicrobial peptide-immobilized nanofiber mat with superior performances than the commercial silver-containing dressing	*Materials Science and Engineering: C*	Yang et al. (2021)
Wound dressing based on electrospun PVA/chitosan/starch nanofibrous mats: Fabrication, antibacterial and cytocompatibility evaluation and in vitro healing assay	*International Journal of Biological Macromolecules*	Adeli et al. (2019)
Red onion skin extract rich in flavonoids encapsulated in ultrafine fibers of sweet potato starch by electrospinning	*Food Chemistry*	da Cruz et al. (2023)

18.4 THERAPEUTIC AND TISSUE ENGINEERING APPLICATIONS

Figure 18.8 shows various biomedical applications of the starch nanofibers.

The utilization of electrospun nanofibers for tissue engineering and delivery of drugs has achieved significant attention during the last few decades. Highly porous nature, high specific surface area, brilliant mechanical functioning, and ease of manufacturing of these nanofibers make it possible to develop perfect delivery systems for growth factors (Sahoo et al. 2009), drugs (Son et al. 2014), bioactive substances (Castro-Muñoz et al. 2023), and even cells (Shabani et al. 2009), DNA (Bellan et al. 2007), and RNA (Stojanov & Berlec 2020), besides scaffolds for cellular adhesion and proliferation (Sill & Von Recum 2008; Diaz-Gomez et al. 2014). The natural carbohydrate polymer starch is inexpensive, readily available, and renewable. Starch is suitable for pharmaceutical applications due to its biodegradability (Cho et al. 2012), biocompatibility (Rodríguez et al. 2012), and bioabsorbility (Kong & Ziegler 2012a; Kong & Ziegler 2014a, 2014b; Wang et al. 2016). The interaction of functional groups within the starch matrix and its constituent components can enhance the capacity to bind and capture diverse hydrophilic

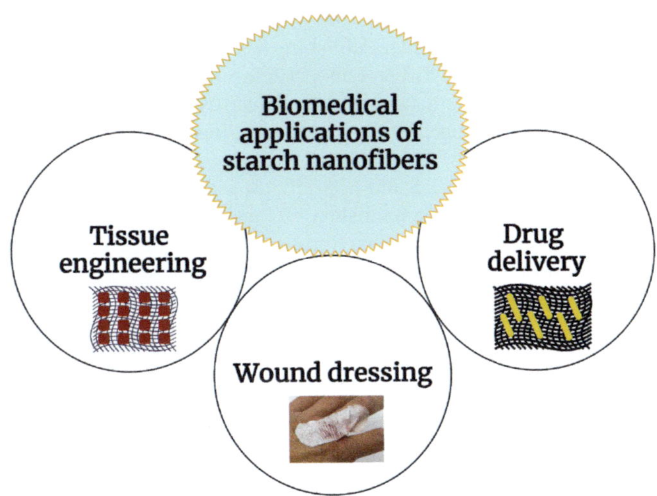

FIGURE 18.8 Various biomedical applications of electrospun starch nanofibers.

and hydrophobic substances. Due to their better thermal durability, starch-based delivery methods have a superior defensive layer for bioactive chemicals under elevated processing temperatures compared to carriers based on lipids or proteins (Fathi et al. 2014). The benefits mentioned above of starch-based substances, along with the adaptability of starch modification and the production of nanofibers, make starch and its byproducts excellent options for application in tissue engineering and drug delivery.

18.4.1 Strach Nanofibers for Delivery of Important Drugs

Based on the encapsulation process, electrospinning can be employed to embed target compounds within starch nanofibers to serve as delivery carriers in pharmaceutical formulations (Kong & Ziegler 2012b; Tang et al. 2016; Lancuški et al. 2017; Fonseca et al. 2022). Electrospun nanofibers are versatile carriers for delivering DNA/RNA (Stojanov & Berlec 2020), cells (Stojanov & Berlec 2020), medications (Nagarajan et al. 2019), proteins (Stojanov & Berlec 2020), antioxidants (Vilchez et al. 2020), viruses (Furuno et al. 2022), and antibacterial agents (Topuz & Uyar 2020) (Sill & Von Recum 2008; Kai et al. 2016). Direct electrospinning subsequent to dissolving or dispersing the necessary components inside the starch nanofiber framework is the primary and typical approach for delivery (Rezaei et al. 2015). To fulfill specific needs, innovative delivery methods based on starch nanofibers have been designed using co-axial electrospinning, emulsion electrospinning, surface-functionalizing electrospun nanofibers, and synergistically combining nanofibers with nanocontainers (Lancuški et al. 2017). Due to its substantial surface area concerning volume, electrospun starch nanofibers enable the discharge of target compounds through a wide surface area. Starch nanofibers' flexibility in dimensions and surface qualities, which are determined by their composition and morphology, enables precise monitoring and adjustments of drug release behavior and bioaccessibility (Sharifi et al. 2016). More aspects of the fiber matrix, for instance,

length and diameter, can be customized in contrast to spherical carriers, where the diameter restricts the ability to modify the specific surface area (Sharifi et al. 2016). Moreover, nanofibers have the potential to transport drugs effectively and efficiently with significant drug loading (Ding et al. 2014; Sánchez et al. 2016). Biodegradable starch nanofibers' drug release mechanism is more intricate than non-biodegradable polymers. Drug diffusion and matrix degradation are linked to drug release characteristics (Yoo et al. 2009). Drug release for non-biodegradable fibers is uniquely regulated by diffusion. However, within a starch fiber system, drug release is primarily dominated by diffusion and degradation processes (Sill & Von Recum 2008; Lu et al. 2016). Because electrospun fibers have a higher surface area than comparable films, drug release typically occurs more quickly. Nanofibers made of starch are hydrophilic and biodegradable. Thus, typical drug release characteristics involve an initial burst pursued by a prolonged and continuous release. Drugs coated on fiber surfaces rapidly diffuse, causing the initial release; however, matrix degradation and drug diffusion regulate the sustained release (Yoo et al. 2009).

18.4.2 Strach Nanofibers for Tissue Engineering

In tissue engineering, electrospun scaffolds have been applied to the bone, cartilage, neurons, tendons, and ligaments. The electrospun starch nanofibers' morphological characteristics, such as their high specific surface area, highly porous nature, permeability, and well-connected pore structure, resemble the extracellular matrix (ECM) present in living things (Tuzlakoglu et al. 2005). To assist cells and give them a place to anchor, nonwoven porous electrospun nanofibers can be used to construct supporting meshwork around the cells. This encourages cell adherence and proliferation (Silva et al. 2013).

Starch has both biocompatibility and biodegradability properties. It is recommended to employ biocompatible materials when producing electrospun scaffolds to reduce the host organism's toxicity, inflammatory reactions, and immunological responses (Sill & Von Recum 2008; Sharifi et al. 2016). The scaffolds' biodegradability could prevent the need for another surgery to remove mats after implantation (Sill & Von Recum 2008). A promising tissue engineering scaffold should degrade at a reasonable rate. If the degradation rate is prolonged, tissue regeneration will be hindered, and if it is too quick, the release of drugs and mechanical stability will be impaired (Huang et al. 2016). Starch nanofibers' porous structure can improve the scaffold's surface area (Li & Xia 2004). The growth of cells, involving adhesion, migration, proliferation, and differentiation, is influenced by both the presence and the size of pores (Sill & Von Recum 2008; Huang et al. 2016). An optimal tissue engineering scaffold must have the proper pore size and degradation rate, as well as regulated porosity, adequate biological response, and mechanical characteristics (Gomes et al. 2002).

18.5 CONCLUSION

Natural materials have been used to create electrospun nanofibers, exhibiting great promise in pharmaceutical applications like drugs, tissue engineering, and wound dressing. These benefits include high porous nature, surface area-to-volume ratio, biocompatibility, biodegradability, and bioabsorbability. The textile, plastic, food, and pharmaceutical sectors have extensively used starch and its derivatives because starch is an affordable, readily available, naturally occurring, biodegradable, and biocompatible polymer. Although electrospinning holds significant potential, the limitations of starch-based materials, such as their inadequate mechanical strength and susceptibility to water, pose considerable challenges in fabricating starch nanofibers at an industrial level. Starch spinnability has been enhanced through the production of blended nanostructures from starch and synthetic polymers and increased prospects for functionalization with various components/molecules. A variety of techniques have been used to create starch fibers through electrospinning, such as choosing the suitable solvents (e.g., DMSO, acetic acid, and chloroform), modifying the starch (acetylated starch, oxidized starch, hydroxypropyl starch, etc.), and utilizing hybrid systems which combine starch with different polymers (PCL, PVA, PEO, etc.). However, several obstacles must be surmounted before electrospinning of starch.

Due to the requirement for modified starches, the choice of a highly efficient solvent, and sophisticated electrospinning methods, the production of electrospun starch nanofibers may encounter substantial challenges. As this polymer holds great promise for application in tissue engineering, regenerative medicine, and other biomedical fields, further research must be conducted to develop native starch nanofibers. Furthermore, well-defined biomolecules like growth hormones and differentiation factors can functionalize starch-based scaffolds to provide new pathways in regenerative medicine. The future challenges in producing starch nanofibers include cost-effectiveness because, for large scale, the electrospinning technique is quite challenging. We must figure out how to reduce solvent waste and safety concerns.

REFERENCES

Adeli, H., Khorasani, M.T. and Parvazinia, M., 2019. Wound dressing based on electrospun PVA/chitosan/starch nanofibrous mats: Fabrication, antibacterial and cytocompatibility evaluation and in vitro healing assay. *International Journal of Biological Macromolecules, 122*, 238–254.

Ainurofiq, A. and Choiri, S., 2015. Drug release mechanism of slightly soluble drug from nanocomposite matrix formulated with zeolite/hydrotalcite as drug carrier. *Tropical Journal of Pharmaceutical Research, 14*(7), 1129–1135.

Alghoraibi, I. and Alomari, S., (eds) 2018. *Handbook of nanofibers*, Springer International Publishing AG.

Alqahtani, T., 2022. Silver nanoparticles impregnated in tapioca starch biofilm made using the electrospinning technique: A cutting-edge material for food packaging. *Biomass Conversion and Biorefinery, 12*(9), 1–8.

Amalorpava Mary, L., Senthilram, T., Suganya, S., Nagarajan, L., Venugopal, J., Ramakrishna, S. and Giri Dev, V.R., 2013. Centrifugal spun ultrafine fibrous web as a potential drug delivery vehicle. *Express Polymer Letters, 7*(3), 238.

Amini, M., Haddadi, S.A., Ghaderi, S., SA, A.R. and Ansarizadeh, M.H., 2018. Preparation and characterization of PVDF/starch nanocomposite nanofibers using electrospinning method. *Materials Today: Proceedings, 5*(7), 15613–15619.

Ashraf, R., Sofi, H.S., Malik, A., Beigh, M.A., Hamid, R. and Sheikh, F.A., 2019. Recent trends in the fabrication of starch nanofibers: Electrospinning and non-electrospinning routes and their applications in biotechnology. *Applied Biochemistry and Biotechnology, 187*, 47–74.

Asl, M.A., Karbasi, S., Beigi-Boroujeni, S., Benisi, S.Z. and Saeed, M., 2021. Evaluation of the effects of starch on polyhydroxybutyrate electrospun scaffolds for bone tissue engineering applications. *International Journal of Biological Macromolecules, 191*, 500–513.

Bailey, E.H.S. and Long, W.S., 1916. On the occurrence of starch in green fruits. *Transactions of the Kansas Academy of Science (1903-), 28*, 153–155.

Balaji, A., Vellayappan, M.V., John, A.A., Subramanian, A.P., Jaganathan, S.K., Supriyanto, E. and Razak, S.I.A., 2015. An insight on electrospun-nanofibers-inspired modern drug delivery system in the treatment of deadly cancers. *RSC Advances, 5*(71), 57984–58004.

Beachley, V. and Wen, X., 2010. Polymer nanofibrous structures: Fabrication, biofunctionalization, and cell interactions. *Progress in Polymer Science, 35*(7), 868–892.

Bellan, L.M., Strychalski, E.A. and Craighead, H.G., 2007. Electrospun DNA nanofibers. *Journal of Vacuum Science & Technology B: Microelectronics and Nanometer Structures Processing, Measurement, and Phenomena, 25*(6), 2255–2257.

Bhardwaj, N. and Kundu, S.C., 2010. Electrospinning: A fascinating fiber fabrication technique. *Biotechnology Advances, 28*(3), 325–347.

Cai, J., Zhang, D., Zhou, R., Zhu, R., Fei, P., Zhu, Z.Z., Cheng, S.Y. and Ding, W.P., 2021. Hydrophobic interface starch nanofibrous film for food packaging: From bioinspired design to self-cleaning action. *Journal of Agricultural and Food Chemistry, 69*(17), 5067–5075.

Cárdenas, W., Gómez-Pachon, E.Y., Muñoz, E. and Vera-Graziano, R., 2016, July. Preparation of potato starch microfibers obtained by electro wet spinning. *IOP Conference Series: Materials Science and Engineering, 138*(1), 012001. IOP Publishing.

Casper, C.L., Stephens, J.S., Tassi, N.G., Chase, D.B. and Rabolt, J.F., 2003. Controlling surface morphology of electrospun polystyrene fibers: Effect of humidity and molecular weight in the electrospinning process. *Macromolecules, 37*(2), 573–578.

Castro-Muñoz, R., Kharazmi, M.S. and Jafari, S.M., 2023. Chitosan-based electrospun nanofibers for encapsulating food bioactive ingredients: A review. *International Journal of Biological Macromolecules, 245*, 125424.

Cho, D., Netravali, A.N. and Joo, Y.L., 2012. Mechanical properties and biodegradability of electrospun soy protein isolate/PVA hybrid nanofibers. *Polymer Degradation and Stability, 97*(5), 747–754.

Cui, H., Lu, J., Li, C., Rashed, M.M. and Lin, L., 2022. Antibacterial and physical effects of cationic starch nanofibers containing carvacrol@ casein nanoparticles against Bacillus cereus in soy products. *International Journal of Food Microbiology*, *364*, 109530.

Cui, J., Li, F., Wang, Y., Zhang, Q., Ma, W. and Huang, C., 2020. Electrospun nanofiber membranes for wastewater treatment applications. *Separation and Purification Technology*, *250*, 117116.

da Cruz, E.P., Jansen, E.T., Fonseca, L.M., dos Santos Hackbart, H.C., Siebeneichler, T.J., Pires, J.B., Gandra, E.A., Rombaldi, C.V., da Rosa Zavareze, E. and Dias, A.R.G., 2023. Red onion skin extract rich in flavonoids encapsulated in ultrafine fibers of sweet potato starch by electrospinning. *Food Chemistry*, *406*, 134954.

De Vrieze, S., Van Camp, T., Nelvig, A., Hagström, B., Westbroek, P. and De Clerck, K., 2009. The effect of temperature and humidity on electrospinning. *Journal of Materials Science*, *44*, 1357–1362.

Deitzel, J.M., Kleinmeyer, J., Harris, D.E.A. and Tan, N.B., 2001. The effect of processing variables on the morphology of electrospun nanofibers and textiles. *Polymer*, *42*(1), 261–272.

Dev, V.G. and Hemamalini, T., 2018. Porous electrospun starch rich polycaprolactone blend nanofibers for severe hemorrhage. *International Journal of Biological Macromolecules*, *118*, 1276–1283.

Diaz-Gomez, L., Alvarez-Lorenzo, C., Concheiro, A., Silva, M., Dominguez, F., Sheikh, F.A., Cantu, T., Desai, R., Garcia, V.L. and Macossay, J., 2014. Biodegradable electrospun nanofibers coated with platelet-rich plasma for cell adhesion and proliferation. *Materials Science and Engineering: C*, *40*, 180–188.

Ding, F., Deng, H., Du, Y., Shi, X. and Wang, Q., 2014. Emerging chitin and chitosan nanofibrous materials for biomedical applications. *Nanoscale*, *6*(16), 9477–9493.

Doi, S., Clark, J.H., Macquarrie, D.J. and Milkowski, K., 2002. New materials based on renewable resources: Chemically modified expanded corn starches as catalysts for liquid phase organic reactions. *Chemical Communications*, *34*(22), 2632–2633.

Donizette Malafatti, J.O., Machado de Oliveira Ruellas, T., Rodrigues Sciena, C. and Paris, E.C., 2023. PLA/starch biodegradable fibers obtained by the electrospinning method for micronutrient mineral release. *AIMS Materials Science*, *10*(2), 200–212.

Draye, J.P., Delaey, B., Van de Voorde, A., Van Den Bulcke, A., De Reu, B. and Schacht, E., 1998. In vitro and in vivo biocompatibility of dextran dialdehyde cross-linked gelatin hydrogel films. *Biomaterials*, *19*(18), 1677–1687.

Drosou, C., Krokida, M. and Biliaderis, C.G., 2018. Composite pullulan-whey protein nanofibers made by electrospinning: Impact of process parameters on fiber morphology and physical properties. *Food Hydrocolloids*, *77*, 726–735.

Du, Z., Lv, H., Wang, C., He, D., Xu, E., Jin, Z., Yuan, C., Guo, L., Wu, Z., Liu, P. and Cui, B., 2023. Organic solvent-free starch-based green electrospun nanofiber mats for curcumin encapsulation and delivery. *International Journal of Biological Macromolecules*, *232*, 123497.

El-Hefnawy, M.E., Alhayyani, S., El-Sherbiny, M.M., Sakran, M.I. and El-Newehy, M.H., 2022. Fabrication of nanofibers based on hydroxypropyl starch/polyurethane loaded with the biosynthesized silver nanoparticles for the treatment of pathogenic microbes in wounds. *Polymers*, *14*(2), 318.

Fathi, M., Martín, Á. and McClements, D.J., 2014. Nanoencapsulation of food ingredients using carbohydrate based delivery systems. *Trends in Food Science & Technology*, *39*(1), 18–39.

Fonseca, L.M., Bona, N.P., Crizel, R.L., Pedra, N.S., Stefanello, F.M., Lim, L.T., Carreño, N.L.V., Dias, A.R.G. and Zavareze, E.D.R., 2022. Electrospun starch nanofibers as a delivery carrier for carvacrol as anti-glioma agent. *Starch-Stärke*, *74*(1–2), 2100115.

Fonseca, L.M., Crizel, R.L., Da Silva, F.T., Fontes, M.R., da Rosa Zavareze, E. and Dias, A.R., 2021a. Starch nanofibers as vehicles for folic acid supplementation: Thermal treatment, UVA irradiation and in vitro simulation of digestion. *Journal of the Science of Food and Agriculture*, *101*(5), 1935–1943.

Fonseca, L.M., da Silva, F.T., Antunes, M.D., Mello El Halal, S.L., Lim, L.T. and Dias, A.R.G., 2019. Aging time of soluble potato starch solutions for ultrafine fibers formation by electrospinning. *Starch-Stärke*, *71*(1–2), 1800089.

Fonseca, L.M., Radünz, M., Crizel, R.L., Camargo, T.M., Gandra, E.A. and Dias, A.R.G., 2021b. Effect of carvacrol encapsulation in starch-based nanofibers: Thermal resistance and antioxidant and antimicrobial properties. *Journal of Food Processing and Preservation*, *45*(5), e15409.

Fonseca, L.M., Radünz, M., dos Santos Hackbart, H.C., da Silva, F.T., Camargo, T.M., Bruni, G.P., Monks, J.L., da Rosa Zavareze, E. and Dias, A.R., 2020. Electrospun potato starch nanofibers for thyme essential oil encapsulation: Antioxidant activity and thermal resistance. *Journal of the Science of Food and Agriculture*, *100*(11), 4263–4271.

Fonseca, L.M., Souza, E.J.D., Radünz, M., Gandra, E.A., da Rosa Zavareze, E. and Dias, A.R.G., 2021c. Suitability of starch/carvacrol nanofibers as biopreservatives for minimizing the fungal spoilage of bread. *Carbohydrate Polymers*, *252*, 117166.

Fürst, W. and Banerjee, A., 2005. Release of glutaraldehyde from an albumin-glutaraldehyde tissue adhesive causes significant in vitro and in vivo toxicity. *The Annals of Thoracic Surgery*, *79*(5), 1522–1528.

Furuno, K., Suzuki, K. and Sakai, S., 2022. Gelatin-based electrospun nanofibers cross-linked using horseradish peroxidase for plasmid DNA delivery. *Biomolecules*, *12*(11), 1638.

Gomes, M.E., Godinho, J.S., Tchalamov, D., Cunha, A.M. and Reis, R.L., 2002. Alternative tissue engineering scaffolds based on starch: processing methodologies, morphology, degradation and mechanical properties. *Materials Science and Engineering: C*, *20*(1–2), 19–26.

Gomez-Caturla, J., Ivorra-Martinez, J., Lascano, D., Balart, R., García-García, D., Dominici, F., Puglia, D. and Torre, L., 2022. Development and evaluation of novel nanofibers based on mango kernel starch obtained by electrospinning. *Polymer Testing*, *106*, 107462.

Gough, J.E., Scotchford, C.A. and Downes, S., 2002. Cytotoxicity of glutaraldehyde crosslinked collagen/poly (vinyl alcohol) films is by the mechanism of apoptosis. *Journal of Biomedical Materials Research*, *61*(1), 121–130.

Hadisi, Z., Nourmohammadi, J. and Nassiri, S.M., 2018. The antibacterial and anti-inflammatory investigation of Lawsonia Inermis-gelatin-starch nano-fibrous dressing in burn wound. *International Journal of Biological Macromolecules*, *107*, 2008–2019.

Haghi, A.K. and Akbari, M., 2007. Trends in electrospinning of natural nanofibers. *Physica Status Solidi (A)*, *204*(6), 1830–1834.

Hekmati, A.H., Rashidi, A., Ghazisaeidi, R. and Drean, J.Y., 2013. Effect of needle length, electrospinning distance, and solution concentration on morphological properties of polyamide-6 electrospun nanowebs. *Textile Research Journal*, *83*(14), 1452–1466.

Hermansson, A.M. and Svegmark, K., 1996. Developments in the understanding of starch functionality. *Trends in Food Science & Technology*, *7*(11), 345–353.

Hu, X., Liu, S., Zhou, G., Huang, Y., Xie, Z. and Jing, X., 2014. Electrospinning of polymeric nanofibers for drug delivery applications. *Journal of Controlled Release*, *185*, 12–21.

Huan, S., Liu, G., Han, G., Cheng, W., Fu, Z., Wu, Q. and Wang, Q., 2015. Effect of experimental parameters on morphological, mechanical and hydrophobic properties of electrospun polystyrene fibers. *Materials*, *8*(5), 2718–2734.

Jaiturong, P., Sirithunyalug, B., Eitsayeam, S., Asawahame, C., Tipduangta, P. and Sirithunyalug, J., 2018. Preparation of glutinous rice starch/polyvinyl alcohol copolymer electrospun fibers for using as a drug delivery carrier. *Asian Journal of Pharmaceutical Sciences*, *13*(3), 239–247.

Jaiturong, P., Sutjarittangtham, K., Eitssayeam, S. and Sirithunyalug, J., 2012. Preparation of glutinous rice starch nanofibers by electrospinning. *Advanced Materials Research*, *506*, 230–233.

Jeong, Y., Lee, K., Kim, K. and Kim, S., 2016. Pore-structure-optimized CNT-carbon nanofibers from starch for rechargeable lithium batteries. *Materials*, *9*(12), 995.

Kai, D., Ren, W., Tian, L., Chee, P.L., Liu, Y., Ramakrishna, S. and Loh, X.J., 2016. Engineering poly (lactide)–lignin nanofibers with antioxidant activity for biomedical application. *ACS Sustainable Chemistry & Engineering*, *4*(10), 5268–5276.

Komur, B., Bayrak, F., Ekren, N., Eroglu, M.S., Oktar, F.N., Sinirlioglu, Z.A., Yucel, S., Guler, O. and Gunduz, O.Ğ.U.Z.H.A.N., 2017. Starch/PCL composite nanofibers by co-axial electrospinning technique for biomedical applications. *Biomedical Engineering Online*, *16*, 1–13.

Kong, L. and Ziegler, G.R., 2012a. Patents on fiber spinning from starches. *Recent Patents on Food, Nutrition & Agriculture*, *4*(3), 210–219.

Kong, L. and Ziegler, G.R., 2012b. Role of molecular entanglements in starch fiber formation by electrospinning. *Biomacromolecules*, *13*(8), 2247–2253.

Kong, L. and Ziegler, G.R., 2013. Quantitative relationship between electrospinning parameters and starch fiber diameter. *Carbohydrate Polymers*, *92*(2), 1416–1422.

Kong, L. and Ziegler, G.R., 2014a. Fabrication of pure starch fibers by electrospinning. *Food Hydrocolloids*, *36*, 20–25.

Kong, L. and Ziegler, G.R., 2014b. Formation of starch-guest inclusion complexes in electrospun starch fibers. *Food Hydrocolloids*, *38*, 211–219.

Kong, Y.Y., Pang, S.C. and Chin, S.F., 2015. Facile synthesis of nickel nanowires with controllable morphology. *Materials Letters*, *142*, 1–3.

Krishnakumar, G.S., Sampath, S., Muthusamy, S. and John, M.A., 2019. Importance of crosslinking strategies in designing smart biomaterials for bone tissue engineering: A systematic review. *Materials Science and Engineering: C*, *96*, 941–954.

Lancuški, A., Ammar, A.A., Avrahami, R., Vilensky, R., Vasilyev, G. and Zussman, E., 2017. Design of starch-formate compound fibers as encapsulation platform for biotherapeutics. *Carbohydrate Polymers*, *158*, 68–76.

Lancuski, A., Vasilyev, G., Putaux, J.L. and Zussman, E., 2015. Rheological properties and electrospinnability of high-amylose starch in formic acid. *Biomacromolecules*, *16*(8), 2529–2536.

Li, D. and Xia, Y., 2004. Electrospinning of nanofibers: Reinventing the wheel? *Advanced Materials*, *16*(14), 1151–1170.

Li, X., Chen, H. and Yang, B., 2016. Centrifugally spun starch-based fibers from amylopectin rich starches. *Carbohydrate Polymers*, *137*, 459–465.

Liang, Q. and Gao, Q., 2023. Effect of amylose content on the preparation for carboxymethyl starch/pullulan electrospun nanofibers and their properties as encapsulants of thymol. *Food Hydrocolloids*, *136*, 108250.

Liao, Y., Loh, C.H., Tian, M., Wang, R. and Fane, A.G., 2018. Progress in electrospun polymeric nanofibrous membranes for water treatment: Fabrication, modification and applications. *Progress in Polymer Science*, *77*, 69–94.

Lineback, D.R., 1986. Current concepts of starch structure and its impact on properties. *Journal of the Japanese Society of Starch Science*, *33*(1), 80–88.

Liu, G., Gu, Z., Hong, Y., Cheng, L. and Li, C., 2017. Electrospun starch nanofibers: Recent advances, challenges, and strategies for potential pharmaceutical applications. *Journal of Controlled Release*, *252*, 95–107.

Liu, Y., He, J.H., Yu, J.Y. and Zeng, H.M., 2008. Controlling numbers and sizes of beads in electrospun nanofibers. *Polymer International*, *57*(4), 632–636.

López de Dicastillo, C., Roa, K., Garrido, L., Pereira, A. and Galotto, M.J., 2017. Novel polyvinyl alcohol/starch electrospun fibers as a strategy to disperse cellulose nanocrystals into poly (lactic acid). *Polymers*, *9*(4), 117.

Lu, Y., Huang, J., Yu, G., Cardenas, R., Wei, S., Wujcik, E.K. and Guo, Z., 2016. Coaxial electrospun fibers: Applications in drug delivery and tissue engineering. *Wiley Interdisciplinary Reviews: Nanomedicine and Nanobiotechnology*, *8*(5), 654–677.

Luan, Z., Zhang, H., Hu, J., Zhang, J. and Liu, Y., 2021. Crosslinked carboxymethyl starch nanofiber mats: Preparation, water resistance and exudates control ability. *European Polymer Journal*, *154*, 110568.

Lv, H., Cui, S., Yang, Q., Song, X., Wang, D., Hu, J., Zhou, Y. and Liu, Y., 2021. AgNPs-incorporated nanofiber mats: Relationship between AgNPs size/content, silver release, cytotoxicity, and antibacterial activity. *Materials Science and Engineering: C*, *118*, 111331.

Lv, H., Cui, S., Zhang, H., Pei, X., Gao, Z., Hu, J., Zhou, Y. and Liu, Y., 2020. Crosslinked starch nanofibers with high mechanical strength and excellent water resistance for biomedical applications. *Biomedical Materials*, *15*(2), 025007.

Macossay, J., Marruffo, A., Rincon, R., Eubanks, T. and Kuang, A., 2007. Effect of needle diameter on nanofiber diameter and thermal properties of electrospun poly (methyl methacrylate). *Polymers for Advanced Technologies*, *18*(3), 180–183.

Meganathan, B. and Panagal, M., 2023. Starch-based electrospun nanofibrous scaffold incorporated with bioactive compound from Cayratia trifolia promotes wound healing by activating VEGF and PDGF signaling pathways. *Applied Biological Research*, *25*(4), 437–446

Megelski, S., Stephens, J.S., Chase, D.B. and Rabolt, J.F., 2002. Micro-and nanostructured surface morphology on electrospun polymer fibers. *Macromolecules*, *35*(22), 8456–8466.

Mistry, P., Chhabra, R., Muke, S., Narvekar, A., Sathaye, S., Jain, R. and Dandekar, P., 2021. Fabrication and characterization of starch-TPU based nanofibers for wound healing applications. *Materials Science and Engineering: C*, *119*, 111316.

Movahedi, M., Asefnejad, A., Rafienia, M. and Khorasani, M.T., 2020. Potential of novel electrospun core-shell structured polyurethane/starch (hyaluronic acid) nanofibers for skin tissue engineering: In vitro and in vivo evaluation. *International Journal of Biological Macromolecules*, *146*, 627–637.

Movahedi, M. and Karbasi, S., 2022. Electrospun halloysite nanotube loaded polyhydroxybutyrate-starch fibers for cartilage tissue engineering. *International Journal of Biological Macromolecules*, *214*, 301–311.

Nagarajan, S., Bechelany, M., Kalkura, N.S., Miele, P., Bohatier, C.P. and Balme, S., 2019. Electrospun nanofibers for drug delivery in regenerative medicine, in *Applications of targeted nano drugs and delivery systems,* eds S.-H. Kim and Y.-K. Kim, Elsevier, pp. 595–625.

Oktay, B., Baştürk, E., Kayaman-Apohan, N. and Kahraman, M.V., 2013. Highly porous starch/poly (ethylene-alt-maleic anhydride) composite nanofiber mesh. *Polymer Composites*, *34*(8), 1321–1324.

Padron, S., Fuentes, A., Caruntu, D. and Lozano, K., 2013. Experimental study of nanofiber production through forcespinning. *Journal of Applied Physics*, *113*(2), 024318.

Pan, W., Liang, Q. and Gao, Q., 2022. Preparation of hydroxypropyl starch/polyvinyl alcohol composite nanofibers films and improvement of hydrophobic properties. *International Journal of Biological Macromolecules*, *223*, 1297–1307.

Pillay, V., Dott, C., Choonara, Y.E., Tyagi, C., Tomar, L., Kumar, P., du Toit, L.C. and Ndesendo, V.M., 2013. A review of the effect of processing variables on the fabrication of electrospun nanofibers for drug delivery applications. *Journal of Nanomaterials*, *2013*(1), 789289.

Rafiq, M., Khan, R.S., Wani, T.U., Rather, A.H., Amna, T., Hassan, M.S., Rather, S.U. and Sheikh, F.A., 2023. Improvisations to electrospinning techniques and ultrasonication process to nanofibers for high porosity: Ideal for cell infiltration and tissue integration. *Materials Today Communications*, *35*, 105695.

Rezaei, A., Nasirpour, A. and Fathi, M., 2015. Application of cellulosic nanofibers in food science using electrospinning and its potential risk. *Comprehensive Reviews in Food Science and Food Safety*, *14*(3), 269–284.

Rodríguez, K., Gatenholm, P. and Renneckar, S., 2012. Electrospinning cellulosic nanofibers for biomedical applications: Structure and in vitro biocompatibility. *Cellulose*, *19*, 1583–1598.

Rutledge, G.C. and Fridrikh, S.V., 2007. Formation of fibers by electrospinning. *Advanced Drug Delivery Reviews*, *59*(14), 1384–1391.

Sahoo, S., Ang, L.T., Goh, J.C.-H., and Toh, S.-L., 2009. Growth factor delivery through electrospun nanofibers in scaffolds for tissue engineering applications. *Journal of Biomedical Materials Research Part A*, *93*(4), 1539–1550.

Sánchez, L.D., Brack, N., Postma, A., Pigram, P.J. and Meagher, L., 2016. Surface modification of electrospun fibres for biomedical applications: A focus on radical polymerization methods. *Biomaterials*, *106*, 24–45.

Shabani, I., Haddadi-Asl, V., Seyedjafari, E., Babaeijandaghi, F. and Soleimani, M., 2009. Improved infiltration of stem cells on electrospun nanofibers. *Biochemical and Biophysical Research Communications*, *382*(1), 129–133.

Sharifi, F., Sooriyarachchi, A.C., Altural, H., Montazami, R., Rylander, M.N. and Hashemi, N., 2016. Fiber based approaches as medicine delivery systems. *ACS Biomaterials Science & Engineering*, *2*(9), 1411–1431.

Sill, T.J. and Von Recum, H.A., 2008. Electrospinning: Applications in drug delivery and tissue engineering. *Biomaterials*, *29*(13), 1989–2006.

Silva, I., Gurruchaga, M., Goni, I., Fernández-Gutiérrez, M., Vázquez, B. and Román, J.S., 2013. Scaffolds based on hydroxypropyl starch: Processing, morphology, characterization, and biological behavior. *Journal of Applied Polymer Science*, *127*(3), 1475–1484.

Son, Y.J., Kim, W.J. and Yoo, H.S., 2014. Therapeutic applications of electrospun nanofibers for drug delivery systems. *Archives of Pharmacal Research*, *37*, 69–78.

Stojanov, S. and Berlec, A., 2020. Electrospun nanofibers as carriers of microorganisms, stem cells, proteins, and nucleic acids in therapeutic and other applications. *Frontiers in Bioengineering and Biotechnology*, *8*, 130.

Sun, B., Long, Y.Z., Zhang, H.D., Li, M.M., Duvail, J.L., Jiang, X.Y. and Yin, H.L., 2014. Advances in three-dimensional nanofibrous macrostructures via electrospinning. *Progress in Polymer Science*, *39*(5), 862–890.

Sundaramurthi, D., Krishnan, U.M. and Sethuraman, S., 2014. Electrospun nanofibers as scaffolds for skin tissue engineering. *Polymer Reviews*, *54*(2), 348–376.

Sutjarittangtham, K., Jaiturong, P., Intatha, U., Pengpat, K., Eitssayeam, S. and Sirithunyalug, J., 2014. Fabrication of natural tapioca starch fibers by a modified electrospinning technique. *Chiang Mai Journal of Science*, *41*(1), 213–223.

Sylvester, M.A., Amini, F. and Tan, C.K., 2020. Electrospun nanofibers in wound healing. *Materials Today: Proceedings*, *29*, 1–6.

Taghavian, H., Ranaei-Siadat, S.O., Kalaee, M.R. and Mazinani, S., 2017. Investigation of the effects of starch on the physical and biological properties of polyacrylamide (PAAm)/starch nanofibers. *Progress in Biomaterials*, *6*, 85–96.

Taghavian, H., Ranaei-Siadat, S.O., Kalaee, M.R., Mazinani, S., Ranaei-Siadat, S.E. and Harati, J., 2015. Optimizing the activity of immobilized phytase on starch blended polyacrylamide nanofibers-nanomembranes by response surface methodology. *Fibers and Polymers*, *16*, 1048–1056.

Tan, S.H., Inai, R., Kotaki, M. and Ramakrishna, S., 2005. Systematic parameter study for ultra-fine fiber fabrication via electrospinning process. *Polymer*, *46*(16), 6128–6134.

Tang, S., Zhao, Z., Chen, G., Su, Y., Lu, L., Li, B., Liang, D. and Jin, R., 2016. Fabrication of ampicillin/starch/polymer composite nanofibers with controlled drug release properties by electrospinning. *Journal of Sol-Gel Science and Technology*, *77*, 594–603.

Topuz, F. and Uyar, T., 2020. Antioxidant, antibacterial and antifungal electrospun nanofibers for food packaging applications. *Food Research International*, *130*, 108927.

Tuah, K.A., Chin, S.F. and Pang, S.C., 2021. Fabrication of drug-loaded starch-based nanofibers via electrospinning technique. *Biointerface Research in Applied Chemistry*, *11*, 10801–10811.

Tuzlakoglu, K., Bolgen, N., Salgado, A.J., Gomes, M.E., Piskin, E. and Reis, R.L., 2005. Nano-and micro-fiber combined scaffolds: A new architecture for bone tissue engineering. *Journal of Materials Science: Materials in Medicine*, *16*, 1099–1104.

Uyar, T. and Besenbacher, F., 2008. Electrospinning of uniform polystyrene fibers: The effect of solvent conductivity. *Polymer*, *49*(24), 5336–5343.

Vilchez, A., Acevedo, F., Cea, M., Seeger, M. and Navia, R., 2020. Applications of electrospun nanofibers with antioxidant properties: A review. *Nanomaterials*, *10*(1), 175.

Waghmare, V.S., Wadke, P.R., Dyawanapelly, S., Deshpande, A., Jain, R. and Dandekar, P., 2018. Starch based nanofibrous scaffolds for wound healing applications. *Bioactive Materials*, *3*(3), 255–266.

Wang, H., Kong, L. and Ziegler, G.R., 2019. Aligned wet-electrospun starch fiber mats. *Food Hydrocolloids*, *90*, 113–117.

Wang, H., Wang, W., Jiang, S., Jiang, S., Zhai, L. and Jiang, Q., 2011. Poly(vinyl alcohol)/oxidized starch fibres via electrospinning technique: fabrication and characterization. *Iranian Polymer Journal*, *20*(7), 551–558.

Wang, H.J., Jin, X., Wang, W.Y., Xiao, C.F. and Tong, L., 2012. Preparation and electrospinning of acidified-oxidized potato starch. *Advanced Materials Research*, *535*, 2340–2344.

Wang, S. and Copeland, L., 2015. Effect of acid hydrolysis on starch structure and functionality: A review. *Critical Reviews in Food Science and Nutrition*, *55*(8), 1081–1097.

Wang, W., Jin, X., Zhu, Y., Zhu, C., Yang, J., Wang, H. and Lin, T., 2016. Effect of vapor-phase glutaraldehyde crosslinking on electrospun starch fibers. *Carbohydrate Polymers*, *140*, 356–361.

Wang, W., Wang, H., Jin, X., Wang, H., Lin, T. and Zhu, Z., 2018. Effects of hydrogen bonding on starch granule dissolution, spinnability of starch solution, and properties of electrospun starch fibers. *Polymer*, *153*, 643–652.

Williams, G.R., Chatterton, N.P., Nazir, T., Yu, D.G., Zhu, L.M. and Branford-White, C.J., 2012. Electrospun nanofibers in drug delivery: Recent developments and perspectives. *Therapeutic Delivery*, *3*(4), 515–533.

Wu, D., Samanta, A., Srivastava, R.K. and Hakkarainen, M., 2017. Starch-derived nanographene oxide paves the way for electrospinnable and bioactive starch scaffolds for bone tissue engineering. *Biomacromolecules*, *18*(5), 1582–1591.

Xu, W., Yang, W. and Yang, Y., 2009. Electrospun starch acetate nanofibers: Development, properties, and potential application in drug delivery. *Biotechnology Progress*, *25*(6), 1788–1795.

Yang, Q., Cui, S., Song, X., Hu, J., Zhou, Y. and Liu, Y., 2021. An antimicrobial peptide-immobilized nanofiber mat with superior performances than the commercial silver-containing dressing. *Materials Science and Engineering: C*, *119*, 111608.

Yoo, H.S., Kim, T.G. and Park, T.G., 2009. Surface-functionalized electrospun nanofibers for tissue engineering and drug delivery. *Advanced Drug Delivery Reviews*, *61*(12), 1033–1042.

Zhang, D., Chen, L., Cai, J., Dong, Q., Din, Z.U., Hu, Z.Z., Wang, G.Z., Ding, W.P., He, J.R. and Cheng, S.Y., 2021. Starch/tea polyphenols nanofibrous films for food packaging application: From facile construction to enhance mechanical, antioxidant and hydrophobic properties. *Food Chemistry*, *360*, 129922.

Zhang, L., Tang, Y. and Tong, L., 2020. Micro-/nanofiber optics: Merging photonics and material science on nanoscale for advanced sensing technology. *iScience*, *23*(1), 100810.

Zhang, Y., Tao, L., Li, S. and Wei, Y., 2011. Synthesis of multiresponsive and dynamic chitosan-based hydrogels for controlled release of bioactive molecules. *Biomacromolecules*, *12*(8), 2894–2901.

Zhu, W., Zhang, D., Liu, X., Ma, T., He, J., Dong, Q., Din, Z.U., Zhou, J., Chen, L., Hu, Z. and Cai, J., 2022. Improving the hydrophobicity and mechanical properties of starch nanofibrous films by electrospinning and cross-linking for food packaging applications. *LWT*, *169*, 114005.

Zobel, H.F., 1988. Molecules to granules: A comprehensive starch review. *Starch-Stärke*, *40*(2), 44–50.

Starch Nanomaterials

19

Edith Agama-Acevedo and Luis A. Bello-Perez

19.1 INTRODUCTION

New and improved applications have been found for nanometric materials. The renewable, biodegradability, the structural characteristics of starch components (amylose and amylopectin), and their arrangement in the granule are desirable for producing nanomaterials. Starch nanomaterials, also known as nanostarches (NSs), are classified into two groups: (1) nanoparticles (SNPs) and (2) nanocrystals (SNCs), which are used interchangeably in the literature (Wang & Zhang 2021). As novel materials, NSs are a type of starch derivative with specific physicochemical and functional characteristics associated with their small size. In contrast, the difference between SNPs and SNCs is their composition; the former includes amorphous or a combination of both amorphous and crystalline areas of the starch granule while the latter comprise only the crystalline one (Figure 19.1).

The interest in starch nanomaterials has increased in recent years due to their unique characteristics such as small size (>600 nm), high surface area per unit volume, good biocompatibility, and eco-friendliness. These qualities are highly desirable for several applications, especially when they interact with molecules, such as lipids, proteins, and bioactive compounds. Then, NSs can be used in drug transport and delivery, dyes, and flavorings as well as biodegradable edible films and filler in biocomposites.

The methods to prepare starch nanomaterials include the top-down and bottom-up approaches. The first can destroy the amorphous regions of the granules and produce a starch derivative with a high crystallinity level by acid hydrolysis, ball milling, and high-pressure homogenization. Meanwhile, bottom-up refers to the self-assembly of nanocrystals by nanoprecipitation, microemulsification, and recrystallization. The crystal structure and morphology of NSs are related to the preparation methods (Gérard et al. 2002). To find new and improved applications, these materials can be obtained from unconventional starch sources (Singh et al. 2024) or modified by chemical reactions, such as cross-linking, oxidation, grafting, acetylation, and esterification. These reactions are based on the presence of hydroxyl groups on the surface of NSs (Zhou et al. 2016).

19.2 STARCH NANOMATERIALS

19.2.1 Nanoparticles

It is well accepted that SNPs are those particles with at least one dimension smaller than 1,000 nm (1 μm); other works consider a stricter size below 100 nm (Liu et al. 2023), and it has been mentioned that at least one dimension should not exceed 300 nm (Sun et al. 2014). As derivatives of a biopolymer, SNPs are non-toxic and can work under different temperatures, pH, light, and other stimuli (Yu et al. 2021). The comparison with its native parental starch showed that SNPs had higher reaction surface, solubility, absorptive capacity, and biological penetration rate (Campelo et al. 2020). Based on these properties, SNPs can be used as carriers (encapsulation) of bioactive compounds and drugs to promote the bioavailability of the substance, biofilm reinforcement (food packing), fat replacers, and Pickering emulsion stabilizers. Early works on SNP preparation used conventional starches, such as corn, potato, cassava, and wheat, since they are commercially available. Recently, due to the interest in the use of regional agronomic resources, other unconventional starch sources, such as taro, quinoa, amaranth, banana, mango, and sago palm, have been proposed for starch isolation and SNP preparation. They likely present characteristics different from those shown by conventional starch sources. The SNPs from unconventional sources can be produced by top-down and bottom-up methods.

19.2.1.1 Preparation methods

The top-down approaches decrease the microscopic size of starch to the nanoscale (<1,000 nm) because the starch granule size is between 70 and 80 μm (potato starch) and 2–3 μm (amaranth starch). For this reason, smaller granule sizes can be more effective in producing SNPs (Figure 19.2). Top-down methods include acid hydrolysis, enzymatic hydrolysis, homogenization, crushing, gamma irradiation, and sonication. Acid hydrolysis is based on the organization of starch components in concentric semi-crystalline rings, and the amorphous regions are attacked first to maintain the crystalline regions of the granule. Hydrochloric acid and sulfuric acid at a temperature of

DOI: 10.1201/9781003464396-19

"

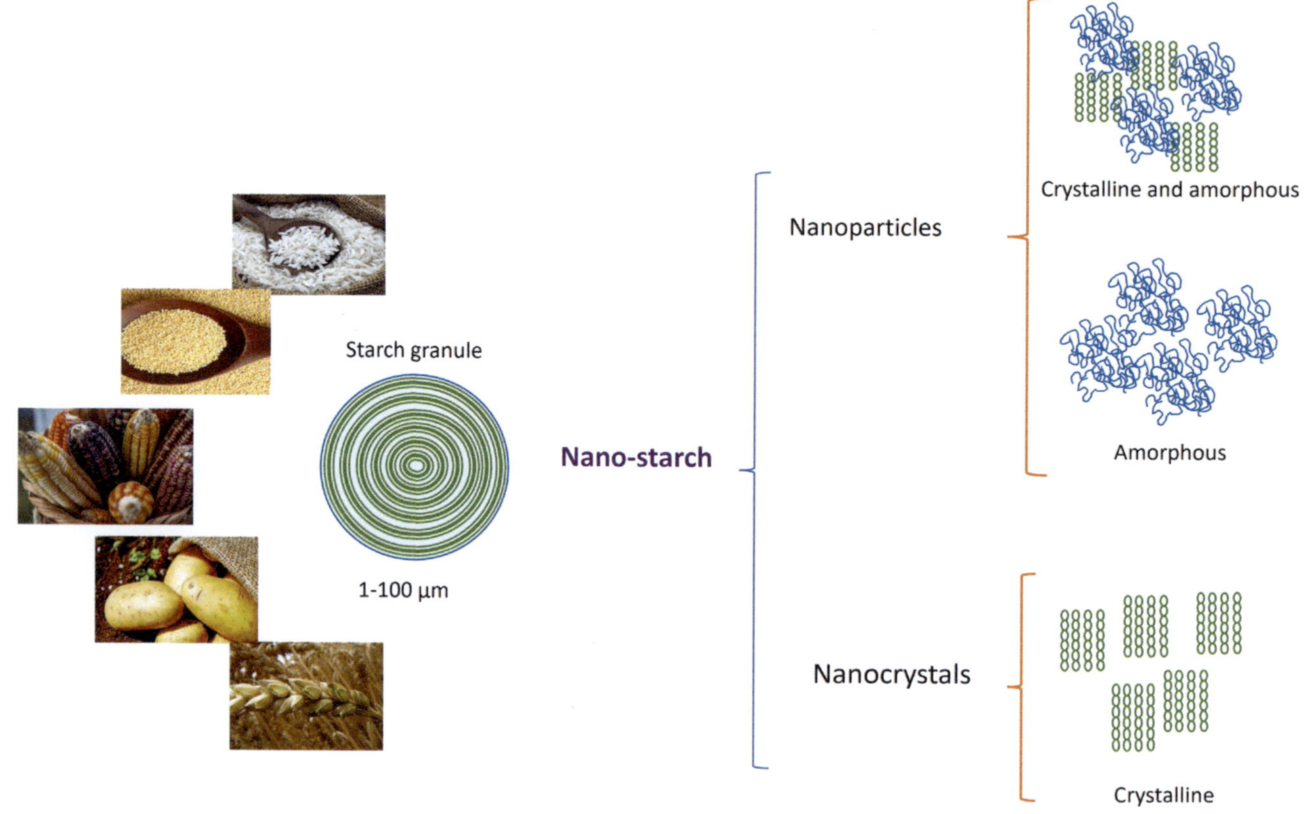

FIGURE 19.1   Main characteristics of starch nanomaterials.

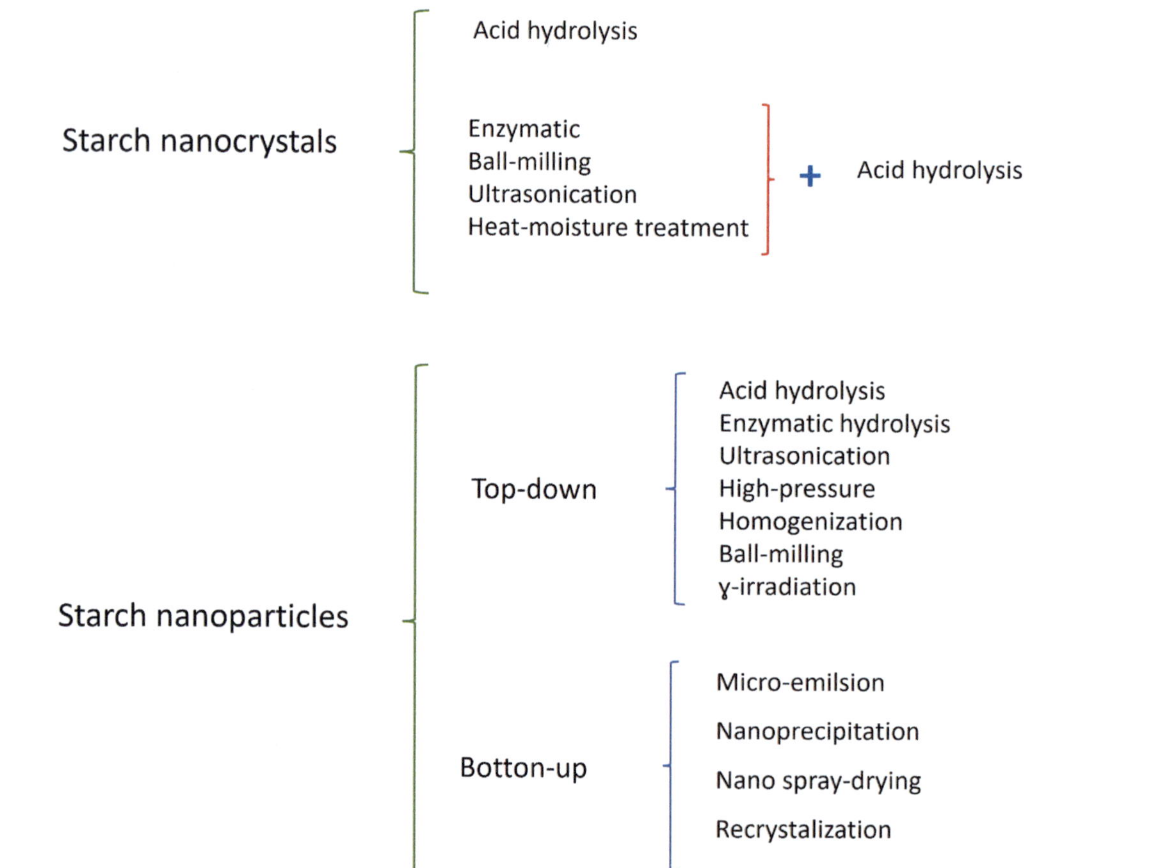

FIGURE 19.2   Preparation methods of starch nanomaterials.

25°C–55°C for extended periods (5–10 days) are used to produce SNPs (Angellier et al. 2004). The acid hydrolysis of starch granule slurry is carried out in three stages: (1) erosion of the granule surface that produces cavities, (2) collapse of the granule by acid radial diffusion, and (3) fragmentation of the growth rings (LeCorre et al. 2012). Both acid concentration and hydrolysis time are determined to obtain a specific size since high acid concentrations and longer times decrease the SNP size. According to reports, the effect of the amylose/amylopectin ratio on SNP production remains unclear. However, the amorphous regions where amylose is present are hydrolyzed to enriched crystalline regions, which are found in higher volumes due to the high amylose concentration (high amylose starch). In that case, the yield of SNPs can be lower than if a waxy starch (high amylopectin content) is used. Systematic studies using waxy starches of different granule sizes, along with varied acid concentrations and temperatures, are necessary to further the knowledge of SNPs.

The enzymatic hydrolysis of starch components is another alternative for SNP production. The branch points of amylopectin are formed by the α-1,6 linkages present in the amorphous areas of the granule. The debranching enzymes, such as pullulanase and isoamylase, hydrolyze those α-1,6 bonds, thus increasing the crystalline areas of the granules. Foresti et al. (2014) reported the use of α-amylase that hydrolyzed α-1,4 linkages joining the glucose monomers in starch chains to produce SNPs with an average diameter of 120–500 nm. The reduction in α-1,6 linkages and the larger amounts of linear chains with α-1,4 linkages may produce SNPs with structures and functionality different from those produced by acid hydrolysis. The combination of methods has produced SNPs with improved functionality and yield. Starches from diverse botanical sources are used to take advantage of the varied size, morphology, amylose/amylopectin ratio, and crystalline arrangement, among others. Several reviews have also reported combined methods, such as acid hydrolysis and nanoprecipitation, acid hydrolysis and ultrasonication, acid hydrolysis, and ball milling, homogenization with pressure and enzymatic hydrolysis, sonication and enzymatic hydrolysis, nanoprecipitation and sonication, and enzymatic hydrolysis and recrystallization (Liu et al. 2023; Marta et al. 2023). Now, green technologies, such as enzymolysis, recrystallization, sonication, enzymatic hydrolysis, and high pressure, are preferred because of the low negative impact on the environment produced when using chemical reagents. When employing a combination of methods to produce SNPs, the first modification for the starch preparation, which increases susceptibility, is essential to the second modification. The latter increases the yield and enhances morphological and structural characteristics that, in turn, generate certain physicochemical and functional properties. Additionally, SNPs have been modified after their production in the search for new and improved applications; still, there are few reports on the application of this strategy. The surface of SNPs can provide specific properties and improve their end-uses. Among the methods for SNP modification are chemical, physical, and enzymatic procedures. The modification of SNPs is based on the numerous polar hydroxyl groups on their surface, which leads to poor dispersibility and compatibility in non-polar solvents. The main goal of SNP modification is improving amphiphilicity, dispersibility, and low thermal stability, among others. The chemical modification of both starch and SNPs is in disuse

due to the pollution linked to the final deposition of the reagents used in the chemical reaction. Novel strategies have been considered for the chemical modification of SNPs, such as grafting of polymer chains directly onto the surface of the reinforcing nanoparticles (Kumari et al. 2020). Three strategies for this modification have been proposed: (1) reaction with small molecules, (2) grafting onto, polymer chain grafting in the presence of coupling agents, and (3) grafting from, grafting of polymer chain by monomer polymerization (Wang et al. 2010). The starch and derivatives have hydrophilic characteristics, a characteristic that restricts their applications. The insertion of amphiphilicity improves the use of SNPs in emulsions, coatings, films, and gel production. To produce this amphiphilic character, the introduction of hydrophobic groups on the surface of SNPs is necessary, and several hydrophobic groups from chemical reagents, organic acids, organic solvents, and fatty acids have been reported. The poor dispersibility of SNPs in non-aqueous solvents hampers applications in hydrophobic systems. To increase dispersibility, the surface of SNPs must be hydrophobic, which can be achieved by esterification with organic acids. Dispersion stability has been improved in SNPs by grafting and cross-linking with sodium tripolyphosphate (Simi & Emilia Abraham 2007). The cross-linking of corn SNPs is carried out with citric acid and used in pea starch nanocomposites. Cross-linked SNPs are dispersed equally in the starch matrix without aggregation, a pattern obtained due to strong interactions between SNPs and the pea starch matrix. The dispersion of SNPs has proven to improve water vapor permeability with an increase in the Young modulus (Ma et al. 2008). Green, physical methods, such as heat moisture treatment (HMT), microwave (MW) treatment, and ultrasonic (U) treatment, have been used to improve SNP functionality. As expected, SNP properties are influenced by the treatment; for instance, the freeze–thaw stability in SNPs has been improved after HMT and MW. Then, modified SNPs acquire specific properties that are useful for certain applications (Das & Sit 2021).

19.2.1.2 Molecular characteristics

Analyzing the morphology, size, and crystalline arrangement is necessary to find the molecular function relationship in SNPs and suggest new applications. The shape and particle size distribution of SNPs has been analyzed by scanning electron microscopy (SEM), transmission electron microscopy (TEM), field emission scanning electron microscopy (FE-SEM), and atomic force microscopy (AFM). The morphology of SNPs determined by TEM and SEM varies depending on the botanical source of the starch, the preparation method, and the modification method when required. The analysis of taro starch nanoparticles by SEM showed a size between 130 and 360 nm, with a fusion and agglomeration (Das & Sit 2021). A similar pattern was reported in sago starch nanoparticles prepared by precipitation with ethanol and a cetyltrimethylammonium bromide solution (Chin et al. 2011). The fusion of SNPs was explained by the gelatinization of the polymer during the preparation that was not separated during precipitation.

The morphology of SNPs has been reported as round, ellipsoidal, flat, platelet, or irregular with cracked and porous surfaces. The preparation modifies the shape and size of SNPs.

Starches from maize with different amylose/amylopectin ratios, potato, and mung bean starch were used for SNPs preparation by acid hydrolysis. Round and oval shapes were found, measuring 41–69.7 nm, while no significant differences in the size of SNPs were observed. The average granule size of regular maize (41 nm) was 20 µm, while that of potato (43.2 nm) was 50 µm (Kim et al. 2012). Potato starch nanoparticles also prepared with acid hydrolysis showed a size between 132.56 and 263.38 nm and an elliptical-polyhedral shape (Torres et al. 2019). Waxy rice starch with a small granule size (2–5 µm) was used to prepare SNPs by acid hydrolysis. The size of rice SNPs ranged between 220 and 279 nm, and the shape was round but irregular. However, the granule size decreased as the acid hydrolysis time was longer (Jeong & Shin 2018). The differences in both the size and shape of SNPs are influenced by the starch botanical source, acid hydrolysis time, and reagents used in precipitation. The combination of acid hydrolysis and precipitation to produce SNPs yielded a smaller size than acid hydrolysis. The precipitation using hydrophobic components (butanol and ethanol) produces an amylose cavity where hydrophobic compounds are accommodated in a single helix (Maryam et al. 2020). Overall, the granule size of starch used for SNP preparation does not affect morphological characteristics, yet more systematic studies on the effect of size and preparation methods are necessary.

The crystallinity of SNPs can be modified as compared with their starch counterpart. It is well known that there are two types of crystalline arrangements of starch: A- and B-type. A third one, the C-type, is the blend of the first two. The V-type crystalline arrangement has been linked to the formation of an amylose–lipid complex and others, such as amylose–polyphenols complexes (Gutiérrez & Tovar 2021; Romero Hernández et al. 2023). Preparation methods, such as gamma irradiation and ultrasonication, produced an amorphous pattern and a decrease in crystallinity (Bel Haaj et al. 2016; Boufi et al. 2018; da Silva et al. 2017; Lamanna et al. 2013; Remanan & Zhu 2021). The change in the X-ray diffraction pattern from A-type to B-type has been reported in high-amylose maize SNPs produced by acid hydrolysis (Kim et al. 2012). Nevertheless, an inverse pattern (B-type to A-type) was found in high-amylose maize SNPs produced by the same method (Perez Herrera & Vasanthan 2018).

19.2.1.3 Thermal and functional characteristics

The thermal properties of SNPs were evaluated by differential scanning calorimetry (DSC) and thermogravimetric analysis (TGA). Both methods provide information on the thermal behavior of the materials and are key to determining applications. The study by DSC in excess water showed a transition phase (low temperatures) associated with disorganization of the crystalline arrangement, which was lower than the starch gelatinization of its native counterpart; contrastingly, the enthalpy associated increased (da Silva et al. 2017; Remanan & Zhu 2021). The conditions used during SNP production altered the thermal characteristics because, at the longest acid hydrolysis time, peak and conclusion temperatures as well as the range of the transition phase increased while the onset

temperature was reduced (Hernández-Jaimes et al. 2013). The combination of enzymatic hydrolysis and recrystallization to produce SNPs resulted in reduced maximum degradation temperature as determined by TGA (Sun et al. 2014).

Some functional properties of SNPs have been reported to be affected by the starch botanical source and preparation method. Swelling and solubility of arrowroot SNPs produced by nanoprecipitation with butanol showed increased swelling volume and solubility when compared with native starch. This pattern is associated with more OH- groups oriented toward the external layers of the material that can bind to more water molecules; they can also be more solubilized due to the disorganization of the starch chains (Winarti et al. 2014). Tuber starch from potatoes and cassava was used to prepare SNPs by a mechanical treatment, and they presented a higher swelling volume vs. native starch (Szymonska et al. 2009). Acid hydrolysis used to produce SNPs resulted in reduced water binding capacity, associated with swelling volume, as compared with native starch; still, no significant change was found as a function of hydrolysis time (Jeong & Shin 2018). The acid hydrolysis produces an increase in the crystalline regions of starch due to a dilution effect; these ordered regions are less prone to binding to water molecules.

19.2.2 Nanocrystals

The starch nanomaterials in the second group, SNCs, are considered eco-friendly organic materials as they are being biodegradable and low cost. It is well known that intact native starch granules are organized by alternating the crystalline and amorphous areas, creating onion-like concentric rings. SNCs maintain a crystalline structure due to the amorphous regions of starch granules that are either removed or destructed during preparation; this structural characteristic provides functionality to the material. However, SNC production is time-consuming and offers low yield (Pinto et al. 2021); the final deposition of sulfuric and hydrochloric acids used in the preparation constitutes another drawback. Starchy sources with high amylopectin content, such as waxy starch, are preferred for SNC production, and waxy maize starch has been reported as raw material for this goal. Non-conventional starch sources like amaranth, classified as showing high amylopectin content (85%–95%) and small granule size, have been proposed to produce SNCs. Studies suggest that a small granule size can increase the yield and produce a higher acid hydrolysis rate due to a larger surface area and short time (Sanchez de la Concha et al. 2018).

19.2.2.1 Preparation methods

The preparation methods to obtain SNCs are similar to those above mentioned for SNPs. The preparation method to produce SNCs needs acid hydrolysis in the first step; the original method from 1886 used hydrochloric acid (7.5%) at 30°C–40°C to hydrolyze potato starch; this process produces a starch derivative with the capacity to be dissolved in hot water producing a clear and transparent solution (Li et al. 2023). Robin (1974) reported the acid hydrolysis of starch to analyze amylopectin structure. Both methods considered the basis to obtain SNCs as reported by Putaux et al. (2003), where SNCs of 20–40 nm in length and 15–30 nm in width were obtained after 6 weeks of acid hydrolysis. This method is considered the first official report of SNC production.

The production of SNCs from 5% waxy maize starch dispersion with HCl (2.2 N) for 6 weeks at 36°C showed a yield of 0.5% based on starch weight (Putaux et al. 2003). This yield was improved to 15.7% by using waxy starch dispersion of around three times (15%), changing sulfuric acid with a concentration of 3.16 M, and a shorter hydrolysis time (5 days) at 40°C (Angellier et al. 2004). However, the long acid hydrolysis time and low yield are drawbacks in SNC production and restrict the end-uses for the development of competitive products (Alves et al. 2021; Ge et al. 2017; Le Corre & Angellier-Coussy 2014). Two strategies have been reported to overcome those shortcomings, and the first one is the addition of substances to prepare (pretreatment) the raw material (native starch) before acid hydrolysis. Second, enzymes such as α-amylase, β-amylase, and glucoamylase hydrolyze the α-1,4 and α-1,6 linkages of the starch components amylose and amylopectin. To obtain an environmentally friendly method for SNC production and improve the yield, an enzymatic treatment with glucoamylase was implemented to hydrolyze both types of bonds (α-1,4 and α-1,6) before the acid hydrolysis. The enzymatic treatment reduced the acid hydrolysis time to 45 hours (less than 2 days) with a similar yield to that obtained in 5 days (LeCorre et al. 2012). The use of glucoamylase for SNC preparation from waxy potato starch was reported; SNCs with minimal size and best dispersion were obtained. Crystallinity increased from 33% to 50.8%, and the original crystalline type changed from B- to A- (Hao et al. 2018). It is important to consider that the addition of substances has disadvantages in the isolation and purification of SNCs. The second strategy suggests the use of instruments to prepare native starch before acid hydrolysis. Ultrasound is simultaneously used with acid hydrolysis (sulfuric acid) to obtain SNCs. This study used normal corn starch dispersion at 15% and 40 w/w in an ultrasound bath at three different temperatures (10°C, 20°C, and 40°C), three times (15, 30, and 45 minutes), and two acid concentrations (3.16 and 4.5 M). The degree of hydrolysis increased with the treatment time and temperature, but it was higher when low starch concentration and acid concentration were used, which produced an increase in SNC yield (21.6%). This procedure has been implemented for the production of SNCs on a large scale (Mohammad Amini & Razavi 2016). Other methods to shorten acid hydrolysis time and increase the yield include wet heat treatment during acid hydrolysis followed by homogenization to obtain higher yields (Kim et al. 2017). The wet ball milling of waxy maize starch was used before acid hydrolysis. Waxy maize starch (10 g) was dispersed with 30 mL ethanol and ball-milled at 300 rpm for different times (15–90 minutes). Afterward, acid hydrolysis of the ball-milled starch was carried out. The increase in ball-milled time produced a reduction in the crystallinity after 30 minutes. The samples ball-milled for 30 and 45 minutes were used in acid hydrolysis; after 3 days of acid hydrolysis, the yield did not change and was higher in 30 minutes ball-milling starch. This strategy could be used to prepare SNCs at a large scale (Dai et al. 2018). The HMT could increase the crystallinity of starch granules before acid hydrolysis to produce SNCs. This treatment uses temperatures up to the starch gelatinization temperature and low moisture content. This treatment produces a rearrangement of starch components in the granule and increases relative crystallinity. The best conditions for HMT were obtained to produce the maximum relative crystallinity. The HMT starch (30% w/v) was hydrolyzed with 3.16 M sulfuric acid at 40°C for 1–5 days under stirring (200 rpm). The SNCs were obtained after hydrolyzing HMT starch for 4 days, and relative crystallinity was increased. The authors reported this as a low-cost, eco-friendly method that does not demand further purification (Dai et al. 2019).

19.2.2.2 Molecular characteristics

The shape and size of SNCs are obtained by TEM, AFM, and FE-SEM. The shape of waxy maize and amaranth SNCs include parallelepiped and square-like, with sizes ranging from 374 to 384 nm (Sanchez de la Concha et al. 2018); spherical-elliptical (Qin et al. 2016), irregular and rough (Dularia et al. 2019), spherical (Minakawa et al. 2019), square (Mukurumbira et al. 2017), and irregular-spherical (Singh et al. 2024; Suriya et al. 2018). The size ranged between 8 and 396 nm based on the starch source and the SNC production method.

19.3 MODIFICATION OF STARCH NANOMATERIALS

Nanostarches or SNs are used to produce new and improved applications. The previous sections show the strategies to produce large-scale SNs in a short time, obtaining high-yield, low-cost, and green technologies. The modification of SNs is a strategy to expand the end-uses of starch nanomaterials as with the modification of native starch (Figure 19.3). Generally, SNs are hydrophilic due to many OH- groups on the surface, show low dispersibility in non-polar solvents, and are poorly compatible with hydrophobic polymers. If the hydrophilic character of SNs is changed to amphiphilic or hydrophobic, the applications of these materials can expand. The common methods for SN modification include esterification, cross-linking, grafting, oxidation, and acylation, where reagents are used in reaction with OH- groups on the surface of the material. The aim of chemical modification of native starch is different from that of SN modification: Starch modification can restrict the swelling of the starch granule and improve gelling properties, consistency, and stability during food processing. However, the goal of SNs is to improve hydrophobicity, dispersibility, and stability in the products where they are added. Acylation and esterification produce hydrophobicity and amphiphilicity.

Esterification is produced by reagents, such as acid anhydride, dodecyl succinic anhydride fatty acid, octenyl succinic anhydride (OSA), and fatty acid chloride. The use of <3% OSA in the reaction was approved by the Food and Drug Administration (FDA) for food and pharmaceutical uses. The esterification with OSA improves emulsifying characteristics and is used in Pickering emulsions. SNs of amaranth starch were esterified with OSA and

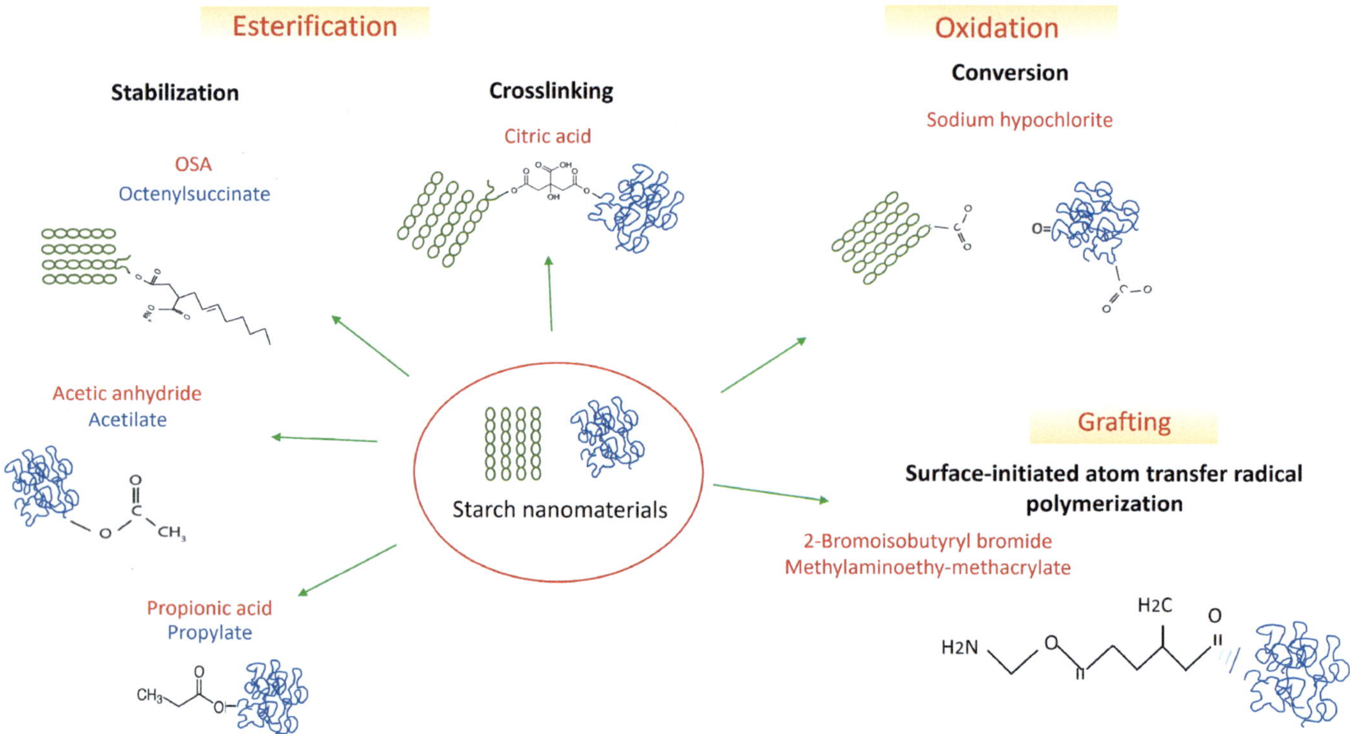

FIGURE 19.3   Strategies for modification of starch nanomaterials.

used to stabilize Pickering emulsion. It was difficult to determine the degree of substitution in the esterified SNs while the functional characteristic (emulsion index) was identified. The emulsification capacity of OSA SNs was enhanced and associated to repulsion forces, as reflected by the ζ-potential. The authors concluded that OSA SNs are viable for use in Pickering emulsions due to their ability to stabilize oil-in-water emulsions (Sánchez de la Concha et al. 2020). The esterification of SNs can produce changes in their morphology from platelet-like shape in unmodified SNs to pseudospherical structure after the modification by 1,4-hexamethylene diisocyanate (Valodkar & Thakore 2010). Cross-linking improves SN functionality, as proven in NSs obtained from banana starch and cross-linked with citric acid. The cross-linking produced SNs with low polarity, higher hydrophobicity, and stability after dispersion. SN cross-linking can be used to encapsulate bioactive substances and reinforce hydrophobic polymer matrices (Ren et al. 2018; Santoyo-Aleman et al. 2019). The oxidation of starch materials introduces carbonyl and carboxyl groups by oxidizing the OH- groups on the surface of SNs into ketone and aldehyde functional groups. The use of sodium hypochlorite for SN oxidation shows that the reaction occurs in the amorphous regions (Liu et al. 2018). The functionalization of SNs by grafting copolymerization can be used. SNs were prepared by acid hydrolysis, and grafting was synthesized through an initiator of 2-Bromoisobutyryl bromide (BIB)-mediated reaction; one halogen of BIB was linked to a carbonyl carbon by acylation, and the other halogen at the end of BIB was grafted with dimethylaminoethyl methacrylate (DMAEMA) to SNs through surface-initiated atom transfer radical polymerization. The grafted SNs were covered by the DMAEMA monomer, and dispersity and adsorption increased (Su et al. 2020). The use of dual or triple modifications

in starches also can be applied in NSs. Green methods have been recommended due to their low impact on the environment. The use of cross-linking and esterification of SNs increased the substitution degree and hydrophobicity, unlike single esterification (Ren et al. 2016). Other methods to modify SNs have been reported in reviews (Li et al. 2023; Wang & Zhang 2021); however, further studies of multiple modifications of NSs are necessary. The modification of native starch and the production of NSs that maintain a level of modification constitute another strategy. Non-conventional starch (banana) was acylated with acetic anhydride, and SNs were obtained and used to encapsulate curcumin (Acevedo-Guevara et al. 2018). The encapsulation and release of curcumin were related to hydrophobic interactions and hydrogen bonding. The esterification of starch changed the hydrophilic character to amphiphilic or hydrophobic.

The modification of SNs depends on the specific applications, and the method(s) of modification should be chosen based on the application and the desirable characteristic of the material where the SNs will be included to provide functionality.

19.4 USES OF STARCH NANOMATERIALS

The applications of SNs and modified SNs are a growing area due to the specific functionality of these starch materials (Figure 19.4). The development of biodegradable packing material has emerged in the food industry given the final deposition of traditional packing material from petroleum

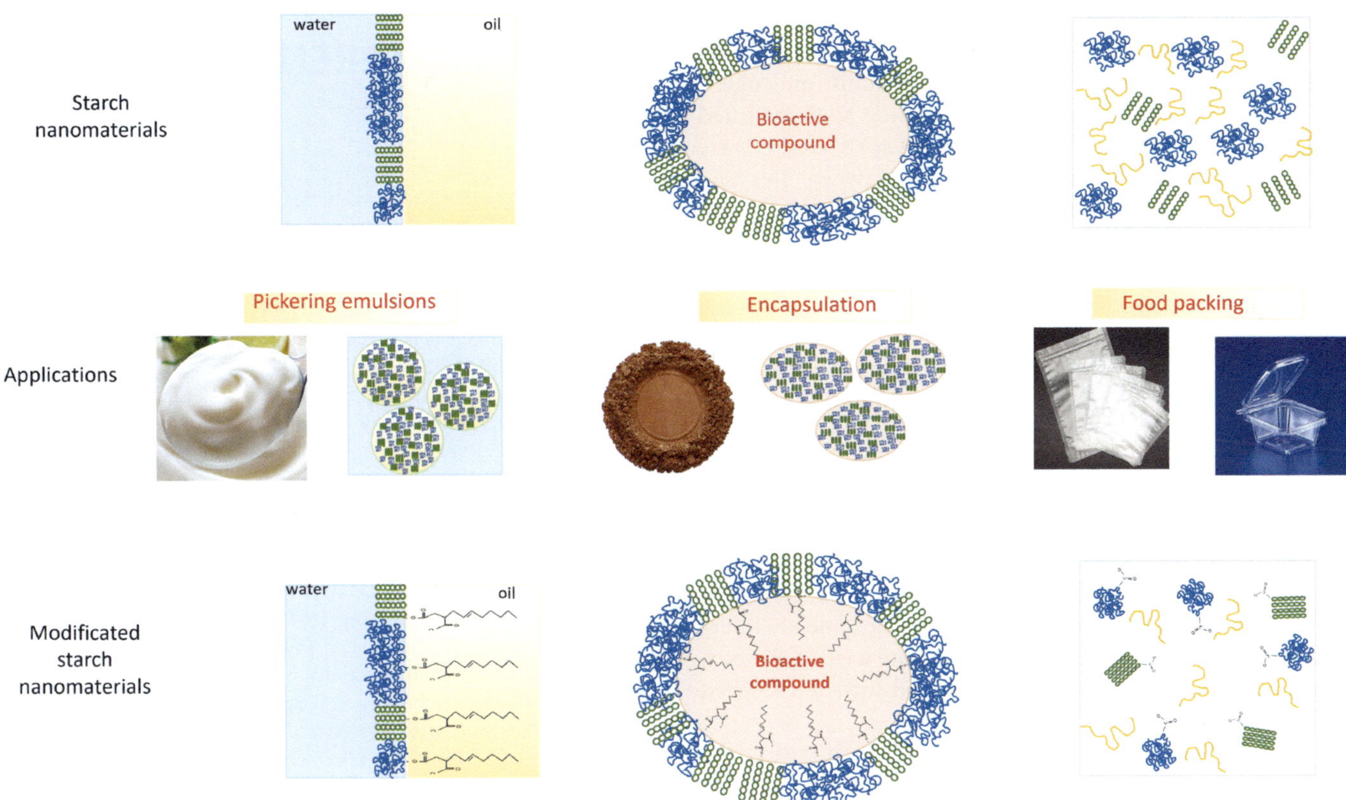

FIGURE 19.4   Application of starch nanomaterials.

derivatives (polyethylene). It is well known that biopolymers such as starch, proteins, polyvinyl alcohol (PVA), polycaprolactone, and chitosan do not meet the functional and barrier properties necessary for packing. The use of nanofillers as SNs is an alternative for overcoming this drawback. SNs are used to reinforce starch films (Fan et al. 2016), especially polymeric matrices; however, poor compatibility of SNs in the matrices hampers the formation of nanocomposites that can be increased by SN modification. Singh et al. (2024) revised the reinforcement of starch films with SNs, and the shielding and resistance of cassava starch films were increased (Velásquez-Castillo et al. 2023). The addition of 5% (w/w) of SNs to the starch solution produced an increase of more than 150% in tensile strength as compared with the control starch film. The inclusion of banana SNs in a banana starch solution improved the mechanical strength and water barrier properties of the film when compared to the control film (Orsuwan & Sothornvit 2017).

The nanometric size of the SNs makes them good candidates for stabilizing Pickering emulsions. Starch of small granule size, such as quinoa, rice, and amaranth, is recommended to stabilize emulsions, and that capacity was improved by esterification with OSA (Timgren et al. 2013). The stability of the emulsion under storage was improved by OSA esterification of SNs. The emulsion stabilized with soluble SNs isolated from sugary-1 maize esterified with OSA did not show phase separation after 30 days; a similar pattern was found when insoluble starch nanoparticles were used (Ye et al. 2017a, b). SNs were prepared from corn starch by acid hydrolysis modified

by acetylation and used to stabilize emulsions. Acetylation improved the emulsifying capacity due to a decrease in SN hydrophilicity and enhanced electrostatic repulsion during emulsification (Qian et al. 2020).

There is a growing interest in the consumption of functional foods; some have bioactive compounds with antioxidant characteristics. The encapsulation of bioactive compounds and drugs with SNs is essential given that those substances show poor water solubility and are sensitive to environmental conditions such as light, pH, and temperature. The protection of those compounds by encapsulation is important to maintain the chemical structure and activity during the processing, and they can be bioaccessible and bioavailable after digestion. NSs are a delivery system for bioactive compounds and drugs so they can be released gradually and be bioavailable (Qiu et al. 2019). Starch SNs were prepared from green banana starch and modified with citric acid by cross-linking. Both unmodified and modified SNs were β-Carotene-loaded, and simulated gastric digestion was used to determine the release of the encapsulated compound. Modified SNs allowed for a more controlled release of the β-Carotene (40%) than when the unmodified counterpart was used (17%). Modified and unmodified SNs are a potential vehicle for the delivery of β-Carotene and can be added to foods to maintain their bioaccessibility (Santoyo-Aleman et al. 2019). Compounds with diverse properties (antimicrobial, antiviral, anti-cancer, antioxidant, and anti-inflammatory), such as curcumin (polyphenol), are useful given the effect associated with disease prevention. Curcumin was encapsulated in SNs prepared from acetylated banana starch, while SNs from native banana starch were used as control. Both SNs showed sizes lower than 250 nm, and the modified SNs presented more encapsulated curcumin molecules. The simulated gastrointestinal study showed that modified SNs loaded with curcumin allowed a more controlled release associated with stronger hydrogen bonds between SNs and curcumin (Acevedo-Guevara et al. 2018).

The encapsulation of different substances included bioactive compounds with antioxidant capacity, such as epigallocatechin gallate, β-carotene, curcumin, bioactive oils (flaxseed, avocado, essential oils, peppermint oil, and clove oil), quercetin, tea polyphenols, catechin, yerba mate extract, vitamin E, folic acid, zinc, magnesium, and prebiotics. Some reviews (Kumari et al. 2020; Liu et al. 2023; Qiu et al. 2019; Rostamabadi et al. 2019; Wang & Zhang 2021) have shown the encapsulation and the delivery of the above-mentioned compounds.

SNs in the food safety field help to detect and quantify microorganisms that produce deterioration of the food and health problems after consumption, as well as some chemical contaminants (metals and pesticides). SNs are used in the production of nanosensors as quantum dots and surface-modified gold nanoparticles.

Although the concentration of SNs is low in diverse applications, their structure is not recognized by digestive enzymes and continues the journey toward the large intestine, where they could become substrate by the microbiota. The contribution of SNs as a source of indigestible carbohydrates (dietary fiber) is marginal; however, this can be considered to expand the use of SNs in diverse products and increase the consumption of non-digestible carbohydrates.

19.5 PERSPECTIVES AND CONCLUSIONS

Nanometric materials show several applications across industries. Starch nanomaterials, such as nanoparticles and nanocrystals, produce specific functional characteristics that are used for the development of new and improved products. The biodegradability and low cost of starch nanomaterials are key to increasing the interest. The methods to produce starch nanomaterials and functionality are of special interest to researchers and industries to increase yield and expand the number of applications. The use of non-conventional starches as raw materials to produce nanomaterials and their modification by green methods constitutes an area of opportunity in this field. Further studies are necessary to properly identify the applications of nanomaterials in nanosensors as a source of resistant starch, packing, encapsulation, Pickering emulsions, stabilizers, and delivery of bioactive and prebiotic compounds and drugs to increase bioaccessibility and bioavailability.

REFERENCES

Acevedo-Guevara, L, Nieto-Suaza, L, Sanchez, LT, Pinzon, MI, & Villa, CC 2018. Development of native and modified banana starch nanoparticles as vehicles for curcumin. *International Journal of Biological Macromolecules*, *111*, 498–504. https://doi.org/10.1016/j.ijbiomac.2018.01.063

Alves, MJDS, Chacon, WD, Gagliardi, TR, Agudelo Henao, AC, Monteiro, AR, & Ayala Valencia, G 2021. Food applications of starch nanomaterials: A review. *Starch/Stärke*, *73*, art. 2100046. https://doi.org/10.1002/star.202100046

Angellier, H, Choisnard, L, Molina-Boisseau, S, Ozil, P, & Dufresne, A 2004. Optimization of the preparation of aqueous suspensions of waxy maize starch nanocrystals using a response surface methodology. *Biomacromolecules*, *5*, 1545–1551. https://doi.org/10.1021/bm049914u

Bel Haaj, S, Thielemans, W, Magnin, A, & Boufi, S 2016. Starch nanocrystals and starch nanoparticles from waxy maize as nanoreinforcement: A comparative study. *Carbohydrate Polymers*, *143*, art. 310–317. https://doi.org/10.1016/j.carbpol.2016.01.061

Boufi, S, Bel Haaj, S, Magnin, A, Pignon, F, Impéror-Clerc, M, & Mortha, G 2018. Ultrasonic assisted production of starch nanoparticles: Structural characterization and mechanism of disintegration. *Ultrasonics Sonochemistry*, *41*, 327–336. https://doi.org/10.1016/j.ultsonch.2017.09.033

Campelo, PH, Sant'Ana, AS, & Pedrosa Silva Clerici, MT 2020. Starch nanoparticles: Production methods, structure, and properties for food applications. *Current Opinion in Food Science*, *33*, 136–140. https://doi.org/10.1016/j.cofs.2020.04.007

Chin, SF, Pang, SC, & Tay, SH 2011. Size controlled synthesis of starch nanoparticles by a simple nanoprecipitation method. *Carbohydrate Polymers*, *86*, 1817–1819. https://doi.org/10.1016/j.carbpol.2011.07.012

da Silva, NMC, Correia, PRC, Druzian, JI, Fakhouri, FM, Fialho, RLL, & de Albuquerque, ECMC 2017. PBAT/TPS composite films reinforced with starch nanoparticles produced by ultrasound. *International Journal of Polymer Science*, 2017, art. 4308261. https://doi.org/10.1155/2017/4308261

Dai, L, Li, C, Zhang, J, & Cheng, F 2018. Preparation and characterization of starch nanocrystals combining ball milling with acid hydrolysis. *Carbohydrate Polymers*, *180*, 122–127. https://doi.org/10.1016/j.carbpol.2017.10.015

Dai, L, Zhang, J, & Cheng, F 2019. Succeeded starch nanocrystals preparation combining heat-moisture treatment with acid hydrolysis. *Food Chemistry*, *278*, 350–356. https://doi.org/10.1016/j.foodchem.2018.11.018

Das, A, & Sit, N 2021. Modification of Taro starch and starch nanoparticles by various physical methods and their characterization. *Starch/Stärke*, *73*, art. 2000227. https://doi.org/10.1002/star.202000227

Dularia, C, Sinhmar, A, Thory, R, Pathera, AK, & Nain, V 2019. Development of starch nanoparticles based composite films from non-conventional source-water chestnut (Trapa bispinosa). *International Journal of Biological Macromolecules*, *136*, 1161–1168. https://doi.org/10.1016/j.ijbiomac.2019.06.169

Fan, H, Ji, N, Zhao, M, Xiong, L, & Sun, Q 2016. Characterization of starch films impregnated with starch nanoparticles prepared by 2,2,6,6-tetramethylpiperidine-1-oxyl (TEMPO)-mediated oxidation. *Food Chemistry*, *192*, 865–872. https://doi.org/10.1016/j.foodchem.2015.07.093

Foresti, ML., Williams, MDP, Martínez-García, R, & Vázquez, A 2014. Analysis of a preferential action of α-amylase from B. licheniformis towards amorphous regions of waxy maize starch. *Carbohydrate Polymers*, *102*, 80–87. https://doi.org/10.1016/j.carbpol.2013.11.013

Ge, S, Xiong, L, Li, M, Liu, J, Yang, J, Chang, R, Liang, CF, & Sun, Q 2017. Characterizations of Pickering emulsions stabilized by starch nanoparticles: Influence of starch variety and particle size. *Food Chemistry*, *234*, 339–347. https://doi.org/10.1016/j.foodchem.2017.04.150

Gérard, C, Colonna, P, Buléon, A, & Planchot, V 2002. Order in maize mutant starches revealed by mild acid hydrolysis. *Carbohydrate Polymers*, *48*, 131–141. https://doi.org/10.1016/S0144-8617(01)00219-3

Gutiérrez, TJ, & Tovar, J 2021. Update of the concept of type 5 resistant starch (RS5): Self-assembled starch V-type complexes. *Trends in Food Science & Technology*, *109*, 711–724. https://doi.org/10.1016/j.tifs.2021.01.078

Hao, Y, Chen, Y, Li, Q, & Gao, Q 2018. Preparation of starch nanocrystals through enzymatic pretreatment from waxy potato starch. *Carbohydrate Polymers*, *184*, 171–177. https://doi.org/10.1016/j.carbpol.2017.12.042

Hernández-Jaimes, C, Bello-Pérez, LA, Vernon-Carter, E, & Alvarez-Ramirez, J 2013. Plantain starch granules morphology, crystallinity, structure transition, and size evolution upon acid hydrolysis. *Carbohydrate Polymers*, *95*, 207–213. https://doi.org/10.1016/j.carbpol.2013.03.017

Jeong, O, & Shin, M 2018. Preparation and stability of resistant starch nanoparticles, using acid hydrolysis and cross-linking of waxy rice starch. *Food Chemistry*, *256*, 77–84. https://doi.org/10.1016/j.foodchem.2018.02.098

Kim, HY, Lee, JH, Kim, JY, Lim, WJ, & Lim, ST 2012. Characterization of nanoparticles prepared by acid hydrolysis of various starches. *Starch/Stärke*, *64*, 367–373. https://doi.org/10.1002/star.201100105

Kim, JH, Park, DH, & Kim, JY 2017. Effect of heat-moisture treatment under mildly acidic condition on fragmentation of waxy maize starch granules into nanoparticles. *Food Hydrocolloids*, *63*, 59–66. https://doi.org/10.1016/j.foodhyd.2016.08.018

Kumari, S, Yadav, BS, & Yadav, RB 2020. Synthesis and modification approaches for starch nanoparticles for their emerging food industrial applications: A review. *Food Research International*, *128*, art. 108765. https://doi.org/10.1016/j.foodres.2019.108765

Lamanna, M, Morales, NJ, García, NL, & Goyanes, S 2013. Development and characterization of starch nanoparticles by gamma radiation: Potential application as starch matrix filler. *Carbohydrate Polymers*, *97*, 90–97. https://doi.org/10.1016/j.carbpol.2013.04.081

Le Corre, D, & Angellier-Coussy, H 2014. Preparation and application of starch nanoparticles for nanocomposites. A review. *Reactive and Functional Polymers*, *85*, 97–120. https://doi.org/10.1016/j.reactfunctpolym.2014.09.020

LeCorre, D, Vahanian, E, Dufresne, A, & Bras, J 2012. Enzymatic pretreatment for preparing starch nanocrystals. *Biomacromolecules*, *13*, 132–137. https://doi.org/10.1021/bm201333k

Li, C, Guo, Y, Chen, M, Wang, S, Gong, H, Zuo, J, Zhang, J, & Dai, L 2023. Recent preparation, modification and application progress of starch nanocrystals: A review. *International Journal of Biological Macromolecules*, *250*, art. 126122. https://doi.org/10.1016/j.ijbiomac.2023.126122

Liu, Q, Li, F, Lu, H, Li, M, Liu, J, Zhang, S, Sun, Q, & Xiong, L 2018. Enhanced dispersion stability and heavy metal ion adsorption capability of oxidized starch nanoparticles. *Food Chemistry*, *242*, 256–263. https://doi.org/10.1016/j.foodchem.2017.09.071

Liu, Y, Qiu, C, Li, X, McClements, DJ, Wang, C, Zhang, Z, Jiao, A, Long, J, Zhu, KF, Wang, J, & Jin, Z 2023. Application of starch-based nanoparticles and cyclodextrin for prebiotics delivery and controlled glucose release in the human gut: A review. *Critical Reviews in Food Science and Nutrition*, *63*, 6126–6137. https://doi.org/10.1080/10408398.2022.2028127

Ma, X, Jian, R, Chang, PR, & Yu, J 2008. Fabrication and characterization of citric acid-modified starch nanoparticles/plasticized-starch composites. *Biomacromolecules*, *9*, 3314–3320. https://doi.org/10.1021/bm800987c

Marta, H, Rizki, DI, Mardawati, E, Djali, M, Mohammad, M, & Cahyana, Y 2023. Starch nanoparticles: Preparation, properties and applications. *Polymers*, *15*, art. 1167.

Maryam, Kasim, A, Novelina, & Emriadi 2020. Preparation and characterization of sago (metroxylon sp.) starch nanoparticles using hydrolysis-precipitation method. *Journal of Physics: Conference Series*, *1481*, art. 012021. https://doi.org/10.1088/1742-6596/1481/1/012021

Minakawa, AFK, Faria-Tischer, PCS, & Mali, S 2019. Simple ultrasound method to obtain starch micro- and nanoparticles from cassava, corn and yam starches. *Food Chemistry*, *283*, 11–18. https://doi.org/10.1016/j.foodchem.2019.01.015

Mohammad Amini, A, & Razavi, SMA 2016. A fast and efficient approach to prepare starch nanocrystals from normal corn starch. *Food Hydrocolloids*, *57*, 132–138. https://doi.org/10.1016/j.foodhyd.2016.01.022

Mukurumbira, AR, Mellem, JJ, & Amonsou, EO 2017. Effects of amadumbe starch nanocrystals on the physicochemical properties of starch biocomposite films. *Carbohydrate Polymers*, *165*, 142–148. https://doi.org/10.1016/j.carbpol.2017.02.041

Orsuwan, A, & Sothornvit, R 2017. Development and characterization of banana flour film incorporated with montmorillonite and banana starch nanoparticles. *Carbohydrate Polymers*, *174*, 235–242. https://doi.org/10.1016/j.carbpol.2017.06.085

Perez Herrera, M, & Vasanthan, T 2018. Rheological characterization of gum and starch nanoparticle blends. *Food Chemistry*, *243*, 43–49. https://doi.org/10.1016/j.foodchem.2017.09.011

Pinto, VZ, Moomand, K, Deon, VG, Biduski, B, Zavareze, EDR, Lenhani, GC, Do Santos, FGH, Lim, LT, & Dias, ARG 2021. Effect of physical pretreatments on the hydrolysis kinetic, structural, and thermal properties of pinhão starch nanocrystals. *Starch/Stärke*, *73*, art. 2000008. https://doi.org/10.1002/star.202000008

Putaux, JL, Molina-Boisseau, S, Momaur, T, & Dufresne, A 2003. Platelet nanocrystals resulting from the disruption of waxy maize starch granules by acid hydrolysis. *Biomacromolecules*, *4*, 1198–1202. https://doi.org/10.1021/bm0340422

Qian, X, Lu, Y, Xie, W, & Wu, D 2020. Viscoelasticity of olive oil/water Pickering emulsions stabilized with starch nanocrystals. *Carbohydrate Polymers*, *230*, art, 115575. https://doi.org/10.1016/j.carbpol.2019.115575.

Qin, Y, Liu, C, Jiang, S, Xiong, L, & Sun, Q 2016. Characterization of starch nanoparticles prepared by nanoprecipitation: Influence of amylose content and starch type. *Industrial Crops and Products*, *87*, 182–190. https://doi.org/10.1016/j.indcrop.2016.04.038

Qiu, C, Hu, Y, Jin, Z, McClements, DJ, Qin, Y, Xu, X, & Wang, J 2019. A review of green techniques for the synthesis of size-controlled starch-based nanoparticles and their applications as nanodelivery systems. *Trends in Food Science & Technology*, *92*, 138–151. https://doi.org/10.1016/j.tifs.2019.08.007

Remanan, MK, & Zhu, F 2021. Encapsulation of rutin using quinoa and maize starch nanoparticles. *Food Chemistry*, *353*, 128534. https://doi.org/10.1016/j.foodchem.2020.128534

Ren, L, Wang, Q, Yan, X, Tong, J, Zhou, J, & Su, X 2016. Dual modification of starch nanocrystals via crosslinking and esterification for enhancing their hydrophobicity. *Food Research International*, *87*, 180–188. https://doi.org/10.1016/j.foodres.2016.07.007

Ren, L, Zhang, Y, Wang, Q, Zhou, J, Tong, J, Chen, D, & Su, X 2018. Convenient method for enhancing hydrophobicity and dispersibility of starch nanocrystals by crosslinking modification with citric acid. *International Journal of Food Engineering*, *14*(4). https://doi.org/10.1515/ijfe-2017-0238

Robin, J 1974. Lint-nerized starches. Gel filtration and enzymatic studies of insoluble residues from prolonged acid treatment of potato starch. *Cereal Chemistry*, *51*, 389–406.

Romero Hernández, HA, Gutiérrez, TJ, Tovar, J, & Bello-Pérez, LA 2023. Complexation of octenyl succinic anhydride-esterified corn starch/polyphenol-rich Roselle (Hibiscus sabdariffa L.) extract: Structural and digestibility features. *Food Hydrocolloids*, *145*, art. 109125. https://doi.org/10.1016/j.foodhyd.2023.109125

Rostamabadi, H, Falsafi, SR, & Jafari, SM 2019. Starch-based nanocarriers as cutting-edge natural cargos for nutraceutical delivery. *Trends in Food Science & Technology*, *88*, 397–415. https://doi.org/10.1016/j.tifs.2019.04.004

Sánchez de la Concha, BB, Agama-Acevedo, E, Agurirre-Cruz, A, Bello-Pérez, LA, & Alvarez-Ramírez, J 2020. OSA esterification of amaranth and maize starch nanocrystals and their use in "pickering" emulsions. *Starch/Stärke*, *72*, art. 1900271. https://doi.org/10.1002/star.201900271

Sanchez de la Concha, BB, Agama-Acevedo, E, Nuñez-Santiago, MC, Bello-Perez, LA, Garcia, HS, & Alvarez-Ramirez, J 2018. Acid hydrolysis of waxy starches with different granule size for nanocrystal production. *Journal of Cereal Science*, *79*, 193–200. https://doi.org/10.1016/j.jcs.2017.10.018

Santoyo-Aleman, D, Sanchez, LT, & Villa, CC 2019. Citric-acid modified banana starch nanoparticles as a novel vehicle for β-carotene delivery. *Journal of the Science of Food and Agriculture*, *99*, 6392–6399. https://doi.org/10.1002/jsfa.9918

Simi, CK, & Emilia Abraham, T 2007. Hydrophobic grafted and crosslinked starch nanoparticles for drug delivery. *Bioprocess and Biosystems Engineering*, *30*, 173–180. https://doi.org/10.1007/s00449-007-0112-5

Singh, AK, Lee, M, Jang, D, & Lee, YS 2024. Non-conventional starch nanoparticles: Novel avenues towards improving sustainability of the food packaging sector. *Trends in Food Science & Technology*, *143*, art. 104273. https://doi.org/10.1016/j.tifs.2023.104273

Su, Q, Wang, Y, Zhao, X, Wang, H, Wang, Z, Wang, N, & Zhang, H 2020. Functionalized nano-starch prepared by surface-initiated atom transfer radical polymerization and quaternization. *Carbohydrate Polymers*, *229*, art. 115390. https://doi.org/10.1016/j.carbpol.2019.115390

Sun, Q, Li, G, Dai, L, Ji, N, & Xiong, L 2014. Green preparation and characterisation of waxy maize starch nanoparticles through enzymolysis and recrystallisation. *Food Chemistry*, *162*, 223–228. https://doi.org/10.1016/j.foodchem.2014.04.068

Suriya, M, Reddy, CK, Haripriya, S, & Harsha, N 2018. Influence of debranching and retrogradation time on behavior changes of Amorphophallus paeoniifolius nanostarch. *International Journal of Biological Macromolecules*, *120*, 230–236. https://doi.org/10.1016/j.ijbiomac.2018.08.059

Szymonska, J, Targosz-Korecka, M, & Krok, F 2009. Characterization of starch nanoparticles. *Journal of Physics: Conference Series, 146*, art. 012027. https://doi.org/10.1088/1742-6596/146/1/012027

Timgren, A, Rayner, M, Dejmek, P, Marku, D, & Sjöö, M 2013. Emulsion stabilizing capacity of intact starch granules modified by heat treatment or octenyl succinic anhydride. *Food Science & Nutrition*, *1*, 157–171. https://doi.org/10.1002/fsn3.17

Torres, FG, Arroyo, J, Tineo, C, & Troncoso, O 2019. Tailoring the properties of native andean potato starch nanoparticles using acid and alkaline treatments. *Starch/ Stärke*, *71*, art. 1800234. https://doi.org/10.1002/star.201800234

Valodkar, M, & Thakore, S 2010. Isocyanate crosslinked reactive starch nanoparticles for thermo-responsive conducting applications. *Carbohydrate Research*, *345*, 2354–2360. https://doi.org/10.1016/j.carres.2010.08.008

Velásquez-Castillo, LE, Leite, MA, Tisnado, VJA, Ditchfield, C, Sobral, PJDA, & Moraes, ICF 2023. Cassava Starch films containing quinoa starch nanocrystals: Physical and surface properties. *Foods*, *12*, art. 576.

Wang, Y, Tian, H, & Zhang, L 2010. Role of starch nanocrystals and cellulose whiskers in synergistic reinforcement of waterborne polyurethane. *Carbohydrate Polymers*, *80*, 665–671. https://doi.org/10.1016/j.carbpol.2009.10.043

Wang, Y, & Zhang, G 2021. The preparation of modified nano-starch and its application in food industry. *Food Research International*, *140*, art. 110009. https://doi.org/10.1016/j.foodres.2020.110009

Winarti, C, Sunarti, T, Mangunwidjaja, D, & Richana, N 2014. Preparation of arrowroot starch nanoparticles by butanol-complex precipitation, and its as bioactive encapsulation matrix. *International Food Research Journal*, *21*, 2207–2213.

Ye, F, Miao, M, Jiang, B, Campanella, OH, Jin, Z, & Zhang, T 2017b. Elucidation of stabilizing oil-in-water Pickering emulsion with different modified maize starch-based nanoparticles. *Food Chemistry, 229*, 152-158. http://dx.doi.org/10.1016/j.foodchem.2017.02.062

Ye, F, Miao, M, Lu, K, Jiang, B, Li, X, & Cui, SW 2017a. Structure and physicochemical properties for modified starch-based nanoparticle from different maize varieties. *Food Hydrocolloids, 67*, 37-44. http://dx.doi.org/10.1016/j.foodhyd.2016.12.041

Yu, M, Ji, N, Wang, Y, Dai, L, Xiong, L, & Sun, Q 2021. Starch-based nanoparticles: Stimuli responsiveness, toxicity, and interactions with food components. *Comprehensive Reviews in Food Science and Food Safety*, *20*, 1075–1100. https://doi.org/10.1111/1541-4337.12677

Zhou, J, Tong, J, Su, X, & Ren, L 2016. Hydrophobic starch nanocrystals preparations through crosslinking modification using citric acid. *International Journal of Biological Macromolecules*, *91*, 1186–1193. https://doi.org/10.1016/j.ijbiomac.2016.06.082

Digestibility and Glycaemic Features of Starches

20

Abayomi Ajala, Elif Turabi Yolacaner, Rosana Colussi, Lovedeep Kaur, and Jaspreet Singh

20.1 INTRODUCTION

Starch is the most common storage carbohydrate in plants and a major source of carbohydrates in human food. It is made up of two types of molecules: amylose, which is a linear polymer of α-D-glucose units linked by a-1,4 glycosidic linkages, and amylopectin, which is a branched polymer of a-D-glucose units linked by α-1,4- and α-1,6-glycosidic linkages. Starch and starchy food products can be categorised based on their digestibility, which is characterised by the rate and duration of the glycaemic response. Predicting and controlling glucose absorption from starchy foods is important for global health concerns. Most starches contain rapidly digesting starch, slowly digesting starch, and resistant starch, which is not broken down by enzymes in the small intestine and passes to the large intestine. A commonly used method to classify starches based on *in vitro* digestion kinetics was proposed by Englyst et al. (1992). This method involves simulating stomach and intestinal conditions to measure glucose release at different times. Based on the method, different starch fractions are defined as:

- *Rapidly digestible starch* (*RDS*): The amount of glucose released after 20 minutes.
- *Slowly digestible starch* (*SDS*): Glucose released between 20 and 120 minutes of *in vitro* digestion.
- *Resistant starch* (*RS*): Total starch minus the amount of glucose released within 120 minutes of *in vitro* digestion, which the following equation can explain:

$$RS = TS - (RDS + SDS)$$

where TS is the total starch.

Resistance starch is recognised as a special type of dietary fibre, which is widely used in some of these foods as a functional ingredient. RS is defined as the type of starches that are not digested in the small intestine but fermented by microorganisms in the colon, leading to the formation of short-chain fatty acids (SCFAs) (Jiali et al. 2023). Therefore, it can serve as a prebiotic with excellent physiological functions such as stabilising postprandial blood glucose levels, preventing type II diabetes, preventing intestinal inflammation, and regulating gut microbiota phenotypes. These physiological effects depend on its structural characteristics (Dongowski et al. 2005).

Resistant starch has also emerged as a crucial ingredient in managing metabolic disorders due to lifestyle changes through its unique digestion and absorption mechanisms (Hemmingsen et al. 2017). It is categorised into five types: RS1 (physically protected), RS2 (native granular), RS3 (retrograded), RS4 (chemically modified), and RS5 (amylose–lipid complex) (Raigond et al. 2015). Each type has unique sources, structures, and digestive properties, affecting its resistance to enzymatic digestion and subsequent physiological impacts. RS1 is a physically encapsulated starch protected by plant cell walls or proteins in grains and beans. It resists digestion due to insufficient milling and coarse particles (Sun et al. 2022). RS2 is the granules with B-type or C-type crystalline structures in bananas, raw rice, and potatoes. Cooking and pulverisation reduce its resistance (Noor et al. 2021). RS3 is a retrograded starch formed by cooling cooked starches, enhancing thermal stability and texture (Li & Hu 2023). RS4 is a chemically modified starch through esterification, phosphorylation, etc., improving textural properties for various food applications. RS5 is an amylose–lipid complex with a V-shaped crystal structure, increasing resistance to digestion and suitable for high-temperature processing (Putseys et al. 2010).

Resistant starch is widely used in food products due to its natural origin, low water holding capacity, high gelatinisation temperature, and good extrusion performance. It is incorporated into various foods to improve health benefits without affecting taste or sensory properties. RS enhances the nutritional profile of baked goods like bread, cakes, and pastries. It can replace a portion of the flour, reducing the glycaemic index (GI) and increasing

DOI: 10.1201/9781003464396-20

dietary fibre content. Studies have shown that bread with 60% RS significantly reduces postprandial glucose and insulin responses (Takahashi et al. 2022; Yamada et al. 2005). In dairy products like yoghurt and cheese, RS improves the texture and viscosity. It serves as a prebiotic, enhancing the survival rate of probiotics such as *Lactobacillus acidophilus*. This improves gut health and provides additional health benefits to consumers (Aryana et al. 2015). RS is added to sausages and other processed meats to improve water retention and texture. It forms a gel with myosin, which helps to retain moisture and flavour. This application also leverages RS's low water-holding capacity, contributing to longer shelf life (Sarteshnizi et al. 2017). Adding RS to pasta and noodles improves their water absorption and reduces viscoelasticity. This not only enhances satiety but also lowers the digestible carbohydrate intake, which is advantageous for weight management (Yaver & Bilgiçli 2021). RS is an excellent thickening and gelling agent in soups, sauces, and gravies, providing a smooth texture without altering the flavour. RS4, in particular, is noted for its ability to maintain gel stability under varying temperatures, making it ideal for a wide range of food products (Patterson et al. 2020). RS is beneficial in formulating low-calorie foods as it provides bulk and texture similar to regular starch but with significantly fewer calories. This application is crucial for developing products for weight management and diabetes control (Higgins 2014).

The precise determination of bioavailable carbohydrates in a given product allows the manufacturer to convey the glycaemic response per serving of a food, particularly in the case of diabetes-therapeutic food, diabetes management, and carbohydrate metabolism disorders. The GI concept has been introduced to categorise foods based on their postprandial blood glucose response. The GI is defined as the postprandial incremental glycaemic area after a test meal, expressed as the percentage of the corresponding area after an equi-carbohydrate portion of a reference food such as glucose or white bread (Goni et al. 1997; Jenkins et al. 1987). Various studies have demonstrated a strong correlation between the rate of *in vitro* digestion and the glycaemic response to food. Such studies can be utilised to identify food that may be beneficial in the diet of individuals with diabetes. The physicochemical and structural characteristics of starch vary among different botanical sources (Singh et al. 2007). Multiple studies have indicated relationships between different starch characteristics and *in vitro* digestibility. Chemically modified starches are commonly employed in food applications as they offer improved functional properties compared to native starches. Several reports suggest that chemically modified starches exhibit greater resistance towards α-amylase digestion (Han & BeMiller 2007). The textural and rheological characteristics of food, the presence of other food components, such as proteins, lipids, and non-starch polysaccharides, and the alterations and interactions occurring in them during food processing significantly impact the enzymatic digestibility of starch. Our chapter provides a comprehensive overview of the factors influencing starch digestibility and glycaemic properties of starch-based foods. It covers all aspects, from starch structure, starch modification, and food processing to digestion, particularly emphasising the effects of other components in the food matrix.

20.2 PHYSICOCHEMICAL CHARACTERISTICS OF STARCH IMPACTING DIGESTION AND GLYCAEMIC PROPERTIES

The starch digestion mechanism tends to follow three phases: (1) Diffusion of enzymes towards the starch, (2) Formation of enzyme–starch complex (adsorption of enzymes onto the starch), and (3) Hydrolytic event (Lehmann & Robin 2007). Based on the starch digestion mechanism, factors that will be considered for limiting the rate of starch digestion would have to either slow down/prevent the access/adsorption of the enzymes (such as granular size and surface properties) or internal starch structure (such as amylose and amylopectin chains length) that prevents the activation of the hydrolytic event (Dhital et al. 2019) (Table 20.1).

TABLE 20.1   Physicochemical characteristics of starch granules that impact digestion properties

STARCH CHARACTERISTICS	STARCH DIGESTIBILITY	REFERENCES
Granular Size and Shape		
Jackfruit starch (~16 μm)	13[a]	Dong et al. (2021)
Potato starch (~50 μm)	44[a]	
Maize starch (~22 μm)	32[a]	
Potato starch (36 μm)	82[a]	Bajaj et al. (2018)
Kidney bean starch (26 μm)	63[a]	
Wheat starch (19 μm)	68[a]	
Sweet potato starch (17 μm)	74[a]	
Low amylose corn starch (16 μm)	71[a]	
Corn starch (15 μm)	75[a]	
Rice starch (5 μm)	78[a]	

(Continued)

TABLE 20.1 (*Continued*)   Physicochemical characteristics of starch granules that impact digestion properties

STARCH CHARACTERISTICS	*STARCH DIGESTIBILITY*	*REFERENCES*
Native potato starch	0.86[b]	Kaur et al. (2007)
Large potato starch granules	0.83[b]	
Medium potato starch granules	1.12[b]	
Small potato starch granules	1.35[b]	
Granular Surface		
Rice starch varieties		Ouyang et al. (2024)
QR	69[c]	
BR50	62[c]	
BR58	65[c]	
(QR and BR58 have more pinholes than BR50)		
Potato starch	9.9[a]	Corgneau et al. (2019)
Cassava starch	19.6[a]	
Maize starch	46.8[a]	
Waxy maize starch	22.7[a]	
Rice starch	48.4[a]	
Waxy rice starch	60[a]	
Wheat starch	36[a]	
(Potato starch has a smoother granular surface)		
Presence of Non-Starchy Substances on the Starch Granules		
Alkali extracted high barley starch (HBS) (0.17, 0.25%)	93.2[b]	Nie et al. (2023)
Ultrasound extracted HBS (0.23, 0.29%)	92.4[b]	
Double enzyme extracted HBS (0.24, 0.30%)	86.7[b]	
Three enzymes extracted HBS (0.21, 0.28%)	88.3[b]	
Ultrasonic-TE extracted HBS (0.21, 0.28%)	90.9[b]	
Complex Internal Structures of the Starch Granules		
High amylose content (%)		Yang et al. (2021)
Heat-treated Japonica rice starch fractions (18.6–26.0)	66.1–70.6[d]	
Heat-treated waxy rice starch fractions (0.3–0.5)	63.1–64.6[d]	
Retrograded rice starch (varieties with higher amount of shorter amylose chains exhibited lower digestibility)	88.5–102.8[b]	Gong et al. (2019)
Corn starch (CS)	82.8[b]	Ouyang et al. (2021)
Potato starch (PTS)	64.7[b]	
Pea starch (PS)	80.4[b]	
Ultrasonic-treated CS	77.1–84.9[b]	
Ultrasonic-treated PTS	66.4–81.6[b]	
Ultrasonic-treated PS	84–90.1[b]	
Normal sorghum starch (raw)	81[a]	Sang et al. (2008)
Waxy sorghum starch (raw)	90[a]	
Waxy rice (cooked)	80[a]	Hu et al. (2004)
Normal rice (cooked)	71[a]	
High amylose maize starch(cooked)	77[a]	Wolf et al. (1999)
Waxy maize starch (cooked)	96[a]	
Normal maize starch (cooked)	86[a]	

[a] Expressed as rapidly digestible starch (%).
[b] Expressed as rate of hydrolysis (%) of starch granules.
[c] Expressed as slowly digestible starch (%).
[d] Expressed as resistant starch (%).

20.2.1 Morphology of Starch Granules (Shape and Size)

The morphological characteristics of starch granules, such as size and shape, vary across biological origins (Singh et al. 2010). Starch granular diameter ranges from 1 to 100 μm while its granular size distribution can be either a single predominant granule type, unimodal (maize, potato, and pea), or with two or more granular types, bimodal (barley, wheat, and oat) (Wang et al. 2022a, 2022b). Many researchers have investigated the relationship between the granular size and shapes of starch granules and their digestibility (Bajaj et al. 2018; Capriles et al. 2008;

Dhital et al. 2010; Kaur et al. 2007; Lal et al. 2021; Langworthy & Deuel 1922; Lindeboom et al. 2004; Tester et al. 2006). The small barley and wheat starch granules have been shown to hydrolyse faster than the large ones (Lindeboom et al. 2004). Kaur et al. (2007) observed that various fractions of potato starches (large, medium, and small) exhibited significant differences in their enzymatic hydrolysis. Separated potato fractions containing a higher percentage of large granules showed lower hydrolysis rates than fractions with medium and small granules. The high specific surface area to the size of the smaller granules may have increased the extent of enzymatic binding/adsorption on the small granules, thus resulting in a higher rate of hydrolysis (Tester et al. 2006). The results from various research suggested that smaller-sized starch granules showed greater susceptibility to enzymes regardless of botanical sources (Bajaj et al. 2018; Dhital et al. 2010; Lal et al. 2021). Dhital et al. (2010) studied the relationship between native potato (15–75 μm) and maize (5–20 μm) starch granule size and their starch digestibility. The results suggested that the rate of digestion of native maize starch was significantly higher than that of potato starch due to the smaller granular size of the maize starch (Cai et al. 2014; Zhang et al. 2018). Similarly, Bajaj et al. (2018) compared the starch granule digestibility from different botanical sources (cereal, tuber, and legumes). Although the results showed that rice starch with the smallest average granular size exhibited a high starch digestibility, the larger potato starch granules were observed to digest more rapidly than all the other sources. This observation indicates that there are other factors besides the size of starch granules that might be responsible for their digestibility.

20.2.2 Surface Characteristics of Starch Granules

The surface characteristics of the starch granules, such as smoothness, have been observed to account for the degree of enzymatic starch digestion across botanical origin, especially in the early stage of hydrolysis (Martens et al. 2018; Warren et al. 2011). Starch granules from legumes and tuber have been reported to have a largely smooth surface lacking pinholes and rough areas than granules from cereals (Singh et al. 2006, 2010; Wani et al. 2016; Wang et al. 2022a, 2022b). These differences in surface characteristics of starch granules subsequently dictate their corresponding mechanism of enzymatic digestibility, which is either by endo-corrosion or exo-corrosion.

Ouyang et al. (2024) studied the correlation between starch structure from three rice varieties and their *in vitro* starch digestibility. The authors observed that rice starch granules with more pinhole surface structure facilitated enzyme–substrate complexes more than the other varieties. Thus, more starch was digested. A similar trend was observed when the digestibility of starch granules from varying botanical sources (potato, corn, maize, waxy maize, and rice) was studied in terms of their surface features (Corgneau et al. 2019). The authors reported that potato starch, which has a smoother granular surface, exhibited a limited extent of starch digestion, while waxy maize and maize starch

granules with more pronounced pores and cracks displayed a high digestion profile.

The surface pores and channels of many cereal starch granules, which are between 0.1–0.3 and 0.07 and 0.1 μm, respectively, are believed to be the initial entry and opening site for the adsorption and activation of the enzymatic digestion (Wang et al. 2022b). The resulting digestion pattern that occurs is termed "endo-corrosion", where the adsorbed α-amylase creates and enlarges pits from the core of the starch granules to the surface (Singh et al. 2010; Wang et al. 2022b).

On the other hand, legumes and potato starch granules, which have a smooth granular surface with no pores or indentations, have been reported to exhibit "exo-corrosion" pattern of enzymatic digestion, i.e., entire granular surface or section of it is eroded during enzymatic digestion (Corgneau et al. 2019; Ma et al. 2017; Singh et al. 2010).

20.2.3 Non-Starch Components of Starch Granules

The presence of some minor components, such as residual proteins and lipids located on the surface or embedded within the native starch granules, may also limit the rate of enzymatic hydrolysis (Baldwin et al. 2015; Singh et al. 2010). They limit the rate of enzymatic hydrolysis by blocking the adsorption sites, therefore preventing the enzymes from binding to the granular surface of the starch (Oates 1997). The composition of residual lipids and protein on starch granules varies significantly across cereal, tubers, and legumes. The literature reports that 0.25%, 0.05%, and 0.01%–1.02% and 1.0%, 0.05%, and 0.06%–0.87% of residual protein and lipids can be found on starch granules from cereals, tubers, and legumes, respectively (Ashogbon et al. 2020; Baldwin et al. 2015).

Nie et al. (2023) studied the morphological features of high barley starch (HBS) extracted by five different methods and investigated their influence on starch digestibility. The authors reported that the alkaline extraction method, which removed most of the granular surface protein and lipids, exhibited the highest starch digestibility, while the least pure HBS produced by the double enzyme extraction method had limited starch digestion. Using a similar approach, endogenous protein, lipids, and β-glucan were removed in single and multiple combinations from HBS (Yang et al. 2021). These authors reported that the rate of starch digestibility increased with an increment in removing the non-starchy components. This further confirms that the non-starch granular surface components can affect the susceptibility of starch towards digestive enzymes (Dhital et al. 2019).

20.2.4 Complex Internal Structure of Starch Granules

The basic component structure in native starch granules comprises amylose and amylopectin. The fine structures and the ratio of amylose to amylopectin have been identified by many studies to influence starch digestion (Bajaj et al. 2018; Chi et al. 2021; Gong et al. 2019; Lin et al. 2018). The biological source of starch granules may dictate their amylose content, which may be attributed to differences in the activities of enzymes involved in the biosynthesis of linear and branched components (Singh et al. 2010). Most starchy foods, such as cereals and tubers, contain about ~30% amylose and 75% amylopectin, while legumes contain ~40% amylose and 70% amylopectin (Hoover & Zhou 2003; Lal et al. 2021).

Kidney bean starch granules with the highest amylose content compared with cereal and roots exhibited a greater limit to starch digestion (Bajaj et al. 2018). Similarly, rice starch granules with the same amylopectin structure and genetic background showed a reduction in the rate of starch digestion as the amylose content increased (Lin et al. 2018). However, research has reported that amylose content, content of shorter amylose chains, and total amylose molecular size correlated negatively with starch digestibility (Cai et al. 2015; Yu et al. 2018). The linear and double helix structures of amylose, which are strongly linked by hydrogen bonds, limit the enzyme action during starch digestion (Chi et al. 2021).

Many studies have indicated that starch digestion correlates with amylopectin chain length (Liu et al. 2018; Martens et al. 2018; Ren et al. 2020). The amylopectin chain lengths are represented as the short:long amylopectin side-chains ratio and the chains are generally classified as long chains above a degree of polymerisation (DP) of 36 glucose residues and as short chains below that (Bertoft 2017). The fine structure of cassava starch and its molecular role in influencing digestibility after cooking were investigated (Liu et al. 2018). The authors reported that amylopectin branch chains with DP 25–36 correlated positively with the RS content. The reason for this positive correlation with starch digestion was that research has shown that DP 25–36 amylopectin chains tend to promote the formation of rigid and crystalline structures, which significantly suppress enzymatic digestion (Zhu 2018).

There are three classes of crystalline structures in native starch granules, namely, A-type (cereals), B-type (tubers), and C-type (legumes). The main differences between these crystalline structures are crystal size, compactness, and crystallinity of the starch (Ouyang et al. 2021). The B-type starch, such as potato starch, has a lower digestion rate and extent compared to the A-type starch due to the well-defined polymorph structures and probably containing more long amylopectin chains (Zhu 2018)

20.3 STARCH MODIFICATION FOR CREATING LOW GLYCAEMIC STARCHES

Low-glycaemic starches may be created by modifying the structure of starch, which can help manage blood glucose levels, making them particularly beneficial for individuals with diabetes and metabolic disorders. Due to increasing health concerns about the effects of high-glycaemic foods, the demand for low glycaemic starches has grown, especially in the food industry, where they are commonly used in functional foods (Maniglia et al. 2021). The limitations in the structure of native starches reduce their suitability for many food applications where stability, texture, and slower digestion are important (Alcázar-Alay & Meireles 2015). Therefore, modification techniques are applied to enhance their properties, such as increasing RS content and reducing digestibility. These modifications, made through genetic (*in planta*), chemical, physical, and enzymatic methods, optimise starch for low-glycaemic applications (Compart et al. 2023).

20.3.1 *In Planta* Starch Modification Applications

In planta modifications are emerging as a pivotal approach for producing starches with reduced glycaemic indices, primarily by enhancing RS content. Genetic engineering techniques enable precise alterations in starch characteristics, addressing limitations in conventional modification methods. By genetically modifying plants, various starch parameters such as granule morphology, granule size, inner structure, and phosphorylation can be tailored to meet industry needs (Table 20.2) (Hebelstrup et al. 2015).

Recent research highlights various strategies for achieving low-glycaemic starches through genetic modifications. Malinova et al. (2017) explored the effects of manipulating starch synthase 4 and plastidal phosphorylase in Arabidopsis, which significantly altered starch granule morphology and phosphorylation. These modifications led to changes in starch digestibility and glycaemic response, demonstrating the potential for genetic interventions to influence starch characteristics effectively. Chia et al. (2020) further contributed by examining the role of carbohydrate-binding proteins in wheat, revealing how modifications in granule size distribution can slow starch digestion and reduce its GI.

Additional studies underscore the broad applicability of genetic modifications in starch production. Bull et al. (2018) utilised CRISPR technology in cassava to create starches with increased RS content, which is known to lower the GI by reducing glucose absorption rates. Similarly, Samodien et al. (2018) investigated starch phosphatase repression in potatoes, finding that increased starch phosphate levels resulted in higher RS content and a lower GI. Miao et al. (2017) supported these findings by demonstrating how genetic edits in starch biosynthesis pathways can enhance RS levels, contributing to a reduced glycaemic response.

While significant progress has been made in research and development of *in planta* modifications, the commercial application of these modified starches is still evolving. Some advancements, particularly those involving CRISPR technology in cassava, are approaching market readiness. However, widespread commercial use of genetically modified starches remains in the early stages, with ongoing research and development paving the way for future applications.

TABLE 20.2   Some *in planta* starch modification studies

PARAMETER	GENE TARGET	SPECIES (GENE CHANGE)	STRUCTURAL CHANGES
Granule morphology	Starch synthase 4 (SS4)	Arabidopsis *Arabidopsis thaliana* (mutation)	Discoid to spherical granules (Malinova et al. 2017)
Granule size	B-granule content 1 (BGC1)	Wheat *Triticum aestivum* (mutation)	B-type granules reduction (Chia et al. 2020)
Amylose content	Granule-bound starch synthase (GBSS)	Cassava *Manihot esculenta* (mutation)	Lower amylose content (Bull et al. 2018)
Amylopectin content	Starch synthase 3 (SS3)	Tomato *Solanum lycopersicum* (overexpression)	Higher amylopectin content (Miao et al. 2017)
Phosphate content	Starch Excess 4 or Like Sex Four 2 (SEX4 or LSF2)	Potato *Solanum tuberosum* (suppression)	Higher phosphate content Altered amylopectin structure Reduced granule size (Samodien et al. 2018)

20.3.2 Chemical Modification Applications

Chemical modification of starches enhances their properties by altering the starch molecule's functional groups without affecting granule size or shape, thereby broadening their applications. Among these techniques, esterification improves starch properties by modifying hydroxyl groups, which enhances hydrophobicity and swelling capacity. For instance, succinylation reduces gelatinisation temperature and improves thickening power, which is beneficial for creating low glycaemic starches used in bakery products and sauces (Singh et al. 2007). Phosphorylation, which affects pasting and thermal stability, has practical implications for products like sauces and soups, contributing to their stability and texture (Helle et al. 2019). Cross-linking, by strengthening starch granules through reactions with multifunctional chemicals, enhances resistance to retrogradation, thus improving viscosity and paste clarity. This has been optimised in recent studies for better performance in bakery products and processed foods (Koo et al. 2010; Wang et al.2020).

Etherification modifies starch to improve its thermal stability and rheological properties, with carboxymethyl starch being a notable example due to its hydrophilicity and cold-water solubility. This makes it suitable for pharmaceutical and food applications (Spychaj et al. 2013). Ozone, an advanced oxidation method, introduces carbonyl and carboxyl groups, which increase hydrophilicity and disrupt crystallinity, benefiting the texture and shelf life of food products. This method is particularly useful in the production of low-GI starches for bakery products, cereals, and functional foods (Pandiselvam et al. 2019). Acid hydrolysis enhances starch solubility and viscosity while preserving granule morphology, making it ideal for use in soups and sauces, contributing to their reduced glycaemic impact (Pratiwi et al. 2018). Grafting, which involves introducing polymer chains into starch, enhances biodegradability and thermal stability, and is being applied to food packaging and agricultural products. Recent advancements such as microwave and UV light-induced grafting are environmentally friendly and improve commercial qualities (Djordjevic et al. 2013; Wang 2020).

Recent research highlights several promising developments for the future of starch modification. Click chemistry, which allows precise starch functionalisation, holds potential for new food applications by enabling the development of tailored low-GI starches (Zhou et al. 2022). Additionally, green chemistry approaches aim to reduce environmental impacts while maintaining starch functionality, which could enhance the sustainability of low-GI starch production. These innovations underscore the potential for future advancements in starch modification, with a focus on low-GI starches and their applications in food products.

20.3.3 Physical Modification Applications

Physical modification for starches offers environmentally friendly alternatives without hazardous chemicals. These techniques can adjust granular structure and packing, improving the starch properties crucial for food applications. Physical modification can be categorised into thermal and non-thermal approaches, with significant interest in food applications for enhancing the functionality of starch-based products.

Physically modified starches play a critical role in enhancing the functionality of food products, particularly through thermal and non-thermal techniques. Heat-moisture treatment (HMT) and annealing, for example, are thermal methods widely used to modify starch properties. HMT alters both the crystalline and amorphous regions of starch granules, improving pasting properties, increasing gelatinisation temperature, and reducing solubility. These characteristics enhance the texture and stability of products like reduced-calorie baked goods and functional snacks, which require controlled digestion and texture (Zavareze & Dias 2011). Annealing, by promoting the reorganisation of the crystalline structure, makes starch more resistant to digestion, aiding in the production of low-GI foods that are beneficial for diabetic-friendly products and meal replacements (Gunaratne 2018).

Another key thermal process, pregelatinisation, involves boiling and drying starch, making it soluble in cold water and enhancing its viscosity and gel stability. This property is particularly valuable for instant food products (Liu et al. 2017).

Non-thermal techniques, such as ultrasonication and micronisation, provide additional avenues for starch modification. Ultrasonication enhances water solubility and rheological properties, making it ideal for improving the mouthfeel of gluten-free products, which often lack elasticity (Bonto et al. 2021). Micronisation, which reduces particle size, results in fine powders that improve dispersibility and smoothness in instant food and powdered beverage applications (Emami et al. 2010). Additionally, cold plasma treatment increases resistant starch content and reduces crystallinity, offering another method for creating low-GI starches, which is critical in health-conscious formulations (Thirumdas et al. 2017).

Physical modification techniques play a crucial role in the production of low-GI starches and in enhancing the functionality of starch-based food products. These methods provide practical and sustainable solutions for the food industry, meeting the growing demand for healthier and more functional ingredients without compromising processing efficiency or product quality.

20.3.4 Enzymatic Modification Applications

Enzymatic modification is a crucial method for altering starch properties in the food industry, offering high specificity and minimal effluent generation. Enzymatic reactions fall into four main categories: endoamylases, exoamylases, debranching enzymes, and transferases, each modifying starch's molecular structure uniquely to suit various food applications.

Endoamylases, such as α-amylases, randomly hydrolyse (1,4)-linkages within amylose and amylopectin, reducing chain length and creating surface pores on starch granules (van der Maarel et al. 2002). This increases solubility while reducing swelling and pasting viscosity, making these starches suitable for baked goods and confectioneries, where product stability and texture are key. These modified starches are particularly beneficial in reduced-calorie baked goods and functional snacks, promoting better starch digestibility and improved glycaemic control. Exoamylases, such as β-amylase, amyloglucosidases, and β-glucosidases, hydrolyse the outer glucose residues of amylose and amylopectin, producing maltose and glucose (Ashok et al. 2000; Dura et al. 2014). These enzymes are essential for the production of maltose and high-glucose syrups, which are widely used in the sweetener industry for products such as candies, jellies, and soft drinks, ensuring sweetness while maintaining texture and moisture retention.

Debranching enzymes, including isoamylase and pullulanase, specifically target α-1,6-glucosidic bonds in amylopectin, producing linear glucans (van der Maarel et al. 2002). Pullulanase is particularly valuable for producing resistant starches, which are gaining popularity in low-GI food products due to their slower digestion rate. Resistant starches are ideal for use in diabetic-friendly foods, meal replacements, and weight management products, as they help regulate postprandial blood glucose levels (Barnett et al. 1999). Transferases, such as cyclodextrin glycosyltransferase and amylomaltase, modify starch by transferring glycosidic residues, creating starches with unique functional properties (Kaur et al. 2012). Cyclodextrins produced through these processes are widely applied for flavour encapsulation and stabilising sensitive food ingredients. Furthermore, cyclodextrins act as fat replacers in the development of low-calorie foods, enhancing texture and mouthfeel without contributing additional calories. This makes them a valuable tool in health-conscious formulations aimed at reducing caloric intake.

Enzymatic modification offers a sustainable alternative to chemical modification with high precision and reduced environmental impact. However, single-enzyme applications often do not achieve the desired starch properties for specific food applications. Therefore, combinations of enzymes or engineered enzymes are increasingly used to enhance catalytic efficiency and expand the range of starch modifications. These approaches hold particular promise for producing low-GI starches, which are in high demand in the development of health-focused food products like snacks, breakfast cereals, and dietary supplements. Further advancements in enzyme engineering and production methods will be crucial for improving the scalability and efficiency of enzymatic starch modifications, making them more widely available for the food industry.

20.4 EFFECT OF FOOD MATRIX ON STARCH DIGESTIBILITY

20.4.1 Composition, Microstructure, and Particle Size

The composition, microstructure, and particle size can significantly affect the starch digestibility and glycaemic response of foods. The type of processing used, formulation, and preparation method also greatly impact the rate of glucose release. The methods for quantifying digestibility and its GI also vary greatly; consequently, the values obtained also vary. Table 20.3 presents the digestible starch or GI content of different starches and starch-based foods, along with the reference for quantification.

The physical structure of foods can affect the digestion of starch and the absorption of its hydrolysis products. For example, in cooked rice, the aleurone layers inhibit the penetration of digestive fluid into the nucleus of the rice grain, which ends up interfering with a lower rate of glucose release. On the other hand, when the particle size is reduced due to the disintegration of the grain that occurs during chewing, there is a greater rate of starch hydrolysis, which in turn will affect the degree of increase in blood sugar in the human body after eating starchy foods (Tamura et al. 2016). Studying the structural differences between white and brown rice and how they affect starch digestion, Kong et al. (2011) reported that fibre helps inhibit the absorption of moisture and acid during cooking and

TABLE 20.3   Digestibility of different starches and starch-based foods

SOURCE	DIGESTIBILITY	REFERENCE AND METHOD
Cereals and Cereal-Based		
Waxy brown rice	92.9[a]	You et al. (2016); *in vitro* digestion by the method of Englyst et al. (1999) using pepsin, pancreatin, and amyloglucosidase
Germinated waxy brown rice	96.9[a]	
White rice	83.8–99.6[a]	You et al. (2015); *in vitro* digestion by the method of Englyst et al. (1999) using pepsin, pancreatin, and amyloglucosidase
Indica white rice	90[b]	Schwanz Goebel et al. (2019); *in vitro* digestion by the method of Dartois et al. (2010) using pepsin, pancreatin, amyloglucosidase, and invertase
Japonica white rice	89[b]	
Gluten-free pasta made with rice and beans	30.2–71.4[a]	Giuberti et al. (2015); *in vitro* digestion by the method of Englyst et al. (1992) using pepsin, pancreatin, and amyloglucosidase
Wheat pasta	60–72[b]	Simonato et al. (2015); *in vitro* digestion by the method of Kim et al. (2008) using pepsin and pancreatic amylase
Extruded maize starch	75–99[a]	Robin et al. (2016); *in vitro* digestion by the method of Englyst et al. (1999) using pepsin, pancreatin, and amyloglucosidase
Wheat grains	55–84[b]	Miranda da Silveira et al. (2020); *in vitro* digestion by the method of Dartois et al. (2010) using pepsin, pancreatin, amyloglucosidase, and invertase
Oat	24[a]	Tang et al. (2019); *in vitro* digestion by the method of Englyst et al. (1992) using pancreatin and amyloglucosidase
Pseudocereals		
Tartary buckwheat flour	69.96[a]	Zhang et al. (2023); *in vitro* digestion by the method of Englyst et al. (1999) using pepsin, pancreatin, and amyloglucosidase
Colored quinoa starch	93.09–98.41[a]	Peng et al. (2022); *in vitro* digestion by the method of Englyst et al. (1999) using pepsin, pancreatin, and amyloglucosidase
Cooked amaranth seed	65[c]	Capriles et al. (2008); *in vitro* digestion by the method of Goni et al. (1997) using pepsin, α-amylase, and amyloglucosidase from *Aspergillus niger*
Extruded amaranth seed	61.06[c]	
Pulses		
Common bean flour	4.40[a]	Chung et al. (2008); *in vitro* digestion by the method of Chung et al. (2008) using pancreatin and amyloglucosidase
Common bean	9.4–24.53[a]	Ovando-Martínez et al. (2011); *in vitro* digestion by the method of Englyst et al. (1992) using pancreatin and amyloglucosidase
kidney beans	56.58[a]	Wang et al. (2023); *in vitro* digestion by the method of Xu et al. (2022) using pepsin, pancreatin, and α-amylase
Cooked pea	~18 to ~28[b]	Ajala et al. (2022); *in vitro* digestion by the method of Schwanz Goebel et al. (2019) using α-amylase, pepsin, pancreatin, amyloglucosidase, and invertase
Navy bean	~80[b]	Berg et al. (2012); *in vitro* digestion by the method of Dartois et al. (2010) using pepsin, pancreatin, amyloglucosidase, and invertase
Roasted chickpea	17.35[a]	Simsek et al. (2016); *in vitro* digestion by the method of Englyst et al. (1999) using pepsin, pancreatin, and amyloglucosidase
Tubers		
Potato	93.1[b]	Chen et al. (2022); *in vitro* digestion by the method of Dartois et al. (2010) using pepsin, pancreatin, amyloglucosidase, and invertase
Potato starch	100[b]	Do et al. (2020); *in vitro* digestion by the method of Dartois et al. (2010) using pepsin, pancreatin, amyloglucosidase, and invertase
Cassava	79.7[a]	Van Hung et al. (2017); *in vitro* digestion by the method of Englyst et al. (1992) using pancreatin and amyloglucosidase
Sweet potato	85.8–94.9[a]	Lu et al. (2020); *in vitro* digestion by the method of Englyst et al. (1992) using pancreatin and amyloglucosidase

(Continued)

TABLE 20.3 (*Continued*)   Digestibility of different starches and starch-based foods

SOURCE	DIGESTIBILITY	REFERENCE AND METHOD
Non-conventional Starch		
Green banana	11.0–19.7[a]	Wang et al. (2024); *in vitro* digestion by the method of Englyst et al. (1992) using pancreatin and amyloglucosidase
Loquat seed starch	13.6–17.6[a]	Kong et al. (2023); *in vitro* digestion by the method of Liu et al. (2023) using pancreatin and amyloglucosidase
Mango kernel starch	20.0–24.4[a]	Sandhu and Lim (2008); *in vitro* digestion by the method of Englyst et al. (1999) using pepsin, pancreatin, and amyloglucosidase
Lotus seed starch	~8.5[d]	Guo et al. (2019); *in vitro* digestion by the method of Wu et al. (2017) using α-amylase, amyloglucosidase, pancreatin, and pepsin and a dynamic rat stomach-duodenum model
Boiled breadfruit	70.1[e]	Chinedum et al. (2018); *in vitro* digestion by the method of Goni et al. (1997) using pepsin, α-amylase, and amyloglucosidase from *Aspergillus niger*

[a] Expressed as rapidly and slowly digestible starch (%).
[b] Expressed as % of starch hydrolysis.
[c] Expressed as % of total starch hydrolysed.
[d] Expressed as glucose concentration (mg/mL).
[e] Expressed as g/100 g of total starch.

gastric digestion in the bran layer of brown rice. The presence of fibres increases the water absorption capacity, enlarging the particle size and favouring greater viscosity of the chyme, thus delaying the disintegration of the rice and its dissolution and, consequently, delaying the emptying of solids.

The influence of the cellular integrity of the cotyledon in navy beans and the influence of reheating on the *in vitro* starch digestibility rate were studied. It was found that the cell walls of navy bean cotyledons impose restrictions on the swelling and gelatinisation of bean starch during cooking, and incomplete gelatinisation reduces the rate and extent of starch hydrolysis (Figure 20.1; Berg et al. 2012). Storage and reheating of grains suggest that the extent of gelatinisation and water availability during cooking may be important factors influencing starch hydrolysis during *in vitro* digestion (Berg et al. 2012). The influence of pea microstructure on hydration kinetics, culinary properties, and starch digestibility was investigated by Ajala et al. (2022), who found that the microstructural characteristics of pea seeds can predict their soaking behaviour and cooking characteristics. This knowledge allows for the prediction of the regulation of starch digestibility and its GI. Understanding this information can be an essential starting point for researchers and product developers in the food industry to create foods with a low GI.

However, when studying the effect of starch entrapment on the structure of the parenchyma in potato cells, no influence on starch digestion *in vitro* was found, despite the need for higher gelatinisation temperatures and lower paste viscosity peaks (Do et al. 2020). This can be explained by the partial disintegration of the cell wall during cooking, facilitating α-amylase access to starch, unlike with beans and peas, as exemplified previously, where the structures remain intact.

The complexity of starch digestibility and the glycaemic response of foods are influenced by various factors, including composition, microstructure, viscosity, and particle size, as well as the type of processing, formulation, and preparation method.

Various studies report the specific effects of the composition of these elements on the release of glucose during digestion. For instance, the presence of aleurone layers in cereals can slow the release of glucose, while reducing particle size during chewing can accelerate starch hydrolysis. Furthermore, the influence of structural characteristics of legumes demonstrates the importance of understanding microstructure in predicting starch digestibility (Berg et al. 2012; Xiong et al. 2019). This knowledge is crucial to guide researchers and product developers in creating foods with a low GI. However, it is important to highlight that the response of starch to cellular structure varies between different foods, as evidenced by the lack of influence of starch entrapment in potato cells on *in vitro* digestion. This contrast highlights the complexity of the role of cellular structure in starch digestibility and underscores the continued need for investigation for a more complete and accurate understanding.

20.4.2 Viscosity of the Digesta

The presence of non-starch polysaccharides from various plant sources can impact the physical characteristics of digesta in all gastrointestinal tract regions. Increased viscosity can slow down several physiological processes related to digestion and absorption of nutrients, thus contributing to the control of glucose intolerance and obesity. The use of agents that alter the viscosity of foods, especially pre-prepared ones, is increasingly common in the food industry. Many of these viscosity agents, such as gums, promote the increase of fibre in formulations. The fibre, particularly soluble dietary fibre, acts like a sponge, absorbing water in the intestine and mixing with the food to form a tangled network. As a result, postprandial glycaemia decreases after ingesting starchy foods (Kaur & Singh 2009).

The use of gums to reduce digestibility has been investigated over the years. Dartois et al. (2010) examined the effect of adding guar gum to a starch-rich food and found that it significantly

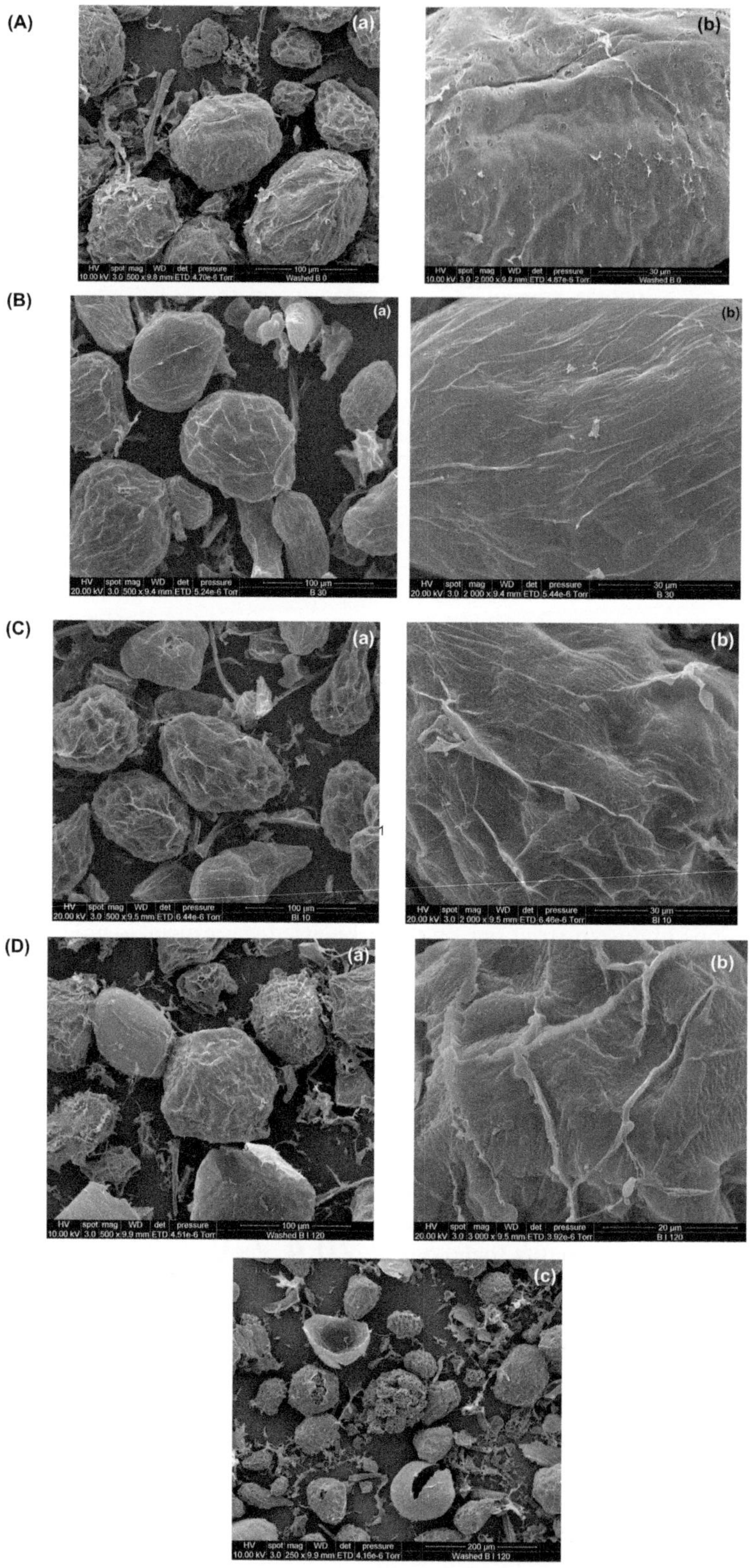

FIGURE 20.1 (A) Scanning electron micrographs collected during *in vitro* gastric and intestinal digestion of starch in cooked navy beans; (a) undigested sample taken at 0 minutes showing cotyledon cells, and (b) magnified view of the cotyledon cell wall. (B) Scanning electron micrographs collected during *in vitro* gastric and intestinal digestion of starch in cooked navy beans: Sample taken after 30 minutes of gastric digestion; (a) cotyledon cells in navy bean digesta and (b) magnified view of the cotyledon cell wall. (C) Scanning electron micrographs collected during *in vitro* gastric and intestinal digestion of starch in cooked navy beans: sample taken after 10 minutes of intestinal digestion; (a) cotyledon cells in navy bean digesta and (b) magnified view of the cotyledon cell wall. (D) Scanning electron micrographs collected during *in vitro* gastric and intestinal digestion of starch in cooked navy beans: sample taken after 120 minutes of intestinal digestion; (a) cotyledon cells in navy bean digesta, (b) magnified view of the cotyledon cell wall, and (c) broken cotyledon cells, picture not representative, see text. (Reproduced with permission from Elsevier from Berg et al. 2012.)

increased the viscosity of the environment. The gum in the food matrix acts to delay and decrease the extent of starch hydrolysis by forming a barrier around the granules, which restricts the transfer of enzymes to the granules and alters the swelling pattern of the starch granules during gelatinisation, resulting in a different distribution of the remaining granule size (Singh et al. 2010). Adding chitosan has also been reported to be effective in reducing starch digestibility (Zheng et al. 2019). Furthermore, Si et al. (2017) reported that complexes formed between corn starch with a high amylose content and chitosan had a synergistic effect in preventing obesity and diabetes in rats fed with a high-fat diet. The complex could control body weight and improve blood lipid composition more efficiently than starch or chitosan alone.

In oats, for example, the presence of β-glucan significantly affects the GI. A study by Tang et al. (2019) demonstrated that removing β-glucan resulted in a significant increase in the estimated GI, from 67.6 to 76.7. Especially in its soluble form, β-glucan can reduce the rate of starch digestion due to increased viscosity. Furthermore, β-glucan can form complexes with adjacent proteins, creating a structure resistant to amylase attack and reducing starch digestibility. When considering starch within the food matrix, the reactions are much broader; therefore, it is impossible to attribute changes to just one constituent.

In addition to the fibre naturally present in the composition of foods, adding certain cereal-based viscous fibre to meals significantly influences the digestion and absorption of carbohydrates (Singh et al. 2010). However, the postprandial glucose response may vary slightly, as these fibres are physically and functionally different from viscous gums. Furthermore, it may be challenging to distinguish between the two effects, whether the decrease in the rate of hydrolysis is due to enzyme inhibition or an increase in the viscosity of the gastrointestinal contents.

20.4.3 Food Composition

20.4.3.1 Protein

The presence of proteins in the food matrix can influence the rate of starch digestion (Singh et al. 2010). Interactions between protein and starch play important roles in the technological, physical, nutritional, and digestibility properties of food products. Proteins can affect starch digestibility in various ways: they may be present on the surface of granules, acting as a physical barrier to enzymes during digestion; they can form gels during processing, increasing the viscosity of the medium and hindering enzyme access. Furthermore, interactions between protein and starch may occur, reducing the assimilation efficiency of amylases and, consequently, the rate of starch digestion.

Even in small concentrations, as is the case with cereals and other food products, proteins alter the functional and digestibility properties of starch (Ezeogu et al. 2008). Protein bodies from albumin, globulins, and glutenins act as barriers, protecting the surface of the granules. Processes such as cooking or processing can reduce starch digestibility due to protein conformational changes, facilitating the formation of disulfide-linked polymers (Oria et al. 1995). The presence of a protein barrier around the starch granule was confirmed by adding alkaline protease to hydrolyse the protein matrix. This showed that removing proteins significantly affects digestibility and, consequently, the estimated GI (Tang et al. 2019).

Ye et al. (2018) investigated the impact of adding proteins and lipids to rice flour on starch digestibility, finding that both components can inhibit starch digestion. Both lipids and proteins slow starch hydrolysis by forming a coating around starch granules, which inhibits their swelling and restricts the access of digestive enzymes to the underlying starch molecules. Remarkably, this protective effect persisted even after cooking rice flour (Ye et al. 2018). The ability of endogenous lipids and proteins to impede starch digestion in rice flour presents an intriguing alternative for mitigating high postprandial blood glucose levels.

Studies examining starch and protein blends have revealed that as the protein ratio increases, the swelling of starch granules during cooking is proportionally restricted, while the retrogradation of amylopectin during cooling is enhanced, leading to a higher content of slowly digestible and resistant starch (Chen et al. 2017; Lu et al. 2016).

20.4.3.2 Lipids

During food processing, starch and lipids combine to form inclusion complexes. This process of formation and stabilisation is characterised by a series of non-covalent interactions, including hydrogen bonds, hydrophobic attractions, and van der Waals forces (Singh et al. 2010). The lipid inclusion complex primarily occurs with amylose. However, Wang et al. (2020) reported that the interaction between amylopectin and lipids can also occur, albeit weaker than that between amylose and lipids. This weaker interaction is attributed to the highly branched structure of amylopectin, which limits the formation of the required single helical glucan conformation.

The elaboration of starch–lipid complexes is of great importance in the food industry, as they positively influence the quality and shelf life of foods. They can improve thermal stability and reduce swelling, retrogradation, and the rate of starch hydrolysis impacting the glycaemic response (Oyeyinka et al. 2021). In addition to the complex naturally formed during food processing, several studies use lipids, mainly SCFAs, to form inclusion complexes with amylose. Most of the methods used are physical, such as high-pressure homogenisation (Guo et al. 2021), ultrasound (Chumsri et al. 2022; Li et al. 2022; Tang et al. 2022), and HMT (Frasson et al. 2024; Yassaroh et al. 2021).

The efficiency of the inclusion complex formation depends on several factors, including the starch source, amylose content, type and amount of lipid or fatty acid used, and process conditions. The literature outlines specific conditions for various starch sources.

Most studies aim to understand the changes in the physical, chemical, and technological characteristics of starch and focus on reducing its digestibility. They suggest that applying and consuming these complexes can help reduce risk factors associated with nutritional problems and health issues, such as cardiovascular diseases and obesity (Raza et al. 2021). According to Guo et al. (2021), the type of lipid used is a critical factor in the efficiency of forming inclusion complexes. They note that the addition of unsaturated fatty acids influences the formation of hydrogen bonds, thereby inducing the formation and dissociation of single and double helices. The authors explain that this helical structure is assembled into ordered aggregates and subsequently aligned into microdomains, ultimately affecting digestibility. Concerning the cis/trans structure, trans unsaturated fatty acids can more easily form the complex with simpler helices, resulting in a more ordered and crystalline structure. In general, the more ordered the domain and the greater the degree of ordering of the complex, the greater the enzymatic resistance it will present, resulting in low glycaemic products.

Chumsri et al. (2022) developed inclusion complexes with rice starch and butyric (C4:0), lauric (C12:0), stearic (C18:0), and linoleic (C18:2) acids. They found that the complexation index was higher when using butyric acid, which has a smaller chain size. The type of fatty acid used and its concentration are important factors to consider in studies aiming to develop products with a low GI. In their study, Chumsri et al. (2022) observed the greatest formation of inclusion complexes with lower digestibility when using 7.5% butyric acid at 20% amplitude for 30 minutes of ultrasonication.

Tang et al. (2022) prepared lotus root starch and myristic acid complexes using an ultrasound-assisted hydrothermal system. This technique improved intermolecular interactions and the order of the crystal structure, leading to increased crystallinity. As a result, the resistant starch content rose significantly from 34.58% to 67.03%, while the proportion of slowly digestible starch decreased from 39.72% to 13.3%. Dietary intervention with 5% and 15% mass fractions of these modified starches yielded notable health benefits in diabetic mice. Specifically, it led to significant reductions in body weight, fasting blood glucose levels, and the regulation of blood lipid levels. Moreover, ingesting the modified starches also increased the concentration of SCFAs, such as acetic and butyric acid, which contribute to restoring intestinal balance.

20.4.3.3 Anti-nutrients and α-Amylase inhibitors

Anti-nutrients are those compounds that have the ability to reduce the absorption and use of nutrients, which can affect human and animal growth. Various food crops, such as legumes – especially beans – and some cereals like rye, wheat, oats, and some fruits contain amylase inhibitors in varying quantities (Singh et al. 2010). An inhibition of amylase enzyme activities can effectively slow down the rate of starch hydrolysis, thus allowing control of postprandial hyperglycaemia. These inhibitors include tannins, phytic acid, saponins, alkaloids, glucosinolates, cyanogenic glycosides, and some oligosaccharides. For example, Li et al. (2018) verified that adding different concentrations (5%, 10%, and 15%) of persimmon tannin to starch proportionally reduced the digestibility rate, decreasing the glycaemic response. The study proved that persimmon tannin can directly interact with starch and strongly inhibit α-amylase and α-glucosidase with IC50 values of 0.35 and 0.24 mg/mL, respectively.

Li et al. (2022) classified the inhibitors against starch digestion into substrate analogues, protein inhibitors based on protein–protein interaction, and fat-soluble components. Substrate analogues are obtained by modifying carbohydrates using physical, chemical, and enzymatic methods to resist digestion and suppress glucose release. Examples include polymers such as rhamnose, arabinose, xylose, mannose, glucose, and galactose extracted from glycyrrhiza residue, which have shown the inhibition of α-glucosidase (Zhang et al. 2020). A significant number of peptides with α-amylase and α-glucosidase inhibitory activities are identified. These peptides can act as competitive inhibitors, which compete with substrates for binding to the active site of carbohydrate digestive enzymes, thus slowing the rate of substrate digestion (Li et al. 2022). Peptides obtained from the most diverse sources are reported, such as camel whey proteins (Baba et al. 2021), germinated soybean (González-Montoya et al. 2018), albumin (Yu et al. 2012), quinoa (Vilcacundo et al. 2017), and rice bran (Uraipong & Zhao 2016). Regarding fat-soluble components, several compounds, such as fatty acids, phytosterols, and fat-soluble vitamins, are reported to be effective in controlling blood sugar (Li et al. 2022).

Studies also aim to reduce postprandial glycaemia through natural means. Some research investigates the effects of teas on the co-digestion of starchy foods such as rice. Aumasa et al. (2023) reported that during *in vitro* digestion, there is a reduction in phytochemicals due to degradation and interaction with α-amylase. Mulberry leaf and Gymnema teas potentially reduce starch digestibility in cooked rice, with phytochemicals playing an important role in their inhibitory effects. Apinanthanuwong et al. (2023) investigated the effect of different types of tea from *Camellia sinensis* (green tea, oolong tea, and black tea) on the stability of tea catechins and their impact on starch digestibility and glycaemic response in cooked rice. The results revealed that green tea exhibited the most potent inhibitory effect on starch digestion, followed by oolong and black teas. The inhibition of starch digestion was attributed to interactions between catechins and digestive enzymes. These findings suggest a potential advantage of herbal teas in reducing postprandial hyperglycaemia and providing antioxidant activities for human health. While the potential benefits of herbal teas in reducing postprandial hyperglycaemia and providing antioxidant activities for human health are clear, both studies suggest quantifying individual phenolic compounds using HPLC techniques to determine which compounds are responsible for the reduction in digestibility.

20.5 EFFECT OF COOKING METHODS AND FOOD PROCESSING ON STARCH DIGESTIBILITY

Cooking methods such as boiling, baking, roasting, and frying and processing techniques like extrusion, milling, and grinding can significantly influence starch digestibility. For example, boiling potatoes gelatinises the starch, which increases its digestibility. However, when gelatinised starch is stored, particularly under refrigeration, its digestibility tends to decrease due to the retrogradation process (Chen et al. 2022; Colussi et al. 2017). The effect of baking on starch digestibility has been extensively studied (Torres et al. 2024). During baking, the breakdown of starch, lipids, and proteins occurs. The proteins surrounding and protecting the starch granules from digestion are also degraded, making the starch more digestible and increasing its GI. Additionally, baking induces interactions between starch and proteins and starch and lipids. While baking significantly increases the content of rapidly digestible starch, it also elevates the content of resistant starch due to these interactions. Moreover, factors such as water availability, oven temperature, and the presence of endogenous macronutrients further affect starch digestibility (Torres et al. 2024).

The controlled roasting process enhances the overall palatability of food by improving flavour, texture, appearance, and colour and also alters starch digestibility. In a study evaluating different roasting methods on chickpeas, Simsek et al. (2016) reported changes in the structural properties of starch and the viscosity of the food matrix. These changes led to significant differences in the starch fractions and the estimated Glycaemic Index (eGI) of the chickpea samples. Specifically, the amounts of rapidly digestible starch (RDS) ranged from 2.80% to 14.43% and decreased with roasting, while the content of resistant starch (RS) increased. Park et al. (2022) reported that steaming followed by roasting could influence the quality and resistant starch levels in brown rice flour with high amylose content by increasing resistant starch and reducing the eGI. The authors suggest that roasting brown rice after steaming the paddy rice to improve flour quality may promote the development of healthier functional foods from rice flour.

Frying starchy foods in oil involves a complex series of interconnected physicochemical reactions and transformations, including starch gelatinisation, water evaporation, protein denaturation, and the formation of a crust (Ding et al. 2018). The frying process reduces the *in vitro* digestibility of starch and protein, as well as the estimated GI levels. This occurs because frying promotes interactions between starch molecules and lipids, leading to the formation of type I and type IIb complexes through fusion and recrystallisation (Wang et al. 2022a, 2022b). According to the same authors, frying can create a honeycomb-like structure that limits continuous contact with digestive enzymes, thereby inhibiting the *in vitro* digestion of instant noodle products. This is due to the formation of starch–lipid complexes, which alter the degree of gelatinisation and transform the protein structure. Tian et al. (2017) studied the microstructure and digestibility of potato strips produced by conventional frying and air-frying. They found that the frying process created a compact crust that partially blocked digestive enzymes from penetrating the starch granules and prevented starch from diffusing out of the disrupted granules. Air-fried potato strips, which contained significantly less oil, had only slightly higher digestibility and sensory attributes compared to conventionally fried strips. Therefore, air frying could be a healthier alternative for preparing fried potato strips while maintaining a similar appeal to customers. Studying the effect of frying on glutinous rice cakes, Ding et al. (2018) found that the process damages the starch structure, impacting digestibility. They observed that different parts of the fried glutinous rice cakes showed small, though not significant, differences. The study reported that the frying conditions used (150°C–190°C for 6 minutes) did not lead to the formation of a RS complex with lipids and amylopectin. Additionally, the fried glutinous rice cakes exhibited a high level of RDS.

Processing techniques like extrusion are common in food production, especially in cereal manufacturing. During extrusion, food macronutrients undergo structural transitions, including starch gelatinisation, molecular breakdown, and retrogradation. These structural changes in starch during extrusion significantly impact its glycaemic response. Small variations in extrusion conditions can alter the structural ordering of starch, leading to differences in digestibility in the final product. Therefore, precise control of process conditions and careful selection of raw materials are crucial for achieving a product with either high glucose release or slow starch digestibility (Huang et al. 2022). Extrusion processing altered the microstructure of broken rice, reduced its relative crystallinity, and transformed the semi-crystalline starch from a typical type A to a mixture of type B and V. Additionally, it significantly decreased the hydrolysis of rice starch (Yang et al. 2020). In a study by Sharma et al. (2015), the extrusion process notably increased the digestibility of corn, pea, and bean starches, presenting a valuable opportunity to incorporate these into easily digestible food products, particularly for weaning foods. In summary, extrusion plays a crucial role in modifying starch properties and food digestibility, offering significant opportunities to optimise the quality and functionality of food products. This process can cater to diverse consumer needs, including those requiring high energy release, such as athletes, and individuals with dietary restrictions, such as diabetes.

Another processing technique that significantly impacts starch digestibility is milling and grinding. Milling or grinding of grains is commonly performed to produce flour, which is used as an ingredient in many food products. Various milling techniques, such as ball milling, hammer milling, and pin milling, utilise mechanical forces, including impact, compression, shear, and friction, to break the grains into smaller fragments or fine particles. During this process, the granular, crystalline, and molecular structures of starch in the grains are also damaged or degraded by these forces (Li et al. 2014). Studies have shown that reducing particle size increases the surface area and promotes damaged starch, allowing greater access to digestive enzymes and consequently improving enzymatic degradation during digestion (Sasaki et al. 2016).

Cooking and processing methods significantly impact starch digestibility and the nutritional composition of food products. Each method brings about distinct changes in the starch structure, affecting factors such as gelatinisation, retrogradation, and resistant starch formation. It is crucial for food scientists and manufacturers to comprehend these effects in order to develop products that meet specific dietary requirements and preferences, catering to those in need of improved digestibility or lower glycaemic responses. Understanding how various processing methods affect starch is imperative for creating healthier, more efficient, and tailored food products to meet diverse consumer needs.

20.6 CONCLUSIONS

The digestion and glycaemic features of starch are influenced by various factors, many of which are interconnected. The efficiency of the enzymes responsible for starch hydrolysis in a lab setting is affected by factors such as starch properties, the enzyme's physical access to the starch, the availability of water for hydrolysis, and the relatively high viscosity-reducing substrate diffusion rates. Additionally, galactomannans like guar gum can lower the post-meal increase in blood glucose by affecting intestinal physiology and directly inhibiting the initial stage of starch breakdown. Other factors, including starch modification, processing methods, and other food components like proteins, lipids, and anti-nutrients, significantly impact starch digestibility. The authors believe further research should explore the interactions between starch and non-starch components in food and their impact on starch digestibility. They also suggest using new plant-based polysaccharides and innovative processing techniques to develop starch-based foods with a low GI.

REFERENCES

Ajala, A., Kaur, L., Lee, S. J., & Singh, J. (2022). Influence of seed microstructure on the hydration kinetics and oral-gastro-small intestinal starch digestion *in vitro* of New Zealand pea varieties. *Food Hydrocolloids*, *129*, 107631. https://doi.org/10.1016/j.foodhyd.2022.107631

Alcázar-Alay, S. C., & Meireles, M. A. A. (2015). Physicochemical properties, modifications and applications of starches from different botanical sources. *Food Science and Technology (Campinas)*, *35*(2), 215–236. https://doi.org/10.1590/1678-457X.6749

Apinanthanuwong, G., Aumasa, T., Ogawa, Y., Singh, J., Panpipat, W., & Donlao, N. (2023). Starch digestibility of cooked rice as influenced by the addition of different tea types (*Camellia sinensis*): An *in vitro* study. *Journal of Functional Foods*, *107*, 105651. https://doi.org/10.1016/j.jff.2023.105651

Aryana, K., Greenway, F., Dhurandhar, N., Tulley, R., Finley, J., Keenan, M., Martin, R., Pelkman, C., Olson, D., & Zheng, J. (2015). A resistant-starch enriched yogurt: Fermentability, sensory characteristics, and a pilot study in children. F1000Research, *4*, 139. https://doi.org/10.12688/f1000research.6451.1

Ashogbon, A. O., Akintayo, E. T., Oladebeye, A. O., Oluwafemi, A. D., Akinsola, A. F., & Imanah, O. E. (2020). Developments in the isolation, composition, and physicochemical properties of legume starches. *Critical Reviews in Food Science and Nutrition*, 61(17), 1–22. https://doi.org/10.1080/10408398.2020.1791048

Ashok, P., Soccol, C. & Thomaz-Soccol, V. T. (2000). Biopotential of immobilized amylases. *Indian Journal of Microbiology*, *40*, 1–14.

Aumasa, T., Ogawa, Y., Singh, J., Panpipat, W., & Donlao, N. (2023). The role of herbal teas in reducing the starch digestibility of cooked rice (*Oryza sativa L.*): An *in vitro* co-digestion study. *NFS Journal*, *33*, 100154. https://doi.org/10.1016/j.nfs.2023.100154

Baba, W. N., Mudgil, P., Kamal, H., Kilari, B. P., Gan, C., & Maqsood, S. (2021). Identification and characterization of novel α-amylase and α-glucosidase inhibitory peptides from camel whey proteins. *Journal of Dairy Science*, *104*(2), 1364–1377. https://doi.org/10.3168/jds.2020-19271

Bajaj, R., Singh, N., Kaur, A., & Inouchi, N. (2018). Structural, morphological, functional and digestibility properties of starches from cereals, tubers and legumes: A comparative study. *Journal of Food Science and Technology*, 55(9), 3799–3808. https://doi.org/10.1007/s13197-018-3342-4

Baldwin, A. J., Egan, D. L., Warren, F. J., Barker, P. D., Dobson, C. M., Butterworth, P. J., & Ellis, P. R. (2015). Investigating the mechanisms of amylolysis of starch granules by solution-state NMR. *Biomacromolecules*, *16*(5), 1614–1621. https://doi.org/10.1021/acs.biomac.5b00190

Barnett, C., Smith, A., Scanlon, B., & Israilides, C. J. (1999). Pullulan production by *Aureobasidium* pullulans growing on hydrolysed potato starch waste. *Carbohydrate Polymers*, *38*(3), 203–209. https://doi.org/10.1016/S0144-8617(98)00092-7

Berg, T., Singh, J., Hardacre, A., & Boland, M. J. (2012). The role of cotyledon cell structure during *in vitro* digestion of starch in navy beans. *Carbohydrate Polymers*, *87*(2), 1678–1688. https://doi.org/10.1016/j.carbpol.2011.09.075

Bertoft, E. (2017). Understanding starch structure: Recent progress. *Agronomy*, *7*(3), 56. https://doi.org/10.3390/agronomy7030056

Bonto, A. P., Tiozon, R. N., Sreenivasulu, N., & Camacho, D. H. (2021). Impact of ultrasonic treatment on rice starch and grain functional properties: A review. *Ultrasonics Sonochemistry*, *71*, 105383. https://doi.org/10.1016/j.ultsonch.2020.105383

Bull, S. E., Seung, D., Chanez, C., Mehta, D., Kuon, J.-E., Truernit, E., Hochmuth, A., Zurkirchen, I., Zeeman, S. C., Gruissem, W., & Vanderschuren, H. (2018). Accelerated ex situ breeding of *GBSS*- and *PTST1*-edited cassava for modified starch. *Science Advances*, 4(9). https://doi.org/10.1126/sciadv.aat6086

Cai, C., Zhao, L., Huang, J., Chen, Y., & Wei, C. (2014). Morphology, structure and gelatinization properties of heterogeneous starch granules from high-amylose maize. *Carbohydrate Polymers*, 102, 606–614. https://doi.org/10.1016/j.carbpol.2013.12.010

Cai, J., Man, J., Huang, J., Liu, Q., Wei, W., & Wei, C. (2015). Relationship between structure and functional properties of normal rice starches with different amylose contents. *Carbohydrate Polymers*, 125, 35–44. https://doi.org/10.1016/j.carbpol.2015.02.067

Capriles, V. D., Coelho, K. D., Guerra-Matias, A. C., & Arêas, A. G. (2008). Effects of processing methods on amaranth starch digestibility and predicted glycemic index. *Journal of Food Science*, *73*(7), H160–H164. https://doi.org/10.1111/j.1750-3841.2008.00869.x

Chen, X., He, X., Zhang, B., Fu, X., Jane, J., & Huang, Q. (2017). Effects of adding corn oil and soy protein to corn starch on the physicochemical and digestive properties of the starch. *International Journal of Biological Macromolecules*, *104*, 481–486. https://doi.org/10.1016/j.ijbiomac.2017.06.024

Chen, Y., Singh, J., Midgley, J., & Archer, R. (2022). Sous vide processed potatoes: Starch retrogradation in tuber and oral-gastro-small intestinal starch digestion *in vitro*. *Food Hydrocolloids*, *124*, 107163. https://doi.org/10.1016/j.foodhyd.2021.107163

Chi, C., Li, X., Huang, S., Chen, L., Zhang, Y., Li, L., & Miao, S. (2021). Basic principles in starch multi-scale structuration to mitigate digestibility: A review. *Trends in Food Science & Technology*, 109, 154–168. https://doi.org/10.1016/j.tifs.2021.01.024

Chia, T., Chirico, M., King, R., Ramirez-Gonzalez, R., Saccomanno, B., Seung, D., Simmonds, J., Trick, M., Uauy, C., Verhoeven, T., & Trafford, K. (2020). A carbohydrate-binding protein,

B-GRANULE CONTENT 1, influences starch granule size distribution in a dose-dependent manner in polyploid wheat. *Journal of Experimental Botany*, *71*(1), 105–115. https://doi.org/10.1093/jxb/erz405

Chinedum, E., Sanni, S., Theressa, N., & Ebere, A. (2018). Effect of domestic cooking on the starch digestibility, predicted glycemic indices, polyphenol contents and alpha amylase inhibitory properties of beans (*Phaseolis vulgaris*) and breadfruit (*Treculia africana*). *International Journal of Biological Macromolecules*, *106*, 200–206. https://doi.org/10.1016/j.ijbiomac.2017.08.005

Chumsri, P., Panpipat, W., Cheong, L., & Chaijan, M. (2022). Formation of intermediate amylose rice starch–lipid complex assisted by ultrasonication. *Foods*, *11*(16), 2430. https://doi.org/10.3390/foods11162430

Chung, J., Liu, Q., Donner, E., Hoover, R., Warkentin, T. D., & Vandenberg, B. (2008). Composition, molecular structure, properties, and *in vitro* digestibility of starches from newly released Canadian pulse cultivars. *Cereal Chemistry*, *85*(4), 471–479. https://doi.org/10.1094/CCHEM-85-4-0471

Colussi, R., Singh, J., Kaur, L., Zavareze, E. D. R., Dias, A. R. G., Stewart, R. B., & Singh, H. (2017). Microstructural characteristics and gastro-small intestinal digestion *in vitro* of potato starch: Effects of refrigerated storage and reheating in microwave. *Food Chemistry*, 226, 171–178. https://doi.org/10.1016/j.foodchem.2017.01.048

Compart, J., Singh, A., Fettke, J., & Apriyanto, A. (2023). Customizing starch properties: A review of starch modifications and their applications. *Polymers*, *15*(16), 3491. https://doi.org/10.3390/polym15163491

Corgneau, M., Gaiani, C., Petit, J., Nikolova, Y., Banon, S., Ritié-Pertusa, L., Le, D. T. L., & Scher, J. (2019). Digestibility of common native starches with reference to starch granule size, shape and surface features towards guidelines for starch-containing food products. *International Journal of Food Science & Technology*, *54*(6), 2132–2140. https://doi.org/10.1111/ijfs.14120

Dartois, A., Singh, J., & Kaur, L. (2010). Influence of guar gum on the *in vitro* starch digestibility—Rheological and microstructural characteristics. *Food Biophysics*, *5*, 149–160. https://doi.org/10.1007/s11483-010-9155-2

Dhital, S., Brennan, C., & Gidley, M. J. (2019). Location and interactions of starches *in planta*: Effects on food and nutritional functionality. *Trends in Food Science & Technology*, *93*, 158–166. https://doi.org/10.1016/j.tifs.2019.09.011

Dhital, S., Shrestha, A. K., & Gidley, M. J. (2010). Relationship between granule size and *in vitro* digestibility of maize and potato starches. *Carbohydrate Polymers*, *82*(2), 480–488. https://doi.org/10.1016/j.carbpol.2010.05.018

Ding, Y., Yang, L., Xia, Y., Wu, Y., Zhou, Y., & Wang, H. (2018). Effects of frying on starch structure and digestibility of glutinous rice cakes. *Journal of Cereal Science*, *83*, 196–203. https://doi.org/10.1016/j.jcs.2018.08.014

Djordjevic, S., Nikolic, L., Kovacevic, S., Miljkovic, M., & Djordjevic, D. (2013). Graft copolymerization of acrylic acid onto hydrolyzed potato starch using various initiators. *Periodica Polytechnica Chemical Engineering*, *57*(1–2), 55. https://doi.org/10.3311/PPch.2171

Do, D. T., Singh, J., Oey, I., & Singh, H. (2020). Isolated potato parenchyma cells: Physico-chemical characteristics and gastro-small intestinal digestion *in vitro*. *Food Hydrocolloids*, *108*, 105972. https://doi.org/10.1016/j.foodhyd.2020.105972

Dong, S., Fang, G., Luo, Z., & Gao, Q. (2021). Effect of granule size on the structure and digestibility of jackfruit seed starch. *Food Hydrocolloids*, *120*, 106964. https://doi.org/10.1016/j.foodhyd.2021.106964

Dongowski, G., Jacobasch, G., & Schmiedl, D. (2005). Structural stability and prebiotic properties of resistant starch type 3 increase bile acid turnover and lower secondary bile acid formation. *Journal of Agricultural and Food Chemistry*, *53*(23), 9257–9267. https://doi.org/10.1021/jf0507792

Dura, A., Błaszczak, W., & Rosell, C. M. (2014). Functionality of porous starch obtained by amylase or amyloglucosidase treatments. *Carbohydrate Polymers*, *101*, 837–845. https://doi.org/10.1016/j.carbpol.2013.10.013

Emami, S., Meda, V., Pickard, M. D., & Tyler, R. T. (2010). Impact of micronization on rapidly digestible, slowly digestible, and resistant starch concentrations in normal, high-amylose, and waxy barley. *Journal of Agricultural and Food Chemistry*, *58*(17), 9793–9799. https://doi.org/10.1021/jf101702e

Englyst, H. N., Kingman, S. M., & Cummings, J. H. (1992). Classification and measurement of nutritionally important starch fractions. *European Journal of Clinical Nutrition*, *46*, S33–S50.

Englyst, K. N., Englyst, H. N., Hudson, G. J., Cole, T. J., & Cummings, J. H. (1999). Rapidly available glucose in foods: An in vitro measurement that reflects the glycemic response. *The American Journal of Clinical Nutrition*, *69*, 448–454.

Ezeogu, L. I., Duodu, K. G., Emmambux, M. N., & Taylor, J. R. N. (2008). Influence of cooking conditions on the protein matrix of sorghum and maize endosperm flours. *Cereal Chemistry*, *85*(3), 397–402. https://doi.org/10.1094/CCHEM-85-3-0397

Frasson, S. F., Colussi, R., Hackbart, H. C. D. S., Borges, C. D., Flores, W. H., & Mendonça, C. R. B. (2024). Rice starch modification by thermal treatments with avocado oil: Autoclave versus microwave methods. *International Journal of Biological Macromolecules*, *267*, 131426. https://doi.org/10.1016/j.ijbiomac.2024.131426

Giuberti, G., Gallo, A., Cerioli, C., Fortunati, P., & Masoero, F. (2015). Cooking quality and starch digestibility of gluten-free pasta using new bean flour. *Food Chemistry*, *175*, 43–49. https://doi.org/10.1016/j.foodchem.2014.11.127

Gong, B., Cheng, L., Gilbert, R. G., & Li, C. (2019). Distribution of short to medium amylose chains are major controllers of *in vitro* digestion of retrograded rice starch. *Food Hydrocolloids*, *96*, 634–643. https://doi.org/10.1016/j.foodhyd.2019.06.003

Goni, I., Garcia-Alonso, A., & Saura-Calixto, F. (1997). A starch hydrolysis procedure to estimate glycemic index. *Nutrition Research*, *17*(3), 427–437. https://doi.org/10.1016/S0271-5317(97)00010-9

González-Montoya, M., Hernández-Ledesma, B., Mora-Escobedo, R., & Martínez-Villaluenga, C. (2018). Bioactive peptides from germinated soybean with anti-diabetic potential by inhibition of dipeptidyl peptidase-IV, α-amylase, and α-glucosidase enzymes. *International Journal of Molecular Sciences*, *19*(10), 2883. https://doi.org/10.3390/ijms19102883

Gunaratne, A. (2018). Heat-moisture treatment of starch. In Z. Sui & X. Kong (Eds.), *Physical Modifications of Starch* (pp. 15–36). Springer.

Guo, T., Hou, H., Liu, Y., Chen, L., & Zheng, B. (2021). In vitro digestibility and structural control of rice starch-unsaturated fatty acid complexes by high-pressure homogenization. *Carbohydrate Polymers*, *256*, 117607. https://doi.org/10.1016/j.carbpol.2020.117607

Guo, Z., Zhao, B., Chen, L., & Zheng, B. (2019). Physicochemical properties and digestion of lotus seed starch under high-pressure homogenization. *Nutrients*, *11*(2), 371. https://doi.org/10.3390/nu11020371

Han, J.-A., & BeMiller, J. N. (2007). Preparation and physical characteristics of slowly digesting modified food starches. *Carbohydrate Polymers, 67*, 366–374.

Hebelstrup, K. H., Sagnelli, D., & Blennow, A. (2015). The future of starch bioengineering: GM microorganisms or GM plants? *Frontiers in Plant Science*, 6. https://doi.org/10.3389/fpls.2015.00247

Helle, S., Bray, F., Putaux, J.-L., Verbeke, J., Flament, S., Rolando, C., D'Hulst, C., & Szydlowski, N. (2019). Intra-sample heterogeneity of potato starch reveals fluctuation of starch-binding proteins according to granule morphology. *Plants, 8*(9), 324. https://doi.org/10.3390/plants8090324

Hemmingsen, B., Gimenez-Perez, G., Mauricio, D., Roqué i Figuls, M., Metzendorf, M.-I., & Richter, B. (2017). Diet, physical activity or both for prevention or delay of type 2 diabetes mellitus and its associated complications in people at increased risk of developing type 2 diabetes mellitus. *Cochrane Database of Systematic Reviews, 2017*(12). https://doi.org/10.1002/14651858.CD003054.pub4

Higgins, J. A. (2014). Resistant Starch and Energy Balance: Impact on Weight Loss and Maintenance. *Critical Reviews in Food Science and Nutrition, 54*(9), 1158–1166. https://doi.org/10.1080/10408398.2011.629352

Hoover, R., & Zhou, Y. (2003). *In vitro* and *in vivo* hydrolysis of legume starches by α-amylase and resistant starch formation in legumes—A review. *Carbohydrate Polymers, 54*(4), 401–417. https://doi.org/10.1016/s0144-8617(03)00180-2

Hu, P., Zhao, H., Duan, Z., Linlin, Z., & Wu, D. (2004). Starch digestibility and the estimated glycemic score of different types of rice differing in amylose contents. *Journal of Cereal Science, 40*(3), 231–237.

Huang, X., Liu, H., Ma, Y., Mai, S., & Li, C. (2022). Effects of extrusion on starch molecular degradation, order–disorder structural transition and digestibility—A review. *Foods, 11*(16). https://doi.org/10.3390/foods11162538

Jenkins, D. J., Thorne, M. J., Wolever, T. M., Jenkins, A. L., Rao, A. V., & Thompson, L. U. (1987). The effect of starch-protein interaction in wheat on the glycaemic response and rate of in vitro digestion. *The American Journal of Clinical Nutrition, 45*, 946–951.

Jiali, L., Wu, Z., Liu, L., Yang, J., Wang, L., Li, Z., & Liu, L. (2023). The research advance of resistant starch: Structural characteristics, modification method, immunomodulatory function, and its delivery systems application. *Critical Reviews in Food Science and Nutrition, 64*, 1–18. https://doi.org/10.1080/10408398.2023.2230287

Kaur, B., Ariffin, F., Bhat, R., & Karim, A. A. (2012). Progress in starch modification in the last decade. *Food Hydrocolloids, 26*(2), 398–404. https://doi.org/10.1016/j.foodhyd.2011.02.016

Kaur, L., & Singh, J. (2009). The role of galactomannan seed gums in diet and health–A review. In J. N. Govil & V. K. Singh (Eds.), *Recent Progress in Medicinal Plants. Standardization of Herbal/Ayurvedic Formulations* (Vol. 24, pp. 429–467). Studium Press, Houston, TX.

Kaur, L., Singh, J., McCarthy, O. J., & Singh, H. (2007). Physico-chemical, rheological and structural properties of fractionated potato starches. *Journal of Food Engineering, 82*(3), 383–394. https://doi.org/10.1016/j.jfoodeng.2007.02.059

Kim, E. H. J., Petrie, J. R., Motoi, L., Morgenstern, M. P., Sutton, K. H., Mishra, S., & Simmons, L. D. (2008). Effect of structural and physicochemical characteristics of the protein matrix in pasta on *in vitro* starch digestibility. *Food Biophysics, 3*, 229–234. https://doi.org/10.1007/s11483-008-9066-7

Kong, F., Oztop, M. H., Singh, R. P., & McCarthy, M. J. (2011). Physical changes in white and brown rice during simulated gastric digestion. *Journal of Food Science, 76*(6), E450–E457. https://doi.org/10.1111/j.1750-3841.2011.02271.x

Kong, X., Yang, W., Zuo, Y., Dawood, M., & He, Z. (2023). Characteristics of physicochemical properties, structure and in vitro digestibility of seed starches from five loquat cultivars. *International Journal of Biological Macromolecules, 253*, 126675. https://doi.org/10.1016/j.ijbiomac.2023.126675

Koo, S. H., Lee, K. Y., & Lee, H. G. (2010). Effect of cross-linking on the physicochemical and physiological properties of corn starch. *Food Hydrocolloids, 24*(6–7), 619–625. https://doi.org/10.1016/j.foodhyd.2010.02.009

Lal, M. K., Singh, B., Sharma, S., Singh, M. P., & Kumar, A. (2021). Glycemic index of starchy crops and factors affecting its digestibility: A review. *Trends in Food Science & Technology, 111*, 741–755. https://doi.org/10.1016/j.tifs.2021.02.067

Langworthy, C. F., & Deuel, H. J. (1922). Digestibility of raw rice, arrowroot, canna, cassava, taro, tree-fern, and potato starches. *Journal of Biological Chemistry, 52*(1), 251–261. https://doi.org/10.1016/s0021-9258(18)85868-9

Lehmann, U., & Robin, F. (2007). Slowly digestible starch – its structure and health implications: A review. *Trends in Food Science & Technology, 18*(7), 346–355. https://doi.org/10.1016/j.tifs.2007.02.009

Li, C., & Hu, Y. (2023). New definition of resistant starch types from the gut microbiota perspectives – A review. *Critical Reviews in Food Science and Nutrition, 63*(23), 6412–6422. https://doi.org/10.1080/10408398.2022.2031101

Li, E., Dhital, S., & Hasjim, J. (2014). Effects of grain milling on starch structures and flour/starch properties. *Starch-Stärke, 66*(1–2), 15–27. https://doi.org/10.1002/star.201200224

Li, K., Yao, F., Du, J., Deng, X., & Li, C. (2018). Persimmon tannin decreased the glycemic response through decreasing the digestibility of starch and inhibiting α-amylase, α-glucosidase, and intestinal glucose uptake. *Journal of Agriculture and Food Chemistry, 66*(7),1629–1637. https://doi.org/10.1021/acs.jafc.7b05833

Li, X., Bai, Y., Jin, Z., & Svensson, B. (2022). Food-derived non-phenolic α-amylase and α-glucosidase inhibitors for controlling starch digestion rate and guiding diabetes-friendly recipes. *LWT, 153*, 112455. https://doi.org/10.1016/j.lwt.2021.112455

Lin, L., Zhang, L., Cai, X., Liu, Q., Zhang, C., & Wei, C. (2018). The relationship between enzyme hydrolysis and the components of rice starches with the same genetic background and amylopectin structure but different amylose contents. *Food Hydrocolloids, 84*, 406–413. https://doi.org/10.1016/j.foodhyd.2018.06.029

Lindeboom, N., Chang, P. R., & Tyler, R. T. (2004). Analytical, biochemical and physicochemical aspects of starch granule size, with emphasis on small granule starches: A review. *Starch-Stärke, 56*(34), 89–99. https://doi.org/10.1002/star.200300218

Liu, K., Zu, Y., Chi, C., Gu, B., Chen, L., & Li, X. (2018). Modulation of the digestibility and multi-scale structure of cassava starch by controlling the cassava growth period. *International Journal of Biological Macromolecules, 120*, 346–353. https://doi.org/10.1016/j.ijbiomac.2018.07.184

Liu, Q., Wang, Y., Yang, Y., Yu, X., Xu, L., Jiao, A., & Jin, Z. (2023). Structure, physicochemical properties and *in vitro* digestibility of extruded starch-lauric acid complexes with different amylose contents. *Food Hydrocolloids, 136*, 108239. https://doi.org/10.1016/j.foodhyd.2022.108239

Liu, Y., Chen, J., Luo, S., Li, C., Ye, J., Liu, C., & Gilbert, R. G. (2017). Physicochemical and structural properties of pregelatinized starch prepared by improved extrusion cooking technology. *Carbohydrate Polymers*, *175*, 265–272. https://doi.org/10.1016/j.carbpol.2017.07.084

Lu, P., Li, X., Janaswamy, S., Chi, C., Chen, L., Wu, Y., & Liang, Y. (2020). Insights on the structure and digestibility of sweet potato starch: Effect of postharvest storage of sweet potato roots. *International Journal of Biological Macromolecules*, *145*, 694–700. https://doi.org/10.1016/j.ijbiomac.2019.12.151

Lu, Z., Donner, E., Yada, R. Y., & Liu, Q. (2016). Physicochemical properties and *in vitro* starch digestibility of potato starch/protein blends. *Carbohydrate Polymers*, *154*, 214–222. https://doi.org/10.1016/j.carbpol.2016.08.055

Ma, M., Wang, Y., Wang, M., Jane, J., & Du, S. (2017). Physicochemical properties and *in vitro* digestibility of legume starches. *Food Hydrocolloids*, *63*, 249–255. https://doi.org/10.1016/j.foodhyd.2016.09.004

Malinova, I., Alseekh, S., Feil, R., Fernie, A. R., Baumann, O., Schöttler, M. A., Lunn, J. E., & Fettke, J. (2017). Starch synthase 4 and plastidal phosphorylase differentially affect starch granule number and morphology. *Plant Physiology*, *174*(1), 73–85. https://doi.org/10.1104/pp.16.01859

Maniglia, B. C., Castanha, N., Le-Bail, P., Le-Bail, A., & Augusto, P. E. D. (2021). Starch modification through environmentally friendly alternatives: A review. *Critical Reviews in Food Science and Nutrition*, *61*(15), 2482–2505. https://doi.org/10.1080/10408398.2020.1778633

Martens, B. M. J., Gerrits, W. J. J., Bruininx, E. M. A. M., & Schols, H. A. (2018). Amylopectin structure and crystallinity explains variation in digestion kinetics of starches across botanic sources in an *in vitro* pig model. *Journal of Animal Science and Biotechnology*, *9*(1). https://doi.org/10.1186/s40104-018-0303-8

Miao, H., Sun, P., Liu, Q., Jia, C., Liu, J., Hu, W., Jin, Z., & Xu, B. (2017). Soluble starch synthase III-1 in amylopectin metabolism of banana fruit: Characterization, expression, enzyme activity, and functional analyses. *Frontiers in Plant Science*, 8. https://doi.org/10.3389/fpls.2017.00454

Miranda da Silveira, M., Lambrecht Dittgen, C., De Souza Batista, C., Biduski, B., Carlos Gutkoski, L., & Levien Vanier, N. (2020). Discrimination of the quality of Brazilian wheat genotypes and their use as whole-grains in human nutrition. *Food Chemistry*, *312*, 126074. https://doi.org/10.1016/j.foodchem.2019.126074

Nie, M., Piao, C., Wang, A., Xi, H., Chen, Z., He, Y., Wang, L., Liu, L., Huang, Y., Wang, F., & Tong, L.-T. (2023). Physicochemical properties and *in vitro* digestibility of highland barley starch with different extraction methods. *Carbohydrate Polymers*, *303*, 120458–120458. https://doi.org/10.1016/j.carbpol.2022.120458

Noor, N., Gani, A., Jhan, F., Jenno, J. L. H., & Arif Dar, M. (2021). Resistant starch type 2 from lotus stem: Ultrasonic effect on physical and nutraceutical properties. *Ultrasonics Sonochemistry*, *76*, 105655. https://doi.org/10.1016/j.ultsonch.2021.105655

Oates, C. G. (1997). Towards an understanding of starch granule structure and hydrolysis. *Trends in Food Science & Technology*, *8*(11), 375–382. https://doi.org/10.1016/s0924-2244(97)01090-x

Oria, M., Hamaker, B., & Schull, J. (1995). *In vitro* protein digestibility of developing and mature sorghum grain in relation to α-, β-, and γ-kafirin disulfide crosslinking. *Journal of Cereal Science*, *22*(1), 85–93. https://doi.org/10.1016/S0733-5210(05)80010-4

Ouyang, J., Wang, C., Huang, Q., Guan, Y., Zhu, Z., He, Y., Jiang, G., Xiong, Y., & Li, X. (2024). Correlation between *in vitro* starch digestibility and starch structure/physicochemical properties in rice. *International Journal of Biological Macromolecules*, *263*, 130316. https://doi.org/10.1016/j.ijbiomac.2024.130316

Ouyang, Q., Wang, X., Xiao, Y., Luo, F., Lin, Q., & Ding, Y. (2021). Structural changes of A-, B- and C-type starches of corn, potato and pea as influenced by sonication temperature and their relationships with digestibility. *Food Chemistry*, *358*, 129858. https://doi.org/10.1016/j.foodchem.2021.129858

Ovando-Martínez, M., Osorio-Díaz, P., Whitney, K., Bello-Pérez, L. A., & Simsek, S. (2011). Effect of the cooking on physicochemical and starch digestibility properties of two varieties of common bean (*Phaseolus vulgaris L.*) grown under different water regimes. *Food Chemistry*, *129*(2), 358–365. https://doi.org/10.1016/j.foodchem.2011.04.084

Oyeyinka, S. A., Singh, S., & Amonsou, E. O. (2021). A review on structural, digestibility and physicochemical properties of legume starch-lipid complexes. *Food Chemistry*, *349*, 129165.

Pandiselvam, R., Manikantan, M. R., Divya, V., Ashokkumar, C., Kaavya, R., Kothakota, A., & Ramesh, S. V. (2019). Ozone: An advanced oxidation technology for starch modification. *Ozone: Science & Engineering*, *41*(6), 491–507. https://doi.org/10.1080/01919512.2019.1577128

Park, J., Oh, S., Chung, H., Shin, D. S., Choi, I., & Park, H. (2022). Effect of steaming and roasting on the quality and resistant starch of brown rice flour with high amylose content. *LWT*, 167, 113801. https://doi.org/10.1016/j.lwt.2022.113801

Patterson, M. A., Maiya, M., & Stewart, M. L. (2020). Resistant starch content in foods commonly consumed in the United States: A narrative review. *Journal of the Academy of Nutrition and Dietetics*, *120*(2), 230–244. https://doi.org/10.1016/j.jand.2019.10.019

Peng, M., Yin, L., Dong, J., Shen, R., & Zhu, Y. (2022). Physicochemical characteristics and in vitro digestibility of starches from colored quinoa (*Chenopodium quinoa*) varieties. *Journal of Food Science*, *87*(5), 2147–2158. https://doi.org/10.1111/1750-3841.16126

Pratiwi, M., Faridah, D. N., & Lioe, H. N. (2018). Structural changes to starch after acid hydrolysis, debranching, autoclaving-cooling cycles, and heat moisture treatment (HMT): A review. *Starch-Stärke*, 70(1–2). https://doi.org/10.1002/star.201700028

Putseys, J. A., Derde, L. J., Lamberts, L., Östman, E., Björck, I. M., & Delcour, J. A. (2010). Functionality of short-chain amylose–lipid complexes in starch–water systems and their impact on in vitro starch degradation. *Journal of Agricultural and Food Chemistry*, *58*(3), 1939–1945. https://doi.org/10.1021/jf903523h

Raigond, P., Ezekiel, R., & Raigond, B. (2015). Resistant starch in food: A review. *Journal of the Science of Food and Agriculture*, *95*(10), 1968–1978. https://doi.org/10.1002/jsfa.6966

Raza, H., Ameer, K., Ren, X., Liang, Q., Chen, X., Chen, H., & Ma, H. (2021). Physicochemical properties and digestion mechanism of starch-linoleic acid complex induced by multi-frequency power ultrasound. *Food Chemistry*, *364*, 130392. https://doi.org/10.1016/j.foodchem.2021.130392

Ren, W., Zhang, H., Chen, Z., & Zhong, Q. (2020). Structural basis for the low digestibility of starches recrystallized from side chains of amylopectin modified by amylosucrase to different chain lengths. *Carbohydrate Polymers*, *241*, 116352–116352. https://doi.org/10.1016/j.carbpol.2020.116352

Robin, F., Heindel, C., Pineau, N., Srichuwong, S., & Lehmann, U. (2016). Effect of maize type and extrusion-cooking conditions on starch digestibility profiles. *International Journal of Food Science & Technology*, *51*(6), 1319–1326. https://doi.org/10.1111/ijfs.13098

Samodien, E., Jewell, J. F., Loedolff, B., Oberlander, K., George, G. M., Zeeman, S. C., Damberger, F. F., van der Vyver, C., Kossmann, J., & Lloyd, J. R. (2018). Repression of *Sex4* and *Like Sex Four2* orthologs in potato increases tuber starch bound phosphate with concomitant alterations in starch physical properties. *Frontiers in Plant Science*, 9. https://doi.org/10.3389/fpls.2018.01044

Sandhu, K. S., & Lim, S. T. (2008). Structural characteristics and *in vitro* digestibility of Mango kernel starches (*Mangifera indica* L.). *Food Chemistry*, *107*, 92–97.

Sang, Y., Bean, S., Seib, P. A., Pedersen, J., & Shi, Y.-C. (2008). Structure and functional properties of sorghum starches differing in amylose content. *Journal of Agricultural and Food Chemistry*, *56*(15), 6680–6685. https://doi.org/10.1021/jf800577x

Sarteshnizi, R. A., Hosseini, H., Khosroshahi, N. K., Shahraz, F., Mousavi Khaneghah, A., Kamran, M., Komeili, R., & Chiavaro, E. (2017). Effect of resistant starch and β-glucan combination on oxidative stability, frying performance, microbial count and shelf life of prebiotic sausage during refrigerated storage. *Food Technology and Biotechnology*, *55*(4). https://doi.org/10.17113/ftb.55.04.17.5479

Sasaki, T., Okunishi, T., Sotome, I., & Okadome, H. (2016). Effects of milling and cooking conditions of rice on *in vitro* starch digestibility and blood glucose response. *Cereal Chemistry*, *93*(3), 242–247. https://doi.org/10.1094/CCHEM-08-15-0155-R

Schwanz Goebel, J. T., Kaur, L., Colussi, R., Elias, M. C., & Singh, J. (2019). Microstructure of indica and japonica rice influences their starch digestibility: A study using a human digestion simulator. *Food Hydrocolloids*, *94*, 191–198. https://doi.org/10.1016/j.foodhyd.2019.02.038

Sharma, S., Singh, N., & Singh, B. (2015). Effect of extrusion on morphology, structural, functional properties and *in vitro* digestibility of corn, field pea and kidney bean starches. *Starch*-Stärke, *67*(9–10), 721–728. https://doi.org/10.1002/star.201500021

Si, X., Strappe, P., Blanchard, C., & Zhou, Z. (2017). Enhanced anti-obesity effects of complex of resistant starch and chitosan in high fat diet fed rats. *Carbohydrate Polymers*, *157*, 834–841. https://doi.org/10.1016/j.carbpol.2016.10.042

Simonato, B., Curioni, A., & Pasini, G. (2015). Digestibility of pasta made with three wheat types: A preliminary study. *Food Chemistry*, *174*, 219–225. https://doi.org/10.1016/j.foodchem.2014.11.023

Simsek, S., Herken, E. N., & Ovando-Martinez, M. (2016). Chemical composition, nutritional value and *in vitro* starch digestibility of roasted chickpeas. *Journal of the Science of Food and Agriculture*, *96*(8), 2896–2905. https://doi.org/10.1002/jsfa.7461

Singh, J., Dartois, A., & Kaur, L. (2010). Starch digestibility in food matrix: A review. *Trends in Food Science & Technology*, *21*(4), 168–180. https://doi.org/10.1016/j.tifs.2009.12.001

Singh, J., Kaur, L., & McCarthy, O. J. (2007). Factors influencing the physico-chemical, morphological, thermal and rheological properties of some chemically modified starches for food applications—A review. *Food Hydrocolloids*, *21*(1), 1–22. https://doi.org/10.1016/j.foodhyd.2006.02.006

Singh, J., McCarthy, O. J., & Singh, H. (2006). Physico-chemical and morphological characteristics of New Zealand Taewa (Maori potato) starches. *Carbohydrate Polymers*, *64*(4), 569–581. https://doi.org/10.1016/j.carbpol.2005.11.013

Spychaj, T., Wilpiszewska, K., & Zdanowicz, M. (2013). Medium and high substituted carboxymethyl starch: Synthesis, characterization and application. *Starch-Stärke*, *65*(1–2), 22–33. https://doi.org/10.1002/star.201200159

Sun, Y., Lv, Y., Qi, S., Zhang, Y., & Wang, Z. (2022). Sensitive colorimetric aptasensor based on stimuli-responsive metal-organic framework nano-container and trivalent DNAzyme for zearalenone determination in food samples. *Food Chemistry*, *371*, 131145. https://doi.org/10.1016/j.foodchem.2021.131145

Takahashi, K., Fujita, H., Fujita, N., Takahashi, Y., Kato, S., Shimizu, T., Suganuma, Y., Sato, T., Waki, H., & Yamada, Y. (2022). A pilot study to assess glucose, insulin, and incretin responses following novel high resistant starch rice ingestion in healthy men. *Diabetes Therapy*, *13*, 1383–1393.

Tamura, M., Singh, J., Kaur, L., & Ogawa, Y. (2016). Impact of structural characteristics on starch digestibility of cooked rice. *Food Chemistry*, *191*, 91–97. https://doi.org/10.1016/j.foodchem.2015.04.019

Tang, J., Liang, Q., Ren, X., Raza, H., & Ma, H. (2022). Insights into ultrasound-induced starch-lipid complexes to understand physicochemical and nutritional interventions. *International Journal of Biological Macromolecules*, *222*, 950–960. https://doi.org/10.1016/j.ijbiomac.2022.09.242

Tang, M., Wang, L., Cheng, X., Wu, Y., & Ouyang, J. (2019). Non-starch constituents influence the *in vitro* digestibility of naked oat (*Avena nuda L.*) starch. *Food Chemistry*, *297*, 124953. https://doi.org/10.1016/j.foodchem.2019.124953

Tester, R. F., Qi, X., & Karkalas, J. (2006). Hydrolysis of native starches with amylases. *Animal Feed Science and Technology*, *130*(1–2), 39–54. https://doi.org/10.1016/j.anifeedsci.2006.01.016

Thirumdas, R., Trimukhe, A., Deshmukh, R. R., & Annapure, U. S. (2017). Functional and rheological properties of cold plasma treated rice starch. *Carbohydrate Polymers*, *157*, 1723–1731. https://doi.org/10.1016/j.carbpol.2016.11.050

Tian, J., Chen, S., Shi, J., Chen, J., Liu, D., Cai, Y., Ogawa, Y., & Ye, X. (2017). Microstructure and digestibility of potato strips produced by conventional frying and air-frying: An *in vitro* study. *Food Structure*, *14*, 30–35. https://doi.org/10.1016/j.foostr.2017.06.001

Torres, J. D., Dueik, V., Contardo, I., Carré, D., & Bouchon, P. (2024). Relationship between microstructure formation and *in vitro* starch digestibility in baked gluten-starch matrices. *Food Chemistry: X*, *22*, 101347. https://doi.org/10.1016/j.fochx.2024.101347

Uraipong, C., & Zhao, J. (2016). Rice bran protein hydrolysates exhibit strong *in vitro* α-amylase, β-glucosidase and ACE-inhibition activities. *Journal of the Science of Food and Agriculture*, *96*(4), 1101–1110. https://doi.org/10.1002/jsfa.7182

van der Maarel, M. J. E. C., van der Veen, B., Uitdehaag, J. C. M., Leemhuis, H., & Dijkhuizen, L. (2002). Properties and applications of starch-converting enzymes of the α-amylase family. *Journal of Biotechnology*, *94*(2), 137–155. https://doi.org/10.1016/S0168-1656(01)00407-2

Van Hung, P., Nguyen, T. H. M., Nguyen, T. L. P., & Nguyen, N. T. T. (2017). Physicochemical characteristics and *in vitro* digestibility of potato and cassava starches under organic acid and heat-moisture treatments. *International Journal of Biological Macromolecules*, *95*, 299–305.

Vilcacundo, R., Martínez-Villaluenga, C., & Hernández-Ledesma, B. (2017). Release of dipeptidyl peptidase IV, α-amylase and α-glucosidase inhibitory peptides from quinoa (*Chenopodium quinoa Willd.*) during *in vitro* simulated gastrointestinal digestion. *Journal of Functional Foods*, *35*, 531–539. https://doi.org/10.1016/j.jff.2017.06.024

Wang, J., Li, A., Hu, J., Zhang, B., Liu, J., Zhang, Y., & Wang, S. (2022a). Effect of frying process on nutritional property, physicochemical quality, and *in vitro* digestibility of commercial instant noodles. *Frontiers in Nutrition*, *9*, 823432. https://doi.org/10.3389/fnut.2022.823432

Wang, J., Li, Y., Ma, W., Zhang, J., Yang, H., Wu, P., Li, J., & Jin, Z. (2024). Physicochemical changes and *in vitro* digestibility of three banana starches at different maturity stages. *Food Chemistry: X*, *21*, 101004. https://doi.org/10.1016/j.fochx.2023.101004

Wang, S., Chao, C., Cai, J., Niu, B., Copeland, L., & Wang, S. (2020). Starch–lipid and starch–lipid–protein complexes: A comprehensive review. *Comprehensive Reviews in Food Science and Food Safety*, *19*(3), 1056–1079. https://doi.org/10.1111/1541-4337.12550

Wang, S., Wang, J., Liu, Y., Liu, X. (2020). Starch Modification and Application In S. Wang (Ed.), *Starch Structure, Functionality and Application in Foods*(131-149), 1st ed. Springer. https://doi.org/10.1007/978-981-15-0622-2_8

Wang, Y., Ral, J.-P., Saulnier, L., & Kansou, K. (2022b). How does starch structure impact amylolysis? Review of current strategies for starch digestibility study. *Foods*, 11(9), 1223. https://doi.org/10.3390/foods11091223

Wang, Z., Fan, M., Hannachi, K., Li, Y., Qian, H., & Wang, L. (2023). Impact of red kidney bean protein on starch digestion and exploring its underlying mechanism. *International Journal of Biological Macromolecules*, 253, 127023. https://doi.org/10.1016/j.ijbiomac.2023.127023

Wani, I. A., Sogi, D. S., Hamdani, A. M., Gani, A., Bhat, N. A., & Shah, A. (2016). Isolation, composition, and physicochemical properties of starch from legumes: A review. *Starch-Stärke*, *68*(9–10), 834–845. https://doi.org/10.1002/star.201600007

Warren, F. J., Royall, P. G., Gaisford, S., Butterworth, P. J., & Ellis, P. R. (2011). Binding interactions of α-amylase with starch granules: The influence of supramolecular structure and surface area. *Carbohydrate Polymers*, *86*(2), 1038–1047. https://doi.org/10.1016/j.carbpol.2011.05.062

Wolf, B. W., Bauer, L. L., & Fahey, G. C. (1999). Effects of chemical modification on *in vitro* rate and extent of food starch digestion: An attempt to discover a slowly digested starch. *Journal of Agricultural and Food Chemistry*, *47*(10), 4178–4183. https://doi.org/10.1021/jf9813900

Wu, P., Bhattarai, R. R., Dhital, S., Deng, R., Chen, X. D., & Gidley, M. J. (2017). *In vitro* digestion of pectin- and mango-enriched diets using a dynamic rat stomach-duodenum model. *Journal of Food Engineering*, *202*, 65–78. https://doi.org/10.1016/j.jfoodeng.2017.01.011

Xiong, W., Zhang, B., Dhital, S., Huang, Q., & Fu, X. (2019). Structural features and starch digestion properties of intact pulse cotyledon cells modified by heat-moisture treatment. *Journal of Functional Foods*, *61*, 103500. https://doi.org/10.1016/j.jff.2019.103500

Xu, X., Bean, S., Wu, X., & Shi, Y. (2022). Effects of protein digestion on in vitro digestibility of starch in sorghum differing in endosperm hardness and flour particle size. *Food Chemistry*, *383*, 132635. https://doi.org/10.1016/j.foodchem.2022.132635

Yamada, Y., Hosoya, S., Nishimura, S., Tanaka, T., Kajimoto, Y., Nishimura, A., & Kajimoto, O. (2005). Effect of bread containing resistant starch on postprandial blood glucose levels in humans. *Bioscience, Biotechnology, and Biochemistry*, *69*(3), 559–566. https://doi.org/10.1271/bbb.69.559

Yang, W., Zheng, Y., Sun, W., Chen, S., Liu, D., Zhang, H., Fang, H., Tian, J., & Ye, X. (2020). Effect of extrusion processing on the microstructure and *in vitro* digestibility of broken rice. *LWT*, *119*, 108835. https://doi.org/10.1016/j.lwt.2019.108835

Yang, Y., Jiao, A., Zhao, S., Liu, Q., Fu, X., & Jin, Z. (2021). Effect of removal of endogenous non-starch components on the structural, physicochemical properties, and *in vitro* digestibility of highland barley starch. *Food Hydrocolloids*, *117*, 106698. https://doi.org/10.1016/j.foodhyd.2021.106698

Yassaroh, Y., Woortman, A. J., & Loos, K. (2021). Physicochemical properties of heat-moisture treated, stearic acid complexed starch: The effect of complexation time and temperature. *International Journal of Biological Macromolecules*, 175, 98–107. https://doi.org/10.1016/j.ijbiomac.2021.01.124

Yaver, E., & Bilgicli, N. (2021). Development of quality characteristics of pasta enriched with Lupin (*Lupinus albus* L.) flour and resistant starch type 4. *Journal of Tekirdag Agricultural Faculty*, *18*, 557–568.

Ye, J., Hu, X., Luo, S., Liu, W., Chen, J., Zeng, Z., & Liu, C. (2018). Effect of endogenous proteins and lipids on starch digestibility in rice flour. *Food Research International*, *106*, 404–409. https://doi.org/10.1016/j.foodres.2018.01.008

You, S. Y., Oh, S. G., Han, H. M., Jun, W., Hong, Y. S., & Chung, H. J. (2016). Impact of germination on the structures and *in vitro* digestibility of starch from waxy brown rice. *International Journal of Biological Macromolecules*, *82*, 863–870. https://doi.org/10.1016/j.ijbiomac.2015.11.023

You, S. Y., Oh, S. K., Kim, H. S., & Chung, H. J. (2015). Influence of molecular structure on physicochemical properties and digestibility of normal rice starches. *International Journal of Biological Macromolecules*, *77*, 375–382. https://doi.org/10.1016/j.ijbiomac.2015.02.054

Yu, W., Tao, K., & Gilbert, R. G. (2018). Improved methodology for analyzing relations between starch digestion kinetics and molecular structure. *Food Chemistry*, *264*, 284–292. https://doi.org/10.1016/j.foodchem.2018.05.049

Yu, Z., Yin, Y., Zhao, W., Liu, J., & Chen, F. (2012). Anti-diabetic activity peptides from albumin against α-glucosidase and α-amylase. *Food Chemistry*, *135*(3), 2078–2085. https://doi.org/10.1016/j.foodchem.2012.06.088

Zavareze, E. D. R., & Dias, A. R. G. (2011). Impact of heat-moisture treatment and annealing in starches: A review. *Carbohydrate Polymers*, *83*(2), 317–328. https://doi.org/10.1016/j.carbpol.2010.08.064

Zhang, L., Zhao, Y., Hu, W., Qian, J.-Y., Ding, X.-L., Guan, C.-R., Lu, Y.-Q., & Cao, Y. (2018). Multi-scale structures of cassava and potato starch fractions varying in granule size. *Carbohydrate Polymers*, *200*, 400–407. https://doi.org/10.1016/j.carbpol.2018.08.022

Zhang, X., Kong, X., Hao, Y., Zhang, X., & Zhu, Z. (2020). Chemical structure and inhibition on α-glucosidase of polysaccharide with alkaline-extracted from glycyrrhiza inflata residue. *International Journal of Biological Macromolecules*, *147*, 1125–1135. https://doi.org/10.1016/j.ijbiomac.2019.10.081

Zhang, Z., Zhu, M., Xing, B., Liang, Y., Zou, L., Li, M., Fan, X., Ren, G., Zhang, L., & Qin, P. (2023). Effects of extrusion on structural properties, physicochemical properties and in vitro starch digestibility of tartary buckwheat flour. *Food Hydrocolloids*, *135*, 108197. https://doi.org/10.1016/j.foodhyd.2022.108197

Zheng, M., Lei, S., Wu, H., Zheng, B., Zhang, Y., & Zeng, H. (2019). Effect of chitosan on the digestibility and molecular structural properties of lotus seed starch. *Food and Chemical Toxicology*, *133*, 110731. https://doi.org/10.1016/j.fct.2019.110731

Zhou, Z., Ye, F., Lei, L., Zhou, S., & Zhao, G. (2022). Fabricating low glycaemic index foods: Enlightened by the impacts of soluble dietary fibre on starch digestibility. *Trends in Food Science & Technology, 122*, 110–122.

Zhu, F. (2018). Relationships between amylopectin internal molecular structure and physicochemical properties of starch. *Trends in Food Science & Technology*, *78*, 234–242. https://doi.org/10.1016/j.tifs.2018.05.024

Clean Label Starch

21

Shinjae Park and Yong-Ro Kim

21.1 INTRODUCTION

Until the early 1980s, many people primarily aimed for economic prosperity and social success (Campbell 1981). However, despite this focus on material wealth, the number of individuals experiencing spiritual satisfaction has declined over time. As a result, mental well-being and personal contentment began to gain significance in people's lives (Furukawa et al. 2019). This societal shift had a profound impact on the food industry, leading to trends such as the "slow food movement" in the mid-1980s and the rise of "well-being foods" in the 2000s (Ares et al. 2014; Chrzan 2004; McMahon, Williams & Tapsell 2010; Pollan 2008). Recently, the global food trend of "clean label" has emerged (Cheung et al. 2016; IFT 2018). The clean label movement, which emerged in the United Kingdom during the 1990s, centered on meeting fundamental standards such as incorporating natural ingredients and reducing processing levels (Baines & Seal 2012). In contrast, the current clean label trend goes beyond simply selecting "healthy foods" to involve thorough examination of ingredient lists and evaluation of the environmental impact of each product (Aschemann-Witzel, Varela & Peschel 2019). The evolution of clean labels has shifted from a narrow emphasis on ingredient choice to a broader consideration of the entire food processing approach.

Go Clean Label™ reflects the clean label movement as emerging from consumers' desire for genuine food, with a focus on transparency and authenticity. Foods following this ideology consist of natural, easily identifiable, and simple components, devoid of any artificial or synthetic additives (Hutt & Sloan 2015). Consequently, to qualify for a clean label, foods must satisfy fundamental requirements such as "no additives", "minimal processing", and "simple ingredient listing" (Aschemann-Witzel, Varela & Peschel 2019; Busken 2015; IFT 2018). Nowadays, consumers with specific preferences increasingly choose clean label products over those based on brand reputation. According to a study conducted by the Center for Food Integrity, 75% of participants rigorously examine nutrition and ingredient labels, while 53% believe that products with clean labels are notably more beneficial for one's health. Furthermore, 54% of American shoppers are prepared to pay extra for products with ingredients they can easily understand (Macdonald 2017). The clean label food market is expected to experience significant expansion, with estimates indicating a rise from $44.8 billion in 2023 to $68.4 billion by 2027, representing a CAGR of 11.1% (The Business Research Company 2023).

Starch, a major dietary carbohydrate source, is naturally plentiful and can be sourced from seeds, tubers, roots, and stems (Wang & Guo 2020). Starch granules exhibit variations in size, structure, shape, and chemical composition depending on their source (Smith 2001). Starch is composed of multiple glucose units connected by glycosidic bonds, giving rise to distinct structures like amylose and amylopectin (Ismail, Irani & Ahmad 2013; Tharanathan 2005). Amylose, which features mainly linear α-1,4-glycosidic bonds, generally accounts for 15%–20% of starch. On the other hand, amylopectin is characterized by its larger size and branching structure, consisting of both α-1,4- and α-1,6-glycosidic linkages, and serves as a significant constituent of starch (Sajilata, Singhal & Kulkarni 2006; Tharanathan 2002). Beyond its role in nutrition, starch demonstrates a range of physical properties, including gelatinization, swelling, and gelation, depending on its composition. The attributes of starch render it a significant component in the food sector, serving as a crucial thickening and stabilizing agent (Karam et al. 2005). These qualities play a critical role in enhancing the performance, consistency, and taste of various food items (Chen, Lai & Lii 2003).

However, native starches are not ideal for industrial use because they are vulnerable to instability when exposed to heat and shear, and they have a propensity to retrograde during cooling or freezing, leading to a decline in food quality (Arocas, Sanz & Fiszman 2009; Karam et al. 2005). To overcome these challenges, extensive research has focused on modifying natural starch with chemical treatments to create modified starches (Zia-ud-Din, Xiong & Fei 2017). The projected growth of the global modified starch market from an estimated $13.7 billion in 2022 to $15.9 billion by 2027 is primarily attributed to the increasing consumer preference for processed and convenience food products (MarketsandMarkets 2021). However, despite this growth, consumers are still cautious about buying products with ambiguous or unclear ingredients, as previously noted. Chemically modified starches are typically labeled

DOI: 10.1201/9781003464396-21

as "modified starch" and receive an E-number under the International Numbering System (INS). On the other hand, starches that have been modified either physically or enzymatically are classified as ingredients rather than additives. As a result, they can be listed on labels simply as "starch" sourced from a variety of origins, without the need for E-numbers (Radeloff & Beck 2016).

This chapter offers an in-depth examination of the latest insights into physically modified "clean label starch". It explores various physical modification techniques, including starch blending, ultrasound, and hydrothermal treatments, along with other methods. Enzymatic modification is also emphasized as a significant process in the production of clean label starch. The content of this chapter is adapted and expanded from the original article by Park and Kim (2021).

21.2 STARCH BLENDING

In order to address the constraints of native starch, the incorporation of small amounts of hydrocolloids, such as gum, into food items is a prevalent practice in the food industry (Oh, Kim & Yoo 2010). Nonetheless, hydrocolloids are often classified as "additives" rather than "ingredients", potentially contradicting clean label standards. An advantageous alternative to this approach is starch blending, which is both economically advantageous and offers a simple technique for modifying the consistency of starch gels without the addition of gums (Oh, Kim & Yoo 2010; Sun 2014). In the field of starch blending, the detection of additive effects allows for the prediction of blend properties through the analysis of individual starch characteristics. However, discrepancies between anticipated and actual properties suggest the presence of non-additive effects, indicating specific interactions between different starch types within the blend (Waterschoot et al. 2015). Various factors, including starch concentration, amylose content, the degree of amylose leaching, swelling power, and the size of starch granule, have been found to significantly impact the properties of starch blends.

21.2.1 Factors Affecting Starch Blending

21.2.1.1 Starch concentration

Hua Liu and Lelievre (1992) noted that when the total starch concentration is below 30 wt.%, the thermal properties of the starch blend closely resemble the sum of the individual starches. However, at higher concentrations, non-additive behavior arises as a result of increased competition for water among the starches as water content diminishes. Waterschoot et al. (2015) also observed that the gelatinization of starches occurs independently when water is sufficient. When water availability is restricted, the starch of the lowest gelatinization point will undergo gelatinization initially, resulting in diminished water availability for the gelatinization of other starches. Consequently, the remaining starches will gelatinize at elevated temperatures, impacting their characteristics.

21.2.1.2 The size of starch granule

Puncha-Arnon et al. (2008) claimed that the difference in the size of starch granules could affect the non-additive behavior of starch blend properties. They discovered that the RVA pasting properties (peak viscosity and setback) and gel hardness observed in blends of canna starch and potato starch, the mean granular diameters of which were similar as $52.3\pm0.0\,\mu m$ and $47.9\pm0.2\,\mu m$, respectively, at different ratios, were linearly influenced by the properties of both starches. However, blends of canna starch with starches of relative smaller granule size, mung bean starch ($\varnothing=24.1\pm0.4\,\mu m$) and rice starch ($\varnothing=6.8\pm1.7\,\mu m$), exhibited those properties that deviated from what was predicted based on the properties of each individual starch, showing more severe non-linearity for rice starch with smaller granule size. These observations might indicate that when there is a substantial disparity in granule size among the starch types in a blend, non-additive effects become more pronounced.

21.2.1.3 Amylose leaching and swelling power

Obanni and Bemiller (1997) reported that amylose begins to leach out of starch granules at an early stage of heating, occurring prior to gelatinization or pasting. This leaching process can lead to the transfer of amylose between different starches, forming molecular interactions that influence pasting characteristics. This phenomenon has also been observed in studies conducted by Zhu and Corke (2011). Chen, Lai and Lii (2003) examined how swelling power affects the pasting behavior of blended starches and found that when starches with low swelling capacity were mixed with starches that swell easily at similar proportions, the overall swelling power decreased considerably and the gelatinization temperature increased. This effect is likely due to the degree of interaction between granules, which is influenced by how much the granules overlap after they come into contact, a factor closely related to swelling power.

21.2.1.4 Amylose content

Extensive research has focused on how differences in amylose content affect the characteristics of blended starches. Novelo-Cen and Betancur-Ancona (2005) conducted a study to examine the chemical and functional characteristics of starch blend combinations derived from lima bean (*Phaseolus lunatus*) and cassava (*Manihot esculenta*). With cassava starch having an amylose content of 17.3% and lima bean starch 32.4%, they found that a blend ratio of 25:75 (lima bean:cassava) provided the best combination of attributes, such as high viscosity

(102 mPa·s) and stability during heating and cooling without retrogradation. The findings of this study indicate that the combination of native starches from different sources can result in the development of novel starches with advantageous properties for a variety of food applications. Juhász and Salgó (2008) conducted an analysis on the pasting characteristics of corn starch blends with varying levels of amylose, including normal corn (27%), waxy corn (0%), and high-amylose corn (70%). The RVA conducted on System I (combinations of normal corn starch with high-amylose or waxy corn starch) and System II (simulated mixtures of waxy corn starch and high-amylose corn starch) demonstrated comparable results, suggesting that alterations in functional characteristics are associated with amylose levels. Nevertheless, disparities in rheological properties observed in the two systems indicate that the properties of starch paste could potentially be influenced by the interplay among starch constituents or granule characteristics (Juhász & Salgó 2008).

21.2.1.5 Blending with specific starch

Introducing starches with distinctive properties, such as potato starch, has a substantial impact on the characteristics of starch mixtures. Potato starch is distinguished by its comparatively large granule size and high phosphorus content within the amylopectin fraction in comparison to other types of starch. Additionally, it exhibits distinctive characteristics such as a high capacity for swelling and the ability to bind metal ions (Tomasik 2009). Park et al. (2009) examined the pasting properties of mixtures composed of potato starch and waxy corn starch. Their findings revealed that peak viscosity and pasting temperature demonstrated a linear relationship in response to differing proportions of the two starches, indicative of additive behavior. However, breakdown and setback exhibited non-additive behavior. Sandhu and Kaur (2010) conducted an experiment that demonstrated that potato starch addition to rice starch had a substantial impact on the properties of noodles. In particular, a mixture of potato and rice starch at a 1:1 ratio resulted in noodles with improved characteristics such as reduced cooking time, enhanced transparency, and a smoother texture. This study highlights the potential benefits of blending potato starch with rice starch in specific proportions to enhance the overall quality of noodles. An overview of the research findings on starch blending, including the types and ratios of starches used and the experimental procedures employed, can be found in Table 21.1.

21.2.2 Applications of Starch Blending

Starch blending plays a crucial role in the production of various food products such as noodles (Noda et al. 2006), wheat substitutes (Oladunmoye et al. 2014), bread (Koca & Anil 2007), cakes (Gómez et al. 2008), and gluten alternatives (Demirkesen et al. 2010; Mancebo, Rodriguez & Gomez 2016; Paucean et al. 2016). Studies have demonstrated the significance of blending techniques in enhancing the quality and characteristics of these food products.

Noda et al. (2006) explored the properties of composite blends comprising wheat flour and potato starch for the

TABLE 21.1   Starch blending conditions and examined properties for clean label starch production (Park & Kim 2021)

STARCHES	RATIO	PROPERTIES	REFERENCES
Normal corn/waxy corn/high amylose corn/potato/wheat/rice	90:10/80:20/70:30/60:40/50:50/ 40:60/30:70/20:80/10:90	Pasting profile/thermal profile/ paste morphology	Obanni and Bemiller (1997)
Wheat/rice	100:0/85:15/75:25/50:50/25:75/ 15:85/0:100	Thermal profile	Liu and Lelievre (1992)
Rice (indica type/waxy type)		Swelling power/pasting profile/ thermal profile/granular morphology	Chen, Lai and Lii (2003)
Corn/cassava/yam	100:0:0/0:100:0/33:34:33/50:50 :0/20:20:60/67:0:33/0:67:33/0: 40:60/40:0:60	Gel morphology/texture	Karam et al. (2005)
Lima bean/cassava	100:0/75:25/50:50/25:75/0:100	Thermal profile/gel profile/pasting profile/texture	Novelo-Cen and Betancur-Ancona (2005)
Canna/potato/mung bean/rice		Pasting profile/thermal profile/gel morphology/texture	Puncha-Arnon et al. (2008)
Corn/waxy corn/high amylose corn	96:4/83:17/66:33/50:50/33:66/1 7:83/4:96	Pasting profile	Juhász and Salgó (2008)
Potato/waxy corn	90:10/80:20/70:30/60:40/50:50/ 40:60/30:70/20:80/10:90	Pasting profile/granular morphology	Park et al. (2009)
Rice (flour) with sweet potato/ potato/cassava/waxy corn/ hydroxypropylated potato/ hydroxypropylated cassava	2:1 (rice:others)	Rheological profile	Oh, Kim and Yoo (2010)
Rice/potato	75:25/50:50/25:75	Pasting profile/texture	Sandhu and Kaur (2010)
Wheat/sweet potato		Pasting profile/thermal profile/ texture	Zhu and Corke (2011)

production of instant noodles. Their results highlighted the importance of phosphate in potato starch modulating the paste viscosity, acting as a critical factor in achieving the desired textural attributes in instant noodles. Similarly, Koca and Anil (2007) studied the effects of blending wheat flour with flaxseed, a functional ingredient, on the rheological and baking properties of the resulting blend. They found that up to a 1:4 ratio with wheat flour, flaxseed enhanced the quality of bread in terms of volume, without negatively impacting its sensory qualities. Also, blending techniques offer significant benefits in the production of gluten-free foods as gluten, a protein found in wheat, can be challenging for certain individuals to digest and is believed to contribute to certain health issues (Holmes et al. 1989). Previous research has demonstrated the effectiveness of utilizing alternative flours such as rice flour (Mancebo, Rodriguez & Gomez 2016; Paucean et al. 2016), buckwheat flour (Dapčević Hadnađev, Torbica & Hadnađev 2013), chestnut flour (Demirkesen et al. 2010), and amaranth (de la Barca et al. 2010) in the production of gluten-free items. Numerous studies have indicated that bakery products with reduced gluten content exhibit similar texture and appearance to their gluten-containing counterparts.

21.3 ULTRASOUND-TREATED STARCH

Ultrasound, also known as ultrasonication, utilizes sound waves with frequencies above the human hearing limit (20 kHz) to generate strong shear forces, high temperatures, and free radicals (Mason & Joyce 2008). This technology is employed to enhance processing efficiency, improve food quality, and ensure safety (Choi & Lee 2017; Knorr et al. 2011). When applied to starch–water systems, ultrasound alters the structure and characteristics of starch by inducing cavitation, which breaks down polymer chains and damages starch granules (Jambrak et al. 2010; Zhu 2015). The extent of these changes is influenced by factors such as ultrasound frequency and intensity, treatment duration, temperature, and the properties of the starch itself (Zhu 2015). This section examines the impacts of ultrasound treatment on starch properties in detail.

21.3.1 Characteristics of Ultrasound-Treated Starch

21.3.1.1 Starch morphology

Following ultrasound treatment, indentations on starch granules can be observed as dark spots when viewed under a light microscope. Zuo et al. (2012) noted that the level of damage to the granules differed based on the intensity of the ultrasound, demonstrating a direct correlation between the extent of damage and the level of energy applied. Starches with larger

granule sizes, such as those found in potato starch, exhibited more significant alterations due to ultrasound treatment, with the severity of damage increasing with prolonged exposure (Carmona-García et al. 2016). Increased severity of granule breaks was observed in close proximity to the Maltese cross region, which may be associated with its fragile surface structure (Carmona-García et al. 2016). Degrois et al. (1974) and Gallant et al. (1972) conducted research on the impact of different surrounding gases on the damage of starch granules under ultrasonication. Their findings revealed that exposure to air resulted in a rough and wrinkled surface of the starch granules, while oxygen caused less damage. In carbon dioxide or vacuum conditions, no erosion was observed. However, when exposed to hydrogen, the starch granules exhibited significantly large and deep pits with a relatively even surface.

21.3.1.2 Swelling power and solubility

A number of researchers have consistently reported that ultrasound treatment increased the swelling power and solubility of starch (Carmona-García et al. 2016; Herceg et al. 2010; Luo et al. 2008; Zheng et al. 2013). Upon ultrasound treatment, an increase in surface fractures and a decrease in intermolecular interactions enhance water permeability into starch, resulting in increased solubility and swelling power. These ultrasound-induced changes in swelling power and solubility became more pronounced at higher ultrasound intensity (Zheng et al. 2013), longer treatment duration (Carmona-García et al. 2016), and higher temperature (Herceg et al. 2010; Luo et al. 2008). Starch with larger granule size showed more appreciable increase in swelling power and solubility than one with smaller size (Carmona-García et al. 2016). From the experiments with waxy, normal, and high amylose corn starches, Luo et al. (2008) claimed that the relative extent of increase in swelling power and solubility upon ultrasound treatment was positively affected by amylose content, since amylose located in the amorphous region of starch granule was more vulnerable to ultrasonic disruption than amylopectin in the crystalline region. Sujka and Jamroz (2013) utilized water and ethanol as solvents in experimental trials to apply ultrasound treatment to starches derived from potato, wheat, corn, and rice. The results indicated a rise in both solubility and swelling power in all starch samples, with a particularly notable increase observed in starches subjected to ultrasonic treatment in water as opposed to those treated in ethanol. This difference is likely due to a greater degree of starch depolymerization occurring when treated in water.

21.3.1.3 Pasting properties

Research by Herceg et al. (2010) has demonstrated that ultrasound treatment induces damage to starch granules through cavitation forces, enhancing their permeability to water upon heating. In a study conducted by Zheng et al. (2013), the effects of dual frequency ultrasound treatment at 25 and 80 kHz on sweet potato starch were investigated, revealing reductions of 6.9% and 9.8% in peak viscosities of starch treated with these frequencies, respectively, as compared to untreated starch. Ultrasound treatment usually results in a reduction in the viscosity of starch paste due to the partial disruption of glycosidic bonds, leading to a weakening of the polymer structure (Herceg et al. 2010; Sujka & Jamroz 2013). However, there are conflicting findings in the literature, as some studies have shown an increase in viscosity following ultrasound treatment. For instance, Chan, Bhat and Karim (2010) and Carmona-García et al. (2016) found that longer ultrasound treatment durations led to a higher peak viscosity of starch. The increase in viscosity of starch paste can be attributed to factors such as prolonged treatment time, repeated cycles, and higher amplitudes during ultrasound processing. These conditions lead to a greater breakdown of starch granules, allowing them to absorb more water and expand further, resulting in a more viscous paste (Sit, Misra & Deka 2014). Additionally, extended ultrasound treatment exceeding 50 minutes can cause a significant breakdown of starch granules, likely due to the accumulation of weakened starch granules (Carmona-García et al. 2016). Consequently, the properties of starch paste vary depending on the specific ultrasound treatment conditions applied.

21.3.1.4 Retrogradation properties

The impact of ultrasound on the retrogradation characteristics of starch is influenced by the ultrasound intensity level and treatment time. Yu et al. (2013) found that high levels of intensity (>500 W power and 10-mm-diameter probe tip) and long treatment duration (>90 minutes) have an appreciable impact on the onset temperature and enthalpy of retrograded rice starch, resulting in an increase in onset temperature and a reduction in retrogradation enthalpy. In a study on the effects of ultrasound treatment to potato starch paste on its retrogradation properties, Nie et al. (2019) observed that the degree of retrogradation measured by the precipitate volume of dilute starch solution increased as ultrasound intensity increased. Also, they found an alteration in the crystalline pattern from B type to V type as well as a reduction of average molecular weight of starch after ultrasonication. They claimed that ultrasound treatment significantly affects both the crystalline pattern and the crystallinity of retrograded potato starch.

21.3.2 Applications of Ultrasound-Treated Starch

There have been a number of laboratory research on the application of ultrasound-treated starch, which can attract industrial interest. These applications include improving enzymatic modifications (Wu et al. 2011), serving as components in encapsulation and delivery systems (Xing et al. 2014; Zhu 2017), acting as emulsifiers (Abbas et al. 2014), and being utilized in the production of edible films (Cheng et al. 2010). Furthermore, ultrasound-treated starch can be involved in the formation of V-type inclusion complexes and is essential in bioethanol

production processes (Khanal et al. 2007; Nitayavardhana et al. 2010).

Wu et al. (2011) examined the impact of incorporating glucoamylase in conjunction with ultrasound treatment on the development of pores and augmentation of the surface area of porous starch granules. Combined ultrasound treatment and glucoamylase digestion emerged as the most efficient method among other approaches for producing microporous starch (Wu et al. 2011). As for an application of porous starch, Xing et al. (2014) have encapsulated *Lactobacillus acidophilus* in porous starch to achieve a significant enhancement of survival rate of *L. acidophilus* in gastrointestinal fluid. Ultrasound treatment also affects the emulsifying properties of starch. Abbas et al. (2014) observed that ultrasound treatment has an impact on the emulsifying characteristics of OSA starch. Specifically, they reported that OSA starch treated with ultrasound at a power density of 1.45 W/mL for 7 minutes exhibited enhanced loading capacity for curcumin in MCT oil, thereby reducing the required amount of emulsifier. Furthermore, Cheng et al. (2010) leveraged the enhanced characteristics of ultrasound-treated starch, including heightened transparency, increased resistance to moisture, and enhanced structural integrity, to develop edible films. In the area of bioethanol production, researchers such as Khanal et al. (2007) and Nitayavardhana et al. (2010) incorporated ultrasound-treated corn and cassava starch, observing a significant decrease in the fermentation duration for ethanol by approximately 24 hours in comparison to the control group.

21.4 HYDROTHERMALLY TREATED STARCH

Heat-moisture treatment (HMT) and annealing (ANN) are commonly used hydrothermal processes to modify the physical and chemical properties of starch while maintaining its granular structure intact (Adebowale et al. 2005; da Rosa Zavareze & Dias 2011; Jacobs & Delcour 1998). HMT is a process that typically involves subjecting starch to low levels of moisture (less than 40% w/w) at temperatures higher than its glass transition temperature (typically exceeding 94°C) for a duration ranging from 1 to 24 hours (Chung, Liu & Hoover 2009). On the contrary, ANN is conducted under conditions of high moisture content (>60% w/w) and temperatures ranging between the glass transition and gelatinization points (typically 5°C–15°C below the onset temperature of gelatinization) (Chung, Liu & Hoover 2009; da Rosa Zavareze & Dias 2011).

In general, HMT is recognized for its ability to increase the gelatinization temperature, thermal stability, and crystallinity of starch. This procedure reduces the swelling of granules by inhibiting the release of amylose, enhancing the bonds within the molecule (Jiranuntakul et al. 2011), and expanding the range of temperatures at which gelatinization occurs in various starch varieties (Chung, Liu & Hoover 2009).

TABLE 21.2   Characteristic change by heat-moisture treatment (HMT) and annealing (ANN) (Park & Kim 2021)

PROPERTIES	HMT	ANN	REFERENCES
Swelling power	–	–	Adebowale et al. (2005), Jacobs and Delcour (1998), da Rosa Zavareze and Dias (2011), Chung, Liu and Hoover (2009), Abraham (1993), Watcharatewinkul et al. (2009), Puncha-Arnon and Uttapap (2013), Gomes, da Silva and Ricardo (2005), Gomes et al. (2004), Chung, Moon and Chun (2000), and Song, Park and Shin (2011)
Solubility	–	–	
Amylose leaching	–	–	
Gelatinization temp.	+	+	
Peak viscosity	–	+/–	
Breakdown	–	–	
Setback	+	+/–	
Gel hardness	++	+	
Crystallinity	+	+	
SDS/RS	++	+	

+, increased; –, decreased

In contrast, ANN typically leads to improvements in granule stability, elevation of gelatinization temperatures, enhancement of crystalline structure, and promotion of interactions within starch chains. However, it can also result in decreased leaching of amylose and reduced swelling of granules, and the effects of these treatments can vary depending on the origin of starch (Chung, Liu & Hoover 2009). Table 21.2 offers a summary of the common changes brought about by HMT and ANN.

21.4.1 Characteristics of HMT/ANN Starch

21.4.1.1 Pasting properties

HMT and ANN both cause notable modifications in the pasting characteristics of starch. Studies have shown that starch treated with HMT exhibits reduced breakdown, suggesting enhanced stability under prolonged heating and mechanical stress (Abraham 1993; Watcharatewinkul et al. 2009). Puncha-Arnon and Uttapap (2013) investigated the impact of moisture content of starch in HMT on starch characteristics, finding that the pasting temperature of starch with 20% moisture slightly increased, while more pronounced increases were observed at moisture levels of 25% and 30% after HMT. Watcharatewinkul et al. (2009) found that subjecting canna starch to HMT led to decreased viscosity, enhanced paste stability, and a higher gelatinization temperature. This indicates that HMT-treated canna starch displayed characteristics similar to those of cross-linked starch.

In contrast, ANN treatment has been shown to result in diverse pasting characteristics based on specific experimental parameters, as demonstrated by da Rosa Zavareze and Dias (2011). Adebowale et al. (2005) observed an increase in peak viscosity in annealed sorghum starch in comparison to native and HMT-treated starch, which they attributed to its enhanced swelling capacity. In addition, findings by Jacobs, Eerlingen and Delcour (1996) indicated that the process of ANN resulted

in elevated levels of both peak and final viscosities in starch obtained from wheat and rice. In contrast, Gomes, da Silva and Ricardo (2005) and Singh et al. (2011) discovered that ANN, similar to HMT, decreased the peak viscosity and swelling capacity of cassava and sorghum starches. Gomes et al. (2004) also noted that ANN resulted in lower peak viscosity and reduced breakdown for cassava starch, with a higher final viscosity compared to the control.

21.4.1.2 Gel texture and rheological properties

Both HMT and ANN methods result in enhanced gel firmness as a consequence of alterations in the gel matrix. These alterations involve enhanced amylose association, optimized crystalline structure, and increased gel density attributed to decreased swelling ability (Cham & Suwannaporn 2010; Eerlingen et al. 1997; Liu, Corke & Ramsden 2000). Eerlingen et al. (1997) examined the rheological characteristics of potato starch subjected to varying moisture content (ranging from 20% to 40%) and processed at temperatures slightly lower than the peak gelatinization temperature. The results of their research demonstrated that hydrothermal treatment led to a decrease in amylose leaching, solubility, and swelling capacity, while improving the packing density of the starch, consequently elevating the storage modulus (G') of the gel. Collado and Corke (1999) employed HMT on sweet potato starch samples varying in amylose content to analyze their gel texture. The findings revealed that gels produced from "Taiwan starch", containing 15.2% amylose, demonstrated notable enhancements in hardness and adhesiveness compared to those from "93-006", which had an amylose content of 28.5% and showed no significant variations in hardness. These results emphasize the influence of the amylose content of the starch on the resulting gel properties obtained through HMT.

Cham and Suwannaporn (2010) conducted an evaluation of the physical attributes of noodles produced using rice flour modified using HMT and ANN. Their findings revealed that the G' of HMT rice flour gel at 25°C, which was correlated to rice noodle quality, exhibited a notably high value of 19,100 Pa, in comparison to that of ANN rice flour gel with a lower G' value of 5,490 Pa. In addition, hydrothermally treated rice flour provided rice noodle with smooth surface due to less amylose leaching. In a study conducted by Jacobs and Delcour (1998), it was observed that gels produced from annealed wheat starches demonstrated elevated G' in comparison to gels made from unprocessed starch, suggesting that the ANN played a role in enhancing gel strength. Likewise, the research conducted by Chung, Moon and Chun (2000) indicated that the ANN aided in the restructuring of starch molecules, leading to decreased solubility and swelling capacity, ultimately resulting in the formation of a stronger gel.

21.4.1.3 Slowly digestible starch/ resistant starch

According to the rate of digestion, starch portions are typically classified into three groups, rapidly digestible starch (RDS), slowly digestible starch (SDS), and resistant starch (RS) (Juansang et al. 2012). Both HMT and ANN have been employed in various studies to generate SDS and RS (Chung, Liu & Hoover 2009; Song, Park & Shin 2011). Chung, Liu and Hoover (2009) employed HMT and ANN techniques to corn, soy, and lentil starches and evaluated their glycemic index based on their rates of digestion. They found that HMT and ANN treatments significantly influenced the starches' vulnerability to enzyme digestion, leading to the production of SDS and RS with high thermal stability. The increase in RS content observed after HMT and ANN suggests an enhancement in starch chain interactions and improvement in crystalline structure. According to the study by Song, Park and Shin (2011), non-waxy rice starch subjected to ANN treatment displayed characteristics comparable to cross-linked starch. This resemblance was believed to be attributed to the restructuring of the external branches of amylopectin within the rice starch granules, resulting in a reduction in the intermolecular gaps between the starch molecules.

21.4.2 Applications of HMT/ ANN Starch

Starches that have undergone hydrothermal treatment could be employed to the canned and frozen food products because of their enhanced thermal resistance and decreased retrogradation (Jayakody & Hoover 2008). Moreover, these treated starches are well-suited for use in various food products such as pasta, bread, and noodles, as indicated by several studies (Bourekoua et al. 2016; Cham & Suwannaporn 2010; Suzuki & Sekiya 1994). Bourekoua et al. (2016) conducted a study to investigate the impact of hydrothermal processing on the properties of rice and corn flours, with a focus on their suitability for gluten-free baking purposes. By subjecting the flours to treatment at 65°C using a flour-to-water ratio of 5:1, significant alterations in bread attributes such as specific volume,

height-to-width ratio, hardness, and chewiness were observed, which were deemed satisfactory based on sensory evaluations when compared to a control group. In their study, Cham and Suwannaporn (2010) effectively developed a method for producing rice noodles using hydrothermal treatment that yielded products with characteristics similar to those made with commercial rice noodle mix. The researchers concluded that ANN was most suitable for producing rice noodles with a more tender consistency, while HMT was found to be more appropriate for semi-dry and dry noodles that necessitate increased tensile strength and enhanced gel firmness. Furthermore, the utilization of rice flour with reduced amylose content and modified by HMT facilitated the development of noodles with a refined texture and appearance. Brumovsky and Thompson (2001) conducted an experiment in which they subjected "Hylon VII" (a high-amylose corn starch) to partial acid hydrolysis for a duration of 6 hours at a temperature of 25°C, followed by annealing treatment for 24 hours at 70°C with a moisture content of 67%. This process led to a 32% augmentation in RS content as determined through total dietary fiber analysis. RS generally has a relatively white color, a micro-particulate structure, and a neutral taste, making it suitable for incorporation into food without altering its texture or appearance (Jayakody & Hoover 2008).

21.5 PREGELATINIZED STARCH

Pregelatinized starch is a special kind of starch that has been cooked and dried in a specific way before being used (Alcázar-Alay & Meireles 2015). The techniques utilized for pregelatinization include drum drying, extrusion cooking, and spray cooking. These methods lead to the breakdown of intermolecular hydrogen bonds between starch molecules, resulting in an irreversible swelling of starch granules (Radeloff & Beck 2016). Such alterations render pregelatinized starch capable of easy melting in cold water. As a result, the paste form can be generated without the necessity of an extra heat processing step, rendering it a viable choice for heat-sensitive food products (Alcázar-Alay & Meireles 2015).

21.5.1 Preparation of Pregelatinized Starch

21.5.1.1 Drum drying

The drum drying technique is commonly utilized in the food sector for the pre-cooking of grain-derived goods like porridge and gruel, as outlined by Björck et al. (1984). This method consists of multiple stages, beginning with the application of a starch slurry onto the heated drum's exterior in a thin layer. Following a single rotation, the starch slurry undergoes prompt dehydration, resulting in the formation of a thin film which are subsequently scraped off with a knife attachment on the dryer (Fritze 1973). Valous et al. (2002)

employed a double drum dryer to produce pregelatinized corn starch and observed intricate interactions among the drum rotation speed, steam pressure, and level of the pregelatinized starch pool during operation. Key variables, such as drum temperature and gap between drums, were significantly affected by steam pressure. The effects of increasing drum speed on responses such as steam pressure were found to have an impact on specific load, mass flow rate, and moisture content of the final dried product (Valous et al. 2002). In their study, Wiriyawattana, Suwonsichon and Suwonsichon (2018) investigated the effects of drum drying on the physical and antioxidant characteristics of pregelatinized riceberry flour. The results indicated that all samples of pregelatinized riceberry flour exhibited significantly greater water absorption index and swelling power in comparison to the control group. Additionally, it was found that drum drying led to a decrease in peak viscosity, pasting temperature, setback, trough viscosity, and final viscosity of riceberry flour. The overall phenolic content and antioxidant capacity experienced a decrease after undergoing drum drying. Therefore, careful consideration of the drum temperature conditions is crucial because such conditions can influence both the antioxidant and physical properties of starches.

21.5.1.2 Extrusion cooking

Extrusion cooking involves subjecting starch granules to a combination of mechanical shear force and elevated temperature for a brief duration within a low humidity setting (Radeloff & Beck 2016). This procedure alters certain functional characteristics of starch. Specifically, the high-temperature, short-time extrusion cooking method is commonly employed by various food manufacturers to create expanded snack products, ready-to-eat cereals, and pet foods (Chinnaswamy & Hanna 1988). In their research, Hirth et al. (2014) examined the impact of processing temperatures and moisture content on the degradation of total anthocyanins in bilberry bioactive compounds, finding that lower temperatures and increased moisture levels helped to minimize this degradation process. Despite the thermomechanical stresses from a higher screw speed, higher flow rates increased anthocyanin retention. The color and stability of the extruded sample's anthocyanin content were maintained over a 40-month storage period at room temperature in a dark environment. Parchure and Kulkarni (1997) conducted a study on the impact of extrusion cooking on RS content in rice and amaranth. The findings revealed a decrease in RS content in the extruded samples compared to their original starch forms, which was believed to be caused by the increased solubilization of starch due to macromolecular degradation during the extrusion cooking process.

21.5.1.3 Spray cooking

Starch with cold-water-swelling properties can be obtained through the process of spray cooking, where starch is cooked and dried simultaneously in a chamber under controlled pressure and temperature conditions (Radeloff & Beck 2016). In a study by Fu et al. (2012), researchers focused on partially gelatinized corn starch achieved through gelatinization and subsequent spray drying. The study revealed that the resulting starch exhibited a primarily amorphous structure. Hence, the partially gelatinized starch exhibited a significantly higher swelling power compared to native corn starch at temperatures below 60°C, suggesting improved hydration at lower temperature levels. Izidoro et al. (2011) examined how spray drying affected green banana starch that had been treated with ultrasound. Simultaneous application of spray drying and ultrasound treatment resulted in a significantly lower RS content, which was attributed to the reduced starch crystallization. The researchers observed that the process of spray drying resulted in heightened swelling power, water absorption capacity, and solubility of green banana starch.

21.5.2 Applications of Pregelatinized Starch

Majzoobi et al. (2011) conducted a study in which pregelatinized starch was manufactured through the utilization of a drum dryer. Their findings revealed that the resulting product displayed reduced crystallinity, heightened viscosity in cold water, enhanced solubility in water, and increased water absorption. Consequently, pregelatinized starch has become a widely utilized component in the food sector due to its efficacy as a thickener and gelling agent. This technology is utilized in a wide range of frozen or instant food products as well as heat-sensitive items including salad dressings, cold desserts, baby foods, and bakery mixes (Alcázar-Alay & Meireles 2015; Majzoobi et al. 2011).

Pregelatinization demonstrates considerable promise within the pharmaceutical sector for its application as diluents in capsules and tablets, as well as disintegrants in capsules. These agents facilitate the absorption of water by starch, allowing for proper disintegration of pills in a controlled release manner (Anwar, Khotimah & Yanuar 2006; Te Wierik et al. 1997). Te Wierik et al. (1997) have used retrograded pregelatinized starch products in compressible tablets designed for controlled-release matrices. The release rate from tablets containing retrograded pregelatinized starch may be adjusted to meet specific criteria, including compression force and tablet geometry, thus demonstrating the capacity to control chemical release using pregelatinized starch with diverse properties.

21.6 HIGH-PRESSURE-TREATED STARCH

Researchers have explored the application of high pressure, also known as high hydrostatic pressure, in the field of food processing and preservation as a potential substitute for conventional methods (Stolt, Oinonen & Autio 2000). The treatment of high pressure is one of the non-thermal processing methods that induces structural changes in food products similar to those caused by heat treatment, but with distinct attributes (Guo et al. 2015a; Oh et al. 2008). This method is known to be suitable for the minimal processing of food products and offers possibilities for creating inventive products with distinct textures or flavors, while causing minimal impact on taste, color, or nutritional value (Pei-Ling, Xiao-Song & Qun 2010). High-pressure treatment can also be promising to alter the characteristics of starch, including granular structure, retrogradation behavior, and chemical reactivity. There have been reports concerning the production of amorphous granular starches (AGSs) through this method (Song et al. 2017; Song et al. 2015). The properties of starch when subjected to high-pressure treatment exhibit considerable changes depending on factors such as treatment pressure level, treatment temperature, and time (Pei-Ling, Xiao-Song & Qun 2010).

21.6.1 Characteristics of High-Pressure-Treated Starch

21.6.1.1 Starch morphology

Stolt, Oinonen and Autio (2000) studied the physical and morphological property changes of barley starch when it is treated with high pressure. They found that the granular integrity of starch was remained with slight swelling even after being treated for 50 minutes at a very high pressure of 600 MPa. Nevertheless, the birefringence completely disappeared after a duration of 30 minutes under a pressure of 450 MPa and immediately at 600 MPa. In another study, Błaszczak, Valverde and Fornal (2005) conducted a similar treatment on potato starch at 600 MPa for 2–3 minutes and observed morphological changes under SEM. Most of starch granules remained granular shape at 2-minute treatment. However, at 3-minute treatment, some of the granules exhibited significant pressure-induced deformation, such that the outer layer appeared compact and condensed with greater resistance to change, whereas the inner part was weaker with some visibly disrupted gel-like network. The loss of birefringence and the development of gel-like structures resulting from water absorption by the amorphous region and crystalline melting suggest that starch undergoes gelatinization when subjected to pressure (Pei-Ling, Xiao-Song & Qun 2010). Song et al. (2015) conducted a study to investigate the potential of producing AGS through non-thermal high-pressure treatment on different types of starches. Their results demonstrated that tapioca, non-waxy rice, and corn starches underwent a transformation into AGS when subjected to specific treatment conditions (550 MPa for 30 minutes), fully gelatinizing while retaining granular structure.

21.6.1.2 Pasting properties

The effects of high-pressure treatment on pasting properties of starch mainly depend on starch source and treatment

pressure. There have been a number of researches on the pasting properties of pressurized starches from difference sources, including lentil (Ahmed et al. 2016), lotus seed (Guo et al. 2015a), sorghum (Liu et al. 2016a), common buckwheat (Liu et al. 2016b), mung bean (Jiang et al. 2015a; Li et al. 2011), rice (Jiang et al. 2015b), quinoa (Li & Zhu 2018), and corn (Li & Zhu 2018). In the reported research experiments, the hydrostatic pressure was applied at levels of 0–600 MPa and the durations of 5–30 minutes. According to the results of above researches, peak viscosity gradually increased (lentil, lotus seed, and mung bean) or decreased (common buckwheat, sorghum, quinoa, and corn) with increasing pressure from the atmospheric pressure up to 400–500 MPa depending on starch sources, and largely decreased at higher pressure (500–600 MPa) to be significantly lower than that of native starch. This trend of peak viscosity was essentially the same as that of final viscosity except for mung bean and rice starch that showed the highest final viscosity values at 600 MPa. A more consistent trend of pasting parameter changes in response to treatment pressures was decreases in breakdown and setback with increasing pressure regardless of starch sources, even though there could be a few exceptions such as mung bean starch that showed the highest setback at 600 MPa (Li et al. 2011). Generally, pasting temperature slightly increased or remained with increasing treatment pressure up to 500 MPa. However, lentil, corn, and quinoa starch exhibited a large decrease in pasting temperature at 600 MPa with the decrease becoming prominent at 500 MPa for quinoa starch. Peak time also consistently increased or remained with treatment pressure with the increase being most prominent at 600 MPa for all of above starch sources. High-pressure treatment causes partial gelatinization of starch with a limited amylose leaching and granule swelling, which leads to the above changes in pasting properties, especially a reduced breakdown and setback, making pressurized starch more stable against heat and shear (Liu et al. 2016a).

21.6.1.3 Retrogradation properties

It is anticipated that the retrogradation characteristics of gels induced by pressure will vary from those induced by heat, as the leaching of amylose is minimal following pressure treatment (Liu et al. 2016b). Specifically, wheat starch gel produced through heat treatment exhibited a more rapid retrogradation compared to starch gel produced by high pressure (Douzals et al. 1996). However, contradictory results were also reported for barley starch gels produced through heat and pressure treatments, showing that retrogradation rate of pressurized barley starch gel did not show a significant difference from that of heat-induced gel (Stolt, Oinonen, & Autio 2000). Guo et al. (2015a) reported that lotus seed starch subjected to elevated pressures ranging from 100 to 500 MPa, followed by heat gelatinization and subsequent storage at 4°C, displayed a notable decrease in melting enthalpy of retrograded starch as compared to native starch. However, similar retrogradation studies conducted for quinoa and corn starch demonstrated that high-pressure treatment did not significantly affect the melting temperature and enthalpy of both retrograded starches (Li & Zhu 2018). Hu et al. (2011) investigated the retrogradation of normal and waxy rice starches that had been gelatinized through high pressure and heat treatments. Their findings revealed that the retrogradation rate of normal rice starch subjected to high-pressure gelatinization was lower compared to heat gelatinization, whereas waxy rice starch did not show the difference in retrogradation rate between heat and pressure tretments. From these results, it was suggested that the retrogradation of high-pressure-treated starch depended on amylose content, starch sources, experimental conditions, and measurement errors (Hu et al. 2011; Li & Zhu 2018).

21.6.1.4 Rheological properties

High-pressure treatment of starch induces disruption of starch crystalline structure, the extent of which depends on treatment pressure and duration. The pressurized starch could rheologically behave as similar as heat-induced partially gelatinized starch (Vallons & Arendt 2009). However, since pressurization causes lower amylose leaching and starch granule swelling due to less water absorption than thermal treatment, the rheological properties of pressurized starch may differ from those of thermally gelatinized starch (Douzals et al. 1996; Stolt, Oinonen & Autio 2000). Numerous studies have conducted to investigate the impact of high-pressure treatment on the rheological characteristics of different types of starches, including rice (Jiang et al. 2015b), mung bean (Jiang et al. 2015a), lotus seed (Guo et al. 2015b), quinoa (Ahmed et al. 2018; Li & Zhu 2018), corn (Li & Zhu 2018), wheat (Douzals et al. 1996), barley (Stolt, Oinonen & Autio 2000), and sorghum (Vallons & Arendt 2009). In many cases of above researches, pressurized starch samples were subject to heating with an excess of water for pasting either in a rheometer or in a water bath and then cooling to form a gel before rheological measurements. With this experimental setup, the resulting gel of pressurized starch showed higher G' and G'' values than those of untreated starch (mung bean, rice, lotus seed, quinoa, and lentil), exhibiting solid-like weak gel characteristics. This result indicated that high-pressure treatment positively affected the formation of rigid gel network during the subsequent pasting and cooling. Among the above studies, both G' and G'' values of rice, quinoa, and lentil starches continuously increased with increasing treatment pressure up to 600 MPa. On the other hand, mung bean, lotus seed, and corn starches showed a decrease in moduli at high pressure of 600 MPa. This phenomenon could be partly explained from the experiment performed by Ahmed et al. (2018). They compared quinoa starch gels produced by pressurization at 450 and 600 MPa and subsequent heating and cooling with starch gels obtained by pressure treatment only. At 450 MPa, the complex viscosity (η^*), representing resistance to deformation and flow, of the pressure and heat-induced gel was higher than that of the pressure-induced gel. However, at 600 MPa,

the pressure-induced gel showed higher rigidity with higher η^* over the gel sample produced by pressure and heat treatment. From these results, they concluded that the pressurization at 600 MPa induced a sufficient gelatinization to form a rigid gel network, which was disturbed by additional heating to decrease gel strength. Stolt, Oinonen and Autio (2000) also studied the pressure induction of barley starch gel. They demonstrated that the barley starch gel obtained by pressure treatment at 550 MPa exhibited higher rigidity with higher G' and lower phase angle than those of the heat-induced gel. However, Douzals et al. (1996) reported opposite results from similar experiments on wheat starch, showing a lower Young's modulus of the pressure-induced starch gel produced at 600 MPa compared to the heat-induced counterpart.

Power law or Herschel–Bulkley model parameters obtained from the steady shear flow curves of the starch paste samples produced by pressure and subsequent heat treatments revealed that consistency coefficient (K) increased and flow behavior index (n) decreased with treatment pressure of up to 600 MPa for lotus seed starch, indicating increases in apparent viscosity and shear thinning behavior with pressure (Guo et al. 2015b). A similar increase in consistency coefficient with treatment pressure was reported for pressurized mung bean starch (Jiang et al. 2015a). However, pressurized quinoa and corn starch exhibited an increase in consistency coefficient and a decrease in flow behavior index at high pressure (500–600 MPa).

21.6.2 Applications of High-Pressure-Treated Starch

The effectiveness and simplicity of high-pressure treatment in modifying granular structures and physicochemical properties of starch have attracted many food researchers to propose diverse opportunities for food industry. Especially, achieving a desired level of gelatinization at ambient temperatures following high-pressure processing indicates a significant advantage for energy saving within various sectors of the food industry. For example, high-pressure-induced granular pregelatinized starch can be used in foods with a smooth and creamy texture such as low-fat spreads, dressings, and dips (Stute et al. 1996). High-pressure-treated corn starch could be used as a carrier for bioactive molecules, such as minerals and antioxidants, due to its increased porosity (Deladino et al. 2017). Also, it could be used to stabilize emulsions produced by homogenization (Villamonte, Jury & de Lamballerie 2016). The pressurized corn starch successfully stabilized emulsions after homogenization step without using surfactants. It is possible to incorporate high-pressure treatment to produce resistant starch, even though pressure treatment alone may not be enough to achieve the high yield of resistant starch (Bauer, Wiehle & Knorr 2005). Along with the aforementioned physicochemical and functional attributes, the influence of this method on the sensory characteristics of the end product needs to be studied further (Pei-Ling, Xiao-Song & Qun 2010).

21.7 PULSED ELECTRIC FIELD-TREATED STARCH

Thermal processing is a commonly employed technique for eliminating microorganisms in food, but it often leads to undesirable effects in taste, color, and nutritional value. Pulsed electric field (PEF) treatment could be one of the alternative non-thermal techniques that have emerged to address these drawbacks (Jeyamkondan, Jayas & Holley 1999). This technology encompasses the application of high-intensity electric pulses (<50 kVcm^{-1}) to pumpable liquid materials for short durations (<40 µs) within a processing chamber (Han et al. 2012). Prior to the PEF treatment, the electrical conductivity of starch suspension needs to be adjusted within a range of 50–200 µs using an ionic salt such as KCl for effective treatment (Castro et al. 2023). Even though, the brief time frame serves to mitigate temperature increase and undesired electrolysis (Zhu 2018), a high electric current flowing through the suspension may generate excessive heat, which needs to be cooled down in a water bath during the treatment (Castro et al. 2023). Primarily owing to its advantages such as suitability for continuous processing, operability at low temperature, uniformity of treatment, and short duration, PEF treatment has gained attention as a potential physical modification method for starch (Han et al. 2009, 2012; Li et al. 2019).

21.7.1 Characteristics of PEF-Treated Starch

21.7.1.1 Starch morphology

Visible microscopic characteristics of starch granules, such as a smooth surface and a distinct Maltese cross pattern, remain at low electric field intensities, even though a few exceptional cases may exist (Castro et al. 2023; Wu et al. 2019). However, damage becomes evident as the intensity exceeds 30 kVcm^{-1}. At 50 kVcm^{-1}, the starch granular morphology is largely deformed, leading to the aggregation of smaller starch fragments that have been postulated to be the lost envelope of starch (Han et al. 2009, 2012; Zeng et al. 2016). The observed changes in morphology indicate that electrical energy has the potential to impact the structural characteristics of starch granules, both internally and externally. The outer layer of a starch granule is typically known for its high density, which provides protection against external physical forces (Błaszczak, Valverde & Fornal 2005). The application of high-intensity PEF results in the notable degradation and fragmentation of the outer layer of starch granule, which accelerates water absorption into the interior structure of starch granules as well as the fragmented parts, finally leading to the aggregation among swollen particles (Han et al. 2009, 2012).

21.7.1.2 Physicochemical properties

Accompanied by the morphological changes of starch granules, PEF, especially when applied at a high electric field intensity from 30 up to 50 kVcm^{-1}, has provoked noticeable changes in the physicochemical properties of starch. In pasting properties, the peak, breakdown, and setback viscosity values of potato and tapioca starches decreased with increasing PEF intensities. The significantly decreased breakdown viscosities for PEF-treated potato and tapioca starches indicated the high stability of hot starch pastes, which could be advantageous for the industrial applications of thermally processed foods (Han et al. 2009, 2012). Changes in the pasting properties of PEF-treated starch were attributed to the partial disruption of starch crystalline structures, which was evidenced by the decreases in the relative crystallinity measured from X-ray diffraction patterns and the gelatinization enthalpy measured using differential scanning calorimeter for both starch samples with increasing PEF intensity (Han et al. 2009, 2012). Zeng et al. (2016) also reported significant decreases in the relative crystallinity, gelatinization enthalpy, and weight average molecular mass of waxy rice starch treated with PEF at 30–50 kVcm^{-1}. They also found that the amounts of RDS, SDS, and RS significantly changed upon high-intensity PEF treatments, with a relatively large increase and decrease in RDS and SDS, respectively, and a slight decrease in RS with increasing PEF intensity. These results were attributed to the increased enzyme susceptibility caused by the PEF-induced disruption of starch granular packaging and semi-crystalline structure. Li et al. (2019) investigated the impacts of PEF processing at varying intensities (ranging from 0 to 8.57 kVcm^{-1}) on the structural properties of wheat starch (type A), potato starch (type B), and pea starch (type C). The results suggested that the application of PEF treatment could lead to alterations in the molecular structure of all three varieties of starch more or less, particularly exhibiting more significant changes in potato starch. Also, even at the relatively low intensity treatment, they found an increase in RDS level, a decrease in SDS level, and a small change in RS level for all three types of starches.

21.7.2 Applications of PEF-Treated Starch

Initially, PEF technology was effectively employed to sterilize a range of low-viscosity, low-electrical-conductivity liquid foods, such as soup, milk, various fruit juices, and liquid eggs (Han et al. 2009; Zeng et al. 2016). Also, PEF techniques have been considered to be highly promising in facilitating chemical modification of starch. For example, at low intensity less than 10 kVcm^{-1}, PEF treatment greatly improved the efficiency of starch acetylation with reduced cost and time (Castro et al. 2023; Hong et al. 2016a, 2016b; Zhu 2018). Starch modification using PEF provides some critical benefits over chemical modification, including significant reductions of processing steps, modification time, and usage of water and waste product

(Castro et al. 2023). However, the industrial application of PEF in producing physically modified clean label starch appears to be challenging at current stage mainly due to lack of scale-up experience with a confirmed technical feasibility. Insufficient understanding of the effects of various factors, including starch types and their molecular interactions in PEF treatment system, also needs to be overcome (Zhu 2018). A quantity and quality of data for more variety of starch sources and experimental setups would be required to differentiate PEF technique from other techniques.

21.8 ENZYMATICALLY MODIFIED STARCH

Enzymes serve as highly selective catalysts, dramatically increasing the rate and specificity of metabolic and biochemical reactions. Enzymatic modification provides a multitude of advantages, including the elimination of undesired byproducts, improved purity levels, cost-effectiveness, and the reliable generation of superior products (Radeloff & Beck 2016). This approach is commonly utilized in starch modification to reproduce the properties of chemically modified starch and introduce novel functions (Patil 2013; Radeloff & Beck 2016). It is essential to note that starch that has been processed using enzymes can be classified as clean label starch due to the absence of chemical additives. This section provides an examination of various types of enzymes utilized in enzymatic procedures and their corresponding uses.

21.8.1 Enzymes Used for Clean Label Starch Production

Among different enzymes that interact with starch, 4-α-glucanotransferases (4αGTases) are widely utilized in the production of clean label starch. These enzymes, part of the glycoside hydrolase (GH) superfamily, specifically cleave the α-1,4-glycosidic bonds in substrates like amylose, amylopectin, glycogen, and maltodextrins. Nevertheless, these transferases do not simply cleave the glycosidic bond; rather, they facilitate the transfer of the excised glucan molecule from the donor to the non-reducing end of another glycosidic acceptor, resulting in the formation of a novel glycosidic linkage. This group of enzymes classified as transferases consists of branching enzyme (BE, EC 2.4.1.18), cyclodextrin glycosyltransferase (CGTase, EC 2.4.1.19), and 4-α-glucanotransferase (4αGTase, EC 2.4.1.25) (Van der Maarel et al. 2002). BE, an enzyme responsible for catalyzing the formation of α-1,6-glucosidic bonds in starch and glycogen, acts by breaking an α-1,4-glycosidic bond and transferring the resulting α-glucan chain to a vacant 6-hydroxyl group on another glucan chain. This process forms an α-1,6-glycosidic bond, creating a branch (Van der Maarel & Leemhuis 2013). This enzyme is responsible for both the branching and cyclization reactions of amylose and amylopectin (Takata et al. 2010). CGTase has the ability to synthesize cyclic oligosaccharides known as cyclodextrins, which consist of 6, 7, or 8 glucose units and have the capability to form inclusion complexes with distinct guest molecules (Uekama, Hirayama & Irie 1998). The enzyme 4αGTase, also referred to as amylomaltase and D-enzyme, facilitates transglycosylation reactions on starch molecules by catalyzing processes including hydrolysis, coupling, disproportionation, and cyclization (Takaha et al. 1996). This enzyme has the ability to generate altered amylopectin clusters and cyclic α-1,4-glucan products, such as cycloamyloses or large-ring cyclodextrins, through both intermolecular and intramolecular transglycosylation processes (Nimpiboon et al. 2020). It is also utilized in the production of thermoreversible starch gels (Kaper et al. 2005; Van der Maarel et al. 2005).

21.8.2 Applications of Enzymatically Modified Starch

Starches treated with 4αGTase have garnered significant interest, resulting in the development of new commercial products (Van der Maarel & Leemhuis 2013). 4αGTase-treated starches include cycloamylose (Takaha et al. 1996; Van der Maarel & Leemhuis 2013), modified amylopectin cluster (Do et al. 2012; Patil 2013; Van der Maarel & Leemhuis 2013), SDS, and RS (Jiang et al. 2014; Yang et al. 2021). The enzyme 4αGTase displays a reaction pattern that bears resemblance to that of CGTase, although distinct differences exist between the two enzymes. The enzyme 4αGTase is known for producing high-molecular-weight products (10^4–10^5 Da) and large-ring cyclic glucans (DP 17–40) from amylose in starch substrates. In contrast, CGTase is characterized by reducing the molecular weight of starch and mainly generating smaller cyclic products with degrees of polymerization of 6, 7, and 8 (Van der Maarel et al. 2002). Starches modified by the enzyme 4αGTase exhibit unique characteristics in forming gels that are reversible with changes in temperature and enhanced stability during freeze–thaw cycles (Do et al. 2012; Lee et al. 2006; Van der Maarel et al. 2005). These starches, when subjected to repeated heating and cooling cycles, can repeatedly melt and solidify, akin to gelatin. A commercially available 4αGTase-treated starch, known as Etenia™, is marketed as clean label potato starch. It can serve as a substitute for gelatin, especially in jelly-type confections, enhance creaminess and mouthfeel in low-fat dairy products, and be used in emulsified low-fat spreads (Patil 2013). 4αGTase-treated starches may contain RS and SDS, with contents that vary depending on the sources of starch and enzyme, as well as the degree of enzyme treatment. Jiang et al. (2014) reported that normal maize starch treated with 4αGTase (10 U/g dry starch) originating from *Acidothermus cellulolyticus* 11B exhibited an increase in RS content from 10.52% to 21.63% after 4 hours of treatment and an increase in SDS content from 9.4% to 20.92% after 12 hours of treatment.

A similar experiment using normal, waxy, and high-amylose maize starches treated with 20 U of 4αGTase from *Azotobacter chroococcum* NCIMB 8003 for 72 hours revealed increases in RS content from 7.9%, 4.6%, and 3.2% to 15.4%, 19.17%, and 17.8%, respectively (Yang et al. 2021).

Highly branched cyclic dextrin (HBCD) is synthesized through the enzymatic treatment of amylopectin-rich waxy starch using *Bacillus stearothermophilus* BE (Takata et al. 2010). The United States Food and Drug Administration (FDA) granted Generally Recognized as Safe (GRAS) status to HBCD in 2011. Therefore, it has been reclassified from a food additive to a food ingredient and is now marketed under the trade name of cyclic cluster dextrin (CCD) (Wilburn, Machek & Ismaeel 2021). The product has been utilized in the formulation of sports drinks and as a spray-drying aid in various industries, and it has also been employed for mitigating undesirable flavors (Takata et al. 2010). HBCD is frequently claimed to possess an ergogenic advantage as a sports nutrition supplement, primarily because of its abilities to promote faster gastric emptying and absorption compared with other carbohydrate sources (e.g., maltodextrin and glucose), which are related to its high molecular weight (average Mw ~160,000 Da) and low osmolality (Furuyashiki et al. 2014; Shiraki et al. 2015; Suzuki et al. 2014; Takii et al. 2007; Wilburn, Machek & Ismaeel 2021). More extensive research with a larger sample size and various control carbohydrates is required to determine the physiological advantages of HBCD (Wilburn, Machek & Ismaeel 2021).

21.9 CONCLUSIONS

This chapter has provided an overview of clean label starches that are currently available. While clean label starch, manufactured without the use of chemicals, is unlikely to completely replace traditional modified starch due to various constraints, there are still numerous possibilities for its utilization. Researchers are consistently enhancing the discipline by providing fresh perspectives on procedures, innovative enzymes, and affordable choices for customers. As a result, despite currently constituting only 7%–8% of the overall modified starch market worldwide, clean label starch is anticipated to experience increased adoption and potentially supplant modified starch over time.

REFERENCES

Abbas, S, Bashari, M, Akhtar, W, Li, WW, & Zhang, X 2014. Process Optimization of Ultrasound-Assisted Curcumin Nanoemulsions Stabilized by Osa-Modified Starch. *Ultrasonics Sonochemistry, 21(4)*, 1265–1274.

Abraham, TE 1993. Stabilization of Paste Viscosity of Cassava Starch by Heat Moisture Treatment. *Starch-Stärke, 45(4)*, 131–135.

Adebowale, KO, Olu-Owolabi, BI, Olayinka, OO, & Lawal, OS 2005. Effect of Heat Moisture Treatment and Annealing on Physicochemical Properties of Red Sorghum Starch. *African Journal of Biotechnology, 4(9)*, 928–933.

Ahmed, J, Thomas, L, Arfat, YA, & Joseph, A 2018. Rheological, Structural and Functional Properties of High-Pressure Treated Quinoa Starch in Dispersions. *Carbohydrate Polymers, 197*, 649–657.

Ahmed, J, Thomas, L, Taher, A, & Joseph, A 2016. Impact of High Pressure Treatment on Functional, Rheological, Pasting, and Structural Properties of Lentil Starch Dispersions. *Carbohydrate Polymers, 152*, 639–647.

Alcázar-Alay, SC, & Meireles, MAA 2015. Physicochemical Properties, Modifications and Applications of Starches from Different Botanical Sources. *Food Science Technology, 35(2)*, 215–236.

Anwar, E, Khotimah, H, & Yanuar, A 2006. An Approach on Pregelatinized Cassava Starch Phosphate Esters as Hydrophilic Polymer Excipient for Controlled Release Tablet. *Journal of Medical Science, 6(6)*, 923–929.

Ares, G, De Saldamando, L, Giménez, A, & Deliza, R 2014. Food and Wellbeing. Towards a Consumer-Based Approach. *Appetite, 74*, 61–69.

Arocas, A, Sanz, T, & Fiszman, S 2009. Clean Label Starches as Thickeners in White Sauces. Shearing, Heating and Freeze/Thaw Stability. *Food Hydrocolloids, 23(8)*, 2031–2037.

Aschemann-Witzel, J, Varela, P, & Peschel, AO 2019. Consumers' Categorization of Food Ingredients: Do Consumers Perceive Them as 'Clean Label' Producers Expect? An Exploration with Projective Mapping. *Food Quality and Preference, 71*, 117–128.

Baines, D, & Seal, R (eds) 2012. *Natural Food Additives, Ingredients and Flavourings*, 1st Ed, Elsevier.

Bauer, BA, Wiehle, T, & Knorr, D 2005. Impact of High Hydrostatic Pressure Treatment on the Resistant Starch Content of Wheat Starch. *Starch-Stärke, 57(3–4)*, 124–133.

Björck, I, Asp, N-G, Birkhed, D, Eliasson, A-C, Sjöberg, L-B, & Lundquist, I 1984. Effects of Processing on Starch Availability *in Vitro* and *in Vivo*. II. Drum-Drying of Wheat Flour. *Journal of Cereal Science, 2(3)*, 165–178.

Błaszczak, W, Valverde, S, & Fornal, J 2005. Effect of High Pressure on the Structure of Potato Starch. *Carbohydrate Polymers, 59(3)*, 377–383.

Bourekoua, H, Benatallah, L, Zidoune, MN, & Rosell, CM 2016. Developing Gluten Free Bakery Improvers by Hydrothermal Treatment of Rice and Corn Flours. *LWT-Food Science and Technology, 73*, 342–350.

Brumovsky, JO, & Thompson, DB 2001. Production of Boiling-Stable Granular Resistant Starch by Partial Acid Hydrolysis and Hydrothermal Treatments of High-Amylose Maize Starch. *Cereal Chemistry, 78(6)*, 680–689.

Busken, DF 2015. Cleaning It Up—What Is a Clean Label Ingredient? *Cereal Foods World, 60*, 112–113.

Campbell, A (eds) 1981. *The Sense of Well-Being in America: Recent Patterns and Trends*, 1st Ed, McGraw-Hill.

Carmona-García, R, Bello-Pérez, L, Aguirre-Cruz, A, Aparicio-Saguilán, A, Hernández-Torres, J, & Alvarez-Ramirez, J 2016. Effect of Ultrasonic Treatment on the Morphological, Physicochemical, Functional, and Rheological Properties of Starches with Different Granule Size. *Starch-Stärke, 68(9–10)*, 972–979.

Castro, LM, Alexandre, EM, Saraiva, JA, & Pintado, M 2023. Starch Extraction and Modification by Pulsed Electric Fields. *Food Reviews International, 39(4)*, 2161–2182.

Cham, S, & Suwannaporn, P 2010. Effect of Hydrothermal Treatment of Rice Flour on Various Rice Noodles Quality. *Journal of Cereal Science, 51(3)*, 284–291.

Chan, H-T, Bhat, R, & Karim, AA 2010. Effects of Sodium Dodecyl Sulphate and Sonication Treatment on Physicochemical Properties of Starch. *Food Chemistry, 120(3)*, 703–709.

Chen, J-J, Lai, VMF, & Lii, C-Y 2003. Effects of Compositional and Granular Properties on the Pasting Viscosity of Rice Starch Blends. *Starch-Stärke, 55(5)*, 203–212.

Cheng, W, Chen, J, Liu, D, Ye, X, & Ke, F 2010. Impact of Ultrasonic Treatment on Properties of Starch Film-Forming Dispersion and the Resulting Films. *Carbohydrate Polymers, 81(3)*, 707–711.

Cheung, T, Junghans, A, Dijksterhuis, G, Kroese, F, Johansson, P, Hall, L, & De Ridder, D 2016. Consumers' Choice-Blindness to Ingredient Information. *Appetite, 106*, 2–12.

Chinnaswamy, R, & Hanna, M 1988. Optimum Extrusion-Cooking Conditions for Maximum Expansion of Corn Starch. *Journal of Food Science, 53(3)*, 834–836.

Choi, E-H, & Lee, J-K 2017. Effects of Sonication on Physicochemical Properties and Pore Formation of Maize Starch. *Korean Journal of Food Science and Technology, 49(5)*, 507–512.

Chrzan, J 2004. Slow Food: What, Why, and to Where? *Food, Culture & Society, 7(2)*, 117–132.

Chung, H, Liu, Q, & Hoover, R 2009. Impact of Annealing and Heat-Moisture Treatment on Rapidly Digestible, Slowly Digestible and Resistant Starch Levels in Native and Gelatinized Corn, Pea and Lentil Starches. *Carbohydrate Polymers, 75(3)*, 436–447.

Chung, K, Moon, T, & Chun, J 2000. Influence of Annealing on Gel Properties of Mung Bean Starch. *Cereal Chemistry, 77(5)*, 567–571.

Collado, LS, & Corke, H 1999. Heat-Moisture Treatment Effects on Sweetpotato Starches Differing in Amylose Content. *Food Chemistry, 65(3)*, 339–346.

da Rosa Zavareze, E, & Dias, ARG 2011. Impact of Heat-Moisture Treatment and Annealing in Starches: A Review. *Carbohydrate Polymers, 83(2)*, 317–328.

Dapčević Hadnađev, TR, Torbica, AM, & Hadnađev, MS 2013. Influence of Buckwheat Flour and Carboxymethyl Cellulose on Rheological Behaviour and Baking Performance of Gluten-Free Cookie Dough. *Food and Bioprocess Technology, 6*, 1770–1781.

de la Barca, AMC, Rojas-Martínez, ME, Islas-Rubio, AR, & Cabrera-Chávez, F 2010. Gluten-Free Breads and Cookies of Raw and Popped Amaranth Flours with Attractive Technological and Nutritional Qualities. *Plant Foods for Human Nutrition, 65*, 241–246.

Degrois, M, Gallant, D, Baldo, P, & Guilbot, A 1974. The Effects of Ultrasound on Starch Grains. *Ultrasonics, 12(3)*, 129–131.

Deladino, L, Teixcira, AS, Plou, FJ, Navarro, AS, & Molina-García, AD 2017. Effect of High Hydrostatic Pressure, Alkaline and Combined Treatments on Corn Starch Granules Metal Binding: Structure, Swelling Behavior and Thermal Properties Assessment. *Food and Bioproducts Processing, 102*, 241–249.

Demirkesen, I, Mert, B, Sumnu, G, & Sahin, S 2010. Utilization of Chestnut Flour in Gluten-Free Bread Formulations. *Journal of food engineering, 101(3)*, 329–336.

Do, HV, Lee, E-J, Park, J-H, Park, K-H, Shim, J-Y, Mun, S, & Kim, Y-R 2012. Structural and Physicochemical Properties of Starch Gels Prepared from Partially Modified Starches Using Thermus Aquaticus 4-α-Glucanotransferase. *Carbohydrate Polymers, 87(4)*, 2455–2463.

Douzals, J-P, Marechal, P-A, Coquille, JC, & Gervais, P 1996. Microscopic Study of Starch Gelatinization under High Hydrostatic Pressure. *Journal of Agricultural and Food Chemistry, 44(6)*, 1403–1408.

Eerlingen, RC, Jacobs, H, Block, K, & Delcour, JA 1997. Effects of Hydrothermal Treatments on the Rheological Properties of Potato Starch. *Carbohydrate Research, 297(4)*, 347–356.

Fritze, H 1973. Dry Gelatinized Starch Produced on Different Types of Drum Dryers. *Industrial & Engineering Chemistry Process Design and Development, 12(2)*, 142–148.

Fu, Z-Q, Wang, L-J, Li, D, & Adhikari, B 2012. Effects of Partial Gelatinization on Structure and Thermal Properties of Corn Starch after Spray Drying. *Carbohydrate Polymers, 88(4)*, 1319–1325.

Furukawa, R, Suto, Y, Ishida, EH, & Yamauchi, T (eds) 2019. *Lifestyle and Nature: Integrating Nature Technology to Sustainable Lifestyles*, 1st Ed, CRC Press.

Furuyashiki, T, Tanimoto, H, Yokoyama, Y, Kitaura, Y, Kuriki, T, & Shimomura, Y 2014. Effects of Ingesting Highly Branched Cyclic Dextrin During Endurance Exercise on Rating of Perceived Exertion and Blood Components Associated with Energy Metabolism. *Bioscience, Biotechnology, and Biochemistry, 78(12)*, 2117–2119.

Gallant, D, Degrois, M, Sterling, C, & Guilbot, A 1972. Microscopic Effects of Ultrasound on the Structure of Potato Starch Preliminary Study. *Starch-Stärke, 24(4)*, 116–123.

Gomes, AM, da Silva, CEM, & Ricardo, NM 2005. Effects of Annealing on the Physicochemical Properties of Fermented Cassava Starch (Polvilho Azedo). *Carbohydrate Polymers, 60(1)*, 1–6.

Gomes, AM, da Silva, CEM, Ricardo, NM, Sasaki, JM, & Germani, R 2004. Impact of Annealing on the Physicochemical Properties of Unfermented Cassava Starch ("Polvilho Doce"). *Starch-Stärke, 56(9)*, 419–423.

Gómez, M, Oliete, B, Rosell, CM, Pando, V, & Fernández, E 2008. Studies on Cake Quality Made of Wheat–Chickpea Flour Blends. *LWT-Food Science and Technology, 41(9)*, 1701–1709.

Guo, Z, Zeng, S, Lu, X, Zhou, M, Zheng, M, & Zheng, B 2015a. Structural and Physicochemical Properties of Lotus Seed Starch Treated with Ultra-High Pressure. *Food Chemistry, 186*, 223–230.

Guo, Z, Zeng, S, Zhang, Y, Lu, X, Tian, Y, & Zheng, B 2015b. The Effects of Ultra-High Pressure on the Structural, Rheological and Retrogradation Properties of Lotus Seed Starch. *Food Hydrocolloids, 44*, 285–291.

Han, Z, Zeng, XA, Yu, SJ, Zhang, BS, & Chen, XD 2009. Effects of Pulsed Electric Fields (PEF) Treatment on Physicochemical Properties of Potato Starch. *Innovative Food Science & Emerging Technologies, 10(4)*, 481–485.

Han, Z, Zeng, XA, Fu, N, Yu, SJ, Chen, XD, & Kennedy, JF 2012. Effects of Pulsed Electric Field Treatments on Some Properties of Tapioca Starch. *Carbohydrate Polymers, 89(4)*, 1012–1017.

Herceg, IL, Jambrak, AR, ŠubArIć, D, Brnčić, M, Brnčić, SR, Badanjak, M, Tripalo, B, Ježek, D, Novotni, D, & Herceg, Z 2010. Texture and Pasting Properties of Ultrasonically Treated Corn Starch. *Czech Journal of Food Sciences, 28(2)*, 83–93.

Hirth, M, Leiter, A, Beck, SM, & Schuchmann, HP 2014. Effect of Extrusion Cooking Process Parameters on the Retention of Bilberry Anthocyanins in Starch Based Food. *Journal of Food Engineering, 125*, 139–146.

Holmes, G, Prior, P, Lane, M, Pope, D, & Allan, R 1989. Malignancy in Coeliac Disease--Effect of a Gluten Free Diet. *Gut, 30(3)*, 333.

Hong, J, Chen, R, Zeng, X-A, & Han, Z 2016a. Effect of Pulsed Electric Fields Assisted Acetylation on Morphological, Structural and Functional Characteristics of Potato Starch. *Food Chemistry, 192*, 15–24.

Hong, J, Zeng, X-A, Buckow, R, Han, Z, & Wang, M-S 2016b. Nanostructure, Morphology and Functionality of Cassava Starch after Pulsed Electric Fields Assisted Acetylation. *Food Hydrocolloids, 54*, 139–150.

Hu, X, Xu, X, Jin, Z, Tian, Y, Bai, Y, & Xie, Z 2011. Retrogradation Properties of Rice Starch Gelatinized by Heat and High Hydrostatic Pressure (HHP). *Journal of Food Engineering, 106(3)*, 262–266.

Hutt, CA, & Sloan, A 2015. Coming Clean: What Clean Label Means for Consumers and Industry, *Global Food Forum-Clean Label Conference*, Itasca, IL.

Ismail, H, Irani, M, & Ahmad, Z 2013. Starch-Based Hydrogels: Present Status and Applications. *International Journal of Polymeric Materials and Polymeric Biomaterials, 62(7)*, 411–420.

Izidoro, DR, Sierakowski, M-R, Haminiuk, CWI, De Souza, CF, & de Paula Scheer, A 2011. Physical and Chemical Properties of Ultrasonically, Spray-Dried Green Banana (Musa cavendish) Starch. *Journal of Food Engineering, 104(4)*, 639–648.

Jacobs, H, & Delcour, JA 1998. Hydrothermal Modifications of Granular Starch, with Retention of the Granular Structure: A Review. *Journal of Agricultural and Food Chemistry, 46(8)*, 2895–2905.

Jacobs, H, Eerlingen, RC, & Delcour, JA 1996. Factors Affecting the Visco-Amylograph and Rapid Visco-Analyzer Evaluation of the Impact of Annealing on Starch Pasting Properties. *Starch-Stärke, 48(7–8)*, 266–270.

Jambrak, AR, Herceg, Z, Šubarić, D, Babić, J, Brnčić, M, Brnčić, SR, Bosiljkov, T, Čvek, D, Tripalo, B, & Gelo, J 2010. Ultrasound Effect on Physical Properties of Corn Starch. *Carbohydrate Polymers, 79(1)*, 91–100.

Jayakody, L, & Hoover, R 2008. Effect of Annealing on the Molecular Structure and Physicochemical Properties of Starches from Different Botanical Origins–A Review. *Carbohydrate Polymers, 74(3)*, 691–703.

Jeyamkondan, S, Jayas, D, & Holley, R 1999. Pulsed Electric Field Processing of Foods: A Review. *Journal of Food Protection, 62(9)*, 1088–1096.

Jiang, B, Li, W, Hu, X, Wu, J, & Shen, Q 2015a. Rheology of Mung Bean Starch Treated by High Hydrostatic Pressure. *International Journal of Food Properties, 18(1)*, 81–92.

Jiang, B, Li, W, Shen, Q, Hu, X, & Wu, J 2015b. Effects of High Hydrostatic Pressure on Rheological Properties of Rice Starch. *International Journal of Food Properties, 18(6)*, 1334–1344.

Jiang, H, Miao, M, Ye, F, Jiang, B, & Zhang, T 2014. Enzymatic Modification of Corn Starch with 4-α-Glucanotransferase Results in Increasing Slow Digestible and Resistant Starch. *International Journal of Biological Macromolecules, 65*, 208–214.

Jiranuntakul, W, Puttanlek, C, Rungsardthong, V, Puncha-Arnon, S, & Uttapap, D 2011. Microstructural and Physicochemical Properties of Heat-Moisture Treated Waxy and Normal Starches. *Journal of Food Engineering, 104(2)*, 246–258.

Juansang, J, Puttanlek, C, Rungsardthong, V, Puncha-Arnon, S, & Uttapap, D 2012. Effect of Gelatinisation on Slowly Digestible Starch and Resistant Starch of Heat-Moisture Treated and Chemically Modified Canna Starches. *Food Chemistry, 131(2)*, 500–507.

Juhász, R, & Salgó, A 2008. Pasting Behavior of Amylose, Amylopectin and Their Mixtures as Determined by RVA Curves and First Derivatives. *Starch-Stärke, 60(2)*, 70–78.

Kaper, T, Talik, B, Ettema, TJ, Bos, H, van der Maarel, MJ, & Dijkhuizen, L 2005. Amylomaltase of *Pyrobaculum Aerophilum* IM2 Produces Thermoreversible Starch Gels. *Applied and Environmental Microbiology, 71(9)*, 5098–5106.

Karam, LB, Grossmann, MVE, Silva, RSS, *Ferrero, C, & Zaritzky*, NE 2005. Gel Textural Characteristics of Corn, Cassava and Yam Starch Blends: A Mixture Surface Response Methodology Approach. *Starch-Stärke, 57(2)*, 62–70.

Khanal, SK, Montalbo, M, Van Leeuwen, J, Srinivasan, G, & Grewell, D 2007. Ultrasound Enhanced Glucose Release from Corn in Ethanol Plants. *Biotechnology Bioengineering, 98(5)*, 978–985.

Knorr, D, Froehling, A, Jaeger, H, Reineke, K, Schlueter, O, & Schoessler, K 2011. Emerging Technologies in Food Processing. *Annual Review of Food Science and Technology, 2*, 203–235.

Koca, AF, & Anil, M 2007. Effect of Flaxseed and Wheat Flour Blends on Dough Rheology and Bread Quality. *Journal of the Science of Food and Agriculture, 87(6)*, 1172–1175.

Lee, KY, Kim, Y-R, Park, KH, & Lee, HG 2006. Effects of A-Glucanotransferase Treatment on the Thermo-Reversibility and Freeze-Thaw Stability of a Rice Starch Gel. *Carbohydrate Polymers, 63(3)*, 347–354.

Li, G, & Zhu, F 2018. Effect of High Pressure on Rheological and Thermal Properties of Quinoa and Maize Starches. *Food Chemistry, 241*, 380–386.

Li, Q, Wu, Q-Y, Jiang, W, Qian, J-Y, Zhang, L, Wu, M, Rao, S-Q, & Wu, C-S 2019. Effect of Pulsed Electric Field on Structural Properties and Digestibility of Starches with Different Crystalline Type in Solid State. *Carbohydrate Polymers, 207*, 362–370.

Li, W, Zhang, F, Liu, P, Bai, Y, Gao, L, & Shen, Q 2011. Effect of High Hydrostatic Pressure on Physicochemical, Thermal and Morphological Properties of Mung Bean (Vigna radiata L.) Starch. *Journal of Food Engineering, 103(4)*, 388–393.

Liu, H, Corke, H, & Ramsden, L 2000. The Effect of Autoclaving on the Acetylation of Ae, Wax, and Normal Maize Starches. *Starch-Stärke, 52(10)*, 353–360.

Liu, H, Fan, H, Cao, R, Blanchard, C, & Wang, M 2016a. Physicochemical Properties and in Vitro Digestibility of Sorghum Starch Altered by High Hydrostatic Pressure. *International Journal of Biological Macromolecules, 92*, 753–760.

Liu, H, & Lelievre, J 1992. A Differential Scanning Calorimetry Study of Melting Transitions in Aqueous Suspensions Containing Blends of Wheat and Rice Starch. *Carbohydrate Polymers, 17(2)*, 145–149.

Liu, H, Wang, L, Cao, R, Fan, H, & Wang, M 2016b. In Vitro Digestibility and Changes in Physicochemical and Structural Properties of Common Buckwheat Starch Affected by High Hydrostatic Pressure. *Carbohydrate Polymers, 144*, 1–8.

Luo, Z, Fu, X, He, X, Luo, F, Gao, Q, & Yu, S 2008. Effect of Ultrasonic Treatment on the Physicochemical Properties of Maize Starches Differing in Amylose Content. *Starch-Stärke, 60(11)*, 646–653.

Macdonald, C 2017. Survey: A Majority of Consumers Are Confused by Food Ingredients. Available from: https://www.fooddive.com/news/survey-a-majority-of-consumers-are-confused-by-food-ingredients/445922/

Majzoobi, M, Radi, M, Farahanaky, A, Jamalian, J, Tongdang, T, & Mesbahi, G 2011. Physicochemical Properties of Pre-Gelatinized Wheat Starch Produced by a Twin Drum Drier. *Journal of Agricultural Science Technology, 13(2)*, 193–202.

Mancebo, CM, Rodriguez, P, & Gomez, M 2016. Assessing Rice Flour-Starch-Protein Mixtures to Produce Gluten Free Sugar-Snap Cookies. *LWT-Food Science and Technology, 67,* 127–132.

MarketsandMarkets 2021. Modified Starch Market: Global Forecast to 2027. Available from: https://www.marketsandmarkets.com/Market-Reports/modified-starch-market-511.html

Mason, T, & Joyce, E 2008. Sonication Used as a Biocide. A Review: Ultrasound a Greener Alternative to Chemical Biocides? *Chemistry Today, 26(6),* 12–15.

McMahon, A-T, Williams, P, & Tapsell, L 2010. Reviewing the Meanings of Wellness and Well-Being and Their Implications for Food Choice. *Perspectives in Public Health, 130(6),* 282–286.

Nie, H, Li, C, Liu, P-H, Lei, C-Y, & Li, J-B 2019. Retrogradation, Gel Texture Properties, Intrinsic Viscosity and Degradation Mechanism of Potato Starch Paste under Ultrasonic Irradiation. *Food Hydrocolloids, 95,* 590–600.

Nimpiboon, P, Tumhom, S, Nakapong, S, & Pongsawasdi, P 2020. Amylomaltase from Thermus Filiformis: Expression in Saccharomyces Cerevisiae and Its Use in Starch Modification. *Journal of Applied Microbiology, 129(5),* 1287–1296.

Nitayavardhana, S, Shrestha, P, Rasmussen, ML, Lamsal, BP, van Leeuwen, J, & Khanal, SK 2010. Ultrasound Improved Ethanol Fermentation from Cassava Chips in Cassava-Based Ethanol Plants. *Bioresource Technology, 101(8),* 2741–2747.

Noda, T, Tsuda, S, Mori, M, Takigawa, S, Matsuura-Endo, C, Kim, SJ, Hashimoto, N, & Yamauchi, H 2006. Effect of Potato Starch Properties on Instant Noodle Quality in Wheat Flour and Potato Starch Blends. *Starch-Stärke, 58(1),* 18–24.

Novelo-Cen, L, & Betancur-Ancona, D 2005. Chemical and Functional Properties of Phaseolus Lunatus and Manihot Esculenta Starch Blends. *Starch-Stärke, 57(9),* 431–441.

Obanni, M, & Bemiller, JN 1997. Properties of Some Starch Blends. *Cereal Chemistry, 74(4),* 431–436.

Oh, HE, Hemar, Y, Anema, SG, Wong, M, & Pinder, DN 2008. Effect of High-Pressure Treatment on Normal Rice and Waxy Rice Starch-in-Water Suspensions. *Carbohydrate Polymers, 73(2),* 332–343.

Oh, JH, Kim, MJ, & Yoo, B 2010. Dynamic Rheological Properties of Rice Flour-Starch Blends. *Starch-Stärke, 62(6),* 321–325.

Oladunmoye, OO, Aworh, OC, Maziya-Dixon, B, Erukainure, OL, & Elemo, GN 2014. Chemical and Functional Properties of Cassava Starch, Durum Wheat Semolina Flour, and Their Blends. *Food Science & Nutrition, 2(2),* 132–138.

Parchure, A, & Kulkarni, P 1997. Effect of Food Processing Treatments on Generation of Resistant Starch. *International Journal of Food Sciences Nutrition, 48(4),* 257–260.

Park, EY, Kim, HN, Kim, JY, & Lim, ST 2009. Pasting Properties of Potato Starch and Waxy Maize Starch Mixtures. *Starch-Stärke, 61(6),* 352–357.

Park, S, & Kim, Y-R 2021. Clean Label Starch: Production, Physicochemical Characteristics, and Industrial Applications. *Food Science and Biotechnology, 30,* 1–17.

Patil, SK 2013. *Emerging and Applied Clean Label Starch Technologies,* Global Food Forum-Clean Label Conference, Oak Brook, IL.

Paucean, A, Man, S, Muste, S, & Pop, A 2016. Development of Gluten Free Cookies from Rice and Coconut Flour Blends. *Bulletin UASVM Food Science and Technology, 73(2),* 164–166.

Pei-Ling, L, Xiao-Song, H, & Qun, S 2010. Effect of High Hydrostatic Pressure on Starches: A Review. *Starch-Stärke, 62(12),* 615–628.

Pollan, M (eds) 2008. In Defense of Food: An Eater's Manifesto, Penguin.

Puncha-Arnon, S, Pathipanawat, W, Puttanlek, C, Rungsardthong, V, & Uttapap, D 2008. Effects of Relative Granule Size and Gelatinization Temperature on Paste and Gel Properties of Starch Blends. *Food Research International, 41(5),* 552–561.

Puncha-Arnon, S, & Uttapap, D 2013. Rice Starch Vs. Rice Flour: Differences in Their Properties When Modified by Heat–Moisture Treatment. *Carbohydrate Polymers, 91(1),* 85–91.

Radeloff, MA, & Beck, RH 2016. "Clean Label"-Starches and Their Functional Diversity. *Sugar Industry/Zuckerindustrie, 141(4),* 209–215.

Sajilata, MG, Singhal, RS, & Kulkarni, PR 2006. Resistant Starch–A Review. *Comprehensive Reviews in Food Science and Food Safety, 5(1),* 1–17.

Sandhu, KS, & Kaur, M 2010. Studies on Noodle Quality of Potato and Rice Starches and Their Blends in Relation to Their Physicochemical, Pasting and Gel Textural Properties. *LWT-Food Science and Technology, 43(8),* 1289–1293.

Shiraki, T, Kometani, T, Yoshitani, K, Takata, H, & Nomura, T 2015. Evaluation of Exercise Performance with the Intake of Highly Branched Cyclic Dextrin in Athletes. *Food science and Technology Research, 21(3),* 499–502.

Singh, H, Chang, YH, Lin, J-H, Singh, N, & Singh, N 2011. Influence of Heat–Moisture Treatment and Annealing on Functional Properties of Sorghum Starch. *Food Research International, 44(9),* 2949–2954.

Sit, N, Misra, S, & Deka, SC 2014. Yield and Functional Properties of Taro Starch as Affected by Ultrasound. *Food Bioprocess Technology, 7(7),* 1950–1958.

Smith, AM 2001. The Biosynthesis of Starch Granules. *Biomacromolecules, 2(2),* 335–341.

Song, JY, Park, JH, & Shin, M 2011. The Effects of Annealing and Acid Hydrolysis on Resistant Starch Level and the Properties of Cross-Linked RS4 Rice Starch. *Starch-Stärke, 63(3),* 147–153.

Song, MR, Choi, SH, Kim, HS, Kim, BY, & Baik, MY 2015. Efficiency of High Hydrostatic Pressure in Preparing Amorphous Granular Starches. *Starch-Stärke, 67(9–10),* 790–801.

Song, MR, Choi, SH, Oh, SM, Kim, HY, Bae, JE, Park, CS, Kim, BY, & Baik, MY 2017. Characterization of Amorphous Granular Starches Prepared by High Hydrostatic Pressure (HHP). *Food Science and Biotechnology, 26,* 671–678.

Stolt, M, Oinonen, S, & Autio, K 2000. Effect of High Pressure on the Physical Properties of Barley Starch. *Innovative Food Science & Emerging Technologies, 1(3),* 167–175.

Stute, R, Heilbronn, Klingler, R, Boguslawski, S, Eshtiaghi, M, & Knorr, D 1996. Effects of High Pressures Treatment on Starches. *Starch-Stärke, 48(11–12),* 399–408.

Sujka, M, & Jamroz, J 2013. Ultrasound-Treated Starch: Sem and Tem Imaging, and Functional Behaviour. *Food Hydrocolloids, 31(2),* 413–419.

Sun, D-S 2014. *Rheological and Thermal Properties of Rice-Starch Blends,* Dongguk University, Goyang, Gyeonggi.

Suzuki, A, & Sekiya, S 1994. Application of Heat-Moisture Treated Starch, "Derica Star" to Food Materials. I: The Application to Pouchpacked and Canned Foods. *Food Chemicals, 10(2),* 67–73.

Suzuki, K, Shiraishi, K, Yoshitani, K, Sugama, K, & Kometani, T 2014. Effect of a Sports Drink Based on Highly-Branched Cyclic Dextrin on Cytokine Responses to Exhaustive Endurance Exercise. *The Journal of Sports Medicine and Physical Fitness, 54(5),* 622–630.

Takaha, T, Yanase, M, Takata, H, Okada, S, & Smith, SM 1996. Potato D-Enzyme Catalyzes the Cyclization of Amylose to Produce Cycloamylose, a Novel Cyclic Glucan (∗). *Journal of Biological Chemistry, 271(6),* 2902–2908.

Takata, H, Akiyama, T, Kajiura, H, Kakutani, R, Furuyashiki, T, Tomioka, E, Kojima, I, & Kuriki, T 2010. Application of Branching Enzyme in Starch Processing. *Biocatalysis and Biotransformation, 28(1),* 60–63.

Takii, H, Kometani, T, Nishimura, T, Kuriki, T, & Fushiki, T 2007. A Sports Drink Based on Highly Branched Cyclic Dextrin Generates Few Gastrointestinal Disorders in Untrained Men During Bicycle Exercise. *Food Science and Technology Research, 10(4),* 428–431.

Te Wierik, GHP, Eissens, AC, Bergsma, J, Arends-Scholte, AW, & Bolhuis, GK 1997. A New Generation Starch Product as Excipient in Pharmaceutical Tablets: III. Parameters Affecting Controlled Drug Release from Tablets Based on High Surface Area Retrograded Pregelatinized Potato Starch. *International Journal of Pharmaceutics, 157(2),* 181–187.

Tharanathan, RN 2002. Food-Derived Carbohydrates—Structural Complexity and Functional Diversity. *Critical Reviews in Biotechnology, 22(1),* 65–84.

Tharanathan, RN 2005. Starch—Value Addition by Modification. *Critical Reviews in Food Science and Nutrition, 45(5),* 371–384.

The Business Research Company 2023. Clean Label Ingredients Global Market Report 2023. Available from: https://www.research-andmarkets.com/reports/5792880/clean-label-ingredients-global-market-report

The Institute of Food Technologists (IFT) 2018. What Is Clean Label. Available from: https://www.ift.org/news-and-publications/blog/2018/november/what-is-clean-label

Tomasik, P 2009. Specific Physical and Chemical Properties of Potato Starch. *Food, 3(Special Issue 1),* 45–56.

Uekama, K, Hirayama, F, & Irie, T 1998. Cyclodextrin Drug Carrier Systems. *Chemical Reviews, 98(5),* 2045–2076.

Vallons, KJ, & Arendt, EK 2009. Effects of High Pressure and Temperature on the Structural and Rheological Properties of Sorghum Starch. *Innovative Food Science & Emerging Technologies, 10(4),* 449–456.

Valous, N, Gavrielidou, M, Karapantsios, T, & Kostoglou, M 2002. Performance of a Double Drum Dryer for Producing Pregelatinized Maize Starches. *Journal of Food Engineering, 51(3),* 171–183.

Van der Maarel, MJ, Capron, I, Euverink, GJW, Bos, HT, Kaper, T, Binnema, DJ, & Steeneken, PA 2005. A Novel Thermoreversible Gelling Product Made by Enzymatic Modification of Starch. *Starch-Stärke, 57(10),* 465–472.

Van der Maarel, MJ, & Leemhuis, H 2013. Starch Modification with Microbial Alpha-Glucanotransferase Enzymes. *Carbohydrate Polymers, 93(1),* 116–121.

Van der Maarel, MJ, Van der Veen, B, Uitdehaag, JC, Leemhuis, H, & Dijkhuizen, L 2002. Properties and Applications of Starch-Converting Enzymes of the α-Amylase Family. *Journal of Biotechnology, 94(2),* 137–155.

Villamonte, G, Jury, V, & de Lamballerie, M 2016. Stabilizing Emulsions Using High-Pressure-Treated Corn Starch. *Food Hydrocolloids, 52,* 581–589.

Wang, S, & Guo, P 2020. Botanical Sources of Starch. In M Sjöö & L Nilsson (eds), *Starch Structure, Functionality and Application in Foods* (pp. 9–27). Woodhead Publishing.

Watcharatewinkul, Y, Puttanlek, C, Rungsardthong, V, & Uttapap, D 2009. Pasting Properties of a Heat-Moisture Treated Canna Starch in Relation to Its Structural Characteristics. *Carbohydrate Polymers, 75(3),* 505–511.

Waterschoot, J, Gomand, SV, Fierens, E, & Delcour, JA 2015. Starch Blends and Their Physicochemical Properties. *Starch-Stärke, 67(1–2),* 1–13.

Wilburn, D, Machek, S, & Ismaeel, A 2021. Highly Branched Cyclic Dextrin and Its Ergogenic Effects in Athletes: A Brief Review. *Journal of Exercise and Nutrition, 4(3).* https://doi.org/10.53520/jen2021.103100

Wiriyawattana, P, Suwonsichon, S, & Suwonsichon, T 2018. Effects of Drum Drying on Physical and Antioxidant Properties of Riceberry Flour. *Agriculture Natural Resources, 52(5),* 445–450.

Wu, C, Wu, Q-Y, Wu, M, Jiang, W, Qian, J-Y, Rao, S-Q, Zhang, L, Li, Q, & Zhang, C 2019. Effect of Pulsed Electric Field on Properties and Multi-Scale Structure of Japonica Rice Starch. *LWT, 116,* 108515.

Wu, Y, Du, X, Ge, H, & Lv, Z 2011. Preparation of Microporous Starch by Glucoamylase and Ultrasound. *Starch-Stärke, 63(4),* 217–225.

Xing, Y, Xu, Q, Ma, Y, Che, Z, Cai, Y, & Jiang, L 2014. Effect of Porous Starch Concentrations on the Microbiological Characteristics of Microencapsulated *Lactobacillus Acidophilus. Food & Function, 5(5),* 972–983.

Yang, Y, Zhao, X, Zhang, T, Hamaker, BR, & Miao, M 2021. Development of a Novel Starch-Based Dietary Fiber Using Glucanotransferase. *Food & Function, 12(13),* 5745–5754.

Yu, S, Zhang, Y, Ge, Y, Zhang, Y, Sun, T, Jiao, Y, & Zheng, XQ 2013. Effects of Ultrasound Processing on the Thermal and Retrogradation Properties of Nonwaxy Rice Starch. *Journal of Food Process Engineering, 36(6),* 793–802.

Zeng, F, Gao, Q-Y, Han, Z, Zeng, X-A, & Yu, S-J 2016. Structural Properties and Digestibility of Pulsed Electric Field Treated Waxy Rice Starch. *Food Chemistry, 194,* 1313–1319.

Zheng, J, Li, Q, Hu, A, Yang, L, Lu, J, Zhang, X, & Lin, Q 2013. Dual-Frequency Ultrasound Effect on Structure and Properties of Sweet Potato Starch. *Starch-Stärke, 65(7–8),* 621–627.

Zhu, F 2015. Impact of Ultrasound on Structure, Physicochemical Properties, Modifications, and Applications of Starch. *Trends in Food Science & Technology, 43(1),* 1–17.

Zhu, F 2017. Encapsulation and Delivery of Food Ingredients Using Starch Based Systems. *Food Chemistry, 229,* 542–552.

Zhu, F 2018. Modifications of Starch by Electric Field Based Techniques. *Trends in Food Science & Technology, 75,* 158–169.

Zhu, F, & Corke, H 2011. Gelatinization, Pasting, and Gelling Properties of Sweetpotato and Wheat Starch Blends. *Cereal Chemistry, 88(3),* 302–309.

Zia-ud-Din, Xiong, H, & Fei, P 2017. Physical and Chemical Modification of Starches: A Review. *Critical Reviews in Food Science and Nutrition, 57(12),* 2691–2705.

Zuo, YYJ, Hébraud, P, Hemar, Y, & Ashokkumar, M 2012. Quantification of High-Power Ultrasound Induced Damage on Potato Starch Granules Using Light Microscopy. *Ultrasonics Sonochemistry, 19(3),* 421–426.

Next Generation of Techniques for Structuring Starch

22

Ming Miao and James N. BeMiller

22.1 STRUCTURE–PROPERTY–PROCESSING STRATEGY FOR STRUCTURING STARCH

It is well understood that the granular nature and composition of starch vary considerably depending on its botanical origin, with the variability being caused by the combination of starch-active enzymes in biosynthetic pathways for an *in planta* or an *in vitro* modification (BeMiller & Whistler 2009; Wang, Henry & Gilbert 2014). Structure parameters of starch include molecular weight distributions, degrees of branching, chain-length distributions, branch points, and amylose:amylopectin ratios; molecular structures determine granular structures and the starch's physicochemical and functional attributes including gelatinization, pasting, retrogradation, organoleptic, and digestion properties (BeMiller & Whistler 2009; Miao & Hamaker 2021; Miao & BeMiller 2023). Thus, an understanding of the structure–functional property–processing relations of starch provide a scientific basis and molecular tools for enzymatically driven strategies of new generations of starch derivatives with precise functionality (Miao et al. 2018; Miao & BeMiller 2023). The state of the art in understanding the causal relationships of a structure–functional property–processing model using bio-approaches to alter the structures of amylose and/or amylopectin molecules, and in turn, their desired attributes are summarized in Figure 22.1 to provide direction for researchers and industry to produce novel derivatives to meet multiple demands. The precise knowledge of starch-active enzymes including glycoside hydrolases and glycosyltransferases for structuring starch to improve functionality has been extensively studied and is summarized below. However, more effort is needed to explore further structure–property–processing relations.

22.1.1 *In Planta* Transformation

Specific structures and contents of amylose and amylopectin molecules are achieved by altering the expression of various enzymes involved in both starch elongation and branching *in vivo*. However, depending on the genetic mechanism involved, the extent of amylose or amylopectin increases and associated changes in functional properties can vary dramatically, even for the same gene in different plant species. Several key starch-synthesizing enzymes, including ADP-glucose pyrophosphorylase (AGPase), granule-bound starch synthase (GBSS), soluble starch synthase (SS), starch branching enzyme (SBE), and debranching enzyme (DBE), have been found to participate in the synthesis of starch in the amyloplast (Chen, Hawkins & Seung 2021). Biosynthesis of amylose is mostly controlled by GBSS and to a lesser extent by SBE, while that of amylopectin is more complex, involving the combined actions of SS, SBE, and DBE (Seung 2020; Wang, Henry & Gilbert 2014). Yu et al. (2019) explored a quantitative data-fitting mathematical model for chain-length distributions of amylose and amylopectin molecules to obtain a mechanistic understanding of the biosynthetic processes of chain growth and termination by means of nonempirical parameterization–simulation–optimization – a process that could be used to predict the biosynthetic pathway required to create a starch with certain desired characteristics (either through genetic modification or targeted breeding programs).

In 1946, Whistler and Kramer reported a corn mutant gene that could make a specialty starch with a higher amylose content (up to 65%) than the normal 25% found in common corn starch (BeMiller & Whistler 2009). There are mainly five mutant genes (amylose-extender (*ae*), dull (*du*), sugary-1 (*su1*), sugary-2 (*su2*), and waxy (*wx*)) affecting the amylose content of grains. High-amylose starches produce high-strength gels and have film-forming ability. They are used primarily for producing and maintaining the shapes of candies, for enhancing the texture of sauces, for keeping the coating on fried products crispy, and for reducing fat uptake

DOI: 10.1201/9781003464396-22

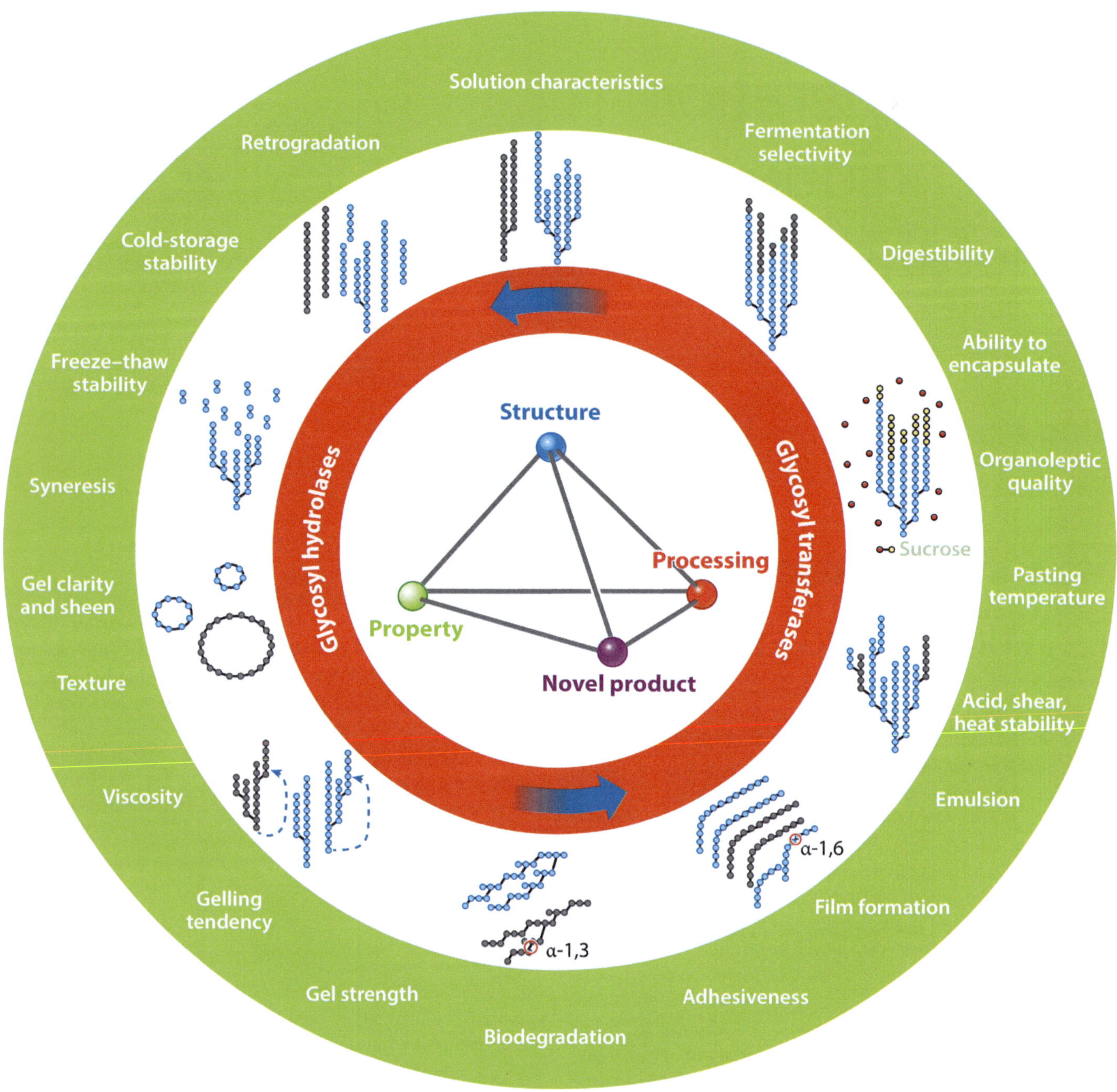

FIGURE 22.1   Overview of structure-property-processing strategy for structuring starch.

of food upon cooking. For example, amylose with a chain length of DP 1,100 has the greatest ability to form intermolecular junctions, a fractal-like network, and large aggregates (Gong et al. 2019). Also, high-amylose starches (with or without further treatment) are a source of resistant starch (RS) via retention of the granular structure (RS2), reassociation of glucan chains (RS3), and formation of lipid–amylose complexes (RS5) – giving high-amylose starches health benefits such as prevention of type 2 diabetes, colon health, glycemic balance, lipid metabolism, appetite, and satiety (Li, Gidley & Dhital 2019).

On the other hand, amylose-free (waxy) starches consist almost exclusively of amylopectin. In starches of waxy mutants, granule-bound starch synthase isoform I (GBSSI) is virtually non-existent, leading to the formation of starch granules without amylose molecules, which greatly reduces retrogradation. Waxy maize/corn starch in a high-yielding hybrid was developed by geneticists at Iowa Agricultural Experiment Station in the 1940s, whereas waxy wheat, waxy potato, and waxy tapioca starches were developed and introduced using traditional breeding methods in the 1990s–2000s (BeMiller & Whistler 2009). Among these, a naturally produced waxy potato starch containing more than 99% amylopectin was marketed in 2005 by Avebe under the brand name ELIANE™. As opposed to amylose-containing starches, waxy starches are usually known to have greater peak viscosities followed by a more noticeable breakdown and a lesser degree of setback during pasting. Waxy starch forms weak gels at relatively high concentrations that exhibit breakdown upon shear (Šárka & Dvořáček 2017). Moreover, waxy potato starch had a greater degree of granule swelling, higher paste viscosities, and better paste clarity than waxy starches from maize, wheat, rice, or tapioca (Hsieh et al. 2019). High proportions of long B chains of amylopectin were negatively correlated with paste breakdown and positively correlated with pasting temperature (Han & Hamaker 2001). For amylopectin molecules, longer A and B1 chains increase the rate of double helix formation during the retrogradation process, leading to a higher gel melting enthalpy and storage modulus (Zhou et al. 2021). Thus, waxy starches are used as thickeners and stabilizers, as well as being clean-label food ingredients. Starch structural features determine its functional properties, and novel specialty starches with tailor-made functionalities can potentially be produced *in planta*, but there is still a considerable gap between our current capabilities and market needs.

22.1.2 *In Vitro* Biotransformation

A variety of starch-converting enzymes, including glycosyl hydrolases and glycosyl transferases, alter amylose and/or amylopectin structures, producing novel products with better behavioral characteristics (Miao & BeMiller 2023). The advantages of enzymatic approaches (as compared to chemical modifications) are faster reactions at lower temperatures, greater assurance of a specific reaction, better process control, high product yields, and no reagents or fewer undesirable by-products in the effluent. These enzymes, which belong to GH13, 57, 70, or 77 families, have been isolated from bacteria, fungi, and archaea and vary widely in their substrate and product specificities (Miao et al. 2018). Glycosyl hydrolases cleave α-1,4- and/or α-1,6-glucosidic bonds present in inner parts of amylose or amylopectin chains, whereas glycosyl transferases cleave an α-1,4-glucosidic bond of the donor molecule and transfer the cleaved part to a glycosyl acceptor with the formation of a new α-1,3, α-1,4, or α-1,6 linkage. Enzymatic modifications have gained attention as tools for producing safer and more healthful starch derivatives to satisfy both environmental and consumer demands.

Branching enzymes, amylomaltases, GH70 α-transglycosylases, amylosucrases, maltogenic amylases, cyclomaltodextrinases, neopullulanases, and other microbial enzymes have been used to produce starches with modified structures. For example, starch-based products with reduced amylose content, more α-1,6 branch points, and high amounts of short chains (DP<13), which are favorable for slowing the starch digestion rate, have been made using such enzymes (Miao et al. 2018). There was a parabolic relationship between the slowly

digestible starch content and amylopectin fine structure, indicating that amylopectin-like molecules with large amounts of either long B chains (DP≥13) or short A chains (DP<13) had higher percentages of slowly digestible starch (Zhang, Ao & Hamaker 2008). For enzymatically treated amylopectin molecules, external chains with chain length of DP>13 form double helices – structures that were positively correlated with gelatinization temperatures and dissociation enthalpy, whereas the amount of fingerprint A-chains (DP 6–8) was negatively correlated with gelatinization peak and conclusion temperatures (Tao et al. 2019). The modified starch had a greater proportion of longer-chain fractions (DP>30) with an A:B chain ratio of 1.5, resulting in higher melting temperatures, enthalpies, and elastic moduli over storage time (Zhou et al. 2021). Although, the use of microbial enzymes for structuring starch to improve functionality has been extensively studied, the desire for using starch-active enzymes working optimally to produce novel starches with ideal performance is still not satisfied, and further effort is needed to establish structure–functional property relations.

22.2 ARTIFICIAL STARCH AND SUGAR SYNTHESIS FROM CARBON DIOXIDE

22.2.1 Chemoenzymatic Starch Synthesis

Starch synthesis in green plants involves about 60 steps and complex regulation. Growing starch-storing plants requires cultivated land and freshwater. Cai et al. (2021) developed a novel technology (using a chemical–biochemical hybrid pathway for cell-free synthesis) that turns carbon dioxide (CO_2) into starch in a highly efficient manner. Their hybrid process that makes starch from CO_2 is an artificial starch anabolic pathway (ASAP). This ASAP could lead to massive savings when it comes to land and water use, as well as moving toward human food security. Using a zinc- and zirconium-based catalyst, the CO_2 is first reduced to methanol, which serves as the building block for the rest of the process. The methanol is then subjected to a select combination of natural and engineered enzymes that turn it into three-carbon sugar molecules, which are then turned into six-carbon sugars, and finally into long-chain polymeric starch molecules. The modular assembly consisting of 11 core reactions (shown in Figure 22.2) was created by computational pathway design, established through modular assembly and substitution, and optimized by protein engineering of three bottleneck-associated enzymes. In a chemoenzymatic system with spatial and temporal segregation, this technique of ASAP, driven by hydrogen, converts CO_2 to starch at a rate of 22 nanomoles of CO_2 per minute per milligram of total catalyst, 8.5 times as efficiently as corn (corn plants require 120 days to make starch). This creative approach opens the way toward future chemo-biohybrid starch synthesis

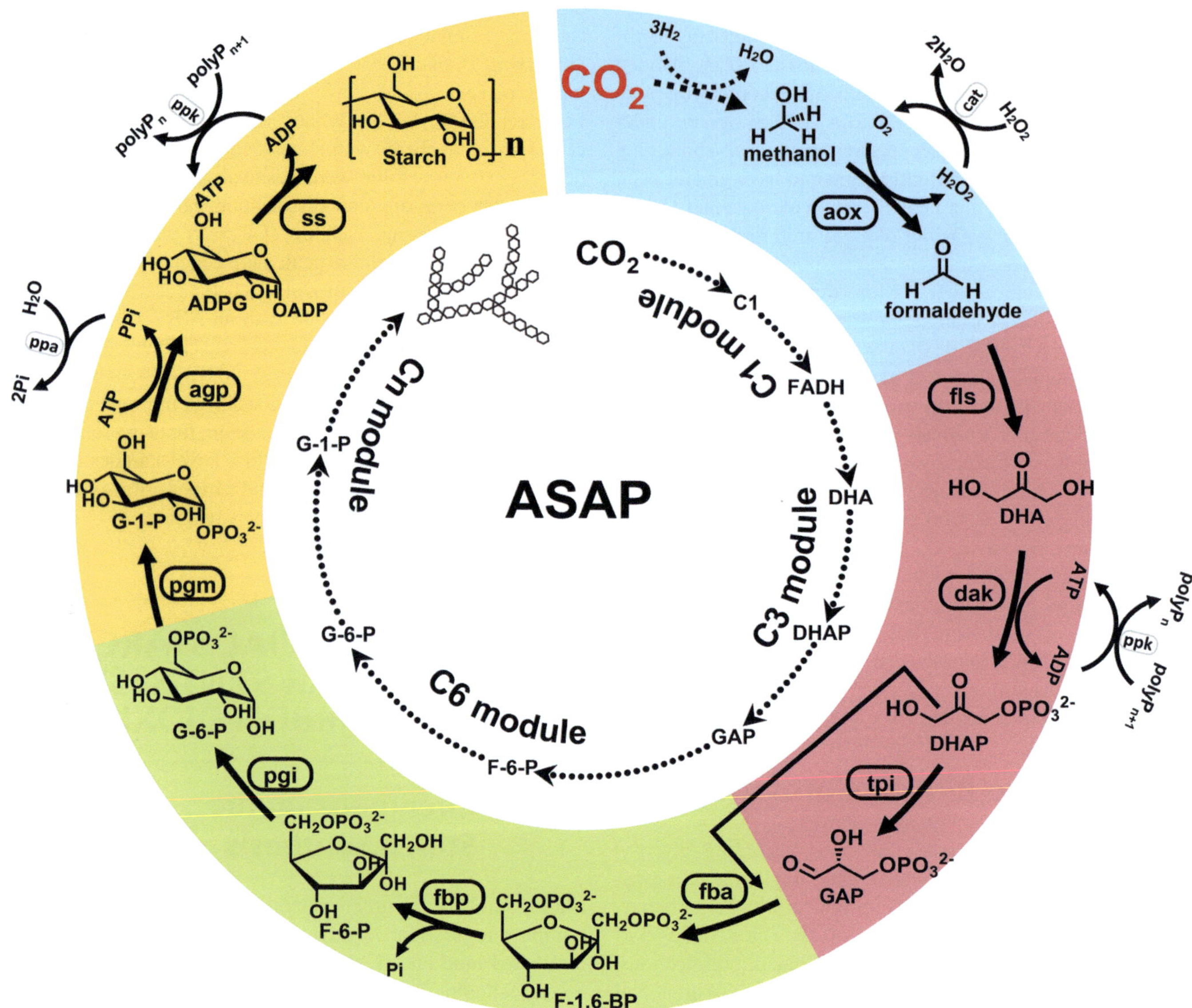

FIGURE 22.2 Starch synthesis via ASAP from CO_2. Inner circle: schematic of the artificial starch pathway drafted by computational pathway design with divided modules. C1 here indicates formic acid and methanol. Outer circle: schematic of artificial starch anabolic pathway (ASAP) 1.0, with individual modules colored. Auxiliary enzymes and chemicals are indicated. ADPG, ADP glucose; aox, alcohol oxidase; FADH, formaldehyde; F-1,6-BP, D-fructose-1,6-bisphosphate; F-6-P, D-fructose-6-phosphate; GAP, D-glyceraldehyde 3-phosphate; pgi, phosphoglucose isomerase; polyP, polyphosphate; pgm, phosphoglucomutase; ppa, pyrophosphatase; ppk, polyphosphate kinase; ss, starch synthase; tpi, triosephosphate isomerase.

from CO_2. However, plenty of room remains to increase the efficiency and productivity of the technique, such as increasing the activity and stability of the stable enzymes on a large scale at low cost, as well as designing and building a proper reactor.

22.2.2 *De Novo* Sugar Synthesis

The conventional method for manufacturing sugars is via extraction from plants (and then conversion in the case of starch). Such manufacturing processes are bioresource dependent with limited energy conversion efficiency. Artificial synthesis of sugars from CO_2 is a promising approach for

addressing challenges posed by land scarcity and climate change. Yang et al. (2023) created an artificial CO_2-to-sugars pathway (ACSP) to facilitate the enantioselective transformation of the carbon atom of CO_2 to specific sugar stereoisomers (Figure 22.3). The versatile chemoenzymatic roadmap of ACSP occurs in four steps: (1) thermochemical reduction of CO_2 to methanol; (2) tandem enzymatic aldol condensation of the one-carbon unit to build a specific stereo-chemical configuration in a phosphorylated sugar; (3) enzymatic iso/epimerization to change the chiral center of this phosphorylated sugar precursor to a defined enantiomer; and (4) dephosphorylation of the generated intermediate "sugar-phosphate" to a specific desired sugar such as D-glucose. Minimum ATP consumption

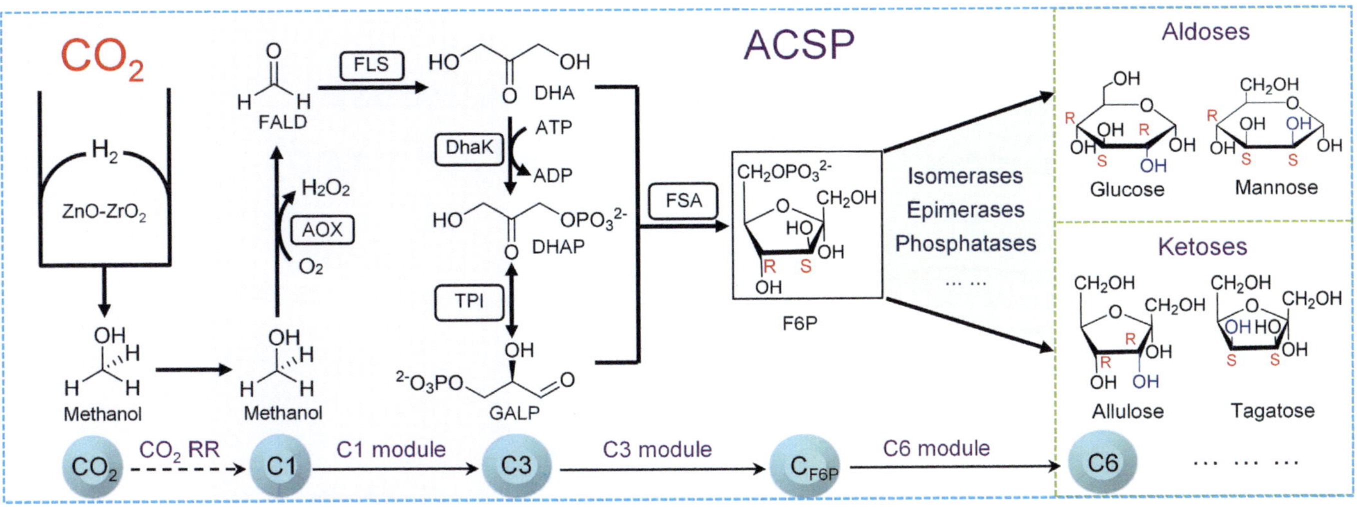

FIGURE 22.3 Sugar synthesis via artificial CO_2-to-sugars route (ACSP). FALD: formaldehyde; DHA: dihydroxyacetone; DHAP: dihydroxyacetone phosphate; GALP: glyceraldehyde 3-phosphate; F6P: fructose-6-phosphate; AOX: alcohol oxidase; FLS: formolase; DhaK: dihydroxyacetone kinase; TPI: triosephosphate isomerase; FSA: fructose 6-phosphate aldolase.

and shortest pathway for bottom-up biosynthesis of D-fructose 6-phosphate (a fundamental precursor) were demonstrated. In comparison to other reported artificial sugar synthesis routes (Antonovsky et al. 2016; Wang et al. 2022; Zheng et al. 2022), the remarkable achievements in terms of minimum ATP consumption (1/6 mol ATP per CO_2 conversion), high CO_2-to-sugar conversion rate (17.8–25.2 mmol·C/L/h), and high specific CO_2-to-sugar conversion rate (59.8 nmol·C/min/mg) were observed for the ACSP platform. In particular, the sugar titer and CO_2-to-sugar conversion rate of the proposed ACSP were sevenfold and nearly twofold higher, respectively, than that of previously reported ASAP for monosaccharide synthesis. Thus, this artificial "CO_2-to-sugar" platform could offer an alternative and bioresource-independent blueprint for precisely synthesizing high-order sugars (such as functional mono-, oligo-, and polysaccharides and their derivatives) via engineering bottleneck-associated enzymes.

(1) partial hydrolysis of cellulose to cellobiose by optimizing the cellobiohydrolase (CBH) and endoglucanase (EG) composition and ratio, and (2) amylose synthesis using cellobiose phosphorylase (CBP) and α-glucan phosphorylase (αGP). For this one-pot enzymatic conversion, CBP reversibly converts cellobiose to glucose-1-phosphate (G-1-P) and glucose in the presence of phosphate ions; αGP adds one glucosyl unit from G-1-P at the non-reducing end of amylose or maltodextrins, and phosphate ions are recycled to maintain nearly constant pH and phosphate levels. Among these enzymes, the special potato αGP was essential to push a partially hydrolyzed intermediate of cellulose forward to the synthesis of amylose. The amylose yield was up to 30% (wt/wt) (i.e., 0.30 g of amylose per gram of cellulose), and the chain length varied from DP 140 to 250, depending on the amount of maltotetraose added and cellobiose availability for amylose synthesis. Also, a bioprocess called simultaneous enzymatic biotransformation and microbial fermentation was developed to coproduce amylose, ethanol, and yeast as single-cell protein in one bioreactor to address the food, biofuels, and environment trilemma.

22.3 ENZYMATIC STARCH SYNTHESIS FROM NON-FOOD BIOMASS

Cellulose (a linear polysaccharide of repeating glucose monomers linked via β-1,4-glucosidic bonds) forms the principal structural component of plant cell walls and is the most abundant biopolymer on Earth. For example, every ton of cereals harvested is usually accompanied by the production of two to three tons of cellulose-rich crop residues, most of which are burned or wasted. To meet the world's future food and sustainability needs, You et al. (2013) designed a non-natural, synthetic, enzymatic pathway that can transform cellulose to amylose with four enzymes originating from bacterial, fungal, and plant sources. As shown in Figure 22.4, this biotransformation pathway has two modules:

22.4 MULTIFUNCTIONAL STARCH-BASED MATERIAL

22.4.1 Shape Memory Material

Shape memory material has the exceptional property of being able to regain its original shape when subjected to an external stimulus, such as temperature, pH, humidity, light, magnetism, or electricity (Chaunier & Lourdin 2009; Chevigny et al. 2016). Such properties have gained considerable interests in both academic and industrial sectors for applications such as drug delivery systems, biomedical devices, self-healing matrices, bio-inks for 4D printing, soft robotics, and flexible electronics (Beilvert

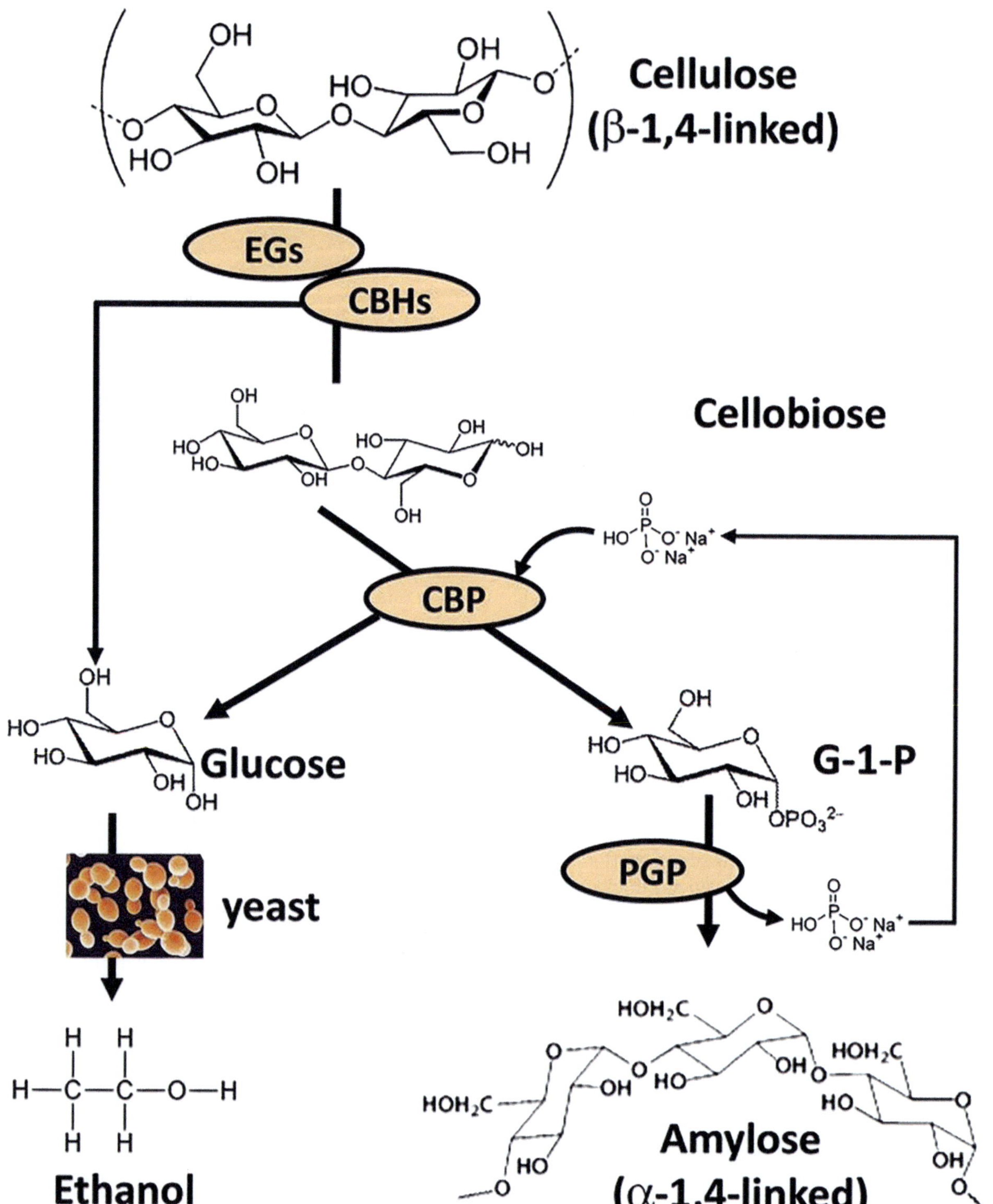

FIGURE 22.4 Enzymatic transformation of cellulose-to-amylose and ethanol. The enzymatic cellulose hydrolysis using endoglucanases (EGs), cellobiohydrolases (CBHs) and beta-glucosidase (BG) in cellulosic ethanol biorefinery versus the synthetic cellulose-to-amylose pathway supplemented with cellobiose phosphorylase (CBP) and potato alpha-glucan phosphorylase (PGP).

et al. 2014; Sessini et al. 2016). The shape changes are achieved throughout the programming and the recovery stages (Chaunier & Lourdin 2009). In the first step of the shape memory cycle, the polymer is deformed under external stress accompanied by heating and fixed in a desired temporary shape by cooling. However, because the temporary shape is more unstable than the original shape, the polymer then returns to the initial shape through the recovery process after a suitable external stimulation. Recently, the shape memory effect has been found in amorphous starch, starch-based blends, and bionanocomposites (Beilvert et al.

2014; Coativy et al. 2015). For example, a shape memory material with a high recovery ratio (>90%) was obtained by extrusion-cooking potato starch at 27% moisture content and 20% glycerol. The shape recovery of the extruded starch could be triggered in media-simulating body humidity and temperature after a subcutaneous implantation in a rat model (Beilvert et al. 2014). The linkage of local orders and entanglements in the amorphous potato starch to the original shape open a path to simple and elegant control of shape memory properties via different treatments (e.g., annealing) and preparation methods (e.g., hot casting and extrusion) (Véchambre et al. 2010, 2011; Chevigny et al. 2016). Coativy et al. (2015) designed shape memory (starch/sodium montmorillonites) bionanocomposites with highly efficient properties by melt processing without additional plasticizer. Maximum relaxation stress significantly improved for the bionanocomposites compared to unreinforced starch. For the multiresponsive shape memory bionanocomposites and blends of ethylene-vinyl acetate and thermoplastic starch, excellent humidity- and thermally activated recovery of the initial shape (with values higher than 80% and 90%, respectively) was obtained (Sessini et al. 2016). The multifunctional starch-based films reinforced with combinations of starch nanocrystals and catechin led to humidity-activated shape memory behavior with a significant increase in the shape-recovery ratio values (Sessini et al. 2018).

22.4.2 Biodegradable Materials

Biodegradable materials have gained more and more attention in the last few decades due to environmental concerns and can be widely used for agricultural, medical, and pharmaceutical applications as well as for food packaging (Cheng et al. 2021; Jiang et al. 2020). Starch is inherently biodegradable. However, it has disadvantages, including its hydrophilic character and the poor barrier and mechanical properties of products made from it. To overcome these defects, different strategies involving physical, chemical, biological, and combined approaches have been investigated to improve performance and decrease the cost of starch-based, biodegradable materials (Jiang et al. 2020; Wang et al. 2024a, b). Extrusion is a popular technique for producing starch-based films, sheets, and foams. Extrusion involves melting-solidification with a multiphysical field of shear, temperature, and pressure input (Bouvier & Campanella 2014). During extrusion processing, starch undergoes multiphase transitions and produces blends and composites (Cheng et al. 2021). Reinforcing starch matrices by addition of natural fillers, nanocrystals, hydrophobic polymers, and other additives have been used in developing functionalized starch-based plastics with excellent specific properties, such as high strength and stiffness, antibacterial and high barrier properties, and low cost, while maintaining biodegradability (Wang et al. 2024a, b). Nevertheless, large-scale economic production of high-performance starch-based biodegradable materials is still a challenge, and more research is required in this area. Moreover, cutting-edge technological approaches to develop innovative starch-based biodegradable materials need to be introduced to stimulate new areas for sustainable industrialization.

22.5 OTHER TECHNOLOGIES FOR STRUCTURING STARCH

Emerging quests for novel starch products with tailor-made structures and functionalities have been attracting more and more academic and industrial interests. As advanced technologies and approaches (including artificial intelligence, synthetic biology, data analysis, precision breeding, and omics technologies) are continuously becoming available, reliable starch derivatives with desired characteristics will be obtained for potential applications in food, agriculture, biomedical, and chemical industries. Also, a comprehensive understanding of quantitative structure–functional property relationships will offer opportunities for plant breeders and food producers to translate precise knowledge for the next generations of structuring starch with targeted structural features with optimal functional properties (Miao & BeMiller 2023).

22.5.1 Conversion of Starch via Polysaccharide Monooxygenases (PMOs)

Many challenges need to be overcome to ensure the successful commercial production and applications of novel starches. However, explorations of multifunctional enzymes as well as processing techniques to better meet the requirements of starch science are already underway. For example, PMOs are industrially important copper-dependent oxidoreductases and found to oxidatively degrade various crystalline polysaccharides, such as starch, cellulose, hemicellulose, and chitin (Lo Leggio et al. 2015; Vu & Marletta 2016; Zhang et al. 2023). PMOs reduce polysaccharides to oligosaccharides with an aldonic acid/aldonolactone unit at the reducing end by means of oxidative cleavage of α-1,4- and β-1,4-glucosidic linkages via hydroxylation (Vu & Marletta 2016). Starch-active PMOs from the auxiliary activity family AA13 possess an uneven active site surface with a shallow groove, as opposed to the flatter surface observed in other PMO families such as AA9-11, which may determine the specificity for starch substrates, including native starch and retrograded starch. These PMOs significantly increase the catalytic efficiency of hydrolytic enzymes under industrially relevant conditions and thus have potential in production of biofuels, sweetener syrups, and other starch-based conversion products.

22.5.2 3D Printing and Pickering Emulsion

Three-dimensional (3D) printing is a rapidly emerging technique known as additive manufacturing based on computer-aided design print paths. 3D printing of a food product involves layer-by-layer deposition of food materials to transform a part model into a complex geometric food (Abedini et al. 2024; Zhang et al. 2022). 3D printing offers potential benefits for food production in terms of modifying texture, for personalized nutrition, and for food customization. Among food materials, starches (with their desirable rheological and gelling properties) have great potential. However, the processing methods and printing parameters are still insufficient for precision design and rapid prototyping with good performance (Abedini et al. 2024). Kierulf et al. (2024) developed a Pickering emulsion platform for building superstructures like 2D cages and 3D hollow sheets using amaranth starch as a LEGO-like building block, a technique which might offer potential application for creation of super-thickeners for food products. As the current knowledge and techniques improve, precise information about structuring starch should be further obtained, offering opportunities for tailor-made structures and corresponding attributes for a variety of applications.

REFERENCES

Abedini, A, Sohrabvandi, S, Sadighara, P, Hosseini, H, Farhoodi, M, Assadpour, E, Sani, MA, Zhang, F, Seyyedi-Mansour, S & Jafari, SM 2024. Personalized nutrition with 3D-printed foods: a systematic review on the impact of different additives. *Advances in Colloid and Interface Science, 328,* 103181.

Antonovsky, N, Gleizer, S, Noor, E, Zohar, Y, Herz, E, Barenholz, U, Zelcbuch, L, Amram, S, Wides, A, Tepper, N, Davidi, D, Bar-On, Y, Bareia, T, Wernick, DG, Shani, I, Malitsky, S, Jona, G, Bar-Even, A & Milo, R 2016. Sugar synthesis from CO_2 in *Escherichia coli*. *Cell, 166,* 115–125.

Beilvert, A, Chaubet, F, Chaunier, L, Guilois, S, Pavon-Djavid, G, Letourneur, D, Meddahi-Pellé, A & Lourdin, D 2014. Shape-memory starch for resorbable biomedical devices. *Carbohydrate Polymers, 99,* 242–248.

BeMiller, JN & Whistler, RL 2009. *Starch: chemistry and technology* (3rd Ed.). Academic Press, San Diego.

Bouvier, J-M & Campanella, OH 2014. *Extrusion processing technology: food and non-food biomaterials*. Hoboken: Wiley-Blackwell Press.

Cai, T, Sun, H, Qiao, J, Zhu, L, Zhang, F, Zhang, J, Tang, Z, Wei, X, Yang, J, Yuan, Q, Wang, W, Yang, X, Chu, H, Wang, Q, You, C, Ma, H, Sun, Y, Li, Y, Li, C, Jiang, H, Wang, Q & Ma, Y 2021. Cell-free chemoenzymatic starch synthesis from carbon dioxide. *Science, 373,* 1523–1527.

Chaunier, L & Lourdin, D 2009. The shape memory of starch. *Starch/Stärke, 61,* 116–118.

Chen, J, Hawkins, E & Seung, D 2021. Towards targeted starch modification in plants. *Current Opinion in Plant Biology, 60,* 102013.

Cheng, H, Chen, L, McClements, DJ, Yang, T, Zhang, Z, Ren, F, Miao, M, Tian, Y & Jin, Z 2021. Starch-based biodegradable packaging materials: a review of their preparation, characterization and diverse applications in the food industry. *Trends in Food Science & Technology, 114,* 70–82.

Chevigny, C, Foucat, L, Rolland-Sabaté, A, Buléon, A & Lourdin, D 2016. Shape-memory effect in amorphous potato starch: the influence of local orders and paracrystallinity. *Carbohydrate Polymers, 146,* 411–419.

Coativy, G, Gautier, N, Pontoire, B, Buléon, A, Lourdin, D & Leroy, E 2015. Shape memory starch-clay bionanocomposites. *Carbohydrate Polymers, 116,* 307–313.

Gong, B, Cheng, L, Gilbert, RG & Li, C 2019. Distribution of short to medium amylose chains are major controllers of *in vitro* digestion of retrograded rice starch. *Food Hydrocolloids, 96,* 634–643.

Han, XZ & Hamaker, BR 2001. Amylopectin fine structure and rice starch paste breakdown. *Journal of Cereal Science, 34,* 279–284.

Hsieh C-F, Liu, W, Whaley, JK & Shi, Y-C 2019. Structure and functional properties of waxy starches. *Food Hydrocolloids, 94,* 238–254.

Jiang, T, Duan, Q, Zhu, J, Liu, H & Yu, L 2020. Starch-based biodegradable materials: challenges and opportunities. *Advanced Industrial and Engineering Polymer Research, 3,* 8–18.

Kierulf, A, Mosleh, I, Li, J, Li, P, Zarei, A, Khazdooz, L, Smoot, J & Abbaspourrad, A 2024. Food LEGO: building hollow cage and sheet superstructures from starch. *Science Advances, 10,* eadi7069.

Li, H, Gidley, MJ & Dhital, S 2019. High-amylose starches to bridge the "fiber gap": development, structure, and nutritional functionality. *Comprehensive Reviews in Food Science and Food Safety, 18,* 362–379.

Lo Leggio, L, Simmons, TJ, Poulsen, JC, Frandsen, KE, Hemsworth, GR, Stringer, M, von Freiesleben, P, Tovborg, M, Johansen, KS, De Maria, L, Harris, PV, Soong, CL, Dupree, P, Tryfona, T, Lenfant, N, Henrissat, B, Davies, GJ & Walton, PH 2015. Structure and boosting activity of a starch-degrading lytic polysaccharide monooxygenase. *Nature Communications, 6,* 5961.

Miao, M & BeMiller, JN 2023. Enzymatic approaches for structuring starch to improve functionality. *Annual Review of Food Science and Technology, 14,* 271–295.

Miao, M & Hamaker, BR 2021. Food matrix effects for modulating starch bioavailability. *Annual Review of Food Science and Technology, 12,* 169–191.

Miao, M, Jiang, B, Jin, Z & BeMiller, JN 2018. Microbial starch-converting enzymes: recent insights and perspectives. *Comprehensive Reviews in Food Science and Food Safety, 17,* 1238–1260.

Šárka, E & Dvořáček, V 2017. New processing and applications of waxy starch (a review). *Journal of Food Engineering, 206,* 77–87.

Sessini, V, Arrieta, MP, Fernández-Torres, A & Peponi, L 2018. Humidity-activated shape memory effect on plasticized starch-based biomaterials. *Carbohydrate Polymers, 179,* 93–99.

Sessini, V, Raquez, JM, Re, GL, Mincheva, R, Kenny, JM, Dubois, P, & Peponi, L 2016. Multiresponsive shape memory blends and nanocomposites based on starch. *ACS Applied Materials & Interfaces, 8,* 19197–19201.

Seung, D 2020. Amylose in starch: towards an understanding of biosynthesis, structure and function. *New Phytologist, 228,* 1490–1504.

Tao, KY, Li, C, Yu, WW, Gilbert, RG & Li, EP 2019. How amylose molecular fine structure of rice starch affects functional properties. *Carbohydrate Polymers, 204,* 24–31.

Véchambre, C, Buléon, A, Chaunier, L, Gauthier, C & Lourdin, D 2011. Understanding the mechanisms involved in shape memory starch: macromolecular orientation, stress recovery and molecular mobility. *Macromolecules, 44,* 9384–9389.

Véchambre, C, Buléon, A, Chaunier, L, Jamme, F & Lourdin, D 2010. Macromolecular orientation in glassy starch materials that exhibit shape memory behavior. *Macromolecules, 43,* 9854–9858.

Vu, VV & Marletta, MA 2016. Starch-degrading polysaccharide monooxygenases. *Cellular and Molecular Life Sciences, 73,* 2809–2819.

Wang, K, Henry, RJ & Gilbert, RG 2014. Causal relations among starch biosynthesis, structure, and properties. *Springer Science Reviews, 2,* 15–33.

Wang, L, Li, D, Ye, L, Zhi, C, Zhang, T & Miao, M 2024a. Development of a self-reinforced starch-derived film with biocompatibility and mechanical properties. *Food Chemistry, 447,* 138974.

Wang, L, Li, D, Ye, L, Zhi, C, Zhang, T & Miao, M 2024b. Starch-based biodegradable composites: effects of in-situ re-extrusion on structure and performance. *International Journal of Biological Macromolecules, 266,* 130869.

Wang, X, Luo, H, Wang, Y, Wang, Y, Tu, T, Qin, X, Su, X, Huang, H, Bai, Y, Yao, B & Zhang, J 2022. Direct conversion of carbon dioxide to glucose using metabolically engineered Cupriavidus necator. *Bioresource Technology, 362,* 127806.

Yang, J, Song, W, Cai, T, Wang, Y, Zhang, X, Wang, W, Chen, P, Zeng, Y, Li, C, Sun, Y & Ma, Y 2023. *De novo* artificial synthesis of hexoses from carbon dioxide. *Science Bulletin, 68,* 2370–2381.

You, C, Chen, H, Myung, S, Sathitsuksanoh, N, Ma, H, Zhang, XZ, Li, J & Zhang, YH 2013. Enzymatic transformation of nonfood biomass to starch. *Proceedings of the National Academy of Sciences, 110,* 7182–7187.

Yu, W, Li, H, Zou, W, Tao, K, Zhu, J & Gilbert, RG 2019. Using starch molecular fine structure to understand biosynthesis-structure-property relations. *Trends in Food Science & Technology, 86,* 530–536.

Zhang, G, Ao, Z & Hamaker, BR 2008. Nutritional property of endosperm starches from maize mutants: a parabolic relationship between slowly digestible starch and amylopectin fine structure. *Journal of Agricultural and Food Chemistry, 56,* 4686–4694.

Zhang, J, Li, Y, Cai, Y, Ahmad, I, Zhang, A, Ding, Y, Qiu, Y, Zhang, G, Tang, W & Lyu, F 2022. Hot extrusion 3D printing technologies based on starchy food: a review. *Carbohydrate Polymers, 294,* 119763.

Zhang, N, Yang, J, Li, Z, Haider, J, Zhou, Y, Ji, Y, Schwaneberg, U & Zhu, L 2023. Influences of the carbohydrate-binding module on a fungal starch-active lytic polysaccharide monooxygenase. *Journal of Agricultural and Food Chemistry, 71,* 18405–18413.

Zheng, T, Zhang, M, Wu, L, Guo, S, Liu, X, Zhao, J, Xue, W, Li, J, Liu, C, Li, X, Jiang, Q, Bao, J, Zeng, J, Yu, T & Xia, C 2022. Upcycling CO_2 into energy-rich long-chain compounds via electrochemical and metabolic engineering. *Nature Catalysis, 5,* 388–396.

Zhou, X, Campanella, OH, Hamaker, BR & Miao, M 2021. Deciphering molecular interaction and digestibility in retrogradation of amylopectin gel network. *Food & Function, 12,* 11460–11468.

Index

Note: **Bold** page numbers refer to tables and *italic* page numbers refer to figures.